P9-AAY-979

QR
82
M8 E36
2009
VAN

Jindrich Kazda · Ivo Pavlik ·
Joseph O. Falkinham III ·
Karel Hruska
Editors

The Ecology of Mycobacteria: Impact on Animal's and Human's Health

Springer

Editors
Prof. Jindrich Kazda
Parkallee 39
23845 Borstel
Germany
j.kazda@t-online.de

Prof. Joseph O. Falkinham III
Virginia Tech
Dept. of Biological Sciences
(MC 406)
2125 Derring Hall
Blacksburg VA 24061-0406
USA
jofiii@vt.edu

Prof. Ivo Pavlik
Veterinary Research Institute
Hudcova 70
621 00 Brno
Czech Republic
pavlik@vri.cz

Prof. Karel Hruska
Veterinary Research Institute
Hudcova 70
621 00 Brno
Czech Republic
hruska@vri.cz

ISBN 978-1-4020-9412-5   e-ISBN 978-1-4020-9413-2
DOI 10.1007/978-1-4020-9413-2
Springer Dordrecht Heidelberg London New York

Library of Congress Control Number: 2008940658

© Springer Science+Business Media B.V. 2009
No part of this work may be reproduced, stored in a retrieval system, or transmitted in any form or by
any means, electronic, mechanical, photocopying, microfilming, recording or otherwise, without written
permission from the Publisher, with the exception of any material supplied specifically for the purpose of
being entered and executed on a computer system, for exclusive use by the purchaser of the work.

Cover image: Mycobacteria (slightly bent, short rods) on the surface of hyalocytes in the grey layer of
Sphognum magellanicum (Photo K. Muller)

Printed on acid-free paper

Springer is part of Springer Science+Business Media (www.springer.com)

Preface to the Second Edition

A decade has passed since the primary literature sources were collected and the first edition of this book written. This period of time seems to be relatively short when one considers that mycobacteria were first reported 100 years ago. On the other hand, the known range of mycobacteria has been greatly extended in recent years. The introduction of molecular biology methods has brought about a remarkable burst in the description of new species. While about 70 mycobacterial species were registered at the time of the first edition, more than 130 of them are known at present. With the discovery of new mycobacterial species, the cases of human and animal immunocompetent and immunosuppressed hosts and the isolation of mycobacteria with the enzymatic potential to cause the degradation of aliphatic organic substances are increasing in numbers almost as rapidly.

In order to be able to cover all of the most significant mycobacterial species, it was necessary to consider the ecology of mycobacteria as a discipline that would not only include the external environment but also the occurrence of mycobacteria in animal and human organisms, where interaction occurs. The environment is neither non-living nor static, but the very opposite. It undergoes periodic and other changes (seasons of the year, changing biotic and abiotic factors), while animal and human organisms have the static tendency towards a status *quo ante*. The classification of mycobacteria into respective disciplines such as epidemiology, epizootiology, immunology and environmental ecology did not contribute to a comprehensive understanding of their significance. Therefore, in this book mycobacteria are presented as a whole, under the general designation of mycobacterial ecology, and without limitation by any particular discipline.

This enabled us to concentrate our attention on the genus *Mycobacterium* in all kinds of environments in which they can live, i.e. in macro-organisms as well as in nature. Special attention was paid to the conditions under which mycobacteria can survive, multiply or exist in a dormant state. Of more than 100 species only a few are obligate pathogens for humans and animals. These are unable to grow in the natural environment, but have developed special strategies for reaching susceptible individuals. Furthermore, potentially pathogenic mycobacteria possess the ability both to multiply in natural environments and to cause diseases. A transitional phenomenon creates mycobacterial species that live in the environment and provoke allergic reactions in animals. The majority of mycobacteria are saprophytic and some of them serve as nutrients for dragonfly larvae.

The phylogeny of mycobacteria indicates that pathogenic species developed from saprophytic ones. There is evidence to suggest that the disturbance of their natural habitats and the overlapping of these biotopes by humans and animals contributed to the spread of mycobacteria and perhaps to their convergence to pathogenicity.

It was not our intention to present a compendium covering all published results, but rather to issue a "readable" book, which is illustrative and thus focused on the principle facts. The increase in the number of Editors has allowed the sharing of original experiences regarding the ecology of mycobacteria, published here for the first time in some cases. The supplemented edition should serve as a guide to these discoveries and also contribute to an understanding of clinically significant species in human and animal medicine.

Borstel, Germany, January 2009   Jindrich Kazda

Editors' Comments

The editors responsible for the chapters are listed under the title of each chapter. Authors are listed under the titles of subchapters.

The references are listed as they appear in the databases Reference Manager (Thomson Reuters, Philadelphia) as imported from Web of Science (Thomson Reuters, Philadelphia) or PubMed (Medline, NLM Bethesda). A few citations, not indexed, were cited according to the reprints or books available. This principle resulted in minor differences in the titles (not all reference titles are in English, some references have capitalized title words, not all species names are according to the contemporary nomenclature and in italics). Some journals are cited with abbreviated titles, some in full, as available in the source databases. These differences were left in the format of the database.

All photos are collected in Chapter 10, with references to the chapter and subchapter where they are quoted.

To keep the structure of the book some information appears in two or more chapters with respect to the chapter's main field. Readers should not consider this as a duplicity.

Acknowledgements

The editors appreciate the invaluable assistance of many colleagues and co-workers, whose comments and enthusiasm in collaborating on the research projects, exemplary library services and administrative support are greatly valued. Special thanks are due to Neysan Donnelly and Catherine Murdoch, University of Aberdeen, UK, for their patient language amendment of the text. The support given by the Ministry of Agriculture of the Czech Republic (project MZE 0002716201) is acknowledged.

Contents

Contributors

Prof. Jiri Damborsky, RNDr. PhD.   Loschmidt Laboratories, Faculty of Science, Masaryk University, Brno, Czech Republic, jiri@chemi.muni.cz

Lenka Dvorska, PhD.   Czech Agriculture and Food Inspection Authority, Brno, Czech Republic, lenka.bartosova@szpi.gov.cz

Prof. Joseph Falkinham III   Virginia Polytechnic Institute and State University, Blacksburg, VA, USA, jofiii@vt.edu

Prof. Karel Hruska, MVDr. CSc.   Veterinary Research Institute, Brno, Czech Republic, hruska@vri.cz

Andrea Jesenska, RNDr. PhD.   Loschmidt Laboratories and National Centre for Biomolecular Research, Faculty of Science, Masaryk University, Brno, Czech Republic, andrea.jesenska@gmail.com

Jarmila Kaustova, MUDr.   National Reference Laboratory for *Mycobacterium kansasii*, Regional Institute of Public Health, Ostrava, Czech Republic, jarmila.kaustova@zuova.cz

Prof. Jindrich Kazda, MVDr. CSc.   University of Kiel, Kiel, Germany, J.Kazda@t-online.de

Lorenz Khol, DVM   Clinic for Ruminants, University of Veterinary Medicine, Vienna, Austria, Johannes.Khol@vu-wien.ac.at

Prof. Jiri Klimes, MVDr. CSc.   Department of Biology and Wildlife Diseases, Faculty of Veterinary Hygiene and Ecology, University of Veterinary and Pharmaceutical Sciences, Brno, Czech Republic, klimesj@vfu.cz

Vojtech Mrlik, MVDr. CSc.   Veterinary Research Institute, Department of Food and Feed Safety, Brno, Czech Republic, mrlik@vri.cz

Prof. Ivo Pavlik, MVDr. CSc.   OIE Reference Laboratories for Paratuberculosis and Avian Tuberculosis; Department of Food and Feed Safety, Veterinary Research Institute, Brno, Czech Republic, pavlik@vri.cz

Jembere Edmealem Shitaye, DVM, PhD.   Veterinary Research Institute, Brno, Czech Republic, *Present address*: Ministry of Agriculture and Rural Development Agricultural Extension Department, Addis Ababa, Ethiopia, edmesh@hotmail.com

Martina Trckova, Ing. PhD.   Veterinary Research Institute, Department of Food and Feed Safety, Brno, Czech Republic, trckova@vri.cz

Authors of Photographs

Takashi Amemori, Veterinary Research Institute, Brno, Czech Republic

Radek Axman, District Veterinary Administration, Chrudim, Czech Republic

Veronika Grymova, Veterinary Clinic AvetuM, Brno, Czech Republic

Roman Halouzka, University of Veterinary and Pharmaceutical Sciences Brno, Czech Republic

Petr Jezek, Regional Hospital, Pribram, Czech Republic

Jarmila Kaustova, Regional Institute of Public Health, Ostrava, Czech Republic

Jindrich Kazda, University of Kiel, Germany

Jiri Lamka, Faculty of Pharmacy in Hradec Kralove, Charles University in Prague, Czech Republic

Lev Mezensky, Regional Institute of Public Health, Brno, Czech Republic

Vojtech Mrlik, Veterinary Research Institute, Brno, Czech Republic

Klaus Muller, University of Kiel, Germany

Ladislav Novotny, University of Veterinary and Pharmaceutical Sciences Brno, Czech Republic

Ivo Pavlik, Veterinary Research Institute, Brno, Czech Republic

Ilona Parmova, State Veterinary Institute, Prague, Czech Republic

Jembere Edmealem Shitaye, Veterinary Research Institute, Brno, Czech Republic

Misa Skoric, University of Veterinary and Pharmaceutical Sciences Brno, Czech Republic

Jiri Stork, The 1st Medical Faculty, Charles University in Prague, Czech Republic

Jana Svobodova, Regional Institute of Public Health, Brno, Czech Republic

Ivo Trcka, Veterinary Research Institute, Brno, Czech Republic

Irena Trneckova, Veterinary Research Institute, Brno, Czech Republic

Richard Tyl, University Hospital, Ostrava, Czech Republic

Vit Ulmann, Regional Institute of Public Health, Ostrava, Czech Republic

Jan Vesely, Bedrich Schwarzenberg's Forestry College (Higher Forestry School) and Secondary Forestry School, Pisek, Czech Republic

Vladimir Vrbas, Veterinary Research Institute, Brno, Czech Republic

Abbreviations

AFB	acid-fast bacilli, acid-fast bacteria
AFLP	amplified fragment length polymorphism
AFR	acid-fast rods
AIDS	Acquired Immunodeficiency Syndrome
ATCC	The American Type Culture Collection
ATP	adenosine triphosphate
BCG	Bacillus Calmette-Guérin or Bacille Calmette-Guérin
BTEX	aromatic hydrocarbons benzene, toluene, ethyltoluene and xylene
CD	Crohn's disease
CFU	colony forming units
CNS	central nervous system
C/N	carbon and nitrogen ratio
CFTR	cystic fibrosis transmembrane conductance regulator
DGGE	denaturing gradient gel electrophoresis
DNA	deoxyribonucleic acid
ESD	endosulfan-degrading
ESM	environmental saprophytic mycobacteria
GC-MS	gas chromatography-mass spectrometry
HIV	human immunodeficiency virus
HP	hypersensitivity pneumonitis
HPLC	high-performance liquid chromatography or high pressure liquid chromatography
M.	*Mycobacterium*
MAC	*M. avium* complex
MAI	*M. avium-intracellulare*
MAIC	*M. avium-intracellulare* complex
MAIS	*M. avium-intracellulare-scrofulaceum* complex
MDT	multi drug therapy
MDP	muramyl dipeptide
MPTR	major polymorphic tandem repeat
MTC	*Mycobacterium tuberculosis* complex

MWF	metal-working fluid
NC AFB	non-cultivable acid-fast bacilli
NCTC	National Collection of Type Cultures
NTM	non-tuberculous mycobacteria
NOD	nucleotide-binding oligomerisation domain
OIE	World Organisation for Animal Health
OPM	obligate pathogenic mycobacteria
PAHs	polycyclic aromatic hydrocarbons
PCR	polymerase chain reaction
PFGE	pulsed field gel electrophoresis
PGL	phenolic glycolipid
PGPR	plant growth-promoting bacteria
PPM	potentially pathogenic mycobacteria
PVC	polyvinylchloride
R.	*Rhodococcus*
rDNA	ribosomal DNA
rRNA	ribosomal RNA
RNA	ribonucleic acid
rep-**PCR**	repetitive-unit-sequence-based PCR
RFLP	restriction fragment length polymorphism
SIV	simian immunodeficiency virus
TCE	trichloroethylene
TLR	toll-like receptor
TMC	Trudeau Mycobacterial Culture Collection
TNF-α	tumor necrosis factor alfa
USA	United States of America
UK	United Kingdom
US EPA	United States Environmental Protection Agency
UV	ultraviolet
WHO	World Health Organization

Chapter 1

The Chronology of Mycobacteria and the Development of Mycobacterial Ecology

J. Kazda

Introduction

In addition to microbiology, three other sciences, epidemiology, biochemistry and molecular biology, have contributed a great deal to recent advances in our knowledge of the genus *Mycobacterium*. Species from this genus are among the most important micro-organisms and include the causative agents of tuberculosis in humans and animals. The ecology of mycobacteria began to develop early after the discovery of the first pathogenic species and its further development differed from mainstream microbiology.

To illustrate the history of mycobacteria, it is useful to describe the contribution of the aforementioned sciences, especially molecular biology, and to mention several important periods in the development of mycobacterial ecology.

1.1 The Microbiology of Mycobacteria

J. Kazda

1.1.1 The First Pathogenic Mycobacteria

The epoch-making discovery was the description of the causative agent of human tuberculosis by Robert Koch (1882; Photo 1.1). He applied and fulfilled the postulate of Henle, who set criteria under which an isolated micro-organism can be regarded to be the causative agent of a disease (commonly known as "Koch's postulates"). A short time after the discovery of *M. tuberculosis*, further pathogenic mycobacteria were discovered: *M. avium* and *M. bovis* as the causative agents of tuberculosis in birds and cattle, respectively (Lehmann and Neumann, 1896). *M. paratuberculosis* was also found in the lesions of intestinal tuberculosis in livestock (Johne and Frothingham, 1895).

In fact, the first known *Mycobacterium* was not *M. tuberculosis* but *M. leprae,* which was discovered as early as 1873 in Bergen, Norway, by Hansen (1875). Due to the lack of suitable staining methods, Hansen identified this bacterium on a native smear by releasing the intracellular bacilli by treatment with a hypotonic solution. A great hindrance to leprosy research is the fact that *M. leprae* is not cultivable on artificial media. Thus, neither Hansen nor his successors were able to fulfil the Henle postulates until the successful multiplication of *M. leprae* in mouse footpads (Shepard, 1960).

1.1.2 Further Progress in Mycobacteriology

Late in the 19th and at the beginning of the 20th century, further attempts were made to identify mycobacteria as a cause of diseases in humans and animals. An improved staining method using carbolic acid and fuchsin introduced by Ziehl (1882) and Neelsen (1883) enabled the differentiation of acid–alcohol–fast rods from other bacteria. This sometimes led

J. Kazda (✉)
University of Kiel, Kiel, Germany
e-mail: J.Kazda@t-online.de

to false conclusions as acid-fast bacilli (AFB) were thought to cause syphilis because of their presence in conglomerates of epithelioid cells in patients. They were reclassified by Alvares and Tavel (1885) as belonging to saprophytic mycobacteria and later designated by Lehmann and Neumann as *M. smegmatis* (Lehmann and Neumann, 1896).

The possibility of the occurrence of mycobacteria in food, especially in butter, was a question that interested Robert Koch and was researched in his laboratory by Rabinowitsch (1897). She did not find any contamination with *M. tuberculosis* but found saprophytic mycobacteria in 39.6% of 80 butter samples collected in Berlin and Philadelphia, most of them identified as *M. smegmatis* (Lehmann and Neumann, 1896). The search for further environmental mycobacteria continued and the findings were evaluated by Courmont and Potet (1903), who reviewed the hitherto published contributions concluding that plants and soil can be a source of non-pathogenic mycobacteria. Water was also examined for mycobacteria. Acid-fast rods found in tap water were described as "morphologically similar to the bacilli of Koch" and named *M. aquae* (Galli-Valerio and Bornand, 1927).

1.1.3 The First Potentially Pathogenic Mycobacteria

M. chelonae was the first such *Mycobacterium* discovered at the beginning of the last century (Friedmann, 1903), followed several years later by *M. lepraemurium* (Marchaux and Sorel, 1912). A spontaneous disease resembling tuberculosis described in a ring snake and later in a carp, that was confirmed morphologically and by the isolation of acid-fast bacilli, was named "*M. piscium*" (Bataillon and Dubart, 1897). Aronson found acid-fast bacilli in the granulomatous lesions of a viper and described this strain as *M. thamnopheos* (Aronson, 1929), after describing in 1926 the causative agents of tuberculosis in saltwater fish, which he named *M. marinum* (Aronson, 1926).

In the following decades, there was no remarkable progress in the field apart from the experiments of Haag (1927), who systematically examined 90 samples of soil, grass, litter and skin smears, finding mycobacteria in 72 samples. He designated these as *M. phlei*,

"*M. lacticola*", "*M. eos*" and "*M. luteum*". Due to the lack of suitable methods for the differentiation of acid-fast micro-organisms, there were then no remarkable developments in the mycobacteriology field for many years.

The attention paid to environmental mycobacteria considerably increased in the late 1930s following the linking of a "pseudo-primary tuberculous complex" of the skin with an abrasion obtained in a swimming pool (Strandberg, 1937). Similar cases which occurred later were described as "water-borne infections of skin in swimming pools" (Hellerstrom, 1952). The increasing interest that followed resulted in the isolation of environmental mycobacteria as the causative agent of granulomatous skin lesions in humans. The source was found to be the water of swimming pools. The disease was described as swimming pool granuloma and the *Mycobacterium* designated as "*M. balnei*" (Linell and Norden, 1954). A comparison revealed that the strain was identical to *M. marinum* described just in 1926 by Aronson (Bojalil, 1959).

At the same time the skin lesions were observed in Sweden, individual cases of lung tuberculosis caused by mycobacteria different from *M. tuberculosis* were reported in the United States (Beavan and Bayne-Jones, 1931). In 1953 two cases of lung tuberculosis were described from which "yellow bacilli" were repeatedly isolated (Buhler and Pollak, 1953). These strains were later designated as *M. kansasii* (Hauduroy, 1955). Avian mycobacteria isolated from human tuberculosis lesions were described in 1939 (Negre, 1939), followed by *M. intracellulare* (Runyon, 1959) and *M. scrofulaceum* (Masson and Prissick, 1956).

1.1.4 The First Methods for the Differentiation of Mycobacteria

The absence of reliable methods for exact differentiation was the main hindrance in the study of mycobacteria. Interest was focused on three pathogenic species: *M. tuberculosis, M. bovis* and *M. avium*. To differentiate between these, a system of "biological methods" was established using inoculation in three kinds of experimental animals: guinea pigs, rabbits and chickens. Heavy lesions in the lungs, liver and spleen of the guinea pigs and rabbits indicated *M. bovis*, while *M. tuberculosis* provoked such lesions only in guinea pigs. *M. avium* was pathogenic only in chickens. This overly simplistic method was favoured by many laboratories until the late 1960s and thus resulted in a "low tide" in mycobacterial taxonomy. Only eight new mycobacterial species were characterised in the first 50 years of the last century (Table 1.1).

One of the first attempts to introduce tests, commonly used in the differentiation of bacteria other than mycobacteria, was made by Gordon and Smith (1953). In rapidly growing mycobacteria they included the hydrolysis of starch, gelatine and casein, acid production from carbohydrates, the utilisation of citrate, succinate and malate, nitrate reductase production and NaCl tolerance. The last two tests have shown a good discriminating value and are still used in taxonomy. On the other hand, an improved test for the production of acid from carbohydrates, developed 17 years later, did not come into general use (Bonicke and Kazda, 1970).

Great progress in the differentiation of mycobacteria began in the early 1960s with the keen observation of Ernest Runyon, who proposed a new and simple concept for the differentiation of mycobacteria associated with disease or found in the environment. Besides the obligate pathogenic species, he used the designation of "atypical mycobacteria", which he divided into four groups: photochromogenic, scotochromogenic, nonchromogenic and rapidly growing (Runyon, 1965). The morphological character of mycobacterial cultures was a simple but very useful tool for further orientation. Thus, the interest of microbiologists became focused on this genus and in a short time a variety of mycobacterial strains were isolated and "grouped" in laboratories all over the world.

1.1.5 Biochemical Methods for the Differentiation of Mycobacteria

Compared with other fields of microbiology, biochemical methods for the differentiation of pathogenic mycobacteria were introduced at a relatively late date. Many clinical microbiologists had great difficulty moving on from the simple scheme of biological trials using experimental animals, as mentioned above.

One of the first methods, known as the "niacin test", was developed by Konno et al. (1958). It is based on the selective production of nicotinic acid by *M. tuberculosis* and was used for its confirmation and the differentiation of this species from *M. bovis* and other slow growers, which gave negative results in this test. Due to the progress being made in the eradication of bovine tuberculosis in livestock, there was an urgent need for a simple method differentiating *M. tuberculosis* from *M. bovis*. Such a test was developed by Virtanen (1960). *M. bovis* did not produce nitrate reductase, which is produced by *M. tuberculosis*. Unlike the niacin test, nitrate reductase is not limited to *M. tuberculosis*, but is widely distributed among mycobacteria. A very important system for the differentiation of mycobacteria was introduced in the early 1960s by Bonicke (1962). This method, known as the "amide row", offered a new way of distinguishing newly isolated mycobacterial strains. The results, obtained by the degradation of 10 different amides, enabled the more precise differentiation of mycobacteria and contributed a great deal to the general acceptance of biochemical methods in their microbiology.

Further tests for the differentiation of mycobacteria soon followed, particularly those based on arylsulfatase activity (Kubica and Vestal, 1961), Tween hydrolysis (Wayne, 1962) and the detection of phosphatase activity (Kappler, 1965). These and other tests enabled great progress to be made in the differentiation of mycobacterial isolates from humans, animals and environmental sources. Between 1951 and 1970, 21 new mycobacterial species were isolated and described, many more than before. This trend resulted in more interest in mycobacterial taxonomy. Acid-fast bacilli, which had formerly been isolated but had exhibited different properties to those of already known species, were collected as stock cultures and were examined from 1967 onwards by the cooperative studies organised by Lawrence Wayne in the International Working Group on Mycobacterial Taxonomy. Nearly all laboratories working on the differentiation of mycobacteria joined this group and took part in the evaluation of taxonomic methods and in the testing of selected mycobacterial strains. Their results have become a valuable guide in mycobacterial taxonomy and more than 50 phenotypic properties have been adopted for the evaluation of new species of mycobacteria (Wayne et al., 1996).

Table 1.1 Survey of the development of the genus *Mycobacterium* during the years 1873–2007

Decade	Species and discovery		Origin
	Name	Year	
1871–1880	*M. leprae*	1873	Humans
1881–1890	*M. tuberculosis*	1882	Humans
	M. smegmatis	1889	Environment
1891–1900	*M. avium*[1] [1]	1891	Birds
	M. paratuberculosis[2] [1]	1895	Cattle
	M. bovis	1896	Cattle
	M. phlei	1899	Environment
1901–1910	*M. chelonae*	1903	Poikilotherms
1911–1920	*M. aquae/M. gordonae*	1912	Environment
	M. lepraemurium	1912	Mice
1921–1930	*M. marinum*	1926	Fish
	M. microti	1927	Animals
1931–1940	*M. fortuitum*	1938	Poikilotherms
1941–1950	*M. intracellulare*	1949	Humans
	M. ulcerans	1950	Humans
1951–1960	*M. abscessus*	1953	Humans
	M. kansasii	1955	Humans
	M. scrofulaceum	1956	Humans
	M. paraffinicum	1956	Environment
	M. farcinogenes	1958	Animals
	M. xenopi	1959	Poikilotherms
	M. salmoniphilum	1960	Fish
1961–1970	*M. flavescens*	1962	Environment
	M. peregrinum	1962	Humans
	M. vaccae	1964	Environment
	M. simiae	1965	Animals
	M. diernhoferi	1965	Environment
	M. nonchromogenicum	1965	Environment
	M. parafortuitum	1965	Environment
	M. terrae	1966	Environment
	M. gastri	1966	Humans
	M. triviale	1966	Humans
	M. aurum	1966	Environment
	M. thermoresistibile	1966	Environment
	M. chitae	1967	Environment
	M. africanum	1969	Humans
1971–1980	*M. agri*	1971	Environment
	M. asiaticum	1971	Humans
	M. duvalii	1971	Humans
	M. gadium	1971	Humans
	M. gilvum	1971	Humans
	M. obuense	1971	Environment
	M. rhodesiae	1971	Humans
	M. neoaurum	1972	Environment
	M. szulgai	1972	Humans
	M. aichiense	1973	Environment
	M. chubuense	1973	Environment
	M. senegalense	1973	Animals
	M. tokaiense	1973	Environment
	M. shimoidei	1975	Humans

Table 1.1 (continued)

Decade	Species and discovery		Origin
	Name	Year	
	M. petroleophilum	1975	Environment
	M. malmoense	1977	Humans
	M. haemophilum	1978	Humans
	M. komossense	1979	Environment
	M. sphagni	1980	Environment
1981–1990	*M. fallax*	1983	Environment
	M. porcinum	1983	Animals
	M. austroafricanum	1983	Environment
	M. pulveris	1983	Environment
	M. moriokaense	1986	Environment
	M. chlorophenolicum	1986	Environment
	M. poriferae	1987	Environment
	M. cookii	1990	Environment
	M. avium subsp. *silvaticum*	1990	Birds
1991–2000	*M. alvei*	1992	Environment
	M. madagascariense	1992	Environment
	M. confluentis	1992	Humans
	M. hiberniae	1993	Environment
	M. brumae	1993	Environment
	M. celatum	1993	Humans
	M. genavense	1993	Humans
	M. intermedium	1993	Humans
	M. interjectum	1993	Humans
	M. mucogenicum	1995	Humans
	M. branderi	1995	Humans
	M. conspicuum	1995	Humans
	M. hodleri	1996	Environment
	M. lentiflavum	1996	Humans
	M. triplex	1996	Humans
	M. mageritense	1997	Humans
	M. heidelbergense	1997	Humans
	M. hassiacum	1997	Humans, environment
	M. novocastrense	1997	Humans
	M. canettii	1997	Humans
	M. bohemicum	1998	Humans, animals, environment
	M. tusciae	1999	Environment, humans
	M. wolinskyi	1999	Humans
	M. caprae	1999	Animals
	M. murale	1999	Environment
	M. goodii	1999	Humans
	M. kubicae	2000	Humans
	M. elephantis	2000	Animals, humans
	M. septicum	2000	Humans
	M. botniense	2000	Environment
2001–2007	*M. immunogenum*	2001	Environment, humans
	M. doricum	2001	Humans
	M. heckeshornense	2001	Humans
	M. frederiksbergense	2001	Environment
	M. palustre	2002	Environment, humans, animals
	M. lacus	2002	Environment, humans

Table 1.1 (continued)

Decade	Species and discovery		Origin
	Name	Year	
	M. holsaticum	2002	Humans
	M. vanbaalenii	2002	Environment
	M. shottsii	2003	Environment
	M. pinnipedii	2003	Animals, humans
	M. montefiorense	2003	Animals
	M. chimaera	2004	Humans
	M. cosmeticum	2004	Environment
	M. boenickei	2004	Humans
	M. brisbanense	2004	Humans
	M. canariasense	2004	Humans
	M. houstonense	2004	Humans
	M. neworleansense	2004	Humans
	M. nebraskense	2004	Humans
	M. parascrofulaceum	2004	Humans
	M. parmense	2004	Humans
	M. psychrotolerans	2004	Environment
	M. pyrenivorans	2004	Environment
	M. saskatchewanense	2004	Humans
	M. pseudoshottsii	2005	Animals
	M. florentinum	2005	Humans
	M. colombiense	2006	Animals
	M. aubagnense	2006	Humans
	M. bollettii	2006	Humans
	M. phocaicum	2006	Humans
	M. arupense	2006	Humans
	M. conceptionense	2006	Humans
	M. fluoranthenivorans	2006	Environment
	M. massiliense	2006	Humans
	M. monacense	2006	Humans
	M. kumamotonense	2007	Humans
	M. seoulense	2007	Humans

[1]*MAA M. avium* subsp. *avium*.
[2]*MAP M. avium* subsp. *paratuberculosis*.
[1] Thorel MF, Krichevsky M, Levy-Frebault VV (1990) Int. J. Syst. Bacteriol. 40:254–260.

1.1.6 Molecular, Genetic and Other Methods in Mycobacterial Taxonomy

After the first phenotypic period, the second major trend in mycobacterial taxonomy was focused on genotypic characteristics. In the early 1980s, a method for the estimation of partial sequences of 16S ribosomal RNA was introduced into the taxonomy of mycobacteria (Stackebrandt and Woese, 1981). The genotypic studies confirmed the validity of previous mycobacterial taxa and were effective in the discrimination of variable regions. The gene encoding the 16S rRNA is still the primary target of molecular taxonomic studies (Tortoli, 2006). These techniques resulted in the splitting of some mycobacterial species and in the description of new taxa, also using previously isolated strains or clusters.

On the other hand, the more sophisticated methods in chemotaxonomy achieved new results in the study of mycolic acids, present in the mycobacterial cell wall. The consequence of both of these techniques was an enormous increase in the description of new mycobacterial species over the last few decades. Compared with the "classical area" beginning with the description of the first *Mycobacterium* species, which lasted until the early 1990s, and during which 64 new species were described, an enormous boom in mycobacterial taxonomy resulted in a total of 67 new species being described between 1991 and 2007. Taking into account the fact that this number also includes about 22 previously isolated strains or clusters, it may represent a source of confusion for diagnostic laboratories and clinicians. These new species of mycobacteria might be associated with specific diseases or occur merely as contaminants. Nearly a quarter of the 67 newly described species originated from the environment or from fauna. The high number of isolates from humans, the most examined tissue samples, does not appear to be in conflict with the claim that human isolates represent only the tip of the iceberg and that the environment is in fact the major source of mycobacteria (Tortoli, 2006). The chronology of the description of the genus *Mycobacterium* is documented in Table 1.1.

1.2 The Ecology of Mycobacteria

J. Kazda

As mentioned above, the ecology of mycobacteria developed parallel to their bacteriology and can be divided into the following stages.

1.2.1 First Steps Towards Mycobacterial Ecology

Occasional examinations of environmental samples, food and poikilotherm, as described above, had an epidemiological or epizootiological character and focused on finding the causative agents responsible for diseases.

The first attempts at ecological studies were made towards the end of the 19th century. Nearly 30 years before the discovery of the causative agent of tuberculosis, new methods of therapy were introduced after the claim was made that the disease was in fact curable (Bremer, 1854). The first sanatorium for the treatment of tuberculosis was founded not in Switzerland but in Germany in Görbersdorf, Schlesien (now Sokolowsko). Its founder, Dr. Bremer, gathered physicians interested in the natural sciences to contribute to his project for the successful treatment of this disease.

One of them was Dr. Alfred Moeller, who was also the chief of a bacteriological laboratory. His conviction was that bacteria have a close relation to floral nature. After his unsuccessful search for *M. tuberculosis* he found acid-fast bacilli colonising the surface of *Phleum pratense* (timothy) when incubated in distilled water (Moeller, 1898). It was the first species of saprophytic mycobacteria found in the environment, designated as *M. phlei* (Lehmann and Neumann, 1896).

Moeller extended his examinations not only to other plants but also to crops and the excrements of farm animals and found a broad variety of acid-fast rods. Some of them exhibited pleomorphism and true branching would probably place them in the genus *Nocardia* (Moeller, 1899). Nevertheless, his efforts to search for the causative agent of tuberculosis outside of patients and especially the large spectrum of environmental samples he examined for the presence of mycobacteria can be regarded as the beginning of mycobacterial ecology.

1.2.2 Descriptive or "Statistical" Mycobacterial Ecology

The frequent occurrence of non-specific skin reactivity to mycobacterial antigens in animals and new mycobacterioses caused by environmental mycobacteria further stimulated research into the ecology of mycobacteria. Extensive examinations were carried out by Beerwerth (1967) who developed decontamination with oxalic acid and sodium hydroxide, which made it possible to cultivate mycobacteria even from highly contaminated samples like animal droppings. In almost all samples of cattle droppings, mycobacteria could be isolated. This indicated that fodder and watering places can be sources of mycobacteria. In the latter, mycobacteria were isolated from 44% of the samples (Gerle, 1972).

In a further contribution to the ecology of mycobacteria, Beerwerth and Schurmann (1969) examined a total of 3434 samples of soil, wastewater, fodder and faeces of domestic animals and found mycobacteria in 86.1 and 70.3% of arable and meadow soil samples, respectively. Mycobacteria were isolated from 70.5% of samples of faeces of grazing cattle. This took place more frequently during the spring when pastures were full of new shoots rather than later in the year when the grass was longer. Furthermore, 36% of arthropods, living in close contact with soil, were positive for mycobacteria (Beerwerth et al., 1979).

Encouraged by more sophisticated methods for the isolation of mycobacteria (Kubica et al., 1963), numerous further attempts were made to find mycobacteria in different kinds of environments. In Japan, several new species of mycobacteria were found in the soil (Tsukamura, 1967). In Australia, the isolation of *M. avium–intracellulare* (*MAI*) from soil, milk and dust and their influence on mycobacterioses in humans were reported (Kovacz, 1962). In South Africa, during an extensive study, the isolation of environmentally derived saprophytic and potentially pathogenic mycobacteria (PPM) was described; they were found in soil, plants and dust, where *M. intracellulare* was particularly prevalent (Kleeberg and Nel, 1973).

1.2.3 Applied Research in Mycobacterial Ecology

Most studies dealing with the isolation of mycobacteria from the environment were based on a high number of examined samples and a statistical evaluation of positive findings. Very little or no attention was paid to biotic and abiotic factors in examined biotopes. This approach often led to a misinterpretation of results, with the result that the impression was created that mycobacteria were ubiquitous in soil and water.

One of the first successful attempts to introduce modern principles to the ecology of mycobacteria can be ascribed to Chapman (1971). To understand the interaction between humans, animals and the environment, he introduced the term "infection as a result of overlapping niches". He showed that environmental mycobacteria are far from being "fastidious" in their nutritional requirements and in their optimum growth temperature and that a variety of them can thrive within a wide pH range (Chapman and Bernard, 1962). They also defined the differences between the niches occupied by obligate pathogenic *M. tuberculosis* and those occupied by potentially pathogenic, environmentally derived mycobacteria and demonstrated that the latter are able to tolerate a variety of metals.

1.2.4 Physiological Mycobacterial Ecology

Detailed studies carried out by J. O. Falkinham and his team included not only the isolation of mycobacteria from environments; their experiments expanded on the conditions under which isolated mycobacteria can thrive. They proposed the term "physiological ecology" to distinguish these experiments from the simple isolation of mycobacteria from the environment. They demonstrated, therefore, that strains of the *M. avium–intracellulare–scrofulaceum* complex (*MAIS*), isolated from water polluted by heavy metals, tolerated higher concentrations of heavy metal salts and oxyanions (Falkinham et al., 1984). The ability to detoxify rather than metabolise metallic and other compounds thus contributes to the colonisation and persistence of mycobacteria in aquatic environments (Falkinham, 1996). This led to an increased interest in the conditions under which mycobacteria can thrive in specific environments.

1.2.5 Molecular Ecology

Similar to epidemiology, the polymerase chain reaction offers a valuable tool for the detection of mycobacteria in the environment, and the DNA fingerprinting technique enabling the distinction of mycobacteria at the single strain level has enabled great progress in the description of their ecology. It has been possible to "follow the routes" of mycobacterial strains not only between habitats but also to determine what kind of vectors contribute to their spreading.

Other sophisticated methods like restriction fragment length polymorphism (RFLP), pulsed field gel electrophoresis (PFGE), repetitive-unit-sequence-based PCR (Rep-PCR) and amplified fragment length polymorphism (AFLP) have allowed the detailed distinction of mycobacteria according to their geographic distribution (Johnson et al., 2000).

1.2.6 The Ecology of Habitat Environments

To understand the dynamics of the distribution of mycobacteria it was necessary to ascertain what types of environments contribute to their multiplication and/or to the spread of mycobacteria. There are many publications which discuss the isolation of mycobacteria from a variety of different environments, but few describe these biotopes in detail. Furthermore, the examination of such biotopes was very seldom repeated. Such results were not even adequate to distinguish particular biotopes as being either sources or merely vectors of mycobacteria. In addition, the biotic and abiotic factors which enable or exclude the multiplication of mycobacteria in these biotopes must be taken into consideration.

The main biotic factors for heterotrophic mycobacteria include those nutrients that are available or released during the decomposition of complex organic substances or are synthesised by autotrophic micro-organisms. Competition from other organisms, especially those belonging to the family *Enterobacteriaceae*, may result in the rapid consumption of nutrients in heavily polluted aquatic environments. The generation time of *Escherichia coli* is 20 min; it is 4.6 h for *M. fortuitum*. Thus, the "strategy" of mycobacteria results in the colonisation of niches which provide conditions suitable for their multiplication and restrict or exclude other competitive micro-organisms from thriving. Such restrictive factors include the acidity of the environment which benefits the selective multiplication of mycobacteria (Kazda, 2000).

Abiotic factors comprise such physical and chemical conditions as temperature, i.e. the accumulation of solar heat, humidity, pH, adhesion to surfaces (especially tubes) and external capillarity in plants. Such biotopes which harbour favourable conditions for the multiplication and longevity of mycobacteria are regarded as habitat environments for mycobacteria. Sphagnum and bryophyte vegetation belong to this group.

1.3 The Classification of Mycobacteria with Regard to Their Ecology

J. Kazda

In the past, several attempts have been made to divide mycobacteria into groups, first into pathogenic and saprophytic ones and later into anonymous, typical or atypical, opportunist, nontuberculous and others. These designations, useful initially, later brought about confusion due to the different ways in which they were interpreted. If it is generally accepted that the term mycobacteriosis is used to define diseases caused by mycobacteria, then the terms atypical, opportunist or nontuberculous cannot be regarded as appropriate for mycobacterial species causing granulomatous tuberculosis. Thus, it cannot be correct to classify *M. avium* as a nontuberculous *Mycobacterium* because of its ability to cause tuberculosis not only in birds but also in humans.

However, such an immensely large genus as *Mycobacterium*, comprising more than 100 species at present, needs at least a guide for initial orientation. For this purpose, three epithets for dividing mycobacteria are currently proposed.

1.3.1 Obligate Pathogenic Mycobacteria

Obligate pathogenic mycobacteria (OPM) are the most specialised species, causing mycobacteriosis in humans and particular species of animals. They possess a high virulence even in the dormant form for a long time in an infected host. Their survival in the environment is very limited, but they can "descend" to feral animals such as the opossum in New Zealand and the badger in Great Britain and Ireland. The control of tuberculosis in feral animals is very difficult and such animals are the source of *M. bovis* infections in cattle. The airborne transmission of OPM to humans is greatly aided by infected particles, which are less than 5 μ in diameter and small enough to remain suspended in the air over a long period of time. When inhaled, their size allows them to enter the respiratory tract and traverse the ciliated epithelium. Their further development depends on the host–parasite relationship. Possessing high virulence, these infected particles are able to cause disease even with a very limited inoculum.

1.3.2 Potentially Pathogenic Mycobacteria

The main source of potentially pathogenic mycobacteria (PPM) is an environment where they are able to multiply, but they can also be found in living hosts, where they colonise suitable niches on the mucous membranes. They possess two main properties: the ability to multiply in favourable types of environments and to provoke mycobacterioses in susceptible hosts. They comprise therefore a transitional group between OPM and environmental saprophytic mycobacteria. Their number is increasing due to more sophisticated methods used in their differentiation and especially as a result of severe mycobacterioses in HIV/AIDS patients.

1.3.3 Environmental Saprophytic Mycobacteria

Environmental saprophytic mycobacteria (ESM) form the largest group of mycobacteria. They were isolated originally from several kinds of environments, in particular from sphagnum and moss vegetation, surface and drinking water and soil-containing organic material. Due to their large spread, they are often found in clinical material and were for a long time regarded as contaminants. Their association with *M. leprae*, however, has shown that their presence supports the pathogenicity of the leprosy bacillus when inoculated simultaneously into the footpads of nude mice. Some ESM, particularly *M. cookii* and *M. hiberniae*, can provoke a non-specific sensitisation to tuberculin, used as the test for bovine tuberculosis. These species belong to the group of slow-growing mycobacteria, similar to OPM, but are not pathogenic for experimental animals (Kazda and Cook, 1988; Cooney et al., 1997).

References

Alvares E, Tavel E (1885) Recherche sur le bacille de Lustgarten. Archives de Physiologie, Normal et. Pathologique. 6:303–321

Aronson JD (1926) Spontaneous tuberculosis in salt water fish. J. Inf. Dis. 39:315–320

Aronson JD (1929) Spontaneous tuberculosis in snakes. *Mycobacterium thamnopheos* n.sp. J. Inf. Dis. 44:215–223

Bataillon E, Dubart L (1897) Un nouveau type de tuberculose. Compt. Rend. Soc. Biol. 49:446–449

Beavan W, Bayne-Jones S (1931) *Mycobacterium* (sp?) Ryan strain, isolated from pleural exudate. J. Inf. Dis. 49:399–419

Beerwerth W (1967) The culture of mycobacteria from feces of domestic animals and their significance for epidemiology and control of tuberculosis (in German). Praxis Pneumol. 21:189–202

Beerwerth W, Eysing B, Kessel U (1979) Mycobacteria in arthropods of different biotopes (in German). Zentralbl. Bakteriol. [Orig A]. 244:50–57

Beerwerth W, Schurmann J (1969) Contribution to the ecology of mycobacteria (in German). Zbl. Bakt. Parasitenk. Infektionskrankh. Hyg. 1. Abt. Orig. 211:58–69

Bojalil LF (1959) Estudio comparativo entre *Mycobacterium marinum* y *Mycobacterium balnei*. Rev. Latinoamer. Microbiol. 2:169–174

Bonicke R (1962) Present status of methods for the routine differentiation of various species of *Mycobacterium* (in German). Ann. Soc. Belg. Med. Trop. 42:403–439

Bonicke R, Kazda J (1970) The occurrence of carbohydrate nitrite reductases in rapid growing *Mycobacterium* species and their importance for the differentiation of these species (in German). Zentralbl. Bakteriol. [Orig.]. 213:68–81

Bremer K (1854) De legibus ad initium atque progressum tuberculosis spectantis. Doctoral Thesis (Cit.: R. Bochalli: Görbersdorf – 100 Jahre). Tuberkulosearzt. 8:696–697

Buhler VB, Pollak A (1953) Human infection with atypical acid-fast organisms; report of two cases with pathologic findings. Am. J. Clin. Pathol. 23:363–374

Chapman JS (1971) The ecology of the atypical mycobacteria. Arch. Environ. Health. 22:41–46

Chapman JS, Bernard JS (1962) The tolerances of unclassified mycobacteria. I. Limits of pH tolerance. Am. Rev. Respir. Dis. 86:582–583

Cooney R, Kazda J, Quinn J, Cook B, Muller K, Monaghan M (1997) Environmental mycobacteria in Ireland as a source of non-specific sensitization to tuberculin. Irish Vet. J. 50:370–373

Courmont P, Potet M (1903) Les bacilles acido-resistants du beurre, du lait et de la nature, compares au bacille du Koch. Archives de medecine experimentale et d'anatomie pathologique. 15:83–128

Falkinham JO (1996) Epidemiology of infection by nontuberculous mycobacteria. Clinical Microbiology Reviews. 9:177–215

Falkinham JO, George KL, Parker BC, Gruft H (1984) *In vitro* Susceptibility of Human and Environmental Isolates of *Mycobacterium-Avium*, *Mycobacterium-Intracellulare*, and *Mycobacterium-Scrofulaceum* to Heavy-Metal Salts and xyanions. Antimicrob. Agents Chemoth. 25:137–139

Friedmann FF (1903) Spontane tuberkulose bei Schildkröten und die Stellung des Tuberkelbacillus im System. Mit einer Übersicht über die Lehre von der Kaltblütlertuberkulose. Ztschr. Tuberk. Leipzig. 4:439–457

Galli-Valerio B, Bornand M (1927) Le *Mycobacterium aquae* Galli-Valerio et son action pathogene. Zentralblatt für Bakteriologie, Parasitenkunde, Infektionskrankheiten und Hygiene, I. Abt. Orig. 101:182–193

Gerle H (1972) Über das Vorkommen atypischer Mykobakterien in Viehtränken, Inaug. Diss. (Freie Univ. Berlin)

Gordon ER, Smith MM (1953) Rapidly growing acid-fast mycobacteria. I. Species description of *Mycobacterium phlei* Lehmann and Neumann and *Mycobacterium smegmatis* (Trevisan) Lehmann and Neumann. J. Bact. 66:41–48

Haag A (1927) Tarnok I.: Soffwechsel in Mykobakterien. In: G. Meissner, A. Schmiedel, A. Nelles, R. Pfaffenberg (Eds.), Mykobakterien und mykobakterielle Krankheiten. Vol. I. 41–244, VEB Gustav Fischer Verlag, Jena, 1980, 377 s.

Hansen GHA (1875) On the aetiology of leprosy. British and Foreign Medico-Chirurgical Review. 55:459–489

Hauduroy P (1955) Derniers aspects du monde des mycobacteries, Mansson et Cie., Paris.

Hellerstrom S (1952) Water-borne tuberculous and similar infections of the skin in swimming pools. Acta Derm. Venereol. 32:449–461

Johne HA, Frothingham L (1895) Ein eigentheumlicher Fall von Tuberculose beim Rind. Dtsch. Ztschr. Tier-Med. 21:438–454

Johnson P, Stinear T, Portaels F, Chamal K, Dubos K, King H (2000) Modern diagnostic methods. In: K. Asieda, R. Sherpbier, M. Raviglione (Eds.), Buruli ulcer. WHO, Geneva

Kappler W (1965) Zur Differenzierung von Mykobakterium mit dem Phosphatase-Test. Beitr. Klin. Tuberk. Spezif. Tuberk. Forsch. 130:223–226

Kazda J (2000) The ecology of mycobacteria. Kluwer Academic Publishers, Dordrecht, Boston, London, 72 pp

Kazda J, Cook BR (1988) Mycobacteria in pond waters as a source of non-specific reactions to bovine tuberculin in New Zealand. N. Z. Vet. J. 36:184–188

Kleeberg HH, Nel EE (1973) Occurrence of environmental atypical mycobacteria in South Africa. Ann. Soc. Belg. Med. Trop. 53:405–418

Koch R (1882) Die Aetiologie der Tuberculose. Berliner Klinische Wochenschrift. 18:221–238

Konno K, Kurzmann R, Bird KT, Sbarra A (1958) Differentiation of human tubercle bacilli from atypical acid-fast bacilli. I. Niacin production of human tubercle bacilli and atypical acid-fast bacilli. Am. Rev. Tuberc. 77:669–680

Kovacz N (1962) Nichtklassifizierte Mykobakterien. Zentralbl. Bakteriol. Parasitenkunde und Infektionskrankheiten. Hyg. I Abt. Orig. 184:46–56

Kubica GP, Beam RE, Palmer JW (1963) A method for the isolation of unclassified acid-fast bacilli from soil and water. Am. Rev. Respir. Dis. 88:718–720

Kubica GP, Vestal AL (1961) The arylsulfatase activity of acid-fast bacilli. I. Investigation of activity of stock cultures of acid-fast bacilli. Am. Rev. Respir. Dis. 83:728–732

Lehmann KB, Neumann R (1896) Atlas und Grundriss der Bakteriologie und Lehrbuch der speziellen bakteriologischen Diagnostik. 1st ed. J.F. Lehmann, Munchen

Linell F, Norden A (1954) *Mycobacterium balnei*. A new acid-fast bacillus occurring in swimming-pools and capable of producing skin lesions in humans. Acta Tub. Scand. 31: Suppl. 33:1–84

Marchaux E, Sorel F (1912) Recherche sur la leprae. Ann. Inst. Pasteur. 26:675–700

Masson AM, Prissick FH (1956) Cervical lymphadenitis in children caused by chromogenic Mycobacteria. Can. Med. Assoc. J. 75:798–803

Moeller A (1898) Über dem Tuberkelbacillus verwandte Mikroorganismen. Ther. Monatshefte. 12:607–613

Moeller A (1899) Ein neuer säure-und alkoholfester Bacillus aus der Tuberkelbacillengruppe, welcher echte Verzweigungsformen bildet. Beitrag zur Pleomorphie der Bakterien. Zentralbl. Bakteriol. Parasitenkunde und Infektionskrankheiten. 25:369–373

Neelsen F (1883) Ein casuistischer Beitrag zur Lehre von der Tuberkulose. Zentralblatt für medizinische Wissenschaften. 28:497–501

Negre L (1939) Caracteres distinctifs du bacille aviare et role de ce germe dans l'infection tuberculeuse de l'homme. Proc. 9th Congres Nationale de la Tuberculose, Lille. 1–26

Rabinowitsch KL (1897) Zur Frage des Vorkommens Tuberkelbacillen in Marktbutter. Zeitschrift für Hygiene und Infektionskrankheiten. 26:90–111

Runyon EH (1959) Anonymous mycobacteria in pulmonary disease. Med. Clin. North Am. 43:273–290

Runyon EH (1965) Pathogenic mycobacteria. Adv Tuberc. Res. 14:235–287

Shepard CC (1960) The experimental disease that follows the injection of human leprosy bacilli into foot-pads of mice. J. Exp. Med. 112:445–454

Stackebrandt E, Woese CR (1981) Towards a phylogeny of the actinomycetes and related organisms. Curr. Microbiol. 5:197–202

Strandberg J (1937) A case of pseudo-primary tuberculous complex of the skin. Acta Derm. Venerol. 18:610–621

Tortoli E (2006) The new mycobacteria: an update. Fems Immunol. Med. Microbiol. 48:159–178

Tsukamura M (1967) Two types of slowly growing, nonchromogenic mycobacteria obtained from soil by the mouse passage method: *Mycobacterium terrae* and *Mycobacterium novum*. Jap. J. Microbiol. 11:163–172

Virtanen S (1960) A study of nitrate reduction by mycobacteria. The use of the nitrate reduction test in the identification of mycobacteria. Acta Tuberc. Scand. Suppl. 48:1–119

Wayne LG (1962) Differentiation of mycobacteria by their effect on Tween 80. Am. Rev. Respir. Dis. 86:579–581

Wayne LG, Good RC, Bottger EC, Butler R, Dorsch M, Ezaki T, Gross W, Jonas V, Kilburn J, Kirschner P, Krichevsky MI, Ridell M, Shinnick TM, Springer B, Stackebrandt E, Tarnok I, Tarnok Z, Tasaka H, Vincent V, Warren NG, Knott CA, Johnson R (1996) Semantide- and chemotaxonomy-based analyses of some problematic phenotypic clusters of slowly growing mycobacteria, a cooperative study of the International Working Group on Mycobacterial Taxonomy. Int. J. Syst. Bacteriol. 46:280–297

Ziehl F (1882) Zur Färbung des Tuberkelbacillus. Deutsche Medizinische Wochenschrift. 8:451–452

Chapter 2

Obligate Pathogenic Mycobacteria

J. Kazda and I. Pavlik (Eds.)

Introduction

Obligate pathogenic mycobacteria (OPM), above all *Mycobacterium tuberculosis* complex *(MTC)* species, are most often spread by transmission between a variety of animals and humans. From an aspect of epidemiology, it is important that they can survive outside the host organism for a long time; this "compensates" for their limited ability to multiply outside the host organism. Even though they do not sporulate, they can still be cultured from a damp environment protected from direct sunlight after the lapse of several months or years.

2.1 *M. tuberculosis* Complex Members

I. Pavlik

At present, eight mycobacterial species are classified as *MTC*, seven of which are obligatory pathogenic mycobacteria; the remaining eighth species is *M. bovis* BCG used for vaccination (van Soolingen et al., 1997; Aranaz et al., 2003; Cousins et al., 2003):

M. tuberculosis: causative agent of human tuberculosis throughout the world (Photo 2.1); infections in animals occasional (Photo 2.2).

J. Kazda (✉)
University of Kiel, Kiel, Germany
J.Kazda@t-online.de

M. africanum: causative agent of human tuberculosis, above all in the inhabitants of Western Africa; infections in animals rare.

M. canettii: causative agent of infections; diagnosed occasionally in people; this species has not been detected in animals yet.

M. bovis: both animals and people are susceptible hosts (Photos 2.3 and 2.4).

M. caprae: the same host spectrum as for *M. bovis* (Photo 2.5).

M. microti: causative agent of tuberculosis in small terrestrial mammals; occasionally found in other animals and people (Photo 2.6).

M. pinnipedii: causative agent of tuberculosis in pinnipeds; transmissible to people, especially from infected animals kept in captivity (Photos 2.7, 6.38 and 6.39).

M. bovis BCG (vaccine strain): occasionally causes post-vaccination infections in immunosuppressed patients; the strain was used in studies of disinfection effectiveness (Omidbakhsh and Sattar, 2006) or survival of mycobacteria in the environment (Vandonsel and Larkin, 1977).

Multiplication outside the host organisms has not been described for any of these eight *MTC* species. These species are most often shed by infected hosts to the environment relative to the localization of infection; it is usually sputum, faeces and urine in the case of humans, milk from dairy animals (especially cattle) and infected body tissues from other domestic and wild animals. The latter, above all from wild animals, become a part of the food chain for other animals or undergo decomposition on the surface of the earth (e.g. wild ruminants), underground (e.g. small terrestrial mammals) or in water (e.g. pinnipeds).

J. Kazda et al. (eds.), *The Ecology of Mycobacteria: Impact on Animal's and Human's Health*,
DOI 10.1007/978-1-4020-9413-2_2, © Springer Science+Business Media B.V. 2009

Accordingly, information regarding seven obligatory pathogenic species of *MTC* is divided into Sections 2.1.1, 2.1.2 and 2.1.3 with regard to their main hosts.

2.1.1 M. tuberculosis, M. africanum and M. canettii

The primary hosts of these three mycobacterial species are people and human-to-human transmission via contaminated aerosol is viewed as the most important (Section 5.9). However, these species can also be shed into the environment through faeces, urine and other excretions or secretions (Section 5.10). From the point of view of ecology, the most important source of these three mycobacterial species is contaminated sewage water; this was mainly investigated in Europe (in Poland and Germany) in the middle of the 20th century. These studies found that *M. tuberculosis* was occasionally detected in sewage water from cities and hospitals (Section 5.2.7).

The available literature provides little information on the occurrence of these three mycobacterial species in sewage water in developing countries with a high occurrence of these infections in humans. *M. africanum* has been isolated from sewage water from hospitals (Nguematcha and Le, 1978); however, other information on the occurrence of *M. africanum* in the environment is almost non-existent. Accordingly, further research in this area should be focused on the investigation of the environment in different niches with a high prevalence and incidence of human infections caused by the above-mentioned mycobacterial species. In-depth knowledge of the circulation of these mycobacterial species in the environment will definitely help to extend control measures in affected countries.

2.1.2 M. bovis and M. caprae

These two species cause tuberculosis in different organs in their hosts, especially in the lungs (Grange and Yates, 1994; Thoen and Steele, 1995; Grange, 1996; Thoen et al., 2006a). The most common sources of infection are ill animals. The intensity of pathogen shedding via secretions or excretions varies depending on the stage of infection; according to the localization of infection, they are shed mainly by sputum, faeces, uterine excretions, urine, semen, etc. Raw milk and milk products from dairy animals (primarily cattle) are important sources of infection for people. The host organism is usually infected by direct contact with an infected animal. Accordingly, at present, tuberculosis caused by these mycobacterial species and paratuberculosis belong to the most serious chronic diseases in animals held in captivity at high densities (e.g. on farms or in zoological gardens).

The sources of infection are usually purchased animals that are in an early stage of the disease and show negative results to intravital diagnostics (especially bovine tuberculin testing). Another risk factor for the spread of causative agents of these diseases is the presence of reservoir animals such as the European badger (*Meles meles*) in Great Britain and Ireland and the brushtail possum (*Trichosurus vulpecula*) in New Zealand. This factor is generally acknowledged and detailed information can be obtained in the recently published books mentioned above (Grange and Yates, 1994; Thoen and Steele, 1995; Grange, 1996; Thoen et al., 2006b). However, it is not the purpose of this book to perform a detailed analysis of all the information but rather to highlight the ecology of these mycobacterial species.

Various components of the environment can become sources of these causative agents of diseases in animals and humans as well. *M. bovis* was detected in 5 (42%) of 12 samples of manure and in 1 (6%) of 18 samples of scrapings from stable walls (Shindler, 1979). Faeces and urine from infected animals on pastures may be sources of contamination of not only soil but also surface water. Kislenko (1972) reported the survival of *M. bovis* in pasture soil from Novosibirsk in Russia for 23 months (entire monitoring period). However, the time of survival of *M. bovis* in soil was shorter in areas of dry climate and high summer temperatures: only 4 weeks in Australia (Duffield and Young, 1985). *M. bovis* has been detected in drinking water for calves in the areas with bovine tuberculosis occurrence in Great Britain (Little et al., 1982).

It is also well known that *M. bovis* can survive in animal carcasses (Thoen et al., 2006b). However, little is known about the risk of *M. bovis* transmission via vegetation on pastures grazed by infected ruminants and studies focused on this topic are scarce. These risks were highlighted by Kislenko (1972). Kislenko performed experiments on guinea pigs, rabbits and cattle fed with mown grass that originated from pastures contaminated with *M. bovis*, either naturally or artificially (the pasture soil was contaminated artificially with *M. bovis* seven months before the grass was fed to animals). Accordingly, the sources of bovine tuberculosis for different animal species in the endemic regions may be various components of the environment, to which little attention has been paid so far.

The importance of contaminated feedstuffs for the transmission of *M. bovis* between different white-tailed deer animals (*Odocoileus virginianus*) and between white-tailed deer and cattle has been demonstrated in the United States (Palmer et al., 2004a,b). However, considering the ecology of mycobacteria, the question concerning the ability of *M. bovis* to survive in different components of such feeds and its potential ability to multiply remains open. It is well known that if processed fodder (especially green fodder) is left standing for a long time, its temperature may rise and fermentation may occur. The temperatures in such fodder do not usually reach values that would reliably kill mycobacteria. On the contrary, the temperatures reached on the margins of the stacks are often about 37 °C (I. Pavlik and P. Miskovic, unpublished data), which can under particular conditions stimulate the growth of *M. bovis*.

2.1.3 M. microti and M. pinnipedii

Information on the quality and occurrence of these two causative agents of both animal and human diseases can be drawn from many review articles (Thoen et al., 2006b; Skoric et al., 2007). The hosts of *M. microti* are especially small terrestrial mammals, their predators and other animals and humans (Section 6.7). It was observed as early as the middle of the last century that insectivores and small rodents can encounter *M. microti* via the consumption of food of vegetable and animal origin (Chitty, 1954). The occurrence of *M. microti* in the environment and its ability to survive in different components of the environment remains obscure.

A comparable situation exists for *M. pinnipedii*. This species causes tuberculosis in water mammals and their breeders in zoological gardens and aquatic parks (Thoen et al., 2006b). It can be supposed that the most important transmission route among animals is direct contact with one another. They gather together at the time of rest, periods of mating and parental care. However, the occurrence and ecology of *M. pinnipedii* in the environment also remains obscure.

2.2 *M. leprae:* Obligate or Potentially Pathogenic *Mycobacterium*?

J. Kazda

2.2.1 General Characteristic

The causative agent of leprosy *M. leprae*, was discovered during 1873 in Norway by Armauer Hansen (1875). At that time, leprosy was not restricted to tropical and subtropical countries but also occurred epidemically in Norway. The number of leprosy cases (very high in the tropics during the last century) decreased with the introduction of multi drug therapy (MDT) from 5.2 million in 1985 to 805 000 in 1995 and to 286 000 at the end of 2004 (WHO, 2005).

The target of eliminating leprosy as a public health problem was defined as a prevalence rate of less than 1 case per 10 000 by the year 2000. This target could not be reached worldwide. In Africa, America, South-East Asia, Eastern Mediterranean and Western Pacific, the registered number of leprosy cases stood at 219 826 cases in 2006. Recently, the effectiveness of the WHO programme of leprosy elimination has been questioned (Fine, 2007), because reduction of leprosy cases to 1 per 10 000 population still leaves a region of high population density, such as India, with more than 100 000 leprosy cases. This cannot be regarded as "elimination". The real eradication of leprosy remains an urgent target in the future.

2.2.2 Ecological Aspects

The "traditional" opinion that the sole source of leprosy infection is an untreated patient cannot be generally accepted. Patients treated with MDT are no longer infectious, but the number of new registered cases is higher than the prevalence. The first hypothesis about the possible role of environmental factors in the spread of leprosy is nearly as old as the discovery of the leprosy bacillus itself. Hansen and Looft (1895) discussed in their book the possibility of the leprosy bacillus occurring outside the human body:

> "Here in Norway where the people often go barefoot, wading in streams and rivers, the back of the feet and the under part of the calves are frequently the seat of the first leprous eruption, not so often in the form of nodules, as of a dense, regular infiltration. ... although as we stated above, the leprosy bacillus has never been found outside of the human body, this might possibly be dependent on insufficient search."

In fact, the first-known examination for mycobacteria in the environment dates back to 1898 (Moeller, 1898).

At almost the same time, an interesting observation was made in a leprosarium in Reitgjaerdet (Norway) by Sand (1910). In his survey of 1221 leprosy patients, he found that the risk of leprosy was higher in men because they have traditionally more environmental contact than women. In contrast, the transmission of leprosy within families was found to be very low. Of 512 married couples, the transmission occurred in only 3.3% of cases. The transmission to children by a leprous father or mother was noted as occurring in 4.9 and 10.5% of cases, respectively. Furthermore, over the 40 years of his investigation, a transmission of leprosy neither within the leprosarium nor between neighbours occurred. The author concluded that the transmission of leprosy did not generally take place directly between humans but indirectly through a medium. He continued questioning as to what kind of medium was needed for the transmission, whether it was a living organism (parasite) or ground containing decomposing material.

The high occurrence of leprosy in the second half of the 19th and the beginning of the 20th century among the population living on isolated farms in coastal Norway gave rise to the question as to how much the environment could influence the epidemiology of this disease. The National Leprosy Registry of Norway documented those farms where leprosy patients were living. The undisturbed environments with deserted houses and a water supply from that time can even be found (Photo 2.8).

During the 1970s and 1980s of the last century, several expeditions were carried out through coastal Norway including the Naustdal district, formerly with the highest leprosy prevalence and incidence rates (Irgens, 1980). To test whether sphagnum moss vegetation could be linked with the occurrence of leprosy in man, the leprosy status of farms was compared with regard to seven environmental variables. The most critical point was the origin of water supply in sphagnum bogs or other close contact to this vegetation at the time of the 1851–1885 leprosy epidemic. These conditions were found to considerably enhance the risk of leprosy. In the same district, leprosy occurred only on farms situated under the southern slopes, characterized by a high accumulation of solar heat beneath the surface of sphagnum vegetation, over 32 °C for a long time, which can enable the growth of mycobacteria. Another important factor was high humidity in summer months. The average incidence rate of leprosy in districts with a relative humidity over 75% in July was 12.4 compared with 0.7 in another district and the ratio of risk was 17.7.

In the former leprosy-endemic coastal areas of Norway, samples of sphagnum and other moss vegetation were collected and inoculated into footpads of mice, at that time (1976) the only suitable method for finding *M. leprae* in the environment. It was possible to examine 759 footpads, originating from 122 samples of which 20.9% contained non-cultivable acid-fast bacilli (NC AFB). These NC AFB continuously multiplied by factors of up to 10^6 in further footpads, although attempts to culture them on media for mycobacteria failed (Photo 2.9).

The NC AFB were positive in the dopa test with a maximum absorption between 480 and 530 nm and could be destained by pyridine, properties shown by *M. leprae*. After inoculation into nine-banded armadillos, antibody production against antigen 7, the presence of AFB in macrophages and pathological findings were similar to those of *M. leprae* (Kazda, 1981). In attempts to identify NC AFB, further techniques have been used: *M. leprae*-specific monoclonal antibodies against the phenolic glycolipid-I (PGL-I) in the indirect immunofluorescence technique (Kolk et al., 1985). It was found that the NC AFB contained PGL-I on their surface, a property characteristic of *M. leprae* (Kazda et al., 1990). In a third study, the polymerase chain reaction technique modified for the testing of samples containing humic acid was used. The results confirmed that the NC AFB present in sphagnum possess the same fragment of the gene encoding superoxide dismutase as *M. leprae* (Mostafa et al., 1995).

The examination of environmental sources of *M. leprae* continued in a recent leprosy-endemic region. In the urban district of Bombay, samples of drinking and surface water and soil were collected and tested for cultivable and non-cultivable AFB. In samples originating from wet soil in a house washing area (Photo 2.10), the footpad technique repeatedly revealed NC AFB, which multiplied in the footpads of white and nude mice causing swelling in the latter. Tests for dopaoxidase and pyridine decolouration were positive. The bacilli contained *M. leprae*-specific phenolic glycolipid I. Biological tests (nerve involvement in nude mice and systemic leprosy in one infected, nine-banded armadillo) were also positive (Kazda et al., 1986).

The isolation of environmentally derived *M. leprae* together with *M. intracellulare* of serotype 19 from the same biotope in Bombay raised the question of whether the accompanying microorganisms could play any role in the development of leprosy. This was supported by the fact that other mycobacteria were frequently isolated from human leproma. Many of these mycobacteria belonged to the *M. avium-intracellulare-scrofulaceum* complex (David, 1984). Furthermore, some authors regarded the "accompanying mycobacteria" as an aetiological factor in leprosy (Kato, 1985). Accordingly, the ecological approach to pathogenicity includes the behaviour of pathogenic species in niches shared with other mycobacteria. Few experiments have been performed until now in which the possible effect of a mixed infection with pathogenic and non-pathogenic mycobacteria has been studied.

In our experiments, *M. leprae* (armadillo-derived) and isolated *M. intracellulare* serotype 19 were inoculated simultaneously into footpads of nude mice. Other nude mice were infected with *M. leprae* and *M. intracellulare* of serotype 19 separately and served as controls (Kazda et al., 1987). It was found that the non-pathogenic *M. intracellulare* of serotype 19 considerably enhanced the pathogenicity of the leprosy bacillus. This supporting effect was demonstrated by an acceleration of footpad swelling beginning just 4 months after inoculation and additionally by the development of cutaneous leproma on dorsal and lateral body sites of nude mice within 6 months. These leproma increased in number and size during the 9 months

they were under observation (Photo 2.11). Their micromorphological characteristics were similar to those of human leproma. In animals inoculated with *M. leprae* alone, a swelling of the inoculated footpads was first observed 12 months after infection. Cutaneous leproma did not develop.

This corresponds with the findings of Lancaster et al. (1984), who described the course of leprosy infection in nude mice. Macroscopic visible cutaneous leproma have not been described in association with the footpad inoculation of *M. leprae* into nude mice and the swelling of footpads has not been previously seen until 9–12 months after inoculation. *M. intracellulare* of serotype 19 inoculated alone neither provoked lesions in nude mice nor showed pathogenic properties when inoculated into rabbits and guinea pigs and can be regarded as non-pathogenic. Thus, this *Mycobacterium*, found together with *M. leprae* in the environment, enhanced pathogenicity in nude mice.

In contrast to past and recent findings, research into the mode of leprosy transmission is still focused on the patient as the sole source of *M. leprae,* although this has often been questioned. It is well known that even in highly endemic areas, contact with a leprosy patient cannot be regarded as the sole source of infection. In a study covering Indonesia, the Philippines, Hawaii and several countries in Africa, no contact could be established in 30–60% of new leprosy cases (Arnold and Fasal, 1973). In the mainland United States, only 25.8% of new detected leprosy cases had had any known contact with a leprosy patient (Enna et al., 1978). As a result of work done by Blake et al. (1987), evidence that environmental non-human sources are critical to human infection with *M. leprae* has been accumulating. These sources include soil, vegetation, water, arthropods and nine-banded armadillos. Recently, a leprosy infection of a woman in Georgia whose only known exposure was to armadillos was described (Lane et al., 2006).

It is generally claimed that the transmission of *M. leprae* takes place in an airborne manner, similar to human tuberculosis, and additionally by direct skin contact from mother to child. However, the presence of viable leprosy bacilli in the environment of endemic regions and systemic leprosy in feral nine-banded armadillos suggests that the leprosy patient is not the sole source of infection and that *M. leprae* can be regarded as a potentially pathogenic *Mycobacterium*. One of the environmental sources is probably polluted surface water, used as drinking water in the tropics, and particularly the contamination of clothes by washing in streams (Photo 2.12). Linen and underwear, which do not properly dry in the humid tropical climate, are used damp. *M. leprae*, if present in surface water, gains access to the skin of humans and to their dwellings. In Indonesia, the use of *M. leprae*-specific probes showed that the prevalence of leprosy among individuals that used contaminated water sources for bathing and washing clothes or dishes was significantly higher than that among the population using water free of *M. leprae* (Matsuoka et al., 1999).

In a recent study aimed at identifying socioeconomic, environmental and behavioural factors associated with the risk of leprosy in an endemic region of Brazil, it has been found that bathing weekly in open water bodies and infrequent changing of bed linen or hammock are significantly associated with a high risk of leprosy. These results support the possibility of an indirect transmission of leprosy and indicate that other reservoirs of *M. leprae* should exist outside of the human body (Kerr-Pontes et al., 2006).

2.2.3 Trends

The ecology of *M. leprae* and PPM will gain more and more importance. In the first edition of this book, it has been shown that the leprosy prevalence in 28 endemic countries from 1985 to 1996 decreased by 78%, but the trend in newly detected cases remained stable (Kazda, 2000). From the WHO Fact Sheet (WHO, 2005) the considerable increase in globally registered new cases of leprosy is evident. A decrease was registered in India between 1993 and 2005; however, the number of new leprosy cases in the other 16 countries reporting 1000 or more cases during 2005 increased by about 20% (from 99 307 in 1993 to 118 207 in 2005). Most of the registered leprosy patients received the multi-drug therapy, which is very effective and prevents interhuman transmission of leprosy in a very short time. Thus, the question of how new leprosy patients could catch this disease arises. The explanations provided by improved case findings, the expansion of health services, an increase in populations at risk or "leprosy

pockets" alone, cannot be sufficient. More likely, other reasons including possible environmental sources of infection must be seriously taken into consideration.

The ecology of *M. leprae*, the first steps of which go back as far as the discovery of this bacillus, must be one of the most important objectives in leprosy research in future. A sophisticated molecular genetic method must be developed and applied to find further sources of *M. leprae* in the environment and to describe more precisely the mechanisms of transmission to humans. The prevention of leprosy, until now somewhat neglected, should be combined with an improvement in social conditions. Successful therapy alone cannot eliminate the stigma of leprosy.

Acknowledgements Section 2.1 was partially supported by grant NPV 1B53009 (Ministry of Agriculture of the Czech Republic).

References

Aranaz A, Cousins D, Mateos A, Dominguez L (2003) Elevation of *Mycobacterium tuberculosis* subsp. *caprae* Aranaz et al. 1999 to species rank as *Mycobacterium caprae* comb. nov., sp. nov. Int. J. Syst. Evol. Microbiol. 53:1785–1789

Arnold HH, Fasal P (1973) Leprosy, diagnosis and management. Charles C. Thomas, Springfield, 2nd Edition

Blake LA, West BC, Lary CH, Todd JR (1987) Environmental nonhuman sources of leprosy. Rev. Infect. Dis. 9:562–577

Chitty D (1954) Tuberculosis Among Wild Voles – with A Discussion of Other Pathological Conditions Among Certain Mammals and Birds. Ecology. 35:227–237

Cousins DV, Bastida R, Cataldi A, Quse V, Redrobe S, Dow S, Duignan P, Murray A, Dupont C, Ahmed N, Collins DM, Butler WR, Dawson D, Rodriguez D, Loureiro J, Romano MI, Alito A, Zumarraga M, Bernardelli A (2003) Tuberculosis in seals caused by a novel member of the *Mycobacterium tuberculosis* complex: *Mycobacterium pinnipedii* sp nov. Int. J. Syst. Evol. Microbiol. 53:1305–1314

David HL (1984) Classification and identification of *Mycobacterium leprae*. Acta Leprol. 2:137–151

Duffield BJ, Young DA (1985) Survival of *Mycobacterium-Bovis* in Defined Environmental-Conditions. Vet. Microbiol. 10:193–197

Enna CD, Jackson RR, Trautman JR, Sturdivant M (1978) Leprosy in the United States, 1967–76. Public Health Rep. 93:468–473

Fine PE (2007) Leprosy: what is being "eliminated"? Bull. World Health Organ. 85:1–84

Grange JM (1996) Mycobacteria and human disease. 2nd ed. London, Arnold, 230 pp

Grange JM, Yates MD (1994) Zoonotic Aspects of *Mycobacterium-Bovis* Infection. Vet. Microbiol. 40: 137–151

Hansen GA, Looft C (1895) Leprosy in its Clinical and Pathological Aspects. Translated by Norman Walker, Bristol, John Wright & Co., London, 1973.

Hansen GHA (1875) On the aetiology of leprosy. British and Foreign Medico-Chirurgical Review. 55:459–489 (cit. Nathan et al., 1990)

Irgens LM (1980) Leprosy in Norway. An epidemiological study based on a national patient registry. Lepr. Rev. 51(Suppl 1):1–130

Kato L (1985) Absence of mycobactin in *Mycobacterium leprae*; probably a microbe dependent microorganism implications. Indian J. Lepr. 57:58–70

Kazda J (1981) Occurrence of non-cultivable acid-fast bacilli in the environment and their relationship to *M. leprae*. Lepr. Rev. 52(Suppl 1):85–91

Kazda J (2000) The ecology of mycobacteria. Kluwer Academic Publishers, Dordrecht, Boston, London, 72 pp

Kazda J, Fasske E, Kolk A, Ganapati R, Schroder KH (1987) The simultaneous inoculation of *Mycobacterium leprae* and *M. intracellulare* into nude mice: development of cutaneous leproma and acceleration of foot pad swelling. Indian J. Lepr. 59:426–434

Kazda J, Ganapati R, Revankar C, Buchanan TM, Young DB, Irgens LM (1986) Isolation of environment-derived *Mycobacterium leprae* from soil in Bombay. Lepr. Rev. 57(Suppl 3):201–208

Kazda J, Irgens LM, Kolk AH (1990) Acid-fast bacilli found in sphagnum vegetation of coastal Norway containing *Mycobacterium leprae*-specific phenolic glycolipid-I. Int. J Lepr. Other Mycobact. Dis. 58:353–357

Kerr-Pontes LRS, Barreto ML, Evangelista CMN, Rodrigues LC, Heukelbach J, Feldmeier H (2006) Socioeconomic, environmental, and behavioural risk factors for leprosy in Northeast Brazil: results of a case-control study. Int. J. Epidemiol. 35:994–1000

Kislenko VN (1972) [Survival of bovine tuberculosis mycobacteria in pasture soils]. Veterinariia. 48:48–51

Kolk AH, Ho ML, Klatser PR, Eggelte TA, Portaels F (1985) Production of monoclonal antibodies against *Mycobacterium leprae* and armadillo-derived mycobacteria. Ann. Inst. Pasteur Microbiol. 136B:217–224

Lancaster RD, McDougall AC, Hilson GR, Colston MJ (1984) Leprosy in the nude mouse. Exp. Cell Biol. 52:154–157

Lane JE, Balagon MV, Dela Cruz EC, Abalos RM, Tan EV, Cellona RV, Sadaya PG, Walsh GP, Walsh DS (2006) *Mycobacterium leprae* in untreated lepromatous leprosy: more than skin deep. Clin. Exp. Dermatol. 31:469–470

Little TW, Naylor PF, Wilesmith JW (1982) Laboratory study of *Mycobacterium bovis* infection in badgers and calves. Vet. Rec. 111:550–557

Matsuoka M, Izumi S, Budiawan T, Nakata N, Saeki K (1999) *Mycobacterium leprae* DNA in daily using water as a possible source of leprosy infection. Indian J. Lepr. 71:61–67

Moeller A (1898) Über dem Tuberkelbacillus verwandte Mikroorganismen. Ther. Monatshefte. 12:607–613

Mostafa HM, Kazda J, Irgens LM, Luesse HG (1995) Acid-fast bacilli from former leprosy regions in coastal Norway showing PCR positivity for *Mycobacterium leprae*. Int. J. Lepr. Other Mycobact. Dis. 63:97–99

Nathan C, Squires K, Griffo W, Levis W, Varghese M, Job CK, Nusrat AR, Sherwin S, Rappoport S, Sanchez E (1990) Widespread intradermal accumulation of mononuclear leukocytes in lepromatous leprosy patients treated systemically with recombinant interferon gamma. J. Exp. Med. 172:1509–1512

Nguematcha R, Le NP (1978) [Detection of pathogenic mycobacteria in the environment of the medical units and of the slaughter-house of an African town (author's transl)]. Med. Trop. (Mars.). 38:59–63

Omidbakhsh N, Sattar SA (2006) Broad-spectrum microbicidal activity, toxicologic assessment, and materials compatibility of a new generation of accelerated hydrogen peroxide-based environmental surface disinfectant. Am. J. Infect. Control. 34:251–257

Palmer MV, Waters WR, Whipple DL (2004a) Investigation of the transmission of *Mycobacterium bovis* from deer to cattle through indirect contact. Am. J. Vet. Res. 65:1483–1489

Palmer MV, Waters WR, Whipple DL (2004b) Shared feed as a means of deer-to-deer transmission of *Mycobacterium bovis*. J. Wildl. Dis. 40:87–91

Sand A (1910) Geschiet die Ansteckung der Lepra durch unmittelbare Übertragung? Mitteilungen und Verhandlungen der 2. Leprakonferenz. G. Fischer, Leipzig. 39–46

Shindler EM (1979) [Epidemiological importance of bovine tuberculosis for rural inhabitants]. Probl. Tuberk. 12–15

Skoric M, Shitaye EJ, Halouzka R, Fictum P, Trcka I, Heroldova M, Tkadlec E, Pavlik I (2007) Tuberculous and tuberculoid lesions in free living small terrestrial mammals and the risk of infection to humans and animals: a review. Veterinarni Medicina 52:144–161

Thoen C, Lobue P, de Kantor I (2006a) The importance of *Mycobacterium bovis* as a zoonosis. Vet. Microbiol. 112:339–345

Thoen CO, Steele JH (1995) *Mycobacterium bovis* infection in animals and humans. Iowa State University Press, 1st ed., 355 pp

Thoen CO, Steele JH, Gilsdorf MJ (2006b) *Mycobacterium bovis* infection in animals and humans. 2nd ed., Blackwell Publishing Professional, Ames, Iowa, USA, 317 pp

van Soolingen D, Hoogenboezem T, de Haas PE, Hermans PW, Koedam MA, Teppema KS, Brennan PJ, Besra GS, Portaels F, Top J, Schouls LM, van Embden JD (1997) A novel pathogenic taxon of the *Mycobacterium tuberculosis* complex, Canetti: characterization of an exceptional isolate from Africa. Int. J. Syst. Bacteriol. 47:1236–1245

Vandonsel DJ, Larkin EP (1977) Persistence of *Mycobacterium-Bovis* BCG in Soil and on Vegetables Spray-Irrigated with Sewage Effluent and Sludge. J. Food Prot. 40:160–163

WHO (2005) Leprosy. Fact Sheet No.101. http://www.who.int/mediacentre/factsheets/fs101/en/index.html

Chapter 3

Potentially Pathogenic Mycobacteria

I. Pavlik, J.O. Falkinham III and J. Kazda (Eds.)

Introduction

Beginning in the middle of the last century, a decreasing incidence of classical tuberculosis was noted in economically developed countries and was attributed to improved nutrition, the consistent use of extensive screening and tracking methods, the introduction of reliable diagnostic methods and the effective antibacterial treatment of tuberculosis. In populations of cattle, national control programmes were adopted, which resulted in the gradual reduction of *Mycobacterium bovis* infection sources and their elimination in many countries. Meat inspection in abattoirs and the obligatory pasteurisation of cow milk during the same period resulted in a substantial decrease or in the elimination of bovine tuberculosis from the human population (Thoen and Steele, 1995; Grange, 1996; Thoen et al., 2006).

3.1 Potentially Pathogenic, Slowly Growing Mycobacteria

I. Pavlik, J. Kaustova, J.O. Falkinham III, J. Kazda, and J.E. Shitaye

The decreased occurrence of infections caused by obligate pathogenic mycobacterial species (e.g.

M. tuberculosis) was complemented by the rise in infections caused by potentially pathogenic mycobacteria (Kazda, 2000). The following factors contributed to this situation:

- The increased prevalence of diseases in humans which decreases their immunocompetence (e.g. HIV/AIDS and cancer).
- The increased use of chemotherapeutic drugs for cancer treatment that result in immunosuppression.
- Increased recognition of the fact that ESM and PPM are important human and animal pathogens.
- Improved techniques for the isolation and identification of ESM and PPM.
- The aging of the human population in the developed world with the accompanying increased susceptibility to disease.
- Lifestyle changes placing humans in habitats occupied by ESM and PPM.
- Improved water treatment, leading to selection for ESM and PPM.
- Changes in the human environment (climatic changes, increasing dustiness in the environment, etc.) affecting the ecology of mycobacteria.

In contrast to classical tuberculosis, human-to-human transmission of infections caused by PPM has not been described. Thus, attention has become focused on identifying environmental sources of infection. It is understood that the human environment includes not only natural environments but also human-engineered environments (e.g. homes, buildings and hospitals) and the materials in those environments (e.g. drinking water, aerosols and instruments). For example, in homes drinking water, shower aerosols, building materials and garden soil can be

I. Pavlik (✉)
Head of OIE Reference Laboratories for Paratuberculosis and Avian Tuberculosis, Department of Food and Feed Safety, Veterinary Research Institute, Brno, Czech Republic
e-mail: pavlik@vri.cz

J. Kazda et al. (eds.), *The Ecology of Mycobacteria: Impact on Animal's and Human's Health*, DOI 10.1007/978-1-4020-9413-2_3, © Springer Science+Business Media B.V. 2009

sources. In hospitals, water, solutions coming into contact with patients, surgical instruments, catheters, dialysis machines and even tissue grafts could be vectors and sources of mycobacterial infections.

Accordingly, the purpose of this chapter is to describe the ecology of potentially pathogenic mycobacteria. The chapter is focused on the more important slowly and rapidly growing species. The most significant and generally applicable information concerning the ecology and sources of infection for people and animals have been summarised (i.e. the most common modes of exposure and occurrence of these micro-organisms in the environment and the localisation of the infection process). Included are those mycobacterial species which are assumed to originate from the environment, even though in some cases they have not been yet detected or the existing data are scarce (*M. malmoense* and *M. haemophilum*). The approach is to identify sources and routes of infection, based on the presentation of infection.

3.1.1 M. kansasii

Acid-fast bacteria that later proved to be *M. kansasii* were first described as "yellow bacilli" (Buhler and Pollak A., 1953) and later renamed *M. kansasii* (Hauduroy et al., 1965 a,b). This species is a slowly growing photochromogenic, potentially pathogenic *Mycobacterium* (Photo 3.1). Five different genotypes of *M. kansasii*, designated I to V, have been distinguished on the basis of restriction fragment length polymorphism patterns with the major polymorphic tandem repeat (MPTR) probe and by restriction fragment patterns of PCR amplicons of the *hsp-65* gene (Picardeau et al., 1997).

Genotype I is the most frequent among *M. kansasii* isolates from human sources worldwide and has not been isolated from environmental sources (Alcaide et al., 1997). It is the only *M. kansasii* type that hybridises to the *M. kansasii* AccuProbe (Gen-Probe, San Diego, CA, USA). Representatives of genotype II have been isolated from both humans and the environment, but this type fails to hybridise to the *M. kansasii* AccuProbe (Alcaide et al., 1997; Picardeau et al., 1997). Genotypes III, IV and V are rarely isolated from humans and have been recovered from environmental samples, most frequently from tap water (Alcaide

et al., 1997; Picardeau et al., 1997). Thus, identification can be problematic, because some hybridisation probes do not react with a number of types (Alcaide et al., 1997; Picardeau et al., 1997). A PCR test based on a species-specific DNA sequence (Yang et al., 1993) does react with all *M. kansasii* types (Picardeau et al., 1997).

M. kansasii can be recovered from tap water, where it can survive for a long time and can most likely multiply. It is possible that the failure to isolate *M. kansasii* of genotype I from water is due to its relative increased susceptibility to decontamination regimens (Engel et al., 1980). The optimum temperature for its *in vitro* isolation ranges between 30 and 33 °C. It causes pulmonary disease in humans and can also be an aetiological agent of extrapulmonary diseases (Table 3.1).

Iatrogenic infections and other modes of transmission (often unexpected) of the pathogen to patients have also been described (Table 3.1; Photo 3.2). In the human population, *M. kansasii* affects individuals from heavily industrialised areas (especially coal mining areas) and urban agglomerations with a high density of building associated with high levels of dust. Increased incidence of the disease has been described in such regions and countries as the USA, Japan, South Africa, Western Australia and also in Europe, above all in England, France and the region on the border between the Czech Republic and Poland (Marks and Jenkins, 1971; Good and Snider, 1982; Tsukamura et al., 1983; Collins et al., 1984; Wayne, 1985; Kaustova et al., 1995).

M. kansasii has also been found to colonise or infect a variety of cold-blooded and warm-blooded animal species. It has been isolated from tissues without gross lesions (above all from lymph nodes) and from lesions in different organ tissues (Table 3.2). Little is known about the infection sources for animals, but considering ecological aspects we can assume that the main reservoir and source of *M. kansasii* is water. It is likely to be the same for humans.

In a bid to identify possible sources of *M. kansasii* infections, different components of the environment have been examined. *M. kansasii* has been recovered from water in regions characterised as having endemic or occasional occurrences of human infections. Publications focused on the investigation of the ecology of this pathogen are scarce because the recovery of *M. kansasii* in drinking water was insufficient

Table 3.1 Human infections caused by *Mycobacterium kansasii*

Mode of transmission	Patients[1]	Infected tissue	Ref.
Infection following steroid infiltration of two plaques of lichen simplex	20/F	Cutaneous infection of East African civil servant living in the UK for 15 years visited doctor in Gujarat (Western India) who infiltrated steroids 4 months before the infection	[5]
Not known	20/M	Fatal septicaemia (primary myelodysplastic syndrome)	[7]
Not known	30/F	Pneumonia and osteomyelitis of the skull (*HIV*+ patient)	[15]
Manipulation with *M. kansasii*-contaminated aquarium sediment	31/M	Isolated skin lesion	[8]
Small wound over the right pretibial area and erythematous swelling after swimming at the beach	38/F	Cellulitis over the left lower leg, which had a poor response to antibiotics (systemic lupus erythematosus)	[6]
Minor leg trauma	44/M	Chronic ulcer of the leg and lateral malleolus (asthmatic steroids treatment)	[12]
Laceration (2 cm) over right elbow in public swimming pool	59/M	Olecranon bursitis	[1]
Bare hand in a sewage pipe by unblocking a toilet on a friend's farm	59/M	18-year history of chronic plague psoriasis and refractory anaemia – lesions on the right hand and wrist	[3]
Tenosynovitis associated with a dog bite	68/M	Accidental puncture wound to the palm of hand while playing with pet dog (adult-onset diabetes mellitus)	[13]
Minor trauma of the skin, unnoticed or intra-operative infections	82/M	6 years after a total knee replacement, necrosis and granuloma formation of periprosthetic tissue	[10]
Not known	?/?	Primary cutaneous infection in an iatrogenically immunosuppressed patient	[14]
Trauma of the joint	5 patients	Following intra-auricular corticoid injections (four *HIV*- and one *HIV*+)	[2]
Not known	46 patients *HIV*+	Infection: pulmonary (36), pulmonary and diarrhoea (5), pulmonary and adenitis (1), adenitis (1), bacteriaemia (1), osteomyelitis (1), pericarditis (1)	[4]
Contaminated tap water in hospital	Hospitalised	Increased frequency of *M. kansasii* isolates from gastric washings	[9]
Not known	10 patients	Blood isolates from immunosuppressed (*HIV*+ or with lymphoma) patients	[11]

[1] Patients ordered according to their age; **M** (male), **F** (female), **?** (not known).

[1] Barham GS, Hargreaves DG (2006) J. Med. Microbiol. 55:1745–1746. [2] Bernard L, Vincent V, Lortholary O, Raskine L, Vettier C, Colaitis D, Mechali D, Bricaire F, Bouvet E, Sadr FB, Lalande V, Perronne C (1999) Clin. Infect. Dis. 29:1455–1460. [3] Breathnach A, Levell N, Munro C, Natarajan S, Pedler S (1995) Clin. Infect. Dis. 20:812–817. [4] Campo RE, Campo CE (1997) Clin. Infect. Dis. 24:1233–1238. [5] Groves RW, Newton JA, Hay RJ (1991) Clin. Exp. Dermatol. 16:300–302. [6] Hsu PY, Yang YH, Hsiao CH, Lee PI, Chiang BL (2002) J. Formos. Med. Assoc. 101:581–584. [7] Kaustova J, Martinek A, Curik R (1993) Eur. J. Clin. Microbiol. Infect. Dis. 12:791–793. [8] Kern W, Vanek E, Jungbluth H (1989) Med. Klin. (Munich) 84:578–583. [9]. Levy-Frebault V, David HL (1983) Rev. Epidemiol. Sante Publique 31:11–20. [10] Neuberger A, Sprecher H, Oren I (2006) J. Clin. Microbiol. 44:2648–2649. [11] Parenti DM, Symington JS, Keiser J, Simon GL (1995) Clin. Infect. Dis. 21:1001–1003. [12] Plaus WJ, Hermann G (1991) Surgery 110:99–103. [13] Southern PM, Jr. (2004) Am. J. Med. Sci. 327:258–261. [14] Stengem J, Grande KK, Hsu S (1999) J. Am. Acad. Dermatol. 41:854–856. [15] Weinroth SE, Pincetl P, Tuazon CU (1994) Clin. Infect. Dis. 18:261–262.

Table 3.2 Animal infections caused by *Mycobacterium kansasii*

Hosts/vectors	Host organism	Infected tissue	Ref.
Poikilothermic	Amoeba (*Acanthamoeba castellanii*)[1]	Body	[14]
	Cockroaches (*Periplaneta americana*)	Body	[10]
	Chinese soft shell turtle (*Pelodiscus sinensis*)	Lung and carapace	[9]
	Cardinal tetra (*Paracheirodon axelrodi*)	Tissues with tuberculoid lesions	[12]
	Siamese fighting fish (*Betta splendens*)	Tissues with tuberculoid lesions	[12]
Homeothermic	Domestic goat (*Capra hircus*)	Healthy lymph nodes	[1]
	Squirrel monkey (*Saimiri sciureus sciureus*)	Healthy lymph nodes	[2]
	Rhesus monkey (*Macaca mulatta*)	Pulmonary lesions	[5]
	Antelope in zoo	Pulmonary lesions (mixed infection caused by *M. a. avium* and *M. kansasii*)	[8]
	Llama	Lesioned mesenteric lymph nodes, liver and lungs	[7]
	Florida manatee (*Trichechus manatus latirostris*)	Tuberculoid nodules in lungs	[13]
	Dog (*Canis familiaris*)	Pleural fluid from persistent pleural effusion (3-year-old spayed female whippet)	[11]
	Cattle (*Bos taurus*)	Lesioned lymph nodes	[6]
	Cattle (*B. taurus*)	Healthy lymph nodes from positive skin-tested animal for bovine tuberculin	[4]
	Black-tailed deer (*Odocoileus hemionus*)	Multiple granulomas in thoracic cavity	[3]

[1] Isolation from the water network was done by the co-culture of amoebae.

[1] Acosta B, Real F, Ferrer O, Deniz S, Poveda B (1998) Vet. Rec. 142:195–196. [2] Brammer DW, O'Rourke CM, Heath LA, Chrisp CE, Peter GK, Hofing GL (1995) J. Med. Primatol. 24:231–235. [3] Hall PB, Bender LC, Garner MM (2005) J. Zoo. Wildl. Med. 36:115–116. [4] Hughes MS, Ball NW, McCarroll J, Erskine M, Taylor MJ, Pollock JM, Skuce RA, Neill SD (2005) Vet. Microbiol. 108:101–112. [5] Jackson RK, Juras RA, Stiefel SM, Hall JE (1989) Lab Anim Sci. 39:425–428. [6] Jarnagin JL, Himes EM, Richards WD, Luchsinger DW, Harrington R, Jr. (1983) Am. J. Vet. Res. 44:1853–1855. [7] Johnson CT, Winkler CE, Boughton E, Penfold JW (1993) Vet. Rec. 133:243–244. [8] Kleeberg HH, Nel EE (1973) Ann. Soc. Belg. Med. Trop. 53:405–418. [9] Oros J, Acosta B, Gaskin JM, Deniz S, Jensen HE (2003) Vet. Rec. 152:474–476. [10] Pai HH, Chen WC, Peng CF (2003) J. Hosp. Infect. 53:224–228. [11] Pressler BM, Hardie EM, Pitulle C, Hopwood RM, Sontakke S, Breitschwerdt EB (2002) J. Am. Vet. Med. Assoc. 220:1336–1334. [12] Rehulka J, Kaustova J, Rehulkova E (2006) Acta Veterinaria Brno 75:251–258. [13] Sato T, Shibuya H, Ohba S, Nojiri T, Shirai W (2003) J. Zoo. Wildl. Med. 34:184–188. [14] Thomas V, Herrera-Rimann K, Blanc DS, Greub G (2006) Appl. Environ. Microbiol. 72:2428–2438.

in most cases. However, the following information can be drawn from the few published results. It has been arranged according to water recycling in nature and then through different users according to respective ecological niches (biomes), starting with soil and dust, and including surface water, service and drinking water and wastewater.

3.1.1.1 *M. kansasii* in Soil and in Dust

M. kansasii is occasionally found in soil (Table 3.3) and according to Joynson (1979) it survives there for only several days, possibly due to the relative increased susceptibility of *M. kansasii* to streptomycin (Chapman, 1971) produced by lower moulds in the soil.

This fact was recognised many years ago but has not been studied further; it may be very interesting to investigate the various activities of antimicrobial substances present in the soil. On the other hand, *M. kansasii* can survive in water all year round and can even propagate under certain conditions (above all at temperatures above 25 °C with sufficient supplies of minerals and organic substances). Table 3.2 and Section 5.5 present the results of attempts to recover *M. kansasii* in soils collected in India, Iran and the USA.

It is worth noting that this species was not found in boreal soil in Finland, in which tens of mycobacterial species in high concentrations were detected. These acid soils, being low in humus found under the forests (particularly coniferous trees), are viewed as important sources of mycobacteria (see more in Section 5.9). The absence of *M. kansasii* in these acidic soils that are rich in other mycobacteria is likely due to its relative sensitivity to acids used in a number of decontamination regimens (Engel et al., 1980). *M. kansasii* was not recovered from soils collected in the region of endemic *M. kansasii* disease in Northern Moravia in the Czech Republic (Table 3.4). In that 1980 study, 93 soil samples from the shores of water basins, brooks and moors were collected and cultured for mycobacteria (Chobot et al., 1997).

M. kansasii is occasionally recovered from dust – for example, from a hospital in Japan (Table 3.2; Section 5.9). Its growth is, however, most likely hindered by the low humidity of concurrent dust present in the majority of niches.

3.1.1.2 *M. kansasii* in Surface Water

M. kansasii is recovered more often from surface water (above all river water) than from either soil or dust (Table 3.3). Even though *M. kansasii* is rarely recovered from surface water, seasonal changes have been recorded in the USA: *M. kansasii* isolates have been

detected in the summer and autumn months (Bland et al., 2005). Its occurrence in this environment is manifested by infections in different animals (Table 3.2). Nevertheless, the frequency of recovery of *M. kansasii* from natural water sources is very low in comparison with other mycobacterial species (Table 3.4). A number of authors (Engel et al., 1980; Powell and Steadham, 1981; Bland et al., 2005) have postulated explanations for the infrequent isolation of *M. kansasii* from surface water samples, including:

- Relative susceptibility to the acidic or basic conditions that are part of almost all decontamination regimens.
- Seasonal changes in *M. kansasii* numbers, which influence the recovery frequency.
- The dilution of *M. kansasii* numbers in lakes and streams which decreases the probability of *M. kansasii* recovery.
- The overgrowth of other micro-organisms in spite of decontamination.
- Culture media may not provide sufficient nutrients or may be toxic for some populations of *M. kansasii*.
- *M. kansasii* cells survive by the formation of cell-wall-deficient, resistant forms (Nyka, 1974).

Table 3.3 *Mycobacterium kansasii* detection in the environment

Country (state)	Locality	Origin of contaminated sample	Ref.
Czech Republic	North Bohemia	Tap water from a patient with mycobacteriosis	[6]
	Pig slaughterhouse	Bristle residue from a splinter	[9]
	Pig slaughterhouse	Meat leftovers from knife and sharpening steel	[9]
	Pig slaughterhouse	Tissue leftovers from a disinfectant container	[9]
	Pig slaughterhouse	Condensed water near a traverse rail	[9]
	Aquarian fish breeder	Sediment from aquarium bottom and filters	[1]
France	Underground coal mining	Dust	[11]
Germany	Bonn	Biofilm of tap water (35 and 45 °C)	[8]
Germany and France	Domestic drinking water	Hot water system (30–69 °C)	[4]
The Netherlands	Rotterdam	Untreated water basins from the central distribution plant	[3]
	Rotterdam	Main pipes from the central distribution plant	[3]
	Bilthoven	Whirlpools (35–40 °C)	[5]
Uganda	Victoria lake	Grass swamps with predominant aquatic plant *Echinochloa pyramidalis*	[10]
USA (Texas)	Rio Grande	River water	[2]
	Different places	Lakes, streams, bayous and city water supplies	[7]

[1] Beran V, Matlova L, Dvorska L, Svastova P, Pavlik I (2006) J. Fish. Dis. 29:383–393. [2] Bland CS, Ireland JM, Lozano E, Alvarez ME, Primm TP (2005) Appl. Environ. Microbiol. 71:5719–5727. [3] Engel HW, Berwald LG, Havelaar AH (1980) Tubercle. 61:21–26. [4] Fischeder R, Schulze-Robbecke R, Weber A (1991) Zentralbl. Hyg. Umweltmed. 192:154–158. [5] Havelaar AH, Berwald LG, Groothuis DG, Baas JG (1985) Zentralbl. Bakteriol. Mikrobiol. Hyg. [B] 180:505–514. [6] Horak Z, Polakova H, Kralova M (1986) J. Hyg. Epidemiol. Microbiol. Immunol. 30:405–409. [7] Powell BL, Jr., Steadham JE (1981) J. Clin. Microbiol. 13:969–975. [8] Schulze-Robbecke R, Fischeder R (1989) Zentralbl. Hyg. Umweltmed. 188:385–390. [9] Shitaye J, Horvathova A, Dvorska-Bartosova L, Moravkova M, Kaevska M, Donnelly N, Pavlik I (2008) Unpublished data. [10] Stanford JL, Paul RC (1973) Ann. Soc. Belg. Med. Trop. 53:389–393. [11] Viallier J, Viallier G, Joubert L (1972) Poumon. Coeur 28:55–65.

Table 3.4 The detection rate of *Mycobacterium kansasii* in water samples collected in the endemic area in North Moravia

Examined water samples			Examined samples			M. kansasii		Ref.
Type[1]	Sample	Origin of sample collection	No.	Posit.[2]	%	No.	%	
Service[3]	Water	Untreated water	103	16	15.5	0	0	[3]
	Biofilm	Showers and pipes in bathrooms	396	Nk	Nk	23	5.8	[2]
	Scrapings	Water duct	201	50	24.9	13	6.5	[3]
	Water	Mines, ironworks and wiring plants	833	Nk	Nk	515	61.8	[2]
Subtotal			1533	Nk	Nk	551	35.9	
Drinking[4]	Surface water	Water treatment plant – before treatment	40	11	27.5	0	0	[3]
	Scrapings	Water treatment plant – scrapings from filters	28	4	14.3	0	0	[3]
	Surface water	Water treatment plant – chlorinated water	46	9	19.6	0	0	[3]
Drinking[5]	Water	Taps	114	26	22.8	0	0	[3]
	Water	Taps in coal mining plants	202	32	15.8	3	1.5	[3]
	Scrapings	Taps	366	145	39.6	18	4.9	[3]
	Scrapings	Taps in coal mining plants	99	34	34.3	2	2.0	[3]
	Water and scrapings	Taps	560	Nk	Nk	39	7.0	[2]
Subtotal			1455	Nk	Nk	62	4.3	
Used	Water and biofilm	Aquariums with drinking water	16	8	50.0	0	0	[1]
	Sediment	Aquariums with drinking water	14	11	78.6	1	7.1	[1]
Waste	Water	Coal mining plants	309	63	20.4	1	0.3	[3]
	Water	Mining water	77	26	33.8	1	1.3	[3]
	Water	Coal mining plants – drainage	88	22	25.0	0	0	[3]
Subtotal			504	130	25.8	3	0.6	

[1] Water from different sources specified in the table. [2] Detection of all mycobacterial species including *M. kansasii*. [3] Water in coal mining plants used for service purposes (water used in coal mining plants). [4] Water in hot water exchangers supplying the whole city with hot water. [5] Water in water treatment stations for public use.

[1] Beran V, Matlova L, Dvorska L, Svastova P, Pavlik I (2006) J. Fish. Dis. 29:383–393. [2] Chobot S, Malis J, Sebakova H, Pelikan M, Zatloukal O, Palicka P, Kocurova D (1997) Cent. Eur. J. Public Health 5:164–173. [3] Kaustova J, Olsovsky Z, Kubin M, Zatloukal O, Pelikan M, Hradil V (1981) J. Hyg. Epidemiol. Microbiol. Immunol. 25:24–30.

3.1.1.3 *M. kansasii* in Industrial Water

For economic reasons, water that is not suitable for drinking can be used for industry, mining and agriculture. Such water is treated according to valid legislation and the regulations of the respective country; however, it fails to meet the requirements for drinking water and therefore, its use for this purpose is not permitted. For better understanding of water recycling in coal mining plants, an example from North Moravia (Karvina and Ostrava) in the Czech Republic will be described.

The dust released by coal extraction, its grinding and breaking up and its subsequent transport and handling, together with the extracted coal, is a significant problem for coal mining plants. Large volumes of water are used to reduce the level of dust. Water for that purpose is usually supplied from surface sources (above all rivers, in some cases from lakes or water reservoirs). The water is treated in either of two ways depending upon use and volumes required because of financial considerations.

Service "technological" water is only treated by sedimentation and subsequent chlorination, but it is not preheated before use; such water is used for dust reduction in collieries and for its transfer and grinding on the surface. This water is again pumped to the surface and following sedimentation in surface tanks, it is returned to the rivers. *M. kansasii* has not been detected in this type of water (Table 3.4). Also, *M. kansasii* has not been recovered from water used in coal mining plants, from wet and dry slack coal and from the organs of rodents trapped in the collieries (Pelikan et al., 1973).

Service water for bathrooms and showers is treated by filtration (open sand filters, stationary fast filters, etc.) with subsequent chlorination and heating:

(a) Waters heated to 42–45 °C for direct use yielded *M. kansasii* from both water and pipe biofilms.
(b) Water heated to 90 °C for use after dilution with cold water yielded *M. kansasii* from both water and biofilms on shower nozzles.

3.1.1.4 *M. kansasii* in Drinking Water

Because waters for drinking are subject to sequential steps designed to remove (filtration and coagulation) and kill (chlorination) micro-organisms, the numbers of micro-organisms present, including mycobacteria, are relatively low. All water treatment steps lead to a decrease in the concentrations of any mycobacteria including *M. kansasii*. It follows that absolute numbers of *M. kansasii* cells are very low and it is necessary to examine a filtrate derived from at least 1–10 litres of water for their successful detection.

The present picture of substantial numbers of patients infected with *M. kansasii* isolates belonging to genotype I stands in contrast to the infrequent recovery of *M. kansasii* from drinking water. Either the source of human infection is something other than water or, as postulated above, there are methodological problems with current methods of isolation. It is important to understand that all methods developed and widely used for mycobacterial isolation were developed and optimised for *M. tuberculosis*, not for environmental mycobacteria. Thus, one should always assume that the tools available are inadequate for the recovery of

any *Mycobacterium*, including *M. kansasii*. Therefore, it would be unwise to propose that a drinking water system posed a low risk of infection simply because the frequency of isolation or number of *M. kansasii* was low.

Both cold and heated drinking water have been reported to contain *M. kansasii*, the details will be listed separately.

M. kansasii in Cold Drinking Water

M. kansasii has infrequently been recovered from cold water samples (10–15 °C) collected from drinking water distribution systems (Table 3.5). The older the water supply system the more biofilm is produced and the risk of *M. kansasii* propagation in this niche increases. Minute amounts of *M. kansasii* were detected in newly built housing estates in Rotterdam; *M. kansasii* was found most frequently in the city's central water distribution systems, which were more than 100 years old (Engel et al., 1980). *M. kansasii* is more frequently recovered from cold water pipelines that run in parallel with warm water pipelines. If the cold water temperature is raised to 18 °C, *M. kansasii* multiplication occurs. Plastic pipes, because their thermal conductivity is lower than metal pipes, may reduce heating and protect against mycobacterial proliferation (Engel et al., 1980). The sites of highest *M. kansasii* (and other mycobacterial) recovery are the dead ends of water mains where chlorine concentrations are low, other microbial competitors are absent and organic matter is available (Engel et al., 1980; Powell and Steadham, 1981).

M. kansasii in Warm Drinking Water

M. kansasii is frequently recovered from warm water distribution systems and warm whirlpool baths (Tables 3.3 and 3.4). The more frequent recovery of *M. kansasii* isolates from warm compared to cold water (Table 3.4) is likely due to the fact that *M. kansasii* grows between 35 and 40 °C. *M. kansasii* has been shown to colonise surfaces and form biofilms in silicone tubing through which warm water (35–45 °C) inoculated with *M. kansasii* flowed (Schulze-Robbecke and Fischeder, 1989). The mycobacteria accumulated in the biofilm after 3 weeks and persisted there

Table 3.5 The detection rate of *Mycobacterium kansasii* in water samples collected in other endemic areas

Endemic area for *M. kansasii* infections in humans			Examined samples		*M. kansasii*			Ref.
Country (locality)	Water type	Origin of samples	No.	Posit.[2]	%	No.	%	
The Netherlands (Rotterdam)	Drinking[1]	Untreated basins, sand filtrates and reservoirs	21	Nk	Nk	1	4.8	[1]
	Drinking[1]	Water from main pipes	112	Nk	Nk	1	0.9	
	Drinking[1]	Water collected six times from 78 taps during 1 year	78	61	78.2	38	48.7	
(Bilthoven)	Drinking[1]	Tap water	69	47	68.1	0	0	[2]
	Pools	Swimming pools (18–25 °C)	132	86	65.2	0	0	
	Pools	Whirlpools (35–40 °C)	73	60	82.2	3	4.1	
USA (Texas)	Surface	Lakes, streams, bayous and city water supplies	180	80	44.4	3	1.7	[3]
	Drinking[1]	Private and public water systems	26	15	57.7	6	23.1	

[1] Drinking water: used after filtration and chlorination. [2] Detection of all mycobacterial species. **Nk** (not known)
[1] Engel HW, Berwald LG, Havelaar AH (1980) Tubercle. 61:21–26. [2] Havelaar AH, Berwald LG, Groothuis DG, Baas JG (1985) Zentralbl. Bakteriol. Mikrobiol. Hyg. [B] 180:505–514. [3] Powell BL, Jr., Steadham JE (1981) J. Clin. Microbiol. 13:969–975.

during the entire experimental period lasting 10 months. Colonisation was intensive, adduced by the fact that 30% of microscopic fields had acid-fast staining bacilli. *M. kansasii* formed biofilms more readily on hydrophobic silicon than on relatively hydrophilic glass surfaces. The formation of biofilms was determined to be highest on paraffin based on the fact that the number of colonies on paraffin after 2 weeks was equal to those formed on silicon after 3 weeks (Schulze-Robbecke and Fischeder, 1989). Although not investigated, biofilm formation would be expected to occur on pipe materials used commonly in households, buildings and hospitals: metal (copper and iron) and plastic (PVC, polyvinylchloride; Photo 3.3). It would be expected that cells of *M. kansasii* may be released from the biofilm and be transported via warm water. In fact, given the slow growth rate of *M. kansasii* and other mycobacteria, biofilm formation may be the only way that mycobacteria can survive and persist in drinking water distribution systems and pipes in households and buildings. The infection of individuals would occur via the formation of aerosols (e.g. showering and washing) or via swallowing. In regions reporting endemic levels of mycobacterial disease or with high numbers of waterborne mycobacteria, including *M. kansasii*, it is logical to recommend individuals to have baths rather than to shower. If showers are the only bathing option, it would be advisable to make sure that adequate ventilation be provided to reduce the concentration of aerosolised mycobacteria.

3.1.1.5 *M. kansasii* in Other Sources

M. kansasii has been recovered from aquaria and water from pig slaughterhouses (Table 3.3). In the first case, the aquarium water may serve as a source of infection of humans via cuts, in the same fashion as *M. marinum* (Photo 3.4). Although it is possible that fish could be infected or may be the source in the aquaria, there are few reports of *M. kansasii* recovery or infection in fish (Beran et al., 2006b).

In the case of slaughterhouse wastewater, the source of *M. kansasii* was most likely the drinking water used for rinsing animal carcasses. This assumption is based on the fact that the slaughterhouse is situated at the boundary of an endemic area of *M. kansasii* occurrence in humans (Northern Moravia; Photo 3.5). It is possible that the processed meat and parenchymatous organs could be contaminated with *M. kansasii*. From the point of view of food safety, it is necessary to ensure that the temperature level in all abattoir premises, where such water is used, is maintained below 10 °C all year around. This does not allow *M. kansasii* propagation.

3.1.1.6 Risk Factors for *M. kansasii* Infections

As is the case in other environmental mycobacteria, human-to-human transmission of *M. kansasii* infections has not been described. Human infections caused by *M. kansasii* are not of an explosive character, as is the case in classical bacterial infections, such as cholera, transmitted via water. *M. kansasii* disease develops in a low proportion of people who are repeatedly exposed to high numbers of *M. kansasii*. Further, most cases involve occupational exposure to dust that, rather than being a source, debilitates lung function, thereby reducing the barriers to infection. Thus, in colliers, the sources of *M. kansasii* were showers and the route of infection was aerosols (Pelikan et al., 1973). Factors other than collier dust are also risk factors because *M. kansasii* infections occur in individuals who are not exposed to an occupational dust.

Other risk factors for *M. kansasii* infection include immunodeficiency or immunosuppression and lung parenchyma damage (pneumoconiosis, chronic bronchitis, a history of tuberculosis or disease due to other environmental mycobacteria). In addition, gastroduodenal ulcers, liver disease, diabetes mellitus, therapy with corticoids or cytostatics, chronic alcoholism and adverse socioeconomic factors are predisposing factors for extrapulmonary diseases (Kaustova et al., 1995). The major risk factor for skin infections is likely to be a break in the skin's integrity, other skin injuries, surgery and some diseases (eczema, psoriasis, etc.).

The number of extrapulmonary disease cases caused by *M. kansasii* in the Czech Republic and in other countries around the world is relatively low compared to those for pulmonary disease (Wayne and Sramek, 1992; Razavi and Cleveland, 2000; del Giudice et al., 2000). Among a cohort of Czech patients investigated between 1984 and 1997, this species was found to be an aetiological agent of extrapulmonary diseases in only 9 (1%) of 785 patients and was found to cause skin lesions, cervical lymphadenitis and uterine infections. A case of fatal septicaemia was also described (Kaustova et al., 1993b).

3.1.1.7 The Prevention of Human Infections Caused by *M. kansasii*

Various measures that have been implemented to eliminate mycobacteria from the water supply system were found to be effective for only a very short period (Pelikan et al., 1973). Between 1990 and 1994 precautions were adopted on a service water supply system used for showering by workers from a coal mining plant in the centre of an endemic region (Northern Moravia). Mechanical cleaning of the service water tank and steaming of the pipeline system and shower nozzles were performed using water heated to 85 °C for 30 min. Significantly decreased counts of colony-forming units/litre (CFU/litre) of *M. kansasii* were found after this treatment. However, gradual re-colonisation of the system with *M. kansasii* (detected in water samples) was detected by weekly inspections. Accordingly, repeated steaming and cleaning was decided upon and was performed every fortnight. This completely eliminated mycobacteria from the shower nozzles (J. Kaustova, unpublished data).

The continuous warming of water to 85 °C and its subsequent cooling to 40 °C performed once a month or continuous water ozone treatment was shown to be equally effective in the elimination of this pathogen. Ozone treatment decreased the CFU/litre counts of *M. kansasii* to 5%. The change of the distributing pipelines was an essential prerequisite before the introduction of this procedure. Subsequently, it was necessary to install ultraviolet (UV) irradiation devices in the hot water distribution system. It was also necessary to slow down the water flow rate to one half of the initial speed (Martinkova et al., 2001).

Chobot et al. (1997) recommended the disinfection of distribution pipelines with hot water of at least 90 °C for 30 min. The reason for the repeated re-colonisation of water supply systems, usually within several weeks, is the presence of *M. kansasii* in biofilm in the dead end of the water supply system where the heat treatment does not fully kill *M. kansasii*.

3.1.1.8 Views and Perspectives on the Research

Even though *M. kansasii* is one of the earliest recognised pathogenic environmental mycobacteria, many issues remain to be clarified:

- The environmental source of the genotype I *M. kansasii* isolates is unknown.
- The basis for regions where *M. kansasii* disease is endemic has yet to be identified.
- The physiochemical conditions of the environment that influence the ecology (niche) for *M. kansasii* remain unidentified and must be measured.
- Because of the difficulty of isolating *M. kansasii* from patients and environmental samples (e.g. decontamination sensitivity), this means that

culture-independent detection and quantification methods (e.g. quantitative real-time PCR) need to be developed.

- The interactions of *M. kansasii* with different matrices (e.g. biofilm) and invertebrate animals (e.g. amoebae) should be investigated and effective techniques for the disinfection of water distributing pipelines should be developed (including effective biofilm elimination).

3.1.2 *M. ulcerans*

3.1.2.1 General Characteristics

A disease with symptoms resembling that of *M. ulcerans* was initially described in 1897 in Uganda, but the causative agent was first isolated more than 50 years later in Australia (Maccallum, 1948). To begin with it was implicated in sporadic cases especially in poor communities of West Africa; however, in the last few decades infection caused by *M. ulcerans* has spread to more than 25 countries (Table 3.6) and is regarded as the third most common mycobacterial disease after tuberculosis and leprosy (Duker et al., 2006). In contrast to humans, there is only one report on *M. ulcerans* infections in animals, such as the koala, possum and alpaca in Australia, but it has not been confirmed in other endemic regions (Merritt et al., 2005).

3.1.2.2 Ecological Aspects

One of the first studies which included ecological methods took place in the early 1970s; it was carried out by the Uganda Buruli Group who showed that contact with the aquatic environment can be regarded as a main risk factor for infection with *M. ulcerans* (Barker, 1972). There is very promising information that aquatic plants, such as *Echinochloa pyramidalis* found in swamps and rivers in endemic areas, can serve as a source of infection. The biofilm obtained from the surface of these plants doubled the growth rate of *M. ulcerans* in laboratory experiments. Furthermore, aquatic snails, small fish and aquatic insects of the orders Hemiptera (water bugs), Odonata (dragonfly larvae) and Coleoptera (beetle larvae) living in

water and marshes in endemic regions have been found to be positive for *M. ulcerans* (Portaels et al., 1999). Recently, molecular methods were able to confirm the presence and multiplication of *M. ulcerans* in the salivary glands of *Naucoris cimicoides*, water bugs common in surface water in the tropics (Marsollier et al., 2005). This biting insect could transfer the infection to mice and probably to humans, but other factors in the aquatic ecosystem, such as an abundant source of *M. ulcerans*, must also contribute and can explain the high endemic occurrence of this infection in such countries as Benin, Ivory Coast, Uganda, etc.

However, this ecologically relevant information is anecdotal and should be viewed in the light of research activity to clarify the competitive advantage of *M. ulcerans* and its increased abundance in the aquatic environment of endemic areas. As mentioned previously, sphagnum and other moss vegetation offers optimal conditions for the growth of mycobacteria. Pronounced multiplication of *M. ulcerans* (type strain ATCC 19423) was observed after inoculation in polyamide capillaries (pore size 100 000), inserted in the grey layer of *Sphagnum magellanicum–S. rubellum* association, when incubated for periods of 12 h at 31 °C by day and 22 °C by night (Kazda, 1978a). Furthermore, in sphagnum vegetation collected from a lake shore in County Cork, Ireland, mycobacteria have been isolated which show a 83% similarity to *M. ulcerans* (Cooney, 1991). This may suggest it has now spread to moderate climate zones. In the tropics, sphagnum vegetation is rare and cannot be regarded as an important source of *M. ulcerans*. In these regions, other moss species are abundant at the edges of streams and cover the shores of bodies of water (Photo 3.6). It was found that such moss vegetation offers favourable conditions for the growth of mycobacteria.

Epidemiological studies have shown higher rates of infection with *M. ulcerans* after flooding. If present in the moss, mycobacteria could be released into water and gain access to the human population. Statistical evaluation revealed a significantly higher incidence of disease in children under 15 years, the group which favour bathing and swimming in rivers and lakes. This activity can resuspend *M. ulcerans* present in biofilm, enhance its concentration in water and allow it to cover the skin; thus, individuals with skin lesions may be infected. Further direct contact can take place with moss and other vegetation, covering the edge and shore

Table 3.6 Countries with reported cases of *Mycobacterium ulcerans* infection[1]

Geographic area	Countries
Africa	Angola, Benin, Burkina Faso, Cameroon, Democratic Republic of Congo, Gabon, Ghana, Guinea, Ivory Coast, Liberia, Nigeria, Sierra Leone, Sudan, Togo, Uganda
Western Pacific	Australia, Papua New Guinea
Asia	China, India, Japan, Malaysia
Americas	Bolivia, French Guyana, Mexico, Peru, Suriname

[1] For more details, see Section 3.1.2.

of water bodies. It has been found that outbreaks of disease are related to drastic man-made disturbances of aquatic environments like impoundments, the extension of agricultural areas and field irrigation (Merritt et al., 2005).

Similar to leprosy, the transmission of infection from the environment to humans is made more likely by washing in contaminated streams. Linen and underwear do not fully dry in climates with high humidity; *M. ulcerans* can, therefore, survive and the infection can gain access to the skin of humans and their dwellings. Recently, the results obtained in a leprosy endemic pocket in Brazil revealed that frequent contact with natural water bodies and infrequent changing of bed linen significantly enhanced the risk of contracting this disease (Kerr-Pontes et al., 2006).

3.1.2.3 Trends

To control endemic disease caused by *M. ulcerans* it will be necessary to intensify the search for this *Mycobacterium* not only in aquatic environments, but especially in moss vegetation surrounding aquatic bodies. The results obtained in Ireland revealed that the highest occurrence of mycobacteria was in these kinds of biotopes. Special attention should be given to the contact between potential sources of infection and humans, to establish effective protection against infection with this environmentally pathogenic *Mycobacterium*.

3.1.3 M. avium Species

The ecological variability and complexity of mycobacteria involved in natural processes can be demonstrated in four subspecies of *M. avium* (*M. a. avium*, *M. a. hominissuis*, *M. a. paratuberculosis* and *M. a. silvaticum*). The range of hosts of each of these subspecies varies as does their growth potential and occurrence in the environment. It is evident from results published so far that an understanding of the ecology of these four subspecies is not at the same level. The ecology of the last-mentioned subspecies *M. a. silvaticum* remains poorly understood. Most information is available about subspecies *M. a. hominissuis*. Major attention has been paid to this subspecies because it is often diagnosed in immunosuppressed people and in pigs.

The subspecies *M. a. avium* was especially a focus of interest with regard to ecology in the 1970s and 1980s. However, new publications have appeared in the last 5–7 years. Attention has been paid to the causative agent of paratuberculosis *M. a. paratuberculosis* from the point of view of ecology, particularly in the past several years, due to the difficulty of its control in infected animal herds and because it is most probably involved in the aetiology of Crohn's disease. For a better understanding of all the above-mentioned relationships, it will be practical to explain the currently accepted taxonomy of *M. avium* subspecies and the *Mycobacterium avium complex (MAC)*.

Taxonomy of *M. avium* species. According to the currently valid taxonomy the *M. avium* species is divided into four subspecies (Thorel et al., 1990; Mijs et al., 2002b; Bartos et al., 2006; Turenne et al., 2007):

    i. *M. a. avium* is the causative agent of avian tuberculosis and can infect a wide range of bird species: domestic poultry (in cages and on commercial farm reared layers), farm raised game birds and pet species and wild birds, as well as domestic pigs, cattle and other mammals including man (Thorel et al., 2001). IS*901* PCR can be used as a rapid method for the identification of this mycobacterial subspecies (Pavlik et al., 2000b) in place of biological trials on pullets or serotyping.

ii. *M. a. hominissuis* causes "avian mycobacteriosis" in pigs, cattle and other animals including man. IS*901*–IS*1245* PCR can be used as an easy and quick method for the identification of this subspecies (Moravkova et al., 2007).

iii. *M. a. paratuberculosis* with IS*900* is the causal agent of paratuberculosis (Johne's disease) in ruminants and other mammals (Ayele et al., 2001; Hasonova and Pavlik, 2006) and is a candidate for one of the causal agents of Crohn's disease (Hermon-Taylor et al., 2000; Hermon-Taylor, 2000; Hermon-Taylor and Bull, 2002).

iv. *M. a. silvaticum*, also called "wood pigeon" strains, is essentially pathogenic for wild and domestic birds (Mijs et al., 2002a).

Currently, all four of these subspecies can be relatively easily identified by PCR (Bartos et al., 2006) and standardised RFLP methods with the following probes: IS*1245* (van Soolingen et al., 1998), IS*900* (Pavlik et al., 1999) and IS*901* (Dvorska et al., 2003).

The growth of *M. a. avium* is known to be mycobactin independent, while *M. a. silvaticum* and *M. a. paratuberculosis* require the presence of mycobactin during primoisolation. However, in the case of *M. a. paratuberculosis* it was confirmed that for some laboratory strains (for example vaccine strains EII and F316) the presence of a growth stimulator in the medium may not be required (Pavlik et al., 1999).

Taxonomy of *M. avium* complex. From the point of view of taxonomy it is important to emphasise that *M. avium* complex (*MAC*) comprises 28 serotypes (Wolinsky and Schaefer, 1973). *M. a. avium* comprises serotypes 1–3 and *M. a. hominissuis* eight serotypes 4–6, 8–11 and 21. The remaining serotypes 7, 12–20 and 22–28 belong to the *M. intracellulare* species, which is the member of the *MAC*, but not a subspecies of *M. avium*. In the environment, *M. intracellulare* is often found in the same compartments as *M. a. hominissuis* (Section 3.1.4).

Two recently described species, *M. chimaera* (Tortoli et al., 2004) and *M. colombiense* (Murcia et al., 2006), are classified as belonging to the *MAC* at present. Neither of these new species has been tested for virulence by biological experiments on pullets and no new serotype has been added to the existing 28 serotypes (Wolinsky and Schaefer, 1973). There

is a lack of information about the ecology of these species. The species *M. colombiense* was described based on the identification of 45 isolates from 43 patients infected with HIV/AIDS in Colombia (Murcia et al., 2006). The species *M. chimaera* was isolated from the lungs of 12 patients in Italy (Tortoli et al., 2004) and from 1 terrestrial rodent in Africa (Gambian rat; *Cricetomys gambianus*; Durnez et al., 2008).

Does *M. scrofulaceum* belong to the *M. avium* complex? Finally, historically speaking, it is necessary to mention the species *M. scrofulaceum*, which was previously classified as belonging to the *M. avium–intracellulare* complex (the commonly used abbreviation is *MAIS* for *M. avium*, *intracellulare*, and *scrofulaceum* complex) based on biochemical similarity. However, this species was later removed from this complex. Based on a collaborative taxonomic study, it was suggested in 1974 to shorten the designation of the *M. avium–intracellulare* complex to *M. avium* complex, later abbreviated to *MAC* (Meissner et al., 1974).

The following chapters will be focused primarily on aspects of the ecology of all the above-mentioned four subspecies of *M. avium*.

3.1.3.1 *M. avium* subsp. *avium*

Avian tuberculosis caused by *M. a. avium* is an insidious, chronic wasting disease with a worldwide distribution affecting many species of birds, domestic and wild animals, and humans (Thorel et al., 1997; Photo 3.7). With regard to domestic mammals, the causative agent of avian tuberculosis has been isolated from, e.g. cattle (Dvorska et al., 2004), pigs (Pavlik et al., 2003, 2005b), horses (Pavlik et al., 2008) and other domestic and wild animals (Thorel et al., 1997; Biet et al., 2005). With this in mind, it is surprising that the causative agent of avian tuberculosis has been detected only occasionally in the environment, e.g. in soil (Sections 5.1 and 5.5), water (Sections 5.2 and 5.3), air and dust (Section 5.9). In water samples from Australia, *M. a. avium* (serotype 1) was detected in only 1 (3.3%) of 30 isolates, in which a serotype was specified (Tuffley and Holbeche, 1980).

Based on these data, it is clear that unlike other members of the *MAC* (*M. a. hominissuis* and *M. intracellulare*), *M. a. avium* is not generally ubiquitous in the environment. The sources of *M. a. avium* in the environment, its ability to survive under defined conditions and to multiply there are puzzles that remain to be solved. Infected birds and mammals are most likely the prime sources of *M. a. avium* as they shed the bacteria via their faeces (Section 5.10). Other potential sources of *M. a. avium* in the environment are tissues from carcasses of infected birds (Sections 6.6 and 6.7).

Birds as a source of environmental contamination. It is surprising that the detection rate of *M. a. avium* in environments where flocks of birds infected with the causative agent of avian tuberculosis have been kept has been very low. For example, in Germany, *M. a. avium* (serotype 2) constituted only 1 (0.07%) of 140 mycobacterial isolates found in the soil of aviaries in winter (Beerwerth and Kessel, 1976). In the Czech Republic, *M. a. avium* (serotype 1) was isolated from 8 (1.6%) of 491 samples from the environment of aviaries where different infected waterfowl species were kept (Dvorska et al., 2007). In one flock of hens infected with avian tuberculosis, *M. a. avium* was detected in 4 (11.4%) of 35 examined samples from the environment; all 4 isolates were of the same IS*901* RFLP type which was detected in the infected birds (Shitaye et al., 2008b).

The shedding of *M. a. avium* into the environment by infected birds depends upon the following factors:

i. **Age of birds.** Although young birds are considered to be more susceptible to the disease, the highest mortality occurs in older age groups (Wilson, 1960).
ii. **Bird species.** Based on the degree of susceptibility of domestic and synanthropic free-living birds to the disease, Hejlicek and Treml (1995) categorised them into four groups:
   Highly susceptible species: domestic fowl (*Gallus gallus* f. *domestica*), sparrows (*Passer domesticus*), pheasants (*Phasianus colchicus*), partridges (*Perdix perdix*) and laughing gulls (*Larus ridibundus*).
   Moderately susceptible: turkeys (*Meleagris gallopavo* f. *domestica*) and guinea fowl (*Numida meleagris* f. *domestica*).
   Moderately resistant: domestic geese (*Anser anser* f. *domestica*) and ducks (*Anas platyrhynchos* f. *domestica*).
   Highly resistant species: turtle doves (*Streptopelia decaocto*) and rooks (*Corvus frugilegus*).

iii. **Stage of the disease.** It has been found for infected domestic fowl that the shedding of *M. a. avium* through faeces is irregular and not as intensive as might be expected. The intensity of contamination of faeces with *M. a. avium* depends on the form of avian tuberculosis in the affected animals. In infected and clinically ill birds, *M. a. avium* was detected in 49 (43.8%) of 112 faecal samples; in the infected but clinically normal birds, *M. a. avium* was isolated from only 10 (17.9%) faecal samples (Shitaye et al., 2008a). However, *M. a. avium* can also be shed via other routes, e.g. eggs (Section 5.7).

Based on this knowledge it can be expected that *M. a. avium* will most often be shed into the environment via the faeces of sparrows and by domestic fowl kept in small flocks around farms (Photos 3.8 and 3.9). The causative agent of avian tuberculosis can also be shed by pheasants that are usually kept in aviaries but are later let out into the wild and hunted (Friend and Franson, 1999).

Exotic birds as a source of environmental contamination. The situation with exotic birds is complicated by the fact that some species are highly resistant to natural infection with *M. a. avium*. Infections can be manifested and/or develop in exotic birds in three forms, in which lesions are observed in different parts of the organs (Gerlach, 1994; Prukner-Radovcic et al., 1998):

i. **The classic form/disseminated form**, i.e. tuberculosis and/or granulomatous lesions are found in various organs, but most notably in parenchymatous organs (liver, spleen, lungs, etc.).
ii. **The paratuberculosis form**: lesions are found in the gastrointestinal tract and it is characterised by the shedding of a high number of micro-organisms.
iii. **The non-tuberculous (atypical) form**, which is difficult to recognise and commonly occurs in finches, canaries and small parrots.

In most avian species, lesions caused by avian tuberculosis have distinct characteristic features, which are frequently seen in parenchymatous organs and in the intestine, though many tissues can be affected. The nature of the lesion caused by a mycobacterial infection can vary according to the genetic heritage of infected avian species and social stress (Gross et al.,

1989). In infected birds, observations of tuberculous lesions mostly in the liver, spleen and intestine have been adequately reported (Photo 3.10). *M. a. avium* is present in very high amounts in these lesions and such tissues are source of *M. a. avium* for other animals in the enclosure and for the soil, water and vegetation (Section 5). Predators and scavengers could be infected by eating such infected birds (Section 6.7). Less frequently, organs such as bone marrow, lungs, kidneys and ovaries also can be affected. In such cases, the bird could be considered as non-infected after the serological and faecal culture examinations and yet be a source of *M. a. avium* (Shitaye et al., 2008a; Shitaye et al., 2008b).

Although the pulmonary form of the infection is rare (Birkness et al., 1999; Marco et al., 2000), granulomas in the trachea have been frequently found (Gerlach, 1994). Despite this, Kul et al. (2005) reported the presence of granulomas in peafowl birds, indicating exposure through the respiratory tract. In geese and ducks, lesions have been found mainly in the liver and spleen and less frequently in other organs such as the intestines and lungs (Cerny, 1982).

Animals other than birds as a source of environmental contamination. Besides birds, other animals are susceptible to infection caused by *M. a. avium*. Friend and Franson (1999) classified the susceptibility of different hosts to *M. a. avium* with regard to their natural resistance as follows:

Highly susceptible (birds, pigs, minks and rabbits).
Moderately susceptible (sheep and deer).
Susceptible (cattle).
Moderately resistant (horses and goats).
Highly resistant (dogs, cats and humans).

Pigs as a source of environmental contamination. It can be seen from the published results that *M. a. avium* is found very rarely in the different components of the environment of domestic animal herds; of 579 mycobacterial isolates from 2412 environmental samples from pig farms, only 13 (2.2%) were identified as *M. a. avium*: 3 isolates were detected in peat fed as a supplement, 7 isolates were from bedding, 1 isolate from scrapings of the stable walls and the remaining 2 isolates were from trapped free-living house sparrows (Matlova et al., 2003). These findings are comparable to the results on a pig farm where a response to avian tuberculin testing occurred (Photo 3.11). Of 140 repeatedly collected samples from the environment over 7 years, none of the 33 isolates were identified as *M. a. avium* (Pavlik et al., 2007) and on 6 pig farms, *M. a. avium* was not found in any of the 102 samples of liquid manure (Fischer et al., 2006).

Other animals as sources of environmental contamination. *M. a. avium* can be shed in the faeces of infected cattle (Pavlik et al., 2005a) or farmed red deer (*Cervus elaphus*; Machackova-Kopecna et al., 2005). When considering other sources of *M. a. avium* that are not included in the above-mentioned list (Friend and Franson, 1999) special mention can be made of faeces and the carcasses of small terrestrial mammals (Fischer et al., 2000). Accordingly, herds of domestic animals and animals held in captivity (especially in zoological gardens) can be protected from the transmission of *M. a. avium* infection by preventing the access of free-living birds and invasions by small terrestrial mammals (Section 7.5).

The survival and propagation of *M. a. avium* in the environment. Long-term survival of *M. a. avium* in the environment has been described. Friend and Franson (1999) reported the survival of *M. a. avium* in poultry litter (for up to 50 months), in poultry carcasses buried 3 ft deep (28 months), *M. a. avium* 5 cm deep in soil (25 months), in body exudates shielded from the sun (11 months), in sawdust at 37 °C (8 months) and at 25 °C (5 months) and in body exudates exposed to the sun (1 month). From these data it could be concluded that *M. a. avium* can resist the following conditions:

High temperatures (Woodley and David, 1976; Merkal et al., 1979; Merkal and Crawford, 1979; Schulze-Robbecke and Buchholtz, 1992).
Low temperatures (Kim and Kubica, 1972, 1973; Portaels et al., 1988a).
pH changes (between pH3 and 7) (Portaels and Pattyn, 1982).
Ultraviolet irradiation (Tsukamura, 1963; McCarthy and Schaefer, 1974; Tsukamura and Dawson, 1981).
Starvation (Archuleta et al., 2002).
Low oxygen levels (Biet et al., 2005).

For growth, however, an optimal pH of between 5.4 and 6.5 has been documented (Portaels and Pattyn, 1982). The multiplication of the causative agent of avian tuberculosis in the environment has also been observed occasionally. For example, Beerwerth (1973) documented the partial multiplication of *M. a. avium* in arable soil, soil from pastures and in peat. It was found under *in vitro* conditions that *M. a. avium* (serotype 1; strain LM1) loses viability at a higher rate when desiccation occurs at higher temperatures; the ability to grow was reduced in the case of a 5–10 day starvation adaptation period (Archuleta et al., 2002). *M. a. avium* (serotype 1; strains 101 and 104) as well as *M. a. hominissuis* (serotype 4; strain 109; Carter et al., 2003) can form biofilm: these can be inhibited by low concentrations of clarithromycin (Carter et al., 2004). The ability of *M. a. avium* (serotype 1; strain 104; Yamazaki et al., 2006b) to form biofilm is associated with the capability to invade and translocate bronchial epithelial cells (Yamazaki et al., 2006a).

When the environment of 4 pig herds was investigated, *M. a. avium* was not detected in any of the 87 samples of liquid manure, 385 larvae, 67 imagoes and 12 syrphid fly exuvia (*Eristalis tenax*). However, it was noteworthy that *M. a. avium* was isolated from the puparium of one of these flies (Fischer et al., 2006). This finding indicated that *M. a. avium* present in the bodies of some animals, including protozoa, could be protected from the devitalising effect of the environment (Sections 6.1 and 6.2 or a review by Thomas and McDonnell, 2007). This effect was observed in murine bone marrow-derived macrophages that protected *M. a. avium* (serotype 2; strain TMC 724) from the destructive effects of clarithromycin (Frehel et al., 1997).

The impact of the environment on the virulence of *M. a. avium*. The virulence of *M. a. avium* is surely determined by many genes, of which the first ones have already been described (strain TMC 724, serotype 2; *ser2* gene cluster) and their activities studied (Belisle et al., 1991; Mills et al., 1994). A decade later, the genes that control the invasion of the intestinal epithelium were identified in two phyla, strains 101 and 104 of serotype 1 (Miltner et al., 2005). Upon infection of *Acanthamoeba castellanii* many genes of *M. a. avium* (serotype 1; strain 104) were upregulated (Tenant and Bermudez, 2006).

The effect of the environment on *M. a. avium* virulence can be evaluated with regard to two aspects:

- *External environment of a host organism* (e.g. soil, water and dust) can adversely affect *M. a. avium* virulence. It is highly probable that besides reduced

temperatures, other at present unrecognised factors can have significant effects on the host's environment: e.g. pH (Section 5.1), the presence of different inhibitory agents including natural and anthropogenic antibiotics (Section 3.1.1), a paucity of water in the soil, etc. (Section 5.5). Ultraviolet irradiation is also a significant factor. Its effects on *M. a. avium* have been described in many studies (see above): it has been shown to hamper *in vitro* capability to grow and stainability according to Ziehl-Neelsen (Murohashi and Yoshida, 1968) and also causes *in vitro* devitalisation (McCarthy and Schaefer, 1974).

The observation of isolates with reduced virulence for pullets suggests that the virulence of *M. a. avium* may be lost in the environment (Photo 3.12). For example, an isolate from peat (Matlova et al., 2005) and one isolate from soil (M. Pavlas, personal communication) failed to cause avian tuberculosis in pullets after intraperitoneal and intramuscular infections. This can be compared with the decreasing virulence for pullets of collection serotype strains stored for a long period of time. It seems that the virulence of the latter was most adversely affected by repeated subculture, freezing and lyophilisation (Pavlik et al., 2000b).

- *The internal environment of a host organism* (ranging from protozoa to vertebrate tissues) can also significantly affect the virulence of *M. a. avium*. For example, Thomas and McDonnell (2007) in their review article report the long-term survival of *M. a. hominissuis*, *M. a. paratuberculosis*, *M. intracellulare* and *M. scrofulaceum* in amoebae and their cysts. Cirillo et al. (1997) provided evidence that a *M. a. avium* strain (serotype 1; strain 101) multiplied in amoeba (*A. castellanii*) and documented its increasing virulence through biological experiments on beige mice.

In contrast, the decreasing virulence of *M. a. avium* for pullets was documented in some naturally highly resistant hosts (Photo 3.13). For example, *M. a. avium* isolates obtained from the faeces of waterfowl resistant to avian tuberculosis (family Threskiornithidae, sacred ibis; *Threskiornis aethiopicus*) were found to not be fully virulent for pullets (Dvorska et al., 2007).

That finding can be compared with the isolates of *M. a. avium* from human patients (Pavlik et al., 2000b), not typical hosts for the causative agent of avian tuberculosis. Other adverse effects on the virulence of *M. a. avium* in humans such as the temperature of their bodies should be considered as well. In any case, the metabolism of the majority of birds is more active, a fact that is manifested by a higher body temperature (up to 42 °C). In mammals (especially cattle and pigs) including humans, mean body temperature varies around the value 37 °C.

The effect of the passage of *M. a. avium* through various living organisms on the virulence of *M. a. avium* was mainly investigated in the 1950s, when different laboratory animals were used for the identification of clinical isolates. Hartwigk and Stottmeier (1963), for example, described an increasing virulence of *M. a. avium* after repeated infections of guinea pigs and a decreasing virulence after repeated infections of rabbits.

It can be seen, therefore, that there is much that remains to be discovered regarding the ecology of *M. a. avium* and rather general preventive measures should be advised. With very few exceptions, most avian species, domestic and wild mammals, and also humans can be naturally infected by *M. a. avium*. The environmental resistance of the causative bacilli combined with their wide distribution indicates how difficult it is to eradicate the disease. Moreover, in particular free scavenging and migratory birds (Section 6.6), apart from being a source, also play a role in spreading the disease over long distances, which complicates control and/or eradication programmes.

Preventive measures. Due to the above-mentioned facts, environmental conditions (an environment contaminated with *M. a. avium*) can pose a risk to host organisms (especially domestic and captive animals). When these animals are under stress it can greatly increase their susceptibility to the causal agent of avian tuberculosis. Domestic and captive animals, which are fed inadequate diets and those which are kept in crowded, wet, cold, poorly ventilated and unhygienic aviaries, may have an increased susceptibility to the causal agent of avian tuberculosis (I. Pavlik, unpublished data).

Disinfection is one of the important steps taken in controlling contamination and the spread of disease. It is, therefore, essential that all water and food containers along with cages and pens be thoroughly cleaned and disinfected daily. Personnel who attend to infected birds should take appropriate precautions

to avoid spreading the causal agent. The placing of a foot bath at a strategic place (Photo 3.14) also helps to reduce the spread of the causal agent of contamination between these workers. Although mycobacteria are known to be resistant to disinfectants, compounds like alcohols, aldehydes, phenolics, etc. have bactericidal activity. However, ordinarily it is impractical to render an infected environment satisfactorily safe by disinfection.

Therefore, the need to prevent the establishment of the infection, particularly in poultry and swine industries, assumes prime importance. With respect to this, early detection of the disease is essential. In poultry farms and/or small aviaries, and also in swine industries, a high standard of all husbandry and/or management practices, hygienic and strict quarantine measures have to be maintained as fundamental procedures. It is equally important that new flocks are introduced into a farm only from infection-free farms and that reliable diagnostic tests are undertaken. Special attention has to be given to small farm aviaries, which represent a potential source of infection for other farms (Photo 3.9).

3.1.3.2 *M. avium* subsp. *hominissuis*

Of all the above-mentioned *M. avium* subspecies, *M. a. hominissuis* is the most successful at growing outside a host organism. It is one of the mycobacteria that have been under investigation longest within the scope of ecological studies (Kazda, 1967, 1973; Kazda and Hoyte, 1972). In contrast to the other three subspecies, *M. a. avium*, *M. a. paratuberculosis* and *M. a. silvaticum*, it has no extra requirements for growth, can grow at varying temperatures (18–45 °C) and is mycobactin independent. With regard to ecology, the components of the environment where *M. a. hominissuis* is found and multiplies are important. *M. a. hominissuis* thus becomes a significant part of the microflora and microfauna involved in the degradation of organic matter.

All sources of *M. a. hominissuis* associated with the environment are documented in different chapters of this book. *M. a. hominissuis* has been recovered from sphagnum vegetation (Section 5.1.1; Photo 3.15), from dust formed by air flow across rivers, agricultural fields and parks, soil (Section 5.5), the air and dust

(Section 5.9), feedstuffs (Section 5.6), foodstuffs (Section 5.7) and from excreta (Section 5.10), among others. Water and biofilms on drinkers were found to be an important reservoir for *M. a. hominissuis*, and surface water in particular, plays a key role in the circulation of this subspecies (Section 5.2). Outside of living organisms, *M. a. hominissuis* has been found in many biotopes including wastewater (Section 5.10.3), sawdust (Section 5.8) and aerosols (Section 5.9.1).

The effects of physical factors on *M. a. hominissuis* as well as on *M. a. avium* have been investigated:

- The effect of chlorine on *M. a. hominissuis* (serotype 4; strain A5; Taylor et al., 2000; Cowan and Falkinham, 2001; Falkinham, 2003; Norton et al., 2004; Steed and Falkinham, 2006).
- The effect of chlorine dioxide and ozone on *M. a. hominissuis* (serotype 4; strain A5; Taylor et al., 2000).
- The effect of temperature on the removal of *M. a. hominissuis* (serotype 4; strain A5) from biofilms: the temperature used for killing must exceed 53 °C (Norton et al., 2004).
- The effect of ultraviolet irradiation: ultraviolet-irradiated monocytes efficiently inhibit the intracellular replication of *M. a. hominissuis* (serotype 4; strains LR 542 and LR 114; Mirando et al., 1992).

Within the scope of investigations of *M. a. hominissuis* ecology, the experiments on the protozoan *A. castellanii* should also be considered as significant. They indicated an ability of a strain of serotype 4 to grow saprozoically and to survive within cyst walls (Steinert et al., 1998). Not surprisingly, given the fact that *M. a. hominissuis* is present in both natural and drinking water, and in soils, it is difficult if not impossible to protect humans, animals and birds from exposure and infection (Section 7.5).

The following factors or actions may put an individual at risk of infection from animals:

Contact with *M. a. hominissuis*-contaminated vehicles: water, soil, dust, excrements from infected animals and contaminated bedding.
Consumption of *M. a. hominissuis*-contaminated food and water.
A decrease of local or general immunity of an organism.

3.1.3.3 *M. avium* subsp. *silvaticum*

M. a. silvaticum has also been isolated from tuberculous lesions in birds; it was isolated for the first time from wood pigeons *(Columba palumbus)*. This subspecies was formerly designated as wood pigeon *Mycobacterium*. It was suggested in 1990 to classify this *Mycobacterium* as a *M. avium* species (Thorel et al., 1990; Steinert et al., 1998). It can be distinguished from subspecies *M. a. avium* and *M. a. hominissuis* by its mycobactin dependence in the first subculture; it differs from subspecies *M. a. paratuberculosis* by the fact that it is virulent for birds. However, as in the case of *M. a. avium*, this is not a constant feature of this subspecies. It was found after repeated testing of the virulence of three collection strains of *M. a. silvaticum* (strains Nos. 6861, 5329 and 3135) that long-term storage resulted in a loss of virulence for pullets (Pavlik et al., 2000b).

By the end of 2007, a total of 26 publications dealing with *M. silvaticum* were listed in the database PubMed. It is noteworthy from the point of view of ecology that none of these isolates originated from the environment; the publications were in large part focused on the detection of this subspecies in different animal species and humans. The study of the ecology of this subspecies is largely hindered by its difficult isolation (the strains are mycobactin dependent and mostly require long-term incubation) and rare occurrence in clinical samples from both animals and humans. Accordingly, the investigation of the ecology of this subspecies can be viewed as a challenging area for scientists.

3.1.3.4 *M. avium* subsp. *paratuberculosis*

Since bovine tuberculosis has now been brought under control, paratuberculosis represents an increasingly large part of the epizootiology of other mycobacterial infections in ruminants in most countries with developed agriculture. According to official and non-official sources of information, paratuberculosis has now spread to all continents, and its prevalence in most countries has been increasing (Ayele et al., 2001; Harris and Barletta, 2001; Rowe and Grant, 2006). Economic losses as a result of infection and growing concerns over food safety are considerable (Hasonova and Pavlik, 2006) and would be significantly higher if the steadily growing suspicion that paratuberculosis may be involved in the development of Crohn's disease in humans be proven (Greenstein, 2003; Hruska et al., 2005; Skovgaard, 2007).

Since the discovery of the causative agent of paratuberculosis in 1895 (Johne and Frothingham, 1895), many studies on *M. a. paratuberculosis* occurrence outside hosts, i.e. in meat and meat products as well as in milk and milk products, have been published. However, this topic is dealt with in other sections of this book.

The first objective of these studies was to primarily highlight sources of *M. a. paratuberculosis* for young animals, above all ruminants (large numbers of young animals get infected through colostrum and milk and then spread the causative agent during adulthood). If the causative agent is spread in this way, the time required for effective sanitation of infected herds of cattle and other ruminants is much longer. The second objective of these studies was to highlight the fact that the human population is exposed to *M. a. paratuberculosis* from different sources including foodstuffs (Section 5.7).

The above-mentioned facts indicate that it is important to study the ecology of the causative agent of paratuberculosis. However, a review of the published data shows that only recently has concentrated attention been paid to the ecology of *M. a. paratuberculosis*. The first studies dealing with the survival of *M. a. paratuberculosis* in different components of the environment were published in the 20th century (Vishnevskii et al., 1940; Lovell et al., 1944; Larsen et al., 1956; Stuart, 1965; Jorgensen, 1977; Olsen et al., 1985). These reports which focused on the survival of this *Mycobacterium* under different conditions were reviewed in the introduction to the study of Whittington et al. (2004). However, with reference to the above-mentioned studies, except for the study of Whittington et al. (2004), it was difficult to define conditions for the investigation of *M. a. paratuberculosis* survival, the amount of the organism constituting an infective dose and methods of its recovery.

The occurrence of live but non-culturable forms largely hinders the investigation of *M. a. paratuberculosis* survival. *M. a. paratuberculosis* can be observed in the following forms:

- With a fully developed cell wall (see below for bacterial characterisation of this causative agent).
- The cell-wall-deficient form; for detailed information see the review of Beran et al. (2006a).

Bacteriological characteristics. *M. a. paratuberculosis* is a slow-growing (only at around 37 °C) mycobactin-dependent bacterium. *M. a. paratuberculosis* contains the specific insertion sequence IS*900* at up to 18 copies per genome (Pavlik et al., 1999). Due to this, the sequence can be used for both the identification of *M. a. paratuberculosis* and in its epidemiological study (Pavlik et al., 1999; Bartos et al., 2006). Using RFLP analysis (Collins et al., 1990) *M. a. paratuberculosis* has been classified into three groups:

i. RFLP type cattle (C).
ii. RFLP type sheep (S).
iii. RFLP type intermediate (I).

At present, a one-copy specific 620 bp long fragment designated f57 is used for the quantification of *M. a. paratuberculosis* by means of quantitative real-time PCR (Poupart et al., 1993).

Interaction with host organism. Infected animals can, according to the intensity of the shedding of *M. a. paratuberculosis*, be categorised as non-shedders, low shedders and high shedders (Toman et al., 2003). Based on this aspect and on the clinical status, the animals in an infected herd may be classified into three groups according to their stage of disease:

i. *Susceptible animals* (clinically normal or ill animals, continuously shedding *M. a. paratuberculosis*, and clinically normal animals, non-intensive and non-regularly shedding *M. a. paratuberculosis*; Photo 3.16).
ii. *Resistant animals* (clinically normal animals, never shedding *M. a. paratuberculosis*).
iii. *Recovered or "super-resistant" animals* (formerly *M. a. paratuberculosis*-shedding animals who have spontaneously recovered).

Growing abilities *in vitro*. The successful multiplication of *M. a. paratuberculosis* was achieved for the first time in a pure culture in Great Britain in 1910 (Twort and Ingram, 1912). The majority of strains of the causative agent of paratuberculosis can be isolated in a medium containing a growth stimulator derived from a soluble component of the bacterial walls of mycobacteria. First of all, an alcohol extract of *M. phlei* designated as mycobactin was used (Francis et al., 1953). The designation of the *Mycobacterium* of origin was later added to the names. Mycobactin from *M. phlei* was renamed as mycobactin P, from *M. tuberculosis* as mycobactin T and from *M. a. paratuberculosis* (formerly designated as *M. johnei*) as mycobactin J (Snow, 1965; Merkal et al., 1968; Merkal and McCullough, 1982). Mycobactin J, which reduces the isolation time by up to 3 weeks and increases the numbers of detected CFU by 10–20%, is used as a growth stimulator at present (McCullough and Merkal, 1982).

Merkal and Curran (1974) discovered that *M. a. paratuberculosis* can be isolated in media enriched with 1% ferric ammonium citrate and sodium pyruvate. The effect of sodium pyruvate was only described later (Jorgensen, 1982). According to growth capability *in vitro* (apparent CFU) at 37 °C, *M. a. paratuberculosis* strains may be classified into the following three groups:

i. *Growing strains*, usually detected in domestic ruminants – cattle and goats – and in wild ruminants – cervids. This group may be divided into two subgroups: (a) relatively fast growing *M. a. paratuberculosis* with growth after 1–3 months and (b) relatively slowly growing *M. a. paratuberculosis* with growth observed after 4 months of incubation (Pavlik et al., 1999).

ii. *Strains that do not grow in subculture* (primary isolations only), usually detected in non-typical hosts such as small terrestrial mammals, carnivores, birds and others, and in the environment or foodstuffs – mostly pasteurised milk (Ayele et al., 2005).

iii. *Non-growing strains*, acid-fast rods (AFR) detected by histology in heavily infected tissues, especially from sheep and moufflons, and occasionally from cattle (Ayele et al., 2004).

RFLP analysis of all these three groups of *M. a. paratuberculosis* strains has not shown a positive association between their genotype and phenotype (i.e. their growth capability *in vitro*). Among domestic ruminants, *M. a. paratuberculosis* strains of RFLP type "C" (cattle) are usually detected in cattle, buffalo and goats, and among wild ruminants, in red deer, fallow deer and roe deer. In contrast, *M. a. paratuberculosis* strains of RFLP types "S" and "I" are usually detected in sheep from among domestic ruminants and from among wild ruminants in wild sheep (moufflons). Nevertheless, the adaptation of *M. a. paratuberculosis* to a specific host has not been confirmed. The PFGE method has revealed a relationship between the genotype of pigmented strains and *M. a. paratuberculosis* strains of the ovine RFLP type.

We can expect that in large part innovation in techniques of culture of the above-mentioned mycobacteria from the environment will significantly contribute to the study of their ecology:

Various decontamination procedures (Gwozdz, 2006).
Various culture media (Sung and Collins, 2003).
Recovery of *M. a. paratuberculosis* from water by the strategy of immune magnetic separation or other techniques before the process of decontamination (Whan et al., 2005a).
Development of techniques using various liquid culture media (Cernicchiaro et al., 2008).

Detection of *M. a. paratuberculosis* in samples from the environment can also be facilitated by the introduction of culture-independent detection techniques based on PCR (Jaravata et al., 2006; Cook and Britt, 2007).

Growing abilities *in vivo*. *M. a. paratuberculosis* mostly enters the host organism through the oral route, less frequently via the uterus or respiratory tract. The intestinal tract is viewed as the most significant entrance gateway for the infection. The rate of *M. a. paratuberculosis* multiplication on the intestinal mucosa is related to the natural immunity of animals and is based on the ability of macrophages to reduce the intracellular multiplication of mycobacteria. A number of experimental studies and in particular four reviews published in the last decade have been focused on these factors (Ayele et al., 2001; Harris and Barletta, 2001; Rowe and Grant, 2006).

Persistence/dormancy in the environment. The causative agent of paratuberculosis is relatively resistant and survives in the environment for a long time: in river water for 270 days, in faeces and black soil for 11 months, in liquid manure at 5 °C for 252 days, but in urine for 7 days only. If frozen to –14 °C, it can survive for at least 1 year (Jorgensen, 1977; Richards, 1981). *M. a. paratuberculosis* can survive inside an infected host organism for a very long time (up to several years). In the environment, the causative agent of paratuberculosis can survive in a state of metabolic quiescence for several months or years (Whittington et al., 2004). Research that might contribute to an explanation underlying these processes should be regarded as being extremely important.

Since 2000, studies dealing with the occurrence and survival of *M. a. paratuberculosis* have been focused on the following two areas:

i. Occurrence of *M. a. paratuberculosis* in surface and drinking water, sewage and in drinking water for animals (Whittington et al., 2003, 2005; Raizman et al., 2004; Pickup et al., 2005; Whan et al., 2005b; Pickup et al., 2006; Berghaus et al., 2006; Norby et al., 2007).
ii. Occurrence of *M. a. paratuberculosis* in liquid manure, dung, lagoons containing animal excrements (Fischer et al., 2003; Lombard et al., 2006; Grewal et al., 2006) and the surroundings of farms (Whittington et al., 2003; Raizman et al., 2004; Berghaus et al., 2006; Lombard et al., 2006).

Generally, we can say that as it is present in various components of the environment *M. a. paratuberculosis* can potentially infect typical and non-typical hosts including humans. However, all the conditions leading to devitalisation of the causative agent of paratuberculosis in the environment or stimulating its growth must be well understood before an assessment of the risks can take place (see below).

Growing abilities outside of a homoiotherm host organism. Only two publications are available which deal with the multiplication of *M. a. paratuberculosis* in an environment outside of warm-blooded host organisms. One of them describes slow growth of the causative agent inside the protozoan *A. polyphaga* that had phagocytised them (Mura et al., 2006). The other publication describes the same phenomenon in the protozoan *A. castellanii* (Whan et al., 2006).

Prevention. Efficacious preventive measures against the occurrence of *M. a. paratuberculosis* in different components of the environment can be adopted after the control of paratuberculosis in infected herds of cattle and free-living animals by the systematic culling of the infected animals. All the subsequent preventive measures aimed at decreasing the risk of infection for animals and humans from the environment are of only very limited effectivity. Various procedures and preparations have been used for the disinfection of different components of the environment such as liquid manure. For example, experiments with the application of radiofrequency power in wastewater management have been performed (Lagunas-Solar et al., 2005) and different, mostly chlorine-based disinfectants have been tested and used (Whan et al., 2001). However, the procedures which have been applied thus far have not been 100% effective and have failed to kill of all *M. a. paratuberculosis* cells in the environment.

With regard to bringing the infectious agent under control in herds of ruminants, especially cattle, a well-managed rearing system of young animals can highly contribute to preventing the spread of the causative agent of paratuberculosis (Sanderson et al., 2000; Wells, 2000; Wells and Wagner, 2000; Pavlik et al., 2000a). However, it is true at present that the most effective method to prevent the spread of the causative agent of paratuberculosis in the environment is the control of this pathogen in the infected herds of ruminants that have become its main source.

3.1.4 *M. intracellulare*

3.1.4.1 Introduction

M. intracellulare has been included as a member of the *M. avium* complex because of its phenotypic and genetic relatedness to *M. avium* and the historic difficulty in separating *M. avium* from *M. intracellulare* isolates by cultural, biochemical and enzymatic tests. In spite of this close relationship, it is important to point out that *M. intracellulare* is a distinct species. Specifically, DNA–DNA hybridisation measurements document that *M. avium* and *M. intracellulare* strains only share approximately 50% similarity (Baess, 1979, 1983). Furthermore, there are differences in the ecology and epidemiology. Today, a variety of different approaches (e.g. HPLC of mycolic acids, DNA probes, RFLP of amplified gene products and gene sequencing) are available to distinguish the two species. In part, the lack of comprehensive knowledge concerning the ecology and epidemiology of *M. intracellulare* is due to the fact that they are considered together with

M. avium and are reported as *MAC* or *MAIC*. Such practice is strongly discouraged because it can lead to the loss of species-specific knowledge.

3.1.4.2 Genetics

Little is known of the genetics and genomics of *M. intracellulare*. The species shares plasmids in common with *M. avium* and *M. scrofulaceum*, demonstrating that horizontal gene transmission occurs between these species (Jucker and Falkinham, 1990). Because *M. avium*, *M. intracellulare* and *M. scrofulaceum* plasmids are large (i.e. >20 kb and as large as 250 kb) and as much as 30% of DNA in *M. avium* and *M. intracellulare* strains can be in plasmids (Meissner and Falkinham, 1986), plasmids are important to members of these three species. For example, plasmids carry genes for mercury (Meissner and Falkinham, 1984) and copper resistance (Erardi et al., 1987), permitting their survival in heavy metal-polluted environments.

Fingerprinting of *M. intracellulare* strains can be performed by the analysis of large restriction fragments separated by pulsed field gel electrophoresis (De Groote et al., 2006). Unfortunately, few *M. intracellulare* strains carry IS*1245* and IS*1311*, the related insertion sequences used in fingerprinting *M. avium* strains (Beggs et al., 2000). To date, no insertion sequence unique to *M. intracellulare* has been identified. In a survey of 92 *M. avium* and 57 *M. intracellulare* isolates from patients, it was shown that all *M. avium* strains carried the gene for the macrophage-induced protein (Plum et al., 1997), whereas none of the *M. intracellulare* strains contained that gene (Beggs et al., 2000).

3.1.4.3 Ecology

M. intracellulare has been recovered from almost every environmental habitat where it has been sought including natural water (Kazda, 1977; Falkinham et al., 1980; von Reyn et al., 1993), drinking water (von Reyn et al., 1993; Falkinham et al., 2001; Vaerewijck et al., 2005), soil (Brooks et al., 1984), sphagnum vegetation (Kazda, 1978a,b), potting soil (De Groote et al., 2006) and aerosols (Wendt et al., 1980). It is of interest to note that neither *M. intracellulare* nor *M. avium* was recovered from soil samples collected in alpine or subalpine habitats that yielded other environmental mycobacteria (Thorel et al., 2004). Although *M. avium* and *M. intracellulare* share the same habitats, namely soil and water, there are some differences. Significantly higher numbers of *M. intracellulare* (i.e. 600 CFU/cm^2) compared to *M. avium* (0.3 CFU/cm^2) were recovered from biofilms in drinking water distribution systems (Falkinham et al., 2001). This suggests a stronger preference for surface and interfacial attachment of *M. intracellulare* over *M. avium*.

3.1.4.4 Physiologic Ecology

Strains of both *M. avium* and *M. intracellulare* can grow in water (George et al., 1980), and their growth is stimulated by humic and fulvic acids (Kirschner et al., 1999). *M. intracellulare* and *M. avium* can grow in the grey layer of sphagnum vegetation (Kazda, 1978a), consistent with their recovery from aerosols of potting soil and from patients with mycobacterial pulmonary disease (De Groote et al., 2006). *M. intracellulare*, like *M. avium*, is readily aerosolised and concentrated in droplets ejected from water (Parker et al., 1983), consistent with its isolation from aerosols (Wendt et al., 1980). Both *M. intracellulare* and *M. avium* grow in the phagocytic, free-living protozoan *Tetrahymena pyriformis* (Strahl et al., 2001) and thereby are capable of surviving predation by protists in water and soil.

M. intracellulare, like *M. avium*, is resistant to disinfectants (Taylor et al., 2000) and antimicrobial compounds (Jones and Falkinham, 2003). Disinfectant resistance coupled with the preference for biofilm formation and the accompanying increased resistance of biofilm-grown cells (Steed and Falkinham, 2006) contribute directly to the persistence of both of these opportunistic pathogens in drinking water distribution systems. *M. intracellulare*, as with a number of other environmental mycobacteria, is relatively heat resistant. For example, at 60 °C, 90% of *M. intracellulare* cells survived for 1.5 min (Schulze-Robbecke and Buchholtz, 1992). Thus, even in water heated to high temperatures in buildings and homes, significant numbers of *M. intracellulare* cells survive.

3.1.5 *M. malmoense*

The first description of *M. malmoense* was made in 1977. Seven isolates from four patients from Malmo (Sweden) with clinical and x-ray evidence of pulmonary mycobacteriosis were described. This nonchromogenic mycobacterial species is very slow growing, dysgenic and exposure to light does not result in colony pigmentation (Photo 3.17; Schroder and Juhlin, 1977). Further cases have primarily taken the form of lung infections and only about half of the patients had predisposing conditions, such as pre-existing disease or other risk factors: rheumatoid arthritis type IIb, lung diseases, tobacco smoking and a history of alcohol abuse. In most cases, isolation from adult patients was consistent with infection, not simply colonisation (Table 3.7). Extrapulmonary cases and disseminated diseases have also been described (Table 3.8). In children and young patients, cervical lymphadenitis or adenitis infections have been reported (Table 3.9).

3.1.5.1 *M. malmoense* in Humans

The majority of reported cases have come from Europe (especially the UK, Sweden, Finland, the Netherlands and Switzerland) and Northern America (Canada and USA) as documented in Tables 3.7, 3.8 and 3.9. The incidence is highest in the Northern European countries such as the UK, Ireland, Sweden and Finland. In the USA the occurrence of infections caused by *M. malmoense* is very low (Buchholz et al., 1998). O'Brien et al. (1987) conducted a survey among state and large city laboratories for the 2-year period and recorded four (0.07%) *M. malmoense* isolates among a total of 5469 non-tuberculous mycobacteria reported.

Three common threads run through clinical reports of *M. malmoense* infections:

- The difficulty in making a diagnosis because of the slow growth (up to 12 weeks at 32–37 °C) of the organism and the possibility of its confusion with other mycobacteria. It is possible to use specially modified medium and hold the cultures for 12 weeks before discarding them (Jenkins and Tsukamura, 1979). In liquid culture systems, the detection of *M. malmoense* is shorter (up to a few weeks).

- Misdiagnoses or late diagnosis leading to the death of patients.
- The poor correlation between *in vitro* drug susceptibility results and clinical response to chemotherapy (Banks et al., 1983, 1985; France et al., 1987).

In some healthy people samples of sputum as well as faeces or urine could be culture positive for *M. malmoense* (Tables 3.7 and 3.10). In HIV/AIDS patients, the patient's pulmonary or other tissue lesions progressed and the patients died with marked cachexia. There does not appear to be a unique geographic focus for *M. malmoense* because patients from southern European countries (such as Italy, Greece or Spain) never travelled abroad. In the absence of any reports of person-to-person transmission, infection must be from environmental sources.

3.1.5.2 Occurrence of *M. malmoense* in Soil and in Water

Published results concerning the detection of *M. malmoense* in the environment are scarce (Table 3.10). Its detection in soil and water in Zaire is associated with the same case studied in detail. Portaels et al. (1988b) showed that the "cellular composition regarding fatty acids, glycolipids and mycolic acids" of the isolates from water, soil and five patients was identical and belonged to type 1 (Table 3.10).

The presence of *M. malmoense* has been confirmed in soil in Japan, France and in stream water in Finland (Table 3.10). From the point of view of ecology, the culture detection of *M. malmoense* in a subalpine region in France not only in soil but also in humus and sphagnum is highly noteworthy. These results indicate that *M. malmoense* micro-organisms circulating in the environment of not yet recognised niches are plentiful. Of 120 mycobacterial isolates from the abovementioned environment in France, 14 (11.7%) were members of this species (Thorel et al., 2004).

3.1.5.3 Clinical Significance of *M. malmoense* from Soil and Water for Animals and People

Exposure of the animal population to *M. malmoense* has been revealed by the occasional detection of this

Table 3.7 *Mycobacterium malmoense* detected in human respiratory tract

Country (state, city)	No.	Patients description[1]	Infected tissue	Ref.
Canada (British Columbia)	1	44/M	Pulmonary disease	[1]
Czech Republic	1	56/M	Pulmonary infection (living in industrial centre with high pollution)	[13]
Denmark	20		Isolation from clinical samples during the 2-year nationwide survey	[18]
France	1	?/?	Chronic obstructive pulmonary disease medico-surgically treated	[12]
Germany	2	*HIV-*	Pulmonary diseases	[5]
Italy (Belluno)	1	33/M (*AIDS+*)	Pulmonary interstitial infiltrate	[19]
Italy (Bergamo)	1	32/M	Pneumonia	[19]
Italy (Bergamo)	1	75/F	Pulmonary cavities	[19]
Italy (Bergamo)	1	81/F	Chronic bronchiectasias (clinical significance uncertain)	[19]
Italy (Brescia)	1	30/M (*AIDS+*)	Pulmonary infiltrate	[19]
Italy (Castelnuovo)	1	64/M	Pulmonary cavities	[19]
Italy (Florence)	1	68/M	Cavitary tuberculosis	[19]
Italy (Florence)	1	74/M	Pulmonary cavities	[19]
Sweden (Malmo)	4		Lung infections	[15]
Sweden	171		Pulmonary infection (170 adult and 1 child)	[10]
Switzerland	1	45/M	Pulmonary diseases (alcoholic and smoker), right upper lobe lobectomy	[14]
The Netherlands	2	64/M, 66/M	Chronic obstructive pulmonary disease treated by inhalation of corticosteroids	[16]
UK (England)	16		Pulmonary infections	[6]
UK (England)	3		Pulmonary diseases with chronic necrotising pulmonary aspergillosis	[9]
UK (England)	1	*HIV+*	Sputum positive smear for AFR pulmonary disease	[17]
UK (England)	1	?/?	Pulmonary disease in disabling myelopathy patient	[21]
UK (Scotland)	41	only one *HIV+*	30 patients had pulmonary diseases	[2]
UK (Scotland)	79		Infected respiratory tract	[4]
UK (Scotland)	20		19 patients had pulmonary diseases	[8]
UK (Wales)	8		Lung infections	[11]
USA (Virginia)	1	43/M	Pulmonary disease – sputum and bronchial washings (first case in the USA)	[20]
USA	1	26/M	Right upper lobe infiltrate and cavitation (sputum and blood cultures)	[7]
USA	70		5 patients pulmonary diseases, 65 patients only isolation from sputum	[3]

[1] Age in years/gender (sex): **M** (male), **F** (female), **?** (not known). **AFR** Acid-fast rods.

[1] Al Moamary MA, Black W, Elwood K (1998) Can. Respir. J. 5:135–138. [2] Bollert FG, Watt B, Greening AP, Crompton GK (1995) Thorax 50:188–190. [3] Buchholz UT, McNeil MM, Keyes LE, Good RC (1998) Clin. Infect. Dis. 27:551–558. [4] Doig, Muckersie L, Watt B, Forbes KJ (2002) J. Clin. Microbiol. 40:1103–1105. [5] Enzensberger R, Hunfeld KP, Krause M, Rusch-Gerdes S, Brade V, Boddinghaus B (1999) Eur. J. Clin. Microbiol. Infect. Dis. 18:579–581. [6] Evans AJ, Crisp AJ, Colville A, Evans SA, Johnston ID (1993) AJR Am. J. Roentgenol. 161:733–737. [7] Fakih M, Chapalamadugu S, Ricart A, Corriere N, Amsterdam D (1996) J. Clin. Microbiol. 34:731–733. [8] France AJ, McLeod DT, Calder MA, Seaton A (1987) Thorax 42:593–595. [9] Hafeez I, Muers MF, Murphy SA, Evans EG, Barton RC, McWhinney P (2000) Thorax 55:717–719. [10] Henriques B, Hoffner SE, Petrini B, Juhlin I, Wahlen P, Kallenius G (1994) Clin. Infect. Dis. 18:596–600. [11] Jenkins PA (1981) Rev. Infect. Dis. 3:1021–1023. [12] Job V, Lacaze O, Carricajo A, Fournel P, Vergnon JM (2004) Rev. Mal Respir. 21:993–996. [13] Kaustova J, Meissner V, Reischl U, Satinska J, Vincent V, Naumann L (2003) Studia Pneumologica Phtiseologica 63:59–63. [14] Leuenberger P, Aubert JD, Beer V (1987) Respiration 51:285–291. [15] Schroder KH, Juhlin I (1977) International Journal of Systematic Bacteriology 27:241–246. [16] Smeenk FW, Klinkhamer PJ, Breed W, Jansz AR, Jansveld CA (1996) Ned. Tijdschr. Geneeskd. 140:94–98. [17] Sullivan AK, Hannan MM, Azadian BS, Easterbrook PJ, Gazzard BG, Nelson MR (1999) Int. J. STD AIDS 10:606–608. [18] Thomsen VO, Andersen AB, Miorner H (2002) Scand. J. Infect. Dis. 34:648–653. [19] Tortoli E., Piersimoni C, Bartoloni A, Burrini C, Callegaro AP, Caroli G, Colombrita D, Goglio A, Mantella A, Tosi CP, Simonetti MT (1997) Eur. J. Epidemiol. 13:341–346. [20] Warren NG, Body BA, Silcox VA, Matthews JH (1984) J. Clin. Microbiol. 20:245–247. [21] White VL, Al Shahi R, Gamble E, Brown P, Davison AG (2001) Thorax 56:158–160.

Table 3.8 *Mycobacterium malmoense* detected in humans with extrapulmonary location (except lymphadenitis – see Table 3.3)

Country (state, city)	No.	Description[1]	Infected tissue	Ref.
Belgium	1	68/M	Sputum and gastric lavage (gastric adenocarcinoma)	[14]
Czech Republic	1	57/M	Pulmonary disease and professional vasoneurosis	[9]
Finland	2	?/?	Chronic skin disorders (no granulomatous lesions found)	[10]
Finland	1	?/?	Infected in skin *prurigo nodularis*	[11]
France	1	75/M	Several similar nodules (10–70 mm) on right arm and papulokeratotic lesions on fingers	[4]
France	1	75/F (*HIV*-)	Cutaneous non-caseating nodules on the back of left hand	[16]
Germany	1	73/?	Granulomatous inflammation of flexor tendon sheaths of wrist	[13]
Greece	1	32/F(*HIV*-)	Pituitary granuloma	[7]
Ireland	1	48/F	Extrapulmonary and disseminated diseases (leukaemia)	[8]
Ireland	1	?/?	Ganglion-like swelling of the flexor (tenosynovitis) of the wrist	[17]
Italy (Bergamo)	1	57/M	Tenosynovitis	[18]
Italy (Bergamo)	1	65/M	Pleurisy	[18]
Spain (Alicante)	1	48/M (*AIDS*+)	Pustular psoriasis and pulmonary infection	[15]
Sweden	1	?/?	Septic cutaneous lesions (culture positivity in skin scrapings and sputum); hairy cell leukaemia	[2]
Switzerland	5	?/?	Extrapulmonary and disseminated infections	[20]
UK (England)	1	?/?	Cold abscess on the hand possibly associated with a hydrocortisone injection (diabetic)	[5]
UK (England)	1	53/F	Right knee intense prepatellar bursitis	[1]
UK (England)	2	Children	Salivary gland masses confirmed by culture	[3]
UK (England)	1	?/F	Middle-aged woman with infected larynx	[12]
UK (England)	1	61/M	Septic arthritis of an interphalangeal joint and osteomyelitis (20-year rheumatoid arthritis treated with steroids)	[19]
USA	1	31/M	Ring-enhancing lesion in the left basal ganglia with slight mass effect and focal enhancement in right thalamic and left parietal areas (blood culture)	[6]

[1] Age in years/gender (sex): **M** (male), **F** (female), **?** (not known).

[1] Callaghan R, Allen M (2003) Ann. Rheum. Dis. 62:1047–1048. [2] Castor B, Juhlin I, Henriques B (1994) Eur. J. Clin. Microbiol. Infect. Dis. 13:145–148. [3] Cox HJ, Brightwell AP, Riordan T (1995) J. Laryngol. Otol. 109:525–530. [4] Doutre MS, Beylot C, Maugein J, Boisseau AM, Long P, Royer P, Roy C (1993) J.R. Soc. Med. 86:110–111. [5] Elston RA (1989) Lancet 1:1144. [6] Fakih M, Chapalamadugu S, Ricart A, Corriere N, Amsterdam D (1996) J. Clin. Microbiol. 34:731–733. [7] Florakis D, Kontogeorgos G, Anapliotou M, Mazarakis N, Richter E, Bruck W, Piaditis G (2002) Clin. Endocrinol. (Oxf) 56:123–126. [8] Gannon M, Otridge B, Hone R, Dervan P, Oloughlin S (1990) International Journal of Dermatology 29:149–150. [9] Kaustova J (2004) Unpublished data. [10] Mattila JO, Katila ML, Vornanen M (1996) Clin. Infect. Dis. 23:1043–1048. [11] Mattila JO, Vornanen M, Vaara J, Katila ML (1996) J. Am. Acad. Dermatol. 34:224–228. [12] McEwan JA, Mohsen AH, Schmid ML, McKendrick MW (2001) J. Laryngol. Otol. 115:920–922. [13] Osterwalder C, Salfinger M, Sulser H (1992) Handchir. Mikrochir. Plast. Chir 24:210–214. [14] Portaels F, Denef M, Larsson L (1991) Tubercle. 72:218–222. [15] Ruiz M, Rodriguez JC, Garcia-Martinez J, Escribano I, Sirvent E, Gutierrez F, Rodriguez-Valera F, Royo G (2002) Res. Microbiol. 153:33–36. [16] Schmoor P, Descamps V, Lebrun-Vignes B, Crickx B, Grossin M, Nouhouayi A, Belaich S (2001) Ann. Dermatol. Venereol. 128:139–140. [17] Syed AA, O'Flanagan J (1998) J. Hand Surg. [Br.] 23:811–812. [18] Tortoli E., Piersimoni C, Bartoloni A, Burrini C, Callegaro AP, Caroli G, Colombrita D, Goglio A, Mantella A, Tosi CP, Simonetti MT (1997) Eur. J. Epidemiol. 13:341–346. [19] Whitehead SE, Allen KD, Abernethy VE, Feldberg L, Ridyard JB (2003) Journal of Infection 46:60–61. [20] Zaugg M, Salfinger M, Opravil M, Luthy R (1993) Clin. Infect. Dis. 16:540–549.

Table 3.9 *Mycobacterium malmoense* lymphadenitis

Country (state, city)	Patients		Infected tissue	Ref.
	No.	Description[1]		
Czech Republic	1	28/F	Cervical lymphadenitis on the left side of the neck	[7]
Finland	2	Children	Cervical adenitis	[6]
Germany	1	1.5/F (*HIV-*)	Craniojugular neck mass (cervical lymphadenitis)	[2]
Germany	1	2.5/M (*HIV-*)	Infra-auricular neck mass (cervical lymphadenitis)	[2]
Germany	1	5/M (*HIV-*)	Unilateral cervical lymphadenitis (in good health)	[3]
Italy (Ancona)	1	5/F	Mycobacterial lymphadenitis	[9]
Italy (Ancona)	1	5/F	Cervical lymphadenopathy	[10]
Italy (Pisa)	1	6/M	Cervical lymphadenopathy	[10]
Italy (Pordenone)	1	6/M	Cervical lymphadenopathy	[10]
Italy (Pisa)	1	9/M	Cervical lymphadenopathy	[10]
Spain (Zaragoza)	1	8/M (*HIV-*)	Cervical lymphadenitis – superficial lymph node in the left cervical area	[8]
Spain (Zaragoza)	1	2/F (*HIV-*)	Cervical lymphadenitis – superficial lymph node in the right cervical area	[8]
Sweden	35	Children	Cervical lymphadenitis	[4]
Sweden	1	Adult	Cervical lymphadenitis	[4]
Sweden	11	Children	Cervical lymphadenitis	[5]
USA	1	?/?	Cervical lymphadenitis	[1]

[1] Age in years/gender (sex): **M** (male), **F** (female), **?** (not known).
[1] Buchholz UT, McNeil MM, Keyes LE, Good RC (1998) Clin. Infect. Dis. 27:551–558. [2] Dunne AA, Kim-Berger HS, Zimmermann S, Moll R, Lippert BM, Werner JA (2003) Otolaryngol. Pol. 57:17–23. [3] Fabbri J, Welgelussen A, Frei R, Zimmerli W (1993) Schweizerische Medizinische Wochenschrift 123:1756–1761. [4] Henriques B, Hoffner SE, Petrini B, Juhlin I, Wahlen P, Kallenius G (1994) Clin. Infect. Dis. 18:596–600. [5] Hoffner SE, Henriques B, Petrini B, Kallenius G (1991) J. Clin. Microbiol. 29:2673–2674. [6] Katila ML, Brander E, Backman A (1987) Tubercle. 68:291–296. [7] Kaustova J (2004) Unpublished data. [8] Lopez-Calleja AI, Lezcano MA, Samper S, de Juan F, Revillo MJ (2004) Eur. J. Clin. Microbiol. Infect. Dis. 23:567–569. [9] Piersimoni C, Felici L, Penati V, Lacchini C (1995) Tuber. Lung Dis. 76:171–172. [10] Tortoli E., Piersimoni C, Bartoloni A, Burrini C, Callegaro AP, Caroli G, Colombrita D, Goglio A, Mantella A, Tosi CP, Simonetti MT (1997) Eur. J. Epidemiol. 13:341–346.

pathogen in skin lesions from cats and from bovine lymph nodes without gross lesions (Table 3.10). The exposure of the human population to *M. malmoense* was confirmed by its detection in sputum (Table 3.7), urine and stool from people with no clinical signs of disease (Table 3.10) and in sputum from individuals with clinical manifestations of disease (Table 3.7). It follows from numerous epidemiological studies analysing hundreds of patients infected with *M. malmoense* as documented by culture (Table 3.10) that

i. The ratio of infected men to women in Sweden (Henriques et al., 1994), Denmark (Thomsen et al., 2002) and USA (Buchholz et al., 1998) are similar.

ii. In *HIV*-negative, healthy children, numbers of cases of cervical lymphadenitis in boys and girls are equal (Table 3.9).

iii. In adult men, both pulmonary (Table 3.7) and extrapulmonary forms of the disease have been found (Table 3.8).

In summary, it would appear that the source of *M. malmoense* is not principally occupational, because both men and women are equally infected. Based on isolation and the pattern of disease, both water and soil are implicated as sources and the routes of infection are either via aerosols or via the gastrointestinal tract.

3.1.5.4 Views and Perspectives on the Research

Knowledge of *M. malmoense* ecology is scarce (Table 3.10) and the assumed routes of infection for people and animals are derived from infected organs (Tables 3.7, 3.8 and 3.9). Accordingly, further research should be focused on the following issues:

i. Application of culture-independent methods, particularly PCR (Hughes et al., 1997; Kauppinen et al., 1999); see also Section 5.5 and systems for rapid detection of *M. malmoense* (Hoffner et al.,

Table 3.10 *Mycobacterium malmoense* detected in animals, human excreta and the environment

Sample description		Anamnestic data about the case		Ref.
Origin	Country (area)	Sample	Tissue (matrix) and notes	
Cat	New Zealand	9-year-old male domestic long hair cat	Multiple cutaneous and subcutaneous lesions on ventrum and dorsum	[1]
		9-year-old male chinchilla cat	Lump on foreleg	[1]
Cattle	UK (Northern Ireland)	5 cattle (*Bos taurus*)	Healthy lymph nodes from positive skin-tested animals to bovine tuberculin (5 positive out of 48 examined)	[2]
Human	Belgium	1 faecal isolate	Stool from 50 healthy European volunteers was examined	[5]
	Italy (Bergamo)	1 urine isolate	Without clinical signs	[8]
	Zaire	5 isolates	Type I	[4]
Water	Zaire	1 isolate	Type II	[4]
	Finland	3 stream water isolates	0.5% of 760 stream water isolates were *M. malmoense*	[3])
Soil	Finland	1 isolate from geese farm		[3]
	Zaire	1 isolate	Type I	[4]
	Japan	1 isolate	First description in Japan environment	[6]
	France	12 isolates	Jura forest and Alpine moor	[7]
Humus	France	1 isolate	Jura and subalpine forest	[7]
Sphagnum	France	1 isolate	Jura forest	[7]

[1]. Hughes MS, Ball NW, Beck LA, de Lisle GW, Skuce RA, Neill SD (1997) J. Clin. Microbiol. 35:2464–2471. [2] Hughes MS, Ball NW, McCarroll J, Erskine M, Taylor MJ, Pollock JM, Skuce RA, Neill SD (15-6-2005) Vet. Microbiol. 108:101–112. [3] Katila ML, Iivanainen E, Torkko P, Kauppinen J, Martikainen P, Vaananen P (1995) Scand. J. Infect. Dis. Suppl 98:9–11. [4] Portaels F, Larsson L, Jenkins PA (1995) Tuber. Lung Dis. 76:160–162. [5] Portaels F, Larsson L, Smeets P (1988) Int. J. Lepr. Other Mycobact. Dis. 56:468–471. [6] Saito H, Tomioka H, Sato K, Tasaka H, Dekio S (1994) Microbiol. Immunol. 38:313–315. [7] Thorel MF, Falkinham JO, Moreau RG (1-9-2004) Fems Microbiology Ecology 49:343–347. [8] Tortoli E., Piersimoni C, Bartoloni A, Burrini C, Callegaro AP, Caroli G, Colombrita D, Goglio A, Mantella A, Tosi CP, Simonetti MT (1997) Eur. J. Epidemiol. 13:341–346.

1991) to study this pathogen's presence in different vehicles for transmission (i.e. water, soil, aerosols and dust).

ii. Identification of predisposing factors for disease development in humans. The first studies (Gelder et al., 2000; Jones et al., 2001) indicate that there exist many other non-recognised factors at present.

iii. Investigation of significance of *M. malmoense* infections in the animal population with regard to non-specific reactivity in tuberculin testing with bovine tuberculin. From the point of view of food safety, the modes of inter-organ transmission of infection in the most significant food animals (cattle and pigs) should be studied; the trends in consumption of non-heat-treated meat and meat products can pose great risks to humans.

iv. Identify sources of *M. malmoense* in the cool geographic regions (Finland, Sweden, UK, Denmark, Holland, etc.) as well as the warmer southern European countries, subtropical and tropical climates.

3.1.6 *M. xenopi*

In the middle of the last century, granulomatous skin lesions were observed in a *Xenopus* toad (*Xenopus laevis*), which was used in pregnancy laboratory testing. *M. xenopi* (Photo 3.18) was described for the first time by Schwabacher (1959) to be a causative agent of human infection with an optimum growth temperature of between 42–45 °C (Marx et al., 1995). *M. xenopi* was found to be more heat resistant than *L. pneumophila*, so thermal measures agains legionellae (60 °C) are expected to be insufficient for the decontamination of water system colonized with M. xenopi (Schulze-Robbecke and Buchholtz, 1992). Thus, from the ecological standpoint, its habitat is predetermined and limited. It has been isolated repeatedly from potable water systems in several northern regions, especially from recirculating, central warm water systems. There is only one report of *M. xenopi* isolation from natural waters (Torkko et al., 2000). Based on the regional localisation of infections and the recovery from natural waters in Finland, it is intriguing to speculate that *M. xenopi* has a unique (among mycobacteria) requirement for a particular compound(s) present in certain waters. It is possible that the compound is found in waters draining from peat lands or boreal forest soils that are rich in humic and fulvic acids.

Consistent with its pattern of isolation, the majority of infections are associated with exposure to warm water systems. In addition, the majority of reports of *M. xenopi* disease are outbreaks or pseudo-outbreaks (Costrini et al., 1981; Slosarek et al., 1993; Sniadack et al., 1993; Kaustova et al., 1993a).

In the tanks and distributing pipelines of exchanger units both organic and inorganic substances accumulate, enabling the clustering and survival of different micro-organisms in the presence of suitable nutrients. Optimum temperature conditions exist here for multiplication of different PPM species, especially *M. xenopi*. The important factors strengthening the resistance of waterborne mycobacteria are the temperature, pH, deposits in the piping, clustering of microorganisms and availability of nutrient factors originating from organic substances (iron, nitrite and sulphite reductants, substances on the reduced chlorine content).

3.1.6.1 Laboratory Diagnosis

Laboratory testing for detection or isolation of any *Mycobacterium* sp. in water is difficult due to the fact that they occur in low numbers (<1000 CFU/ml). Before culture examination, it is necessary to concentrate mycobacteria either by centrifugation or filtration using large volumes of water. Culture examination is commonly performed at 37 °C to isolate other mycobacteria, even though higher incubation temperatures (42–45 °C) are rather selective. Upon primary isolation and subculture at temperatures above 37 °C, *M. xenopi* forms distinctive X- or star-shaped species-specific colonies in the first 10 days of incubation. Thus, it is important, if *M. xenopi* is suspected, to examine colonies within the first 2 weeks. Experience with *M. xenopi* appears to be a factor in its detection and isolation which is dependant upon the culture method employed (Donnabella et al., 2000).

Because of the difficulty in detecting and isolating *M. xenopi*, as is the case for all mycobacteria, rapid methods using PCR coupled to growth in liquid

or on solid media have been developed (Fauville-dufaux et al., 1995; Nieminen et al., 2006). Further, a species-specific probe for *M. xenopi* has been described (Picardeau and Vincent, 1995). Rapid detection of *M. xenopi* by means of gas chromatography–mass spectrometry (GC-MS) was compared with parallel culture results (Alugupalli et al., 1992). It was found that the results obtained with GC-MS were in agreement with the culture results, but whereas *M. xenopi* requires an incubation time of several weeks, the GC-MS method enables the rapid detection of this species in drinking water. Results can be obtained within 2 days of receipt of the sample (Alugupalli et al., 1992).

At present, other techniques based on PCR are being introduced; these will substantially accelerate and facilitate detection of mycobacteria in the environment. For epidemiologic studies, two methods for DNA fingerprinting by RFLP have been described. The first uses IS*1395*, which is present in high copy number in *M. xenopi* strains, though not unique to that species (Picardeau et al., 1996). The second method employs RFLP analysis using IS*1081* as a probe (Collins, 1994). The utility of IS*1081*, a *M. tuberculosis* complex insertion sequence, is based on its sequence similarity (89%) relatedness to IS*1356* (Picardeau et al., 1996). It is important to point out that the designation of an isolate as *M. xenopi* must be certain before fingerprinting can be initiated, because of the cross-reactivity of the probes with DNA sequences in other mycobacteria.

3.1.6.2 Sources of *M. xenopi* Infection in Humans

M. xenopi can cause pulmonary and extrapulmonary diseases (Wayne and Sramek, 1992; Danesh-Clough et al., 2000; Katoch, 2004). Various organs and tissues can be infected with *M. xenopi*, especially in immunosuppressed patients (Table 3.11).

Isolation of *M. xenopi* from water samples collected from hospitals, dental units and households has been described by a number of authors in Europe, USA and Australia (Bullin et al., 1970; McSwiggan and Collins, 1974; Collins and Yates, 1984; Wright et al., 1985; Oga et al., 1993). Recovery of *M. xenopi* from water samples has most commonly been associated with outbreaks of pulmonary disease (Costrini et al., 1981; Slosarek et al., 1993; Sniadack et al., 1993; Kaustova et al., 1993a). In addition, pseudo-infections

Table 3.11 Human infections caused by *Mycobacterium xenopi*

Mode of transmission	Patients[1]	Infected tissue	Ref.
Bronchoscopy-associated pseudoinfections	3 ill patients and 17 infected	Bronchoscopes were disinfected in a 0.13% glutaraldehyde-phenate and tap water bath and were rinsed in tap water	[1]
Cellulitis on the right side of face (perhaps caused by contaminated water)	42/M	T7 to T8 spondylitis – isolation from the vertebral biopsy sample	[4]
Perhaps related to invasive procedures	7 ill patients	After invasive surgical procedure on affected joint, arthritis was diagnosed	[5]
Not known	7/M	Lymphoblastic leukaemia	[3]
Not known	29/M	Chest radiograph: bilateral diffuse infiltration and bone marrow infection (*HIV+*)	[6]
Not known	34/M	Night sweats, anorexia and weight loss	[2]
Not known	36/M	Chest radiograph: bilateral nodular abnormality (*HIV+*)	[6]
Not known	50/M	Chest radiograph: bilateral infiltration (*HIV+*)	[6]

[1] Age in years/gender (sex): **M** (male).

[1] Bennett SN, Peterson DE, Johnson DR, Hall WN, Robinson-Dunn B, Dietrich S (1994) Am. J. Respir. Crit Care Med. 150:245–250. [2] el Helou P, Rachlis A, Fong I, Walmsley S, Phillips A, Salit I, Simor AE (1997) Clin. Infect. Dis. 25:206–210. [3] Levendoglu-Tugal O, Munoz J, Brudnicki A, Fevzi OM, Sandoval C, Jayabose S (1998) Clin. Infect. Dis. 27:1227–1230. [4] Meybeck A, Fortin C, Abgrall S, Adle-Biassette H, Hayem G, Ruimy R, Yeni P (2005) J. Clin. Microbiol. 43:1465–1466. [5] Salliot C, Desplaces N, Boisrenoult P, Koeger AC, Beaufils P, Vincent V, Mamoudy P, Ziza JM (15-10-2006) Clin. Infect. Dis. 43:987–993. [6] Szlavik J, Sarvari C (2003) Eur. J. Clin. Microbiol. Infect. Dis. 22:701–703.

have been described caused by contamination of biological materials which were being collected or processed. Infections linked to the presence of *M. xenopi* in disinfectant solutions have also been described (Table 3.12).

The presence of *M. xenopi* in water supply systems in buildings is due to its ability to propagate in warm water piping or in water heaters. This leads to contamination of different auxiliary devices with *M. xenopi* (e.g. shower nozzles, taps, water containers, etc.), which in turn leads to infection.

Between 1980 and 1987, *M. xenopi* was isolated during an outbreak in which 14% of 78 patients devel-

oped pulmonary mycobacteriosis. *M. xenopi* was isolated from 15% of 48 samples of water and swabs from water supply nozzles and taps collected in the patients' households. *M. xenopi* was also isolated from 1 of 12 swabs taken from water taps in the patients' health care units. In contrast, *M. xenopi* was not found in 83 water samples and swabs collected in waterworks, suggesting that the likely source of infection was the hot water systems in either the households or medical facilities (Horak et al., 1986, 1991; Fuchsova et al., 1990).

Similarly, *M. xenopi* was isolated from sputum samples from 21 patients in Prague in 1990. This species was isolated repeatedly from 13 men and was

Table 3.12 Numbers of persons from whom *Mycobacterium xenopi* was isolated in the Czech Republic in 1990–1996

Region No.	Name	Year 1990	1991	1992	1993	1994	1995	1996
1[1]	North Moravia	21	24	18	11	26	81	94
2[1]	North Bohemia	33	29	17	8	5	10	12
3[1]	Prague (Capital)	5	25	27	11	12	9	6
Subtotal		59	78	62	30	43	100	112
%		78	80	82	81	69	90	82
4–8	Five other regions	17	19	14	7	19	11	25
%		22	20	18	19	31	10	18
1–8	Czech Republic	76	97	76	37	62	111	137

[1] Region with intensively developed industry.

identified as an aetiological agent of pulmonary disease. Water samples were collected in the households of these patients and *M. xenopi* was isolated from 12 of them (Slosarek et al., 1993). *M. xenopi* was detected in warm water pipelines, taps and shower nozzles in the households of infected patients.

3.1.6.3 Regional Distribution of *M. xenopi*

Studies in the Czech Republic point to a limited geographic range of *M. xenopi*, suggesting a unique requirement for some waterborne nutrient. Between 1990 and 1996, 69–90% of people, infected with *M. xenopi* were living in three of a total of eight regions in the Czech Republic. It is noteworthy that these three regions are the most significant industrial areas: North Moravia, North Bohemia and the capital city Prague (Table 3.12). Subsequent analysis of anamnestic data from the region of North Moravia revealed that the majority of the inhabitants lived in the city of Ostrava metropolis. *M. xenopi* detection in these inhabitants was analysed chronologically.

1982–1989: *M. xenopi* was isolated from only five men (*M. xenopi* was the aetiological agent of pulmonary disease in two of them). During investigation of the infection sources, *M. xenopi* was not isolated from any sample of water or scrapings from shower baths or taps in the residences of these five persons.

1990–1996: Increased isolation rates of *M. xenopi* were noted during this period. The majority of the infected men lived in Sector I (Table 3.13). *M. xenopi* was isolated largely from sputum samples: among other samples, it was isolated from urine and pulmonary tissue from five and two patients, respectively. Investigations gradually began to show that shower nozzles were the sources of *M. xenopi* infection. Nevertheless, the detection of *M. xenopi* in the warm water exchanger units was viewed as most significant. This water was distributed into the flats of the majority of the infected patients.

1997–2005: The remaining sources of *M. xenopi* were investigated in both Sectors I and II (patients' residences) supplied with warm water of the same origin. *M. xenopi* constituted 36.8% of all mycobacterial isolates obtained from water and scrapings from shower nozzles in both Sectors I and II (Table 3.14).

Over this 9-year period, *M. xenopi* was isolated from 224 inhabitants of Ostrava (Table 3.13).

Water sources in the regions: In the early 1990s, Sector I was supplied from a water tower situated in a large-scale agricultural area. Organic pollution and the contamination of water with *M. xenopi* was most likely more intensive here than in Sector II. The latter was supplied from another water tower situated in an area of less intensive agriculture (mostly forests) and drinking water originated from local ground sources. However, the drinking water supply underwent substantial changes in the mid-1990s. Both Sectors I and II began to be supplied from both water towers. Consequently, equal pressure of *M. xenopi* infection was exerted on the inhabitants of both Sectors I and II. Since 2000, the detection rates of *M. xenopi* in water samples from patients' residences in Sector II have increased (Table 3.14).

The clinical impact of *M. xenopi* on the inhabitants of the city of Ostrava became significant. The incidence of the disease between 2000 and 2005 was 5.3–16.0 times higher there compared to the whole Czech Republic (Table 3.13).

3.1.6.4 Sources of *M. xenopi* Infection in Animals

This route of infection most commonly runs through other animals, above all domestic and wild pigs. *M. xenopi* was detected on farms keeping domestic pigs, e.g. in peat fed as a supplement and in dust and spider webs in stables (Photo 3.19). *M. xenopi* was also detected in tuberculous lesions, particularly in head and intestinal lymph nodes (Table 3.15). Warm water, which is the most common source of *M. xenopi*, is not usually used on animal farms for financial reasons (Photo 3.20).

However, in the late 1990s, the technology of feeding liquid diets distributed through pipelines started to be introduced to large-scale pig production farms. Powdered mixtures are mixed with water and liquid feed supplements such as yeasts, whey and molasses can be added. In one herd, a long-term effort failed to find sources of mycobacteria causing frequent tuberculous lesions in the intestinal lymph nodes from market pigs after slaughter. After the identification of *M. xenopi* in almost one half of the infected animals, the epidemiologic investigation focused on warm

Table 3.13 *Mycobacterium xenopi* isolated from persons living in Ostrava and in the whole Czech Republic

Year	Ostrava city					Czech Republic		Ostrava/Czech Republic[1]
	Bacteriologically confirmed			Clinically ill		Clinically ill		
	Sector I	Sector II	Total	Abs.	Rel.	Abs.	Rel.	
1990	13[2]	0	13	Nk	Nk	Nk	Nk	Nk
1991	19	1	20	Nk	Nk	Nk	Nk	Nk
1992	10	1	11	Nk	Nk	Nk	Nk	Nk
1993	6	1	7	Nk	Nk	Nk	Nk	Nk
1994	22	1	23	Nk	Nk	Nk	Nk	Nk
1995	46	12	58	Nk	Nk	Nk	Nk	Nk
1996	50	9	59	Nk	Nk	Nk	Nk	Nk
1997	33	9	42	2	0.62	Nk	Nk	Nk
1998	4	7	11	2	0.62	Nk	Nk	Nk
1999	24	7	31	2	0.62	Nk	Nk	Nk
2000	26	18	44	3	0.93	10	0.10	9.3
2001	18	16	34	7	2.19	15	0.15	14.6
2002	16	9	25	2	0.62	9	0.09	6.9
2003	6	7	13	2	0.64	14	0.14	4.6
2004	6	8	14	2[3]	0.64	4	0.04	16.0
2005	5	5	10	3	0.96	18	0.18	5.3

[1] Ratio of the number of clinically ill patients per 100 000 inhabitants of the city of Ostrava to the number of clinically ill patients per 100 000 inhabitants in the whole Czech Republic. [2] Repeated detections in several persons. [3] One patient had a relapse.
Abs. Absolute numbers of bacteriologically confirmed cases of *M. xenopi* in clinically ill patients.
Rel. Relative numbers of bacteriologically confirmed cases of *M. xenopi* in clinically ill patients per 100 000 inhabitants.
Nk Not known (not available from the Register of Tuberculosis of the Czech Republic).

Table 3.14 Results of examination of water samples and scrapings collected from exchanger units of two Sectors I and II in Ostrava between 1997 and 1998

Sector	Water and scrapings			Isolated PPM species		
	Total	Posit.	%	*M. kansasii*	*M. xenopi*	*M. gordonae*
I	26	19	73.1	1	7	11
II	19	19	100	2	7	10
Total	45	38	84.4	3	14	21
%		100		7.9	36.8	55.3

Table 3.15 Animal infections caused by *Mycobacterium xenopi*

Host organism	Infected tissue	Ref.
Domestic pig (*Sus scrofa* f. *domestica*)	Lesioned lymph nodes	[3], [7], [9], [6]
Wild boar (*Sus scrofa*)	Lesioned lymph nodes	[2]
Domestic cat (*Felis catus*)	Skin lesions	[8]
	Lymphadenitis and peritonitis	[5]
Cattle (*Bos taurus*)	Not known	[4]
Atlantic bottlenose dolphins (*Tursiops truncatus*)	Serum antibodies	[1]

[1] Beck BM, Rice CD (2003) Mar. Environ. Res. 55:161–179. [2] Corner LA, Barrett RH, Lepper AW, Lewis V, Pearson CW (1981) Aust. Vet. J. 57:537–542. [3] Jarnagin JL, Richards WD, Muhm RL, Ellis EM (1971) American Review of Respiratory Disease 104:763-&. [4] Lazovskaia AL, Slinina KN, Vorobéva ZG, Druchkova MV, Poluektov EI (2006) Probl. Tuberk. Bolezn. Legk. 50–52. [5] MacWilliams PS, Whitley N, Moore F (1998) Vet. Clin. Pathol. 27:50–53. [6] Shitaye JE, Parmova I, Matlova L, Dvorska L, Horvathova A, Vrbas V, Pavlik I (2006) Veterinarni Medicina 51:497–511. [7] Thoen CO, Jarnagin JL, Richards WD (1975) Am. J. Vet. Res. 36:1383–1386. [8] Tomasovic AA, Rac R, Purcell DA (1976) Aust. Vet. J. 52:103. [9] Windsor RS, Durrant DS, Burn KJ (1984) Vet. Rec. 114:497–500.

water sources. The result was unexpected: the breeder used warm water from several old water heaters for the dilution of powder mixtures for the fattening of younger animals (Photo 3.20). Subsequent examination by microscopy revealed AFB in the sediment of the water heaters. *M. xenopi* was cultured from one third of the water and from one half of the sediment samples (I. Pavlik, unpublished data).

A disease caused by *M. xenopi* is detected occasionally in other domestic animals (cattle or domestic cats). Serum antibodies against antigen prepared from *M. xenopi* have been detected in sea mammals, i.e. dolphins (Table 3.15). This finding is not surprising when one considers the detection of this agent in poikilothermic animals (Section 6.5).

3.1.6.5 Other Sources of *M. xenopi* Infection

M. xenopi enters the environment through wastewater (Photo 3.21). Very little is known about its further fate within the entire ecosystem. Therefore, a study published in 1982 in Hungary investigating the occurrence and sources of *M. xenopi* for 97 inhabitants of Pecs is noteworthy (Szabo et al., 1982). All inhabitants were living in the vicinity of a slurry container where organic components were gradually fermented. The content of the container was partly used for fertilisation in gardens and partly as earth for ornamental plants. The above-mentioned people grew vegetables on this substrate in adjacent gardens and in their households.

The slurry container situated 500–1000 m from peoples' residences dried up in summer and became dusty. The prevailing wind blew in the direction of buildings where the people lived. *M. xenopi* was isolated from sewage inflow and from samples collected at different places and layers of the slurry container. *M. fortuitum*, *M. terrae*, *M. chelonae* and *M. smegmatis* were also detected. *M. xenopi* and *MAC* were detected in 73 and 24 persons, respectively, of 97 subjects examined due to suspected infection. Pulmonary mycobacteriosis was detected in 24 people (21 men and 3 women) and *M. xenopi* was isolated from 21 of them. Some persons who exhibited bacteriological positivity did not experience any clinical problems at the time of the detection of PPM; some of them had problems concurrent to lung findings outlasting for 1–4 years. A history of alcohol abuse was noted in many patients (Szabo et al., 1982).

3.1.6.6 Preventive Measures

It would appear that the principal source of *M. xenopi* infection in humans and animals is warm water from domestic systems, not natural waters. The combination of higher water temperatures and disinfectant selects for the survival and proliferation of *M. xenopi* in warm water systems. It is likely that the same measures that would result in the reduction of numbers of *M. avium* and *M. intracellulare* in drinking water distribution systems (e.g. ultraviolet irradiation) would also reduce *M. xenopi* numbers. Based on the observation that a pseudo-outbreak of *M. xenopi* infection followed a decrease in hot water temperature from 130 to 120 °F (Sniadack et al., 1993), increasing the temperature of water in households and recirculating hot water systems in hospitals, buildings and apartments may reduce numbers of *M. xenopi*, other mycobacteria (e.g. *M. avium* and *M. intracellulare*) and other bacteria (e.g. *Legionella* sp.).

According to Szabo et al. (1982) it is possible to prevent an infection from sludge by heat treatment before using it as fertiliser. The time and temperature necessary for *M. xenopi* devitalisation have not been reported in available literature. Due to the thermoresistance of *M. xenopi* and its ability to survive in soil, possibly even for many years, it is necessary to perform detailed experiments focused on this issue.

3.1.7 M. marinum

The photochromogenic *M. marinum* was one of the first described slow-growing species from the PPM (Photo 3.22). Aronson (1926) reported an acid-fast micro-organism to be the causative agent of a fish disease. A skin disease caused by mycobacteria was later observed in children from Sweden (Linell and Norden, 1954). The morphological and biochemical characteristics of mycobacterial isolates from the children and a swimming pool, where the children went swimming, were comparable with those described by Aronson. The newly described mycobacterial species was named *M. balnei* and later renamed *M. marinum*, according to Aronson's original designation. After several decades of research in this field *M. marinum* was detected to be the causative agent of skin diseases resulting from skin damage (Bhatty et al., 2000).

Accordingly, human medicine is highly interested in this causative agent of skin diseases that also causes skin and organ diseases in cold-blooded animals, especially aquarium fish.

3.1.7.1 What is the Difference Between "Swimming Pool" and "Fish Tank" Granuloma?

Almost until the end of the last century, contact with water in swimming pools was in most cases a shared characteristic of people infected with *M. marinum*. These people became infected with *M. marinum* in water reservoirs and swimming pools through contact of their damaged skin with contaminated fresh or saltwater and therefore this disease was originally designated as "swimming pool" granuloma (Leoni et al., 1999). However, fishermen and aquarium fish breeders were later found to be often infected via injured skin. This disease was newly designated as "fish tank granuloma" according to its origin. It is, therefore, necessary to note that with regard to the causative agent (*M. marinum*), these two diseases are identical (Photos 3.23 and 3.24).

Contact with aquarium fish and various components of the aquarium environment have prevailed in the history of infected patients over the last two decades (Table 3.16), possibly due to the introduction of more effective disinfection and cleaning procedures in swimming pools. These procedures are highly effective in removing biofilms and organic deposits often contaminated with *M. marinum*.

Dermatologists usually discover that patients suffering from skin mycobacteriosis caused by *M. marinum* keep exotic fish in their households (Photos 3.25 and 3.26). Due to the fact that the incubation time is very long – up to several weeks or months, many of them do not recall minor skin wounds that could become the route of infection by *M. marinum* from the environment of their aquaria or from the fish. Therefore, if dermatologists now find out that a patient is a fish breeder, it leads them to carry out diagnostic steps (biopsy of damaged tissue for laboratory testing), which makes the detection of the causative agent of the disease easier.

3.1.7.2 What Are the Main Sources of Human Infections?

The sources of *M. marinum* infections can be categorised as follows:

i. Water from swimming pools.
ii. Aquarium fish and all components of their life environment (water, plants, stones, sediment, invertebrate animals, biofilm on the walls, etc.).
iii. Coal mine water.
iv. Surface water (e.g. river water in Southeast Asia).
v. Other contaminated water sources (e.g. sewage, seawater, etc.).

Intact human skin provides sufficient protection against *M. marinum* penetration. Any skin damage (Photo 3.27) and concurrent immunodeficiency can, however, allow the infection of a patient (Photos 3.25 and 3.26). It is necessary to point out that besides the above-mentioned sources, contact with infected aquarium walls and contact with infected water or sediment can also represent routes of infection (Tables 3.16, 3.17, 3.18 and 3.19).

Mycobacteria can propagate in organic and inorganic sediments formed in different parts of water distribution systems. The number of viable mycobacteria isolated from water depend, besides other factors, on the amount of water flow and the water flow rate. In aquaria, there exists a suitable environment for the propagation of different PPM including *M. marinum*, i.e. stagnant, relatively warm water (multiplication of mycobacteria begins at 18–20 °C in the environment), plant surface, tank deposit consisting of sludge rich in organic substances, etc. A number of water animals such as snails and fish feed (sludge worms) described in Section 6.3 can participate in the spread of *M. marinum* throughout the environment of water tanks.

The most significant component of the aquarium environment infected with *M. marinum* are exotic (tropical) fish, because temperatures of 20 °C and higher are favourable for their breeding. Due to the fact that this is also the optimum indoor ambient temperature in most people's residences, fish from subtropical and tropical regions are most commonly bred indoors. *M. marinum* is adapted to such temperatures and can easily multiply at 20 °C. However, water for fish requiring lower temperatures (e.g. trout require 10–15 °C) must be cooled, which requires powerful equipment.

3.1.7.3 What Are the Most Frequent Modes of Human Infection?

The penetration of a pathogen is facilitated above all by chronic skin diseases or diseases that weaken the

Table 3.16 *Mycobacterium marinum* infection in fish fanciers without other predisposition factors

Mode of transmission[1]	Patients[2]	Infected tissue	Ref.
Manipulation with aquarium fish, water and sediment	19/?	Sporotrichoid skin inflammation	[8]
Manipulation with aquarium fish, water and sediment	20/?	Sporotrichoid skin inflammation	[8]
Manipulation with aquarium fish, water and sediment	21/?	Sporotrichoid skin inflammation	[8]
Fish fanciers	23/M	Skin infection	[9]
Fish fanciers	23/F	Skin infection	[9]
Agriculture student studying mycobacterial infections in fish	25/M	Skin infection	[9]
Fish tank owner	25/M	Sporotrichoid skin lesions	[10]
Agriculture student studying mycobacterial infections in fish	27/F	Skin infection	[9]
Aquarist	33/M	Skin infection	[6]
Infection from cleaning infected aquarium	37/M	Injured skin on hand from broken glass when cleaning an aquarium	[3]
Fish tank owner	40/M	Sporotrichoid skin lesions	[10]
Cleaned aquarium with tropical fish	42/M	Livid verrucous, painless nodules on right upper extremity	[1]
Fish fancier	43/M	Multiple granulomas presented with sporotrichoid-like skin infection of the right arm with multiple papulonodular lesions along lymphatic drainage	[2]
Manipulation with tropical fish	45/M	Painless nodular plagues on both hands	[5]
Owner of an aquarium shop	48/F	Painful ulcer on the dorsum of right hand and tender nodules on right forearm	[7]
Manipulation with aquarium fish, water and sediment	52/?	Sporotrichoid skin inflammation	[8]
Fish tank owner	57/M	Sporotrichoid skin lesions	[10]
Fish tank owner	59/F	Sporotrichoid skin lesions	[10]
Kept tropical fish as pets	60/F	Extensive subcutaneous abscesses developing on all limbs	[4]
Fish fancier	67/F	Skin infection	[9]

[1] Anamnestic data from the published paper. [2] Age in years/gender (sex): **M** (male), **F** (female), **?** (not known); patients are ordered according to their age.
[1] Belic M, Miljkovic J, Marko PB (2006) Acta Dermatovenerol. Alp Panonica. Adriat. 15:135–139. [2] Borradori L, Baudraz-Rosselet F, Beer V, Monnier M, Frenk E (1991) Schweiz. Med. Wochenschr. 121:1340–1344. [3] Engbaek HC, Vergmann B, Baess I (1970) Acta Pathol. Microbiol. Scand.[B] Microbiol. Immunol. 78:619–631. [4] Enzensberger R, Hunfeld KP, Elshorst-Schmidt T, Boer A, Brade V (2002) Infection 30:393–395. [5] Guarda R, Gubelin W, Gajardo J, Rohmann I, Valenzuela MT (1992) Rev. Med. Chil. 120:1027–1032. [6] Huminer D, Pitlik SD, Block C, Kaufman L, Amit S, Rosenfeld JB (1986) Arch. Dermatol. 122:698–703. [7] Kasick JM, Bergfeld WF, Taylor JS (1982) Archives of Dermatology 118:949. [8] Kern W, Vanek E, Jungbluth H (1989) Med. Klin. (Munich) 84:578–583. [9] Kullavanijaya P, Sirimachan S, Bhuddhavudhikrai P (1993) Int. J. Dermatol. 32:504–507. [10] Lewis FM, Marsh BJ, von Reyn CF (2003) Clin. Infect. Dis. 37:390–397.

natural defence system (Table 3.17). When investigating the history of a patient, the doctors must be very cautious and thoroughly evaluate all anamnestic data. It is sometimes very difficult to determine the direct contact with *M. marinum* that the patient could have had through a contaminated environment. In some cases, the modes of spread of infection and patient contact with a contaminated aquarium environment or its components are very odd, e.g. the stones from a contaminated aquarium, which the father of the family threw into the garden. His child was infected with *M. marinum* (Table 3.18)

after playing with them. Another example is the infection of a 14-month-old girl; her father put the cleaned aquarium content into the bathtub in which the child was bathed (Speight and Williams, 1997; Table 3.18).

If the skin damage is severe or the infectious dose of *M. marinum* is high, incubation takes only several days or weeks. In such cases, the patients usually still recall a skin injury, e.g. by broken glass or another pointed object or fish fin while cleaning a fish tank. One female patient mentioned an open lesion on her finger from a recent wart removal (Table 3.19).

Table 3.17 *Mycobacterium marinum* infection in fish fanciers with predisposition factors

Mode of transmission[1]	Patients[2]	Predisposition factors	Infected tissue	Ref.
Fish tank owner	36/F	Patient with rheumatoid arthritis, corticoid therapy	Granulomatous skin lesion	[3]
Fish tank owner	50/F	Patient with diabetes mellitus	Tenosynovitis of the right index finger	[3]
Fish tank owner	55/M	Patient with psoriasis, melanoma and prednisone therapy	Spread sporotrichoid skin lesions and an extensive deep infection including osteomyelitis	[3]
Keeper of aquarium regularly handling the fish with bare hands during the routine maintenance of the aquarium	84/M	Patient with a past history of diabetes mellitus, congestive cardiac failure and hypertension	Abscess in the ring finger metacarpophalangeal joint which spread into the finger and dorsum of the hand (there was no history of trauma or foreign body in the hand)	[2]
Fish fancier infected while cleaning his aquarium with bare hands	?/?	Patient suffered from chronic hand eczema	Bilateral, symmetric, sporotrichoid granulomas involving the dorsa of fingers and wrists	[1]

[1] Anamnestic data from the published paper. [2] Age in years/gender (sex): **M** (male), **F** (female), **?** (not known); patients are ordered according to their age.
[1] Alinovi A, Vecchini F, Bassissi P (1993) Acta Derm. Venereol. 73:146–147. [2] Bhatty MA, Turner DPJ, Chamberlain ST (2000) Br. J. Plast. Surg. 53:161–165. [3] Lewis FM, Marsh BJ, von Reyn CF (2003) Clin. Infect. Dis. 37:390–397.

Cases of skin infection among workers from coal mines in the Czech Republic are occasional; nevertheless, such infections caused by *M. marinum* with an endemic occurrence in a particular area in South Moravia were waterborne and were transmitted via coal mine water (Ulicna et al., 1968; Horacek and Ulicna, 1973).

3.1.7.4 Clinical Manifestations in Humans

The first signs of a skin disease are a rash, tubercles and reddish scabs, which can appear within 2–4 weeks after infection (Jernigan and Farr, 2000). After an additional 3–5 weeks, the typical diameter of the lesions is 1–2.5 cm. The lesions are either isolated or in clusters,

Table 3.18 Indirect contact with fish or other cold-blooded animals and not known sources of *Mycobacterium marinum* infection

Mode of transmission	Patients[1]	Infected tissue	Ref.
Infected aquarium was cleaned in the bathtub in which the child was bathed	Child/?	Disseminated dermatitis, osteomyelitis and bacteriaemia in immunocompromised child	[6]
Father had tanks with tropical fish	1.1/F	Sporotrichoid skin lesions on right leg	[7]
Fish tank was cleaned in the bathtub in which the child was bathed	1.4/F	"Pimple-like" skin lesions on lower extremities	[4]
Stones from infected aquarium were thrown into the garden, where the boy had played with them	4/M	Nontender erythematous nodule at the base of the dorsal aspect on index finger	[2]
Labourer who swam in the river	17/M	Skin infection on foot	[5]
Husband kept tropical fish	50/F	She recalled splinter going into the dorsum of the index finger while gardening 3 weeks prior to admission	[1]
Tool and die maker: no history of trauma on the infected finger, kept fish tank	56/M	Septic arthritis of the proximal interphalangeal joint of the ring finger on the left hand	[3]

[1] Age in years/gender (sex): **M** (male), **F** (female), **?** (not known); patients are ordered according to their age.
[1] Bhatty MA, Turner DP, Chamberlain ST (2000) Br. J. Plast. Surg. 53:161–165. [2] Bleiker TO, Bourke JE, Burns DA (1996) Br. J. Dermatol. 135:863–864. [3] Harth M, Ralph ED, Faraawi R (1994) J. Rheumatol. 21:957–960. [4] King AJ, Fairley JA, Rasmussen JE (1983) Arch. Dermatol. 119:268–270. [5] Kullavanijaya P, Sirimachan S, Bhuddhavudhikrai P (1993) Int. J. Dermatol. 32:504–507. [6] Parent LJ, Salam MM, Appelbaum PC, Dossett JH (1995) Clin. Infect. Dis. 21:1325–1327. [7] Speight EL, Williams HC (1997) Pediatr. Dermatol. 14:209–212.

Table 3.19 *Mycobacterium marinum* infection in injured aquarists

Mode of transmission	Patients[1]	Infected tissue	Ref.
Cut himself on the little finger while cleaning a fish tank	24/M	Two granulomatous lesions over the dorsum of the right hand	[3]
Bitten by one of the exotic fish	25/M	Nodular lymphangitis on the right hand of an exotic fish seller	[6]
Cut himself on the third finger of the right hand while cleaning a fish tank	34/M	Blue red, granulomatous skin infection	[2]
Nurse working on a medical ward cleaned a fish tank while she had an open lesion on her finger from a recent wart removal	38/F	Sporotrichoid skin lesions	[4]
Injured middle finger while handling sick catfish in a tropical aquarium	46/M	Swollen and painful granulomatous inflammation on middle finger on the right hand	[5]
Accidentally stabbed right thumb with a fork while cleaning a fresh water aquarium	46/F	Erythema, swelling and lymphangitis of the right thumb	[5]
Cut while cleaning an aquarium	51/F	Nontender swelling, sporotrichoid spread to forearm	[8]
Fish dealer, whose right middle finger had been punctured by a fish 4 months before the visit to hospital	52/F	Two month history of progressive pain, numbness, tenderness and erythematous swelling of the right middle finger	[7]
Cut while cleaning aquarium	61/F	Linear, indurated and subcutaneous nodular ulcers on the hand and forearm	[8]
Aquarist, minimal erosion on the hand	65/F	Nodular plaque on the left hand	[1]

[1] Age in years/gender (sex): **M** (male), **F** (female); patients are ordered according to their age.
[1] Escalonilla P, Esteban J, Soriano ML, Farina MC, Piqu E, Grilli R, Ramirez JR, Barat A, Martin L, Requena L (1998) Clin. Exp. Dermatol. 23:214–221. [2] Junger H, Witzani R (1981) Z. Hautkr. 56:16–18. [3] Laing RB, Flegg PJ, Watt B, Leen CL (1997) J. Hand Surg. [Br.] 22:135–137. [4] Lewis FM, Marsh BJ, von Reyn CF (2003) Clin. Infect. Dis. 37:390–397. [5] Paul D, Gulick P (1993) J. Fam. Pract. 36:336–338. [6] Seiberras S, Jarnier D, Guez S, Series C (2000) Presse Med. 29:2094–2095. [7] Shih JY, Hsueh PR, Chang YL, Chen MT, Yang PC, Luh KT (1997) J. Formos. Med. Assoc. 96:913–916. [8] Street ML, Umbert-Millet IJ, Roberts GD, Su WP (1991) J. Am. Acad. Dermatol. 24:208–215.

Table 3.20 *Mycobacterium marinum* infection in injured patients others than aquarists

Mode of transmission	Patients[1]	Infected tissue	Ref.
Abraded knee on an Astroturf field and had exposed the wound 5 days later during a fishing trip in Chesapeake Bay	23/M	Solitary plaque on the lateral surface of the right knee subsequently evolved into a verrucous plaque with some areas of atrophy	[3]
Scratched by a water trough while "dipping" cattle	44/M	Tendosynovitis and abscess on finger adjacent to scratch	[1]
Habitual nail biting of care worker with cutaneous sarcoidosis	46/M	Sporotrichoid skin infection on right hand	[4]
Insect bite on dorsum of the right hand	49/F	Verrucous plaque in the place of biting	[5]
Injury at abattoir	51/M	Inflamed finger joint following injury	[1]
Employed in timber industry	62/M	Plaque on knee, no known injury	[1]
Scraped hands on barnacles while cleaning the underside of a boat	67/M	Solitary nontender nodules on hands and forearm	[6]
Deep inoculation (2 patients)	?/?	Tendosynovitis of the abductors	[2]

[1] Age in years/gender (sex): **M** (male), **F** (female), **?** (not known); patients are ordered according to their age.
[1] Blacklock ZM, Dawson DJ (1979) Pathology 11:283–287. [2] Contios S, Roguedas AM, Genestet M, Volant A, Le Nen D (2005) Presse Med. 34:587–588. [3] Cummis DLC, Delacerda D, Tausk FA (2005) Int. J. Dermatol. 44:518–520. [4] Gudit VS, Campbell SM, Gould D, Marshall R, Winterton MC (2000) J. Eur. Acad. Dermatol. Venereol. 14:296–297. [5] Hunfeld KP, Enzensberger R, Brade V (1997) Dtsch. Med. Wochenschr. 122:917. [6] Johnson RP, Xia Y, Cho S, Burroughs RF, Krivda SJ (2007) Cutis 79:33–36.

often painless, but some patients may experience considerable pain. Ulcerous or warty scab-covered lesions can gradually develop (Photo 3.27).

The penetration of mycobacteria from contaminated water or biofilm on the walls of water pools through skin with disturbed integrity is a precondition for development of the swimming pool disease. Typical locations of skin lesions associated with swimming are elbows, knees and lower legs and in professional and amateur fishermen and aquarists especially the hands and fingers (Tables 3.16, 3.17, 3.18 and 3.19).

Occasional cases of infection disseminated to different body organs in immunodeficient patients or in relation to HIV/AIDS disease have been described (Gombert et al., 1981; Hanau et al., 1994; Parent et al., 1995). The time of treatment of this disease with antituberculosis drugs ranges between 6 and 12 months. Prognosis of this disease is infaust in immunocompromised patients.

3.1.7.5 Diagnosis of the Disease

Histological detection of granulomatous inflammation and culture detection of the causative agent can facilitate treatment. Due to the adaptation of *M. marinum* to the aquatic environment in subtropical and tropical areas, the optimum culture temperature is below 37 °C. The optimum growth temperature of 37 °C is suitable for the isolation of obligatory pathogenic mycobacteria (such as *M. tuberculosis* or *M. bovis*). For the *in vitro* isolation of *M. marinum*, the optimum growth temperatures range between 30 and 33 °C. Therefore, we assume that acral parts of patients' bodies with a temperature lower than 37 °C are usually affected.

When performing culture examination of infection sources for human patients, it is necessary to be aware of the fact that the environment in aquaria is suitable for the life of both *M. marinum* and other species of PPM or saprophytic mycobacteria. Likewise, the environment in the aquaria may often be favourable for *M. chelonae*, *M. fortuitum*, *M. abscessus* and other PPM that can cause skin infections in humans. For more information see Section 6.5.

3.1.7.6 Preventive Measures and Human Health Issues

Aquarists should be informed about the various risk factors associated with their hobby. *M. marinum* infection poses a risk not only to the fish breeders but also to all members of their families and to the customers. Accordingly, the aquarists should be completely aware of these risks and limit their exposure as much as possible.

Many dermatologists admit that they rarely take into consideration the possibility of an infection of mycobacterial aetiology for differential diagnosis of skin diseases. If there is no mention about the risk associated with the breeding of aquarium fish in the patient's history, the questions concerning the patient's anamnesis do not touch on other risk factors. At present, these factors include frequent travel to holiday resorts located in subtropical and tropical zones, the increasing number of aquatic sports resulting in potential skin injury, fishing and the ever more common practice of consuming non-heat-treated fish.

The processing of fish in the kitchen also poses a risk because careless individuals can be injured with bones, teeth, spiked fins, or knives. It follows from Sections 5, 6 and 7 that PPM are present in different components of the environment including locations where its occurrence does not seem very probable at first sight. Therefore, many dermatologists assume that this disease is "underdiagnosed" and that the actual number of infected patients will rise as knowledge of this disease increases.

3.1.8 Other Potentially Pathogenic and Saprophytic Slowly Growing Species

I. Pavlik, J. Kaustova

A number of slowly growing potentially pathogenic mycobacteria causing infection have been described in the literature. They are responsible for a relatively low number of infections and consequently little is known of their epidemiology and ecology. It is presumed that all are of environmental origin, although that is debatable. In the absence of systematic studies of their environmental distribution, it is only possible to speculate, on the basis of the types of infections, on their sources and routes of infection.

3.1.8.1 *M. haemophilum*

M. haemophilum was described in 1978 as a new slowly growing non-photochromogenic mycobacterial species in a patient suffering from Hodgkin's disease in Israel (Sompolinsky et al., 1978). *M. haemophilum* requires special conditions for growth: a source of ferric ions (chocolate agar, hemin, ferric ammonium citrate X-factor; Photo 3.28), a culture temperature of between 30 and 32 °C and an environment with an increased concentration of CO_2 (La Bombardi and Nord, 1998). *M. haemophilum* was originally thought to be a typical causative agent of tropical and subtropical diseases similarly to *M. ulcerans* (Section 3.1.2). However, its occurrence across all climatic zones on all continents, except Antarctica, has been demonstrated by the more than 150 described human clinical cases caused by *M. haemophilum* (Kiehn and White, 1994; Tables 3.21 and 3.22).

There has only been one report of the isolation of *M. haemophilum* from an environmental sample: water from a drinking water system (Falkinham et al., 2001). In spite of the many surveys conducted to isolate and enumerate mycobacteria in water and soil, there have been no reports of *M. haemophilum* isolation. There are several reasons for this. First, *M. haemophilum* requires a complexed source of iron, which is usually not added to media used for the isolation of mycobacteria. Second, the optimal temperature for the growth of *M. haemophilum* is between 30 and 32 °C and in many instances, 37 °C is used for the primary isolation of mycobacteria.

Clearly then, the habitats occupied by *M. haemophilum* would include those containing a source of free or uncomplexed ferric iron. Because iron is in the reduced ferrous state in most water systems, the search for habitats for *M. haemophilum* must take that into account. It is likely that *M. haemophilum*, like other mycobacteria, is capable of forming biofilms, surviving and growing in protozoa and amoebae. The latter environment would provide the iron needed for its growth.

M. haemophilum has not been detected in the households or workplaces of infected patients in the Czech Republic (J. Kaustova, unpublished data). Gouby et al. (1988) failed to isolate *M. haemophilum* from 20 examined water samples from a haemodialysis unit in a hospital, where two patients were infected with this mycobacterial species. Investigating the molecular epidemiology with the use of the RFLP method, it was found that the RFLP types of isolates from patients in respective hospitals were identical (Kikuchi et al., 1994).

In spite of the fact that sources of *M. haemophilum* have not been uncovered, the symptoms of a number of disease presentations suggest that it is a water inhabitant. First, cervical lymphadenitis caused by *M. haemophilum* has been reported in children throughout the world (Table 3.21). The presentation of infection is consistent with water or soil as the source of infection. In immunocompetent, immunodeficient and immunosuppressed patients, the infection is usually found on the skin, often at joints, suggesting that the route of infection is through trauma or abrasions (Table 3.22). Here again the source could be soil or water.

The capability of this pathogen to grow at 25 °C (Dawson and Jennis, 1980) and its optimum growth temperature which ranges between 30 and 32 °C explain its frequent detection, above all at joints and tendons where the temperature is rather low. Its occurrence in fish (Kent et al., 2004), insects (Pai et al., 2003) and reptiles (Hernandez-Divers and Shearer, 2002) and its rare occurrence in warm-blooded animals is consistent with its restricted growth range. The recovery of *M. haemophilum* from infected zebrafish (Kent et al., 2004) certainly suggests that *M. haemophilum*, like *M. marinum*, is present in aquaria and that working with aquaria may be a risk factor for disease (Tables 3.23 and 3.24).

An epidemiological study of children with cervical lymphadenitis was performed in The Netherlands. Two groups of children were compared with regard to the infection risk posed by the environment: 32 and 94 children were infected with *M. haemophilum* and members of the *MAC*, respectively. By "univariate analysis" no significant differences between the groups of children were detected throughout the year: most clinical symptoms (enlarged and painful lymph nodes) were detected in autumn (29%), less in winter (25%) and in spring (25%) and the least in summer (21%). The children infected with the *MAC* species were statistically more often infected, provided they

Table 3.21 Children's infections caused by *Mycobacterium haemophilum*

Country (state, city)[1]	Patient description Age/sex[2]	Infected tissue (predisposition, mode of transmission, if it is known)	Ref.
Australia (Brisbane)	5 children	Perihilar or cervical lymphadenitis (immunocompetent)	[1]
Czech Republic	12/M	Cervical lymphadenitis	[4]
Israel	12 children	Of 29 children with cervical lymphadenitis the following species were detected: *M. haemophilum* (12 children), *MAC* (14 children), *M. simiae* (2 children) and *M. scrofulaceum* (1 child)	[3]
	5 children	Biopsies from infected submandibular lymph nodes with lymphadenopathies (immunocompetent)	[8]
	10 children	9 children with cervical lymphadenitis, 1 child with right preauricular lymphadenitis (all were *HIV*- and of 1–10 years of age)	[8]
The Netherlands	32 children	Cervicofacial lymphadenitis (immunocompetent)	[7]
	21 children	Cervicofacial lymphadenitis (immunocompetent)	[6]
	5/F	Inguinal lymphadenitis diagnosed in fine needle biopsy of lymph nodes (small indolent wound on the dorsum of left foot, immunocompetent)	[5]
UK (England)	16/M	Nodules on the palmar surface of the distal interphalangeal joint of the two fingers and middle phalanx of the right hand (immunosuppressed: renal transplantation 5 years ago)	[2]

[1] Countries (state, city) are listed alphabetically. [2] Age in years/gender (sex): **M** (male), **F** (female).

[1] Armstrong KL, James RW, Dawson DJ, Francis PW, Masters B (1992) J. Pediatr. 121:202–205. [2] Campbell LB, Maroon M, Pride H, Adams DC, Tyler WB (2006) Pediatr. Dermatol. 23:481–483. [3] Haimi-Cohen Y, Zeharia A, Mimouni M, Soukhman M, Amir J (2001) Clin. Infect. Dis. 33:1786–1788. [4] Kaustova J, Boznsky J, Svobodova J, Wendrinska J, Zima P, Zichacek R, Rosinska D, Blazkova H, Chocholac D, Pacola R, Velart D, Palion J, Reischl U, Naumann L (2005) Klin. Mikrobiol. Infekc. Lek. 11:105–108. [5] Lindeboom JA, Kuijper CF, van Furth M (2007) Pediatr. Infect. Dis. J. 26:84–86. [6] Lindeboom JA, Kuijper EJ, Prins JM, Bruijnesteijn van Coppenraet ES, Lindeboom R (2006) Clin. Infect. Dis. 43:1547–1551. [7] Lindeboom JA, Prins JM, Bruijnesteijn van Coppenraet ES, Lindeboom R, Kuijper EJ (2005) Clin. Infect. Dis. 41:1569–1575. [8] Samra Z, Kaufmann L, Zeharia A, Ashkenazi S, Amir J, Bahar J, Reischl U, Naumann L (1999) J. Clin. Microbiol. 37:832–834.

had played in sandpits, compared to children infected with *M. haemophilum* with a history of contact with water (swimming water). No effect on the occurrence of cervical lymphadenitis was noted for the frequency of their visits to the children's farms (Lindeboom et al., 2005).

3.1.8.2 Views and Perspectives on the Research

For a better understanding of *M. haemophilum* ecology, it will be necessary to use not only culture but also culture-independent methods for further studies of the environment and water supply systems in hospitals and households of patients (van Coppenraet et al., 2005). It is assumed that the primary sites of *M. haemophilum* occurrence are biofilms and deposits.

The research should be focused on the ecology of a group of five recently described mycobacterial species: *M. celatum* discovered in 1993 (Photo 3.29), *M. mucogenicum* discovered in 1995 (Photo

3.30), *M. lentiflavum* discovered in 1996 (Photo 3.31), *M. hassiacum* discovered in 1997 and *M. bohemicum* discovered in 1998 (Chapter 1). Little is known about their ecology and preventive measures cannot, therefore, be focused on sources of infection (Table 3.25). The development of culture-independent methods can enable further studies of the environment and explain the ecology of these species.

3.2 Rapidly Growing Potentially Pathogenic Mycobacteria

J.O. Falkinham III

The source of infection with potentially pathogenic mycobacteria is the environment. *M. abscessus*, *M. chelonae* and *M. fortuitum* (Photo 3.32) are the species most commonly linked to disease (Wallace and Brown, 1998). Many of the infections that have been

Table 3.22 HIV/AIDS+ patients infected by *Mycobacterium haemophilum*

Country (state, city)[1]	Age/sex[2]	Infected tissue (predisposition, mode of transmission, if it is known)	Ref.
Brazil (Bahia)	30/M	Osteomyelitis in elbow (*HIV+*)	[9]
Brazil (Sao Paulo)	43/M	Biopsy specimen from nasal ulcer (*HIV+*)	[9]
Germany	53/M	Osteomyelitis of the right medial head of the tibia (*HIV/AIDS+*)	[3]
	7 patients	4 patients, skin lesions; 1 patient, lymphonodular infection; and 2 patients, bone infections (*HIV/AIDS+*)	[8]
Ghana	30/F	Two indurated, ulcerating and secreting skin nodules near the right elbow (*HIV+*)	[8]
Italy	51/M	Cutaneous erythematous papules on extremities (*HIV/AIDS+*)	[6]
Japan	51/M	Fell and had been injured near the right knee, ulcerated lesion followed by right inguinal lymphadenopathy (*Pneumocystis carinii* pneumonia, *HIV+*)	[1]
USA (Georgia)	25/F	Left knee pretibial oedema and retropatellar tenderness; no history of trauma was elicited, isolation from synovial aspiration (*HIV/AIDS+*)	[7]
USA (Illinois)	37/M	Diffuse, tender, pruritic, nodular skin lesions on the arm and thigh followed by fatal bilateral pneumonia (*HIV/AIDS+*)	[4]
USA (Illinois)	37/M	Erythematous swollen, tender left knee; synovial fluid of the knee was positive (*HIV/AIDS+* and cutaneous Kaposi's sarcoma)	[4]
USA (New York)	36/M	Osteomyelitis in the right distal lateral femoral condyle and multiple ulcerating skin abscesses on the left hand and both shoulders (*HIV+*)	[5]
USA (Washington)	40/M	Pain and swelling in multiple interphalangeal and metacarpophalangeal joints and both knees (*HIV+*)	[2]

[1] Countries (state, city) are listed alphabetically. [2] Age in years/gender (sex): **M** (male), **F** (female).
[1] Endo T, Takahashi T, Suzuki M, Minamoto F, Goto M, Okuzumi K, Oyaizu N, Nakamura T, Iwamoto A (2001) J. Infect. Chemother. 7:186–190. [2] Geisler WM, Harrington RD, Wallis CK, Harnisch JP, Liles WC (2002) Arch. Dermatol. 138:229–230. [3] Gruschke A, Enzensberger R, Brade V (2002) Dtsch. Med. Wochenschr. 127:1947–1950. [4] Kiehn TE, White M, Pursell KJ, Boone N, Tsvitis M, Brown AE, Polsky B, Armstrong D (1993) Eur. J. Clin. Microbiol. Infect. Dis. 12:114–118. [5] Lefkowitz RA, Singson RD (1998) Skeletal Radiol. 27:334–336. [6] Martinelli C, Farese A, Carocci A, Giorgini S, Tortoli E, Leoncini F (2004) J. Eur. Acad. Dermatol. Venereol. 18:83–85. [**7**] Olsen RJ, Cernoch PL, Land GA (2006) Arch. Pathol. Lab. Med. 130:783–791. [**8**] Paech V, Lorenzen T, von Krosigk A, von Stemm A, Meigel WM, Stoehr A, Rusch-Gerdes S, Richter E, Plettenberg A (2002) Clin. Infect. Dis. 34:1017–1019. [**9**] Sampaio JL, Alves VA, Leao SC, De Magalhaes VD, Martino MD, Mendes CM, Misiara AC, Miyashiro K, Pasternak J, Rodrigues E, Rozenbaum R, Filho CA, Teixeira SR, Xavier AC, Figueiredo MS, Leite JP (2002) Emerg. Infect. Dis. 8:1359–1360.

reported in the literature are nosocomial, recognised as outbreaks among a population of patients. Sporadic nosocomial outbreaks have been reported following cardiac bypass surgery, dialysis and augmentation mammoplasty (Wallace et al., 1998). In a number of reports, the sources of these outbreaks have been traced to water or contaminated instruments or fluids (Wallace and Brown, 1998). Pseudo-infections or outbreaks have also been linked to the contamination of fluids or instruments by rapidly growing mycobacteria (Wallace et al., 1998).

Because of the emergence of the rapidly growing mycobacteria as important agents of nosocomial infections, a great deal of effort has been made in developing tools for fingerprinting the species. To date, patterns of large restriction fragments separated by pulsed field gel electrophoresis (PFGE) have proven useful for all isolates from which DNA can be isolated (Burns et al., 1991; Hector et al., 1992; Wallace et al., 1993). It was observed that genomic DNA could be isolated from approximately 50% of *M. abscessus* strains (Wallace et al., 1993; Zhang et al., 1997). Along with the analysis of large restriction fragments separated by PFGE, arbitrary primed PCR analysis (Zhang et al., 1997) appears to be superior to plasmid or antibiotic susceptibility profiling (Yew et al., 1993) for fingerprint analysis of rapidly growing mycobacteria.

DNA fingerprinting studies have clearly established that the rapidly growing mycobacteria are normal inhabitants of drinking water. Among the sites sampled, hospital water supplies were rich in rapidly growing mycobacteria (Carson et al., 1988) and the

Table 3.23 Naturally infected animals with *Mycobacterium haemophilum*

Country (state)[1]	Case description (animal species, tissue examined, pathological findings)		Ref.
	Age/sex[2]	Infected tissue (mode of transmission, if it is known)	
Italy	1 case	Zebrafish (*Danio rerio*): numerous bacterial colonies in the spinal cord in absence of inflammation	[3]
Taiwan; Republic of China	1 case	Surface of adult cockroach (*Periplaneta americana*) trapped in supply room in one hospital	[4]
USA (Georgia)	Adult/F	Royal python (*Python regius*): 18-month history of chronic respiratory tract disease, lesions in cranial part of the lungs[3]	[1]
USA (Kansas)	16/F	American bison (*Bison bison*): ataxia and hind limb paresis; there was a 1.5 cm × 1.0 cm, well-delineated, pale tan, firm intradural mass (granuloma) found in the caudal lumbar region of the spinal cord	[2]
USA (Oregon)	1 case	Zebrafish (*D. rerio*): numerous bacterial colonies in the spinal cord in absence of inflammation	[3]

[1] Countries (state) are listed alphabetically. [2] Age in years/gender (sex): **F** (female). [3] Mixed infection caused by *M. marinum* and *M. haemophilum*.

[1] Hernandez-Divers SJ, Shearer D (2002) J. Am. Vet. Med. Assoc. 220:1661–3, 1650. [2] Jacob B, Debey BM, Bradway D (2006) Vet. Pathol. 43:998–1000. [3] Kent ML, Whipps CM, Matthews JL, Florio D, Watral V, Bishop-Stewart JK, Poort M, Bermudez L (2004) Comp Biochem. Physiol C. Toxicol. Pharmacol. 138:383–390. [4] Pai HH, Chen WC, Peng CF (2003) J. Hosp. Infect. 53:224–228.

two tested species, *M. abscessus* and *M. mucogenicum*, were capable of growing in water (Carson et al., 1978). Furthermore, pulmonary infection can occur via showers (Burns et al., 1991) or the consumption of ice (Laussucq et al., 1988), disinfecting or marking solutions (Safranek et al., 1987) and contaminated endoscopes (Wallace et al., 1993).

Although the rapidly growing mycobacteria are listed as members of the genus *Mycobacterium*, the length and sequence of the 16S rDNA gene are different between the rapid- and slow-growing members of the genus (Stahl and Urbance, 1990). Those differences occur with the helical region between positions 451 and 482 in the *Escherichia coli* 16S rRNA

Table 3.24 Experimentally infected animals with *Mycobacterium haemophilum*

Animal's model used description			
Species	Experimental details	Infected tissue (mode of transmission)	Ref.
Mice (*Mus musculus*)	Conventional breed	Intravenously, intramuscularly and subcutaneously resistant; however, a large number of AFB were found in smears of the liver, kidney and spleen	[3]
	Prednisolone-treated mice	In 12 of 30 intravenously infected mice skin lesions on the ears, *M. haemophilum* was isolated	[1]
	Prednisolone-nontreated mice	In 30 intravenously infected mice no lesions were observed	[1]
	ICR outbred mice	Immunocompetent: subcutaneous infection-resistant	[2]
	BALB/c nu/+ inbred mice	Immunocompetent: subcutaneous infection-resistant	[2]
	BALB/c nu/nu inbred mice	Immunodeficient: subcutaneous infection-sensitive (histopathological lesions in the skin and spleen observed in all mice)	[2]
Guinea pig (*Cavia aperea*)	Conventional breed	Intravenously, intramuscularly and subcutaneously resistant; however, a large number of AFB were found in smears of the liver, kidney and spleen	[3]
Frog	Conventional breed (reared at room temperature)	Intramuscularly to the thighs: resistant	[3]
	Conventional breed (reared at 30 °C)	Intramuscularly to the thighs: sensitive (died within 8–20 days and clumps of AST were found in smears of the liver and kidney	[3]

[1] Abbott MR, Smith DD (1980) J. Med. Microbiol. 13:535–540. [2] Atkinson BA, Bocanegra R, Graybill JR (1995) Antimicrob. Agents Chemother. 39:2316–2319. [3] Sompolinsky D, Lagziel A, Naveh D, Yankilevitz T (1978) Int. J. Syst. Bacteriol. 28:67–75.

Table 3.25 Detection of five recently described mycobacterial species

Mycobacterial species (year of discovery)	Case description (animal species, tissue examined, pathological findings)		Ref.
	Host[1]	Infected tissue (mode of transmission, if it is known)	
M. celatum (1993)	Human	Mycobacterial species description	[6]
	Human	Interstitial infiltrations in the lungs of seven patients	[14]
	28/F	Blood samples (partner was *HIV*+)	[20]
	39/M	Blood sample (fever, anaemia and diarrhoea present)	[20]
	48/M	Sputum (*HIV*+ with cytomegalovirus infection)	[2]
	Ferret	4-year-old male ferret (*Mustela putorius f. furo*), infected lungs	[23]
	Bird	White-tailed trogon (*Trogon viridis*)	[4]
	Aquarium	Biofilm	[3]
M. mucogenicum (1995)	23/M	CSF (*HIV*-, meningeal syndrome for 2 weeks)	[1]
	82/M	CSF (*HIV*-, diabetes, arterial hypertension, permanent pacemaker)	[1]
	Water	Drinking water from distribution systems that use surface water	[7]
	Water	City water, hot water, old and new hose, shower, salt and soda	[11]
M. lentiflavum (1996)	85/F	Intervertebral disc (synostosis of thoracic vertebrae Th9 and Th10)	[18]
	38/F	Ascites (*HIV*+)	[5]
	36/F	Pus and one of two sputum samples (multiple cutaneous abscesses on both legs and in the neck)	[5]
	35/F	Pleural effusion (cough and left pleural effusion)	[5]
	3.5/M	Pus of the left cervical lymph node (lymphadenitis)	[9]
	2.8/M	Enlarged right submandibular lymph node	[9]
	45/F	Left submandibular lymph node (severe periodontal disease recently treated by oral surgery)	[13]
	Human	12 immunosuppressed patients (6 sputum, 2 urine, 4 BAL)	[16]
	Human	76 patients from Finland (1996–2003)	[22]
	Water	Water distribution system from Finland	[22]
M. hassiacum (1997)	Human	Urine of patient without clinical disease (species description)	[17]
	Human	Urine of 45-year-old female with recurrent cystitis	[21]
	Cattle	Faeces of cattle with detected *M. hassiacum/M. buckleii*	[8]
M. bohemicum (1998)	Human	Three consecutive sputum samples (species description)	[15]
	2/F	Cervical lymphadenitis	[12]
	Cattle	Lymph nodes	[10]
	85/F	Ulcerative dermatitis	[19]
	Goat/F	2-year-old Rocky Mountain goat (*Oreamnos americanus*) in zoo (mesenteric lymph node)	[19]
	Stream	Four isolates were detected	[19]

[1] Age in years/gender (sex) of human patient: **M** (male), **F** (female). **BAL** Bronchoalveolar lavage in three patients, bronchial wash in one patient. **CSF** Cerebrospinal fluid.

[1] Adekambi T, Foucault C, La Scola B, Drancourt M (2006) J. Clin. Microbiol. 44:837–840. [2] Bell HC, Heath CH, French MA (2005) AIDS 19:2047–2049. [3] Beran V, Matlova L, Dvorska L, Svastova P, Pavlik I (2006) J. Fish. Dis. 29:383–393. [4] Bertelsen MF, Grondahl C, Giese SB (2006) Avian Pathol. 35:316–319. [5] Buijtels PC, Petit PL, Verbrugh HA, van Belkum A, van Soolingen D (2005) J. Clin. Microbiol. 43:6020–6026. [6] Butler WR, O'Connor SP, Yakrus MA, Smithwick RW, Plikaytis BB, Moss CW, Floyd MM, Woodley CL, Kilburn JO, Vadney FS (1993) Int. J. Syst. Bacteriol. 43:539–548. [7] Covert TC, Rodgers MR, Reyes AL, Stelma GN, Jr. (1999) Appl. Environ. Microbiol. 65:2492–2496. [8] Glanemann B, Hoelzle LE, Wittenbrink MM (2002) Dtsch. Tierarztl. Wochenschr. 109:528–529. [9] Haase G, Kentrup H, Skopnik H, Springer B, Bottger EC (1997) Clin. Infect. Dis. 25:1245–1246. [10] Hughes MS, Ball NW, McCarroll J, Erskine M, Taylor MJ, Pollock JM, Skuce RA, Neill SD (2005) Vet. Microbiol. 108:101–112. [11] Kline S, Cameron S, Streifel A, Yakrus MA, Kairis F, Peacock K, Besser J, Cooksey RC (2004) Infect. Control Hosp. Epidemiol. 25:1042–1049. [12] Palca A, Aebi C, Weimann R, Bodmer T (2002) Pediatr. Infect. Dis. J. 21:982–984. [13] Piersimoni C, Goteri G, Nista D, Mariottini A, Mazzarelli G, Bornigia S (2004) J. Clin. Microbiol. 42:3894–3897. [14] Piersimoni C, Tortoli E, de Lalla F, Nista D, Donato D, Bornigia S, De Sio G (1997) Clin. Infect. Dis. 24:144–147.

numbering system (Stahl and Urbance, 1990). It is debatable whether the lack of a defining sequence is sufficient to suggest that rapidly and slowly growing mycobacteria belong to a different genus. Such a proposal has not gained support, perhaps because the rapidly and slowly growing species share so many similarities in structure, physiology, ecology and epidemiology.

In addition to differing in the 16S rDNA gene sequence, most rapidly growing mycobacteria have an average of two rDNA cistrons, compared to only one among the slowly growing mycobacteria (Bercovier et al., 1986). The exceptions are *M. abscessus* and *M. chelonae* that have only a single rDNA cistron (Tortoli, 2003). Furthermore, the mycolic acids that make up the outer membrane of the rapidly growing mycobacteria are different from those in the slowly growing mycobacteria and these may impart increased permeation to the membrane (Brennan and Nikaido, 1995). It is likely that both these differences contribute to the relatively increased growth rate and even susceptibility to antimicrobial agents of the rapidly growing mycobacteria. However, it is important that even the rapidly growing mycobacteria grow slower than many environmental opportunistic pathogens that are useful models for the epidemiology of mycobacteria (e.g. *Pseudomonas aeruginosa* and *Legionella pneumophila*).

As is the case for the genus *Mycobacterium* in general (Tortoli, 2003), there has been substantial taxonomic revision of species and reassignment of isolates to species among the rapidly growing mycobacteria (Levy-frebault et al., 1986; Kusunoki and Ezaki, 1992; Pitulle et al., 1992; Wallace, 1994). The use of subspecies or biovariants has been discarded (Wallace, 1994). Thus, care must be taken in drawing conclusions from reports predating 1995, because the species assignment for rapidly growing mycobacteria may be incorrect. Accurate identification has been demonstrated using patterns of mycolic acids separated by high-performance liquid chromatography (Glickman et al., 1994), although there are difficulties in separating *M. abscessus* and *M. chelonae* (Wallace, 1994).

Sequencing the 16S rRNA genes (Springer et al., 1996; Tortoli et al., 2001, 2003) is becoming the gold standard. Both sequencing and HPLC of mycolic acids are preferred over classical biochemical, enzymatic and cultural tests for identification (Springer et al., 1996). It is important to note that the species associated with the majority of infections, namely *M. abscessus, M. chelonae* and *M. fortuitum*, are distinct species (Levy-Frebault et al., 1986; Wallace, 1994; Domenech et al., 1994) and hybrid names (e.g. *M. chelonae–abscessus*) should be avoided. Finally, a significant proportion of mycobacteria remain unidentifiable in spite of the use of biochemical, enzymatic and cultural tests; HPLC of mycolic acids; and 16S rDNA sequencing (Springer et al., 1996; Tortoli et al., 2001).

3.2.1 *M. abscessus, M. chelonae* and *M. fortuitum*

M. abscessus, M. chelonae and *M. fortuitum* (Photo 3.32) have been recovered repeatedly from drinking water samples throughout the world. In Germany, *M. fortuitum* was isolated from a drinking water system and one hot water system yielded *M. chelonae* (Fischeder et al., 1991). In Berlin, 70% of home and hospital waters yielded mycobacteria of which 14% were *M. chelonae* and 3% were *M. fortuitum* (Peters et al., 1995). In the USA, *M. fortuitum* was isolated from 3% of samples of drinking water, 83% of hospital ice samples and *M. chelonae* was recovered

Table 3.25 (continued) [15] Reischl U, Emler S, Horak Z, Kaustova J, Kroppenstedt RM, Lehn N, Naumann L (1998) Int. J. Syst. Bacteriol. 48 Pt 4:1349–1355. [16] Safdar A, Han XY (2005) Eur. J. Clin. Microbiol. Infect. Dis. 24:554–558. [17] Schroder KH, Naumann L, Kroppenstedt RM, Reischl U (1997) Int. J. Syst. Bacteriol. 47:86–91. [18] Springer B, Wu WK, Bodmer T, Haase G, Pfyffer GE, Kroppenstedt RM, Schroder KH, Emler S, Kilburn JO, Kirschner P, Telenti A, Coyle MB, Bottger EC (1996) J. Clin. Microbiol. 34:1100–1107. [19] Torkko P, Suomalainen S, Iivanainen E, Suutari M, Paulin L, Rudback E, Tortoli E, Vincent V, Mattila R, Katila ML (2001) J. Clin. Microbiol. 39:207–211. [20] Tortoli E, Piersimoni C, Bacosi D, Bartoloni A, Betti F, Bono L, Burrini C, De Sio G, Lacchini C, Mantella A (1995) J. Clin. Microbiol. 33:137–140. [21] Tortoli E, Reischl U, Besozzi G, Emler S (1998) Diagn. Microbiol. Infect. Dis. 30:193–196. [22] Tsitko I, Rahkila R, Priha O, Ali-Vehmas T, Terefework Z, Soini H, Salkinoja-Salonen MS (2006) FEMS Microbiol. Lett. 256:236–243. [23] Valheim M, Djonne B, Heiene R, Caugant DA (2001) Vet. Pathol. 38:460–463.

from a single membrane filter effluent sample (Covert et al., 1999). In Greece, 4% of drinking water samples yielded *M. chelonae* and 3% *M. fortuitum* and in spite of replacement of the system, 2% of water samples still yielded *M. chelonae* (Tsintzou et al., 2000). In Paris, *M. chelonae* and *M. fortuitum* were recovered from groundwater (7 and 3%, respectively) or treated surface water (3 and 9%, respectively; Le Dantec et al., 2002).

Because there have been no systematic studies of the ecology, as opposed to epidemiology, of any of the rapidly growing mycobacteria, comments concerning the ecology of these three major rapidly growing mycobacteria will be considered together. For the most part, the ecological studies have come in response to outbreaks of infection, most of which have been nosocomial. The outbreaks have involved cardiac surgery, plastic surgery, postinjection abscesses and dialysis (Wallace et al., 1998). The outbreaks have been traced to hospital water (Carson et al., 1988), ice (Laussucq et al., 1988), inadequately sterilised water or solutions (Camargo et al., 1996), inadequately sterilised needles or injectors (Wenger et al., 1990), contaminated skin marking or disinfecting solutions (Safranek et al., 1987) and water used for rinsing dialysers (Bolan et al., 1985).

Following the discovery that hospital water used for rinsing dialysers contained *M. chelonae*, a survey of haemodialysis centres showed that the prevalence of mycobacteria in water supplies was high (Carson et al., 1988). Environmental mycobacteria were recovered from 95 samples of water (83%) collected from 115 dialysis centres and 65% of the total numbers of isolates were rapidly growing mycobacteria (Carson et al., 1988). Samples collected from the hospital water system (2.8×10^4 CFU/100 ml) and water used to rinse dialysers and prepare disinfectant solutions (2.2×10^4 CFU/100 ml) were significantly higher than the municipal water entering the hospital (2.5×10^3 CFU/100 ml), documenting the concentration of mycobacteria in the hospitals' water systems.

The factors leading to the presence of high numbers of rapidly growing mycobacteria in water systems include the ability to grow in water (Carson et al., 1978), the ability to form biofilms (Schulze-Robbecke and Fischeder, 1989) and disinfectant resistance (Carson et al., 1978; Hayes et al., 1982; Collins, 1986; Safranek et al., 1987; Russell, 1999; Walsh

et al., 2001). The fact that the optima temperature for growth of *M. chelonae* and some *M. abscessus* strains is between 28 and 30 °C is certainly consistent with the amplification of their numbers in hospital water systems. Strains of *M. fortuitum*, *M. chelonae* and *M. flavescens* have all been shown to form biofilms (Schulze-Robbecke and Fischeder, 1989). Cells in biofilms in bronchoscopes, dialysers and rinsing instruments cannot be dislodged by rinsing and are more resistant to disinfectants than cells in suspension. Furthermore, if bronchoscopes and other instruments are not thoroughly dried following disinfection, mycobacteria surviving in biofilms can grow in the remaining water before the instrument is used. It may be the case that once a bronchoscope is contaminated with mycobacteria, it will always be a source of mycobacteria.

In addition to cold water, rapidly growing mycobacteria would be expected to be present in warm water. Measurements of heat susceptibility of mycobacteria demonstrated that *M. chelonae*, *M. fortuitum* and *M. phlei* strains were relatively resistant to heat. For example, to kill at 55 °C 90% of cells of *M. chelonae*, *M. fortuitum* and *M. phlei* took 23, 25 and 70 min, respectively (Schulze-Robbecke and Buchholtz, 1992). Although it would be expected that only thermotolerant mycobacteria would be able to grow in warm or hot water, they could survive for substantial periods.

Because of the use of tetracycline for the treatment of infections by rapidly growing mycobacteria, a study was undertaken to determine whether any tetracycline-resistant isolates of *M. fortuitum* harboured resistance genes. Two of seven tetracycline-resistant isolates carried genes identified in Gram-positive microbes (*tetK* and *tetL*) that result in resistance (Pang et al., 1994). Not only does this suggest that the rapidly growing mycobacteria could transmit antibiotic resistance to other mycobacteria, but it also suggests that the genus *Mycobacterium* is not shielded from horizontal gene transmission from outside the genus.

3.2.2 Other Potentially Pathogenic Rapidly Growing Mycobacteria

A number of rapidly growing mycobacteria not belonging to the three species listed above have been recovered from water or soil. *M. flavescens* is frequently isolated among this group (Fischeder et al., 1991; Peters et al., 1995; Tsintzou et al., 2000). A number of these species are characterised as being thermotolerant: *M. phlei*, *M. flavescens* and *M. thermoresistibile* (Pitulle et al., 1992). The recovery of *M. flavescens* from a domestic hot water system (30–69 °C) demonstrates that thermotolerance is not just a laboratory phenomenon (Fischeder et al., 1991). Furthermore, the persistence of these mycobacteria is likely due to their ability to form biofilms and the accompanying resistance to antimicrobial agents, as demonstrated by *M. flavescens* (Schulze-Robbecke and Fischeder, 1989), *M. fortuitum* (Hall-Stoodley and Lappin-Scott, 1998), *M. chelonae* (Hall-Stoodley et al., 1999) and *M. phlei* (Bardouniotis et al., 2001). Among this group, *M. neoaurum* stands out as being frequently isolated from indwelling Hickman catheters (George and Schlesinger, 1999).

Within the list of rapidly growing mycobacteria isolated from the environment, three groups stand out.

First, it is striking that a number of species (four to date) have been recovered from soil or water contaminated with chlorinated hydrocarbons, including *M. chlorophenolicum* (Hagglblom et al., 1994), *M. frederiksbergense* (Willumsen et al., 2001), *M. hodleri* (Kleespies et al., 1996) and *M. vanbaalenii* (Khan et al., 2002). The capability to degrade chlorinated hydrocarbons is certainly consistent with the high surface hydrophobicity that would allow for partitioning into hydrocarbon-containing environments.

Whether the capacity of these species to degrade is due to horizontal gene transmission or to the evolution of fatty acid gene synthesis remains to be tested. However, their presence and metabolic capabilities coupled with the relative resistance of mycobacteria to a variety of compounds and their ability to grow in natural habitats suggest that these mycobacteria could be used for *in situ* remediation of polluted habitats.

Second, a number of the species were isolated from sphagnum vegetation, including *M. cookii* (Kazda et al., 1990), *M. hiberniae* (Kazda et al., 1993) and *M. madagascariense* (Kazda et al., 1992). This too is consistent with other data concerning the mycobacteria, their preference for acidic, humic and fulvic acid-rich habitats where competition with other micro-organisms is reduced.

Third, a number of mycobacteria including slowly growing (e.g. *M. avium*, *M. intracellulare*, *M. scrofulaceum*, *M. gordonae* and *M. terrae*), rapidly growing (e.g. *M. chelonae*, *M. flavescens*, *M. fortuitum* and *M. mucogenicum*) and one of the novel rapidly growing species, *M. murale*, have been recovered from the walls of a water-damaged building (Vuorio et al., 1999; Huttunen et al., 2000; Torvinen et al., 2006). These mycobacteria may all be contributing to the respiratory problems experienced by individuals who work or reside in water-damaged buildings.

3.2.3 Conclusions

It may be a general characteristic of rapidly growing mycobacteria and perhaps all representatives of the genus *Mycobacterium* that they are capable of invading and proliferating in marginal habitats where competition with other micro-organisms is reduced because of the presence of inhibitory pollutants or novel, difficult-to-degrade nutrients.

3.3 Nosocomial Infections

I. Pavlik, J. Kaustova, and J.O Falkinham

More than 100 PPM species are known at present. While in the middle of the last century their species range was considerably limited, we have witnessed an explosion of new PPM species "description" in the last decade. Current methods of molecular biology (above all sequencing) allow us not only to discover but also to detect and study new species relatively quickly and reliably. However, the limited availability of these techniques for common clinical practice and routine laboratory diagnosis remains a problem. Accordingly, the purpose of the present section is not to describe all cases of newly detected clinically significant and often clinically insignificant PPM species. We have tried to analyse PPM with regard to current knowledge on the ecology of mycobacteria and to highlight and describe the main risk factors. In particular, PPM that cause infections in humans (as presented in the text and tables) should not be neglected.

From the point of view of mycobacterial ecology, information on PPM can be summarised as follows:

1. PPM are natural inhabitants of different environmental habitats that are shared by mycobacteria, humans and animals.
2. PPM can be transmitted to humans and animals through a variety of different vehicles (dust, water, aerosol, plants, etc.) where they can grow or survive.
3. PPM colonise various organs of clinically normal people and animals.
4. PPM can, under certain conditions, affect human and animal health.

Although hospitals impose conditions to reduce nosocomial infections (e.g. reduced humidity and surface sterilisation), many of the measures (e.g. disinfection) lead to selection for mycobacteria. For example, environmental mycobacteria which are present in drinking water and form biofilms on surfaces are resistant to disinfectants that kill other micro-organisms. Thus, it is not surprising that there has recently been an increased occurrence of nosocomial infections on many continents.

A number of different sources linked to nosocomial infections caused by environmental mycobacteria have been identified (Table 3.26; Photo 8.33):

i. Cold and warm piped water systems in hospitals or rainwater penetrating through open hospital windows in the form of an aerosol.
ii. Ice that comes into contact with patients.
iii. Disinfectant solutions used for surface sterilisation (e.g. gentian violet).
iv. Dust particles transmitted on different subjects including the clothes of patients, visitors, physicians and auxiliary hospital staff.
v. Non-heat-treated foodstuffs (above all fruits and vegetables).
vi. Soil in flower pots and freshly cut flowers used as decoration in hospitals or as gifts to patients, ponds with fish or other cold-blooded animals.
vii. Invertebrates in the subtropical and tropical zones and mites and other microscopic invertebrates in the temperate zone.

The multiplication of PPM occurs mainly in organic and inorganic sediment from different parts of the hospital environment, most frequently in the piped water system. The practice of reducing the temperature of hot water to prevent burns may be partly responsible for increasing the frequency of nosocomial PPM diseases (Table 3.26). The potting soil in flower pots and the plants and water of decorative aquaria can harbour high numbers of PPM. Due to the fact that today a number of hospitals try to make the ward environment pleas-

Table 3.26 Human infections in hospitals caused by conditionally pathogenic mycobacteria contaminating water

Mode of transmission	No. of patients/samples	Infected tissue, clinical signs etc	Isolates[1]	Ref.
Contaminated ice machine	47 patients	Contamination of sputum samples on *HIV* ward	*M. fortuitum*	[3]
Contaminated ice machine	30 patients	Colonies in sputum	*M. fortuitum*	[4]
Pseudo-outbreak due to laboratory contamination	21 patients – clinical samples	Contamination of 18 of 21 samples occurred in one laboratory the same day	*M. gordonae*	[2]
Contaminated hospital water supply	62 patients	Nosocomial pseudo-outbreak	*M. simiae*	[1]
Contaminated hospital potable water supply	117 sputum samples, 33 urine samples, 10 bronchoscopy samples, 3 gastric aspirates	Control measures: hyperchlorination of the potable water system of the hospital	*M. terrae*	[5]

[1] Mycobacterial species are listed in alphabetical order.

[1] El Sahly HM, Septimus E, Soini H, Septimus J, Wallace RJ, Pan X, Williams-Bouyer N, Musser JM, Graviss EA (2002) Clin. Infect. Dis. 35:802–807. [2] Esteban J, Fernandez-Roblas R, Ortiz A, Garcia-Cia JI (2006) Clin. Microbiol. Infect. 12:677–679. [3] Gebo KA, Srinivasan A, Perl TM, Ross T, Groth A, Merz WG (2002) Clin. Infect. Dis. 35:32–38. [4] Laussucq S, Baltch AL, Smith RP, Smithwick RW, Davis BJ, Desjardin EK, Silcox VA, Spellacy AB, Zeimis RT, Gruft HM (1988) Am. Rev. Respir. Dis. 138:891–894. [5] Lockwood WW, Friedman C, Bus N, Pierson C, Gaynes R (1989) Am. Rev. Respir. Dis. 140:1614–1617.

ant for patients and sometimes imitate the domestic environment (e.g. in paediatric oncology clinic), live flower decorations, small fish ponds or winter gardens can also be encountered.

3.3.1 Modes of Host Infection with Potentially Pathogenic Mycobacteria

How can a patient acquire a PPM infection? Mycobacteria can penetrate the host organism either before coming to a hospital (e.g. contamination of open fracture with soil, water or dust) or during their stay there. Because of the long incubation time between exposure and disease for PPM, it is difficult to reliably differentiate between prior exposure or hospital exposure modes of infection. The PPM infection may be detected during the hospital stay, but can also manifest itself after discharge. To better understand the host response to PPM, we will explain the following terms that will allow us to estimate the clinical significance of PPM for animals and humans. The clinical presentation depends on the PPM species. The isolation of PPM from a clinical specimen may represent an infection in one of the following stages (Phillips and von Reyn, 2001):

- Colonisation, defined as the establishment of PPM within the patient's microflora, without evidence of disease or tissue invasion.
- Pseudo-infection, defined as a positive culture result from a patient without evidence of a true infection or colonisation, which is typically caused by contamination during the handling of specimens.
- Infection, manifested as diseases of different organs or tissues.

It can be seen from the available literature that immunodeficient patients, in particular, are infected in hospitals through two modes of transmission:

i. Natural openings, the integrity of mucous membranes is not disturbed, particularly through respiratory and urogenital systems, less frequently through the eyes or ears. PPM-contaminated catheters or bronchoscopes are the major sources of infection. The highest rates of PPM colonisation in potable water systems are found in haemodialysis and transplantation centres; cases of nosocomial findings in bronchoscopy rooms have been described as well. *M. gordonae*, an occasional aetiological agent of human disease, has been reported as the most frequent contaminating agent of clinical specimens collected from human airways (Table 3.27).

Table 3.27 Iatrogenic infections in hospitals caused by contaminated catheters and bronchoscope with potentially pathogenic mycobacteria

Mode of transmission	Patients[1]	Infected tissue, clinical signs etc.	Isolates[2]	Ref.
Catheter-related bloodstream infection	54/F	After a modified radical mastectomy and subsequent plastic surgery skin flap complication	*M. brumae*	[4]
Catheter-related infections	15 patients	Not known	*M. fortuitum* complex	[5]
Bronchoscopy	6 patients	Contaminated tap water system and contamination during the sample examination	*M. gordonae*	[6]
Contamination of a drain during the previous operation	76/M	Tonsillitis on the left side and cervical lymphadenitis	*M. gordonae*	[2]
Catheter-related bloodstream infection	23/F pregnant	A tunnelled central venous catheter had been placed due to hyperemesis	*M. mageritense*	[1]
Catheter-related sepsis	32/F	Blood samples	*M. mageritense*	[7]
Contaminated vascular catheter	46/M	Hickman catheter placed 10 months ago	*M. neoaurum*	[3]

[1] Age in years/gender (sex): **M** (male), **F** (female). [2] Mycobacterial species are listed in alphabetical order.
[1] Ali S, Khan FA, Fisher M (2007) Journal of Clinical Microbiology 45:273. [2] Fleisch F, Pfyffer GE, Thuring C, Luthy R, Weber R (1997) Dtsch. Med. Wochenschr. 122:51–53. [3] George SL, Schlesinger LS (1999) Clin. Infect. Dis. 28:682–683. [4] Lee SA, Raad II, Adachi JA, Han XY (2004) J. Clin. Microbiol. 42:5429–5431. [5] Raad II, Vartivarian S, Khan A, Bodey GP (1991) Rev. Infect. Dis. 13:1120–1125. [6] Stine TM, Harris AA, Levin S, Rivera N, Kaplan RL (1987) JAMA 258:809–811. [7] Wallace RJ, Jr., Brown-Elliott BA, Hall L, Roberts G, Wilson RW, Mann LB, Crist CJ, Chiu SH, Dunlap R, Garcia MJ, Bagwell JT, Jost KC, Jr. (2002) J. Clin. Microbiol. 40:2930–2935.

Table 3.28 Iatrogenic infection in humans after surgery by potentially pathogenic mycobacteria

Mode of transmission	Patients[1]	Infected tissue, clinical signs, etc.	Isolates[2]	Ref.
Kidney transplantation	66/M	Reddish-brown, circumscribed, infiltrative papules on extremities	*M. abscessus*	[5]
Elective cosmetic surgery	30/F	Abscess in breast	*M. chelonae*	[1]
Hernia repair mesh-associated infection of wound	65/M	Erythematous and purulent inflammation at the right inguinal herniorrhaphy site with cellulitic lesions from the right thigh into the groin, scrotum and abscess	*M. goodii*	[7]
Severe periodontal disease treated with oral surgery	45/F	Cervical lymphadenitis in the left submandibular area	*M. lentiflavum*	[4]
Surgical wound, knee effusion	25/M	Wound infection, possible osteomyelitis	*M. mageritense*	[8]
Infected wound after subcutaneous implantation	74/M	Automatic implantable cardio-defibrillator followed by mycobacteraemia	*M. peregrinum*	[6]
Dental procedure in immunocompetent patient	?/F	Localised chronic infection at the left mandible after tooth extraction – died with overwhelming infection	*M. szulgai* and *M. terrae*	[3]
Infection following knee replacement surgery	73/F	Bilateral total knee replacement, 2 months after surgery material draining from the incision was infected	*M. thermoresistibile*	[2]

[1] Age in years/gender (sex): **M** (male), **F** (female), **?** (not known). [2]Mycobacterial species are listed in alphabetical order.
[1]. Blacklock ZM, Dawson DJ (1979) Pathology 11:283–287. [2]. LaBombardi VJ, Shastry L, Tischler H (2005) J. Clin. Microbiol. 43:5393–5394. [3]. Mahaisavariya P, Chaiprasert A, Khemngern S, Manonukul J, Gengviniij N, Ubol PN, Pinitugsorn S (2003) J. Med. Assoc. Thai. 86:52–60. [4]. Piersimoni C, Goteri G, Nista D, Mariottini A, Mazzarelli G, Bornigia S (2004) J. Clin. Microbiol. 42:3894–3897. [5]. Scholze A, Loddenkemper C, Grunbaum M, Moosmayer I, Offermann G, Tepel M (2005) Nephrol. Dial. Transplant. 20:1764–1765. [6]. Short WR, Emery C, Bhandary M, O'Donnell JA (2005) J. Clin. Microbiol. 43:2015–2017. [7]. Sohail MR, Smilack JD (2004) J. Clin. Microbiol. 42:2858–2860. [8]. Wallace RJ, Jr., Brown-Elliott BA, Hall L, Roberts G, Wilson RW, Mann LB, Crist CJ, Chiu SH, Dunlap R, Garcia MJ, Bagwell JT, Jost KC, Jr. (2002) J. Clin. Microbiol. 40:2930–2935.

Table 3.29 Human infection after the injection and iatrogenic infection in hospitals after the dialysis

Mode of transmission	Patients[1]	Infected tissue, clinical signs etc.	Isolates[2]	Ref.
Postinjection joint infection from cotton balls soaked in benzalkonium chloride	10 patients	Infected knee, heel, elbow, gluteus and shoulder in different patients	*M. abscessus*	[6]
Accidental self-inoculation of 0.1 mg of semi-moist isolate during experimental animal infection	38/M	Painful abscess in the inoculated distal phalanx of the forefinger of the left hand	*M. a. hominissuis*	[4]
Haemodialysis	?/?	Localised infection on the left upper limb, downstream from the arteriovenous fistula	*M. chelonae*	[1]
Recent extensive tattoos on his skin – intradermal inoculation	27/M	Enlarged and excised left axillary lymph node	*M. elephantis*	[7]
Suspected infection through intrabursal infection or subsequent postoperative infection	60/M	History of hypertension, osteoarthritis, type II diabetes mellitus – bursitis on right olecranon	*M. goodii*	[2]
Renal dialysis unit infection	49/M	Painful swelling of the left middle finger and nodular lesions on the left ankle and the face	*M. haemophilum*	[3]
Renal dialysis unit infection	51/M	Pain in the left knee (infected pus obtained by aspiration)	*M. haemophilum*	[3]
Contaminated bottle of veterinary anabolic steroid (Stanozol)	31/M 27/F	Thigh abscesses after the injection of semi-professional weight lifter	*M. smegmatis*	[5]

[1] Age in years/gender (sex): **M** (male), **F** (female), **?** (not known). [2]Mycobacterial species are listed in alphabetical order.
[1] Drouineau O, Rivault O, Le Roy F, Martin-Passos E, Young P, Godin M (2006) Nephrol. Ther. 2:136–139. [2] Friedman ND, Sexton DJ (2001) J. Clin. Microbiol. 39:404–405. [3] Gouby A, Branger B, Oules R, Ramuz M (1988) J. Med. Microbiol. 25:299–300. [4] Kazda J, Vrubel F, Dornetzhuber V (1967) Am. Rev. Respir. Dis. 95:848–853. [5] Plaus WJ, Hermann G (1991) Surgery 110:99–103. [6] Tiwari TS, Ray B, Jost KC, Jr., Rathod MK, Zhang Y, Brown-Elliott BA, Hendricks K, Wallace RJ, Jr. (2003) Clin. Infect. Dis. 36:954–962. [7] Turenne C, Chedore P, Wolfe J, Jamieson F, May K, Kabani A (2002) J. Clin. Microbiol. 40:1230–1236.

Table 3.30 Human infection after cosmetic procedures

Mode of transmission	Patients[1]	Infected tissue, clinical signs, etc.	Isolates[2]	Ref.
Mesotherapy of obese patients	22 patients	Soft tissue infections following mesotherapy	*M. abscessus*	[4]
Infection following liposuction – contaminated biofilm in the piped water system in one office	12 patients	Skin lesions (abscesses) in the areas of liposuction	*M. chelonae*	[1]
Mesotherapy of obese patients	4 patients	Soft tissue infections following mesotherapy	*M. chelonae*	[4]
Furunculosis after pedicure	27/F	Multiple, tender, violaceous nodules on lower legs	*M. chelonae/fortuitum*	[3]
Mesotherapy of obese patient	?/?	Soft tissue infections following mesotherapy	*M. cosmeticum*	[4]
Mesotherapy of obese patients	10 patients	Soft tissue infections following mesotherapy	*M. fortuitum*	[4]
Infection after localised microinjection (mesotherapy) treatment	24/F, 27/F, 44/F	Painful nodules at the site of mesotherapy	*M. fortuitum*	[2]
Contamination of a footbath at a nail salon	110 patients	Mycobacterial furunculosis	*M. fortuitum*	[6]
Furunculosis after pedicure	31/F	Solitary, nontender, erythematous to violaceous, perifollicular, smooth, firm nodule on the left lower leg	*M. fortuitum*	[3]
Furunculosis after pedicure	28/F	Multiple nontender, perifollicular, erythematous to violaceous nodules and plaques on both anterior lower legs	*M. fortuitum/peregrinum*	[3]
Furunculosis after pedicure	22/F	Front and sides of lower legs	*M. fortuitum/peregrinum*	[3]
Wound infection following liposuction	37/F	Surgical wound (thigh)	*M. mageritense*	[5]
Mesotherapy of obese patients	2 patients	Soft tissue infections following mesotherapy	*M. peregrinum*	[4]
Mesotherapy of obese patient	?/?	Soft tissue infections following mesotherapy	*M. simiae*	[4]

[1] Age in years/gender (sex): **F** (female), **?** (not known). [2]Mycobacterial species are listed in alphabetical order.
[1] Meyers H, Brown-Elliott BA, Moore D, Curry J, Truong C, Zhang Y, Wallace Jr RJ (2002) Clin. Infect. Dis. 34:1500–1507. [2] Nagore E, Ramos P, Botella-Estrada R, Ramos-Niguez JA, Sanmartin O, Castejon P (2001) Acta Derm. Venereol. 81:291–293. [3] Redbord KP, Shearer DA, Gloster H, Younger B, Connelly BL, Kindel SE, Lucky AW (2006) J. Am. Acad. Dermatol. 54:520–524. [4] Rivera-Olivero IA, Guevara A, Escalona A, Oliver M, Perez-Alfonzo R, Piquero J, Zerpa O, de Waard JH (2006) Enferm. Infecc. Microbiol. Clin. 24:302–306. [5] Wallace RJ, Jr., Brown-Elliott BA, Hall L, Roberts G, Wilson RW, Mann LB, Crist CJ, Chiu SH, Dunlap R, Garcia MJ, Bagwell JT, Jost KC, Jr. (2002) J. Clin. Microbiol. 40:2930–2935. [6] Winthrop KL, Albridge K, South D, Albrecht P, Abrams M, Samuel MC, Leonard W, Wagner J, Vugia DJ (2004) Clin. Infect. Dis. 38:38–44.

ii. Parenterally, above all through skin lesions or by contamination of a surgical wound (Tables 3.28 and 3.29); apart from the cases associated with, e.g. transplantations of various organs, we can encounter other diseases (Table 3.30). Plastic surgery as well as acupuncture (Woo et al., 2002) are also found to be associated with PPM infection.

3.3.2 The Most Important Risk Factors for the Spread of Potentially Pathogenic Mycobacterial Infection in Hospitals

Contaminated hospital water systems, medical technology and medical aids are the main vectors of the above-mentioned PPM. It is highly risky if safety precautions concerning sterilisation, preparation of solutions, the use of permanent catheters, cleaning, etc. are not observed. Endoscopes, especially bronchoscopes and automated endoscope washers, are frequently the sources of pseudo-outbreaks of PPM (Pankhurst et al., 1998; Phillips and von Reyn, 2001).

Another critical risk factor is the "immunological condition" of the host organism. In human populations from economically developed countries, the number of immunodeficient patients with suppressed cell immunity has increased due to the successful control of human tuberculosis and many other diseases. In contrast, in economically less developed countries, the immunocompetence of the population has been basically devastated by the HIV/AIDS pandemic that can not yet be stopped. PPM affects various organs or tissues in these immunosuppressed individuals.

3.3.3 Clinical, Therapeutic and Preventive Aspects

From the clinical, therapeutic and preventive points of view, the problems of nosocomial infections caused by both fast and slowly growing PPM species can be summarised in the following three points (Khooshabeh et al., 1994; Kelley et al., 1995; Sastry and Brennan, 1995; Smith et al., 2001; Goldblatt and Ribes, 2002; De Groote and Huitt, 2006):

i. Diagnosing cases caused by PPM is difficult and often delayed.
ii. A high degree of PPM resistance not only to a number of classic antituberculotics but also to a series of antibiotics or chemotherapeutics with an antibacterial effect used for the treatment of non-specific infections constitutes a great problem.
iii. It is often difficult or even impossible to impose effective control measures to prevent the spread of infection (e.g. contaminated piped water supply in hospitals, introduction of air-conditioning in hospitals in subtropical and tropical climate zones, etc.).

As PPM cannot be easily eliminated from the hospital environment, careful surveillance must be used to identify potential outbreaks. PPM may also contaminate microbiological specimens, which lead to harmful diagnostic procedures and sometimes to unnecessary therapy with antituberculotics and also other antibacterial drugs, respectively.

For a better understanding of the different and often odd and incredible modes of PPM transmission together with the clinical significance of a variety of PPM species, the cases of PPM reported as an aetiological agent of nosocomial infections and the cases of colonisation under different conditions have been included.

Acknowledgements Section 3.1.3 was partially supported by the European Commission (PathogenCombat FOOD-CT-2005-007081 and ParaTBTools FP6-2004-FOOD-3B-023106) and by the Ministry of Agriculture of the Czech Republic (NAZV QH81065) and Sections 3.1 and 3.3 by the Ministry of Agriculture of the Czech Republic (NPV 1B53009).

References

Alcaide F, Richter I, Bernasconi C, Springer B, Hagenau C, Schulze-Robbecke R, Tortoli E, Martin R, Bottger EC, Telenti A (1997) Heterogeneity and clonality among isolates of *Mycobacterium kansasii*: implications for epidemiological and pathogenicity studies. J. Clin. Microbiol. 35:1959–1964

Alugupalli S, Larsson L, Slosarek M, Jaresova M (1992) Application of Gas-Chromatography Mass-Spectrometry for Rapid Detection of *Mycobacterium-Xenopi* in Drinking-Water. Appl. Environ. Microbiol. 58:3538–3541

Archuleta RJ, Mullens P, Primm TP (2002) The relationship of temperature to desiccation and starvation tolerance of the *Mycobacterium avium* complex. Arch. Microbiol. 178:311–314

Aronson JD (1926) Spontaneous tuberculosis in salt water fish. J. Inf. Dis. 39:315–320

Ayele WY, Bartos M, Svastova P, Pavlik I (2004) Distribution of *Mycobacterium avium* subsp *paratuberculosis* in organs of naturally infected bull-calves and breeding bulls. Vet. Microbiol. 103:209–217

Ayele WY, Machackova M, Pavlik I (2001) The transmission and impact of paratuberculosis infection in domestic and wild ruminants. Veterinarni Medicina 46:205–224

Ayele WY, Svastova P, Roubal P, Bartos M, Pavlik I (2005) *Mycobacterium avium* subspecies *paratuberculosis* cultured from locally and commercially pasteurized cow's milk in the Czech Republic. Appl. Environ. Microbiol. 71:1210–1214

Baess I (1979) Deoxyribonucleic acid relatedness among species of slowly-growing mycobacteria. Acta Pathol. Microbiol. Scand. [B]. 87:221–226

Baess I (1983) Deoxyribonucleic acid relationships between different serovars of *Mycobacterium avium*, *Mycobacterium intracellulare* and *Mycobacterium scrofulaceum*. Acta Pathol. Microbiol. Immunol. Scand. [B]. 91:201–203

Banks J, Jenkins PA, Smith AP (1985) Pulmonary Infection with *Mycobacterium-Malmoense* – A Review of Treatment and Response. Tubercle. 66:197–203

Banks J, Smith AP, Jenkins PA (1983) *Mycobacterium-Malmoense* – Problems with Treatment and Diagnosis – A Case-Report. Tubercle. 64:217–219

Bardouniotis E, Huddleston W, Ceri H, Olson ME (2001) Characterization of biofilm growth and biocide susceptibility testing of *Mycobacterium phlei* using the MBEC (TM) assay system. Fems Microbiol. Lett. 203:263–267

Barker DJP (1972) Distribution of Buruli Disease in Uganda. Trans. R. Soc. Trop. Med. Hyg. 66:867–874

Bartos M, Hlozek P, Svastova P, Dvorska L, Bull T, Matlova L, Parmova I, Kuhn I, Stubbs J, Moravkova M, Kintr J, Beran V, Melicharek I, Ocepek M, Pavlik I (2006) Identification of members of *Mycobacterium avium* species by Accu-Probes, serotyping, and single IS*900*, IS*901*, IS*1245* and IS*901*-flanking region PCR with internal standards. J. Microbiol. Methods. 64:333–345

Beerwerth W (1973) [The use of natural substrates as culture media for mycobacteria]. Ann. Soc. Belg. Med. Trop. 53:355–360

Beerwerth W, Kessel U (1976) [Mycobacteria in the environment of man and animal (proceedings)]. Zentralbl. Bakteriol. [Orig. A]. 235:177–183

Beggs ML, Stevanova R, Eisenach KD (2000) Species identification of *Mycobacterium avium* complex isolates by a variety of molecular techniques. J. Clin. Microbiol. 38: 508–512

Belisle JT, Pascopella L, Inamine JM, Brennan PJ, Jacobs WR, Jr. (1991) Isolation and expression of a gene cluster responsible for biosynthesis of the glycopeptidolipid antigens of *Mycobacterium avium*. J. Bacteriol. 173: 6991–6997

Beran V, Havelkova M, Kaustova J, Dvorska L, Pavlik I (2006a) Cell wall deficient forms of mycobacteria: a review. Veterinarni Medicina 51:365–389

Beran V, Matlova L, Dvorska L, Svastova P, Pavlik I (2006b) Distribution of mycobacteria in clinically healthy ornamental fish and their aquarium environment. J. Fish. Dis. 29:383–393

Bercovier H, Kafri O, Sela S (1986) Mycobacteria Possess A Surprisingly Small Number of Ribosomal-RNA Genes in Relation to the Size of Their Genome. Biochem. Biophys. Res. Commun. 136:1136–1141

Berghaus RD, Farver TB, Anderson RJ, Jaravata CC, Gardner IA (2006) Environmental sampling for detection of *Mycobacterium avium* ssp. *paratuberculosis* on large California dairies. J. Dairy Sci. 89:963–970

Bhatty MA, Turner DPJ, Chamberlain ST (2000) *Mycobacterium marinum* hand infection: case reports and review of literature. Br. J. Plast. Surg. 53:161–165

Biet F, Boschiroli ML, Thorel MF, Guilloteau LA (2005) Zoonotic aspects of *Mycobacterium bovis* and *Mycobacterium avium*-intracellulare complex (MAC). Vet. Res. 36:411–436

Birkness KA, Swords WE, Huang PH, White EH, Dezzutti CS, Lal RB, Quinn FD (1999) Observed differences in virulence-associated phenotypes between a human clinical isolate and a veterinary isolate of *Mycobacterium avium*. Infect. Immun. 67:4895–4901

Bland CS, Ireland JM, Lozano E, Alvarez ME, Primm TP (2005) Mycobacterial ecology of the Rio Grande. Appl. Environ. Microbiol. 71:5719–5727

Bolan G, Reingold AL, Carson LA, Silcox VA, Woodley CL, Hayes PS, Hightower AW, McFarland L, Brown JW, III, Petersen NJ (1985) Infections with *Mycobacterium chelonei* in patients receiving dialysis and using processed hemodialyzers. J. Infect. Dis. 152:1013–1019

Brennan PJ, Nikaido H (1995) The Envelope of Mycobacteria. Ann. Rev. Biochem. 64:29–63

Brooks RW, Parker BC, Gruft H, Falkinham JO, III (1984) Epidemiology of infection by nontuberculous mycobacteria. V. Numbers in eastern United States soils and correlation with soil characteristics. Am. Rev. Respir. Dis. 130:630–633

Buchholz UT, Mcneil MM, Keyes LE, Good RC (1998) *Mycobacterium malmoense* infections in the United States, January 1993 through June 1995. Clin. Infect. Dis. 27: 551–558

Buhler VB, Pollak. A. (1953) Human infection with atypical acid-fast organisms; report of two cases with pathologic findings. Am. J. Clin. Pathol. 23:363–374

Bullin CH, Tanner EI, Collins CH (1970) Isolation of *Mycobacterium-Xenopei* from Water Taps. J. Hyg. Camb. 68:97–100

Burns DN, Wallace RJ, Schultz ME, Zhang YS, Zubairi SQ, Pang YJ, Gibert CL, Brown BA, Noel ES, Gordin FM (1991) Nosocomial Outbreak of Respiratory-Tract Colonization with *Mycobacterium-Fortuitum* – Demonstration of the Usefulness of Pulsed-Field Gel-Electrophoresis in An Epidemiologic Investigation. Am. Rev. Respir. Dis. 144: 1153–1159

Camargo D, Saad C, Ruiz F, Ramirez ME, Lineros M, Rodriguez G, Navarro E, Pulido B, Orozco LC (1996) Iatrogenic outbreak of *M-chelonae* skin abscesses. Epidemiol. Infect. 117:113–119

Carson LA, Bland LA, Cusick LB, Favero MS, Bolan GA, Reingold AL, Good RC (1988) Prevalence of nontuberculous mycobacteria in water supplies of hemodialysis centers. Appl. Environ. Microbiol. 54:3122–3125

Carson LA, Petersen NJ, Favero MS, Aguero SM (1978) Growth-Characteristics of Atypical Mycobacteria in Water and Their Comparative Resistance to Disinfectants. Appl. Environ. Microbiol. 36:839–846

Carter G, Wu M, Drummond DC, Bermudez LE (2003) Characterization of biofilm formation by clinical isolates of *Mycobacterium avium*. J. Med. Microbiol. 52:747–752

Carter G, Young LS, Bermudez LE (2004) A subinhibitory concentration of clarithromycin inhibits *Mycobacterium avium* biofilm formation. Antimicrob. Agents Chemother. 48: 4907–4910

Cernicchiaro N, Wells SJ, Janagama H, Sreevatsan S (2008) Influence of type of culture medium on characterization of *Mycobacterium avium* subsp *paratuberculosis* subtypes. J. Clin. Microbiol. 46:145–149

Cerny L (1982) [Development of morphological changes after experimental *Mycobacterium avium* infection in ducks and geese]. Veterinarni Medicina 27:95–100

Chapman JS (1971) The ecology of the atypical mycobacteria. Arch. Environ. Health. 22:41–46

Chobot S, Malis J, Sebakova H, Pelikan M, Zatloukal O, Palicka P, Kocurova D (1997) Endemic incidence of infections caused by *Mycobacterium kansasii* in the Karvina district in 1968–1995 (analysis of epidemiological data – review). Cent. Eur. J. Public Health. 5:164–173

Cirillo JD, Falkow S, Tompkins LS, Bermudez LE (1997) Interaction of *Mycobacterium avium* with environmental amoebae enhances virulence. Infect. Immun. 65: 3759–3767

Collins CH, Grange JM, Yates MD (1984) Mycobacteria in water. J. Appl. Bacteriol. 57:193–211

Collins CH, Yates MD (1984) Infection and Colonization by *Mycobacterium-Kansasii* and *Mycobacterium-Xenopi* – Aerosols As A Possible Source. J. Infect. 8:178–179

Collins DM (1994) DNA-Fingerprinting of *Mycobacterium-Xenopi* Strains. Lett. Appl. Microbiol. 18:234–235

Collins DM, Gabric DM, de Lisle GW (1990) Identification of two groups of *Mycobacterium paratuberculosis* strains by restriction endonuclease analysis and DNA hybridization. J. Clin. Microbiol. 28:1591–1596

Collins FM (1986) Bactericidal Activity of Alkaline Glutaraldehyde Solution Against A Number of Atypical Mycobacterial Species. J. Appl. Bacteriol. 61:247–251

Cook KL, Britt JS (2007) Optimization of methods for detecting *Mycobacterium avium* subsp. *paratuberculosis* in environmental samples using quantitative, real-time PCR. J. Microbiol. Methods. 69:154–160

Cooney RP (1991) A study of environmental mycobacteria in Ireland. M.A. Thesis, National University of Ireland, Dublin.

Costrini AM, Mahler DA, Gross WM, Hawkins JE, Yesner R, Desopo ND (1981) Clinical and Roentgenographic Features of Nosocomial Pulmonary-Disease Due to *Mycobacterium-Xenopi*. Am. Rev. Respir. Dis. 123:104–109

Covert TC, Rodgers MR, Reyes AL, Stelma GN, Jr. (1999) Occurrence of nontuberculous mycobacteria in environmental samples. Appl. Environ. Microbiol. 65:2492–2496

Cowan HE, Falkinham JO, III (2001) A luciferase-based method for assessing chlorine-susceptibility of *Mycobacterium avium*. J. Microbiol. Methods. 46:209–215

Danesh-Clough R, Theis JC, van der Linden A (2000) *Mycobacterium xenopi* infection of the spine – A case report and literature review. Spine. 25:626–628

Dawson DJ, Jennis F (1980) Mycobacteria with a growth requirement for ferric ammonium citrate, identified as *Mycobacterium haemophilum*. J. Clin. Microbiol. 11:190–192

De Groote MA, Huitt G (2006) Infections due to rapidly growing mycobacteria. Clin. Infect. Dis. 42:1756–1763

De Groote MA, Pace NR, Fulton K, Falkinham JO, III (2006) Relationships between *Mycobacterium* isolates from patients with pulmonary mycobacterial infection and potting soils. Appl. Environ. Microbiol. 72:7602–7606

del Giudice P, Bernard E, Perrin C, Bernardin G, Fouche R, Boissy C, Durant J, Dellamonica P (2000) Unusual cutaneous manifestations of miliary tuberculosis. Clin. Infect. Dis. 30:201–204

Domenech P, Menendez MC, Garcia MJ (1994) Restriction-Fragment-Length-Polymorphisms of 16S Ribosomal-RNA Genes in the Differentiation of Fast-Growing Mycobacterial Species. Fems Microbiol. Lett. 116:19–24

Donnabella V, Salazar-Schicchi J, Bonk S, Hanna B, Rom WN (2000) Increasing incidence of *Mycobacterium xenopi* at

Bellevue Hospital – An emerging pathogen or a product of improved laboratory methods? Chest. 118:1365–1370

Duker AA, Portaels F, Hale M (2006) Pathways of *Mycobacterium ulcerans* infection: A review. Environ. Int. 32:567–573

Durnez L, Eddyani M, Mgode GF, Katakweba A, Katholi CR, Machang'u RR, Kazwala RR, Portaels F, Leirs H (2008) First detection of mycobacteria in African rodents and insectivores, using stratified pool screening. Appl. Environ. Microbiol. 74:768–773

Dvorska L, Bull TJ, Bartos M, Matlova L, Svastova P, Weston RT, Kintr J, Parmova I, van Soolingen D, Pavlik I (2003) A standardised restriction fragment length polymorphism (RFLP) method for typing *Mycobacterium avium* isolates links IS*901* with virulence for birds. J. Microbiol. Methods. 55:11–27

Dvorska L, Matlova L, Ayele WY, Fischer OA, Amemori T, Weston RT, Alvarez J, Beran V, Moravkova M, Pavlik I (2007) Avian tuberculosis in naturally infected captive water birds of the Ardeideae and Threskiornithidae families studied by serotyping, IS*901* RFLP typing, and virulence for poultry. Vet. Microbiol. 119:366–374

Dvorska L, Matlova L, Bartos M, Parmova I, Bartl J, Svastova P, Bull TJ, Pavlik I (2004) Study of *Mycobacterium avium* complex strains isolated from cattle in the Czech Republic between 1996 and 2000. Veterinary Microbiology. 99: 239–250

Engel HW, Berwald LG, Havelaar AH (1980) The occurrence of *Mycobacterium kansasii* in tapwater. Tubercle. 61: 21–26

Erardi FX, Failla ML, Falkinham JO, III (1987) Plasmid-encoded copper resistance and precipitation by *Mycobacterium scrofulaceum*. Appl. Environ. Microbiol. 53: 1951–1954

Falkinham JO (2003) Factors influencing the chlorine susceptibility of *Mycobacterium avium*, *Mycobacterium intracellulare*, and *Mycobacterium scrofulaceum*. Appl. Environ. Microbiol. 69:5685–5689

Falkinham JO, III, Norton CD, LeChevallier MW (2001) Factors influencing numbers of *Mycobacterium avium*, *Mycobacterium intracellulare*, and other Mycobacteria in drinking water distribution systems. Appl. Environ. Microbiol. 67:1225–1231

Falkinham JO, III, Parker BC, Gruft H (1980) Epidemiology of infection by nontuberculous mycobacteria. I. Geographic distribution in the eastern United States. Am. Rev. Respir. Dis. 121:931–937

Fauvilledufaux M, Maes N, Severin E, Farin C, Serruys E, Struelens M, Younes N, Vincke JP, Devos MJ, Bollen A, Godfroid E (1995) Rapid identification of *Mycobacterium xenopi* from bacterial colonies or Bactec culture by the polymerase chain-reaction and a luminescent sandwich hybridization assay. Res. Microbiol. 146:349–356

Fischeder R, Schulze-Robbecke R, Weber A (1991) Occurrence of mycobacteria in drinking water samples. Zentralbl. Hyg. Umweltmed. 192:154–158

Fischer O, Matlova L, Bartl J, Dvorska L, Melicharek I, Pavlik I (2000) Findings of mycobacteria in insectivores and small rodents. Folia Microbiol. (Praha). 45:147–152

Fischer OA, Matlova L, Bartl J, Dvorska L, Svastova P, du Maine R, Melicharek I, Bartos M, Pavlik I (2003) Earthworms (Oligochaeta, Lumbricidae) and mycobacteria. Vet. Microbiol. 91:325–338

Fischer OA, Matlova L, Dvorska L, Svastova P, Bartos M, Weston RT, Pavlik I (2006) Various stages in the life cycle of syrphid flies (*Eristalis tenax*; Diptera: Syrphidae) as potential mechanical vectors of pathogens causing mycobacterial infections in pig herds. Folia Microbiol. (Praha). 51:147–153

France AJ, Mcleod DT, Calder MA, Seaton A (1987) *Mycobacterium-Malmoense* Infections in Scotland – An Increasing Problem. Thorax. 42:593–595

Francis J, Macturk HM, Madinaveitia J, Snow GA (1953) Mycobactin, A Growth Factor for *Mycobacterium-Johnei*. 1. Isolation from *Mycobacterium-Phlei*. Biochem. J. 55: 596–607

Frehel C, Offredo C, de Chastellier C (1997) The phagosomal environment protects virulent *Mycobacterium avium* from killing and destruction by clarithromycin. Infect. Immun. 65:2792–2802

Friend M, Franson JC (1999) Field manual of wildlife diseases; general field procedures and diseases of birds. USGS-National Wildlife Health Center, http://www.nwhc.usgs.gov/publications/field_manual.

Fuchsova M, Zima Z, Horak Z, Kubin M (1990) Nosocomial occurrence of *M. xenopi* among hospitalized patients. (In Czech). Stud. Pneumol. Phtiseol. Cechoslov. 50: 557–562

Gelder CM, Hart KW, Williams OM, Lyons E, Welsh KI, Campbell IA, Marshall SE (2000) Vitamin D receptor gene polymorphisms and susceptibility to *Mycobacterium malmoense* pulmonary disease. J. Infect. Dis. 181:2099–2102

George KL, Parker BC, Gruft H, Falkinham JO, III (1980) Epidemiology of infection by nontuberculous mycobacteria. II. Growth and survival in natural waters. Am. Rev. Respir. Dis. 122:89–94

George SL, Schlesinger LS (1999) *Mycobacterium neoaurum* – An unusual cause of infection of vascular catheters: Case report and review. Clin. Infect. Dis. 28: 682–683

Gerlach H (1994) *Mycobacterium*. In: B.W. Ritchie, G.J. Harrison, L.R. Harrison (Eds.), Avian Medicine: Principles and Applications. Lake Worth, Florida, Wingers Publishing. 971–975

Glickman SE, Kilburn JO, Butler WR, Ramos LS (1994) Rapid Identification of Mycolic Acid Patterns of Mycobacteria by High-Performance Liquid-Chromatography Using Pattern-Recognition Software and A Mycobacterium Library. J. Clin. Microbiol. 32:740–745

Goldblatt MR, Ribes JA (2002) *Mycobacterium mucogenicum* isolated from a patient with granulomatous hepatitis. Arch. of Pathol. Lab. Med. 126:73–75

Gombert ME, Goldstein EJC, Corrado ML, Stein AJ, Butt KMH (1981) Disseminated *Mycobacterium-Marinum* Infection After Renal-Transplantation. Ann. Int Med. 94:486–487

Good RC, Snider DE, Jr. (1982) Isolation of nontuberculous mycobacteria in the United States, 1980. J. Infect. Dis. 146:829–833

Gouby A, Branger B, Oules R, Ramuz M (1988) Two cases of *Mycobacterium haemophilum* infection in a renal-dialysis unit. J. Med. Microbiol. 25:299–300

Grange JM (1996) Mycobacteria and human disease. 2nd ed. London, Arnold, 230 pp

Greenstein RJ (2003) Is Crohn's disease caused by a mycobacterium? Comparisons with leprosy, tuberculosis, and Johne's disease. Lancet Infect. Dis. 3:507–514

Grewal SK, Rajeev S, Sreevatsan S, Michel FC, Jr. (2006) Persistence of *Mycobacterium avium* subsp. *paratuberculosis* and other zoonotic pathogens during simulated composting, manure packing, and liquid storage of dairy manure. Appl. Environ. Microbiol. 72:565–574

Gross WB, Falkinham JO, III, Payeur JB (1989) Effect of environmental-genetic interactions on *Mycobacterium avium* challenge infection. Avian Dis. 33:411–415

Gwozdz JM (2006) Comparative evaluation of two decontamination methods for the isolation of *Mycobacterium avium* subspecies *paratuberculosis* from faecal slurry and sewage. Vet. Microbiol. 115:358–363

Hagglblom MM, Nohynek LJ, Palleroni NJ, Kronqvist K, Nurmiaholassila EL, Salkinojasalonen MS, Klatte S, Kroppenstedt RM (1994) Transfer of Polychlorophenol-Degrading Rhodococcus-Chlorophenolicus (Apajalahti Et-Al 1986) to the Genus *Mycobacterium* as *Mycobacterium-Chlorophenolicum* Comb-Nov. Int. J. Syst. Bacteriol. 44:485–493

Hall-Stoodley L, Keevil CW, Lappin-Scott HM (1999) *Mycobacterium fortuitum* and *Mycobacterium chelonae* biofilm formation under high and low nutrient conditions. J. Appl. Microbiol. 85:60S–69S

Hall-Stoodley L, Lappin-Scott H (1998) Biofilm formation by the rapidly growing mycobacterial species *Mycobacterium fortuitum*. FEMS Microbiol. Lett. 168:77–84

Hanau LH, Leaf A, Soeiro R, Weiss LM, Pollack SS (1994) *Mycobacterium-Marinum* Infection in A Patient with the Acquired-Immunodeficiency-Syndrome. Cutis. 54:103–105

Harris NB, Barletta RG (2001) *Mycobacterium avium* subsp. *paratuberculosis* in veterinary medicine. Clin. Microbiol. Rev. 14:489-+

Hartwig H, Stottmeier D (1963) On the variability of mycobacteria. Zentralbl. Bakteriol. [Orig.]. 189:430–453

Hasonova L, Pavlik I (2006) Economic impact of paratuberculosis in dairy cattle herds: a review. Veterinarni Medicina 51:193–211

Hauduroy P, Hovanessian A, Roussianos D (1965a) [Instability of chromogeneity in strains of *Mycobacterium kansasii*]. Ann. Inst. Pasteur. (Paris). 109:142–144

Hauduroy P, Hovanessian A, Roussianos D (1965b) [Study on some characteristics of scotochromogenic variants of *Mycobacterium kansasii*, and especially of their pathogenicity for hamsters]. Ann. Inst. Pasteur. (Paris). 109:138–141

Hayes PS, Mcgiboney DL, Band JD, Feeley JC (1982) Resistance of *Mycobacterium-Chelonei*-Like Organisms to Formaldehyde. Appl. Environ. Microbiol. 43: 722–724

Hector JSR, Pang YJ, Mazurek GH, Zhang YS, Brown BA, Wallace RJ (1992) Large Restriction Fragment Patterns of Genomic *Mycobacterium-Fortuitum* DNA As Strain-Specific Markers and Their Use in Epidemiologic Investigation of 4 Nosocomial Outbreaks. J. Clin. Microbiol. 30:1250–1255

Hejlicek K, Treml F (1995) [Comparison of the pathogenesis and epizootiologic importance of avian mycobacteriosis in various types of domestic and free-living syntropic birds]. Veterinarni Medicina 40:187–194

Henriques B, Hoffner SE, Petrini B, Juhlin I, Wahlen P, Kallenius G (1994) Infection with *Mycobacterium-Malmoense* in Sweden – Report of 221 Cases. Clin. Infect. Dis. 18: 546–600

Hermon-Taylor J (2000) *Mycobacterium avium* subspecies *paratuberculosis* in the causation of Crohn's disease. World J. Gastroenterol. 6:630–632

Hermon-Taylor J, Bull T (2002) Crohn's disease caused by *Mycobacterium avium* subspecies *paratuberculosis*: a public health tragedy whose resolution is long overdue. J. Med. Microbiol. 51:3–6

Hermon-Taylor J, Bull TJ, Sheridan JM, Cheng J, Stellakis ML, Sumar N (2000) Causation of Crohn's disease by *Mycobacterium avium* subspecies *paratuberculosis*. Can. J. Gastroenterol. 14:521–539

Hernandez-Divers SJ, Shearer D (2002) Pulmonary mycobacteriosis caused by *Mycobacterium haemophilum* and *M marinum* in a royal python. J. Am. Vet. Med. Assoc. 220:1661–1663

Hoffner SE, Henriques B, Petrini B, Kallenius G (1991) *Mycobacterium-Malmoense* – An Easily Missed Pathogen. J. Clin. Microbiol. 29:2673–2674

Horacek J, Ulicna L (1973) *M. balnei* in colliery waters – infection agents of verrucous skin disease. Ceskoslovenska. dermatologie. 48:97–99

Horak Z, Janasova V, Polakova H, Kralova M, Rychlikova E (1991) Occurrence of pulmonary mycobacterioses due to *M. xenopi* in urban population. (In Czech). Stud. Pneumol. Phtiseol. Cechoslov. 51:51–56

Horak Z, Polakova H, Kralova M (1986) Water-borne *Mycobacterium xenopi*–a possible cause of pulmonary mycobacteriosis in man. J. Hyg. Epidemiol. Microbiol. Immunol. 30:405–409

Hruska K, Bartos M, Kralik P, Pavlik I (2005) *Mycobacterium avium* subsp. *paratuberculosis* in powdered infant milk: paratuberculosis in cattle – the public health problem to be solved. Veterinarni Medicina 50:327–335

Hughes MS, Ball NW, Beck LA, deLisle GW, Skuce RA, Neill SD (1997) Determination of the etiology of presumptive feline leprosy by 16S rRNA gene analysis. J. Clin. Microbiol. 35:2464–2471

Huttunen K, Ruotsalainen M, Iivanainen E, Torkko P, Katila ML, Hirvonen MR (2000) Inflammatory responses in RAW264.7 macrophages caused by mycobacteria isolated from moldy houses. Environ. Toxicol. Pharmacol. 8:237–244

Jaravata CV, Smith WL, Rensen GJ, Ruzante JM, Cullor JS (2006) Detection of *Mycobacterium avium* subsp. *paratuberculosis* in bovine manure using Whatman FTA card technology and Lightcycler real-time PCR. Foodborne. Pathog. Dis. 3:212–215

Jenkins PA, Tsukamura M (1979) Infections with *Mycobacterium malmoense* in England and Wales. Tubercle 60:71–76

Jernigan JA, Farr BM (2000) Incubation period and sources of exposure for cutaneous *Mycobacterium marinum* infection: Case report and review of the literature. Clin. Infect. Dis. 31:439–443

Johne HA, Frothingham L (1895) Ein eigentheumlicher Fall von Tuberculose beim Rind. Dtsch. Ztschr. Tier-Med. 21:438–454

Jones DC, Gelder CM, Ahmad T, Campbell IA, Barnardo MCNM, Welsh KI, Marshall SE, Bunce M (2001) CD1 genotyping of patients with *Mycobacterium malmoense* pulmonary disease. Tissue Antigens. 58:19–23

Jones JJ, Falkinham JO, III (2003) Decolorization of malachite green and crystal violet by waterborne pathogenic mycobacteria. Antimicrob. Agents Chemother. 47:2323–2326

Jorgensen JB (1977) Survival of *Mycobacterium paratuberculosis* in slurry. Nord. Vet. Med. 29:267–270

Jorgensen JB (1982) An improved medium for culture of *Mycobacterium paratuberculosis* from bovine faeces. Acta Vet. Scand. 23:325–335

Joynson DHM (1979) Water – Natural Habitat of *Mycobacterium-Kansasii*. Tubercle. 60:77–81

Jucker MT, Falkinham JO, III (1990) Epidemiology of infection by nontuberculous mycobacteria IX. Evidence for two DNA homology groups among small plasmids in *Mycobacterium avium, Mycobacterium intracellulare*, and *Mycobacterium scrofulaceum*. Am. Rev. Respir. Dis. 142: 858–862

Katoch VM (2004) Infections due to non-tuberculous mycobacteria (NTM). Indian J. Med. Res. 120:290–304

Kauppinen J, Mantyjarvi R, Katila ML (1999) *Mycobacterium malmoense*-specific nested PCR based on a conserved sequence detected in random amplified polymorphic DNA fingerprints. J. Clin. Microbiol. 37:1454–1458

Kaustova J, Charvat B, Mudra R, Holendova E (1993a) Ostrava – a new endemic focus of *Mycobacteria xenopi* in the Czech Republic. Cent. Eur. J Public Health. 1:35–37

Kaustova J, Chmelik M, Ettlova D, Hudec V, Lazarova H, Richtrova S (1995) Disease due to *Mycobacterium kansasii* in the Czech Republic: 1984–89. Tuber. Lung Dis. 76:205–209

Kaustova J, Martinek A, Curik R (1993b) A case of fatal septicemia due to *Mycobacterium kansasii*. Eur. J. Clin. Microbiol. Infect. Dis. 12:791–793

Kazda J (1967) [Atypical mycobacteria in drinking water–the cause of para-allergies against tuberculin in animals]. Z. Tuberk. Erkr. Thoraxorg. 127:111–113

Kazda J (1973) [The importance of water for the spread of potentially pathogenic Mycobacteria. I. Possibilities for the multiplication of Mycobacteria (author's transl)]. Zentralbl. Bakteriol. [Orig. B]. 158:161–169

Kazda J (1977) [The importance of sphagnum bogs in the ecology of Mycobacteria (author's transl)]. Zentralbl. Bakteriol. [Orig. B]. 165:323–334

Kazda J (1978a) [Multiplication of mycobacteria in the gray layer of sphagnum vegetation (author's transl)]. Zentralbl. Bakteriol. [Orig. B]. 166:463–469

Kazda J (1978b) [The behaviour of *Mycobacterium intracellulare* serotype Davis and *Mycobacterium avium* in the head region of sphagnum moss vegetation after experimental inoculation (author's transl)]. Zentralbl. Bakteriol. [Orig. B]. 166:454–462

Kazda J (2000) The ecology of mycobacteria. Kluwer Academic Publishers, Dordrecht, Boston, London, 72 pp.

Kazda J, Cooney R, Monaghan M, Quinn PJ, Stackebrandt E, Dorsch M, Daffe M, Muller K, Cook BR, Tarnok ZS (1993) *Mycobacterium-Hiberniae* Sp-Nov. Int. J. Syst. Bacteriol. 43:352–357

Kazda J, Hoyte R (1972) Concerning Ecology of *Mycobacterium Intracellular* Serotype Davis. Zentralblatt fur Bakteriologie Mikrobiologie und Hygiene Series A-Medical

Microbiology Infectious Diseases Virology Parasitology. 222:506–509

Kazda J, Muller HJ, Stackebrandt E, Daffe M, Muller K, Pitulle C (1992) *Mycobacterium-Madagascariense* Sp-Nov. Int. J. Syst. Bacteriol. 42:524–528

Kazda J, Stackebrandt E, Smida J, Minnikin DE, Daffe M, Parlett JH, Pitulle C (1990) *Mycobacterium-Cookii* Sp-Nov. Int. J. Syst. Bacteriol. 40:217–223

Kelley LC, Deering KC, Kaye ET (1995) Cutaneous *Mycobacterium chelonei* presenting in an immunocompetent host: case report and review of the literature. Cutis. 56: 293–295

Kent ML, Whipps CM, Matthews JL, Florio D, Watral V, Bishop-Stewart JK, Poort M, Bermudez L (2004) Mycobacteriosis in zebrafish (*Danio rerio*) research facilities. Comp. Biochem. Physiol. C, Toxicol. Pharmacol. 138:383–390

Kerr-Pontes LRS, Barreto ML, Evangelista CMN, Rodrigues LC, Heukelbach J, Feldmeier H (2006) Socioeconomic, environmental, and behavioural risk factors for leprosy in Northeast Brazil: results of a case-control study. Int. J Epidemiol. 35:994–1000

Khan AA, Kim SJ, Paine DD, Cerniglia CE (2002) Classification of a polycyclic aromatic hydrocarbon-metabolizing bacterium, *Mycobacterium* sp strain PYR-1, as *Mycobacterium vanbaalenii* sp nov. Int. J. Syst. Evol. Microbiol. 52:1997–2002

Khooshabeh R, Grange JM, Yates MD, Mccartney ACE, Casey TA (1994) A Case-Report of *Mycobacterium-Chelonae* Keratitis and A Review of Mycobacterial Infections of the Eye and Orbit. Tuber. Lung Dis. 75:377–382

Kiehn TE, White M (1994) *Mycobacterium haemophilum*: an emerging pathogen. Eur. J. Clin. Microbiol. Infect. Dis. 13:925–931

Kikuchi K, Bernard EM, Kiehn TE, Armstrong D, Riley LW (1994) Restriction-Fragment-Length-Polymorphism Analysis of Clinical Isolates of *Mycobacterium-Haemophilum*. J. Clin. Microbiol. 32:1763–1767

Kim TH, Kubica GP (1972) Long-term preservation and storage of mycobacteria. Appl. Microbiol. 24:311–317

Kim TH, Kubica GP (1973) Preservation of mycobacteria: 100 percent viability of suspensions stored at –70 °C. Appl. Microbiol. 25:956–960

Kirschner RA, Parker BC, Falkinham JO (1999) Humic and fulvic acids stimulate the growth of *Mycobacterium avium*. FEMS Microbiol. Ecol. 30:327–332

Kleespies M, Kroppenstedt RM, Rainey FA, Webb LE, Stackebrandt E (1996) *Mycobacterium hodleri* sp nov, a new member of the fast-growing mycobacteria capable of degrading polycyclic aromatic hydrocarbons. Int. J. Syst. Bacteriol. 46:683–687

Kul O, Tunca R, Haziroglu R, Diker KS, Karahan S (2005) An outbreak of avian tuberculosis in peafowl (*Pavo cristatus*) and pheasants (*Phasianus colchicus*) in a zoological aviary in Turkey. Veterinarni Medicina 50:446–450

Kusunoki S, Ezaki T (1992) Proposal of *Mycobacterium-Peregrinum* Sp-Nov, Nom Rev, and Elevation of *Mycobacterium-Chelonae* Subsp *Abscessus* (Kubica Et-Al) to Species Status – *Mycobacterium-Abscessus* Comb-Nov. Int. J. Syst. Bacteriol. 42:240–245

Lagunas-Solar MC, Cullor JS, Zeng NX, Truong TD, Essert TK, Smith WL, Pina C (2005) Disinfection of dairy and animal farm wastewater with radiofrequency power. J. Dairy Sci. 88:4120–4131

Larsen AB, Merkal RS, Vardaman TH (1956) Survival Time of *Mycobacterium-Paratuberculosis*. Am. J. Vet. Res. 17:549–551

Laussucq S, Baltch AL, Smith RP, Smithwick RW, Davis BJ, Desjardin EK, Silcox VA, Spellacy AB, Zeimis RT, Gruft HM, Good RC, Cohen ML (1988) Nosocomial *Mycobacterium-Fortuitum* Colonization from A Contaminated Ice Machine. Am. Rev. Respir. Dis. 138: 891–894

Le Dantec C, Duguet JP, Montiel A, Dumoutier N, Dubrou S, Vincent V (2002) Occurrence of mycobacteria in water treatment lines and in water distribution systems. Appl. Environ. Microbiol. 68:5318–5325

Leoni E, Legnani P, Mucci MT, Pirani R (1999) Prevalence of mycobacteria in a swimming pool environment. J. Appl. Microbiol. 87:683–688

Levy-frebault X, Grimont F, Grimont PAD, David HL (1986) Deoxyribonucleic-Acid Relatedness Study of the *Mycobacterium-Fortuitum-Mycobacterium-Chelonae* Complex. Int. J. Syst. Bacteriol. 36:458–460

Lindeboom JA, Prins JM, Bruijnesteijn van Coppenraet ES, Lindeboom R, Kuijper EJ (2005) Cervicofacial lymphadenitis in children caused by *Mycobacterium haemophilum*. Clin. Infect. Dis. 41:1569–1575

Linell L, Norden A (1954) *Mycobacterium balnei*, a new acid-fast bacillus occurring in swimming pools and capable of producing skin lesions in humans. Acta Tuberc. Scand. Suppl. 33:1–84

Lombard JE, Wagner BA, Smith RL, McCluskey BJ, Harris BN, Payeur JB, Garry FB, Salman MD (2006) Evaluation of environmental sampling and culture to determine *Mycobacterium avium* subspecies *paratuberculosis* distribution and herd infection status on US dairy operations. J. Dairy Sci. 89:4163–4171

Lovell R, Levi M, Francis J (1944) Studies on the survival of Johne's bacilli. J. Comp. Pathol. 54:120–129

Maccallum P (1948) A New Mycobacterial Infection in Man. 1. Clinical Aspects. J. Pathol. Bacteriol. 60:93–102

Machackova-Kopecna M, Bartos M, Straka M, Ludvik V, Svastova P, Alvarez J, Lamka J, Trcka I, Treml F, Parmova I, Pavlik I (2005) Paratuberculosis and avian tuberculosis infections in one red deer farm studied by IS*900* and IS*901* RFLP analysis. Vet. Microbiol. 105:261–268

Marco I, Domingo M, Lavin S (2000) *Mycobacterium* infection in a captive-reared capercaillie (*Tetrao urogallus*). Avian Dis. 44:227–230

Marks J, Jenkins PA (1971) The opportunist mycobacteria–a 20-year retrospect. Postgrad. Med. J. 47:705–709

Marsollier L, Aubry J, Coutanceau E, Andre JPS, Small PL, Milon G, Legras P, Guadagnini S, Carbonnelle B, Cole ST (2005) Colonization of the salivary glands of *Naucoris cimicoides* by *Mycobacterium ulcerans*requires host plasmatocytes and a macrolide toxin, mycolactone. Cell. Microbiol. 7:935–943

Martinkova I, Sebakova H, Pelikan M, Zatloukal O (2001) [Endemic incidence of *Mycobacterium kansasii* infection in Karvina District 1968–1999; overview of the descriptive characteristics]. Epidemiol. Microbiol. Immunol. 50: 165–180

Marx CE, Fan KL, Morris AJ, Wilson ML, Damiani A, Weinstein MP (1995) Laboratory and Clinical-Evaluation of *Mycobacterium-Xenopi* Isolates. Diagnostic Microbiology and Infections Disease 21:195–202

Matlova L, Dvorska L, Ayele WY, Bartos M, Amemori T, Pavlik I (2005) Distribution of *Mycobacterium avium* complex isolates in tissue samples of pigs fed peat naturally contaminated with mycobacteria as a supplement. J. Clin. Microbiol. 43:1261–1268

Matlova L, Dvorska L, Bartl J, Bartos M, Ayele WY, Alexa M, Pavlik I (2003) Mycobacteria isolated from the environment of pig farms in the Czech Republic during the years 1996 to 2002. Veterinarni Medicina 48:343–357

McCarthy CM, Schaefer JO (1974) Response of *Mycobacterium avium* to ultraviolet irradiation. Appl. Microbiol. 28:151–153

McCullough WG, Merkal RS (1982) Structure of mycobactin. J. Curr. Microbiol. 7:337–341

McSwiggan DA, Collins CH (1974) The isolation of *M. kansasii* and *M. xenopi* from water systems. Tubercle. 55:291–297

Meissner G, Schroder KH, Amadio GE, Anz W, Chaparas S, Engel HW, Jenkins PA, Kappler W, Kleeberg HH, Kubala E, Kubin M, Lauterbach D, Lind A, Magnusson M, Mikova Z, Pattyn SR, Schaefer WB, Stanford JL, Tsukamura M, Wayne LG, Willers I, Wolinsky E (1974) A co-operative numerical analysis of nonscoto- and nonphotochromogenic slowly growing mycobacteria. J. Gen. Microbiol. 83:207–235

Meissner PS, Falkinham JO, III (1984) Plasmid-encoded mercuric reductase in *Mycobacterium scrofulaceum*. J. Bacteriol. 157:669–672

Meissner PS, Falkinham JO, III (1986) Plasmid DNA profiles as epidemiological markers for clinical and environmental isolates of *Mycobacterium avium, Mycobacterium intracellulare*, and *Mycobacterium scrofulaceum*. J. Infect. Dis. 153:325–331

Merkal RS, Crawford JA (1979) Heat inactivation of *Mycobacterium avium-Mycobacterium intracellulare* complex organisms in aqueous suspension. Appl. Environ. Microbiol. 38:827–830

Merkal RS, Crawford JA, Whipple DL (1979) Heat inactivation of *Mycobacterium avium-Mycobacterium intracellulare* complex organisms in meat products. Appl. Environ. Microbiol. 38:831–835

Merkal RS, Curran BJ (1974) Growth and metabolic characteristics of *Mycobacterium paratuberculosis*. Appl. Microbiol. 28:276–279

Merkal RS, Larsen AB, Kopecky KE, Ness RD (1968) Comparison of Examination and Test Methods for Early Detection of Paratuberculous Cattle. Am. J. Vet. Res. 29:1533–1538

Merkal RS, McCullough WG (1982) A new mycobactin, mycobactin J, from *Mycobacterium paratuberculosis*. Current Microbiol. 7:333–335

Merritt RW, Benbow ME, Small PLC (2005) Unraveling an emerging disease associated with disturbed aquatic environments: the case of Buruli ulcer. Front Ecol. Environ. 3:323–331

Mijs W, de Haas P, Rossau R, Van der Laan T, Rigouts L, Portaels F, van Soolingen D (2002a) Molecular evidence to support a proposal to reserve the designation *Mycobacterium avium* subsp *avium* for bird-type isolates and '*M-avium* subsp *hominissuis*' for the human/porcine type of *M-avium*. Int. J. Syst. Evol. Microbiol. 52:1505–1518

Mijs W, De Vreese K, Devos A, Pottel H, Valgaeren A, Evans C, Norton J, Parker D, Rigouts L, Portaels F, Reischl U, Watterson S, Pfyffer G, Rossau R (2002b) Evaluation of a commercial line probe assay for identification of Mycobacterium species from liquid and solid culture. Eur. J. Clin. Microbiol. Infect. Dis. 21:794–802

Mills JA, McNeil MR, Belisle JT, Jacobs WR, Jr., Brennan PJ (1994) Loci of *Mycobacterium avium* ser2 gene cluster and their functions. J. Bacteriol. 176:4803–4808

Miltner E, Daroogheh K, Mehta PK, Cirillo SL, Cirillo JD, Bermudez LE (2005) Identification of *Mycobacterium avium* genes that affect invasion of the intestinal epithelium. Infect. Immun. 73:4214–4221

Mirando WS, Shiratsuchi H, Tubesing K, Toba H, Ellner JJ, Elmets CA (1992) Ultraviolet-Irradiated Monocytes Efficiently Inhibit the Intracellular Replication of *Mycobacterium-Avium-Intracellulare*. J. Clin. Invest. 89:1282–1287

Moravkova M, Hlozek P, Beran V, Pavlik I, Preziuso S, Cuteri V, Bartos M (2007) Strategy for the detection and differentiation of *Mycobacterium avium* species in isolates and heavily infected tissues. Research in Veterinary Science (Article in press, available online at www.sciencedirect.com)

Mura M, Bull TJ, Evans H, Sidi-Boumedine K, McMinn L, Rhodes G, Pickup R, Hermon-Taylor J (2006) Replication and long-term persistence of bovine and human strains of *Mycobacterium avium* subsp. *paratuberculosis* within *Acanthamoeba polyphaga*. Appl. Environ. Microbiol. 72:854–859

Murcia MI, Tortoli E, Menendez MC, Palenque E, Garcia MJ (2006) *Mycobacterium colombiense* sp. nov., a novel member of the *Mycobacterium avium* complex and description of MAC-X as a new ITS genetic variant. Int. J. Syst. Evol. Microbiol. 56:2049–2054

Murohashi T, Yoshida K (1968) Effect of ultraviolet irradiation on the acid-fastness of difficult to culture and unculturable mycobacteria. Am. Rev. Respir. Dis. 97:306–310

Nieminen T, Pakarinen J, Tsitko I, Salkinoja-Salonen M, Breitenstein A, Ali-Vehmas T, Neubauer P (2006) 16S rRNA targeted sandwich hybridization method for direct quantification of mycobacteria in soils. J. Microbiol. Methods. 67:44–55

Norby B, Fosgate GT, Manning EJ, Collins MT, Roussel AJ (2007) Environmental mycobacteria in soil and water on beef ranches: Association between presence of cultivable mycobacteria and soil and water physicochemical characteristics. Vet. Microbiol. 124:153–159

Norton CD, LeChevallier MW, Falkinham JO, III (2004) Survival of *Mycobacterium avium* in a model distribution system. Water Res. 38:1457–1466

Nyka W (1974) Studies on the effect of starvation on mycobacteria. Infect. Immun. 9:843–850

O'Brien RJ, Geiter LJ, Snider DE (1987) The Epidemiology of Nontuberculous Mycobacterial Diseases in the United-States – Results from A National Survey. Am. Rev. Respir. Dis. 135:1007–1014

Oga M, Arizono T, Takasita M, Sugioka Y (1993) Evaluation of the Risk of Instrumentation As A Foreign-Body in Spinal Tuberculosis – Clinical and Biologic Study. Spine. 18:1890–1894

Olsen JE, Jorgensen JB, Nansen P (1985) On the Reduction of *Mycobacterium-Paratuberculosis* in Bovine Slurry Subjected to Batch Mesophilic Or Thermophilic Anaerobic-Digestion. Agric. Wastes. 13:273–280

Pai HH, Chen WC, Peng CF (2003) Isolation of non-tuberculous mycobacteria from hospital cockroaches (*Periplaneta americana*). J. Hosp. Infect. 53:224–228

Pang Y, Brown BA, Steingrube VA, Wallace RJ, Roberts MC (1994) Tetracycline Resistance Determinants in *Mycobacterium and Streptomyces* Species. Antimicrob. Agents Chemother. 38:1408–1412

Pankhurst CL, Johnson NW, Woods RG (1998) Microbial contamination of dental unit waterlines: the scientific argument. Int. Dent. J. 48:359–368

Parent LJ, Salam MM, Appelbaum PC, Dossett JH (1995) Disseminated *Mycobacterium-Marinum* Infection and Bacteremia in A Child with Severe Combined Immunodeficiency. Clin. Infect. Dis. 21:1325–1327

Parker BC, Ford MA, Gruft H, Falkinham JO, III (1983) Epidemiology of infection by nontuberculous mycobacteria. IV. Preferential aerosolization of *Mycobacterium intracellulare* from natural waters. Am. Rev. Respir. Dis. 128: 652–656

Pavlik I, Horvathova A, Dvorska L, Bartl J, Svastova P, du Maine R, Rychlik I (1999) Standardisation of restriction fragment length polymorphism analysis for *Mycobacterium avium* subspecies *paratuberculosis*. J. Microbiol. Methods. 38:155–167

Pavlik I, Jahn P, Moravkova M, Matlova L, Treml F, Cizek A, Nesnalova E, Dvorska-Bartosova L, Halouzka R (2008) Lung tuberculosis in a horse caused by *Mycobacterium avium* subsp *avium* of serotype 2: a case report. Veterinarni Medicina 53:111–116

Pavlik I, Matlova L, Dvorska L, Bartl J, Oktabcova L, Docekal J, Parmova I (2003) Tuberculous lesions in pigs in the Czech Republic during 1990–1999: occurrence, causal factors and economic losses. Veterinarni Medicina 48:113–125

Pavlik I, Matlova L, Dvorska L, Shitaye JE, Parmova I (2005a) Mycobacterial infections in cattle and pigs caused by *Mycobacterium avium* complex members and atypical mycobacteria in the Czech Republic during 2000–2004. Veterinarni Medicina 50:281–290

Pavlik I, Matlova L, Gilar M, Bartl J, Parmova I, Lysak F, Alexa M, Dvorska-Bartosova L, Svec V, Vrbas V, Horvathova A (2007) Isolation of conditionally pathogenic mycobacteria from the environment of one pig farm and the effectiveness of preventive measures between 1997 and 2003. Veterinarni Medicina 52:392–404

Pavlik I, Rozsypalova Z, Vesely T, Bartl J, Matlova L, Vrbas V, Valent L, Rajsky D, Mracko I, Hirko M, Miskovic P (2000a) Control of paratuberculosis in five cattle farms by serological tests and faecal culture during the period 1990–1999. Veterinarni Medicina 45:61–70

Pavlik I, Svastova P, Bartl J, Dvorska L, Rychlik I (2000b) Relationship between IS*901* in the *Mycobacterium avium* complex strains isolated from birds, animals, humans, and the environment and virulence for poultry. Clin. Diagn. Lab. Immunol. 7:212–217

Pavlik I, Trcka I, Parmova I, Svobodova J, Melicharek I, Nagy G, Cvetnic Z, Ocepek M, Pate M, Lipiec M (2005b) Detection of bovine and human tuberculosis in cattle and other animals in six Central European countries during the years 2000–2004. Veterinarni Medicina 50:291–299

Pelikan M, Mikova Z, Kaustova J, Kubin M (1973) Aerosol from water for industrial purposes – a possible transmission factor for infection by atypical mycobacteria (in Czech:Aerosol z užitkové vody jako pravděpodobný faktor přenosu při infekci atypickými mykobakteriemi). Ceskoslovenska hygiena. 6:316–323

Peters M, Muller C, Ruschgerdes S, Seidel C, Gobel U, Pohle HD, Ruf B (1995) Isolation of Atypical Mycobacteria from Tap Water in Hospitals and Homes – Is This A Possible Source of Disseminated Mac Infection in Aids Patients. J. Infect. 31:39–44

Phillips MS, von Reyn CF (2001) Nosocomial infections due to nontuberculous mycobacteria. Clin. Infect. Dis. 33:1363–1374

Picardeau M, Prod'Hom G, Raskine L, LePennec MP, Vincent V (1997) Genotypic characterization of five subspecies of *Mycobacterium kansasii*. J. Clin. Microbiol. 35:25–32

Picardeau M, Varnerot A, Rauzier J, Gicquel B, Vincent V (1996) *Mycobacterium xenopi* IS*1395*, a novel insertion sequence expanding the IS*256* family. Microbiology-Uk. 142:2453–2461

Picardeau M, Vincent V (1995) Development of A Species-Specific Probe for *Mycobacterium-Xenopi*. Res. Microbiol. 146:237–243

Pickup RW, Rhodes G, Arnott S, Sidi-Boumedine K, Bull TJ, Weightman A, Hurley M, Hermon-Taylor J (2005) *Mycobacterium avium* subsp *paratuberculosis* in the catchment area and water of the river Taff in South Wales, United Kingdom, an its potential relationship to clustering of Crohn's disease cases in the city of Cardiff. Appl. Environ. Microbiol. 71:2130–2139

Pickup RW, Rhodes G, Bull TJ, Arnott S, Sidi-Boumedine K, Hurley M, Hermon-Taylor J (2006) *Mycobacterium avium* subsp. *paratuberculosis* in lake catchments, in river water abstracted for domestic use, and in effluent from domestic sewage treatment works: diverse opportunities for environmental cycling and human exposure. Appl. Environ. Microbiol. 72:4067–4077

Pitulle C, Dorsch M, Kazda J, Wolters J, Stackebrandt E (1992) Phylogeny of Rapidly Growing Members of the Genus *Mycobacterium*. Int. J. Syst. Bacteriol. 42:337–343

Plum G, Brenden M, Clark-Curtiss JE, Pulverer G (1997) Cloning, sequencing, and expression of the mig gene of *Mycobacterium avium*, which codes for a secreted macrophage-induced protein. Infect. Immun. 65: 4548–4557

Portaels F, Elsen P, Guimaraes-Peres A, Fonteyne PA, Meyers WM (1999) Insects in the transmission of *Mycobacterium ulcerans* infection. Lancet. 353:986–986

Portaels F, Fissette K, De Ridder K, Macedo PM, De Muynck A, Silva MT (1988a) Effects of freezing and thawing on the viability and the ultrastructure of *in vivo* grown mycobacteria. Int. J. Lepr. Other Mycobact. Dis. 56:580–587

Portaels F, Larsson L, Smeets P (1988b) Isolation of Mycobacteria from Healthy-Persons Stools. Int. J. Lepr. Other Mycobact. Dis. 56:468–471

Portaels F, Pattyn SR (1982) Growth of mycobacteria in relation to the pH of the medium. Ann. Microbiol. (Paris). 133:213–221

Poupart P, Coene M, Vanheuverswyn H, Cocito C (1993) Preparation of A Specific Rna Probe for Detection of

Mycobacterium-Paratuberculosis and Diagnosis of Johnes Disease. J. Clin. Microbiol. 31:1601–1605

Powell BL, Steadham JE (1981) Improved technique for isolation of *Mycobacterium kansasii* from water. J. Clin. Microbiol. 13:969–975

Prukner-Radovcic E, Culjak K, Sostaric B, Mazija H, Sabocanec R (1998) Generalised tuberculosis in pheasants at a commercial breeding farm. Zeitschrift fur Jagdwissenschaft. 44:33–39

Raizman EA, Wells SJ, Godden SM, Bey RF, Oakes MJ, Bentley DC, Olsen KE (2004) The distribution of *Mycobacterium avium* ssp. *paratuberculosis* in the environment surrounding Minnesota dairy farms. J. Dairy Sci. 87:2959–2966

Razavi B, Cleveland MG (2000) Cutaneous infection due to *Mycobacterium kansasii*. Diagn. Microbiol. Infect. Dis. 38:173–175

Richards WD (1981) Effects of physiol and chemical factors on the viability of *Mycobacterium paratuberculosis*. J. Clin. Microbiol. 14:587–588

Rowe MT, Grant IR (2006) *Mycobacterium avium* ssp. *paratuberculosis* and its potential survival tactics. Lett. Appl. Microbiol. 42:305–311

Russell AD (1999) Bacterial resistance to disinfectants: present knowledge and future problems. J. Hosp. Infect. 43:S57-S68

Safranek TJ, Jarvis WR, Carson LA, Cusick LB, Bland LA, Swenson JM, Silcox VA (1987) *Mycobacterium-Chelonae* Wound Infections After Plastic-Surgery Employing Contaminated Gentian-Violet Skin-Marking Solution. N. Engl. J. Med. 317:197–201

Sanderson MW, Dargatz DA, Garry FB (2000) Biosecurity practices of beef cow-calf producers. J. Am. Vet. Med. Assoc. 217:185–189

Sastry V, Brennan PJ (1995) Cutaneous Infections with Rapidly Growing Mycobacteria. Clin. Dermatol. 13:265–271

Schroder KH, Juhlin I (1977) *Mycobacterium-Malmoense* Sp-Nov. Int. J. Syst. Bacteriol. 27:241–246

Schulze-Robbecke R, Buchholtz K (1992) Heat susceptibility of aquatic mycobacteria. Appl. Environ. Microbiol. 58:1869–1873

Schulze-Robbecke R, Fischeder R (1989) Mycobacteria in biofilms. Zentralbl. Hyg. Umweltmed. 188:385–390

Schwabacher H (1959) A strain of *Mycobacterium* isolated from skin lesions of a cold-blooded animal, *Xenopus laevis*, and its relation to atypical acid-fast bacilli occurring in man. J. Hyg. (Lond). 57:57–67

Shitaye JE, Matlova L, Horvathova A, Moravkova M, Dvorska-Bartosova L, Trcka I, Lamka J, Treml F, Vrbas V, Pavlik I (2008a) Diagnostic testing of different stages of avian tuberculosis in naturally infected hens (*Gallus domesticus*) by the tuberculin skin and rapid agglutination tests, faecal and egg examinations. Veterinarni Medicina 53:101–110

Shitaye JE, Matlova L, Horvathova A, Moravkova M, Dvorska-Bartosova L, Treml F, Lamka J, Pavlik I (2008b) *Mycobacterium avium* subsp. *avium* distribution studied in a naturally infected hen flock and in the environment by culture, serotyping and IS*901* RFLP methods. Vet. Microbiol. 127:155–164

Skovgaard N (2007) New trends in emerging pathogens. Int. J. Food Microbiol. 120:217–224

Slosarek M, Kubin M, Jaresova M (1993) Water-borne household infections due to *Mycobacterium xenopi*. Cent. Eur. J. Public Health. 1:78–80

Smith MB, Schnadig VJ, Boyars MC, Woods GL (2001) Clinical and pathologic features of *Mycobacterium fortuitum* infections. An emerging pathogen in patients with AIDS. Am. J. Clin. Pathol. 116:225–232

Sniadack DH, Ostroff SM, Karlix MA, Smithwick RW, Schwartz B, Sprauer MA, Silcox VA, Good RC (1993) A Nosocomial Pseudo-Outbreak of *Mycobacterium-Xenopi* Due to A Contaminated Potable Water-Supply – Lessons in Prevention. Infect. Control Hosp. Epidemiol. 14:636–641

Snow GA (1965) Structure of Mycobactin P A Growth Factor for *Mycobacterium Johnei* and Significance of Its Iron Complex. Biochem. J. 94:160–165

Sompolinsky D, Lagziel A, Naveh D, Yankilevitz T (1978) *Mycobacterium-Haemophilum* Sp-Nov, A New Pathogen of Humans. Int. J. Syst. Bacteriol. 28:67–75

Speight EL, Williams HC (1997) Fish tank granuloma in a 14-month-old girl. Pediatr. Dermatol. 14:209–212

Springer B, Stockman L, Teschner K, Roberts GD, Bottger EC (1996) Two-laboratory collaborative study on identification of mycobacteria: Molecular versus phenotypic methods. J. Clin. Microbiol. 34:296–303

Stahl DA, Urbance JW (1990) The division between fast- and slow-growing species corresponds to natural relationships among the mycobacteria. J. Bacteriol. 172:116–124

Steed KA, Falkinham JO, III (2006) Effect of growth in biofilms on chlorine susceptibility of *Mycobacterium avium* and *Mycobacterium intracellulare*. Appl. Environ. Microbiol. 72:4007–4011

Steinert M, Birkness K, White E, Fields B, Quinn F (1998) *Mycobacterium avium* bacilli grow saprozoically in coculture with *Acanthamoeba polyphaga* and survive within cyst walls. Appl. Environ. Microbiol. 64:2256–2261

Strahl ED, Gillaspy GE, Falkinham JO, III (2001) Fluorescent acid-fast microscopy for measuring phagocytosis of *Mycobacterium avium*, *Mycobacterium intracellulare*, and *Mycobacterium scrofulaceum* by *Tetrahymena pyriformis* and their intracellular growth. Appl. Environ. Microbiol. 67:4432–4439

Stuart P (1965) Vaccination Against Johnes Disease in Cattle Exposed to Experimental Infection. Br. Vet. J. 121:289–318

Sung NM, Collins MT (2003) Variation in resistance of *Mycobacterium paratuberculosis* to acid environments as a function of culture medium. Appl. Environ. Microbiol. 69:6833–6840

Szabo I, Kiss KK, Varnai I (1982) Epidemic pulmonary infection associated with *Mycobacterium xenopi* indigenous in sewage-sludge. Acta Microbiol. Acad. Sci. Hung. 29:263–266

Taylor RH, Falkinham JO, III, Norton CD, LeChevallier MW (2000) Chlorine, chloramine, chlorine dioxide, and ozone susceptibility of *Mycobacterium avium*. Appl. Environ. Microbiol. 66:1702–1705

Tenant R, Bermudez LE (2006) *Mycobacterium avium* genes upregulated upon infection of *Acanthamoeba castellanii* demonstrate a common response to the intracellular environment. Curr. Microbiol. 52:128–133

Thoen CO, Steele JH (1995) *Mycobacterium bovis* infection in animals and humans. Iowa State University Press, 1st ed., 355 pp

Thoen CO, Steele JH, Gilsdorf MJ (2006) *Mycobacterium bovis* infection in animals and humans. 2nd ed., Blackwell Publishing Professional, Ames, Iowa, USA, 317 pp

Thomas V, McDonnell G (2007) Relationship between mycobacteria and amoebae: ecological and epidemiological concerns. Lett. Appl. Microbiol. 45:349–357

Thomsen VO, Andersen AB, Miorner H (2002) Incidence and clinical significance of non-tuberculous mycobacteria isolated from clinical specimens during a 2-y nationwide survey. Scand. J. Infect. Dis. 34:648–653

Thorel MF, Falkinham JO, Moreau RG (2004) Environmental mycobacteria from alpine and subalpine habitats. Fems Microbiol. Ecol. 49:343–347

Thorel MF, Huchzermeyer H, Weiss R, Fontaine JJ (1997) *Mycobacterium avium* infections in animals. Literature review. Vet. Res. 28:439–447

Thorel MF, Huchzermeyer HF, Michel AL (2001) *Mycobacterium avium* and *Mycobacterium intracellulare* infection in mammals. Rev. Sci. Tech. 20:204–218

Thorel MF, Krichevsky M, Levy-Frebault VV (1990) Numerical taxonomy of mycobactin-dependent mycobacteria, emended description of *Mycobacterium avium*, and description of *Mycobacterium avium* subsp. *avium* subsp. nov., *Mycobacterium avium* subsp. *paratuberculosis* subsp. nov., and *Mycobacterium avium* subsp. *silvaticum* subsp. nov. Int. J. Syst. Bacteriol. 40:254–260

Toman M, Faldyna M, Pavlik I (2003) Immunological characteristics of cattle with *Mycobacterium avium* subsp *paratuberculosis* infection. Veterinarni Medicina 48:147–154

Torkko P, Suomalainen S, Iivanainen E, Suutari M, Tortoli E, Paulin L, Katila ML (2000) *Mycobacterium xenopi* and related organisms isolated from stream waters in Finland and description of *Mycobacterium botniense* sp nov. Int. J. Syst. Evol. Microbiol. 50:283–289

Tortoli E (2003) Impact of genotypic studies on mycobacterial taxonomy: the new mycobacteria of the 1990s. Clin. Microbiol. Rev. 16:319–354

Tortoli E, Bartoloni A, Bottger EC, Emler S, Garzelli C, Magliano E, Mantella A, Rastogi N, Rindi L, Scarparo C, Urbano P (2001) Burden of unidentifiable mycobacteria in a reference laboratory. J. Clin. Microbiol. 39:4058–4065

Tortoli E, Rindi L, Garcia MJ, Chiaradonna P, Dei R, Garzelli C, Kroppenstedt RM, Lari N, Mattei R, Mariottini A, Mazzarelli G, Murcia MI, Nanetti A, Piccoli P, Scarparo C (2004) Proposal to elevate the genetic variant MAC-A, included in the *Mycobacterium avium* complex, to species rank as *Mycobacterium chimaera* sp. nov. Int. J. Syst. Evol. Microbiol. 54:1277–1285

Torvinen E, Meklin T, Torkko P, Suomalainen S, Reiman M, Katila ML, Paulin L, Nevalainen A (2006) Mycobacteria and fungi in moisture-damaged building materials. Appl. Environ. Microbiol. 72:6822–6824

Tsintzou A, Vantarakis A, Pagonopoulou O, Athanassiadou A, Papapetropoulou M (2000) Environmental mycobacteria in drinking water before and after replacement of the water distribution network. Water Air Soil Pollut. 120:273–282

Tsukamura M (1963) Modification of ultraviolet-induced mutation frequency and survival in *Mycobacterium avium* by pre-irradiation incubation in phosphorus-deficient medium. Jpn. J. Microbiol. 11:97–104

Tsukamura M, Dawson DJ (1981) An attempt to induce *Mycobacterium intracellulare* from *Mycobacterium scrofulaceum* by ultraviolet irradiation. Microbiol. Immunol. 25:531–535

Tsukamura M, Kita N, Otsuka W, Shimoide H (1983) A study of the taxonomy of the *Mycobacterium nonchromogenicum* complex and report of six cases of lung infection due to *Mycobacterium nonchromogenicum*. Microbiol. Immunol. 27:219–236

Tuffley RE, Holbeche JD (1980) Isolation of the *Mycobacterium avium-M. intracellulare-M. scrofulaceum* complex from tank water in Queensland, Australia. Appl. Environ. Microbiol. 39:48–53

Turenne CY, Wallace R, Jr., Behr MA (2007) *Mycobacterium avium* in the postgenomic era. Clin. Microbiol. Rev. 20:205–229

Twort FW, Ingram GLY (1912) A method for isolating and cultivating the *Mycobacterium enteritidis chronicae pseudotuberculosae bovis* and some experiments on the preparation of a diagnostic vaccine for pseudotuberculosis enteritis of bovine. Proc. Roy. Soc. 84:517–543

Ulicna L, Sytarova J, Kazda J (1968) *Mycobacterium marinum (balnei)*- agents of TB cutis verrucosa in coal workers. Rozhl. Tuberk. 28:695–698

Vaerewijck MJ, Huys G, Palomino JC, Swings J, Portaels F (2005) Mycobacteria in drinking water distribution systems: ecology and significance for human health. FEMS Microbiol. Rev. 29:911–934

van Coppenraet LES, Kuijper EJ, Lindeboom JA, Prins JM, Claas ECJ (2005) *Mycobacterium haemophilum* and lymphadenitis in children. Emerg. Infect. Dis. 11: 62–68

van Soolingen D, Bauer J, Ritacco V, Leao SC, Pavlik I, Vincent V, Rastogi N, Gori A, Bodmer T, Garzelli C, Garcia MJ (1998) IS*1245* restriction fragment length polymorphism typing of *Mycobacterium avium* isolates: proposal for standardization. J. Clin. Microbiol. 36:3051–3054

Vishnevskii PP, Mamatsev EG, Chernyshev VV, Chernyshev NS (1940) The viability of the bacillus of Johne's disease (in Russian). Sovyet Vet. 11–12:89–93

von Reyn CF, Waddell RD, Eaton T, Arbeit RD, Maslow JN, Barber TW, Brindle RJ, Gilks CF, Lumio J, Lahdevirta J (1993) Isolation of *Mycobacterium avium* complex from water in the United States, Finland, Zaire, and Kenya. J. Clin. Microbiol. 31:3227–3230

Vuorio R, Andersson MA, Rainey FA, Kroppenstedt RM, Kampfer P, Busse HJ, Viljanen M, Salkinoja-Salonen M (1999) A new rapidly growing mycobacterial species, *Mycobacterium morale* sp. nov., isolated from the indoor walls of a children's day care centre. Int. J. Syst. Bacteriol. 49:25–35

Wallace RJ (1994) Recent Changes in Taxonomy and Disease Manifestations of the Rapidly Growing Mycobacteria. Eur. J. Clin. Microbiol. Infect. Dis. 13:953–960

Wallace RJ, Brown BA (1998) Catheter sepsis due to *Mycobacterium chelonae*. J. Clin. Microbiol. 36:3444–3444

Wallace RJ, Brown BA, Griffith DE (1998) Nosocomial outbreaks pseudo-outbreaks caused by nontuberculous mycobacteria. Ann. Rev. Microbiol. 52:453–490

Wallace RJ, Zhang YS, Brown BA, Fraser V, Mazurek GH, Maloney S (1993) Dna Large Restriction Fragment Patterns of Sporadic and Epidemic Nosocomial Strains of *Mycobacterium-Chelonae* and *Mycobacterium-Abscessus*. J. Clin. Microbiol. 31:2697–2701

Walsh SE, Maillard JY, Russell AD, Hann AC (2001) Possible mechanisms for the relative efficacies of ortho-phthalaldehyde and glutaraldehyde against glutaraldehyde-resistant *Mycobacterium chelonae*. J. Appl. Microbiol. 91:80–92

Wayne LG (1985) The "atypical" mycobacteria: recognition and disease association. Crit. Rev. Microbiol. 12:185–222

Wayne LG, Sramek HA (1992) Agents of newly recognized or infrequently encountered mycobacterial diseases. Clin. Microbiol. Rev. 5:1–25

Wells SJ (2000) Biosecurity on dairy operations: hazards and risks. J. Dairy Sci. 83:2380–2386

Wells SJ, Wagner BA (2000) Herd-level risk factors for infection with *Mycobacterium paratuberculosis* in US dairies and association between familiarity of the herd manager with the disease or prior diagnosis of the disease in that herd and use of preventive measures. J. Am. Vet. Med. Assoc. 216:1450–1457

Wendt SL, George KL, Parker BC, Gruft H, Falkinham JO, III (1980) Epidemiology of infection by nontuberculous Mycobacteria. III. Isolation of potentially pathogenic mycobacteria from aerosols. Am. Rev. Respir. Dis. 122:259–263

Wenger JD, Spika JS, Smithwick RW, Pryor V, Dodson DW, Carden GA, Klontz KC (1990) Outbreak of *Mycobacterium-Chelonae* Infection Associated with Use of Jet Injectors. JAMA. 264:373–376

Whan L, Ball HJ, Grant IR, Rowe MT (2005a) Development of an IMS-PCR assay for the detection of *Mycobacterium avium* ssp *paratuberculosis* in water. Lett. Appl. Microbiol. 40:269–273

Whan L, Ball HJ, Grant IR, Rowe MT (2005b) Occurrence of *Mycobacterium avium* subsp. *paratuberculosis* in untreated water in Northern Ireland. Appl. Environ. Microbiol. 71:7107–7112

Whan L, Grant IR, Rowe MT (2006) Interaction between *Mycobacterium avium* subsp. *paratuberculosis* and environmental protozoa. BMC. Microbiol. 6:63-

Whan LB, Grant IR, Ball HJ, Scott R, Rowe MT (2001) Bactericidal effect of chlorine on *Mycobacterium paratuberculosis* in drinking water. Lett. Appl. Microbiol. 33:227–231

Whittington RJ, Marsh IB, Reddacliff LA (2005) Survival of *Mycobacterium avium* subsp. *paratuberculosis* in dam water and sediment. Appl. Environ. Microbiol. 71:5304–5308

Whittington RJ, Marsh IB, Taylor PJ, Marshall DJ, Taragel C, Reddacliff LA (2003) Isolation of *Mycobacterium avium* subsp. *paratuberculosis* from environmen-

tal samples collected from farms before and after destocking sheep with paratuberculosis. Aust. Vet. J. 81:559–563

Whittington RJ, Marshall DJ, Nicholls PJ, Marsh IB, Reddacliff LA (2004) Survival and dormancy of *Mycobacterium avium* subsp. *paratuberculosis* in the environment. Appl. Environ. Microbiol. 70:2989–3004

Willumsen P, Karlson U, Stackebrandt E, Kroppenstedt RM (2001) *Mycobacterium frederiksbergense* sp nov., a novel polycyclic aromatic hydrocarbon-degrading *Mycobacterium* species. Int. J. Syst. Evol. Microbiol. 51:1715–1722

Wilson J (1960) Avian tuberculosis. An account of the disease in poultry, captive birds and wild birds. British Veterinary Journal. 116:380–393

Wolinsky E, Schaefer WB (1973) Proposed Numbering Scheme for Mycobacterial Serotypes by Agglutination. Int. J. Syst. Bacteriol. 23:182–183

Woo PC, Leung KW, Wong SS, Chong KT, Cheung EY, Yuen KY (2002) Relatively alcohol-resistant mycobacteria are emerging pathogens in patients receiving acupuncture treatment. J. Clin. Microbiol. 40:1219–1224

Woodley CL, David HL (1976) Effect of temperature on the rate of the transparent to opaque colony type transition in *Mycobacterium avium*. Antimicrob. Agents Chemother. 9:113–119

Wright EP, Collins CH, Yates MD (1985) *Mycobacterium-Xenopi* and *Mycobacterium-Kansasii* in A Hospital Water-Supply. J. Hosp. Infect. 6:175–178

Yamazaki Y, Danelishvili L, Wu M, Hidaka E, Katsuyama T, Stang B, Petrofsky M, Bildfell R, Bermudez LE (2006a) The ability to form biofilm influences *Mycobacterium avium* invasion and translocation of bronchial epithelial cells. Cell Microbiol. 8:806–814

Yamazaki Y, Danelishvili L, Wu M, MacNab M, Bermudez LE (2006b) *Mycobacterium avium* genes associated with the ability to form a biofilm. Appl. Environ. Microbiol. 72:819–825

Yang M, Ross BC, Dwyer B (1993) Isolation of a DNA probe for identification of *Mycobacterium kansasii*, including the genetic subgroup. J. Clin. Microbiol. 31:2769–2772

Yew WW, Wong PC, Woo HS, Yip CW, Chan CY, Cheng FB (1993) Characterization of *Mycobacterium-Fortuitum* Isolates from Sternotomy Wounds by Antimicrobial Susceptibilities, Plasmid Profiles, and Ribosomal Ribonucleic-Acid Gene Restriction Patterns. Diag. Microbiol. Infect. Dis. 17:111–117

Zhang YS, Rajagopalan M, Brown BA, Wallace RJ (1997) Randomly amplified polymorphic DNA PCR for comparison of *Mycobacterium abscessus* strains from nosocomial outbreaks. J. Clin. Microbiol. 35:3132–3139

Chapter 4

Physiological Ecology of Environmental Saprophytic and Potentially Pathogenic Mycobacteria

J.O. Falkinham III

Introduction

Physiological ecology refers to the study and identification of the physiological features of the environmental saprophytic and potentially pathogenic mycobacteria that are determinants of its regional and local distribution, its ecology. Although mycobacteria are slow growing relative to most other environmental micro-organisms, they persist in a number of habitats where suspended micro-organisms would be washed away. Furthermore, mycobacteria are found in habitats where there is reduced competition for nutrients and other micro-organisms, because of the physiochemical conditions and absence of readily degraded nutrients.

Although research to date has primarily focused on *Mycobacterium tuberculosis* because of its global health impact, lessons learned from that microbe do not necessarily hold true for the ESM and PPM. For example, the ESM and PPM, although relatives of members of the *M. tuberculosis* complex, are capable of survival and growth in natural and engineered environments. Furthermore, they do not have the fastidious nutrient requirements of *M. tuberculosis*. They can be grown on defined minimal medium (e.g. glycerol, ammonia, phosphate and minerals).

4.1 Cell Envelope Structure and Hydrophobicity

The most striking feature of the ESM, PPM and all members of the genus *Mycobacterium* is the presence of a lipid- and wax-rich outer membrane (Brennan and Nikaido, 1995). Long chain lipids (C_{60}–C_{80} in some species) are arranged perpendicularly with respect to the cell wall. Freeze fracture studies have shown that there is a fracture plane, indicating that the outermost lipid-rich layer acts like a bilayer (a membrane). As a consequence of its presence and composition, the outer mycobacterial membrane results in cells that are hydrophobic and impermeable to hydrophilic compounds. For example, hydrophobicity likely contributes to the reduced uptake of hydrophilic nutrients and antibiotics. As a consequence, mycobacteria are resistant to a variety of hydrophilic antibiotics and disinfectants (e.g. chlorine).

It is important to point out that the extreme hydrophobicity of mycobacterial cells has beneficial as well as detrimental effects. For example, although the hydrophobicity reduces permeation to nutrients, it provides a barrier to antimicrobial agents. Furthermore, even though hydrophobicity contributes to the reduced growth rate of mycobacteria, it contributes to surface attachment and the ability to persist in flowing systems (e.g. rivers and pipes).

Mycobacterial cells are significantly more hydrophobic than cells of other micro-organisms leading to a predilection to attach to surfaces or air–water interfaces. Furthermore, transport through the mycobacterial outer membrane is reduced compared to trans-membrane transport exhibited by other micro-organisms (Brennan and Nikaido, 1995).

J.O. Falkinham III (✉)
Virginia Polytechnic Institute and State University, Blacksburg
Virginia USA
e-mail: jofiii@vt.edu

J. Kazda et al. (eds.), *The Ecology of Mycobacteria: Impact on Animal's and Human's Health*,
DOI 10.1007/978-1-4020-9413-2_4, © Springer Science+Business Media B.V. 2009

Elevated hydrophobicity also leads to preferential phagocytosis of mycobacterial cells by macrophages (Van Oss et al., 1975). As transparent and opaque colonial variants of *M. avium* differ in cell surface hydrophobicity (Stormer and Falkinham, 1989), it is likely that colonial variants of other ESM and PPM also differ in hydrophobicity as well.

4.2 Slow Growth

The ESM and PPM grow substantially slower than other environmental micro-organisms; even those members of the genus described as "rapidly growing" require 3–7 days for colony formation. In spite of their slow growth, mycobacteria do not have a slow metabolism. Their consumption of oxygen and their metabolism is equal to that of other micro-organisms growing at the same temperature. Two characteristics of the mycobacteria likely contribute to their reduced rates of growth: the presence of the lipid-rich outer membrane and the low number of ribosomal RNA genes.

First, the lipid-rich outer membrane is impermeable to hydrophilic compounds, many of which are essential nutrients (Brennan and Nikaido, 1995). Although the effect of reduced permeation on the growth rate of mycobacteria has not been independently assessed, it likely contributes to the slow growth rate of mycobacteria. In support of that hypothesis is the observation that the less hydrophobic opaque colonial variants of *M. avium* grow more rapidly in laboratory culture medium than the hydrophobic, transparent colonial variants (Stormer and Falkinham, 1989). It is important to ensure that laboratory cultivation does not lead to the selection of opaque variants from transparent types that are commonly isolated from patients. Growth rate may also be reduced because a substantial proportion of energy and carbon are channelled into the synthesis of the long chain fatty acids, lipids and waxes in the outer membrane.

Second, members of the genus *Mycobacterium* have only one or, at most, two copies of the ribosomal RNA (rRNA) genes (Bercovier et al., 1986). Multiple rDNA cistrons are the rule in most micro-organisms and in as much as the rate of protein synthesis and consequently growth is dependent upon the number of ribosomes (Maaloe and Kjeldgaard, 1966), mycobacteria would be expected to grow slowly.

As was illustrated for hydrophobicity, the slow growth of mycobacteria has advantages and disadvantages. Slow growth means they can be crowded out of certain environments. However, slow growth goes hand-in-hand with slow death. Rapidly growing cells are killed by antimicrobial agents because metabolism and growth can be unbalanced (e.g. absence of protein synthesis) more readily. Slowly growing mycobacterial cells are less susceptible to such imbalances, in part because they can adapt to changing conditions before irreversible events leading to death occur. For example, cells of *M. avium* growing at reduced growth rates compared to cells growing at faster rates were significantly more resistant to chlorine (Falkinham, 2003).

4.3 Surface and Interface Adherence

As a consequence of the very high surface hydrophobicity of mycobacteria, they prefer attachment to surfaces or localisation at air–water interfaces, rather than remaining suspended in water. In water, they may prefer attachment to particulate matter (e.g. soil). To illustrate the extent of the hydrophobicity, the addition of 0.1 ml of hexadecane to 3.0 ml of a turbid suspension (10^8 CFU/ml) of *M. avium* in water leads to their complete removal from the aqueous suspension and concentration in the hexadecane (Stormer and Falkinham, 1989).

4.3.1 Surface Adherence

In drinking water distribution systems, ESM and PPM are predominantly found in biofilms on pipe surfaces. For example, *M. avium* cells in drinking water systems throughout the world were found more frequently and in higher numbers in biofilms on pipe surfaces compared to the water (Falkinham et al., 2001; Torvinen et al., 2004). Other mycobacteria, specifically *M. kansasii* (Schulze-Robbecke and Fischeder, 1989), *M. chelonae* (Schulze-Robbecke et al., 1992), *M. fortuitum* (Hall-Stoodley and Lappin-Scott, 1998) and *M. phlei* (Bardouniotis et al., 2001) also form biofilms on surfaces including high-density polyethylene and silastic rubber.

It is likely that the high cell surface hydrophobicity of mycobacteria (Van Oss et al., 1975) contributes to biofilm formation. Individuals interested in the detection or recovery of ESM and PPM from drinking water distribution systems, household, hospital or building plumbing should collect biofilms (e.g. showerheads or water tap surfaces) rather than simply collect water samples. It is likely that the mycobacteria in water have simply been released from biofilms.

The mycobacterial predilection for surface attachment also leads to the colonisation of catheter surfaces. Catheter-associated infections have been reported for *M. avium* (Schelonka et al., 1994), *M. neoaurum* (Davison et al., 1988; Holland et al., 1994; George and Schlesinger, 1999; Woo et al., 2000), the *M. neoaurum*-like species *M. lacticola* (Kiska et al., 2004), *M. fortuitum* and *M. chelonae* (Flynn et al., 1988; Raad et al., 1991). These reports all established that the isolation of either slowly or rapidly growing mycobacteria from catheters should not be dismissed as laboratory contamination, but that catheter-associated infection should be considered.

4.3.2 Interface Adherence and Concentration

Because of their elevated cell surface hydrophobicity, mycobacterial cells are concentrated at air–water interfaces (Wendt et al., 1980). The mechanism of concentration appears to be the collection of mycobacterial cells in the water by air bubbles rising in the water column (Parker et al., 1983). Water-borne hydrophobic mycobacterial cells coming in contact with air bubbles will adhere to the bubbles and be brought to the surface. At this point, the air bubble bursts, forms a crater and there is the ejection of drops (called "jet drops") that are approximately one-tenth the diameter of the air bubbles. The drops are ejected to heights of 10 cm above the air–water interface and droplets small enough can be carried elsewhere by wind currents (Wendt et al., 1980).

These ejected droplets can be collected by inverting a Petri dish with microbiological media 10 cm from the surface. The droplets can be spread on the medium surface and the number of mycobacteria counted. Such studies established the fact that mycobacterial cells were enriched in the droplets relative to their concen-

tration in the water by factors as high as 10 000 (Parker et al., 1983). The concentration of mycobacteria at air–water interfaces can lead to the transmission of a substantial proportion of water-borne mycobacterial cells from water to air.

4.4 Formation and Consequences of Biofilm Formation

Biofilm formation by other micro-organisms leads to an increased resistance to antimicrobial agents (Nickel et al., 1985; Bardouniotis et al., 2001, 2003; Donlan, 2001). As expected, cells of *M. avium* grown in biofilms are more resistant to chlorine (Steed and Falkinham, 2006) or antibiotics (Falkinham, 2007) compared to cells grown in a suspension in the same medium. What was unexpected was the fact that *M. avium* cells grown in biofilms, although harvested, washed and suspended, were of intermediate susceptibility (Steed and Falkinham, 2006; Falkinham, 2007). The cells were neither as resistant as the cells grown and exposed to the antimicrobial agent in biofilms nor as susceptible as those grown and exposed to antimicrobial agent in suspension. The resistance was transient, because growth of the biofilm-grown and suspended cells in medium for 24 h at 37 °C led to cell populations that were as susceptible as were cells grown from low inocula in suspension.

This is an example of the fact that the slow growth of the ESM and PPM is not always detrimental to survival, but provides time for cells to adapt to changing conditions through the induction of genes leading to the appearance of novel proteins of adaptive benefit.

4.5 Physiological and Metabolic Features

In addition to the major determinants of ESM and PPM, i.e. cell surface hydrophobicity, limited permeation and slow growth, there are a number of metabolic and physiologic features that are determinants of mycobacterial ecology. They include the following factors:

- The relative temperature resistance.
- The relatively wide range of pH and temperature tolerated by growing bacteria.
- The ability to grow under micro-aerobic conditions and survive anaerobic conditions.
- The ability of mycobacteria to metabolise compounds that are not substrates for other micro-organisms and to grow under conditions which limit the growth of many other micro-organisms.

Isolates of many ESM and PPM species grow over a relatively wide range of temperatures and pH. For some species, growth temperature is restricted to temperatures below 30–35 °C. Specifically, the optimal temperature for growth of *M. marinum* is 27–28 °C, and for *M. ulcerans* it is 32–33 °C. Other species are capable of growing at higher temperatures. For example, *M. avium* and *M. xenopi* strains are characteristically capable of growing at 45 °C, which likely contributes to their presence in hot water tanks and hot water distribution systems (du Moulin et al., 1988). Furthermore, a variety of mycobacteria were isolated from different alpine habitats (Thorel et al., 2004). Although no alpine isolates were recovered at 37 °C, it is not known whether those alpine isolates are capable of adaptation to grow at 37 °C.

The pH range for the growth of ESM and PPM is relatively wide, usually between 4.5 and 7.0. In contrast, the pH range for growth of *M. tuberculosis* is restricted to 5.8–6.5 (Chapman and Bernard, 1962; Portaels and Pattyn, 1982; Piddington et al., 2000), further underscoring the distinctiveness of that *Mycobacterium*. None of the ESM and PPM species have been reported to grow well below pH 4 or above 7 (Chapman and Bernard, 1962; Portaels and Pattyn, 1982). Many ESM and PPM, for example, *M. avium, M. intracellulare* and *M. scrofulaceum*, grow best between pH 5 and 5.5 (Portaels and Pattyn, 1982; George and Falkinham, 1986). Furthermore, *M. avium* has been shown to be resistant to exposure to the acidity of the human stomach (Bodmer et al., 2000).

The fact that the pH optimum for growth of most ESM and PPM is below 6.5 suggests that standard mycobacterial media (i.e. Middlebrook 7H9 and 7H10) whose pH is 6.6 may not be optimal for recovery and growth. An interesting study showed that a relationship existed between the growth of mycobacteria, including *M. tuberculosis*, ESM and PPM, at an acidic pH and different Mg^{2+} concentrations (Piddington et al.,

2000). In medium adjusted to pH 6.25, higher levels of Mg^{2+} were required for the growth of both *M. tuberculosis* and *M. kansasii*.

In contrast, growth of *M. marinum, M. avium, M. scrofulaceum, M. fortuitum* and *M. chelonae* occurred at pH 6.0 without any increase in the Mg^{2+} demand (Piddington et al., 2000). These results lend support to the notion that the apparent absence of *M. kansasii* group I isolates from water is not due to their non-presence but their susceptibility to a low pH.

Measurements of heat susceptibility of ESM and PPM showed a 10-fold range amongst ESM and PPM species. Isolates of *M. avium, M. intracellulare, M. scrofulaceum* and *M. xenopi* were the most resistant. For example, at 60 °C, approximately 4–30 min would be required to reduce the populations by 90%. Based on that data, it is not surprising that those species are found in hot water systems. In contrast, isolates of *M. chelonae, M. fortuitum, M. kansasii* and *M. marinum* exhibited greater susceptibility (Schulze-Robbecke and Buchholtz, 1992).

The growth of strains of *M. avium* was shown to be stimulated by humic and fulvic acids (Kirschner et al., 1999), which explains, in part, the high numbers of *M. avium* and other mycobacteria in humic acid-rich environments such as coastal swamps in the eastern United States (Kirschner et al., 1992), boreal forest soil in Finland (Iivanainen et al., 1997) and sphagnum-rich potting soil (De Groote et al., 2006).

Furthermore, ESM and PPM belonging to a variety of different slowly and rapidly growing species are capable of degrading and utilising hydrocarbons and chlorinated hydrocarbons for growth. For example, a variety of ESM and PPM, including *M. kansasii, M. avium, M. intracellulare* and *M. fortuitum*, can grow on paraffin (Ollar et al., 1990). An *M. fortuitum* strain capable of degrading polychlorinated phenolics has been described (Nohynek et al., 1993), a strain of *M. aurum* degrades the antimicrobial agent morpholine (Poupin et al., 1998) and a *Mycobacterium* sp. isolate was capable of the degradation of phenanthrene, fluorine, fluoranthene and pyrene (Boldrin et al., 1993).

A particularly troubling report documented the *Mycobacterium* sp.-catalysed reduction and activation of compounds to mutagens from nitroaromatic compounds that are widely distributed by pollutants (Rafii et al., 1994). We predict that ESM and PPM may be important agents of the transformation of pollutants throughout the world.

It is important at this point to take note of the fact that the measurements of the hydrophobicity of cells, the composition of the outer membrane and pH and temperature ranges for growth are measured on cells growing in laboratory medium. It is possible and in a number of instances likely that the results from laboratory measurements may not be replicated in the natural habitats occupied by mycobacteria. For example, cells grown in rich laboratory medium at high growth rates are considerably more sensitive to chlorine compared to cells grown under conditions that mimic the habitat in drinking water distribution systems (Falkinham, 2003). Furthermore, some enzymatic activities are subject to regulation, such as the repression of nitrate reductase activity as a consequence of growth in ammonia or amino acids. Therefore, the reader is urged to be cautious in extrapolating from laboratory data to ESM and PPM habitats.

4.6 Mycobacterial Adaptation (Dormancy)

The slow growth rate of mycobacteria is responsible, in part, for the ability of these environmental micro-organisms to adapt to changing conditions. Slow growth means slow death; the action of antimicrobial agents is not followed by an immediate and deterministic pathway of death because before the irreversible events occur, mycobacteria can induce protective or survival strategies. Although slow growing, the metabolism of mycobacteria is not slow; its rate of ATP formation equals that of *Escherichia coli* under the same conditions of temperature and nutrients.

Slow growth is due, in part, to the utilisation of ATP for synthesis of the long chain mycolic acids. Perhaps the best known example of mycobacterial adaptation is the acquisition of resistance to anaerobic conditions. Gradual reduction in the oxygen concentration in cultures does not affect the survival of cells by virtue of their ability to enter a dormant stage (Wayne and Hayes, 1996; Dick et al., 1998). Such gradual decreases in oxygen would be expected to occur in the distal portions of drinking water

distribution systems due to microbial-and chemical-catalysed oxygen consumption.

4.7 Formation of Biofilms

In spite of their slow growth compared to other environmental micro-organisms, mycobacteria have impacts on the habitats they occupy. Because of their hydrophobic envelope and surface (Brennan and Nikaido, 1995), mycobacteria are concentrated at interfaces and are likely "pioneers" in the formation of biofilms. In the latter role, mycobacteria form the conditioned surface required for the adherence of other micro-organisms. The environmental mycobacteria do not have fastidious growth requirements, but are actually oligotrophs capable of growing on a wide variety of compounds. In fact, their nutritional diversity coupled with their hydrophobicity likely contributes to their ability to metabolise hydrocarbons that, like mycobacteria, are concentrated at air–water interfaces.

Mycobacterial persistence in a number of human-engineered habitats (e.g. drinking water distribution systems) is due to their oligotrophy, nutritional diversity, biofilm formation and disinfectant resistance. The use of disinfectants in drinking water systems or the introduction of hydrocarbon pollutants in soils and water appear to be selective agents, leading to the proliferation and ultimate dominance of mycobacteria in those habitats. It is likely that human activities (e.g. disinfection and pollution) are leading to selection for mycobacteria at a time when the aging of populations and the increased prevalence of immunodeficient individuals are predisposing more individuals to infection and disease.

4.7.1 Mycobacteria as Biofilm "Pioneers"

The extreme cell surface hydrophobicity of environmental mycobacteria is a major determinant of their distribution in the environment. Hydrophobicity drives their enrichment in aerosols (Parker et al., 1983), their concentration at air–water interfaces (Wendt et al., 1980) and their attachment to surfaces (Steed and Falkinham, 2006; Falkinham, 2007). Attachment to non-conditioned, freshly washed, sterile surfaces including glass, copper, iron and plastics is rapid and numbers per square centimetre can reach thousands (glass) within 2 hours (Steed and Falkinham, 2006). The attachment of *M. avium* to catheter tubing was slower (9 CFU/cm^2 in 2 h) but there was reduced flow in the catheter tubing, thereby reducing agitation and contact of cells with the surface (Falkinham, 2007).

Comparison values for other micro-organisms, including mycobacteria, are difficult to obtain because the methods employed did not measure adherence separate from adherence and growth (i.e. accumulation). That is especially a problem for assessing the adherence of rapidly growing bacteria. Adherent mycobacteria serve as a conditioning agent, like proteins, and their presence increases the adherence of hydrophilic gram-positive and gram-negative bacteria (J. O. Falkinham III, unpublished data). Mycobacteria may be important agents accelerating biofilm formation in natural and human-engineered environments.

4.7.2 Biofilms in Hospitals

Evidence for adaptation has also come from studies of *M. avium* growth in protozoa and amoebae and in biofilms. *M. avium* recovered from *Acanthamoeba castellanii* are more antibiotic resistant compared to cells grown in medium (Cirillo et al., 1997). Likewise, cells of *M. avium* grown in biofilms on either glass beads or catheter tubing are more resistant to disinfection by chlorine (Steed and Falkinham, 2006) or antibiotics (Falkinham, 2007). The increased resistance is not due to antimicrobial-resistant "persisters" or mutants, because cultivation of the amoebae- or biofilm-grown cells in medium in suspension results in its loss.

References

Bardouniotis E, Ceri H, Olson ME (2003) Biofilm formation and biocide susceptibility testing of *Mycobacterium fortuitum* and *Mycobacterium marinum*. Curr. Microbiol. 46:28–32

Bardouniotis E, Huddleston W, Ceri H, Olson ME (2001) Characterization of biofilm growth and biocide susceptibility testing of *Mycobacterium phlei* using the MBEC (TM) assay system. Fems Microbiol. Lett. 203:263–267

Bercovier H, Kafri O, Sela S (1986) Mycobacteria Possess A Surprisingly Small Number of Ribosomal-RNA Genes in Relation to the Size of Their Genome. Biochem. Biophys. Res. Commun. 136:1136–1141

Bodmer T, Miltner E, Bermudez LE (2000) *Mycobacterium avium* resists exposure to the acidic conditions of the stomach. FEMS Microbiol. Lett. 182:45–49

Boldrin B, Tiehm A, Fritzsche C (1993) Degradation of phenanthrene, fluorene, fluoranthene, and pyrene by a *Mycobacterium* sp. Appl. Environ. Microbiol. 59:1927–1930

Brennan PJ, Nikaido H (1995) The Envelope of Mycobacteria. Ann. Rev. Biochem.. 64:29–63

Chapman JS, Bernard JS (1962) The tolerances of unclassified mycobacteria. I. Limits of pH tolerance. Am. Rev. Respir. Dis. 86:582–583

Cirillo JD, Falkow S, Tompkins LS, Bermudez LE (1997) Interaction of *Mycobacterium avium* with environmental amoebae enhances virulence. Infect. Immun. 65:3759–3767

Davison MB, McCormack JG, Blacklock ZM, Dawson DJ, Tilse MH, Crimmins FB (1988) Bacteremia caused by *Mycobacterium neoaurum*. J. Clin. Microbiol. 26:762–764

De Groote MA, Pace NR, Fulton K, Falkinham JO, III (2006) Relationships between *Mycobacterium* isolates from patients with pulmonary mycobacterial infection and potting soils. Appl. Environ. Microbiol. 72:7602–7606

Dick T, Lee BH, Murugasu-Oei B (1998) Oxygen depletion induced dormancy in *Mycobacterium smegmatis*. FEMS Microbiol. Lett. 163:159–164

Donlan RM (2001) Biofilms and device-associated infections. Emerg. Infect. Dis. 7:277–281

du Moulin GC, Stottmeier KD, Pelletier PA, Tsang AY, Hedley-Whyte J (1988) Concentration of *Mycobacterium avium* by hospital hot water systems. JAMA. 260:1599–1601

Falkinham JO (2003) Factors influencing the chlorine susceptibility of *Mycobacterium avium*, *Mycobacterium intracellulare*, and *Mycobacterium scrofulaceum*. Appl. Environ. Microbiol. 69:5685–5689

Falkinham JO, III (2007) Growth in catheter biofilms and antibiotic resistance of *Mycobacterium avium*. J. Med. Microbiol. 56:250–254

Falkinham JO, III, Norton CD, LeChevallier MW (2001) Factors influencing numbers of *Mycobacterium avium*, *Mycobacterium intracellulare*, and other Mycobacteria in drinking water distribution systems. Appl. Environ. Microbiol. 67:1225–1231

Flynn PM, Van HB, Gigliotti F (1988) Atypical mycobacterial infections of Hickman catheter exit sites. Pediatr. Infect. Dis. J. 7:510–513

George KL, Falkinham JO, III (1986) Selective medium for the isolation and enumeration of *Mycobacterium avium-intracellulare* and *M. scrofulaceum*. Can. J. Microbiol. 32:10–14

George SL, Schlesinger LS (1999) *Mycobacterium neoaurum* - An unusual cause of infection of vascular catheters: Case report and review. Clin. Infect. Dis. 28:682–683

Hall-Stoodley L, Lappin-Scott H (1998) Biofilm formation by the rapidly growing mycobacterial species *Mycobacterium fortuitum*. FEMS Microbiol. Lett. 168:77–84

Holland DJ, Chen SC, Chew WW, Gilbert GL (1994) *Mycobacterium neoaurum* infection of a Hickman catheter in an immunosuppressed patient. Clin. Infect. Dis. 18:1002–1003

Iivanainen EK, Martikainen PJ, Raisanen ML, Katila ML (1997) Mycobacteria in boreal coniferous forest soils. Fems Microbiol. Ecol. 23:325–332

Kirschner RA, Parker BC, Falkinham JO, III (1992) Epidemiology of infection by nontuberculous mycobacteria. *Mycobacterium avium*, *Mycobacterium intracellulare*, and *Mycobacterium scrofulaceum* in acid, brown-water swamps of the southeastern United States and their association with environmental variables. Am. Rev. Respir. Dis. 145:271–275

Kirschner RA, Parker BC, Falkinham JO (1999) Humic and fulvic acids stimulate the growth of *Mycobacterium avium*. Fems Microbiol. Ecol. 30:327–332

Kiska DL, Turenne CY, Dubansky AS, Domachowske JB (2004) First case report of catheter-related bacteremia due to "*Mycobacterium lacticola*". J. Clin. Microbiol. 42:2855–2857

Maaloe O, Kjeldgaard N (1966) Control of macromolecular synthesis; a study of DNA, RNA, and protein synthesis in bacteria (Ed.: B. D. Davis). W.A. Benjamin, New York

Nickel JC, Ruseska I, Wright JB, Costerton JW (1985) Tobramycin resistance of *Pseudomonas aeruginosa* cells growing as a biofilm on urinary catheter material. Antimicrob. Agents Chemother. 27:619–624

Nohynek LJ, Haggblom MM, Palleroni NJ, Kronqvist K, Nurmiaho Lassila EL, Salkinojasalonen M (1993) Characterization of A *Mycobacterium-Fortuitum* Strain Capable of Degrading Polychlorinated Phenolic-Compounds. Syst. Appl. Microbiol. 16:126–134

Ollar RA, Dale JW, Felder MS, Favate A (1990) The use of paraffin wax metabolism in the speciation of *Mycobacterium avium-intracellulare*. Tubercle. 71:23–28

Parker BC, Ford MA, Gruft H, Falkinham JO, III (1983) Epidemiology of infection by nontuberculous mycobacteria. IV. Preferential aerosolization of *Mycobacterium intracellulare* from natural waters. Am. Rev. Respir. Dis. 128:652–656

Piddington DL, Kashkouli A, Buchmeier NA (2000) Growth of *Mycobacterium tuberculosis* in a defined medium is very restricted by acid pH and Mg(2+) levels. Infect. Immun. 68:4518–4522

Portaels F, Pattyn SR (1982) Growth of mycobacteria in relation to the pH of the medium. Ann. Microbiol. (Paris). 133:213–221

Poupin P, Truffaut N, Combourieu B, Besse P, Sancelme M, Veschambre H, Delort AM (1998) Degradation of morpholine by an environmental *Mycobacterium* strain involves a cytochrome P-450. Appl. Environ. Microbiol. 64:159–165

Raad II, Vartivarian S, Khan A, Bodey GP (1991) Catheter-related infections caused by the *Mycobacterium fortuitum* complex: 15 cases and review. Rev. Infect. Dis. 13:1120–1125

Rafii F, Selby AL, Newton RK, Cerniglia CE (1994) Reduction and mutagenic activation of nitroaromatic compounds by a *Mycobacterium* sp. Appl. Environ. Microbiol. 60:4263–4267

Schelonka RL, Ascher DP, McMahon DP, Drehner DM, Kuskie MR (1994) Catheter-related sepsis caused by *Mycobacterium avium* complex. Pediatr. Infect. Dis. J. 13:236–238

Schulze-Robbecke R, Buchholtz K (1992) Heat susceptibility of aquatic mycobacteria. Appl. Environ. Microbiol. 58:1869–1873

Schulze-Robbecke R, Fischeder R (1989) Mycobacteria in biofilms. Zentralbl. Hyg. Umweltmed. 188:385–390

Schulze-Robbecke R, Janning B, Fischeder R (1992) Occurrence of mycobacteria in biofilm samples. Tuber. Lung Dis. 73:141–144

Steed KA, Falkinham JO, III (2006) Effect of growth in biofilms on chlorine susceptibility of *Mycobacterium avium* and *Mycobacterium intracellulare*. Appl. Environ. Microbiol. 72:4007–4011

Stormer RS, Falkinham JO, III (1989) Differences in antimicrobial susceptibility of pigmented and unpigmented colonial variants of *Mycobacterium avium*. J. Clin. Microbiol. 27:2459–2465

Thorel MF, Falkinham JO, Moreau RG (2004) Environmental mycobacteria from alpine and subalpine habitats. Fems Microbiol. Ecol. 49:343–347

Torvinen E, Suomalainen S, Lehtola MJ, Miettinen IT, Zacheus O, Paulin L, Katila ML, Martikainen PJ (2004) Mycobacteria in water and loose deposits of drinking water distribution systems in Finland. Appl. Environ. Microbiol. 70:1973–1981

Van Oss C, Gillman C, Neumann A (1975) Phagocytic Engulfment and Cell Adhesiveness. Marcel Dekker, Inc., New York.

Wayne LG, Hayes LG (1996) An in vitro model for sequential study of shiftdown of *Mycobacterium tuberculosis* through two stages of nonreplicating persistence. Infect. Immun. 64:2062–2069

Wendt SL, George KL, Parker BC, Gruft H, Falkinham JO, III (1980) Epidemiology of infection by nontuberculous Mycobacteria. III. Isolation of potentially pathogenic mycobacteria from aerosols. Am. Rev. Respir. Dis. 122:259–263

Woo PC, Tsoi HW, Leung KW, Lum PN, Leung AS, Ma CH, Kam KM, Yuen KY (2000) Identification of *Mycobacterium neoaurum* isolated from a neutropenic patient with catheter-related bacteremia by 16S rRNA sequencing. J. Clin. Microbiol. 38:3515–3517

Chapter 5

Environments Providing Favourable Conditions for the Multiplication and Transmission of Mycobacteria

I. Pavlik, J.O. Falkinham III and J. Kazda (Eds.)

Introduction

As a result of a number of systematic surveys to detect or enumerate ESM and PPM in water and soil throughout the world, it has been possible to identify those habitats that harbour the highest numbers of ESM and PPM. It would be expected that those sources should be linked to the highest frequencies of disease, provided that there are routes for human or animal transmission.

Mycobacteria either constitute a natural part of the ecosystem or penetrate the environment through human activities (Costallat et al., 1977; Masaki et al., 1989; Kazda, 2000). With respect to their occurrence in the environment, the substrate seems to play a dominant role in their survival. In most cases, the initial amount of mycobacteria present in a substrate is very low. Their further fate in the environment depends upon the conditions and proceeds according to one of the following scenarios:

- Mycobacteria persist in a substrate, which only serves as a transport medium.
- Mycobacteria propagate in a substrate, which serves as a source of mycobacteria for host organisms under certain conditions.

5.1 Mycobacteria in Sphagnum, Peats and Potting Soils

J. Kazda and J.O. Falkinham III

A variety of mycobacteria, including *Mycobacterium intracellulare*, *M. avium*, *M. xenopi*, *M. marinum* and *M. simiae*, have been recovered from sphagnum vegetation from around the world (Kazda, 1977; Kazda et al., 1979; Schroder et al., 1992). Further, isolates of *M. avium* and *M. intracellulare* have been shown to grow in sphagnum vegetation (Kazda, 1978a, b).

Thus, it is not surprising to find that peat lands, boreal forest soil (Iivanainen et al., 1997) and water draining from peat lands and moors have high numbers of mycobacteria (Iivanainen et al., 1999a). That in turn leads to higher numbers of *M. avium* and other mycobacteria in Finnish surface (Katila et al., 1995) and drinking water (von Reyn et al., 1993) and the high frequency of *M. avium* infection in Finnish HIV/AIDS patients (Ristola et al., 1999). It is expected that drinking water supplies that draw surface water from peat lands and boreal (pine) forests across Northern Europe, the USA and Canada ought to have higher levels of mycobacteria. This hypothesis is supported by evidence of high numbers of *M. avium* and other mycobacteria in the drinking water distribution system in the Boston, Massachusetts, area in the northeast of the USA (du Moulin and Stottmeier, 1986).

Conditions similar to those in peatlands are found in the acid, brown water swamps of the eastern coast of the USA (Kirschner et al., 1992). The high numbers of ESM and PPM, particularly *M. avium*, *M. intracellulare* and *M. scrofulaceum*, in this water

I. Pavlik (✉)
Head of OIE Reference Laboratories for Paratuberculosis and Avian Tuberculosis, Department of Food and Feed Safety, Veterinary Research Institute, Brno, Czech Republic
e-mail: pavlik@vri.cz

J. Kazda et al. (eds.), *The Ecology of Mycobacteria: Impact on Animal's and Human's Health*, DOI 10.1007/978-1-4020-9413-2_5, © Springer Science+Business Media B.V. 2009

were correlated with low pH, high humic and fulvic acid concentrations, high zinc content and low oxygen content (Kirschner et al., 1992).

Peat, drainage water from boreal forest soil and peatlands and acid, brown water swamps all have high concentrations of humic and fulvic acids. Further studies have led to the discovery that the growth of *M. avium* was stimulated by humic and fulvic acids (Kirschner et al., 1999). Thus, the conclusion is that the regional and geographic distribution of ESM and PPM is dictated by the physiological conditions tolerated and conducive to their growth. It is likely that the habitats occupied by ESM and PPM are rather selective for mycobacteria and restrict the growth of many rapidly growing micro-organisms.

5.1.1 Sphagnum Vegetation: The Habitat of Mycobacteria

Sphagnum vegetation is one of the best studied environments from the point of view of mycobacteria. Sphagnum or peat moss belongs taxonomically to the class *Bryopsida*, subclass *Sphagnidae*, containing one genus *Sphagnum*. Its distribution comprises a variety of sphagnum bogs in Europe, Asia, Africa, North and South America and New Zealand. The vaulted surface of sphagnum bogs shows a relief of relatively dry hummocks created by moderately hygrophilous sphagnum species and the shallow wet part occupied by highly hygrophilous sphagnum species (Photo 5.1).

5.1.1.1 General Characteristics

In contrast to vascular plants, sphagnum has developed an external transport system for nutrients between the capillary spaces of branches and stems in their closely compact vegetation. This microstream of nutrients moves upwards in dry weather and downwards during rain, supporting the actively growing top region of sphagnum. The ion exchanger localised in the cell wall of sphagnum binds cations and releases an equivalent amount of H^+ ions, which thus creates an acidic environment. Intact sphagnum vegetation develops three different strata: the actively growing green region on the top, with the highest acidity (pH 2.0–2.5); the

deeper grey layer where the first stage of peat decay takes place (pH 4.5–5.0); and the brown stratum, containing peat, situated at the bottom (pH about 4.0; Photo 5.2).

During the growth period from early spring to late autumn, a new green stratum develops on the surface, reaching a thickness of 3–5 cm in moderately hygrophilous species and 20–30 cm in highly hygrophilous sphagnum species (Photo 5.3). The overgrown, formerly green region suffers from a lack of light and decays into peat. This decay is caused by the pectinase enzymes of mycorrhiza fungi, which decompose the more deeply situated parts of sphagnum into brown melanin–humin compounds (Reuther, 1957).

Products from both the decomposition of sphagnum and fungi metabolism, mainly carbohydrates and amino acids, are released and circulate in a liquid form on the surface of the sphagnum stems. The first indication that sphagnum biotopes could be a possible source of mycobacteria was the frequent positive results obtained when these micro-organisms were isolated from acidic water streams originating in sphagnum bogs (Kazda, 1973a).

5.1.1.2 Ecological Aspects

To localise the stratum of the sphagnum vegetation where growth of mycobacteria can take place, compact samples of highly and moderately hygrophilous sphagnum association were collected and exposed to temperatures of 30 °C during the day and 22 °C at night. Later, suspensions of selected mycobacterial strains were spread onto the head region of the sphagnum and CFU followed. The number of mycobacteria present in the green region at the beginning dropped very rapidly. In the highly hygrophilous associations, mycobacteria disappeared totally within 10–13 days and in moderately hygrophilous ones within 2–3 weeks after inoculation.

To examine the condition in the grey layer *in situ*, experiments were performed in which polyamide hollow fibres, pore size 100 000, were inserted into the grey layer of a moderately hygrophilous *Sphagnum magellanicum* association. The hollow fibres were filled with suspensions of 19 mycobacterial species, and the sphagnum vegetation was exposed to temperatures of 31 °C by day and 22 °C by night. All

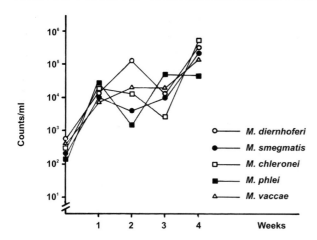

Fig. 5.1 Pronounced multiplication of mycobacteria in hollow fibres (pore size 100000) inserted into the grey layer of a moderately hygrophilous sphagnum association

Table 5.1 The spectrum of amino acids in the fluid film obtained from the surface of the grey layer of *Sphagnum magellanicum* and *S. rubellum* association. Concentrations are shown before inoculation with *Mycobacterium sphagni* and after 3 weeks of incubation at 31 °C

Amino acid	Concentration in nano mol per mg%	
	Before inoculation	After incubation
Asparagine	130.1	0
Glutamic acid	60.2	0
Serine and threonine	60.0	40.2
Glutamine	41.4	0
Alanine	19.1	16.2
Glycine	18.4	10.0
Gamma-aminobutyric acid	18.1	0
Arginine	14.4	0
Ornithine and tryptophane	11.1	0
Valine	9.4	0
L-Methylhistidine	8.8	0
Isoleucine	3.8	0
Leucine	3.4	0
Phenylalanine	3.3	0
Lysine	3.2	0
Histidine	2.8	0
3-Methylhistidine	2.4	0
Tyrosine	2.2	0
DL-Cystathionine	1.7	0

inoculated mycobacteria showed pronounced multiplication; the dynamics of some selected strains are displayed in Fig. 5.1.

The analysis of the fluid film gained from the grey layer of *S. magellanicum* and *S. rubellum* association revealed 19 amino acids with asparagine and glutamic acid constituting the highest proportion. When inoculated with *M. sphagni*, uptake of amino acids occurred (Table 5.1). On agar medium, enriched with the sphagnum fluid, colonies of *M. sphagni* and other rapidly growing mycobacteria developed, their growth being similar to that on conventional media.

Sphagnum is also rich in carbohydrates, fructose, glucose, galactose, arabinose, xylose, mannose and rhamnose, which account for about 90% of their organic material. *M. sphagni* and *M. komossense* can utilise glucose, fructose and rhamnose as a source of carbon. Nearly all mycobacteria can utilise aspartic and glutamic acids as a source of nitrogen. These compounds belong to the standard ingredients of conventional media for the cultivation of mycobacteria.

It was found that the fluid film covering the surface of the grey layer of sphagnum offers unique conditions for mycobacteria to thrive. They follow as successive micro-organisms in the course of the first stage of peat decay, caused by fungi. It is an "open system" in contrast to the vascular plants with the inner nutrient transport. These niches are accessible for mycobacteria which are adapted to the prevailing acidity of the grey layer. To find conditions which are relevant to the multiplication of mycobacteria, a field project was started in the largest intact sphagnum bog in Komosse,

South Sweden, and in the Fördefjord region in coastal Norway.

First, the association between the botanical species of sphagnum and the occurrence of the mycobacteria was examined. From the Komosse sphagnum bog, 20 specimens, each of 11 different sphagnum species, were taken and the yield for mycobacteria was monitored. In this biotope, the highest occurrence of mycobacteria was found in the highly hygrophilous species *S. balticum*, *S. recurvum* and *S. tenellum* with 65, 50 and 45%, respectively. Here, the lowest yield of mycobacteria was found in *S. rubellum* (15%). Compared with the sphagnum vegetation collected in coastal Norway where the moderately hygrophilous *S. rubellum* creates large carpets due to high precipitation and increased humidity in summer, the highest yield of 48% was found in this species. Thus, not the sphagnum species but the microclimate in the grey layer of sphagnum is important for the growth of the mycobacteria.

Further factors which promote high colonisation with mycobacteria are found in more southernly situated sphagnum carpets, when pH values lie between

4.0 and 5.2, where the coverage with higher plants forms up to 30% of the surface, in the presence of Ericaceae and in high concentrations of organic substances in sphagnum fluid (Kazda, 2000).

Based on these results, a worldwide search for mycobacteria has been started. To date, 86 larger sphagnum biotopes from 5 continents, Europe, North America, South America, Africa and New Zealand, have been examined. Mycobacteria were found in all of them. The frequency of mycobacterial isolations varied from 31.9% in Sweden to 79.0% in Ireland, with an average of 41.2% (Table 5.2). Twenty six different species of *mycobacteria* have been isolated from sphagnum vegetation, including five new species: *M. komossense*, *M. sphagni*, *M. cookii*, *M. madagascariense* and *M. hiberniae*. This is the largest spectrum of mycobacteria found in the same kind of biotope (Kazda, 2000). After being described as a new species, *M. sphagni* has been identified in samples collected in sphagnum bogs worldwide and thus can be regarded as the habitat micro-organism of this vegetation.

On the other hand, the *MAC*, *M. simiae* and *M. madagascariense* have been isolated only from sphagnum growing in the tropical climate of Madagascar. The isolated position of New Zealand may be the reason why *M. cookii* is limited to these islands and can reach a very high density in sphagnum there. In contrast, no explanation can be found as to why *M. komossense* is common only in the boreal and subboreal climates of Scandinavia and North America.

Table 5.2 Survey of sphagnum biotopes examined for mycobacteria

Continent	Country	Number of		Isolated in % of	
		Biotopes	Samples	Bogs	Samples
Europe	Czech Republic	6	18	100	55.5
	France	2	30	100	66.6
	Germany	9	217	100	45.2
	Great Britain	4	32	100	34.4
	Ireland	3	19	100	79.0
	Norway	15	192	100	34.4
	Sweden	2	310	100	31.9
Americas	Colombia	2	32	100	40.6
	USA	6	29	100	34.5
Africa	Madagascar	5	18	100	55.5
Western Pacific	New Zealand	32	54	100	77.8
Total		86	951		

The high affinity of mycobacteria for sphagnum vegetation raises the question of how long mycobacteria have colonised sphagnum. In one sphagnum bog, "Weisses Moor", in Northwest Germany, samples of the whole peat profile were aseptically collected and examined. This relatively young sphagnum bog evolved about 1000 years ago on sedge peat. In the grey layer, a high occurrence of mycobacteria was found, with a viability of 60–90%. In the brown layer (10–15 cm deep) the mycobacterial population diminished, with a viability of 20–60%. In the deeper part of the peat profile up to 100 cm deep, the smears still showed acid-fast rods but the cells were no longer viable.

This indicates that acid-fast micro-organisms resembling mycobacteria colonised the sphagnum vegetation about 1000 years ago, when this sphagnum bog began to develop (Kazda, 2000). It is evident that this "long-term" colonisation of the same biotope with relatively slow-growing mycobacteria required special constant conditions, which benefit these micro-organisms and restrict or prevent the growth of others. The ion exchanger in the grey layer of the sphagnum cell wall maintains a pH of about 4.5, favourable for ESM but adverse for most other micro-organisms.

The acidic environment of sphagnum bogs makes them hostile to higher plants. It promotes the development of large sphagnum carpets and their open surface offers the best conditions for the accumulation of heat from the sun under the surface in the grey layer. When the sun shines, the temperature can reach values of between 16 and 26.8 °C higher than that of the air. A temperature of over 30 °C can persist for a long period of time. In the moderate climate of North Germany values of 30 °C and higher were recorded for 10–16 h a day from the beginning of June until the middle of August (Kazda, 1978a). In experiments with *M. fortuitum*, incubated at a temperature alternating at 12 h intervals between 31 and 22 °C, the generation time was not as expected (50%), but only 19% longer compared with the continuous incubation at 31 °C. This booster effect, provoked by the changing temperature, might favour the growth of mycobacteria under field conditions.

The fact that sphagnum bogs are a rich source of mycobacteria is of great importance to their ecology. Hygrophilous sphagnum species have close contact to surface water and release mycobacteria into small streams, particularly during flooding. Thus, streams originating in sphagnum bogs serve as a main distributor of mycobacteria (Photo 5.4).

5.1.1.3 Trends

To isolate mycobacteria from sphagnum it is necessary to use a technique which is different from that which is used for other environmental sources. The optimal yield of mycobacteria can be found in the grey layer of sphagnum vegetation. For the cultivations it is necessary to squeeze the fluid (directly or after addition of sterile distilled water) and to use the centrifuged sediment. Instead of treatment, dilution on agar-enriched plates (Kazda, 2000) and incubation at 31 °C is recommended. Only this method can show the whole spectrum of mycobacteria.

5.1.2 Potting Soil

Based on the fact that high numbers of mycobacteria have been found to be present in sphagnum vegetation and peat (see above) and that a high frequency of soil samples in potted plants collected from the homes of HIV/AIDS patients with disseminated *M. avium* infections have yielded *M. avium* (Yajko et al., 1995), the hypothesis that commercial potting soil constitutes a source of *M. avium* infection was tested (De Groote et al., 2006). Specifically, the experiment sought to determine whether dust aerosols of potting soil, generated by transferring potting soil to containers from packages, contained mycobacteria that were associated with particles able to enter the bronchioles and alveoli of the human lung.

Because most commercial potting soil mixes contain a substantial proportion of peat, it was thought that the potting soil either contained mycobacteria or that mycobacteria introduced in drinking water proliferated in the potting soil mix. The investigation demonstrated that potting soil mixes taken directly from containers had substantial numbers of mycobacteria (e.g. 10^6/g).

A group of elderly patients with mycobacterial pulmonary infections but lacking the classic risk factors for mycobacterial pulmonary disease (e.g. lung damage, smoking, occupation involving dust exposure) provided samples of potting soil. Almost all samples of potting soil yielded mycobacteria (e.g. *M. avium, M. intracellulare, M. kansasii, M. simiae, M. chelonae* and *M. fortuitum*), many of which were associated with particle sizes able to enter the human lung as far as the bronchioles and alveoli. Importantly, in a substantial proportion of patients, the species in their potting soil aerosols matched their lung isolate (De Groote et al., 2006).

Furthermore, in a number of instances (e.g. *M. avium* and *M. intracellulare*), the pattern of restriction endonuclease digestion fragments separated by pulsed field gel electrophoresis of the patient and their potting soil isolate was almost identical (De Groote et al., 2006). Thus, not only do the high numbers of mycobacteria in peat lead to high numbers in water which becomes a vector for infection, but potting soil aerosols can also serve as a vector for mycobacterial infection.

A recent report documented an elevated incidence of tuberculous lesions in the lymph nodes of slaughtered pigs in the Czech Republic. An important factor was thought to be the feeding of peat to the pigs as a supplement. A retrospective study of farms using peat demonstrated that a high frequency of peat samples yielded mycobacteria, particularly *M. a. hominissuis*, and that the IS*1245* restriction fragment length polymorphism patterns of peat and lymph node isolates were identical (Matlova et al., 2005).

5.1.3 Moss Vegetation (Bryopsida, Bryidae) as a Source of Mycobacteria

Aside from sphagnum, there are other kinds of moss which offer favourable conditions for the growth of mycobacteria. They belong to the class Bryopsida, of which the subclass Bryidae alone contains over 95% of these moss species.

5.1.3.1 General Characteristics

Unlike sphagnum which requires special conditions for growth, other mosses can colonise a variety of different biotopes worldwide. It is more resistant to desiccation and pollution and can be found in different soil types in a wide range of climates. In favourable conditions, this moss creates hummocks, a life form very similar to that of sphagnum (Photo 5.5). It often possesses an ion exchanger in the cell wall, resulting in a rapid acidification of the immediate environment and nutrients are transported by external conduction as adult plants lack roots and inner capillarity.

5.1.3.2 Ecological Aspects

Little is known about the distribution of mycobacteria in these moss species, although some, collected occasionally in Norway and Russia, have been found to be positive. *Agrobacterium* sp. have been isolated from moss collected in different geographical locations (Spiess and Lippincott, 1981). One of the first systematic studies dealing with the occurrence of mycobacteria in moss vegetation (without sphagnum) was carried out in Ireland. From July through September of 1990 and in April 1991, 32 samples of *Hylocomium splendens*, 23 of *Thuidium tamariscinum*, 21 of *Breutelia chrysocoma*, 21 of *Ctenidium molluscum*, 16 of *Acrocladium cuspidatum* and a small number of *Cratoneuron filicinum*, *Rhytidiadelphus squarrosus* and *Entodon schreberi* were collected and examined for mycobacteria. The grey layer of these mosses showed an unusually high density of mycobacteria by direct microscopic enumeration (Photo 5.6). A total of 59 strains of rapidly and slowly growing mycobacteria

were isolated, most frequently *M. terrae*, *M. aichiense*, *M. sphagni*, *M. gordonae*, *M. aurum* and occasionally *M. chubuense*, *M. nonchromogenicum* and the *MAC* (Cooney et al., 1997). The highest density of mycobacteria was found in moss growing at the edges of small water ponds, temporarily floating.

To find ecological conditions relevant to the multiplication of mycobacteria, the three most frequently growing species of *B. chrysocoma*, *T. tamariscinum* and *H. splendens* were collected in the Burren, Ireland. The pH values and the density of AFB in the green, grey and brown layers were monitored. The highest occurrence was found in the grey layer, but – in contrast to sphagnum – the green head region contained a relatively high number of AFB. These findings correspond to the values between pH 5.4 and 5.9 in the head region, pH 5.4 and 5.6 in the grey and pH 6.5 and 6.8 in the brown layer. However, the higher pH values probably resulted from a chemical interaction with the underlying alkaline limestone common in this region.

To examine the solar heat accumulation in early spring (1–7 April), the probe of a rotating barrel thermograph was inserted in the grey layer of a hummock of *H. splendens* association and the temperature values were compared with the air and ground temperature measured in the nearest meteorological station. For approximately 6 h each day, the moss temperature exceeded 25 °C with a maximum of 40 °C. The air and ground temperatures in the same period reached only 8–10 °C (Fig. 5.2). At a temperature of

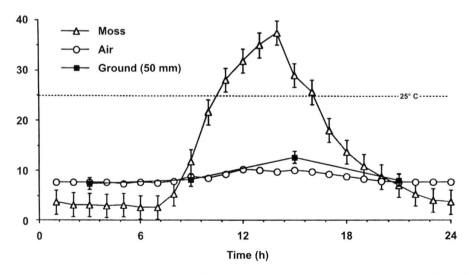

Fig. 5.2 The temperature variation of *Hylocomium splendens* association compared to the temperature variations of air and ground (50 mm deep)

over 25 °C, the mesophilic mycobacteria can thrive and their growth can probably be enhanced by the changing temperatures, as mentioned in the previous paragraph dealing with sphagnum. To confirm the availability of nutrients for the growth of mycobacteria, pressed fluid from the grey layer of *H. splendens* vegetation (Photo 5.7) was tested after sterile filtration and incubation at 31 °C. The multiplication of *M. aichiense*, *M. aurum*, *M. sphagni* and *M. avium* was observed (Kazda et al., 1997).

Apart from sphagnum which requires the wet conditions of bogs, other moss is widespread in nature. The presence of moss on grassland in Ireland is particularly evident in the early spring following the outdoor wintering of livestock (O'Donovan, 1987). Winter grazing reduces the vegetation cover to a low mat, 3–5 cm in height. This low canopy cover, combined with increasing temperature and regular rainfall in the spring, favours moss growth that accounts for nearly one quarter of the total biomass. As grazing is scarce, the ingestion of moss by animals takes place instead of grass (Photo 5.8). In this period, AFB have been found to be very abundant in faeces of the wintering livestock. Their numbers diminished rapidly in late spring when the higher grass canopy had developed. The ingestion of mycobacteria together with moss provoked a non-specific tuberculin reaction in grazing livestock. *M. hiberniae*, a new mycobacterial species isolated from moss in Ireland, was found to be the causative agent of this reaction (Cooney et al., 1997).

The results showed that a variety of moss species provide favourable conditions for the growth of mycobacteria. This is not restricted to Ireland. In Germany, twenty six randomly collected samples of *R. squarrosus*, *A. cuspidatum* and *Brachythecium nutabulum* contained a very high number of AFB. Similar results were obtained with samples from New Zealand. It is evident that the affinity for mycobacteria is not restricted to sphagnum, but is widely distributed in other kinds of moss vegetation.

5.1.3.3 Trends

Acid rain and air pollution benefit the spread of moss which continuously displace the grass vegetation in garden lawns, meadows and pastures. The mycobacteria present in moss can have easy access to animals and humans. As mentioned previously, the environ-

ment can be the source of further new mycobacterial species with unknown interactions with living organisms.

The isolation of two slowly growing mycobacterial species, *M. cookii* in New Zealand and *M. hiberniae* in Ireland, provoking non-specific tuberculin reactions, can illustrate how to find sources of PPM in the environment. Until now, the search for the source of *M. ulcerans* was concentrated only on fauna found in tropical water, although the possible connection of mycobacteria with plant sources was first hinted at more than 100 years ago. Further research activity should be focused on the occurrence and living conditions of diverse moss vegetation growing at the edges of water bodies in *M. ulcerans* endemic regions, to localise the possible reservoir of this infection.

5.2 Water as a Habitat and Vector for Transmission

I. Pavlik and J.O. Falkinham III

Following the recognition that there was no evidence of person-to-person transmission of ESM and PPM (summarised in Wolinsky, 1979), environmental sources, primarily water, were investigated. This led some to propose water as a possible source of infection based on the isolation or detection of ESM and PPM in natural and drinking waters. Proof that water was a source of infection was provided by the evidence that DNA fingerprints of water and patient isolates were identical (von Reyn et al., 1994).

In spite of the relatively high numbers of ESM and PPM in water (Covert et al., 1999; Falkinham et al., 2001; Vaerewijck et al., 2005) and the identification of patients with disease (both HIV positive and HIV negative), there have been surprisingly few reports of matches between patient and environmental isolates. Several factors may be responsible:

i. Decontamination regimens reduce the numbers of the infecting *Mycobacterium*. For example, *M. kansasii* is relatively sensitive to decontamination, especially acid or base (Engel et al., 1980). It is important to point out that decontamination is usually not necessary for drinking water samples.

ii. The true source of the infecting *Mycobacterium* may not have been sampled. Samples collected from different taps in a single house with a single input water source differ in the species distribution and number of mycobacteria. Some taps may not yield mycobacteria and others large numbers. Thus, comprehensive sampling is required. Water samples are probably the least likely to yield mycobacteria because most mycobacteria in drinking water are attached to pipe walls in biofilms. The only cells in water are those released from the biofilm. Sources yielding the most mycobacteria are biofilms in taps and shower heads and in-line water filter materials.

iii. The expense and effort required for such fingerprinting studies are high. A real effort by the patient, physician and laboratory is needed.

iv. Evidence of disease may come long after exposure (e.g. 6 months) and from the individual patient's perspective, unless there is concern of repeated infection, there may be no strong motivation to identify the source.

v. Even if fingerprinting is performed, the generation of clonal diversity in both the patient and the source (water) will lead to the absence of identical or even close matches (De Groote et al., 2006). This is quite common because only single patient isolates are saved, a holdover perhaps from *M. tuberculosis* diagnostic microbiology. In order to ensure an adequate number of isolates for fingerprinting analysis, at least 10 isolates from both the patient and their suspected environmental source should be collected and analysed. In that way, even if there is clonal diversity of isolates within a particular species, the probability of identifying a matching patient and environmental isolate can be increased.

5.2.1 Natural Water as a Habitat and Vector for Transmission

Natural waters throughout the world have been shown to yield ESM and PPM (Falkinham et al., 1980; Schroder et al., 1992; Iivanainen et al., 1993, 1997, 1999a; von Reyn et al., 1993; Katila et al., 1995). The most valuable studies have included not only isolation and identification but methods to provide estimates of numbers of mycobacteria. The word "estimates" is used deliberately, because many researchers decontaminate water samples, as is done for sputum samples. Because decontamination reduces the numbers of mycobacteria (fortunately to a lesser extent than for most other micro-organisms), not all mycobacteria are isolated (Brooks et al., 1984). The most valuable studies have included measurements of physiochemical variables that have led to the identification of characteristics of the mycobacteria that are determinants of their geographic distribution and numbers.

Studies of the natural waters in Finland not only provided estimates of the frequency of recovery of different mycobacterial species, but also showed that high mycobacterial counts were significantly associated with low pH, low levels of dissolved oxygen, high levels of organic matter, high levels of the metals Fe, Al, Cu, Co, Cr and anions and the presence of peatlands (Iivanainen et al., 1993; Katila et al., 1995). Such conditions are also met in the acid, brown water swamps of the eastern coast of the USA where numbers of mycobacteria are quite high (Kirschner et al., 1992).

ESM and PPM are not just present in natural waters because they have been leached from soil or peat, but because they can grow there. A variety of ESM and PPM species, such as *M. avium*, *M. intracellulare* and *M. scrofulaceum*, have been shown to grow in natural waters (Kazda, 1973a, b; George et al., 1980). Thus, the conclusion is that ESM and PPM are natural inhabitants of waters and capable of survival, growth and persistence; they are not contaminants of natural waters.

5.2.2 Drinking Water Distribution Systems as a Habitat and Vector for Transmission

There is little doubt that one of the major sources of infection by ESM and PPM is drinking water distribution systems (Katila et al., 1995; Covert et al., 1999; Falkinham et al., 2001; Vaerewijck et al., 2005). This is not only because so many individuals are exposed to drinking water, but also because the systems select for mycobacteria. Mycobacteria are introduced into systems on particulates, because mycobacterial numbers are correlated with source water turbidity. ESM and

PPM are normal inhabitants of drinking water systems, not just contaminants.

Mycobacteria are hydrophobic and form biofilms (Schulze-Robbecke and Fischeder, 1989; September et al., 2004; Steed and Falkinham, 2006), so they are not washed out, in spite of their slow growth. Evidence that biofilms contribute directly to the presence of mycobacteria in drinking water systems is provided by data showing that replacement of the water network reduced the percentage of drinking water samples with mycobacteria from 21% before to 2% after the replacement (Tsintzou et al., 2000). The biofilms are resistant to heat (Schulze-Robbecke and Buchholtz, 1992) and disinfectants (Taylor et al., 2000) and can grow on many of the degradation-resistant compounds in water, e.g. humic acids (Kirschner et al., 1999).

Following disinfection, mycobacteria lack competitors and are free to utilise available carbon sources for growth, as long as the concentration is above $50 \mu g$ assimilable organic carbon per litre (Norton et al., 2004). The choice of disinfectant also influences the frequency and numbers of mycobacteria in drinking water systems. Although a change in disinfectant from chlorine to chloramine led to an increase in the frequency and number of mycobacteria (Pryor et al., 2004), the combination of ozonisation and filtration in a surface drinking water system failed to reduce the persistence of *M. avium* (Hilborn et al., 2006). Furthermore, some species like *M. avium* and *M. intracellulare* can grow at the low oxygen concentrations found at the end points of systems. In fact, numbers of *M. avium* and *M. intracellulare* are higher at the ends compared to samples coming from the treatment plant (Falkinham et al., 2001).

The ESM and PPM grow in hot water heaters and piping in homes, buildings and hospitals. *M. avium* numbers have been shown to be significantly higher in hospital hot water (recirculating) systems than in the input water (du Moulin et al., 1988), and *M. xenopi* numbers in hot water systems in buildings reporting outbreaks of pulmonary disease were higher than those of the input water (Slosarek et al., 1994). Water samples collected from hospitals throughout the world have a substantial number and variety of mycobacteria (Carson et al., 1988; Graham, Jr. et al., 1988; Vantarakis et al., 1998; Kusnetsov et al., 2003; Tobin-D'Angelo et al., 2004). In fact, an *M. avium* pseudobacteriuria was linked to the presence of *M. avium* in the hospital's water supply (Graham, Jr. et al., 1988).

5.2.3 Spas and Hot Tubs as Habitats and Vectors for Transmission

Spas and hot tubs have been shown to be sources of ESM and PPM infections in different countries throughout the world (Aubuchon et al., 1986; Kahana et al., 1997; Embil and Warren, 1997; Glazer et al., 2008; Mangione et al., 2001; Khoor et al., 2001; Lumb et al., 2004). Primarily, the infections have been pulmonary and caused by *M. avium*, which has been isolated from both patients and spas or hot tubs to which they were exposed. In a number of instances, it has been shown that the DNA fingerprint patterns of large restriction fragments separated by pulsed field gel electrophoresis for the patient and some of the isolates recovered from the spa or hot tub were identical (Kahana et al., 1997; Embil and Warren, 1997; Mangione et al., 2001; Lumb et al., 2004).

In one case, the exposure of a healing wound in a hot tub led to a soft tissue infection caused by a rapidly growing *Mycobacterium* (Aubuchon et al., 1986). As is the case for drinking water distribution systems, conditions in the spa or hot tub select for ESM and PPM, for example, *M. avium*. The water contains organic matter, is heated at least periodically and is disinfected (e.g. chlorine or bromine). Thus, competitors of the mycobacteria are killed (e.g. skin flora) and the mycobacteria can proliferate. Eradication of the mycobacteria in the spa or hot tub is difficult because of their ability to form biofilms, making removal by cleaning difficult and providing even higher levels of disinfectant resistance (Muilenberg et al., 1998; Lumb et al., 2004).

5.2.4 Footbaths as Habitats and Sources of Mycobacterial Infection

Like spas and hot tubs, the environment provided by footbaths in podiatrist's offices and nail salons appears to be a habitat for mycobacteria. In May 2002, an outbreak of furunculosis caused by *M. fortuitum* in over 100 healthy female patrons of a nail salon in Northern California was described (Winthrop et al., 2002). Patterns of large restriction fragments of DNA of *M. fortuitum* isolates from the patients and recovered from the footbaths in the nail salon separated by pulsed field gel electrophoresis were identical (Winthrop et al., 2002). Other reports of nail salon-associated furunculosis have identified *M. chelonae*, *M. fortuitum* (Sniezek et al., 2003) and *M. mageritense* (Gira et al., 2004) as causative agents.

M. mageritense was isolated from both the patients and the footbaths and PFGE fingerprints of the patient and footbath isolates matched (Gira et al., 2004). One common factor shared by many of the infected women was that they had shaved their legs before patronising the nail salons (Sniezek et al., 2003; Gira et al., 2004). In a follow-up study to the Northern California outbreak, it was shown that 29 of 30 (97%) different footbaths in 5 Californian counties yielded mycobacteria. Isolated mycobacteria included *M. fortuitum* (47%), *M. mucogenicum* (23%), *M. mageritense* (20%), the *MAC* (17%) and others (Vugia et al., 2005). Thus, there can be no doubt that footbaths are habitats for both rapidly and slowly growing mycobacteria and proprietors and patrons of nail salons should take special care to avoid infection (e.g. avoid shaving).

5.2.5 Surface Water

Water is the main vector for the spread/transmission of most ESM and PPM species on Earth. Surface water plays an important role in the circulation of mycobacteria throughout the environment. Surface water constitutes the most important part of the hydrosphere (water envelope) surrounding the Earth. Mycobacteria, however, have also been isolated from atmospheric and underground water.

5.2.5.1 Mycobacteria in the Atmosphere

Water in the atmosphere can exist in three forms: as a gas (water vapour), as a liquid (rain drops) and as a solid (snow flakes). The occurrence of mycobacteria as an aerosol is relatively rare. Nevertheless, PPM can affect the responses to both human and avian tuberculin testing (Section 5.9.1; Eilertsen, 1969), especially in coastal areas. In intensively exploited inland areas, water used for irrigation can become an important source of a wide variety of mycobacterial species present in aerosol (Table 5.3; Photo 5.9).

ESM and PPM have also been found in rainwater and snow (Table 5.3). In the ecology of mycobacteria, water in these forms and states can become their vectors, spreading them over great distances under favourable weather conditions. PPM found in the rainwater of Australia may be consequential to health; in arid regions, water harvested from the reservoirs during periods of rain is an important source not only for animals but also for humans (Photo 5.10). The water vapour can best condensate as droplets of rain around the condensation centres; these can be the dust particles contaminated with mycobacteria, or mycobacteria themselves. Dust contaminated with mycobacteria and subsequently rainwater or snow contaminated with

Table 5.3 Examples of mycobacterial detection in different water sources from the atmosphere and irrigation water

Water origin	Country	Isolates	Ref.
Aerosol[1]	USA	*MAIS*	[7], [1]
Irrigation water	Israel	*M. nonchromogenicum, M. terrae, M. gordonae, M. chelonae, M. fortuitum, M. flavescens*	[3]
	Australia	*M. ulcerans*	[4]
Rain water	USA	*MI*	[7]
	Australia	*MAA* (1), *MAH* (6, 8), *MI* (7, 14, 18, 19, 23), *M. gordonae, M. terrae–triviale–nonchromogenicum* complex, RGM	[6]
		M. fortuitum, M. intracellulare, RGM, photochromogenic mycobacteria	[5]
Snow	USA	*Mycobacterium* sp.	[2]

[1] See also Sections 5.3.2, 5.9.1, 5.9.2, 7.2 and 7.3.

MAIS M. avium–intracellulare–scrofulaceum complex. **RGM** Rapidly growing mycobacterial species. *MI M. intracellulare.* *MAA M. a. avium.* *MAH M. a. hominissuis.* (**1**) Serotype 1.

[1]Brooks RW, Parker BC, Gruft H, Falkinham JO, III (1984) Am. Rev. Respir. Dis. 130:630–633. [2] Goslee S, Wolinsky E (1976) Am. Rev. Respir. Dis. 113:287–292. [3] Haas H, Fattal B (1990) Water Research 24:1233–1235. [4] Ross BC, Johnson PD, Oppedisano F, Marino L, Sievers A, Stinear T, Hayman JA, Veitch MG, Robins-Browne RM (1997) Appl. Environ. Microbiol. 63:4135–4138. [5] Singer E, Rodda GM (1965) Tubercle. 46:209–213. [6] Tuffley RE, Holbeche JD (1980) Appl. Environ. Microbiol. 39:48–53. [7] Wendt SL, George KL, Parker BC, Gruft H, Falkinham JO, III (1980) Am. Rev. Respir. Dis. 122:259–263.

mycobacteria can thus become significant sources of these pathogens.

The detection of tuberculous lesions and subsequently mycobacteria in the pulmonary (tracheobronchial) lymph nodes of sows that were kept in stables cooled with aerosols in the hot summer months was unexpected. This was generated from drinking water using a special apparatus equipped with filters for the prevention of mechanical contamination. The aerosol in stables was emitted from nozzles above the heads of sows. Due to the fact that *M. fortuitum* and *M. a. hominissuis* were detected in drinking water on this farm, we can suppose that the aspirated contaminated aerosol was most likely the source of mycobacteria for the sows (I. Pavlik, unpublished data; Photo 5.11).

The detection of *MAC* members in condensed water on the space station "Mir" (Table 5.5) is a serious finding with potential implications for the health of the astronauts. These samples were formed in the condensate of steam inside the space station. Astronauts may be under severe stress during long-term endurance of weightlessness; this can then be further complicated by a potential mycobacterial infection (Kawamura et al., 2001).

5.2.5.2 Mycobacteria in Saltwater

Saltwater is mainly found on the Earth's surface, i.e. in the oceans and seas. The occurrence of mycobacteria in saltwater has largely been studied in sea aquariums (Section 6.5.1). However, the occurrence of mycobacteria here cannot be compared with the occurrence under natural conditions in the oceans and seas. The occurrence of mycobacteria has been studied at the seashore of the USA in the Atlantic Ocean (Table 5.4). Water in the oceans and seas has the following natural reducing effects on mycobacteria:

i. *Temperature* is a significant factor, which limits the growth of mycobacteria and reduces their survival. The growth of different *MAC* members was observed at 17.8, 20.0 and 22.5 °C. The growth was stopped at 15.5 °C and a gradual reduction of the original amount of *MAC* was seen at 4.8 and 9.4 °C by 30 days (George et al., 1980).

ii. *NaCl* at a concentration of 1.5% and higher exerts inhibiting effects on the *MAC* members

and this increases with increasing NaCl concentrations (George et al., 1980). The survival of *M. tuberculosis* is affected by saltwater. At concentrations of 5, 20 and 35% at 4, 25 and 37 °C they survive for only 8 days (Jamieson et al., 1976).

iii. *Intensive and rapid dilution* of mycobacteria occurs due to the continuous flow of water in the oceans and seas. Mycobacteria are usually transferred into saltwater through river water and water draining from large agglomerations and industrial centres.

5.2.5.3 Mycobacteria in Fresh Underground Water

Under the Earth's surface, water is generally bound to humus and other components of the soil. Mostly free water is found in underground lakes and springs. ESM and PPM have been isolated from both spring and well water (Table 5.4). The following ESM species *M. gordonae* and *M. nonchromogenicum* and PPM species *M. fortuitum*, *M. peregrinum*, *M. chelonae* and *M. intracellulare* have been isolated from ground water used as drinking water in Paris (Le Dantec et al., 2002). The frequent colonisation of biofilms in the water piping systems can be explained by this finding. Mycobacteria are subsequently released from the biofilm into drinking water (Table 5.4; Photo 5.12).

Only water obtained from artesian wells has been found not to be contaminated with mycobacteria (I. Pavlik, unpublished data; Photo 5.13). A low content or absence of mycobacteria has also been noted in well water in Africa, a water source important for both people and animals (Kleeberg and Nel, 1973). This factor probably contributes to the relatively low detection rate of non-specific responses to bovine and avian tuberculin testing in different countries of Africa (J. Zinsstag, personal communication).

5.2.5.4 Mycobacteria in Fresh Surface Water

On the other hand, mycobacteria have been isolated from different sources of fresh surface water (burns, rivers, lakes, water reservoirs, etc.). Table 5.4 shows that mycobacteria have been isolated from the water of springs, burns, rivers and estuaries. Mycobacteria have not only been isolated from water (water column "pelagus") but also from sediment and biofilms on different

Table 5.4 Examples of mycobacterial detection in surface water

Water type	Water origin	Ref.
Saltwater	Beaches	[5, 4, 3]
	Beaches-sand	[17]
	Atlantic Ocean (East Coast, USA)	[6]
	Atlantic Ocean	[4]
	Port	[3]
Fresh underground water	Wells and springs	[5]
	Ground water	[11]
Fresh surface water	Brook	[10, 9, 23]
	River	[1, 5, 4, 21, 19, 15, 20, 2, 16]
	Dam-sediment and water	[23, 22]
	Lakes	[5, 4, 3, 20, 12, 16]
	Ponds	[5, 4, 3, 7]
	Camping	[8]
	Holiday resort parks	[8]
	Mud	[17, 18]
	River catchments	[19]
	Trough water (surface) for animals	[14, 13]

[1] Ancusa M, Terbancea W (1970) Z. Gesamte Hyg. 16:913–916. [2] Bland CS, Ireland JM, Lozano E, Alvarez ME, Primm TP (2005) Appl. Environ. Microbiol. 71:5719–5727. [3] Brooks RW, George KL, Parker BC, Falkinham JO, III, Gruff H (1984) Can. J. Microbiol. 30:1112–1117. [4] George KL, Parker BC, Gruft H, Falkinham JO, III (1980) Am. Rev. Respir. Dis. 122:89–94. [5] Goslee S, Wolinsky E (1976) Am. Rev. Respir. Dis. 113:287–292. [6] Gruft H, Falkinham JO, III, Parker BC (1981) Rev. Infect. Dis. 3:990–996. [7] Haas H, Fattal B (1990) Water Research 24:1233–1235. [8] Havelaar AH, Berwald LG, Groothuis DG, Baas JG (1985) Zentralbl. Bakteriol. Mikrobiol. Hyg.[B] 180:505–514. [9] Iivanainen E, Martikainen PJ, Vaananen P, Katila ML (1999) J. Appl. Microbiol. 86:673–681. [10] Iivanainen EK, Martikainen PJ, Vaananen PK, Katila ML (1993) Appl. Environ. Microbiol. 59:398–404. [11] Le Dantec C, Duguet JP, Montiel A, Dumoutier N, Dubrou S, Vincent V (2002) Appl. Environ. Microbiol. 68:5318–5325. [12] Niva M, Hernesmaa A, Haahtela K, Salkinoja-Salonen M, Sivonen K, Haukka K (2006) Boreal Environment Research 11:45–53. [13] Norby B, Fosgate GT, Manning EJ, Collins MT, Roussel AJ (2007) Vet. Microbiol. 124:153–159. [14] Pearson CW, Corner LA, Lepper AW (1977) Aust. Vet. J. 53:67–71. [15] Pelletier PA, du Moulin GC, Stottmeier KD (1988) Microbiol. Sci. 5:147–148. [16] Pickup RW, Rhodes G, Bull TJ, Arnott S, Sidi-Boumedine K, Hurley M, Hermon-Taylor J (2006) Appl. Environ. Microbiol. 72:4067–4077. [17] Singer E, Rodda GM (1965) Tubercle. 46:209–213. [18] Stanford JL, Paul RC (1973) Ann. Soc. Belg. Med. Trop. 53:389–393. [19] Tuffley RE, Holbeche JD (1980) Appl. Environ. Microbiol. 39:48–53. [20] von Reyn CF, Barber TW, Arbeit RD, Sox CH, O'Connor GT, Brindle RJ, Gilks CF, Hakkarainen K, Ranki A, Bartholomew C (1993) J. Infect. Dis. 168:1553–1558. [21] Wendt SL, George KL, Parker BC, Gruft H, Falkinham JO, III (1980) Am. Rev. Respir. Dis. 122:259–263. [22] Whittington RJ, Marsh IB, Reddacliff LA (2005) Appl. Environ. Microbiol. 71:5304–5308. [23] Whittington RJ, Marsh IB, Taylor PJ, Marshall DJ, Taragel C, Reddacliff LA (2003) Aust. Vet. J. 81:559–563.

surfaces. Mycobacteria can survive in biofilms and can also multiply there. In biofilms, mycobacteria are more effectively protected from various factors causing their death (Hall-Stoodley et al., 1999; Carter et al., 2003; Parsek and Fuqua, 2004; Section 4.7).

A perusal of different reviews reveals that tens of mycobacterial species have been isolated from surface water; among ESM, the following were most often detected: *M. gordonae*, *M. flavescens*, *M. gastri*, *M. nonchromogenicum*, *M. terrae* and *M. triviale*, and among PPM, the following have sometimes been isolated from water: *M. kansasii*, *M. xenopi*, *MAC* members, *M. fortuitum* and *M. chelonae* (Collins and Yates, 1984; Wayne and Kubica, 1986; Pedley et al., 2004).

5.2.5.5 Mycobacteria in Water Treated by Distillation and Deionisation

It is noteworthy that mycobacteria have also been isolated from distilled and deionised water (Table 5.5).

Table 5.5 Examples of mycobacterial detection in different treated water sources

Water treatment	Water origin	Ref.
Deionisation	Tap water	[24, 10, 21]
Distillation	Tap water	[9, 2]
Space station	"Mir"	[14]
Drinking	Water plants	[12]
	Sand filter of reservoir	[6]
	Treated surface water	[18, 19]
	Reservoirs	[9, 25, 1]
	Hospitals	[10, 4, 2, 1, 8], Section 3.3
	Commercial buildings	[1]
	Laboratories	[9]
	Fountains	[9]
	Cold and hot water	[21, 7, 26, 27]
	Bottled water	[27]
	Ice making machine	[9]
	Swabs and scrapings	[15, 16, 17]
Others	Fitness clubs	[11], Section 3.3
	Holiday resort parks	[11]
	Public bath	[22]
	Sauna	[11]
	Swimming pool	[23, 5, 3, 13, 20]
	Swimming pool – hotels	[11]
	Whirlpools	[11]
	Zoo and aquariums	[9], Section 6.5

[1] Aronson T, Holtzman A, Glover N, Boian M, Froman S, Berlin OG, Hill H, Stelma G, Jr. (1999) J. Clin. Microbiol. 37:1008–1012. [2] Carson LA, Bland LA, Cusick LB, Favero MS, Bolan GA, Reingold AL, Good RC (1988) Appl. Environ. Microbiol. 54:3122–3125. [3] Dailloux M, Hartemann P, Beurey J (1980) Zentralbl. Bakteriol. Mikrobiol. Hyg.[B] 171:473–486. [4] du Moulin GC, Stottmeier KD, Pelletier PA, Tsang AY, Hedley-Whyte J (1988) JAMA 260:1599–1601. [5] Eilertsen E (1969) Scand. J. Respir. Dis. Suppl 69:85–88. [6] Engel HW, Berwald LG, Havelaar AH (1980) Tubercle. 61:21–26. [7] Fischeder R, Schulze-Robbecke R, Weber A (1991) Zentralbl. Hyg. Umweltmed. 192:154–158. [8] Gomila M, Ramirez A, Lalucat J (2007) Appl. Environ. Microbiol. 73:3787–3797. [9] Goslee S, Wolinsky E (1976) Am. Rev. Respir. Dis. 113:287–292. [10] Graham L, Jr., Warren NG, Tsang AY, Dalton HP (1988) J. Clin. Microbiol. 26:1034–1036. [11] Havelaar AH, Berwald LG, Groothuis DG, Baas JG (1985) Zentralbl. Bakteriol. Mikrobiol. Hyg.[B] 180:505–514. [12] Hilborn ED, Covert TC, Yakrus MA, Harris SI, Donnelly SF, Rice EW, Toney S, Bailey SA, Stelma GN, Jr. (2006) Appl. Environ. Microbiol. 72:5864–5869. [13] Iivanainen E, Northrup J, Arbeit RD, Ristola M, Katila ML, von Reyn CF (1999) APMIS 107:193–200. [14] Kawamura Y, Li Y, Liu H, Huang X, Li Z, Ezaki T (2001) Microbiol. Immunol. 45:819–828. [15] Kubalek I, Komenda S (1995) APMIS 103:327–330. [16] Kubalek I, Mysak J (1995) Cent. Eur. J. Public Health 3:39–41. [17] Kubalek I, Mysak J (1996) Eur. J. Epidemiol. 12:471–474. [18] Le Dantec C, Duguet JP, Montiel A, Dumoutier N, Dubrou S, Vincent V (2002) Appl. Environ. Microbiol. 68:1025–1032. [19] Le Dantec C, Duguet JP, Montiel A, Dumoutier N, Dubrou S, Vincent V (2002) Appl. Environ. Microbiol. 68:5318–5325. [20] Leoni E, Legnani P, Mucci MT, Pirani R (1999) J. Appl. Microbiol. 87:683–688. [21] Pelletier PA, du Moulin GC, Stottmeier KD (1988) Microbiol. Sci. 5:147–148. [22] Saito H, Tsukamura M (1976) Jpn. J. Microbiol. 20:561–563. [23] Singer E, Rodda GM (1965) Tubercle. 46:209–213. [24] Stine TM, Harris AA, Levin S, Rivera N, Kaplan RL (1987) JAMA 258:809–811. [25] Tuffley RE, Holbeche JD (1980) Appl. Environ. Microbiol. 39:48–53. [26] von Reyn CF, Barber TW, Arbeit RD, Sox CH, O'Connor GT, Brindle RJ, Gilks CF, Hakkarainen K, Ranki A, Bartholomew C (1993) J. Infect. Dis. 168:1553–1558. [27] Yajko DM, Chin DP, Gonzalez PC, Nassos PS, Hopewell PC, Reingold AL, Horsburgh CR, Jr., Yakrus MA, Ostroff SM, Hadley WK (1995) J. Acquir. Immune. Defic. Syndr. Hum. Retrovirol. 9:176–182.

Some residual growth has also been observed in distilled water under laboratory conditions. Residual growth was more intensive in autoclaved water than in sterile water obtained by filtration according to George et al. (1980).

5.2.6 Water for Animals

Drinking water from water piping systems as well as water from wells and surface water is commonly given to farm animals. Water from all these sources may be contaminated with mycobacteria (Tables 5.4 and 5.5). On farms, expansion tanks (usually placed in the upper part of the stables) can constitute a great risk (Photo 5.12). Mycobacteria from water can accumulate (settle on the bottom) as well as multiply in the tanks under the following favourable conditions:

i. Temperatures above 18–20 °C can be found especially in expansion tanks or along the entire water piping system in the summer months (Photo 5.14). The warmth accumulates in the upper parts of buildings and under the ceiling, where these tanks are often situated.
ii. A sufficient amount of iron from the water piping system, which settles on the bottom of expansion tanks and where a deep layer of biofilm is often formed (Photos 5.12 and 5.15).
iii. Intensive organic pollution, predominantly found on farms where the expansion tanks are not covered or otherwise protected from the penetration of dust, arthropods and vertebrates, above all birds.

At present, animals in stables, especially pigs, are neither commonly given surface water to drink nor is surface water used for feed preparation (Photos 5.16 and 5.17). Nevertheless, under certain conditions (mostly due to technical problems – e.g. a dried well on the farm), surface water is used. That is usually – especially in the hot months – massively contaminated with ESM and PPM that are highly immunogenic and can cause sensitisation of humans and animals to various tuberculins (Section 7.4; Lepper, 1977; Norby et al., 2007). High concentrations of ESM and PPM in drinking water can cause the formation of tuberculous lesions in mesenteric lymph nodes, mainly in pigs (I. Pavlik and L. Matlova, unpublished data).

On pastures, farm animals are only given drinking water under exceptional circumstances because it is expensive. Surface water from different sources is most often used for watering the animals. Stationary surface water is most dangerous for them (Photo 5.18). Thus, the access of animals on pastures to different shallow pools, celestial pools, morasses, peat bogs and other common sources of PPM should be prevented (Photos 5.19–5.21). These water sources also often become sources of causative agents of parasitic and other diseases.

5.2.7 Wastewater

Mycobacteria have also been found in wastewater (Table 5.6). They were repeatedly isolated from wastewater sediment with concentrations reaching 1×10^5/ml (Brooks et al., 1984). Accordingly, we can assume that wastewater and its sediment is a suitable substrate for the survival and multiplication of mycobacteria owing to the high content of nutrients. After all, in peat bogs, mycobacteria are a natural part of the ecosystem and participate in mineralisation of the organic matter present (Sections 5.1.1 and 5.4). The conditions in wastewater and the sediment are most likely less suitable for the growth of mycobacteria than in burn water (Table 5.7). *In vitro* experiments have shown that after a 3-month incubation in wastewater and its sediment, the following species could grow: among slowly growing mycobacteria only *M. terrae* and *M. triviale*, and among fast-growing mycobacteria *M. fortuitum*, *M. vaccae*, *M. diernhoferi* and *M. flavescens* could grow out of the 20 tested mycobacterial species from the four groups according to Runyon's classification. No growth of members of Runyon's two remaining groups was detected in wastewater or the sediment (Table 5.7; Photo 5.22).

For a better understanding of the sources of mycobacteria and their fate in wastewater, these are categorised according to their origin (source) of wastewater contamination as follows:

i. *Communal wastewater* from cities and villages (city agglomerations are the biggest sources of this category of wastewater).
ii. *Industrial wastewater* constitutes the largest source of the contamination of life environments

Table 5.6 Examples of mycobacterial detection in wastewater sources

Wastewater origin	Isolates	Country	Ref.
Kitchen sink	*M. kansasii*	The Netherlands	[7]
River sediment (wastewater present)	*M. tuberculosis*	Romania	[1]
Fresh sewage used for pastures and fields	*M. tuberculosis*, RGM, *M.* sp.	Germany	[3]
Raw sewage, sewage effluent	*M. tuberculosis*	Poland	[10]
Sewage water around tuberculous sanatoria	*M. bovis*, *M. tuberculosis*, *M.* sp.	Kazakhstan	[4]
Sewage from tuberculous sanatorium and hospitals, towns and sewage purification plants	*M. bovis*, *M.* sp., *M. tuberculosis*	Poland	[5], [6]
Sewage in different areas	*M. chelonae*, *M. flavescens*, *M. fortuitum*, *M. gordonae*, *M. phlei*, *M. scrofulaceum M. smegmatis*, *M. terrae complex*, *M. vaccae*, *M.* sp.	Korea	[8]
Sewage sediment	*M. tuberculosis*	Poland	[2]
Settled sewage, raw sludge, drying bed sludge, centrifuged sludge cake, final effluent	*MAA* (2), *MAP*, *M. chelonae*, *M. fortuitum*, *M. gordonae*, *M. peregrinum*, *M. scrofulaceum*, *M.* sp.	UK	[9]
Wash-off water from wearing apparel, crockery, household utensils, etc.	*M. tuberculosis*	Russia	[12]
Water from poultry slaughterhouse	*M. gordonae*	Brazil	[13]
Sanatorium sewage: inlet, settling tank and outlet	*M. tuberculosis*	India	[14]
Sewage from cattle farm used for pastures	*M. bovis*, *M. tuberculosis*	Poland	[15], [17]
Fresh and drying sludge near Pecs	*M. xenopi*	Hungary	[16]
Corroded concrete sewer pipes	*M. sydneyiensis*, *M.* sp.	Belgium	[18]
Sewage treatment works (domestic)	*MAP*	UK	[11]

MAA (2) M. a. avium (serotype 2). **RGM** Rapidly growing mycobacteria. ***MAP** M. a. paratuberculosis.* **MAH** *M. a. hominissuis.*
[1] Ancusa M, Terbancea W (1970) Z. Gesamte Hyg. 16:913–916. [2] Bedrynska-Dobek M (1966) Pol. Med. J. 5:1058–1064. [3] Beerwerth W, Schurmann J (1969) Zentralbl. Bakteriol. [Orig.] 211:58–69. [4] Blagodarnyi I, Vaksov VM (1972) Probl. Tuberk. 50:8–12. [5] Buczowska Z (1965) Biul. Inst. Med. Morsk. Gdansk. 16:49–56. [6] Buraczewski O, Osinski J (1966) Pol. Med. J. 5:1065–1072. [7] Engel HW, Berwald LG, Havelaar AH (1980) Tubercle. 61:21–26. [8] Jin BW, Saito H, Yoshii Z (1984) Microbiol. Immunol. 28:667–677. [9] Jones PW, Rennison LM, Matthews PR, Collins P, Brown A (1981) J. Hyg. (Lond) 86:129–137. [10] Ogielski L, Zawadzki Z (1961) Acta Microbiol. Pol. 10:433–437. [11] Pickup RW, Rhodes G, Bull TJ, Arnott S, Sidi-Boumedine K, Hurley M, Hermon-Taylor J (2006) Appl. Environ. Microbiol. 72:4067–4077. [12] Poptsova NV (1974) Probl. Tuberk. 17–20. [13] Prince KA, Costa AR, Malaspina AC, Luis AF, Leite CQ (2005) Rev. Argent Microbiol. 37:106–108. [14] Saldanha FL, Sayyid SN, Kulkarn I, Sr. (1964) Indian J. Med. Res. 52:1051–1056. [15] Skurski A, Szulga T, Wachnik Z, Madra J, Kowalczyk H (1965) Arch. Immunol. Ther. Exp.(Warsz.) 13:189–196. [16] Szabo I, Kiss KK, Varnai I (1982) Acta Microbiol. Acad. Sci. Hung. 29:263–266. [17] Szulga T, Szaro A, Madra J, Kowalczyk H (1965) Arch. Immunol. Ther. Exp. (Warsz.) 13:331–335. [18] Vincke E, Boon N, Verstraete W (2001) Appl. Microbiol. Biotechnol. 57:776–785.

in some regions (wastewater from slaughterhouses may be included in this category).

iii. *Wastewater from agriculture* includes animal excrement, silage juices, fertilisers used in excess on fields with a low content of humus, residues of pesticides and other components (Photo 5.23).

5.2.7.1 Communal Wastewater

Hospitals and sanatoria for the treatment of human tuberculosis are the primary sources of OPM in communal wastewater. At present, in economically developed countries, decontamination of wastewater is ensured by numerous preventive measures including continuously operating purification plants. However, wastewater is a serious hygiene problem in developing countries and in less economically developed countries; its analysis would lie outside the purview of this book.

By examination of 13 samples of wastewater from cities, tuberculosis hospitals and sanatoria, 9 isolates were obtained and classified as *MTC* (Buraczewski and Osinski, 1966). *M. tuberculosis* was isolated more often than *M. bovis* from hospital wastewater: 21 iso-

Table 5.7 The growth of mycobacterial species at 22 and 37 °C in media prepared from autoclaved water, wastewater and sediment within 3 months ([1]; modified)

Runyons' group[1]	Mycobacterial species	Water from pond	Wastewater	Wastewater sediment
	M. tuberculosis	−	−	−
	M. bovis	−	−	−
I	*M. marinum*	++	−	−
II	*M. gordonae*	++	−	−
	M. scrofulaceum	++	−	−
III	*M. a. avium* (serotype 2)[2]	+	−	−
	M. a. avium (serotype 3)[2]	+	−	−
	M. a. hominissuis (serotype 4)	+	−	−
	M. a. hominissuis (serotype 8)	++	−	−
	M. gastri	+	−	−
	M. nonchromogenicum	++	−	−
	M. terrae	++	−	++
	M. triviale	−	−	+
	M. xenopi	+	−	−
IV	*M. smegmatis*	−	−	−
	M. phlei	−	−	−
	M. fortuitum	+	−	+
	M. vaccae	+	+++	+++
	M. diernhoferi	+	+++	+++
	M. flavescens	+++	+++	+++
pH of substrate used		5.4	6.0	5.6

[1] Divided according to Runyons' groups (**I** Photochromogenic species, **II** Scotochromogenic species, **III** Non-photochromogenic species and **IV** Rapid growers).
[2] Growth only at 37 °C. Intensity of growth: − no growth, + separated acid-fast rods and microscopic detection of few colony forming units (CFU), ++ microscopic detection of many CFU, +++ CFU strongly detected by the naked eye.
[1] Beerwerth W (1973) Ann. Soc. Belg. Med. Trop. 53:355–360.

lates of *M. tuberculosis* and 11 isolates of *M. bovis* were isolated from 131 samples from three hospitals and one sanatorium. Their most likely sources were sputum, stool and urine from ill patients. The causative agent of human tuberculosis was detected 0.5 km downstream from the waste discharge (Buczowska, 1965). In Romania, *M. tuberculosis* was swept with wastewater up to 5 km downriver from its source – a tuberculosis sanatorium (Ancusa and Terbancea, 1970).

The survival of *M. tuberculosis* in wastewater sediment was studied in Poland. *M. tuberculosis* fully virulent for guinea pigs was isolated from depths of 20 cm. In experiments with artificially infected sediment, *M. tuberculosis* fully virulent for guinea pigs was isolated after 9 months. Investigation of the effect of sediment depths on the virulence of *M. tuberculosis* for guinea pigs showed that sediment from depths of 7 m

destroyed *M. tuberculosis*. The results show that sediment from wastewater cannot be used for the fertilisation of vegetables, but can be used for the fertilisation of root crops, cereals or flowers (Bedrynska-Dobek, 1966; Beerwerth and Schurmann, 1969).

An extensive study of 1400 wastewater samples collected as far as several kilometres away from tuberculosis sanatoria in Kazakhstan was performed. Of 22 mycobacterial isolates, 15 were identified as *M. tuberculosis*, 5 as *M. bovis* and 2 isolates were classed as "atypical" mycobacteria after biological experiments. Subsequent tuberculin testing showed a higher positivity in the inhabitants from the "microdistrict" supplied with surface drinking water than in the inhabitants from a microdistrict supplied with "water of good quality" (Blagodarnyi and Vaksov, 1972).

Besides human excrement present in communal wastewater (Section 5.10.1), mycobacteria-contaminated water itself can be the source of ESM and PPM (Section 5.2). Water from households used for washing root vegetables and other foodstuffs can also be contaminated with ESM and PPM and become a source of mycobacteria (Section 5.7).

5.2.7.2 Industrial Wastewater

M. bovis has more frequently been isolated from slaughterhouse wastewater than *M. tuberculosis*: of 36 samples, 4 and 2 isolates of *M. bovis* and *M. tuberculosis* were isolated, respectively. Slaughtered cattle affected by bovine tuberculosis were the source of *M. bovis* (Buczowska, 1965). The most likely source of *M. tuberculosis* was sputum, stool and urine from ill personnel in the abattoir because cattle and other animals are more resistant to *M. tuberculosis* infection (Section 2.1.1). In the same Polish town, where the above-mentioned wastewater from the slaughterhouse was examined, only 4 isolates of *M. bovis* were obtained from 36 samples of communal wastewater (Buczowska, 1965). This result can be viewed from two aspects. First, the source of the causative agent of bovine tuberculosis could originate from both the slaughter wastewater and the wastewater from hospitals and households where infected patients were present. Second, with regard to the ecology of both *MTC* members (*M. tuberculosis* and *M. bovis*), the ability of *M. bovis* to survive in wastewater may be higher.

M. a. hominissius (serotype 8), which is most often found in sawdust, was isolated from water in the surroundings of a wood-processing plant (Photo 5.24). If the animals drink such contaminated river water containing mycobacteria, extensive gross lesions can develop in their lymph nodes with concurrent positive tuberculin responses.

5.2.7.3 Agricultural Wastewater

Cattle are the primary source of OPM (particularly *M. bovis*) in agriculture. Wastewater from farms can contain high concentrations of *M. bovis* that can infect both the people handling cattle and other animals. In Poland, 12 (16.0%) of 75 mycobacterial isolates from wastewater transported from the infected herds to pastures were identified as *M. bovis* and *M. tuberculosis* (Skurski et al., 1965; Szulga et al., 1965). Nevertheless,

communal wastewater used sometimes for the irrigation of pastures can become a source of mycobacterial infection with PPM, *M. bovis* and *M. tuberculosis* (Sobiech and Wachnik, 1966).

Floods should also be viewed as highly risky because mycobacteria from wastewater can gain access to arable soil or pastures. Sensitisation to bovine tuberculin PPM was observed in one insemination station. In spring, the PPM were regularly transmitted to the pastures and the bull paddocks with water from a burn containing effluent from the surrounding villages (I. Pavlik, unpublished data).

5.2.7.4 Spontaneous Sewage Purification (Self-Purification of Wastewater)

Organically polluted wastewater can clean itself, without human intervention; this is known as the self-cleaning (self-purification) effect. The process lasts quite a long period of time and is risky, due to possible eutrophisation of the surface water and the spread of causative agents of different diseases including mycobacteria (Table 5.6). The principle of this process is the gradual sedimentation of particles that are denser than water on the bottom of burns, rivers, ponds, lakes or seas. The particles, which are less dense than water, float on the surface (e.g. oil) and are trapped in different biofilms on the subjects present at the water surface. This mechanical (first, or primary) cleaning is not very important for the concentration of mycobacteria in wastewater. For instance, *M. tuberculosis* and *M. bovis* have been detected in both the "water column" and in the sediment in wastewater (Buczowska, 1965; Buraczewski and Osinski, 1966; Ancusa and Terbancea, 1970; Blagodarnyi and Vaksov, 1972).

The subsequent step is biochemical (second or secondary) cleaning; this involves fermentation of organic matter. According to the character of the contaminated surface water efflux, these processes occur under either aerobic conditions (e.g. burns or fast rivers) or anaerobic conditions (e.g. slow rivers, lakes or ponds; Photo 5.25 and 5.26). With regard to the fact that mycobacteria have been isolated from both the biofilm and the sediment of fresh surface water, the killing effect of these processes on, e.g., *M. tuberculosis* is not high. The long-term presence of *M. tuberculosis* in surface water that is undergoing a cleaning process for a long time poses a risk, even if its concentrations are low.

Therefore, such surface water should not be used for the irrigation of vegetables for raw consumption (Buczowska, 1965; Buraczewski and Osinski, 1966; Blagodarnyi and Vaksov, 1972).

When aerosols containing mycobacteria are formed, people and animals can become infected through aspiration (inhalation). Buraczewski and Osinski (1966) referred to the risk posed by *M. tuberculosis*, *M. bovis* and PPM from wastewater and Szabo et al. (1982) studied the risk constituted by *M. xenopi* from drying sludge tanks (Section 3.1.6).

5.2.7.5 Wastewater Treatment Plants (Sewage Plants)

In the wastewater treatment plants, the above-described self-cleaning process occurring in nature is "artificially" accelerated and the entire process can be professionally (in a sophisticated manner) controlled. Thus, the natural process of self-purification becomes more effective and the by-products can be utilised; e.g. biogas can be used for the generation of electricity, or the fermented sediment can be used as an excellent fertiliser. With respect to ecology and health, the risks of these cleaning plants can be seen in the following facts.

Wastewater Passes Through Sewer Pipes (Most Often Constructed as Enclosed Systems Situated Underground) to Wastewater Treatment Plants

In economically developed countries, open sewers and ditches were formerly used to dispose of wastewater; these are still used in economically developing countries (Photo 5.27). This way of discharging of wastewater poses a great risk, especially during floods or when wastewater seeps into ground or river water (if sewer pipes are located higher than the river surface is). The causative agent of human tuberculosis can reach this water in large part from households with infected patients. In Russia, 1210 samples of wastewater (wash-off water from clothing, household utensils, etc.) from 188 households with patients suffering from various forms of tuberculosis were examined; on average 12–15 samples from every household were tested. *M. tuberculosis* was isolated from wastewater from 15.2% of the households

(Poptsova, 1974). OPM (mostly *M. tuberculosis* and *M. bovis*) were detected in the effluent from hospitals and sanatoria with tuberculosis patients and from slaughterhouses for cattle affected by tuberculosis (Table 5.6).

Mycobacteria present in drinking water or in water from different premises also enter the wastewater. For example, in a poultry slaughterhouse in Brazil, *M. gordonae* of the same genotype was found in water used for rinsing the slaughtered poultry and in wastewater (Prince et al., 2005).

Concrete sewer pipes which were being corroded were investigated. It was found that mycobacteria participated in the corrosion. Of 22 sequenced bacterial isolates, 5 isolates were mycobacteria; one of them was identified as *M. sydneyiensis* (Vincke et al., 2001).

Wastewater Treatment Plants Constitute a Risk for the Spread of Mycobacteria, Above All if Their Capacity Is Insufficient or Their Function Is Impaired

An examination of 19 samples of wastewater from different parts of the wastewater treatment plants situated near the tuberculosis sanatorium resulted in the isolation of 11 members of the *MTC*, of which 4 isolates were identified as *M. tuberculosis* and 4 isolates as *M. bovis* (Buraczewski and Osinski, 1966). The fermented (digested) sediment (sludge) can also significantly participate in the spread of mycobacteria and constitutes a health hazard both for people and animals. The mycobacteria can reach high concentrations. In Texas, USA, the following concentrations of CFU of mycobacteria per 1 g were detected, 4.0×10^7 in sludge, which had undergone digestion in a two-stage high-rate anaerobic digestion system, 2.4×10^9 in lagooned sludge after anaerobic digestion for 11 days and 4.1×10^7 in sludge after standard anaerobic digestion for 20–30 days with no mixing or heating (Dudley et al., 1980). In India, *M. tuberculosis* was detected in water leaking from a settling tank with wastewater (Saldanha et al., 1964).

The distribution of *M. tuberculosis* in wastewater treatment plants remains obscure. In Poland, for example, *M. tuberculosis* was isolated from 6 (13.0%) of 46 samples of fresh "sewage". Culture examination of 46 sediment samples for the presence of *M. tuberculosis* was negative; however, *M. tuberculosis* was isolated from 3 (6.5%) of 46 samples of sewage effluent from irrigation pipes. Little is known at present about what is responsible for the absence of *M. tuberculosis* in samples of tank sediment. The testing of 12 water samples from a river flowing near the fields irrigated with this waste effluent was also negative for *M. tuberculosis*. The authors thought that the results may be attributable to an effective "filtration capability" of the soil, through which the contaminated water was seeping into the riverbed (Ogielski and Zawadzki, 1961).

It seems that only partial killing of other different pathogenic mycobacteria (e.g. *M. a. avium* serotype 2 and *M. a. paratuberculosis*) or PPM (e.g. *M. fortuitum* and *M. scrofulaceum*) occurs during fermentation. These species have been isolated from different stages of the cleaning process from both aerobic and anaerobic wastewater treatment plants (Jones et al., 1981). *M. a. paratuberculosis* was detected in water effluent from domestic waste treatment works, lake and river water (Pickup et al., 2006).

With regard to ecology, we will also mention the occurrence of mycobacterial species, e.g. *M. austroafricanum*, that can acquire carbon from, for instance, methyl tert-butyl ether added to "gasoline storage tanks" (Ferreira et al., 2006). Detailed information about this topic is presented in Section 8.3. The role of vertebrates in the spread of mycobacteria in areas irrigated with wastewater is relatively small (Fertig, 1961); for detailed information see Section 6.7.3.

Handling Liquid Waste (e.g. Aeration) in Wastewater Treatment Plants Poses a Risk of the Spread of Pathogens Including Mycobacteria Through Aerosols

High amounts of aerosol containing various pathogenic bacteria (e.g. *Klebsiella* sp. and *Streptococcus* sp.) are formed during the handling of effluent in wastewater treatment plants (Photo 5.28). Even though mycobacteria were not detected in aerosols, the ozonation of air failed to destroy these bacteria, which were detected at a distance of 300 m away from the wastewater treatment plant building (Pereira and Benjaminson, 1975). With regard to the fact that no

mycobacterial species were found to be present in the wastewater or in the sediment piles, similar testing should be performed with the known species of OPM (e.g. *M. tuberculosis* or *M. bovis*) and PPM (above all *MAC* members) because *MAC* members have been detected in the air during soil handling while repotting plants (Sections 5.1.2 and 5.9.3.1).

5.2.7.6 Reducing Factors for Mycobacteria in Wastewater and Sediments

Not only bacteria and fungi but also protozoa play a role in the fermentation of organic matter. Trunova (1971) discovered that protozoa can participate in reducing the amount of *M. tuberculosis* in settling tanks. From the point of view of the prevention of the occurrence of OPM in wastewater, possibilities of effective disinfection have been studied by a number of authors. It was found during "ultraviolet and chlorine disinfection" that *M. terrae* could aggregate, thereby reducing the effects of these disinfection systems (Bohrerova and Linden, 2006). The effect of gamma radiation, e.g. on *M. fortuitum*, in artificially infected effluents was also reduced (Garcia et al., 1987).

Of all the above-mentioned water sources, mycobacteria have most often been found in wastewater. The following factors may help explain this fact:

i. The most intensively contaminated water from the warm water supply system is highly diluted with cold water that is not as heavily contaminated with mycobacteria.

ii. Antagonistic properties of other prokaryotes towards mycobacteria, by as yet unrecognised mechanisms, can be exerted both in the soil and in wastewater. As mentioned above (Section 3.1.1), the effects of antibiotics produced by lower moulds are assumed (Chapman, 1971). However, various mechanisms of inhibition surely exist, which may be revealed in the future. These can be different detergent-derived enzymes, other contaminating chemicals, etc.

iii. Temperature levels of wastewater (usually lower than 20 °C) are not favourable for the growth of mycobacteria.

iv. Mineralisation processes occurring in the wastewater destroy various pathogenic or conditionally pathogenic micro-organisms. They may also cause the devitalisation of mycobacteria.

5.2.8 Views and Perspectives on the Research

Even though the above-mentioned studies and a review article by Peccia and Hernandez (2006) attest to the recent attention that has been concentrated on mycobacteria in aerosol, further research in this field will be necessary. The potential presence of mycobacteria as well as of other bacteria in different strata of the atmosphere has not been thoroughly studied yet.

With respect to the fact that wastewater constitutes a serious ecological problem in countries with rapidly developing industry (more specifically in Asia and South America), it is necessary to investigate this sphere as well.

Novel culture-independent detection and identification methods have already been used in some laboratories (Kim et al., 2004). A wider application of these methods resulting in the identification of different mycobacterial species will allow a more thorough study of their ecology. The potential identification of mycobacterial species that have not yet been detected in wastewater would be another qualitative step.

5.3 Occupational Habitats and Vectors for Transmission

J.O. Falkinham III

5.3.1 Swimming Pools as Habitats for Transmission

Attendants in indoor swimming pools are at risk of hypersensitivity pneumonitis (HP), so-called "swimming pool lung" (Rose et al., 1998; Martyny and Rose, 1999). One risk factor is the duration of exposure to the aerosols produced in indoor swimming facilities (Rose et al., 1998). Initial reports focused on the measurement of aerosol endotoxin levels; however, there was no correlation between endotoxin levels and the prevalence of HP (Rose, 1999). Although there has been

no study designed to determine whether there is a link between the presence of ESM and PPM in swimming pools and the aerosols generated in indoor swimming pools, it is reasonable to propose that mycobacteria are the causative agents.

First, mycobacteria have been recovered from swimming pools, with high numbers correlating with low chlorine, high temperature, low pH and low redox potential (Kaffka et al., 1979; Dailloux et al., 1980; Havelaar et al., 1985; Emde et al., 1992; Leoni et al., 1999; Iivanainen et al., 1999a).

Second, the appearance of HP in individuals is consistent with the presentation of disease conditions following exposure to mycobacterial antigens. For example, it has been shown that the exposure of mice to *M. terrae* (Huttunen et al., 2000) and to mycobacterial heat-shock proteins resulted in hypersensitivity reactions (Rao et al., 2002).

Third, the appearance of "swimming pool lung", mycobacterial infections associated with exposure to spas and hot tubs and HP in workers exposed to metalworking fluid, has always followed disinfection of the water or metalworking fluid.

In addition, the route of either the mycobacteria or antigens resulting in the appearance of granulomatous lesions in the lungs is via aerosols, certainly consistent with the ready aerosolisation and concentration of cells of mycobacteria (Wendt et al., 1980; Parker et al., 1983).

Consistent with the presence of ESM and PPM in swimming pools are reports of the presence of *M. intracellulare* in a public bath in Japan (Saito and Tsukamura, 1976) and the identification of the source of sporotrichoid dermatitis caused by *M. abscessus* at a public bath in Korea (Lee et al., 2000).

5.3.2 Metalworking Fluids and Aerosols as Sources of Mycobacterial Exposure

Outbreaks of hypersensitivity pneumonitis have been reported in mechanics exposed to aerosols generated by machining operations, such as grinding and milling (Bernstein et al., 1995; Kreiss and CoxGanser, 1997; Shelton et al., 1999). These operations use metalworking fluids to cool the cutting face in the cutting/grinding of the tool and to carry off metallic particles. Because the metalworking fluids are sprayed on the metal-cutting tool face, aerosols are generated in a large volume and particle (droplet) density.

Although fresh metalworking fluids (diluted in water from the neat fluid) lack a microbial flora (e.g. *Acinetobacter*), when used the fluids begin to accumulate a population of micro-organisms which decrease the efficacy of their activity necessitating the addition of biocides. Reported cases of HP occur in spite of the addition of biocides. Recovery of mycobacteria, particularly a novel species, *M. immunogenum* (Moore et al., 2000; Wilson et al., 2001), has led to the hypothesis that mycobacteria are responsible for HP in the mechanics exposed to aerosols generated during cutting, grinding and milling operations. This hypothesis is consistent with:

i. The ability of mycobacterial cells (Huttunen et al., 2000) or proteins (Rha et al., 2002) to induce hypersensitivity reactions.
ii. The high relative resistance to antimicrobial agents (biocides) of the ESM and PPM.
iii. The ability of mycobacterial isolates to degrade some of the components of metalworking fluids (Krulwich and Pelliccione, 1979; Guerin and Jones, 1988; Combourieu et al., 1998a,b). It would follow that in metalworking fluids, killing the primary (rapidly growing) microbial flora with biocides is followed by the growth of mycobacteria (slowly growing) on the available carbon compounds in the absence of competitors.

Note that the appearance of mycobacteria (and an associated public health problem) is preceded by the selective step of biocide (antimicrobial) addition, as is the case in drinking water systems, spas and hot tubs and swimming pools.

5.3.3 *Damp and* Water-Damaged Houses *and Buildings*

A variety of micro-organisms, including mycobacteria, have been recovered from building material in damp or moisture-damaged buildings (Andersson et al., 1997; Torvinen et al., 2006). Occupants of such structures,

whether workers or residents, have been known to report respiratory problems. In a study of moisture-damaged building materials, including ceramic plasterboard and wood (88 samples), ESM and PPM were isolated from 23% of samples. A wide variety of mycobacterial species were isolated including *M. intracellulare*, *M. chelonae*, *M. fortuitum* and *M. flavescens* (Torvinen et al., 2006).

Based on the documented recovery of ESM and PPM from moisture-damaged building materials and the ability of mycobacterial cells (Huttunen et al., 2000) or proteins (Rha et al., 2002) to induce hypersensitivity reactions, it is possible that respiratory symptoms reported by workers or occupants of water-damaged or damp buildings are due, in part, to the presence of mycobacteria. This might be a major problem in houses and buildings in the New Orleans area of the USA that suffered extensive water damage as a result of Hurricane Katrina in 2005. That area has high numbers of ESM and PPM in waters (Falkinham et al., 1980) and soil (Brooks et al., 1984).

5.3.4 Instruments

Catheters and bronchoscopes have been implicated as sources of mycobacterial infections. Catheter-related infections by *M. avium* (Schelonka et al., 1994) and *M. chelonae* (Hsueh et al., 1998; Hsueh and Luh, 1998) have been reported. *M. avium* has been shown to form biofilms in catheters and cells from such biofilms are more resistant to antibiotics, even when released from the biofilm (Falkinham, 2007).

It is likely that the reported difficulty in treating catheter-related infections is due to the antibiotic resistance of biofilm-grown mycobacteria. In the cases of catheter-related infections, the original source of the *Mycobacterium* was not identified. Quite possibly it could be disinfecting solutions used to sterilise the skin surface before the catheter emplacement. Both rapidly and slowly growing mycobacterial species are resistant to disinfectants (e.g. gentian violet) used in skin sterilisation and marking (Jones and Kubica, 1965; Safranek et al., 1987; Jones and Falkinham et al., 2008).

Reports of bronchoscope- or endoscope-transmitted mycobacterial infections appear regularly in the literature. A variety of different ESM and PPM species have been isolated from the devices following isolation from patients not suspected of having mycobacterial infections. These include *M. intracellulare* (Dawson et al., 1982), *M. xenopi* (Bennett et al., 1994) and *M. chelonae* (Wang et al., 1995). In addition to bronchoscopes, bronchoscope cleaning machines have also been implicated as the source of bronchoscope contamination and hence patient infection (Gubler et al., 1992; Takigawa et al., 1995). In addition to the difficulty in cleaning and sterilising all surfaces of bronchoscopes and their cleaning machines, a number of other problems lead to the potential contamination of these instruments:

i. Bronchoscopes and cleaning equipments accumulate organic matter. Organic matter binds and consumes disinfectant solutions, so there is no disinfectant left. For example, after mixing 1 ppm of chlorine and as few as 1 000 000 mycobacterial cells (it takes 100 times more to make a turbid suspension of cells), the chlorine is gone (Cowan and Falkinham, 2001).

ii. It is likely that mycobacteria forming biofilms on bronchoscope surfaces and cells in biofilms are more resistant to disinfectants (Steed and Falkinham, 2006). Biofilms are difficult to remove, especially in joints and it is difficult to reach some portions of the bronchoscopes.

iii. Following cleaning and disinfection, surviving mycobacteria likely grow and form biofilms in the remaining moisture. Thorough drying after cleaning and disinfection may prevent the re-growth and renewed formation of biofilm.

iv. Mycobacteria are susceptible to disinfectants, including chlorine, chloramine, chlorine dioxide and ozone. However, mycobacterial cells are at least 100 times more resistant compared to other bacteria (Taylor et al., 2000). Therefore, if the guidance for disinfection of normal gram positive (*Staphylococcus aureus*) or gram negative (*Pseudomonas aeruginosa*) directs use of a concentration of 10 ppm for 10 min (dosage equal to the product = 100 ppm min), a 100-fold higher dose is required.

5.4 Peatland Runoff Waters

J. Kazda

Extensive studies of peatland runoff water under different climatic conditions and of the sediment in burns for the presence of mycobacteria have been carried out in Finland. Iivanainen et al. (1993) examined burn water collected from 53 drainages of a peatland area, which was ditched about 30 years before for forestry. The number of mycobacteria ranged from 10 to 2200 CFU/litre. These counts correlated positively with the proportion of peatlands in the drainage area, high precipitation, brown water colour and concentration of Fe, Al, Cu, Co and Cr.

For comparison, samples of runoff water from natural and drained peatland were collected from May to October. The highest CFU count was found in August with 7.3×10^3/litre. The runoff water from the natural peatlands had similar median counts compared with the drained water. The counts of mycobacteria correlated with precipitation but not with air temperature, indicating that mycobacterial growth took place in vegetation and soil but not in the runoff water (Iivanainen et al., 1999b).

In the burn sediment, the CFU varied from 110 to 150 000 CFU/g and were related to climatic conditions, the peat characteristic of the drained region, the content of C and Pb in sediment and the chemical oxygen demand of the water. Furthermore, the increase in acidity positively influenced the count of mycobacteria and decreased the counts and activity of other heterotrophic bacteria. Compared with forest soil, the mycobacterial count in burn sediment was more than 100 times lower than in forest soils (Iivanainen et al., 1999a).

About 60% of the peatland in Finland has been ditched and drained, especially for forestry. These anthropogenic changes have influenced the quality and mycobacterial content of burn water. The number of mycobacterial isolations from the human respiratory tract in Finland increased more then 5-fold from 1975 to 1990 – the period of intensive draining of peatlands. It has been emphasised that such widespread environmental manipulation of peatland has contributed to the increase in the mycobacterial colonisation of the respiratory tract of humans (Iivanainen et al., 1993).

5.5 Soil

I. Pavlik

Soil is viewed as a natural reservoir of mycobacteria, because of their detection in various types of soil from different places (Table 5.8). The occurrence of particular mycobacterial species in soil is ubiquitous and, e.g., the *M. avium*, *M. terrae*, *M. fortuitum* and *M. flavescens* complex members have been detected in soil from a number of different niches in countries on all continents (Table 5.9). In the biologically active soil layer, mycobacteria are often found directly by microscopic examination after Ziehl–Neelsen staining and their proportion is estimated as 100–1 00 000 AFB/g of soil (Beerwerth and Kessel, 1976). Nutrient content in soil is a key factor affecting the quantity and concentration of respective mycobacterial species.

Various studies have reported different dominating mycobacterial species in different soil types (Table 5.9). It remains unclear why, e.g., *M. nonchromogenicum*, *M. triviale*, *M. gordonae*, *M. scrofulaceum* and other species have only been found in some countries and particular soil types. It has likely been caused by the use of differing culture techniques in respective laboratories (varying decontamination processes, different culture media, etc.). These can fundamentally alter both the detection rate of mycobacteria in soil samples and the range of isolated mycobacterial species.

5.5.1 Soil Composition and Preconditions for Mycobacterial Colonisation

Soil is the outermost layer of the Earth (pedosphere). Soil formation is the combined effect of degradation and weathering processes. It is formed mostly through the action of water, air and live organisms including mycobacteria that are involved in the change of the outermost layer of compounds into soil. Analysis of the published data shows that mycobacteria have been isolated from 3122 (67.4%) of 4801 examined soil samples of different origin (Table 5.8). Soil differs substantially in its materials of origin, bedrock, composition, colour and especially in its fertility (Photo 5.29). Soil is a complex system composed of living and nonliving materials that vary considerably between soil types and niches. These factors (especially pH and the availability of organic matters and trace elements) are also assumed to determine the species of mycobacteria present (Table 5.9).

5.5.1.1 Nonliving Materials in Soil and Physical Factors

The nonliving materials in soil which affect water and air permeability are the amount of clay, earth, sand and gravel. These are vulnerable to the natural processes of erosion and decay. The mechanical activity of moving particles in the environment (mainly due to wind, water in different states and the motion of completely loose sediments) is the cause of erosion. Erosion has always existed as a natural process; however, it has increased due to human activity in many places. These include the deforestation of landscapes, dewatering of soil, the overgrazing of grassland without allowing for its natural renewal, etc. (Photo 5.30).

Erosion, as a natural phenomenon, can be beneficial to the entire ecosystem to some degree. Excessive erosion, however, can lead to ecosystem damage and loss of its functionality. The presence of mycobacteria in these biomes has not yet been studied and in some respects (especially ecological) investigation may yield useful information.

Soil is classified into three standard types according to the proportion of solid particles:

   i. Sandy soil.
  ii. Earthy soil.
 iii. Clay-like soil.

Not much attention has been concentrated on these three soil types when the occurrence of mycobacteria in soil has been studied. Only Beerwerth and Schurmann (1969) reported lower amounts of mycobacteria in sandy soils than in other soil types. From the point of view of seasonal changes and its effect on the occurrence of mycobacteria, no significant differences in the amounts and species composition have been found in Germany, Finland and other countries in Europe (Beerwerth, 1971). It is obvious that one of the most important components is water, which affects both the quantity and variety of living organisms. This was

Table 5.8 Mycobacteria detection rates in soil of different origin

Soil samples		Examined samples			Ref.
Origin	Country	No.	Pos.	%	
Boreal soil	Finland	47	47	100	[10]
Forest	Germany	631	188	29.8	[3]
Forest	Germany	198	78	39.4	[4]
Forest	Germany	299	63	21.1	[2]
Subtotal		1175	376	32.0	
Meadows and pastures	Germany	658	518	78.7	[3]
Meadows and pastures	Germany	175	123	70.3	[4]
Pastures	Bulgaria	40	15	37.5	[8]
Pastures	Germany	299	275	92.0	[2]
Subtotal		1172	931	79.4	
Arable	Germany	299	290	97.0	[2]
Arable	Germany	585	543	92.8	[3]
Arable	Germany	158	138	87.3	[4]
Arable	Iran	220	91	41.4	[9]
Arable, mud and clay	USA	72	62	86.1	[17]
Farm yards	Korea	54	41	75.9	[12]
Subtotal		1388	1165	83.9	
Potting	USA	79	47	59.5	[7]
Irrigated	Germany	99	99	100	[4]
Recreation parks	Thailand	100	60	60.0	[11]
Subtotal		278	206	74.1	
Paddocks for pigs	Brazil	100	83	83.0	[6]
Paddocks for pigs	Bulgaria	40	15	37.5	[8]
Paddocks for pigs	Czech Rep.	19	3	15.8	[14]
Soil and bedding		22	19	86.4	[16]
Subtotal		181	120	66.3	
Suburb of Adelaide	Australia	52	16	30.8	[15]
Villages in Malawi	Malawi	148	76	51.4	[5]
City of Cordoba	Argentina	97	20	20.6	[1]
Villages in India	India	300	207	69.0	[13]
Subtotal		597	319	53.4	
Total		4791	3117	65.1	

[1] Ballarino GJ, Eseverri MV, Salas AV, Giayetto VO, Gonzalez S, Wolff L, Pessah O (2002) Rev. Fac. Cien. Med. Univ Nac. Cordoba 59:39–44. [2] Beerwerth W (1971) Prax. Pneumol. 25:661–668. [3] Beerwerth W, Kessel U (1976) Zentralbl. Bakteriol. [Orig. A] 235:177–183. [4] Beerwerth W, Schurmann J (1969) Zentralbl. Bakteriol. [Orig.] 211:58–69. [5] Chilima BZ, Clark IM, Floyd S, Fine PE, Hirsch PR (2006) Appl. Environ. Microbiol. 72:2343–2350. [6] Costallat LF, Pestana de Castro AF, Rodrigues AC, Rodrigues FM (1977) Aust. Vet. J. 53:349–350. [7] De Groote MA, Pace NR, Fulton K, Falkinham JO, III (2006) Appl. Environ. Microbiol. 72:7602–7606. [8] Dimov I, Gonzales E (1986) Vet. Med. Nauki, Sofia 23: 47–52. [9] Ghaemi E, Ghazisaidi K, Koohsari H, Khodabakhshi B, Mansoorian A (2006) East Mediterr. Health J. 12:280–285. [10] Iivanainen EK, Martikainen PJ, Raisanen ML, Katila ML (1997) Fems Microbiology Ecology 23:325–332. [11] Imwidthaya P, Suthiravitayavaniz K, Phongpanich S (1989) J. Med. Assoc. Thai. 72:317–320. [12] Jin BW, Saito H, Yoshii Z (1984) Microbiol. Immunol. 28:667–677. [13] Kamala T, Paramasivan CN, Herbert D, Venkatesan P, Prabhakar R (1994) Appl. Environ. Microbiol. 60:2180–2183. [14] Matlova L, Dvorska L, Bartl J, Bartos M, Ayele WY, Alexa M, Pavlik I (2003) Veterinarni Medicina 48:343–357. [15] Reznikov M, Dawson DJ (1980) Pathology 12:525–528. [16] Sanchez I, Rosell R (1983) Revista Cubana de Ciencias Veterinarias 14:29–33. [17] Wolinsky E, Rynearson TK (1968) Am. Rev. Respir. Dis. 97:1032–1037.

Table 5.9 Mycobacterial species isolated from soil

Country[1]	Origin of soil	Isolates[2]	Ref.
Australia	Roots of wheat	MAIS, M. aichiense, M. bohemicum, M. cookii, M. flavescens, M. heidelbergense, M. palustre	[4]
Australia	Paddocks for pigs	M. terrae	[5]
Bulgaria	Pastures	M. fortuitum, M. gordonae, M. phlei, M. terrae	[7]
Colombia	Cages with primates	M. abscessus, M. fortuitum, M. intracellulare, M. nonchromogenicum	[1]
Finland	Boreal soil	M. cookie, M. elephantis, M. farcinogenes, M. flavescens, M. fortuitum, M. manitobense, M. porcinum, M. senegalense, M. septicum, M. terrae	[15]
Germany	Irrigated soil with sewage from city	M. tuberculosis	[2]
India	Villages	MAIS, M. diernhoferi, M. flavescens, M. fortuitum, M. gadium, M. gastri, M. kansasii, M. nonchromogenicum, M. parafortuitum, M. phlei, M. smegmatis, M. szulgai, M. terrae, M. thermoresistibile, M. vaccae	[12]
India	Arable	MAC, M. chelonae, M. fortuitum, M. terrae	[16]
Iran	Arable soil	M. fortuitum, M. flavescens, M. chelonae, M. thermoresistibile, M. phlei, M. triviale, M. terrae, M. gordonae, M. fallax, M. gastri, M. marinum, M. kansasii	[8]
Japan	Nk	M. agri, M. aurum, M. chelonae, M. fortuitum, M. gordonae, M. nonchromogenicum, M. phlei, M. scrofulaceum, M. smegmatis, M. terrae, M. thermoresistibile	[13]
Japan	Soil, mud, wastes	MAC, M. gordonae, M. nonchromogenicum, M. scrofulaceum, M. terrae, M. triviale	[9]
Korea	Nk	M. agri, M. aurum, M. fortuitum, M. gordonae, M. nonchromogenicum, M. phlei, M. scrofulaceum, M. smegmatis, M. terrae, M. thermoresistibile	[13]
Korea	Farm yards	MAC, M. chelonae, M. flavescens, M. fortuitum, M. gordonae, M. phlei, M. scrofulaceum, M. smegmatis, M. terrae complex, M. vaccae	[11]
Malawi	Different parts of Karonga district	M. canariense, M. celatum, M. chlorophenolicum, M. chubuense, M. cosmeticum, M. duvali, M. farcinogenes, M. fortuitum, M. goodie, M. lactiocola, M. mageritense, M. moriokaense, M. peregrinum, M. poriferae, M. psychrotolerans, M. pulveris, M. senegalense, M. septicum, M. smegmatis, M. tokaiense, M. wolinskyi	[3]
Spain	Nk	M. brumae	[14]
Thailand	Recreation parks	M. chelonae, M. fortuitum, M. gordonae	[10]
USA	Potting	MAC, M. asiaticum, M. chelonae, M. flavescens, M. fortuitum, M. gordonae, M. interjectum, M. kansasii, M. peregrinum, M. porcinum, M. simiae, M. smegmatis, M. szulgai, M. terrae, M. triviale	[6]
USA	Arable soil, mud, clay	MAIS, M. fortuitum, M. kansasii, M. smegmatis	[18]
Zaire	Soil	M. malmoense	[17]

[1] Countries are ordered alphabetically.

[2] Mycobacterial species are ordered alphabetically; unidentified mycobacterial isolates are not mentioned.

MAIS *Mycobacterium avium–intracellulare–scrofulaceum* complex (consisting of *M. avium* subsp. *avium*, *M. avium* subsp. *hominissuis*, *M. intracellulare* and *M. scrofulaceum*). **MAC** *Mycobacterium avium* complex (consisting of *M. avium* subsp. *avium*, *M. avium* subsp. *hominissuis* and *M. intracellulare*). **Nk** Not known.

[1] Alfonso R, Romero RE, Diaz A, Calderon MN, Urdaneta G, Arce J, Patarroyo ME, Patarroyo MA (2004) Vet. Microbiol. 98:285–295. [2] Beerwerth W, Schurmann J (1969) Zentralbl. Bakteriol. [Orig.] 211:58–69. [3] Chilima BZ, Clark IM, Floyd S, Fine PE, Hirsch PR (2006) Appl. Environ. Microbiol. 72:2343–2350. [4] Conn VM, Franco CM (2004) Appl. Environ. Microbiol. 70:1787–1794. [5] Costallat LF, Pestana de Castro AF, Rodrigues AC, Rodrigues FM (1977) Aust. Vet. J. 53:349–350. [6] De Groote MA, Pace NR, Fulton K, Falkinham JO, III (2006) Appl. Environ. Microbiol. 72:7602–7606. [7] Dimov I, Gonzales E (1986) Vet. Med. Nauki, Sofia 23:47–52. [8] Ghaemi E, Ghazisaidi K, Koohsari H, Khodabakhshi B, Mansoorian A (2006) East Mediterr. Health J. 12:280–285. [9] Ichiyama S, Shimokata K, Tsukamura M (1988) Microbiol. Immunol. 32:733–739. [10] Imwidthaya P, Suthiravitayavaniz K, Phongpanich S (1989) J. Med. Assoc. Thai. 72:317–320. [11] Jin BW, Saito H, Yoshii Z (1984) Microbiol. Immunol. 28:667–677. [12] Kamala T, Paramasivan CN, Herbert D, Venkatesan P, Prabhakar R (1994) Appl. Environ. Microbiol. 60:2180–2183. [13] Kim SK (1981) Yonsei Med. J. 22:1–20. [14] Luquin M, Ausina V, Vincentlevyfrebault V, Laneelle MA, Belda F, Garciabarcelo M, Prats G, Daffe M (1993) International Journal of Systematic Bacteriology 43:405–413. [15] Niva M, Hernesmaa A, Haahtela K, Salkinoja-Salonen M, Sivonen K, Haukka K (2006) Boreal Environment Research 11:45–53. [16] Parashar D, Chauhan DS, Sharma VD, Chauhan A, Chauhan SV, Katoch VM (2004) Appl. Environ. Microbiol. 70:3751–3753. [17] Portaels F, Larsson L, Jenkins PA (1995) Tuber. Lung Dis. 76:160–162. [18] Wolinsky E, Rynearson TK (1968) Am. Rev. Respir. Dis. 97:1032–1037.

confirmed by the highest detection rate of mycobacteria occurring in watered arable soils (Beerwerth and Schurmann, 1969; Photo 5.9).

5.5.1.2 Living Components of Soil

The interrelationships between the method of soil cultivation, fertilisation and other factors have not been studied recently. However, it was found that the presence of easily utilisable organic matters in soil can be a significant precondition for the occurrence of mycobacteria there (Beerwerth, 1971). The studies published in late 1960s and 1970s (Tables 5.8 and 5.9) can provide underlying information for further investigation of the ecology and significance of mycobacteria for soil fertility. The processes of organic matter accumulation and conversion in soil can be divided into the following categories:

i. *Humification* is the conversion of organic matter in soil into humus.
ii. *Mineralisation* is the release of inorganic compounds, such as CO_2 and NH_4, etc. during decomposition.
iii. *Accumulation of organic matter* is a process which leads to the accumulation of plant and animal residues in soil and depends on bioclimatic conditions.
iv. *Decomposition* is caused by microbial processes and by the zooedaphone in soil.
v. *Transformation of organic matter in soil* is fragmentation by means of the zooedaphone, accumulation of transformed products, humification and mineralisation, humus stabilisation and the production of organo-mineral complexes.
vi. *Peat formation* is the accumulation of organic matter under anaerobic conditions.

Living components in soil (bacteria, fungi, invertebrate animals, etc.) vary depending on the soil composition and their natural habitat. They form a complex that determines soil fertility and the ability of soil to provide plants with sufficient amounts of nutrients, water and air in permeable soil. People purposely increase soil fertility by appropriate cultivation, fertilisation and irrigation or drainage according to the area and the conditions. Soil, which contains a lot of humus, is almost black in colour; accordingly, it is called black earth. It is very rich in nutrients and thus the most fertile among soils. If the amount of humus decreases, its black colour gradually turns brown; soil of brown colour is designated brown earth. All of these factors determine not only the presence of mycobacteria (Table 5.8) but also their species composition (Table 5.9).

The presence of mycobacteria in these types of soil has not been specially studied in association with the amounts of solid particles (Tables 5.8 and 5.9). However, it follows from the previously published data (Table 5.8) that higher amounts of mycobacteria will be found in soil with higher water and organic compound content. As of now, several studies dealing with the symbiotic activities of mycobacteria in root systems of different plants have been published: wheat (*Triticum aestivum*; Conn and Franco, 2004), cotton (*Gossypium hirsutum*) and pea (*Pisum sativum*; Egamberdiyeva and Hoflich, 2004).

As mentioned above, besides water, the presence of mycobacteria is primarily affected by the availability of organic matter. Accordingly, humus (the most fertile layer of soil) is the most significant component in this regard. Humus is dead organic matter of vegetable and animal origin that is in different stages of conversion. In the first phase, living organisms, e.g. bacteria and fungi, and inorganic factors such as water, temperature, etc. participate in this conversion. This process of decomposition is called mineralisation and proceeds up to the release of simple inorganic compounds such as H_2O_2 CO_2, H_2S and NH_4. Subsequently, in the second phase, due to biological activity, very complicated organic compounds such as humic acids (humification stage) are formed. These two phases are closely interrelated.

5.5.2 Soil Types and the Occurrence of Mycobacteria

Soil characteristics differ very much in different parts of the world and depend upon the mother rock, climate and localisation. Soil formation in hot and damp torrid zones is most intensive due to chemical weathering. However, the fertility of these soils is low due to heavy rainfalls which leach out many nutrients. On the other hand, physical weathering prevails in the cold Polar regions; therefore, soils with a low content of clay minerals, composed mainly of grit, are present there. Soils in the temperate zone are among the most fertile soils in the world. Many of them have been continuously cultivated for many thousands of years without depletion. Agricultural soil is exploited by people for the production of crops and the largest part is arable soil. The topmost layer of cultivated land is arable land; it is hoed by ploughing and can therefore be easily blown away by the rainwater or wind together with mycobacteria.

In arable soil, *M. terrae*, *M. nonchromogenicum* and *M. triviale* species have predominated compared to meadow soils, pastures and forests in which *M. gordonae* and *M. scrofulaceum* have been detected most often (Beerwerth and Kessel, 1976). The species of mycobacteria present are most likely affected by soil composition. According to different proportions of respective components present in different soils, these are classified into the following 10 types and ordered according to decreasing fertility.

5.5.2.1 Black Soils and Brown Soils

Slow decomposition of the vegetation residues gives rise to humus that is not washed off to the deeper layers of soil. Due to that fact its colour is brown-black to black and this type of soil was named accordingly. They are usually found in lowlands and the lower reaches of hilly regions up to 300 m above sea level where the annual total rainfall ranges between 400 and 500 mm. Due to the fact that humus is rich in nutrients for plants, this type of soil is the most fertile. Brown soils are found on the margins of black soil regions, mostly up to a height of 450 m above sea level. They originate in the zone of deciduous trees on the margins of black soil regions. They were likely formed by the degradation of black soil due to the damp climate and forest vegetation. The content of nutrients is lower than in black soils but they are suitable for agriculture if fertilisation is sufficient.

According to the study of Conn and Franco (2004) the profile of symbiotic bacteria in the root system of plants (endophyte) also influences the composition of arable soil. They detected differences in the proportions of *M. aichiense*, *M. bohemicum*, *M. cookii*, *M. flavescens*, *M. heidelbergense* and *M. palustre* in wheat (*T. aestivum*) endophyte within three localities. Perhaps this phenomenon will be explained by future discoveries.

5.5.2.2 Brown Forest Soils

Brown forest soils originate on non-calcic rocks below forest vegetation. This type of soil is typical of hilly areas with an annual temperature ranging between 6.5 and 8.3 °C and a total annual rainfall of between 600 and 800 mm.

Boreal soils. Boreal soils are found in the northern regions between the Polar circle and the parallel at 50° northern latitude, where coniferous trees prevail. The majority of such forests are found in Canada and especially in Siberia, where the name of "taiga" originated. The heavy rainfalls and subsequent chemical reactions with fallen needles give rise to podzol soils. Acid water causes the elution of iron compounds; these flow down into lower levels and form a non-permeable layer, which allows the formation of morass, swamps and peat bogs that are typical of Siberian taiga. Iron and aluminium oxides accumulate there and are responsible for low water permeability.

The amount of mycobacteria in boreal soil positively correlates with the amounts of Ca and Mn. Factor analysis showed that among others the concentrations of Ca, Mn, K and Zn were associated with the soil pH, which ranged between 3.0 and 4.0. On the other hand, together with decreasing concentrations of mycobacteria in boreal soils, increasing concentrations of Fe and Al, associated with decreasing pH and decreasing Ca concentrations, were noted (Iivanainen et al., 1997).

Organic matter is concentrated both in live plants and in dead biomass. When considering the ecology of mycobacteria, this biome seems to be the most important source of mycobacteria in the Earth's northernmost regions (Iivanainen et al., 1997, 1999a,b). This further supports the observation that ESM and PPM are able to multiply at temperatures of between 5 and 10 °C (George et al., 1980; Ermolenko et al., 1997).

Boreal soils under coniferous trees in Finland are quite acidic (pH 3.0–4.0), poor in organic oxygen (23–48%), total nitrogen (0.88–1.49%) and living heterotrophic bacteria: 1.6×10^6 to 4.1×10^7 CFU/g dry soil (Iivanainen, 1995). However, the occurrence of mycobacteria is based on the fact that they were found in 100% of the examined samples in concentrations of between 4.5×10^4 and 1.2×10^6 CFU/g of dry soils. The concentration of mycobacteria in boreal soils is highly associated with the amount and composition of organic matter. Due to unfavourable conditions (rapid washing off and non-permeable bed) organic matter is rapidly washed off into water streams, which take it from these regions to lakes. With respect to the results obtained in Finland, which indicated the presence of mycobacteria in all examined samples of boreal soils, this biome should be viewed as a significant source of ESM and PPM (Iivanainen et al., 1997).

This assumption was confirmed by the following publication. A lower concentration of mycobacteria was detected in burn or stream sediments than in boreal soil, where they originated (Iivanainen et al., 1999a,b). It was reported in a previous study that the numbers of mycobacteria, low pH and high content of organic matter in this burn or stream water correlated with the high proportion of peat bogs, from which water originated (Iivanainen et al., 1993). However, percentages of mycobacteria and other members of Actinomycetes in lakes fluctuate and mycobacteria are not the most significant component there in comparison with boreal soil. The species of mycobacteria in boreal soils are also different than in surface water, brooks and lakes (Niva et al., 2006).

Tundra. The treeless tundra landscape is found in the north of the boreal region; it is situated along the shore of northern Europe from Trondheim stretching eastwards, in Iceland and in Europe's northernmost islands. Tundra is characterised by boggy ground due to abundant moisture. The main vegetation is lichen and mosses, in warmer localities dwarf trees and shrubs, especially birches. Under the tundra vegetation there is a shallow profile of tundra soils poor in humus with a typical gley horizon. The presence of mycobacteria in these locations has not been studied yet. Nevertheless, we can suppose that mycobacteria can be transferred into this biome through surface waters (in particularly rainwater) and dust particles.

5.5.2.3 Calcic Soils, Rankers, Gley and Illimerised Soils

Calcic soils (rendsins) are analogous to brown forest soils, but in contrast to those, they arose on a calcic base. They are abundant in limestone detritus of various sizes, designated as the skeleton. They are rich in nutrients, but very shallow, which decreases their agricultural value. These soils arise from rocks under high temperatures; the base slowly undergoes chemi-

cal weathering and thus only a few fine particles are formed. Calcic soils are usually a transient stage in the development of other soil types. They can also be formed in lower altitudes, in places where fine particles are washed away with water, or through the action of the wind. They are found on the slopes of mountains and in vale notches and are the components of soil of the lowest fertility.

Gley soils originate in depressions or in (alluvial) plains with a consistently high groundwater level. Below the humic horizon, the gley horizon is formed, which is of griseous colour owing to reducing processes. The uppermost part of the horizon is often spotted because oxygen access has been intermittent due to the changing water level. These soils are the heaviest type of soil and if they are not drained they are constantly moist. Therefore, they are common in meadows.

Illimerised soils, which go to a depth of more than 2 m, originate in lowland margins and hilly areas. Clay particles are transferred here; they permeate together with rainwater into deeper layers of soil.

5.5.2.4 Podzols

Podzols are the most acidic soils in mountain regions. They are formed on the acidic non-calcic base under coniferous trees in cool and moist conditions at higher altitudes. Plant residues undergo the process of slow decomposition and accumulate in the horizon as raw humus. That produces various organic acids, which cause the dissociation of minerals to iron and aluminium oxides; these are transferred to the deepest soil layer along with leaking rainwater.

These oxides accumulate in this layer to form a distinctly coloured layer which is 70–100 cm high. Its hardness restricts the growth of tree roots (i.e. iron band or placic horizon). On the other hand, the deepest part is depleted of clay minerals, it has a typical bleached ashy colour (zola = ash in Russian, podsole = ash like). Podzolisation is a complex process supported by coniferous trees due to the fact that the decomposition of needles is very slow.

5.5.2.5 Salty Soils

Salty soils are found in warm and dry regions, where evaporation prevails over rainfall. Water rich in minerals in the dry and warm summer season rises through capillary attraction in gaps between rock particles to the surface, where it vaporises. During water vaporisation, salts are excluded and form salt layers of various heights, which are designated "solonetz" and "solonchak" or white alkaline soils. The low rainfall in these dry and warm regions fails to wash the precipitated salts off back to the soil – mostly alkaline in these regions.

5.5.2.6 Soil in Slag Heaps, Overburdens and Removed Soil

These soils arise through the actions of humans. They are one of the most devastated soils, which have completely or partly lost their original characteristics. Such soils are designated as anthropogenic, or also anthropic. They include soils affected by air pollutants from industry and soils that arose in consequence of the heavy mining of raw materials. Extensive and expensive recultivations are necessary before their incorporation into a biologically balanced environment.

A series of new mycobacterial species have been isolated from such soils in Japan (*M. vanbaalenii, M. mageritense, M. frederiksbergense, M. vanbaalenii/ M. austroafricanum* and *M. chubuense/M. chlorophenolicum*). Some other isolates could not be more closely identified and some were isolated only after co-culture with protozoa (Wang et al., 2006).

5.5.3 The Influence of Different Factors on the Presence of Mycobacteria in Soil

Beerwerth and Schurmann (1969) failed to find significant differences in the occurrence of mycobacteria in soil containing different amounts of calcium, sand, clay or peat. However, the highest recovery of mycobacteria was from the soil of meadows and pastures and in particular from arable soil, which can be explained by the high content of humus. However, Donoghue et al. (1997) did not confirm these results in Great Britain; they isolated less mycobacteria from arable soil than from forest soil.

The discrepancy can be explained by different decontaminating procedures and media used for the isolation of mycobacteria (Kazda, 2000; Thorel et al., 2004). For example, the highest counts of mycobacte-

ria from boreal coniferous forest soils and a low rate of contamination were obtained when decontamination with NaOH–malachite green–cycloheximide was combined with culture on glycerol and cycloheximide-supplemented egg medium at pH 6.5 (Iivanainen, 1995).

Accordingly, the next analysis will be focused on some of the most important factors determining the amounts and different species of mycobacteria found in soil.

5.5.3.1 Soil Composition

The amount of soil bacteria including mycobacteria is highly affected by the presence of clay which is found in soil as three dominant minerals: kaolinite, illite and montmorillonite. Besides the different chemical composition, these also differ in "plasticity, cohesion and cation adsorption; kaolinite being the lowest in each case and montmorillonite highest" (Garcia and McKay, 1970).

Among macro elements in soil, nitrogen appears to be important for the nutrition of some mycobacteria species; arable soil has been systematically supplemented with nitrogen over centuries or even millennia in some locations. Species composition has also been affected by a change to organic farming; the artificial application of the nitrogen in the form of ammonium salts has been replaced by nitrogen present in inorganic compounds from rotten manure (Donoghue et al., 1997).

5.5.3.2 Vegetation

Vegetables grown on arable land or continual forest vegetation also influence the levels of mycobacteria and species composition in soil. Rhizosphere with various root exudates markedly stimulates the growth of soil bacteria and the same can also be supposed for mycobacteria. In studies of boreal soils in Finland, no difference in the recovery of mycobacteria was found between deciduous and coniferous forests with trees of between 40 and 60 years old. The concentration of the isolated mycobacteria ranged between 4.5×10^4 and 1.2×10^6 CFU/g dry soil (Iivanainen et al., 1997).

Nevertheless, the species composition of mycobacteria in arable soil can be affected, in particular, by the grown crops as explained in detail in Section 5.7.

The crops can also play a significant role in the spread of mycobacteria in soil; a theory of "motorways" was recently advanced. They are represented by liquid films produced around fungal hyphae through which not only bacteria but also mycobacteria can spread (Kohlmeier et al., 2005).

5.5.3.3 Water Quantity

Water is essential not only for the survival of mycobacteria but also for their multiplication. Soil bacteria including mycobacteria are mainly present in the pores of aggregated soil and they are also aggregated on complexes composed of clay particles and humic compounds. Water together with well-decomposed organic material (humus) forms the most important colloidal soil system. Favourable conditions for the growth of soil microflora, including mycobacteria in the presence of the appropriate concentrations of dissolved salts and other compounds, are formed in this system (Garcia and McKay, 1970; Thorel et al., 2004).

Beerwerth and Schurmann (1969) detected the highest amounts of mycobacteria in watered soils. The recovery from forest soils was lower, with no difference between deciduous and coniferous forests. Donoghue et al. (1997) isolated most mycobacteria from pastures for cattle along burns or streams and watering places for cattle in the summer months, when moist weather was accompanied by increased temperatures.

On the other hand, the recovery of mycobacteria from soil samples from Iran was considerably lower due to a lack of soil moisture (Ghaemi et al., 2006). The same results were obtained by analysis of soil and other components from the environment of farm animals in South Africa. The recovery of mycobacteria from soil (mostly dry) was much lower in contrast to deep bedding, which was moistened by animal urine and faeces. Soil in this area seems to be the main reservoir of mycobacteria over the rainy winter season. Subsequently, plant leaves which later become food for animals are contaminated through the rising dust (Kleeberg and Nel, 1973).

5.5.3.4 Temperature

Besides the organic and inorganic nutrients available in soil, moisture, vegetation type and temperature sig-

nificantly affect the amount of mycobacteria in soil (Kubalek and Komenda, 1995; Donoghue et al., 1997; Kazda, 2000). During a 15-month investigation in Great Britain it was found that the mean temperature of forest soil in different niches was 16 °C in the summer months. This mean temperature was significantly lower ($p < 0.025$) than the mean temperatures in burn or stream water (19 °C) and arable soil (21.4 °C). The mean temperature of pasture soil was 19.5 °C and ranged between 14 and 25 °C (Donoghue et al., 1997).

It is evident that the temperature of 37 °C which is optimum for these mycobacterial species was not found in any of the soils. Nevertheless, it is necessary to emphasise that mycobacteria do not require temperatures between 18 and 20 °C for their growth. Their growth, e.g. of species *M. flavescens*, has been observed at 4 °C (Ermolenko et al., 1997) and *MAC* members (*M. a. avium* and *M. intracellulare*) and *M. scrofulaceum* have been observed to grow slowly at 10 °C (George et al., 1980).

During studies on the occurrence of mycobacteria in boreal soils in Finland, it was detected by factor analysis that an ambient temperature negatively correlated with soil moisture, which influenced the amount of mycobacteria present in soil (Iivanainen et al., 1997). However, all of these factors should be considered with respect to local specific conditions. These findings cannot be generalised due to a lack of similar studies in other soil types at present. In the Southeastern US, Kirschner, Jr. et al. (1992) detected the majority of the *MAC* members and *M. scrofulaceum* found in the study in soils from the vicinity of acidic, brown, low oxygen-containing swamps in spring and in summer, when the temperatures are usually the highest. The seasonal occurrence of mycobacteria in soil and in surface water was compared in South India. A significantly higher recovery from soil ($p = 0.2$) and water ($p = 0.01$), respectively, was observed for samples collected in the warm summer months than those taken in colder winter months (Kamala et al., 1994).

Data published in the early 1970s can be explained with respect to the above-mentioned results published over the last two decades. In this work Beerwerth (1971) failed to detect a relationship between the concentration of mycobacteria in soil and the season. The most likely cause was the decontamination process used and agars employed for the isolation of ESM and PPM.

5.5.4 Transmission of Mycobacteria from Soil to Other Components of the Environment and Its Impact on the Health of Animals and People

The main vector for the transmission of mycobacteria from soil is surface water. It washes them off during heavy rain into water streams (rills, burns or streams and rivers), where sediment with mycobacteria is then carried by the flow of water to lower altitudes. Another important vector is the air (Photo 5.31). During strong winds, especially in dry weather, soil materials containing mycobacteria are picked up and carried far (Section 5.9). Various parts of plants (roots, stems, blossoms, seeds, etc.) and trunks are less significant vectors of soil mycobacteria (Sections 5.7 and 5.8). These vectors can, however, be transported very far by various vehicles including humans (transport of crop or trunks). Animals can spread mycobacteria mainly through their surfaces (hair, wool, feathers or skin; Photo 5.32) or in their intestinal tract (Section 5.10) and less often through urine, sputum and other secretory and excretory products.

Soil is also considered as a primary source of *MAC* members. It is assumed that all *MAC* members are washed off from soils into rivers and then transported into the ocean. Aerosol is formed during surf; that also contains *MAC* members to which animals and people are exposed (Codias and Reinhardt, 1979; Brooks et al., 1984). A higher occurrence of infections caused by the *MAC* in a village population in the USA can be explained by their higher exposure to *MAC* members (in particular *M. a. hominissuis*) from agricultural soil and dust released from the soil (Codias and Reinhardt, 1979).

Nevertheless, it is also necessary to take into consideration the results concerning the composition of mycobacterial species in soil and water in South India where different species were found in the following two different ecological biomes: *M. fortuitum* complex members prevailed in soil and *MAC* members and *M. scrofulaceum* prevailed in water (Kamala et al., 1994). These results implicate soil as a significant source of mycobacteria found in water; however, the water environment is also rather specific and some mycobacterial species are found in this biome more often than in soil (Section 5.2).

5.5.4.1 Invertebrates

Soil is not only an important habitat for invertebrate animals, but organic components in soil are very important food for them (Photo 5.33). However, invertebrate animals can also become contaminated with mycobacteria from soil in the following ways (Section 6.3):

i. The body surfaces of all invertebrate animals that are in contact with soil are contaminated with various hydrophobic components of both soil and mycobacteria.
ii. Unicellular organisms (protozoa and others) "graze" on soil microflora including mycobacteria (Ronn et al., 2001).
iii. Invertebrate animals (such as earthworms) feed on organic remnants together with soil contaminated with mycobacteria.
iv. Predators of unicellular and higher invertebrate animals can be infected with mycobacteria contaminating the body surface and internal organs of their victims.

The circulation of mycobacteria through soil is then completed by shedding in the faeces of these invertebrate animals or in the faeces of their predators (Sections 6.3 and 8.3).

5.5.4.2 Vertebrates

Above all geophagia, which is the deliberate consumption of earthy substances by higher organisms, helps spread mycobacteria from soil. Due to their resistance, mycobacteria can survive passage through different sections of the intestinal tract of various animals (Section 5.10) and are returned through the shed excrements to the environment:

i. Rumen (environment with numerous protozoa "grazing" on cellulose including mycobacteria).
ii. Stomach (acid and highly saccharose and proteolytic environment).
iii. Small intestines (environment rich in lytic enzymes).
iv. Blind intestines (environment rich in lytic enzymes produced by other micro-organisms).

v. Large intestines (environment characterised by the activities of specialised proteo-, lipo- and fibrolytic bacteria).

vi. Rectum (environment with a strong dehydrating ability).

Geophagia is the deliberate eating of earth (Photo 5.34) containing mostly indigestible clayish earthy matters (Trckova et al., 2004). Kaolinite is the most common soil mineral, which together with other clay minerals and with organic matter forms the humic complex in soil. Geophagia has been observed not only in animals, but also in humans, especially in children (Johns and Duquette, 1991a,b; Mahaney et al., 1996a,b; Knezevich, 1998). There exist several hypotheses which seek to explain this phenomenon (Wilson, 2003):

i. Detoxication of harmful or indigestible matters in food.

ii. Treatment of gastrointestinal disorders, most often diarrhoea.

iii. Supplementation with mineral substances (such as iron in newborn piglets).

iv. Treatment of excessive acidity in the digestivetract.

Geophagia in the human population is viewed as abnormal behaviour or a sign of metabolic dysfunction. The most acceptable explanation of geophagia in humans is gastrointestinal adsorption of plant metabolites and enterotoxins (Dominy et al., 2004). In South Africa (especially among black women living in Northern Transvaal), pregnant women often take soil into their mouth (mostly clay), which they spit out after some time. The contamination of this clayey soil with mycobacteria is sufficient for the infection of the oral cavity. The following investigation showed that *M. fortuitum*, *M. smegmatis* and *M. vaccae* species were detected in the sputum of these women; these probably contaminated the sputum in the oral cavity. To make the picture complete, it is necessary to say that these mycobacterial species were also found in the soil of this area as well as in Asia, the USA and Europe (Table 5.9).

Geophagia was observed in 76% of subjects from the free-living group of monkeys *Macaca mulatta*; even though high numbers of them were affected by one or more types of endoparasite (89% of animals), the prevalence of diarrhoeal diseases was very low and

due to geophagia reached only 2% (Knezevich, 1998). The consumption of soil with a higher content of kaolin ($<20\%$) by cattle caused the adsorption of toxic and indigestible components from the diet and thus likely prevented diarrhoea (Mahaney et al., 1996a). Geophagia in free-living birds leads to a reduction in the biological adsorption of toxic substances (especially alkaloids and tannins) present in particular in seeds and different fruit (Diamond et al., 1999; Gilardi et al., 1999).

Vertebrates subsequently shed mycobacteria from soil through their faeces (Section 6.3) and thus the circulation of mycobacteria between soil and vertebrates is completed.

5.5.4.3 Soil and Soil Mycobacteria in Relation to Animal Health

Geophagia from the point of view of mycobacteria has been studied extensively to date. The experiments on rats orally infected with a mineral oil suspension containing dead *M. butyricum* (*M. smegmatis* according to currently accepted taxonomy) are noteworthy. It was shown that in contrast to non-infected rats, the consumption of kaolin by infected animals was significantly increased (Burchfield et al., 1977). On animal farms, especially those keeping domestic pigs and cattle, ESM and PPM are quite often found in soil (rich in organic substances) in the paddocks and on unpaved roads (Table 5.8).

With respect to common geophagia in pigs and the indiscriminate intake of feed from the ground by cattle, these mycobacteria can be ingested. On such farms, animals are repeatedly in contact with soil contaminated with urine and faeces; the permeability of water through this soil is decreased during rain. Consequently, the conditions are beneficial for the survival and multiplication of ESM and PPM. Accordingly, an increased occurrence of non-specific responses to bovine and avian tuberculin was noted. In cattle experimentally infected with mycobacterial isolates from soil, non-specific reactions to bovine tuberculin were observed for 4 weeks and disappeared 10 weeks later (Corner and Pearson, 1979). In pig herds, a higher occurrence of tuberculous lesions in head and intestinal lymph nodes was documented (Corner and Pearson, 1979; Dvorska et al., 1999; Matlova et al., 2003).

It is noteworthy that the factor of age plays a role in defence mechanisms. Tuberculous lesions in lymph nodes are usually formed in pigs that have been infected up to the age of about 2 months. If they are sensitised to mycobacteria from the environment and then encounter this infectious agent again under the age of 2 months, granulomatous inflammation with concurrent necrosis (caseification) begins to develop in their lymph nodes. After the infection of an older pig with mycobacteria from soil (especially *MAC* members), the stimulation of their cellular immunity occurs. It is manifested by non-specific responses in sows and boars to tuberculin testing with bovine and avian tuberculin. After this stage, the infection is mostly overcome without the occurrence of necrotic lesions associated with caseification (Kleeberg and Nel, 1973).

Detections of comparable mycobacterial species from gastric lavages and from the blood of New World primates, from soil and pigs in Bulgaria, or the species composition of mycobacteria from soil in Japan, Korea and Argentina provide evidence of the source of infection for animals from soil (Table 5.10). Mycobacteria

found in soil are also in contact with the skin of different animals living in soil. In Perth, Australia, skin infections in male and female numbats (*Myrmecobius fasciatus*) were caused by *MAC* members (serotypes 1, 6 and 8) and *M. fortuitum* (Gaynor et al., 1990). Numerous mycobacterial species commonly found in soil were also found in small terrestrial mammals and most probably infected them (Fischer et al., 2000; Skoric et al., 2007).

5.5.4.4 Soil and Soil Mycobacteria in Relation to the Health of People

With reference to the frequent occurrence of *MAC* members in soil, it is evident from Table 5.11 that almost all serotypes found in human patients were also detected in soil. In addition to wounds (Table 5.12), the route of the infection in cases of pulmonary mycobacteriosis was most likely inhalation (Photo 5.35). Handling earth for plant cultivation in flower pots is considered as highly risky at present (Table 5.12; Section 5.6).

Table 5.10 Mycobacterial species cultured from soil from different countries

Mycobacterial species	Colombia[1] Soil[a]	Gastric lavage[b]	Blood[c]	Japan[3] Soil[d]	Korea[3] Soil[e]	Bulgaria[2] Soil	Pigs	Argentina[4] Soil
M. abscessus	4	21	1	0	0	0	0	0
M. agri	0	0	0	10	8	0	0	0
M. aurum	0	0	0	4	10	0	0	0
M. chelonae	0	3	0	3	0	0	0	0
M. fortuitum	1	1	1	69	78	1	8	17
M. gastri	0	0	0	0	0	0	8	0
M. gordonae	0	0	0	6	1	5	32	0
M. intracellulare	1	3	0	0	0	0	0	0
M. kansasii	0	0	0	0	0	0	0	3
M. nonchromogenicum	2	1	0	7	12	0	0	0
M. phlei	0	1	0	13	3	3	0	5
M. scrofulaceum	0	0	0	11	19	0	4	0
M. simiae	0	1	0	0	0	0	0	0
M. smegmatis	0	0	1	13	4	0	8	0
M. thermoresistibile	0	0	0	18	5	0	0	0
M. triviale	0	1	0	0	0	0	6	0
M. terrae	0	1	0	1	2	6	0	0
M. vaccae	0	0	0	0	0	0	7	0
M. sp.	0	2	0	17	21	0	11	2
Total	8	35	3	172	163	15	84	27

[a] Examined soil samples from cages in a zoological garden with New World primates. [b] Gastric lavages from New World primates.
[c] Culture examinations from blood samples from New World primates. [d] Samples from Hiroshima, [e] Samples from Seoul.
[1] Alfonso R, Romero RE, Diaz A, Calderon MN, Urdaneta G, Arce J, Patarroyo ME, Patarroyo MA (2004) Vet. Microbiol. 98:285–295. [2] Dimov I, Gonzales E (1986) Vet. Med. Nauki, Sofia 23:47–52. [3] Kim SK (1981) Yonsei Med. J. 22:1–20. [4] Oriani DS, Sagardoy MA (2002) Rev. Argent Microbiol. 34:132–137.

Table 5.11 Serotype detection of *M. avium* complex members from soil, plants and different hosts (humans and pigs)

Country		USA			Australia			Brazil[2]	South Africa[5]				Japan[8]
Member/serotype		Soil[3]	Humans[4]	Humans[1]	Soil[7]	Soil[6]	Humans[6]	Soil^a	Soil^b	Plants^c	Pigs^d	Humans^e	Humans
M. a. avium	1	4	0	13	0	0	31	1	1	0	19	2	14
	2	5	0	0	0	0	24	0	0	2	4	2	0
	3	0	1	0	0	0	11	0	0	0	1	1	0
M. a. hominissuis	4	0	2	4	0	0	51	0	0	14	30	3	0
	5	0	0	0	0	0	1	0	0	0	0	11	0
	6	0	0	2	1	0	26	7	0	0	0	9	0
	8	0	2	2	0	0	92	4	0	10	32	1	5
	9	1	0	4	2	0	50	0	0	1	0	10	2
	10	0	0	0	0	0	14	1	2	1	16	1	0
	11	0	0	0	0	0	2	0	0	0	0	1	0
	21	0	1	0	0	0	3	0	0	0	0	0	3
M. intracellulare	7	0	0	4	8	3	19	1	0	0	0	38	1
	12	0	0	6	0	1	30	0	0	0	1	12	3
	13	1	0	4	3	0	20	0	1	0	0	17	1
	14	0	0	9	15	2	28	0	1	1	0	31	6
	15	0	0	4	0	0	10	2	0	0	0	8	0
	16	0	0	26	4	0	66	0	0	1	0	10	15
	17	0	1	4	2	0	13	0	0	3	0	9	0
	18	0	0	3	3	2	14	0	0	0	1	7	0
	19	0	1	5	3	3	50	3	0	0	0	26	0
	20	0	0	0	0	0	4	1	0	0	0	8	0
	22–28	0	1	0	0	0	47	0	0	0	0	19	0
Mixed or unknown		4	12	26	29	18	Nk	15	8	6	67	339	0
Total		15	21	116	70	29	606	35	13	39	171	565	50

^a Soil from the pig farms and surrounding environment. ^b Soil and dust isolates. ^c Plants, animal feed and bedding. ^d Swine lymph nodes with lesions. ^e Sputa of healthy rural humans. **Nk** Not known.

[1] Codias EK, Reinhardt DJ (1979) Am. Rev. Respir. Dis. 119:965–970. [2] Costallat LF, Pestana de Castro AF, Rodrigues AC, Rodrigues FM (1977) Aust. Vet. J. 53:349–350. [3] Matthews PR, Brown A, Collins P (1979) J. Appl. Bacteriol. 46:425–430. [4] McClatchy JK (1981) Rev. Infect. Dis. 3:867–870. [5] Nel EE (1981) Rev. Infect. Dis. 3:1013–1020. [6] Reznikov M, Dawson DJ (1980) Pathology 12:525–528. [7] Reznikov M, Leggo JH (1974) Pathology 6:269–273. [8] Saito H, Kai M, Kobayashi K (1998) Kekkaku 73:379–383.

ESM and PPM found in soil can also penetrate through the dust particles or aerosol into deeper parts of the lungs and can cause pulmonary mycobacteriosis in immunocompromised adult patients and cervical lymphadenitis in children whose immunity is not yet developed (but not immunosuppressed; Sections 3.1 and 5.9). However, on the one hand, ESM and PPM can (especially in developing countries, in which people live in close contact with soil) enhance the immunity of people against human tuberculosis and leprosy in some regions and on the other hand, they can only "mask" their protection during BCG vaccination (Chapter 7; Fine, 1995).

In Finland, the constantly increasing occurrence of mycobacteriosis in humans caused by *MAC* members and *M. malmoense* was recorded between 1978 and 1987. The main cause was likely the dewatering of peat bogs with the aim of forestation. This caused an increased water outflow accompanied by an increasing concentration of mycobacteria in soil. During windy weather, mycobacteria with dust particles from the dried peat and soil were spread into populated areas (Katila et al., 1995).

5.5.5 Gross Lesions in Cattle and Pigs Caused by Rhodococcus equi and Its Relation to Soil

Gross lesions in lymph nodes resembling tuberculosis (tuberculoid lesions) that have been described in pigs and cattle (Dvorska et al., 1999; Pavlik et al., 2005; Shitaye et al., 2006) can also be caused by bacteria

Table 5.12   Soil as a contaminated vehicle infecting wounded skin followed by mycobacterial infections in humans

Mode of mycobacteria transmission	Patients[1]	Infected tissue	Isolates[2]	Ref.
Farmer stepped on a nail while wearing tennis shoes	33/M	Erythema over the dorsal surface of the right third metatarsophalangeal joint followed by drained pus	*M. abscessus*	[6]
A laceration on the left elbow caused 1 year earlier while travelling in northern Queensland, Australia	24/M	Low-grade olecranon bursitis of the left elbow of army officer	*M. asiaticum*	[3]
Bicycle accident	7/M	Chronic abscess at the site of the wound	*MAIS*	[1]
Fall from horse	63/F	Chronic arthritis in wrist	*MAIS*	[1]
Minor leg trauma	73/F	Chronic lateral leg ulcer (asthmatic steroid treatment)	*M. chelonae*	[7]
Injured by garden fork	7/M	Chronic abscess at the site of the wound	*M. fortuitum*	[1]
Bicycle accident	10/F	Chronic ulcer on thigh at the site of the wound	*M. fortuitum*	[1]
Puncture wound	27/F	Chronic ulcer on foot at the site of the puncture	*M. fortuitum*	[1]
Injury by garden fork	45/F	Osteomyelitis in big toe	*M. fortuitum*	[1]
Injury by garden fork	60/F	Chronic ulcer on ankle at site of puncture	*M. fortuitum*	[1]
Sitting on the ground and selling fresh chicken meat	16/M	Large annular plaque on the buttocks	*M. fortuitum*	[4]
Minor leg trauma	44/M	Chronic ulcer of leg and lateral malleolus (asthmatic steroid treatment)	*M. kansasii*	[7]
Cut by slate	38/M	Plaques, indurated pustules on dorsum of the left hand	*M. marinum*	[8]
Stepped on gardening shears when spreading mulch containing an unknown type of manure on his lawn	?/M	Violaceous indurated plaque on foot growth at 52 °C	*M. thermoresistibile*	[2]
Injury to right ankle while working	33/M	Oedema and erythema of the right ankle followed by osteomyelitis	*M. tuberculosis* and *M. fortuitum*	[5]

[1] Age in years/gender (sex): **M** (male), **F** (female), **?** (not known).
[2] Mycobacterial species are in alphabetical order.

[1] Blacklock ZM, Dawson DJ (1979) Pathology 11:283–287. [2] Cummings GH, Natarajan S, Dewitt CC, Gardner TL, Garces MC (2000) Clin. Infect. Dis. 31:816–817. [3] Dawson DJ, Blacklock ZM, Ashdown LR, Bottger EC (1995) J. Clin. Microbiol. 33:1042–1043. [4] Kullavanijaya P (1999) Clin. Dermatol. 17:153–158. [5] Lazzarini L, Amina S, Wang J, Calhoun JH, Mader JT (2002) Eur. J. Clin. Microbiol. Infect. Dis. 21:468–470. [6] Meredith FT, Sexton DJ (1996) Clin. Infect. Dis. 23:651–653. [7] Plaus WJ, Hermann G (1991) Surgery 110:99–103. [8] Street ML, Umbert-Millet IJ, Roberts GD, Su WP (1991) J. Am. Acad. Dermatol. 24:208–215.

(e.g. *R. equi*) other than mycobacteria. However, it can greatly complicate the assessment of meat edibility by the naked eye at meat inspections in abattoirs due to the fact that gross lesions often resemble tuberculous lesions. The presence of the gross lesions often leads to the condemnation of whole animals. In some pig herds, the identification of *R. equi* predominates after culture examination of the gross lesions from the lymph nodes.

Farmers are usually glad to hear that their herd is not affected by a mycobacterial infection. However, they soon find out that the consequent losses due to the finding of gross lesions in slaughtered animals are almost as high as those due to mycobacterial infections. Accordingly, they ask about the source of this intracellular pathogen. With respect to the fact that the infections caused by *MAC* members and *R. equi* are often found together and that soil is their significant source, we will try to explain the present understanding of this problem to the readers of this book.

5.5.5.1 The Occurrence of *R. equi* in Horses, Cattle, Pigs and Soil

Farms with a long history of keeping horses have traditionally been at the highest risk of *R. equi* infection. In the 1980s, the discovery of *R. equi* in faeces from healthy horses led to the assumption that *R. equi* is a component of normal horse faeces (Woolcock et al., 1980; Nakazawa et al., 1983). Foals bred in such farms are steadily exposed to *R. equi* from adult animals and soil (Prescott, 1991; Takai et al., 1991). *R. equi* is a soil micro-organism with minimal requirements for growth. They mainly multiply in manure and in soil with a high content of organic matter. Accordingly, manure contaminated with *R. equi* can become a significant reservoir of infection if left in the pasture or if used for the fertilisation of soil with growing biomass intended for feeding as green fodder (Prescott et al., 1984).

The highest number of *R. equi* was found on the soil surface, while hardly any *R. equi* was found at a depth of 30 cm under the surface (Takai et al., 1991). *R. equi* was isolated from 100% of tested samples from soil in a feeding place for cattle; it was comparable with the samples from soil in the feeding places for pigs and horses (Photo 5.36). The occurrence of *R. equi* in soil collected from other places (such as street margins,

riversides and cultivated fields) was much lower (Takai and Tsubaki, 1985).

Nevertheless, *R. equi* was detected in 50 and 35% of faeces of cattle and pigs, respectively, and in the cervical lymph nodes of 35% of pigs. However, *R. equi* was not isolated from any of 54 bovine retropharyngeal and submaxillary lymph nodes (Mutimer and Woolcock, 1980). Due to the fact that *R. equi* is an obligate aerobic micro-organism, it cannot multiply in the large intestines of adult herbivores and thus cannot be a constituent of their normal intestinal microflora. Its presence in the faeces of these animals only reflects its ingestion on pastures or from contaminated fodder (Prescott, 1991).

5.5.5.2 The Health Implications of *R. equi* and Its Occurrence in Pigs Together with Potentially Pathogenic Mycobacteria

What are the principles of the intracellular parasitism of *R. equi*? At the level of the cell the most important manifestation of *R. equi* pathogenicity is the ability to survive in macrophages and to subsequently destroy them (Prescott, 1991). This characteristic feature of intracellular parasitism is pre-conditioned by its ability to prevent fusion of phagosomes with lysosomes. Induced non-specific lysosome degranulation leads to the destruction of not only the infected macrophage but also of the surrounding tissue and to the influx of neutrophils to the original granulomatous infectious lesion (Hietala and Ardans, 1987).

The problem with the simultaneous occurrence of *R. equi* and PPM in submaxillary and mesenteric lymph nodes from healthy pigs remains obscure. Mesenteric lymph nodes represent a more suitable environment for the survival of ESM and PPM (19% of 90 samples); however, they do not provide convenient conditions for *R. equi* (0%). The occurrence of *R. equi* (7.4% of 148 samples) in the submaxillary lymph nodes of pigs was not found to be markedly dependent upon the presence of atypical mycobacteria (5% of 148 samples) and vice versa. *R. equi* and PPM seem to be "cooperating", or simultaneously participating in the development of caseous lymphadenitis of pigs. The mycobacterial isolates detected in pigs together with *R. equi* were identified as *M. chelonae*, *M. fortuitum* and *M. xenopi*. The presence of *R. equi* and atypical mycobacteria in the digestive tract of pigs is the result of ingestion (deglutition) of these bacteria with food. The organisms are then transmitted from the intestines – likely by macrophages – to the submaxillary and mesenteric lymph nodes (Takai et al., 1986).

5.5.5.3 Prevention of Infections Caused by *R. equi*

Due to the above-mentioned facts, it is necessary to limit the contact of animals, especially pigs, with arable soil, which is commonly contaminated with *R. equi*. It is also necessary to minimise the supply of the feed contaminated with earth (raw potatoes, carrots, etc.). The surfaces of paddocks for pigs (best covered with concrete or another material) should be kept clean (regular removal of animal faeces).

5.5.6 The Role of Soil in Spreading Obligatory Pathogenic Mycobacteria

Soil, similar to sewage, can be contaminated with highly specialised pathogenic mycobacterial species (*M. tuberculosis*, *M. bovis*, *M. a. paratuberculosis,* etc.) through animal and human faeces. They can be present there for several weeks or months (Garcia and McKay, 1970) and thus survive in soil in slowly decreasing amounts. These pathogenic species with highly specialised host requirements need a temperature of about 37 °C for their multiplication, which is not commonly reached in soil or sewage sediments (Section 5.5.3.3; Beerwerth and Kessel, 1976).

With regard to epidemiological and epizootiological aspects, it is important that the temperature in these substrates does not usually reach 37 °C. However, a number of authors working in human or veterinary mycobacteriology do not share this opinion (Chiodini et al., 1984; Chiodini, 1989). They view the circulation of, e.g., *M. a. paratuberculosis* in the environment as epizootiologically and epidemiologically essential.

Some mycobacterial species such as *M. kansasii* specialised to "water niches" can contaminate soil but cannot survive there for a long time in contrast to the above-mentioned species (Section 3.1.1).

5.5.7 Prevention

Due to the considerably long survival time of mycobacteria in soil, impervious, easily cleanable surfaces in paddocks without sawdust (Section 5.8) and with correctly stored straw used as bedding (Section 5.7.4) should be recommended to farmers (above all for pig breeding). It is recommended to control the contact of children with soil or sand, due to potential geophagia. Adult men, especially immunocompromised patients, should be well informed about the risks posed by frequent contact with soil (high probability of contamination with *MAC* members).

5.5.8 Views and Perspectives on the Research

Researchers have struggled with culture techniques for the detection of mycobacteria in soil for more than 100 years. It is possible that the relatively low interest in this niche, which is likely an important in the ecology of mycobacteria, is due to this factor. Efforts over several decades to modify the decontamination process using different culture media, the incubation conditions (various temperatures, oxygen or carbon dioxide environments) and others have not brought the expected results. This may, at least in part, be due to the occurrence of mycobacteria as different cell-wall-deficient forms in soil, which hinder culture detection: they are living, but "non-culturable strains" (Garcia and McKay, 1970; Beran et al., 2006).

With the passage of time and the development of culture-independent methods (above all PCR techniques) new possibilities have appeared in the field of research into the ecology of mycobacteria in soil. Today, it is becoming evident that the range of mycobacterial species in soil was considerably underestimated using the culture techniques of the past (Cheung and Kinkle, 2001). Accordingly, it is necessary to continue the already initiated efforts to develop culture-independent methods based, in particular, on PCR (Mendum et al., 2000; Nieminen et al., 2006).

Using these culture-independent methods, we can expect that the results obtained in warm and dry regions (e.g. northern parts of India) will be confirmed. Mycobacteria exposed to high temperatures in soil (up to 47 °C) and a dry atmosphere (with only 28% humidity) can become more sensitive to current decontaminating methods (Parashar et al., 2004), which cause the loss of their ability to grow *in vitro* and hence the results of the culture examination are not representative.

From the point of view of human and animal health, the role of the soil mycobacteria and soil on its own is not clear. According to the assumption presented by two Finnish authors (von Hertzen and Haahtela, 2006), the limited contact of the human population with soil is due to urbanisation. This condition leads to the increased occurrence of asthma and atopy cases – hypersensitivity based on genetic predisposition. The research dealing with the ecology of mycobacteria should therefore be focused primarily on ESM and PPM occurrence in soil and their complex effect (concurrently with various components of soil) on the health of people with different genetic predispositions.

5.6 Feedstuffs

I. Pavlik

The animal population also comes into contact with mycobacteria through feedstuffs. The intensity of animal exposure to mycobacteria depends on both the degree of feed contamination and the ability of mycobacteria to propagate in some of the feed components. When considering the ecology of mycobacteria in feedstuffs, the origin of the feedstuffs and their composition is important. According to their origin, feedstuffs can be classified as follows:

i. *Farm feedstuffs* are produced on the same farms on which they are also fed to animals (e.g. silage, mixed feeds, cereals and haylage).
ii. *Industrial feedstuffs* mostly originate from vegetable waste from industrial production (e.g. rapeseed oil cake).

According to their composition, feedstuffs can be classified as follows:

i. *Vegetable feedstuffs* (fodder, root crops, cereals, straw, hay, etc.).

ii. *Animal feedstuffs* (milk, abattoir waste including blood, bone- and fish meal, etc.).
iii. *Synthetic feedstuffs* (protein concentrates, biofactor supplements, etc.).
iv. *Inorganic supplements* (mineral supplements – salts, trace elements, zinc compounds, kaolin, bentonites, wood coal, peat, etc.).

From the point of view of the ecology of mycobacteria, their occurrence is most probable in farm feedstuffs of vegetable origin. It is evident that conditions for the growth of ESM and PPM are more favourable in arable soil and meadows than in forest soil or soil collected from a depth of more than 1 m underground. The conditions for the growth of mycobacteria in surface waters of ponds are better than in sludge or decayed mud (Table 5.13; Beerwerth, 1973). Accordingly, the occurrence of mycobacteria in feedstuffs of vegetable origin much depends upon the techniques used in the harvesting and handling of these feedstuffs and on their further storage. Feedstuffs can be contaminated with mycobacteria at all these stages and the main source of their contamination may sometimes remain hidden.

Mycobacteria are most likely more frequently present in the faeces of herbivores compared to other animals (Section 5.10). Owing to their nutrition (they mainly ingest vegetable feedstuffs) their potential intake of mycobacteria is higher than that of animals fed other feedstuffs (e.g. feedstuffs of animal origin). Nevertheless, both feedstuffs of animal and synthetic origin and inorganic feed supplements may be contaminated with mycobacteria. The ecology of mycobacteria in different feedstuffs according to their composition will be described in the following sections.

5.6.1 Plants as a Feedstuff for Animals

Based on the above-mentioned facts concerning the occurrence of mycobacteria in soil (Section 5.5), water (Section 5.2) and dust (Section 5.9), the relatively frequent isolation of mycobacteria from non-preserved (Table 5.14a) and preserved (Table 5.14b) feedstuffs of vegetable origin is not surprising. Some mycobacteria most likely live in symbiosis with the root system of particular plant species. For example, non-specified

Table 5.13 The growth of mycobacterial species at 22 and 37 °C in media prepared from different types of autoclaved material ([1]; modified)

Runyon's group [1]	Mycobacterial species	Type of soil				Surface water		Decayed mud
		Arable	Meadow	Forest	Underground	Pond	Sludge	
	M. tuberculosis	–	–	–	–	–	–	–
	M. bovis	–	–	–	–	–	–	–
I	*M. marinum*	++	+++	–	–	++	–	+
II	*M. gordonae*	++	+++	–	+	++	–	+
	M. scrofulaceum	+++	+++	++	+	++	–	+
III	*MAA* (2)[2]	++	++	+	+	+	–	–
	MAA (3)[2]	++	++	+	–	+	–	–
	MAH (4)	+++	+++	+	+	+	–	–
	MAH (8)	+++	+	–	–	++	–	–
	M. gastri	++	++	–	+	+	–	–
	M. nonchromogenicum	+	++	–	–	++	–	–
	M. terrae	++	++	+	++	++	–	++
	M. triviale	++	+++	–	++	–	–	+
	M. xenopi	+	+	–	–	+	–	–
IV	*M. smegmatis*	+++	+++	–	–	–	–	–
	M. phlei	+++	+++	–	–	–	–	–
	M. fortuitum	+++	+++	+	+++	+	–	+
	M. vaccae	+++	+++	–	++	+	+++	+++
	M. diernhoferi	+++	+++	–	+	+	+++	+++
	M. flavescens	+++	+++	–	+	+++	+++	+++
pH of substrate used		5.7	5.7	4.0	5.4	5.4	6.0	5.6

[1] Classified according to Runyon's groups (**I** Photochromogenic mycobacterial species, **II** Scotochromogenic mycobacterial species, **III** Non-photochromogenic mycobacterial species and **IV** Rapid growers).
[2] Growth only at 37 °C.
Intensity of growth: – no growth, + separated acid-fast rods and microscopically detected few colony forming units (CFU), ++ microscopic detection of many CFU, +++ CFU strongly detected by the naked eye. *MAA M. a. avium. MAH M. a. hominissuis.* (2) Serotype 2.
[1] Beerwerth W (1973) Ann. Soc. Belg. Med. Trop. 53:355–360.

mycobacterial isolates were obtained along with other bacteria from the underground roots of the terrestrial orchid *Calanthe vestita* var. *rubrooculata* (Tsavkelova et al., 2001). *M. phlei* present in soil and in the roots of the cotton plant (*G. hirsutum*) and pea (*P. sativum*) stimulates their growth (Egamberdiyeva and Hoflich, 2004). Hence, *M. phlei* can be classified among the plant growth-promoting bacteria (PGPR). These PGPR bacteria do not only stimulate the growth of some plants but also protect the plants from the effects of external sources of stress through stimulating the growth of their root system. Mycobacteria can even inhibit soil and plant pathogens.

The occurrence of mycobacteria in moss vegetation on pastures should also be expected. These pose a risk, especially at times when there is a lack of suitable pasture and when the grazing animals, particularly cattle, consume moss due to hunger (Section 5.1.3; Photo 5.8; Cooney et al., 1997).

5.6.1.1 Green Fodder as the Main Unfermented Feed Source

Non-preserved (non-fermented) feedstuffs of vegetable origin can be contaminated with numerous mycobacterial species as follows (Table 5.15):

i. *The root systems of plants are* found in soil, which is one of the substrates containing the highest numbers of mycobacteria (Section 5.5). Therefore, mycobacteria contaminate vegetable feedstuffs via

Table 5.14a Mycobacterial contamination rate in feeding stuffs

Examined feeding stuffs		Examined samples			
Group	Sample specification	No.	Pos.	%	Ref.
Grass, ensilage and haylage	Green fodder	221	67	30.3	[2]
	Green fodder from pastures	120	48	40.0	[2]
	Green fodder from meadows	55	7	12.7	[2]
	Green fodder	94	40	42.6	[5]
	Maize silage	20	3	15.0	[3]
	Fresh green lucerne	2	2	100	[7]
Mash rations, mixed concentrate, lucerne, etc.		52	38	73.1	[8]
Root crop	Root crop	46	12	26.1	[2]
Cereals and their products	Feeding concentrate	511	17	3.3	[1]
	Feeding concentrate	400	13	3.3	[2]
	Feeding concentrate	94	9	9.6	[5]
	Feeding concentrate	270	28	10.4	[9]
	Feeding concentrate for pigs	113	19	16.8	[3]
	Feeding concentrate for pigs	50	2	4.0	[4]
	Unripe cereals	111	4	3.6	[2]
	Cereal powder	15	3	20.0	[6]
	Grains of maize	8	2	25.0	[3]
	Waste of baked goods (pasta)	435	201	46.2	[4]

[1] Beerwerth W, Kessel U (1976) Zentralbl. Bakteriol. [Orig. A] 235:177–183. [2] Beerwerth W, Schurmann J (1969) Zentralbl. Bakteriol. [Orig.] 211:58–69. [3] Cvetnic Z, Kovacic H, Ocepek M (1998) Wiener Tierarztliche Monatsschrift 85:18–21. [4] Dalchow W (1988) Tietärztliche Umschau 43:62–74. [5] Dimov I, Gonzales E (1986) Veterinarni Medicina Nauki, Sofia 23:47–52. [6] Kauker E, Rheinwald W (1972) Berl Munch. Tierarztl. Wochenschr. 85:384–387. [7] Kleeberg HH, Nel EE (1973) Ann. Soc. Belg. Med. Trop. 53:405–418. [8] Kleeberg H, Nel E (1969) J S Afr Vet Med Assoc 40:233–250. [9] Matlova L, Dvorska L, Bartl J, Bartos M, Ayele WY, Alexa M, Pavlik I (2003) Veterinarni Medicina 48:343–357.

Table 5.14b Mycobacterial contamination rate in feeding stuffs

Examined feeding stuffs		Examined samples			
Group	Sample specification	No.	Pos.	%	Ref.
Hay, straw, dry lucerne	Straw	53	27	50.9	[5]
	Straw	19	3	15.8	[1]
	Hay	8	4	50.0	[5]
	Dry and milled lucerne	8	7	87.5	[3]
Feed supplements	Kaolin[a]	88	6	6.8	[4]
	Peat[a]	327	213	65.1	[4]
	Charcoal[a]	28	4	14.3	[4]
Other feedstuffs	Bone- and fishmeal	25	9	36.0	[2]
	Bone- and fishmeal	8	3	37.5	[1]
	Fishmeal	5	3	60.0	[3]

[a] Used as a feed supplement for piglets under two months of age.

[1] Cvetnic Z, Kovacic H, Ocepek M (1998) Wiener Tierarztliche Monatsschrift 85: 18–21. [2] Kauker E, Rheinwald W (1972) Berl Munch. Tierarztl. Wochenschr. 85:384–387. [3] Kleeberg HH, Nel EE (1973) Ann. Soc. Belg. Med. Trop. 53:405–418. [4] Matlova L, Dvorska L, Bartl J, Bartos M, Ayele WY, Alexa M, Pavlik I (2003) Veterinarni Medicina 48:343–357. [5] Pavlas M, Patlokova V (1985) Acta Vet. Brno. 54:85–90.

Table 5.15 Mycobacterial species detected in feedstuffs

Examined material	Mycobacterial species in alphabetical order	Ref.
Plants	*M. gastri, M. intracellulare, M. terrae, M. xenopi,* M. sp.	[6]
Grasses	*M. gordonae, M. kansasii, MAC, M. nonchromogenicum, M. vaccae*	[8]
Green fodder	*M. fortuitum, M. gordonae, M. phlei, M. terrae*	[3]
Plant tissue culture	*M. scrofulaceum*	[9]
Feeding concentrates	*M. fortuitum, M. gordonae, M. smegmatis, M. triviale, M. vaccae*	[3]
Feeding concentrate for pigs	*M. chelonae, M. fortuitum, M. nonchromogenicum*	[1]
Feeding concentrate for pigs	*M. fortuitum, M. terrae*	[2]
Waste of kohlrabi	Rapidly growing mycobacterial species	[4]
Silage and haylage	*M. terrae*	[4]
Silage for deer	*M. chelonae*	[7]
Fishmeal	*MAC, MAA* (2, 3)	[5]
Fishmeal	*M. gordonae*	[1]
Bone meal	*MAC, MAA* (2)	[5]
Protein concentrate from eggs	*MAC, MAA* (2)	[5]
Plant material for pigs	*MAC*[1], *M. terrae, M. gastri, M. xenopi*	[6]
Grains of maize	*M. nonchromogenicum*	[1]
Waste of baked goods (pasta)	*MAH* (8)	[2]
Barleycorn	*MAA*	[5]
Vitamin and mineral supplement for pigs	*M. gordonae*	[1]

[1] Based on the biological trial on chicken, virulent *M. a avium* was not detected.

MAC *M. avium* complex. ***MAA*** *M. a. avium.* ***MAH*** *M. a. hominissuis.* (**2, 3**) Serotypes 2 and 3.

[1] Cvetnic Z, Ocepek M, Kovacic M, Mitak M (1996) Zb. Vet. Fak. Univ. Ljubljana 33: 181–189. [2] Dalchow W (1988) Tietärztliche Umschau 43:62–74. [3] Dimov I, Gonzales E (1986) Vet. Med. Nauki, Sofia 23:47–52. [4] Fodstad FH (1977) Acta Vet. Scand. 18:374–383. [5] Kauker E, Rheinwald W (1972) Berl Munch. Tierarztl. Wochenschr. 85:384–387. [6] Kleeberg H, Nel E (1969) J S Afr Vet Med Assoc 40:233–250. [7] Prosser BA (1989) J. Appl. Bacteriol. 66:219–226. [8] Stanford JL, Paul RC (1973) Ann. Soc. Belg. Med. Trop. 53:389–393. [9] Taber RA, Thielen MA, Falkinham JO, Smith RH (1991) Plant Sci. 78:231–236.

the roots through carelessness during harvesting or via soil present on the surface of, e.g., wheels of tractors when silage is produced.

ii. *Plant stalks* may contain mycobacteria in their vascular bundles. The mycobacteria can penetrate the stalks with water when they are injured (Kauker and Rheinwald, 1972). Another potential source of mycobacterial contamination of plants is their surface, which can get contaminated in many ways. The most frequent way is soil-derived dust which moves with the rain and subsequently sticks to the surface of stalks or other parts of plants.

iii. *The leaves of plants* constitute the largest surface, which can be contaminated (the upper side) with mycobacteria present in raindrops or in fertilisers if applied on leaves. From the underside, leaves are most often contaminated with dust particles of soil-containing mycobacteria. As mentioned above, during rain, large drops of water cause the soil to disperse and the soil particles then mainly stick to the underside of the leaves (Beerwerth and Schurmann, 1969). *M. a. hominissuis* serotype 8/9 was isolated from the upper part of a nettle growing in an aviary for water birds (L.

Dvorska and I. Pavlik, unpublished data; Dvorska et al., 2007).

iv. *Root crops*, especially potatoes, carrots, parsley, etc., may become highly contaminated with soil on their surface if harvested during moist weather (Beerwerth and Schurmann, 1969). Little is known at present about the transmission of mycobacteria to the bulb if, e.g., its surface is injured.

Animals, mainly ruminants and gallinaceous fowl, can ingest mycobacteria at pasture. At the highest risk are the small ruminants due to their modified upper lip which they use to ingest grass along with parts of roots. They also swallow ESM and PPM from soil; these can subsequently affect the tuberculin (e.g. diagnosis of bovine tuberculosis) or serological testing results (e.g. diagnosis of paratuberculosis; Photo 5.30).

The technique of harvesting also affects the occurrence of mycobacteria in green fodder. Due to the relatively frequent occurrence of mycobacteria in soil, these can be much more frequently found in fodder plants, which have been mowed close to the ground.

The handling of soil, plants or their parts in gardens may also become a source of mycobacterial infection for people with small hand injuries. When diagnosing such cases, a mycobacterial origin of the skin (or also lymph node) infection is not usually considered. In many cases, it takes several months before the disease is diagnosed and causal therapy can begin (Holland et al., 1997).

5.6.1.2 Ensilage and Haylage as the Main Fermented Feed Source

Little is known about the effect of green matter fermentation on mycobacteria. It can be supposed that the viability of mycobacteria is not completely destroyed by this process and that living non-culturable strains may be present there. This is likely due to the acid–alcohol fast character of the mycobacterial wall which can protect them for some time. When mown green matter is dumped in silage pits, mycobacteria can access this matter through three basic routes (Photo 5.37):

i. Based on the above-mentioned facts concerning the occurrence of mycobacteria in soil, it is not sur-

prising that mycobacteria are spread together with contaminated plants.

ii. Green matter, which is dumped in a pit and packed, is contaminated with soil present on the wheels of tractors, caterpillar vehicles, etc.

iii. Mycobacteria are brought to silage pits, haylage towers and other facilities together with rain, dust, the faeces of flying birds, etc. (Photo 5.38).

During fermentation, pH changes occur and different acids are produced. All these factors participate in the killing of *M. a. paratuberculosis*, which does not survive longer than 14 days in well-made silage under experimental conditions at 30 and 37 °C. When the tolerance of *M. a. paratuberculosis* to different acids was investigated it was found that they were tolerant to 0.05% (w/v) formic and acetic acids, 0.10% (w/v) lactic and propionic acids and 0.30% (w/v) butyric acid (Katayama et al., 2000).

5.6.1.3 Cereals and Their Products

With regard to the ecology of mycobacteria, it is necessary to view information from Australia concerning the occurrence of mycobacteria in the root system of wheat (*T. aestivum*) as highly noteworthy. It is highly probable that the isolated species (*MAIS* members, *M. aichiense*, *M. bohemicum*, *M. cookii*, *M. flavescens*, *M. heidelbergense*, *M. palustre* and *M.* sp.) not only are soil components but also have symbiotic relationships exclusively with wheat (Conn and Franco, 2004). The question arises which mycobacterial species can occur in the root systems of other plants and which of them are in symbiotic relationship with these plants. In any case, these parts of plants (cereals) also represent a source of mycobacteria, e.g. for straw (see below) or grains.

Nassal et al. (1974) analysed corn products purchased in different bakeries in Heidelberg (oatmeal, groats, flour, etc.). They did not find mycobacteria in any of the 40 examined samples. Beerwerth and Schurmann (1969) isolated mycobacteria from 3.6% of 111 samples of unripe cereals and from 3.3% of 400 samples of corn feedstuff supplied to farmers. The above-mentioned authors explained this relatively low detection rate by the fact that a long distance (usually more than 1 m) separated the corn ears from soil. This helped to prevent contamination of the cereal ear surface with soil particles containing mycobacteria (Beerwerth and Schurmann, 1969; Nassal et al., 1974).

The amounts of mycobacteria in mixed feeds reported in a number of studies were very low; when considering the possible development of gross lesions in the mesenteric lymph nodes of pigs, it is subminimal (Table 5.14a). The contamination of mixed feeds with ESM and PPM may increase through inappropriate handling and storage: e.g. the continual dumping of mixed feeds in feedstuff bins and storage in moist storehouses. The warm season is particularly critical in this regard because significant propagation of mycobacteria may occur as a consequence of increased humidity (Section 7.5; Photo 5.39).

The Importance of Mycobacteria-Contaminated Cereal Products Fed to Pigs

Mycobacteria have also been found in the waste of baked goods, which were fed to pigs (Dalchow, 1988) or cattle (I. Pavlik, unpublished data; Photos 5.40 and 5.41). A study performed on a market pig farm in Germany (Dalchow, 1988) is especially noteworthy as regards ecology; it can serve as an illustration of the above-mentioned risk factors. *M. a. hominissuis* (serotype 8) was isolated from the lymph nodes of pigs that both had and lacked tuberculous lesions. Mycobacteria, however, were not detected either in the waste of baked goods – paste from a factory – or in the components for their production. However, *M. a. hominissuis* (serotype 8) was isolated from 46.2% of feed samples prepared from this waste and subsequently from pig faeces.

This waste was highly contaminated while it was stored on the floor because the lower layer was heated and fermented. According to the author's opinion, it was a consequence of the "compost effect", which most likely provided the optimum conditions for the multiplication of *M. a. hominissuis* (serotype 8) by its warmth, dampness and convenient substrate. This theory was also supported by the subsequent soaking of the waste of baked goods in drinking water for 48 h in which *M. gordonae* was subsequently detected; however, this species was not isolated from either any pig lymph nodes or faeces (Dalchow, 1988). This was likely due to its decreased virulence, which is manifested in particular in immunosuppressed hosts (Section 3.2).

Nevertheless, *M. terrae* and *M. fortuitum*, of which just *M. fortuitum* was detected in the lymph nodes without tuberculous lesions, were found in commercial concentrated diet fed to pigs (Dalchow, 1988). With regard to the ecology of mycobacteria, it is evident that especially *M. a. hominissuis* (serotype 8) can become part of different components of the environment (soil, pig faeces, etc.) under particular conditions and can very soon and to a large extent interfere with the health of animals, particularly pigs (Section 3.1.3; Matlova et al., 2003).

The infection of pigs with PPM (especially *M. a. hominissuis* of serotypes 4, 6, 8 and 9) from contaminated sawdust (Section 5.8) is frequent in cases where pigs are kept on deep bedding. In some herds, sawdust is used for sprinkling the floor in pens for piglets and their broods. The newborn piglets then dry off faster as the floor in the pens is not slippery for them. When alerted to the health risks posed by the use of sawdust and other wood products, some farmers started to use bran for sprinkling the floor (Photo 5.42). However, it was shown that bran was inappropriately stored in mills or bakeries for a long time where it became contaminated with mycobacteria (in particular *M. a. hominissuis*). This resulted in the frequent diagnosis of tuberculous lesions in the lymph nodes of fattened pigs (Section 7.5; Fischer et al., 2004).

5.6.2 Hay and Straw

Pigs kept on dry and clean chopped straw as bedding do not usually develop tuberculous lesions in their intestinal lymph nodes. However, hay and straw mainly become contaminated with soil and dust particles during their harvest, transportation, storage and subsequent handling (Photo 5.43). If hay and straw are stored under unsuitable conditions, the detection rate of mycobacteria may reach up to several tens of per cent of the examined samples (Tables 5.14b and 5.15). Pavlas and Patlokova (1985) isolated *M. terrae*, *M. gordonae*, *M. triviale*, *M. flavescens* and *M. intracellulare* from hay and straw samples.

On one pig farm, *M. a. hominissuis* (serotype 8) and *M. fortuitum* were detected in the head and mesenteric lymph nodes of slaughtered sows, which gave positive responses to avian tuberculin. It was found then that both these mycobacterial species circulated either together or separately in different components of stables and in the surroundings of the farm. In any case, these mycobacterial species were detected in old straw stored in a field stack. The straw was then transported to the loft where these mycobacterial species were detected as well; in the subsequent study, their presence was also revealed in the stables – in scrapings from troughs in the pens and on the floors, in straw residues swept from floors in the halls and pens for pigs. Other places from which mycobacteria were isolated were spider's webs on the windows in the farrowing house, pig faeces mixed with straw in the pens, bedding in stables for non-pregnant sows, and the larvae and adults of Diptera under the feeding places for piglets in the farrowing house (Fischer et al., 2000; Pavlik et al., 2007).

In Table 5.16, an analysis of the literature data concerning the detection of serotypes of different *MAC* members in feedstuffs and the lymph nodes of pigs and cattle in the same country (South Africa and the Czech Republic) is presented. The results indicate that the feedstuffs may be sources of all three *MAC* members. From the point of view of prevention, it is therefore suitable to protect feeds from bird faeces, which may contain the causative agent of avian tuberculosis *M. a. avium* (Section 6.6).

5.6.3 Feeding Supplements

Feeding supplements can be considered as non-typical sources of mycobacteria for animals (especially pigs). Over the last few decades, in particular, peat (Trckova et al., 2005), kaolin and bentonites (Trckova et al., 2004) and wood coal (Matlova et al., 2003) have been used due to their favourable dietetic characteristics. They prevent or alleviate the course of diarrhoea improving the health of the intestinal tract and owing to their absorptive ability they eliminate or reduce the adverse consequences of different toxins. The feeding of supplements to piglets begins most often after birth and continues till weaning. Recently, the feeding of supplements has been practiced much more extensively in some herds; they are given to all animals on some farms. With respect to their frequent contamination with mycobacteria, the most common feeding supplements will be described in the following chapters,

with particular emphasis on the ecology of mycobacteria and their possible health hazard.

5.6.3.1 Peat as a Feed Supplement

Besides its excellent effects on soil as fertiliser (Abbes et al., 1993) peat has favourable effects on the organisms of both people and animals. Namely due to these characteristics, peat has started to be used for both humans and in animal herds for several purposes, which can be often combined:

i. **Peat as an absorbing material.** The absorptive qualities of peat are used in biofilters for the removal of ammonia (Choi et al., 2003), carbon disulphide and hydrogen sulphide (Hartikainen et al., 2001) and leachate from landfill sites (Heavey, 2003). Deep bedding from chopped straw (40%) and peat (60%) have been shown to release significantly less ammonia to the environment in both pig herds (Jeppsson, 1998; Rizzuti et al., 1999; Sutton et al., 1999) and cattle herds (Jeppsson, 1999) and in anaerobic lagoons (Picot et al., 2001).

ii. **Peat as bedding and its favourable effects on skin.** Due to the fact that peat is a soft and highly absorptive material (Abbes et al., 1993), it is especially used as bedding for cattle and pigs (Danilova et al., 1968). It has been found through experiments with various types of bedding used for 11-week-old piglets that peat was preferred by the animals. Other bedding materials are listed according to decreasing popularity as follows: mushroom compost, sawdust, sand, wood bark and straw (Beattie et al., 1998). Owing to its structure and soft character, peat prevents limb abrasions in piglets (Lysons, 1996; Durrling et al., 1998). These characteristics of peat may be due to its water-soluble components (fulvic and ulmic acids), which can penetrate through skin and affect it favourably (Beer et al., 2003a,b).

iii. **Peat as a source of an extract with favourable effects on the respiratory tract.** In Poland, the Tolpa Torf Preparation (TTP) has been tested; it has been shown to be an effective drug in the treatment of recurrent respiratory tract infections (Jankowski et al., 1993).

Table 5.16 Serotype detection of *M. avium* complex members from the environment (including feeding stuffs), feeding concentrates, peat and animals' tissues

Member/serotype		South Africa [3]			Czech Republic			Denmark [2]					
		Environment^a	Pigs	Cattle	Feeding concentrates[4]	Pigs[5]	Cattle[1]	Peat	Pigs	Deer	Cattle	Birds	Humans
M. a. avium	1	1	18	4	0	0	0	0	1	0	0	1	0
	2	2	4	1	1	421	17	0	18	1	1	0	3
	3	0	0	2	0	16	0	0	0	0	0	0	4
M. a. hominissuis	4	10	18	1	2	1	1	5	1	0	0	0	23
	5	0	0	0	0	0	0	0	0	0	0	0	0
	6	0	0	0	1	0	0	2	5	0	0	0	16
	8	10	18	1	12	643	1	2	0	2	0	0	3
	9	0	0	1	3	74	0	3	1	0	0	0	3
	10	0	11	0	0	0	0	0	0	0	0	0	0
	11	0	0	0	0	0	0	0	0	0	0	0	0
	21	0	0	0	1	0	0	0	0	0	0	0	0
M. intracellulare	7	1	0	0	0	0	0	0	0	0	0	0	0
	12	0	1	0	1	0	0	0	0	0	0	0	0
	13	0	0	0	0	0	0	0	0	0	0	0	0
	14	0	0	0	0	0	0	0	0	0	0	0	0
	15	0	0	1	0	0	0	0	0	0	0	0	0
	16	1	0	0	0	0	0	0	0	0	0	0	0
	17	3	0	0	0	0	0	0	0	0	0	0	2
	18	0	1	0	0	0	0	0	0	0	0	0	0
	19	0	0	0	0	0	0	0	0	0	0	0	0
	20	0	0	0	0	0	0	0	0	0	0	0	0
	22–28	0	0	0	0	0	0	0	0	0	0	0	2
Mixed/unknown		7	43	0	7	338	0	5	5	3	0	0	30
Total		35	114	11	28	1 493	19	17	31	6	1	1	86

^a Environmental isolates included feeding stuff isolates as well.

[1] Dvorska L, Matlova L, Bartos M, Parmova I, Bartl J, Svastova P, Bull TJ, Pavlik I (2004) Vet. Microbiol. 99:239–250. [2] Klausen J, Giese SB, Fuursted K, Ahrens P (1997) APMIS 105:277–282. [3] Kleeberg HH, Nel EE (1973) Ann. Soc. Belg. Med. Trop. 53:405–418. [4] Pavlik I, Matlova L, Dvorska L (2008) Unpublished data. [5] Shitaye JE, Parmova I, Matlova L, Dvorska L, Horvathova A, Vrbas V, Pavlik I (2006) Veterinarni Medicina 51:497–511.

iv. **Peat as a feed supplement (Photo 5.44).** Fulvic and ulmic acids also favourably affect smooth musculature, including intestinal tract muscles (Beer et al., 2000; Beer et al., 2002) and some components of the immune system (Bellometti et al., 1997). Their compounds enhance iron uptake by the organism, which is highly important in suckling piglets (Fuchs et al., 1990).

The relatively low pH of peat (4.0–4.5) has bactericidal effects on both pathogenic coliform bacteria and other species of pathogenic bacteria of intestinal microflora. Thus, it can alleviate the course of or completely prevent diarrhoea, especially in piglets (Lenk and Benda, 1989; Framstad et al., 2001). High fibre content supports water absorption from food. The enlargement of the volume of the piglet intestinal system which follows results in the ingestion of higher amounts of feed with a consequent growth acceleration and increased weight gains (Roost et al., 1990; Fuchs et al., 1995).

In the second half of the 1990s, peat started to be used more often for feeding as a supplement for neonatal and growing piglets. Owing to the good experiences of farmers with the preventive feeding of piglets with peat, this raw material started to be generally fed to all newborn piglets in some herds. However, veterinary meat inspections revealed more tuberculous lesions in their lymph nodes after slaughter at the age of 6–7 months (Pavlik et al., 2003).

Favourable conditions for the survival of different ESM and PPM species are formed in peat after extraction. These can even propagate at temperatures between 18 and 20 °C (Kazda, 2000). When different feedstuffs, feed supplements, drinking water and environmental samples from stables were examined,

AFB were most often detected by direct microscopy and culture in peat; the culture positivity was up to 68.2%. Of eight identified mycobacterial species, 81.2% were *M. a. hominissuis* and 1.4% *M. a. avium* (Matlova et al., 2003). Both those *MAC* members can cause tuberculous lesions in the lymph nodes of animals, generally pigs. Primarily in herds where farmers started to use peat to feed all newborn piglets, tuberculous lesions appeared in increased numbers of cases in the intestinal and head lymph nodes of market pigs after slaughter (Pavlik et al., 2003; Matlova et al., 2005).

M. sphagni has been the most often isolated species from peat from natural localities; however, *M. fortuitum*, *M. terrae*, *M. chelonae*, *M. gordonae*, *M. xenopi*, *M. phlei*, *M. marinum*, *M. flavescens*, *M. farcinogenes* and *M. scrofulaceum* have also been detected (Kazda, 2000). These mycobacterial species have also been found in pig organs (Bercovier and Vincent, 2001). It remains obscure why *M. a. hominissuis* predominated in peat from the Czech Republic (Matlova et al., 2003, 2005). In Denmark, *M. a. hominissuis* was also detected in peat and its serotypes were compared with serotypes of the other different *MAC* isolates from different hosts, i.e. pigs, red deer, cattle, birds and people (Table 5.16).

In the Czech Republic, a high similarity between the isolates from peat fed as a supplement and from the lymph nodes of infected pigs was detected by IS*1245* RFLP analysis of *M. a. hominissuis* isolates (Matlova et al., 2005). The originally sterile underground peat (according to the producer), which was most likely contaminated after extraction, was probably the source of *M. a. hominissuis* for pigs. Due to the fact that according to the producer it was transported and further handled "in the open" for several months, its contamination with mycobacteria from dust, water, faeces of animals and other sources from environment is highly likely (I. Pavlik, unpublished data; Photo 5.30).

However, *M. a. avium* was also detected in peat fed to the pigs as a supplement. Even though it was not possible to find its source, it can be supposed that it could be free-living birds or domestic gallinaceous fowl. However, this isolate was not virulent in biological trials on pullets in contrast to isolates from birds with avian tuberculosis (Matlova et al., 2005). Long-term stored isolates of *M. a. avium* in collections or isolates from non-typical hosts (possibly also man) lose

virulence for birds (Photo 3.12; Pavlik et al., 2000). Therefore, it is possible to assume the loss of virulence of this isolate after long survival in contaminated peat (Matlova et al., 2005).

Peat contaminated with mycobacteria may also be a source of ESM and PPM for other products, to which it is merely added. This was most likely the case in the feeding of "pig-compost" as a supplement to pigs in Holland and where peat was added into the communal compost (Section 5.10.3; Engel et al., 1978). High amounts of ESM and PPM have also been isolated from different kinds of soil for the repotting of plants to which peat is also added (I. Pavlik and L. Matlova, unpublished data).

5.6.3.2 Kaolin as a Feed Supplement

Another feed supplement, kaolin, is used in some herds of piglets to prevent diarrhoea (Photo 5.45). Similarly to peat, kaolin has good adsorption properties and is not generally toxic. It is used for the treatment of diarrhoeal and digestive disorders in humans (Heimann, 1984; Kasi et al., 1995; Gebesh et al., 1999). Other characteristics and possibilities for using kaolin have been described in detail in a review article of Trckova et al. (2004). However, it may contain mycobacteria; 6 (6.8%) of 88 examined samples were shown to contain mycobacteria and *M. a. hominissuis* was isolated from two of the positive samples (Matlova et al., 2003).

Kaolin may most likely be contaminated with mycobacteria during its processing, because the extracted raw material (white earth) is usually sterile. An investigation in a kaolin mine found that mycobacteria which contaminated the extracted kaolin came from surface water used for its levigation; mycobacteria were detected in 3 (23.1%) of 13 examined samples of water from a pond in the mine. On the other hand, no mycobacteria were detected in kaolin after extraction. They appeared in the final product after kaolin levigation; mycobacteria including *M. a. hominissuis* were detected in 3 (1.3%) of 230 samples. When kaolin produced in this way was fed to the animals for 6 months, tuberculous lesions in the lymph nodes were detected in 1723 (16.1%) of 10 673 slaughtered market pigs. During one of these months, tuberculous lesions were even detected in 612 (33.1%) of 1851 slaughtered pigs. Two months before the initiation of the feeding of peat as a supplement and 3 months after its ceasing, the occurrence of tuberculous lesions was significantly lower; tuberculous lesions were detected in only 182 (2.4%) of 7675 pigs (Matlova et al., 2004a).

5.6.3.3 Bentonite, Zeolites and Charcoal as Feed Supplements

After the above-mentioned negative experiences with increased incidence of tuberculous lesions in pigs fed peat and kaolin as supplements (see above) some farmers have started to use bentonites and charcoal.

Bentonites are clayey rocks with the prevailing clay mineral montmorillonite from the group of smectites. Zeolites are natural or synthetic crystalline aluminosilicates with ion exchanging properties; clinoptilolite is the best known among the high number of naturally occurring zeolites. For more information on their composition, mining and use see the review article of Trckova et al. (2004). With regard to the ecology of mycobacteria, it is necessary to mention their detection in a commercially available bentonite preparation for animals, especially pigs. Through the examination of samples from an original package, mycobacterial species *M. fortuitum*, *M. a. hominissuis* and *M.* species were isolated from three of them (L. Matlova, I. Pavlik, Z. Zraly, unpublished data).

Some farmers have started to use charcoal as a feed supplement for piglets. Good absorption properties are its main advantage. However, the surface of inappropriately stored charcoal may be contaminated with dust, fungi and mycobacteria from the environment. When handling charcoal, its high absorption capacity must be considered. High numbers of various fungal species have been detected in samples of crushed charcoal used in different pig herds (I. Pavlik, M. Trckova, Z. Zraly, unpublished data). Mycobacteria were detected in 4 (14.3%) of 28 examined samples of charcoal (Matlova et al., 2003).

Accordingly, it is necessary to view all the above-mentioned raw materials as risky from the point of view of feed safety owing to their potential contamination with mycobacteria.

5.6.4 Conclusions, Views and Perspectives on Research

As mentioned above, Beerwerth (1973) performed an *in vitro* study on the effect of various substrates on the growth of different mycobacterial species (Table 5.13). These experiments provided new information, but their importance should not be overstated because all substrates had been autoclaved before use. Therefore, it will be necessary to evaluate the importance of different substrates for mycobacterial growth under natural conditions; it would be illuminating to assess the roles played by different components and subsequently their synergistic or antagonistic effects under natural conditions.

The potential contamination of green fodder, mixed feeds and feed supplements with mycobacteria present in the faeces of free-living birds, in stable dust and in the building for feed preparation or other contaminated facilities used for the preparation of animal feeds should be always considered.

The problem of the evaluation of the quality of peat used in both human medicine and the prevention of animal diseases remains to be solved. For example, in Germany it has been suggested to perform microbiological analysis of peat and "marine mud or Fargo" (peloids) for the bacteria *Escherichia coli*, Coliform organisms, *S. aureus*, *P. aeruginosa* and *Candida albicans* according to the "Definitions of the German Health Resorts Associations" (Eichelsdorfer, 1992). However, mycobacteria are not listed among the monitored bacteria and contamination of these products with ESM and PPM is highly probable in many cases.

5.7 Foodstuffs

I. Pavlik

Mycobacteria enter the human food chain through a number of routes. Foodstuffs intended for human consumption may be contaminated with mycobacteria; for example, vegetables contaminated inside their tissues (Section 5.6) and meat from animals infected by various causative agents such as *MTC* members (especially beef and pork), *MAC* (especially pork and poultry meat) and *M. marinum* and *M. fortuitum* (especially sea food). However, mycobacteria can contaminate the surface of foodstuffs; the most common routes or vehicles of their transmission are contaminated air/aerosols and dust (Section 5.9), water (Section 5.2), soil (Section 5.5), etc. The foodstuffs can be contaminated secondarily not only during their storage (vegetables in cellars, bowls on tables) but also during their processing or preparation.

However, as regards heat treatment, there are basic differences between human food and feed for animals which usually ingest feed in its natural form. People usually eat heat-treated foodstuffs and a smaller proportion of raw food. When people consume fruit and vegetables it is always mechanically cleaned from soil and other organic remnants and washed with water (Nassal et al., 1974). In contrast to people, animals ingest much higher amounts of vegetables that are mostly contaminated with earth (Sections 5.5 and 5.6). This fact is reflected in the lower occurrence of mycobacteria in the stools of people in comparison with, e.g., faeces from the majority of domestic animals (Section 5.10).

As far as the ecology of mycobacteria is concerned, further handling of the foodstuffs and their remnants is crucial. Mycobacteria from foodstuffs are not involved in ecology after the heat treatment of foodstuffs at temperatures that destroy mycobacteria. Insufficiently heat-treated (e.g. cold-smoked salmon or various special sausages and salamis) or raw foodstuffs can also become sources of mycobacteria in the environment. They enter the environment having passed through and being shed from the intestinal tract of people (Section 5.10).

Organic remnants and leftovers of foodstuffs can play an important role in the ecology of mycobacteria. They are directly discarded in waste dumps and effluents, or are composted or otherwise fermented (Section 5.10.3). This is how ESM can return to the environment (their original habitat) together with PPM and OPM from foodstuffs of vegetable and animal origin. All these routes of the spread of mycobacteria from contaminated foodstuffs and their remnants have only been partly studied so far, even though some information can be found in the available literature. One of the aims of the present chapter is therefore to describe current knowledge of the importance of foodstuffs in the ecology of mycobacteria.

5.7.1 Fruit and Vegetables

Mycobacteria have also been detected in fruit and vegetables of different origin; mycobacteria were found in 21 (17.4%) of 121 samples (Tables 5.17 and 5.18; Argueta et al., 2000).

Some specialists assume that mycobacteria from soil are transported to plants via osmotic pressure through the root system of plants. Particular attention has been concentrated on the *MAC* members. Rheinwald (1972) isolated *M. a. hominissuis* from aseptically collected sawdust from the wood of spruce, oak and beech. Thus, the author concluded that mycobacteria were transported from soil via osmotic pressure through roots to the tissues of trees. This hypothesis was confirmed by experiments on nettles (*genus Urtica*); grown nettle stalks were cut and submerged in water containing mycobacteria. Culture examination revealed their subsequent penetration via osmotic pressure to the capillary net of the nettles (Kauker and Rheinwald, 1972). Based on the above-mentioned observations, Nassal et al. (1974) focused on the occurrence of mycobacteria in 1000 samples of 24 species or types of fruit and 26 species or types of vegetables. They examined 20 samples from every commodity, of which 10 samples were collected before and 10 samples after washing.

5.7.1.1 Naturally Contaminated Fruit

Non-photochromogenic mycobacteria from fruit (group III according to Runyon) were isolated from 5 (1.0%) of 480 samples from 2 (8.3%) of 24 species. These two species, i.e. garden strawberries and rhubarb, grow very close to the ground, from where the mycobacteria most likely originated. In fruit, which grows at greater distances from the soil surface (e.g. on shrubs or trees), no mycobacteria were found (Table 5.17; Nassal et al., 1974). The five above-mentioned positive samples were collected from unwashed fruit (Nassal et al., 1974); with respect to food safety, thoroughly washing fruit with tap water is sufficient (fruit examined in this study was washed five times with water). Even though small amounts of mycobacteria are also found in drinking water (Section 5.2), their role is most likely small if fruit is washed well.

In the study of Argueta et al. (2000) *M. scrofulaceum* and *M. simiae* were isolated from non-heat-treated juices and ciders from one manufacturer (Table 5.18). With regard to the fact that both

Table 5.17 Mycobacterial contamination rate in foodstuff

Examined feeding stuffs		Examined samples			No. of isolates	Ref.
Group	Samples specification	No.	Pos.	%		
Vegetable	Fresh vegetable, different kinds	121	7	5.8	7	[9]
	Fresh mushrooms, salads, lettuce, cider, juice	121	21	17.4	29	[1]
Milk	Raw bulk cows' milk – winter	351	175	49.9	175	[2]
	Raw bulk cows' milk – spring	207	51	24.6	51	[2]
	Raw bulk cows' milk – summer	212	35	16.5	35	[2]
	Raw milk	193	77	39.9	77	[5]
	Farm vats	88	6	6.8	6	[3]
	Farm pick-up tankers	97	26	26.8	26	[3]
	Storage tanks and transport tankers	103	45	43.7	45	[3]
	Pasteurised milk	76	0	0	0	[3]
	Pasteurised milk	43	0	0	0	[4]
	Raw milk	51	35	68.6	64	[4]
	Cows with mastitis	2035	46	2.3	46	[6]
	Dairy creamery effluent sludge	63	27	42.9	32	[7]
	Industrial centrifugation deposits of cows' milk	84	40	47.6	40	[8]
Total		3845	591	15.4	633	

[1] Argueta C, Yoder S, Holtzman AE, Aronson TW, Glover N, Berlin OG, Stelma GN, [Jr., Froman S, Tomasek P (2000) J. Food Prot. 63:930–933. [2] Chapman JS, Bernard JS, Speight M (1965) Am. Rev. Respir. Dis. 91:351–355. [3] Dunn BL, Hodgson DJ (1982) J. Appl. Bacteriol. 52:373–376. [4] Hosty TS, McDurmont CI (1975) Health Lab Sci. 12:16–19. [5] Jones RJ, Jenkins DE, Hsu KH (1966) Can. J. Microbiol. 12:979–984. [6] Koehne G, Maddux R, Britt J (1981) Am. J. Vet. Res. 42:1238–1239. [7] Matthews PR, Collins P, Jones PW (1976) J. Hyg. (Lond) 76:407–413. [8] Tacquet A, Tison F, Devulder B, Roos P (1966) Ann. Inst. Pasteur Lille. 17:173–179. [9] Yajko DM, Chin DP, Gonzalez PC, Nassos PS, Hopewell PC, Reingold AL, Horsburgh CR, Jr., Yakrus MA, Ostroff SM, Hadley WK (1995) J. Acquir. Immune. Defic. Syndr. Hum. Retrovirol. 9:176–182.

these species cause human diseases, this finding can be viewed as serious.

From the point of view of epidemiology, the fact that coprophagous flies sit on plants and above all on ripening fruit is serious. They seek the sugars and vitamins there, which they need for the developing ova, often laid into the faeces (Bejsovec, 1962). Among fruit, they prefer tomatoes, pears, peaches and bananas, less so cherries, sour cherries, grapes, raspberries and red currants. Vegetables are also attractive for them (O. A. Fischer, personal communication; Gawaad et al., 1997). Thus, they can become vectors for ESM, PPM or OPM (Section 6.3.5).

5.7.1.2 Naturally Contaminated Vegetables

Nassal et al. (1974) isolated non-photochromogenic mycobacteria from fresh vegetables from 9 (1.7%) of 520 samples, from 5 (19.2%) of 26 kinds of vegetables (Table 5.17) growing close to the ground or directly in soil: potatoes, radishes, chicory, lettuce and parsley (Table 5.17; Nassal et al., 1974). With regard to

food safety, it is important that mycobacteria were also isolated from the same vegetables washed five times. Washing the vegetables did have a certain effect, however; lower amounts of mycobacteria were detected in washed compared to unwashed vegetables (Nassal et al., 1974).

A later publication (Yajko et al., 1995) described a relatively low occurrence of mycobacteria in fresh vegetables. Mycobacteria were isolated from 7 (5.8%) of 121 samples of fresh vegetables; only one (0.8%) of the isolates was classified as belonging to the *MAC*. However, in the study of Argueta et al. (2000) different species of mycobacteria were isolated from two kinds of lettuce and from leeks, parsley and sweet basil. This indicates that vegetables may pose a risk, especially if consumed raw.

When mycobacteria were investigated in the household of a long-term oncology patient in the Czech Republic (17-year-old boy), it was found in the anamnesis that he was given only heat-treated food according to the doctor's recommendations. Only the scrambled eggs were seasoned with green parsley all year around. That parsley was planted in the garden in an earthen pot. *M. engbaekii* was isolated from the soil

Table 5.18 Mycobacterial species detection in fruits, vegetables, cereals and mushrooms

Examined samples		Isolates	Ref.
Vegetables	Fresh unspecified kinds	*MAC*	[4]
	Lettuce	*MAH*	[3]
	Potatoes, white charlock, chicory, parsley	Non-photochromogenic (III Runyon)	[2]
	Carrots, garlic, kohlrabi, celery, onion, lettuce salad, garden cress, melon, red pepper, rhubarb, red cabbage, cucumber, chive, white cabbage	Negative culture for mycobacteria	[2]
	Packed salad and packed broccoli salad	*MAC*,RGM	[1]
	Romaine lettuce, red leaf lettuce, spinach	*MAC*,RGM, *M. gordonae* V	[1]
	Leeks	*MAC*, RGM, *M. gordonae* VI	[1]
	Parsley, rhubarb Swiss chard	RGM	[1]
	Basil	*M. genavense, M. gordonae* III	[1]
	Apple pomegranate juice[1]	*M. scrofulaceum*	[1]
	Apple cider, apple berry juice, apple cherry juice[1]	*M. simiae*	[1]
Fruits	Strawberries, rhubarb	Non-photochromogenic (III Runyon)	[2]
	Cucumber, bilberries, raspberries, red currant, black currant, gooseberries, blackberry, tomato, white grapes, blue grapes, orange, grapefruit, lemon, banana, elderberries, pear, apricot, sweet cherry, Mirabelle, peach, plum, apple	Negative culture for mycobacteria	[2]
Cereals	Oat flakes, rye flour, wheat flour, feeding concentrate	Negative culture for mycobacteria	[2]
Mushrooms	Portobella and Monterey mushrooms	*MAC, M. flavescens* II	[1]
	Daikon sprouts, Italian brown mushrooms	*MAC*	[1]

[1] Unpasteurized juice from the same manufacturer.
RGM non-specified rapidly growing mycobacteria. *MAC M. avium* complex member not specified. *MAH M. a. hominissuis.*
[1] Argueta C, Yoder S, Holtzman AE, Aronson TW, Glover N, Berlin OG, Stelma GN, Jr., Froman S, Tomasek P (2000) J. Food Prot. 63:930–933. [2] Nassal J, Breunig W, Schnedelbach U (1974) Prax. Pneumol. 28:667–674. [3] Sequeira PC, Fonseca LS, Silva MG, Saad MH (2005) Mem. Inst. Oswaldo Cruz 100:743–748. [4] Yajko DM, Chin DP, Gonzalez PC, Nassos PS, Hopewell PC, Reingold AL, Horsburgh CR, Jr., Yakrus MA, Ostroff SM, Hadley WK (1995) J. Acquir. Immune. Defic. Syndr. Hum. Retrovirol. 9:176–182.

and from the green parsley haulm (V. Mrlik, J. Sterba, I. Pavlik I., unpublished data; Photo 5.46).

5.7.1.3 The Impact of Experimentally Contaminated Soil with *M. a. paratuberculosis* on the Contamination of Planted Vegetables

Three groups of plants were used (lettuce, radish and tomato) which were grown for 2–3 months in conventional soil. Subsequently, *M. a. paratuberculosis* was added to one and *M. a. hominissuis* of serotype 8 to another group of plants. All biological sam-

ples from the control group (the group of plants which were not contaminated with mycobacteria) of plants were negative for *M. a. paratuberculosis* and *M. a. hominissuis*. From the group of plants contaminated with *M. a. paratuberculosis*, the same agent of identical IS*900* RFLP type B-C1 (Pavlik et al., 1999) was isolated from the soil, roots, leafs, "head" (surface) and "stalk" (inside) of all plants. Similar results were observed in the last group of plants contaminated with *M. a. hominissuis*. This causative agent was isolated from all biological samples examined (Table 5.19).

The risks of planting vegetables on substrates, which contain high concentrations of mycobacteria,

Table 5.19 Impact of mycobacterial contamination (CFU) of soil on mycobacteria detection in three kinds of vegetables [1]

| Examined samples | | Lettuce[1] | | | Radish[2] | | | Tomato[3] | | |
Material	Location	C	MAP	MAH	C	MAP	MAH	C	MAP	MAH
Soil	Surface	0	3	20	0	2	1	0	1	20
	Middle (10 cm)	0	200	200	0	100	100	0	100	100
	Bottom (20 cm)	0	2	20	0	200	20	0	1	10
Roots		0	10	20	0	20	30	0	1	10
Leafs	Lower	0	1	50	0	5	20	0	0	10
	Middle	0	1	30	0	Nt	Nt	0	0	20
	Upper	0	2	10	0	1	60	0	0	50
"Head"	Surface	0	5	30	0	10	50	0	0	20
"Stalk"	Inside	0	0	100	0	2	20	0	1	20

[1] *Lactuca sativa.*
[2] *Raphanus sativus.*
[3] *Solanum lycopersicum.* **Nt** Not tested.
CFU Colony forming units. **C** Control group No. 1 with planted plants in soil without *M. a. paratuberculosis (MAP)* and *M. a. hominissuis (MAH)*. **MAP** Group No. 2 with planted plants in soil with **MAP**. **MAH** Group No. 3 with planted plants in soil with **MAH** (serotype 8). [1] Pavlik I, Rozsypalova Z, Moravkova M (2008) Unpublished data.

can be assessed from the obtained results. In such cases, thorough washing of vegetables before consumption cannot protect the consumers from infection. These results are also in accordance with Nassal et al. (1974) who detected mycobacteria in potato and radish pulp and Laukkanen et al. (2000) who isolated mycobacteria from aseptically collected tissues of Scots pine (*Pinus sylvestris*).

5.7.2 The Relationship of Plants and Their Products with Mycobacteria

The relationship between plants, especially their roots, and mycobacteria was described previously (Section 5.6.1). However, it has been found in different parts of the world that extracts from some plants (that have not been identified yet) have different relationships with mycobacteria. These components can both stimulate and destroy the growth of mycobacteria.

5.7.2.1 The Stimulating Effect of Some Plants and Their Products on the Growth of Mycobacteria

The growth of *M. ulcerans* was stimulated *in vitro* by an extraction from two green algae *Rhizoclonium* sp. and *Hydrodictyon reticulatum* (Marsollier et al.,

2004). "Sugar cane mud" formed during the production of sugar cane is named "cachaza" in Cuba. This is an important contaminant, especially in areas with intensive cultivation systems and production of sugar cane. It was found that mycobacteria are present in "cachaza" that can bioconvert the "sugar cane mud" phytosterols (Perez et al., 2005).

It is, however, necessary to admit that the stimulatory/inhibitory effects of higher plants on a variety of mycobacteria, i.e. ESM, PPM and OPM remain obscure. Accordingly, further research should be focused on this issue. In particular, the simulating effect on mycobacteria of different agricultural plants used as feed for animals should be investigated. Concerning the human population, it will be important to investigate the relationships between mycobacteria and various kinds of fruit and vegetables with respect to composting, which today is an important ecological way of waste disposal from cities and agglomerations.

Their effect on mycobacteria found in human organisms should also be studied. This issue is timely, above all due to the fact that the consumption of raw fruit and vegetables may cause complications, e.g. during the treatment of human tuberculosis. For example, it was found that after the consumption of orange juice, the bioavailability of clofazimine antibiotics was decreased in human patients (Nix et al., 2004).

The relationship between a vegetarian diet and the incidence of tuberculosis in humans was investigated by Chanarin and Stephenson (1988) in England. They detected, by means of a survey of 1187 inhabitants who came from India (Asiatic Indians), that the incidence of tuberculosis in vegetarians was 133 of 1000 men compared with people eating mixed diets in whom the incidence was only 48 of 1000 men. They assumed that the frequent deficiency of cobalamin in the vegetarian food primarily influenced the increased incidence of tuberculosis in vegetarians.

5.7.2.2 The Antimycobacterial Effect of Some Plants and Their Products on the Growth of Mycobacteria

Some plant components exert an antimicrobial effect on various bacterial species, including mycobacteria. An inhibitory effect of garlic (*Allium sativum*) extract on the growth of different isolates of *M. tuberculosis* (Delaha and Garagusi, 1985) and on the growth of *M. smegmatis* (Naganawa et al., 1996) has been described. The antimycobacterial effect of phorbol esters in fruit of *Sapium indicum* has also been detected (Chumkaew et al., 2003).

The antimycobacterial effects of different commercially tested juices (cordials) from raspberries, black currants, cranberries and blackberries on *M. phlei* (Cavanagh et al., 2003) are worthy of mention. They were found to reduce the contamination of these products with ESM or PPM to a certain degree; however, this should be tested on other mycobacterial species.

Honey is considered as a dietetically valuable food. Its antimycobacterial effects have been studied *in vitro*. It was found that honey adversely affected the growth of the following species of mycobacteria: *M. tuberculosis*, *M. bovis* BCG, the *MAC*, *M. kansasii*, *M. gordonae* and *M. marinum* (Ulker, 1967). The effect of different concentrations of honey on the growth capability of different isolates of *M. tuberculosis* on the surface of Lowenstein–Jensen agar was tested. No inhibition of the growth was observed after the addition of 1.0, 2.5 and 5.0% of honey (Asadi-Pooya et al., 2003).

5.7.3 Cereals and Their Products

The results of culture examination of cereals and their products have been reviewed in the previous chapters and in Tables 5.17 and 5.18. The negative result from testing cereal products (oatmeal, cornmeal, flour, etc.), purchased from different bakeries in Heidelberg (Nassal et al., 1974), is noteworthy. Even though properly heat-treated products (most often baked and boiled) are usually made of these raw materials, their importance in the transmission of mycobacteria is limited. It is the subsequent careless handling of these products which can lead to additional contamination with mycobacteria (Section 5.6; Dalchow, 1988).

However, new risks may be posed by various specialities and dishes that are prepared from insufficiently heat-treated or non-heat-treated cereals or legumes. The consumption of various sprouted cereals is also popular. A study dealing with survival (maintenance of virulence) of the avian tuberculosis causative agent, *M. a. avium*, in the following artificially contaminated cereals and legumes (Abdullin et al., 1969) revealed that they survived in barley grains for 1398 days (virulence was maintained during the entire period), in wheat grains for 976 days (virulence for 883 days), in pea grains for 976 days (virulence for 883 days) and in all of these grains that had been ground and mixed together for 793 days (virulence for 700 days).

The ability of *M. tuberculosis* to survive in the traditional dish "poi" was investigated. It is made of starch from a large underground rootstock of a perennial plant "elephant ear". Before artificial infection, the crushed starch was mixed with water and stored at room and refrigerator temperature. *M. tuberculosis* survived for only 3 days at room temperature, but for up to 11 days at refrigerator temperatures. The main cause of *M. tuberculosis* killing was the decreasing pH during the fermentation of "poi", which reaches the value of 3.5 after only 3 days at room temperature. With respect to the fact that "poi" is consumed after cooking or steam stewing, the hazard of the spread of mycobacteria is relatively low (Ichiriu and Bushnell, 1950).

5.7.4 The Spread of Mycobacteria by Milk and Milk Products

Ruminant milk is regarded, especially in economically less developed countries, as the primary source of protein and is sometimes referred to as "meat for children". Also, milk is often consumed raw immediately after milking because cooling it is difficult without available instrumentation and electric power, above all in the tropics and subtropics. The risk of transmission of *M. bovis* through milk from infected ruminants has been described (Thoen and Steele, 1995; Thoen et al., 2006). Nevertheless, the purpose of this book is not to deal with this issue in detail.

5.7.4.1 Milk

As far as the ecology of mycobacteria is concerned, the contamination of milk, primarily with ESM and PPM, can be assumed (Photo 5.47). Different species may originate from both the environment (mostly ESM) and from infected cows (mostly PPM).

The effect of the seasons of the year has been described. In the cool months (i.e. in winter in the Northern hemisphere) when vegetation is sparse, animals usually gather in the feeding places, or are kept in the stables. Based on the examination of 770 samples of milk during the winter, spring and summer seasons in Australia, it was found that the mycobacteria content of milk was significantly higher in the hot summer months (Chapman et al., 1965). This finding can be explained by the multiplication of ESM and PPM in the environment, from which they most likely easily contaminated the milk.

The fact that PPM species are more resistant to increased temperatures than is *M. bovis* poses a problem for food safety. For instance, Harrington, Jr. and Karlson (1965) found that *M. tuberculosis*, *M. bovis*, *M. avium*, *M. fortuitum* and *M. bovis* BCG did not survive at 62.8 °C for 30 min and at 71.7 °C for 15 min. On the other hand, some photochromogenic and fast-growing mycobacterial isolates survived 62.8 °C for 30 min and some scotochromogenic and non-chromogenic mycobacteria survived 71.7 °C for 15 s. The highest thermoresistance at 71 °C was detected in *M. phlei*, which survived during the entire period of investigation, i.e. 80 s. All the remaining tested species *M. borstelense* (*M. chelonae*), *M. diernhoferi*, *M. fortuitum*, *M. smegmatis* and *M. vaccae* survived this temperature for only 5 s (Schliesser et al., 1972).

Grant et al. (1996) obtained similar results: *M. bovis* was killed after 20 min and *M. fortuitum* after 10 min at 63.5 °C; on the other hand, *M. a. avium* (strain NCTC 8552 of serotype 2), *M. intracellulare* and *M. kansasii* survived this temperature for 30 min.

Both clinically normal females and females affected by mycobacterial mastitis can shed ESM or PPM through milk via the blood or lymphatic system (Tables 5.17 and 5.20), even though milk may lack concurrent visible changes. In advanced stages of mastitis, pus or blood flakes may be seen in milk. Mastitis caused by ESM or PPM occurs less commonly and its diagnosis is difficult because the mycobacterial origin of mastitis is often not considered. However, when mycobacteria are massively shed, the diagnosis is usually mycobacterial mastitis (Schultze et al., 1985; Schultze and Brasso, 1987).

5.7.4.2 Human Breast Milk

The shedding of mycobacteria through milk has also been detected in women. In the late 1960s and in the late 1990s, the shedding of *M. leprae* (Pedley, 1967) and *M. a. paratuberculosis* (Naser et al., 2000) through milk was documented, respectively.

5.7.4.3 Dairies

Dairy creamery effluent sludge can be a significant source of mycobacteria in the environment, e.g. in fields (Matthews et al., 1976). Among the ESM and PPM species, causative agents of, e.g., skin diseases of people and animals, *M. marinum* and *M. fortuitum*, have been detected: (Tables 5.17 and 5.18).

5.7.4.4 Dairy Products

Mycobacteria can be spread by milk products. In the middle of the last century, research was focused on the survival of *M. bovis* in cheese as seen in the review of Keogh (1971). At the beginning of this century, attention was concentrated on the investigation of the occurrence of *M. a. paratuberculosis* in different types of cheese. After artificial contamination of cow's milk with *M. bovis*, which was used for the production of three types of cheese, significant differences were found in the experiments on guinea pigs. Only in Emmental cheese, was *M. bovis* immediately killed after reaching the temperature of 55 °C at the beginning of production. In Camembert cheese, however, *M. bovis* was killed after 60 days

Table 5.20 Mycobacterial species detected in cows' milk

Examined samples	Mycobacterial species (available numbers of isolates)	Ref.
Fresh bulk milk	*M. fortuitum, M. kansasii, M. sp.*	[2]
Farm vats	*M. flavescens, M. fortuitum, M. intracellulare*	[3]
Farm pick-up tankers	*M. flavescens, M. fortuitum, M. gordonae, MI, M. scrofulaceum, M. terrae, M. thermoresistibile*	[3]
Storage tanks and transport tankers	*M. flavescens, M. fortuitum, M. gastri, M. gordonae, MI, M. scrofulaceum, M. smegmatis, M. terrae*	[3]
Raw milk	*MAC, M. fortuitum, M. flavescens, M. gordonae, M. marinum, M. parafortuitum*complex, *M. scrofulaceum, M. terrae* complex	[4]
Raw milk	*M. fortuitum* (14 isolates), *M. phlei* (45 isolates), *M. smegmatis* (1 isolate), scotochromogens (14 isolates), Non-chromogenes (9 isolates)	[6]
Raw milk	*M. a. paratuberculosis*	[1]
Baby formula	*M. a. paratuberculosis*	[5]
Dairy creamery effluent sludge	*M. fortuitum* (13 isolates), *M. gordonae* (5 isolates), *M. peregrinum* (6 isolates), *M. marinum/M. scrofulaceum* (4 isolates), *M.* sp. (4 isolates)	[8]
Industrial centrifugation sediments of cows' milk	*MAA* (8 isolates), *M. bovis* (20 isolates), *M. gordonae* II (2 isolates), *M.* sp. pigment (10 isolates)	[11]
Cows with mastitis	*M. fortuitum* (21 isolates), *M. flavescens* (1 isolate), *M. vaccae* (5 isolates), RGM (19 isolates)	[7]
Cows with mastitis	*M. smegmatis, M.* sp.	[9]
Cows with mastitis	*M. smegmatis*	[10]
Cows with mastitis	*M. smegmatis*	[12]
Cows with mastitis	*M. chelonae, M. fortuitum*	[13]

RGM non-specified rapidly growing mycobacteria. ***MAC*** *M. avium* complex. ***MI*** *M. intracellulare.* ***MAA*** *M. a. avium.*

[1] Ayele WY, Svastova P, Roubal P, Bartos M, Pavlik I (2005) Applied and Environmental Microbiology 71:1210–1214. [2] Chapman JS, Bernard JS, Speight M (1965) Am. Rev. Respir. Dis. 91:351–355. [3] Dunn BL, Hodgson DJ (1982) J. Appl. Bacteriol. 52:373–376. [4] Hosty TS, McDurmont CI (1975) Health Lab Sci. 12:16–19. [5] Hruska K, Bartos M, Kralik P, Pavlik I (2005) Veterinarni Medicina 50:327–335. [6] Jones RJ, Jenkins DE, Hsu KH (1966) Can. J. Microbiol. 12:979–984. [7] Koehne G, Maddux R, Britt J (1981) Am. J. Vet. Res. 42:1238–1239. [8] Matthews PR, Collins P, Jones PW (1976) J. Hyg. (Lond) 76:407–413. [9] Richardson A (1970) Vet. Rec. 86:497–498. [10] Schultze WD, Brasso WB (1987) Am. J. Vet. Res. 48:739–742. [11] Tacquet A, Tison F, Devulder B, Roos P (1966) Ann. Inst. Pasteur Lille. 17:173–179. [12] Thomson JR, Mollison N, Matthews KP (1988) Vet. Rec. 122:271–274. [13] Wetzstein M, Greenfield J (1992) Can. Vet. J. 33:826.

and in Edamer cheese, *M. bovis* was fully virulent for guinea pigs during the entire experimental period of 80 days (Frahm, 1959). In Germany *M. bovis* was not detected in 100 samples of Emmental cheese; however, if the cheese was produced from artificially contaminated milk, the *M. bovis* was still virulent 3 months after infection of the guinea pigs (Hahn, 1959).

The causative agent of paratuberculosis, both killed (Ikonomopoulos et al., 2005; Clark, Jr. et al., 2006) and live (Ikonomopoulos et al., 2005), was detected in

milk using the IS*900* PCR. A review by Grant (2005) describes the occurrence of *M. a. paratuberculosis* not only in non-pasteurised but also in pasteurised milk and in cheese made from them. The IS*900* element, specific for *M. a. paratuberculosis*, has also been detected in "semi-hard Swiss cheeses", even though subsequent culture did not reveal the presence of live cells of the causative agent of paratuberculosis (Stephan et al., 2007). This fact may be associated with the presence of live, but "non-culturable" bacteria among others, related to potential occurrence of L forms of mycobacteria (Beran et al., 2006).

An investigation of the effect of cheese ripening on the survival of *M. a. paratuberculosis* found that increasing concentrations of NaCl did not significantly influence the survival of the causative agent of paratuberculosis. However, an initial pasteurisation of milk caused more rapid killing of *M. a. paratuberculosis* with decreasing pH. The devitalisation of *M. a. paratuberculosis* during the 60-day period of ripening cheese made from pasteurised milk (concentrations lower than 10^3 CFU/ml) was documented (Sung and Collins, 2000). However, in hard and semi-hard types of cheese *M. a. paratuberculosis* with initial concentrations of between 10^4 and 10^5 CFU/ml was not killed even after 120 days of ripening (Spahr and Schafroth, 2001).

After the experimental infection of yogurt, it was found that *M. tuberculosis*, *M. bovis* and *M. bovis* BCG were killed after 18–24 h, provided that the pH decreased during fermentation. On the other hand, in neutralised yogurt, *M. tuberculosis* (strain H37R$_v$) survived during the entire period of 6-day monitoring (Tacquet et al., 1961). The causative agent of paratuberculosis, *M. a. paratuberculosis*, was found to survive in yogurt for an entire 20-day period of an investigation (I. Pavlik and Z. Rozsypalova, unpublished data).

5.7.5 Meat, Parenchymatous Organs and Their Products as a Source of Mycobacteria

Meat can be involved in the ecology of mycobacteria in different ways. As described previously (Grange and Yates, 1994; Thoen and Steele, 1995; Grange, 1996; Thoen et al., 2006), meat can become infected with mycobacteria during different mycobacterial infections (e.g. bovine tuberculosis).

The purpose of this book is not to describe the courses of different mycobacterial diseases and the localisation of infection in meat and other parts of the bodies of slaughtered animals that are sources of meat products. Rather, this section will describe how mycobacteria from meat and meat products can be involved in ecology. The three main ways of their spread in the environment and the routes of contamination of meat and meat products will be discussed.

5.7.5.1 Slaughterhouses and Meat Factories

It seems that slaughterhouses and to a lesser extent meat-processing plants are sources of mycobacteria in the environment. On the one hand, these plants produce meat (most often beef, pork, mutton/lamb, venison and poultry) and meat products for human consumption; these are usually heat treated which can significantly reduce or completely devitalise all mycobacterial species (preservation of meat by freezing does not destroy mycobacteria). On the other hand, the slaughterhouses and meat-processing plants also produce large amounts of:

i. Organic remnants (the content of the intestines and forestomachs of ruminants, intestines of other animals including blood, urine, etc.).
ii. Wastewater (containing – besides contaminated rinsing water – small pieces of organic remnants, hair, bristles and various bodily fluids, i.e. most often haemolysed blood, urine, the content of the intestines and bile).

In developed countries, organic remnants are usually safely liquidated in rendering plants. In less economically developed countries, waste is discarded into rivers, lakes or seas. Their storage in sewage lagoons and lakes, from where they are subsequently drawn as fertilisers for fields, is not unusual. In some countries with fish farming industry (especially in countries of Southeast Asia), organic remnants are used as feed for the fish.

In first and second world countries, wastewater passes to the wastewater treatment plants. The most important parts of sewage are the "splash" waters, which may sometimes constitute a great technology problem. These occur if the capacity of the wastewater treatment plants is insufficient to ensure complete purification. The wastewater from the slaughterhouses can also become a source of mycobacteria during transport and during its application in the environment, e.g. in fields or pastures (Section 5.10.3).

With respect to the ecology of mycobacteria, biofilms formed in the slaughterhouses and meat-processing plants are most important; these have been described previously (Section 4.7). Biofilms are formed on different surfaces in these premises, including working surfaces (e.g. tables and production lines), protective clothing of personnel and tools (e.g. saws and knives). Contaminated knives used for the collection of samples for laboratory analysis in slaughterhouses may under certain conditions be significant sources of mycobacteria (Norton et al., 1984). Mycobacteria present in biofilms can in return contaminate meat, parenchymatous organs and the final meat products (Table 5.21; J. E. Shitaye, A. Horvathova and I. Pavlik, unpublished data).

Accordingly, it is necessary to ensure that surfaces in abattoirs are as much as possible free of organic remnants and biofilms, in which mycobacteria owing to their hydrophobicity can be easily retained (Table 5.21) and multiply under favourable conditions (Section 4.7).

Table 5.21 Examples of mycobacterial detection in meat (flash) and parenchymatous without tuberculous or tuberculoid lesions in pigs[1] and the environmental samples from pig slaughterhouses

Tissue	Mycobacterial species	Ref.
Meat	*MAA*, *M. bovis*	[2]
	MAA, *M. fortuitum*, *M.* sp.	[4]
	MAA	[5]
Liver	*MAH* (3, 8, 9)	[1]
	MAA	[5]
	MAA	[5]
Spleen	*MAH* (3, 8, 9)	[1]
Kidney	*MAA* (2)	[3]
Curded blood mixed with wastewater	*M. intermedium*, *M. szulgai*	[6]
Bristle residue from splinter	*M. kansasii*, *M.* sp.	[6]
Meat leftovers from laser needle for fat control	*M. intermedium*, *M. szulgai*	[6]
Meat leftovers from knife and sharpening steel	*M. kansasii*	[6]
Tissue leftovers from disinfectant container	*M. kansasii*	[6]
Condensed water near traverse rail	*M. kansasii*	[6]
Condensed water from pipeline near the live weight balance	*M.* sp.	[6]

[1] In most animals tuberculous lesions were detected in lymph nodes only. *MAA M. a. avium*. *MAH M. a. hominissuis*. **(3, 8, 9)** Serotypes 3, 8, 9.

[1] Dalchow W (1988) Tietärztliche Umschau 43:62–74. [2] Gotze U (1967) Berl. Munch. Tierarztl. Wochenschr. 80:47–49. [3] Jorgensen JB, Engbaek HC, Dam A (1972) Acta Vet. Scand. 13:68–86. [4] Nassal J (1965) Berl Munch. Tierarztl. Wochenschr. 78:273–275. [5] Pavlas M, Patlokova V (1977) Veterinarni Medicina 22:1–8. [6] Shitaye J, Horvathova A, Kaevska M, Donnelly N, Pavlik I (2008) Unpublished data.

5.7.5.2 Slaughterhouses as a Source of Mycobacteria for Workers

Mycobacteria present in the environment of slaughterhouses and meat-processing plants can also cause infection in the workers who are in contact with them. The transmission of *M. bovis* in slaughterhouses where cattle affected with bovine tuberculosis were slaughtered is well known. Analysis of five cases of *M. bovis* infection in workers from a slaughterhouse in Australia revealed lung and kidney infections in four and one of them, respectively (Robinson et al., 1988). Tongue infection with *M. bovis*, though rare, has also been described (Pande et al., 1995).

People handling meat and meat products can become infected with mycobacteria both in slaughterhouses and in their households. The wild animal hunters who can become infected by the non-professional handling of the bodies, meat and parenchymatous organs of the caught animals are categorised in a special group. This risk is acute in the areas where free-living animals infected by, e.g., *M. bovis* are present and where intensive hunting takes place. Such areas are found in, e.g., some states of the USA, where *M. bovis* still occurs in free-living red deer (VanTiem, 1997) and in the Republic of South Africa where *M. bovis* is found in water buffalos and other free-living ruminants, particularly in the Kruger National Park (Weyer et al., 1999).

Good health education and observing basic rules of hygiene can ensure that the risk of infection for hunters and other men handling the infected animals is relatively low.

5.7.5.3 Meat as a Source of Mycobacteria

As far as the ecology of mycobacteria is concerned, the importance of meat itself is relatively low. This is mainly due to the fact that mycobacteria are intracellular parasites and are usually present in the blood of newly infected animals for only a very short time. The meat of animals with gross lesions in their lymph nodes and/or in parenchymatous organs is condemned by the meat inspectors. Consumers are thus protected from mycobacterial infections that cause infections in animals, especially bovine tuberculosis. All the risk factors posed by the presence of OPM or PPM in the blood and meat are, however, described in detail in many books dealing with, e.g., bovine tuberculosis (Grange and Yates, 1994; Thoen and Steele, 1995; Thoen et al., 2006).

Mycobacteria are usually isolated from animals with advanced tuberculous processes (Gotze, 1967; Jorgensen et al., 1972; Francis, 1973; Pavlas and Patlokova, 1977; Windsor et al., 1984; Hancox, 2002; Gutierrez Garcia, 2006). However, *M. fortuitum* and other species have also been isolated from meat and parenchymatous organs without gross lesions (Table 5.21).

Mycobacteria from condemned animal bodies, organs and meat can be involved in ecology. If these wastes are processed by canning as feed for dogs and cats, or safely disposed in rendering plants, they do not become part of the ecosystem. However, this may occur in less economically developed countries where dead animal cadavers are buried in carrion places and animal burial sites or carcasses and organic remnants are left free in the environment (Photos 5.48 and 5.49). In the countries of Asia, "disposal in the air" of animal carcasses or organic remnants from slaughterhouses is performed. They deposit them in free locations (meadows and fields in lowlands; rocks in the mountain regions), where they are disposed of by scavengers (above all vultures, beasts of prey and invertebrates).

In economically developed countries, the hazard consists mainly in leaving the dead bodies or parts of the dead bodies of caught animals as carrion for free-living animals. Another great risk is the insufficient interring of the dead or killed animal bodies and their organic remnants. Various carrion animals can easily unearth them and these may then pose a great risk. Mycobacteria in this environment can rapidly become involved in ecological systems and can also be spread quickly through water. Dead animals are also sometimes thrown into rivers (particularly in various countries in Asia and Africa) or the sea (I. Pavlik, unpublished data; Photo 5.50).

Under such conditions, mycobacteria including OPM can be spread quite quickly. The so far unexplained ecology of *M. pinnipedii* which causes tuberculosis in water mammals in the southern hemisphere is a good example (Thoen et al., 2006).

5.7.5.4 Sources of Mycobacterial Contamination for Meat Products

As described above, mycobacteria can contaminate meat and parenchymatous organs of animals during their slaughter in slaughterhouses and/or during their subsequent processing in abattoirs. The majority of common processing procedures for meat products ensure their safety with regard to potential infection with *M. bovis*. For example, during the production of seven heat-treated frankfurters and sausages, *M. bovis* was killed at 70–82 °C, which was reached inside the products. During production by "roasting", 90–95 °C was reached for a period of 50–150 min and during "cooking in a steam chamber", 80–95 °C was reached for a period of 45–180 min, which was sufficient for the killing of *M. bovis* (Kozhemiakin et al., 1967).

If meat products are insufficiently heat treated during their production, the risk with regard to ESM and PPM can be seen in these two facts:

i. The products are made from intestines, meat and other components massively infected with mycobacteria.
ii. The product treatment (fermentation, curing, drying, etc.) does not ensure the death of mycobacteria.

Intestines used for the production of different meat products may also be contaminated with ESM or PPM from their contents, in which mycobacteria were also detected (Section 5.10). Due to the fact that in, e.g., pigs, two *MAC* members (*M. a. avium* and *M. a. hominissuis*) of PPM are most often found (M. Pavlas, personal communication), their capability to survive in salted and non-salted "*ex vivo*" intestines intended for further use was studied. The results showed that *M. a. avium* survived in the non-salted and salted intestines for 5 and 30 weeks, respectively. *M. a. hominissuis* was more resistant and survived in the non-salted and salted intestines for 13 weeks and 12 months, respectively (Palasek et al., 1991).

5.7.5.5 Meat Products as a Source of Mycobacteria

It was found during the production of Sicilian salami from non-heat-treated meat that *M. bovis* survival depended on the origin of the infection. In salami made from naturally infected beef from cows with miliary bovine tuberculosis, *M. bovis* was detected on day 87 of culture (it was virulent for guinea pigs only up to day 72). However, after artificial contamination,

M. bovis was still alive on day 117 (it was virulent for guinea pigs only before day 102). The differences were explained by, in particular, a higher concentration of *M. bovis* used for artificial contamination of the meat emulsion (Iannuzzi, 1968).

In Bulgaria, the survival of three species of mycobacteria (*M. tuberculosis*, *M. bovis* and *M. a. avium*) added at different concentrations to three different kinds of salami made from raw dried meat (Koprivshtenska and Troyanska Loukanka sausages) and fumigated raw meat (Gornooryahovska Nadenitsa sausage) was investigated. The survival varied between 150 and 180 days depending on the mycobacterial species and salami type (Table 5.22; Savov, 1975).

The thermoresistance of ESM and PPM is the same or higher than that of OPM, which can result in the potential contamination of meat products. Studies on the survival of, e.g., PPM in meat products revealed that the two most important *MAC* members (*M. a. avium* and *M. a. hominissuis*) are more resistant to higher temperatures than is the OPM *M. bovis* (Merkal and Whipple, 1980). These facts are consistent with the in vitro study of Pavlas (1990).

Table 5.22 Mycobacterial species survivance and virulence in respected animal model in different meat products (modified published data by Savov [1])

Mycobacterial species	Meat products	Positivity (days) by	
		Culture	Guinea pig
M. tuberculosis	Koprivshtenska sausage	180	180
	Troyanska Loukanka sausage	160	170
	Gornooryahovska Nadenitsa sausage	160	170
M. bovis	Koprivshtenska sausage	160	170
	Troyanska Loukanka sausage	150	160
	Gornooryahovska Nadenitsa sausage	170	180
M. a. avium	Koprivshtenska sausage	180	180
	Troyanska Loukanka sausage	150	150
	Gornooryahovska Nadenitsa sausage	180	170

[1] Savov D (1975) Vet. Med. Nauki 12:39–43.

With respect to the ecology of mycobacteria, concentrated attention was paid to the survival of two *MAC* members in Wiener sausages. Merkal et al. (1979) found that the thermoresistance of *M. a. avium* (serotype 2) and *M. a. hominissuis* (serotype 10) in Wiener sausages was not changed by the addition of either sodium nitrite or by the smoking temperature of 50 °C for 30 min. It was found that *M. a. hominissuis* (serotype 10) was immediately devitalised by a temperature of 68.3 °C in a water bath and after 18–20 min by smoking at 60–65 °C. It was completely devitalised in the sausage meat emulsion at 65.5 °C after 10 min.

On the other hand, the following noteworthy results concerning the second *MAC* member that most often causes avian tuberculosis (Section 3.1.3), *M. a. avium* serotype 2, were also observed in work with the same sausages:

First: It was more difficult to kill an *in vitro* isolated strain than *M. a. avium* of identical serotype, originating from naturally infected meat.

Second: *M. a. avium* serotype 2 was killed in Wiener sausages at 60 °C in either case (Merkal et al., 1981). This was lower than the temperature at which the above-mentioned *M. a. hominissuis* serotype 10 (Merkal et al., 1979) was killed.

Third: This *MAC* member (*M. a. hominissuis*) is one of the most common PPM in herds of domestic animals, especially in pigs and in the environment (dust, water, etc.); therefore, it can contaminate products from external sources.

5.7.6 Fish and Fish Products as a Source of Mycobacteria

The occurrence of mycobacteria in fish has been described in a separate chapter (Section 6.5.1). Fish, processing plants (including fishing boats) and fish products can also be viewed as sources of ESM and PPM owing to a high proportion of fish in human food. Various reviews deal with the transmission of infection from fish to men (Greenlees et al., 1998; Durborow, 1999; Ghittino et al., 2003; Novotny et al., 2004).

Effluent from processing plants directly running into river or sea water should be viewed as risky with respect to ecology. It leads to the relatively fast spread of different mycobacterial species (especially *M. marinum* and *M. fortuitum*) in aquatic environments. Under certain conditions, it can cause serious problems to organisms that live their whole life in water (such as fish and other water vertebrates), or organisms including men that only occasionally inhabit aquatic environments (Sections 3.1.7, 3.2 and 6.5).

The increasing popularity of raw fish dishes (e.g. sushi) and other non-heat-treated fish products is also a matter for concern (Suppin et al., 2007). Insufficiently heat-treated fish or fish products can likewise be dangerous. For example, *M. chelonae* survived the temperature of 55 °C for 15 min, 60 °C for 2.5 min and 65 °C for 0.5 min; it was completely killed at 70 °C. In fish silage, prepared from whole fish or from their parts, *M. chelonae* survived for 90 min (Whipple and Rohovec, 1994).

The occurrence of mycobacteria in frozen fish (at temperatures of between -18 and -22 °C) of different species was also investigated (Mediel et al., 2000). These whole-frozen fish came from Denmark (*Solea solea*), Spain (*Merluccius merluccius*), Iceland (*Gadus morhua*), Argentina (*Genypterus blacodes*) and Ireland (*Lophius piscatorius*). The following mycobacterial species were isolated from 29 of 100 samples: the *M. terrae* complex, *M. peregrinum*, *M. nonchromogenicum*, *M. chelonae*, *M. fortuitum* and *M. gordonae*. The absence of *M. marinum* can be explained by its more frequent occurrence in fish from subtropical and tropical regions.

In fish and products made from them and other aquatic animals, it will be necessary to focus primarily on the following three spheres concerning the occurrence and survival of mycobacteria:

i. Different parts of processing plants (including boats, etc.).
ii. Non-heat-treated products (consumed raw, smoke dried, brine canned, etc.).
iii. Waste and effluent.

5.7.7 Domestic Poultry as a Source of Mycobacteria

The ecology of mycobacteria in poultry can be analysed with regard to different aspects. The first is the occurrence of mycobacteria in eggs, meat and products of these. The second aspect is the importance of waterfowl and gallinaceous poultry in the ecology of mycobacteria. These play an important role in the ecology of mycobacteria, both while alive (they can spread, e.g., *M. a. avium*; Section 3.1.3) and after slaughter in the slaughterhouses.

5.7.7.1 Poultry Slaughterhouses as a Source of Mycobacteria

If poultry from a flock affected with avian tuberculosis is slaughtered, various localisations of infection including bone marrow should be considered (Gonzalez et al., 2002). The bodies of dead birds (especially poultry from small-scale production systems) as well as kitchen and abattoir waste, for example effluent, can become an important source of *M. a. avium* in the environment.

Different components of the environment in the slaughterhouses for poultry can become sources of mycobacteria, similarly as in the slaughterhouses for cattle, pigs and other animals. In Brazil, *M. gordonae* of an identical genotype was isolated from water (untreated water, treated water and served water) piped to a slaughterhouse. In the Czech Republic, *M. chelonae* was isolated from frozen mixed chicken meat for soup and frozen duck hearts (J. E. Shitaye, A. Horvathova, I. Pavlik, unpublished data).

5.7.7.2 Poultry Products as a Source of Mycobacteria

Shedding *M. a. avium* through eggs is rather important for food safety. Such infected eggs can also become sources of *M. a. avium* for martens, weasels, hedgehogs and other animals which can further disseminate the pathogen.

Nevertheless, infected free range hens on farms and in villages (in less economically developed countries) become important sources of *M. a. avium* for the surrounding environment. Despite this, the shedding of *M. a. avium* is irregular even in massively infected birds (Shitaye et al., 2008a,b). These birds commonly move over quite wide areas. It is likely that the occurrence of avian tuberculosis in free range birds has fallen owing to decreased numbers of these small flocks of hens in central European countries (I. Pavlik, I. Trcka, J. Lamka, unpublished data). The frequency of *M. a. avium* detection in slaughtered domestic pigs has also decreased (Shitaye et al., 2006).

The detection of *M. a. avium* in the mucosa of a 17-year-old boy suffering from leukaemia is an unusual and alarming finding. After cytostatic treatment, a desquamated piece of intestinal mucosa was collected from his stool, in which granulomatous inflammation was detected by histology. The PCR method revealed the presence of the *M. a. avium*-specific IS*901* fragment in these samples. Local investigation showed that the boy consumed eggs delivered by three small flock breeders; the flock consisted of several hens, in which the occurrence of *M. a. avium* could with high probability be assumed (B. Gabalec, J. Sterba, I. Pavlik, V. Mrlik, I. Tesinska, P. Kralik, unpublished data; Photo 5.46).

Attention should, therefore, be paid to the capability of all *MAC* members to survive in poultry meat and eggs even after heat treatment applied according to current recipes. These are above all insufficiently heat-treated steaks and "quick" grilled chicken that is not fully cooked. *M. a. avium* can also be present in bone marrow, half-raw scrambled eggs or creams made of raw yolk used to fill cakes, etc.

5.7.8 Other Non-traditional Foodstuffs

Almost nothing is known about the importance of algae in ecology and the risk they pose in the spread of mycobacteria. Due to the fact that they are used in different recipes and their increasing popularity as human food, future research would do well to focus on them. Various invertebrate animals, the human consumption of which is increasing, should be regarded as at least worthy of mention with regard to mycobacteria.

Fungi are a special group of foodstuffs. Some kinds of these, e.g. mushrooms, are cultivated on different substrates. From the point of view of mycobacteria the faeces of horses and other domestic animals should, for instance, be viewed as risky. Nevertheless, the origin of mycobacteria in the investigated mushrooms has not been satisfactorily explained yet (Table 5.18; Argueta et al., 2000). In some parts of the world (e.g. central European countries) mushrooms growing in the wild are a popular food. Nothing is known about the

occurrence of mycobacteria in the tens of mushroom species found in different niches in nature or about the capabilities of mycobacteria to survive the different treatments applied to mushrooms before consumption.

5.7.9 Raw Food Product Consumption and the Risk It Poses to Humans

Mycobacteria present in non-heat-treated food can significantly affect, in particular, the following groups of inhabitants:

i. *Children*: mycobacterial infections caused by OPM (particularly *M. bovis* causing tuberculosis of cervical lymph nodes), PPM (above all *MAC* infection of different organs) and ESM species (in combination with all of the above-mentioned mycobacteria) can convert the response to various tuberculin.

ii. *HIV/AIDS patient*: mycobacterial infections by the *MAC* members are manifested by extrapulmonary symptoms, e.g. intestinal, hepatic and renal (Horsburgh, Jr. et al., 1994).

iii. *Oncological and other immunosuppressed patients*: various forms of mycobacterial diseases with extrapulmonary localisation.

The above-mentioned groups of inhabitants from different parts of the world are exposed to the risk factors, which are associated with potential mycobacterial infection from foodstuffs and economic, social and cultural conditions. For instance, the consumption of the raw meat of sea animals is more traditional in coastal countries. On the other hand, raw meat and organs from ruminants, e.g. "tartar steak", were originally consumed in inland countries. Differences in eating habits are disappearing as a result of globalisation, but the insufficient hygiene in some restaurants employing staff and technologies which do not meet the required food safety standards poses new risks (Suppin et al., 2007).

This range of risks has been extended recently in association with migration of the inhabitants of various cultures together with their different eating habits. The spread of mycobacterial infections from foodstuffs will likely thus change as the habits of migrants will differ from the habits common among "original" inhabitants. Travel abroad may also constitute a certain risk factor (Besser et al., 2001) because foreign travel (especially to exotic countries) is usually associated with the tasting of local food. Accordingly, it is always necessary to ascertain whether food has come from a trustworthy source and that the raw materials for its preparation have been heat treated.

5.7.9.1 Food as a Risk Factor for Bovine Tuberculosis in Humans

The fact that bovine tuberculosis has not been brought under control in almost all economically underdeveloped countries is an important risk factor. The alimentary route seems to be one of the main routes in the spread of causative agents of the disease. With respect to the fact that this disease occurs in less economically developed countries at present, the availability of credible data is considerably limited. Nevertheless, it is possible to estimate the impact of consumption of non-heat-treated ruminant milk containing the causative agent of bovine tuberculosis on people from review articles (Cosivi et al., 1995, 1998; Bonsu et al., 2000; Ayele et al., 2004; Thoen et al., 2006; Shitaye et al., 2007).

However, bovine tuberculosis also poses a threat to the population of economically developed countries, in which the causative agent has once again been spread among cattle by free-living reservoir animals. New Zealand is an example of a country where transmission of the causative agent of bovine tuberculosis to inhabitants still continues (Baker et al., 2006). A comparable situation can also be seen in the Southern USA where cheese is made of non-pasteurised milk and the occurrence of bovine tuberculosis in children of Hispanic origin is rising (Besser et al., 2001).

The relationship between the occurrence of bovine tuberculosis in caught wild animals in, e.g., Africa and the USA and the occurrence of bovine tuberculosis in humans (especially hunters, veterinary workers and consumers) is currently under analysis (VanTiem, 1997; Michel, 2002). The raw liver of caught animals and foodstuffs made from their meat are usually consumed by adults. Because of the long incubation time necessary for the development of clinical bovine tuberculosis in humans, instructive data will likely only be available after several years, or even longer.

5.7.9.2 Food as a Risk Factor for *M. avium* Complex Infections in Humans

All three *MAC* members are mycobacterial species found quite often in animals, the environment and foodstuffs as mentioned in many chapters of this book. *M. a. avium* serotypes 1–3 cause avian tuberculosis in birds and animals. People become infected with *MAC* members most likely and most frequently by eating insufficiently heat-treated food from infected animals or their products (eggs, milk, etc.). The consumption of eggs from small-scale poultry farms should be considered as particularly risky. Contact between hens (captive for several years) and free-living birds increases the possibility of their infection with *M. a. avium* and its shedding via eggs is highly probable. Accordingly, the consumption of raw or insufficiently heat-treated eggs of such origin (semi-boiled eggs) is not advised.

The most common *MAC* members found in the environment and concurrently in foodstuffs is *M. a. hominissuis* of serotypes 4–6, 8–11 and 21. Previously, these serotypes were classified as species *M. intracellulare* and were designated as intermediary serotypes (Section 3.1.3). The new name of this subspecies *M. a. hominissuis* proposed by Mijs et al. (2002) refers to its hosts (men and pigs); it has most often been isolated from these. However, the name does not say anything about their origin, specifically that they are an important constituent of different niches and that with regard to ecology, are one of the most regularly isolated mycobacteria.

The third *MAC* member is the *M. intracellulare* species, comprising the remaining serotypes 7, 12–20 and 22–28. Its detection in the environment and hosts is less frequent than from the two above-mentioned members. Nevertheless, *M. intracellulare* also causes serious diseases in humans (especially in children; Section 3.1.4).

5.7.9.3 Pork Meat as a Risk Factor for *M. avium* Complex Infections in Humans

With respect to the frequent detection of these three *MAC* members in tuberculous lesions in the lymph nodes of pigs, the relationship between the consumption of pork and the transmission of *MAC* members to the human population has been discussed for many years. This issue is, however, only relevant in countries where pork is eaten. Originally, the consumption of insufficiently heat-treated pork was considered to be one of the risk factors for the development of infections in humans (Mallmann, 1972). In particular, the introduction of serotyping methods (Wolinsky and Schaefer, 1973) and their subsequent application for the serotyping of isolates from the environment, animals and people allowed the identification of the origin of different *MAC* members causing infections in humans (Codias and Reinhardt, 1979; Reznikov and Dawson, 1980; Nel, 1981; McClatchy, 1981).

These results led some to suggest that contaminated environment was the source of *MAC* members for humans and captive pigs (Brown and Tollison, 1979; Groothuis, 1985). In areas where high amounts of pork are traditionally consumed, all known recipes employ heat treatment. However, eating habits are beginning to change. In some countries, the consumption of "red steaks" made of pork is beginning to expand; it is also eaten raw as "tartar steaks" which were originally prepared from beef.

At present, molecular methods (above all RFLP analyses) targeting appropriate specific fragments for different *MAC* members allow an explanation of infection sources for both pigs and humans (Section 3.1.3; Matlova et al., 2005; Bartos et al., 2006; Moravkova et al., 2007).

5.7.9.4 Food as a Risk Factor for *M. avium* Complex Infections in HIV/AIDS Patients

HIV/AIDS patients constitute a special group. In these patients, the relationship between the occurrence of *MAC* in the environment, the consumption of raw foodstuffs and *MAC*-caused infections has been documented (Yajko et al., 1995; Ristola et al., 1999; Yoder et al., 1999). With respect to the immunosuppression of these patients, it is necessary that they be educated on the risks associated with the consumption of non-heat-treated foodstuffs (Kaplan et al., 2002).

Water for drinking and washing various foodstuffs (e.g. salads) and for everyday personal hygiene is another highly risky factor for exposure to *MAC* members (Section 5.2; Rice et al., 2005). Their detection in the stomach content reveals the ingestion of mycobacteria with food or water; analysis of the stomach content for the presence of AFB can make the diagnosis of

human mycobacterial diseases easier (Body and Boyd, 1988).

The microwave treatment of milk may be one of the preventive measures for nurslings and other consumers (Kindle et al., 1996; George, 1997).

5.8 Sawdust, Wood Shavings, Wood and Bark

I. Pavlik and M. Trckova

Sawdust, wood shavings, bark and other wooden raw materials are the by-products of wood processing. The are cheap and hence are easily available and widely used in agriculture, the construction industry, landscape architecture and other areas.

In agriculture, wood wastes (especially sawdust and shavings) are largely used for bedding, particularly in pig herds (Photo 5.51). Their beneficial role in animal welfare, above all in sustaining the temperature in pens (Marlier et al., 1994) and the reduction of humidity and malodour in the stable environment (Kao, 1993; Marlier et al., 1994; Chan et al., 1994; Nicks et al., 2003, 2004; Nicks, 2004), has been documented in the literature. Many experiments confirm a statistically significant decrease in the release of ammonia (NH_3), methane (CH_4), N_2O and CO_2 while using sawdust for bedding instead of straw (Nicks et al., 2003, 2004). A mixture of straw and sawdust can also be used for bedding; however, pure sawdust brings better results (Nicks et al., 1998).

Some authors have recommended sawdust for the bedding of pregnant sows because of its stimulating effect on the activity of sows in the pre-parturition stage, which can contribute to a rapid progress of parturition and a reduction in the numbers of piglets that become crushed intra-partum (Cronin et al., 1993; Pavicic et al., 2006).

Recently, the practice of pig fattening on composted or fermented deep bedding systems has been used on numerous farms (Kao, 1993; Ong et al., 1993; Chan et al., 1994; Lavoie et al., 1995). The key to this technology is the application of special enzymatic environmentally friendly preparations to the sawdust bedding in a stable. Bedding containing faeces and urine is naturally fermented during the course of fattening and the end product serves as a good quality fertiliser. The differences in animal efficiency and carcass qualities obtained with this technology are not statistically significant in comparison with conventional pig fattening (Ong et al., 1993).

Aside from the factors of temperature, moisture, microbial populations, enzymatic activity and others (Tiquia, 2002), the C/N ratio of the composted materials affects the course of the used sawdust composting in pile. If the C/N ratio is low due to an increased proportion of liquid manure in the composted material, the composting time can be prolonged to over 63 days. If using compost based on bedding with a high proportion of faeces which increase its salinity, the growth of plants may be inhibited (Huang et al., 2004). Compost based on fermented sawdust bedding (sawdust swine waste compost) with a pH of 8.0 (Kao, 1993) can enhance the physical, chemical and biological qualities of soil (Lavoie et al., 1995). However, due to a strong possibility of mycobacterial contamination (Table 5.23), the use of composts is considerably risky, restrictions have been imposed and farmers should be warned of the risk.

An analysis of published results has partly revealed the risks posed by respective parts of trees and raw materials from the point of view of their potential mycobacterial contamination (Table 5.23):

i. Timber without bast fibre and bark (processed trunks or planks) is not contaminated with mycobacteria.
ii. Bast fibre may contain mycobacteria in vessels, which transport fluids and nutrients together with mycobacteria (if they are present there) through the roots from soil.
iii. Mycobacterial contamination of the bark of living trees is very low and is due in large part to rainwater and dust contaminated with mycobacteria (Photo 5.52).

5.8.1 Mycobacteria in Living Trees

Mycobacteria seem to be able to penetrate injured roots, the vessels that supply trees with water and nutrients. Kauker and Rheinwald (1972) described mycobacterial penetration into the vessels of young nettles (*Urtica colous*), which were unrooted and left standing in a jar containing water contaminated with mycobacteria. Water was continuously stirred with a magnetic stirrer to prevent the sedimentation of *M. a. hominisuis*, which was later detected by culture in vessels from all parts of the plants. Most of the mycobacteria were present in the lower two thirds of the stalks. When the hazard analysis of mycobacterial occurrence in such raw materials is performed, the bast fibre emerges as the most risky. The water supply to woody matter is very low and thus, penetration of mycobacteria is considerably limited.

5.8.2 Mycobacterial Contamination of Wood Products

The substantial proportion of mycobacterial contamination of all wood products occurs after the extraction and handling of trees. Mycobacteria spread to all wood products during the processing of trunks at a sawmill (Table 5.23):

i. Soil mycobacteria largely contaminate bark on stacked tree trunks.
ii. Mycobacteria contaminate sawdust, which is a by-product of cutting whole trunks with bark and bast fibre.
iii. Sawdust and wood shavings, which are the by-products of cutting logs that are free of bark and bast fibre, are almost sterile and may be contaminated with mycobacteria from the environment (particularly from dust).
iv. During the transport and further storage of wood products in an environment favourable for pathogens (if there is no protection from rain), their contamination with PPM occurs (most commonly with *M. a. hominissuis* of serotypes 4, 8 and 9, *M. fortuitum* and *M. terrae*).
v. If stored within reach of free-living animals (particularly birds) these products can be contaminated with avian tuberculosis pathogens (*M. a. avium* of serotypes 1 and 2).

Accordingly, it is recommended to use fresh wood shavings for agriculture or landscape architecture due

Table 5.23 Mycobacteria detection rate in wood products

Examined wood material		Culture examination			
Origin	Specification	No.	Posit.	%	Ref.
Bark	Living deciduous and coniferous trees	100	3	3.0	[2]
	Trees from stores contaminated with soil	108	62	57.4	[2]
Sawdust	Obtained sterile from wood trunks	110	0	0	[1]
	Obtained sterile from wood trunks	164	0	0	[2]
	Fresh from sawmill	225	139	61.8	[2]
	Not specified	13	1	7.7	[7]
	From the pig farms with mycobacteriosis	78	34	43.6	[4]
	From the pig farms without mycobacteriosis	40	21	52.5	[2]
	Used as bedding	375	228	60.8	[1]
	From the pig farms with mycobacteriosis	110	68	61.8	[2]
	Not specified	9	8	88.9	[1]
	Sawdust used as bedding	12	11	91.7	[5]
	Fresh and used sawdust as bedding	56	19	33.9	[3]
Wood shavings	Fresh	20	0	0	[4]
Mixed	Sawdust and wood shavings	29	11	37.9	[1]
	Sawdust and wood shavings	94	39	41.5	[6]

[1] Beerwerth W, Kessel U (1976) Zentralbl. Bakteriol. [Orig. A] 235:177–183. [2] Beerwerth W, Popp K (1971) Zentralbl. Veterinarmed. B 18:634–645. [3] Dalchow W (1988) Tietärztliche Umschau 43:62–74. [4] Matlova L, Dvorska L, Bartl J, Bartos M, Ayele WY, Alexa M, Pavlik I (2003) Veterinarni Medicina 48:343–357. [5] Matlova L, Dvorska L, Palecek K, Maurenc L, Bartos M, Pavlik I (2004) Veterinary Microbiology 102:227–236. [6] Pavlas M, Patlokova V (1985) Acta Veterinaria Brno 54:85–90. [7] Saitanu K, Holmgaard P (1977) Nord. Vet. Med. 29:221–226.

to the fact that mycobacteria have not been detected in fresh wood shavings, or their content was very low (Table 5.23). On the other hand, if any wood product is stored under unsuitable conditions, such as high humidity for a long time, its contamination with mycobacteria can increase. Mycobacteria can propagate in sawdust, particularly in the summer season when the outside temperature rises above 18 °C (Kazda, 2000). The risk of the use of sawdust after its storage under inadequate conditions on pig farms has been described (Matlova et al., 2004b)

Uhlemann et al. (1975) described a case of wet sawdust delivered to a pig farm and stored 3 m high in a roofed shed. In summer, the temperature reached 63 °C at a depth of 1 m and later decreased to between 40 and 60 °C. They described how the propagation of potentially pathogenic bacteria could be rapid at temperatures of above 40 °C. PPM of various species were found in all layers, except at a depth of 25 cm.

It was found under laboratory conditions that a laboratory strain of *M. a. avium* can survive temperatures of between 18 and 22 °C for 7 months (Schliesser and Weber, 1973). *M. a. avium* has also been observed to

survive in wood shavings at temperatures of between −20 and +30 °C for more than 1 year (Charette et al., 1989). It follows that it is not advisable to use wood products even after the lapse of a period of time sufficient for the natural devitalisation of mycobacteria.

The analysis of the mycobacterial species profile in wood shavings (Table 5.24a) and other wood products (Table 5.24b) has highlighted various sources of contamination. *MAC* members have most frequently been isolated from fresh sawdust. The causative agent of avian tuberculosis, *M. a. avium* of serotype 1, can be considered as most significant. Among other *MAC* members, *M. a. hominissuis* of serotypes 4, 6, 8 and 9 were isolated; these have also been detected in tuberculous lesions on the lymph nodes and parenchymatous organs (above all in the liver and spleen) of pigs. Besides these two significant *MAC* members, *M. fortuitum* (one of the PPM) has been detected; it also causes tuberculous lesions in pigs (Table 5.24a). Besides the above-mentioned causative agents of tuberculous lesions in pigs, other species of PPM have been isolated from wood shavings combined with sawdust, or sawdust alone, and also from tuberculous lesions in pigs (Table 5.24b).

Table 5.24a Mycobacterial species detected in sawdust and pig tissues

Sawdust		Pig samples		
Origin	Mycobacterial species	Origin	Isolates	Ref.
Fresh from sawmill	MAH (4, 6, 8, 9), M. fortuitum	Lnn	MAH (4, 6, 8, 9), M. fortuitum, M. chelonae	[4]
Fresh from sawmill	MAA (1), MAH (4)	Lnn	MAH (4, 6), M. xenopi	[6]
Used as bedding	MAH (8, 9), M. fortuitum	Lnn	MAH (3, 8, 9), M. fortuitum	[2]
Used as bedding	MAH	Lnn	MAH, M. fortuitum, M. chelonae	[4]
Used as bedding	MAA (1), M. fortuitum	Lnn	MAH (4, 6), M. xenopi	[6]
Pig farm	M. aquae, M. triviale, M. diernhoferi, M. smegmatis, M. fortuitum, M. triviale, MAH (4, 6, 8, 9)	Faeces	M. aquae, M. triviale, M. diernhoferi, M. fortuitum, M. triviale, M. vaccae, MAH (8), MAA (2), M. phlei, M. terrae, M. nonchromogenicum	[1]
Pig farm	RGM	Lnn	RGM	[3]
Pig farm	MAA (1, 2), MAH (4, 6, 8, 9 and mixed)	Lnn	MAA (2, 3), MAH (4, 8, 9 and mixed)	[5]

RGM Rapidly growing mycobacteria. **Lnn** Lymph nodes. **MAA** *M. a. avium.* **MAH** *M. a. hominissuis.* **(4)** Serotype 4.
[1] Beerwerth W, Popp K (1971) Zentralbl. Veterinarmed. B 18:634–645. [2] Dalchow W (1988) Tietärztliche Umschau 43:62–74. [3] Fodstad FH (1977) Acta Vet. Scand. 18:374–383. [4] Matlova L, Dvorska L, Palecek K, Maurenc L, Bartos M, Pavlik I (2004) Veterinary Microbiology 102:227–236. [5] Piening C, Anz W, Meissner G (1972) Dtsch. Tierarztl. Wochenschr. 79:316–321. [6] Windsor RS, Durrant DS, Burn KJ (1984) Vet. Rec. 114:497–500.

5.8.3 The Pathogenesis of the Mycobacterial Infections Caused by Contaminated Wood Products

Analysis of the occurrence of tuberculous lesions in the lymph nodes of pigs reveals that piglets aged less than 2 months are most frequently infected (Brooks, 1971; Carpenter and Hird, 1986; Matlova et al., 2004b). Mycobacteria from wood products have been responsible for an increase in positive reactions to avian tuberculin testing (Fodstad, 1977; Pavlas and Patlokova, 1985). The main problem during the investigation of infection sources is caused by the long incubation time. Sawdust is mostly used in boxes for weaned piglets. These are sold at the age of 2–3 months to other farms,

Table 5.24b Mycobacterial species detection from wood materials other than sawdust and pig tissues

Sawdust		Pig samples		
Origin	Isolates	Origin	Isolates	Ref.
Wood shavings	MAH (4), M. fortuitum	Lnn	MAC	[1]
Wood shavings	MAC[1], M. xenopi, M. terrae, M. gastri	Lnn	M. tuberculosis, MAC	[2]
Sawdust and wood shavings	MAA (1), MAH (8), M. terrae, M. fortuitum, M. abscessus	Lnn	MAH (5, 8)	[5]
Sawdust and wood shavings	Negative culture examination done later	Lnn	MAA (1, 2), MAH (4, 5, 8)	[4]
Sawdust and dust	MAH (8)	Lnn	MAH (8)	[3]

Lnn Lymph nodes. **MAC** *M. avium* complex. **MAA** *M. a. avium.* **MAH** *M. a. hominissuis.* **(4)** Serotype 4.
[1] Charette R, Martineau GP, Pigeon C, Turcotte C, Higgins R (1989) Can. Vet. J. 30:675–678. [2] Kleeberg H, Nel E (1969) J S Afr Vet Med Assoc 40:233–250. [3] Saitanu K, Holmgaard P (1977) Nord. Vet. Med. 29:221–226. [4] Songer JG, Bicknell EJ, Thoen CO (1980) Can. J. Comp. Med. 44:115–120. [5] Uhlemann J, Held R, Müller K, Jahn H, Dürrling H (1975) Monatshefte für Veterinärmedizin 30:175–180.

where their fattening continues for the following 4–5 months before slaughter (Saitanu and Holmgaard, 1977; Matlova et al., 2004b). The incubation time from infection to the appearance of gross tuberculous lesions in lymph nodes varies from 4–6 months (Carpenter and Hird, 1986; Matlova et al., 2004b).

In the past authors have taken a different view as to the route of mycobacterial infections to pigs from wood products. While Szabo et al. (1975) only described injuries to the mucous membrane of the oral cavity, Engel et al. (1978) described the ingestion of sawdust by piglets. We observed an intensive intake of sawdust by piglets several hours after weaning on one farm where a 2–5 cm layer of sawdust was used as bedding (Photo 5.51). It is possible that piglets alleviated the stress caused by their separation from dams at the age of 28 days by a search for feed and the consumption of sawdust (I. Pavlik, unpublished observation).

A similar behaviour was observed in piglets aged 2–3 months during transportation to a fattening facility by truck. The bottom of the truck was covered with a 5–10 cm layer of sawdust and wood shavings mixed with straw to prevent sliding on the smooth surface. The truck driver observed intensive consumption of the bedding by the piglets. Due to the fact that the bedding was contaminated with PPM, tuberculous lesions were then observed in the pig head and mesenteric lymph nodes at slaughter after 4–5 months of fattening. Due to the fact that the occurrence of PPM was not increased in samples from the environment on the investigated farms, we supposed that the transportation time (30–50 min) of pigs (piglets) was sufficient for their infection under the above-mentioned conditions (I. Pavlik, unpublished data).

However, the wood used in pig pens can also be a source of *M. a. hominissuis*. On one farm we observed pigs to bite decaying and rotten planks used in a pen barrier. After staining according to Ziehl–Neelsen, AFB were detected by microscopic examination mostly in clusters in the wet parts of the decaying wood. Non-specific responses to avian tuberculin were detected in sows; numerous antibodies against all three *MAC* members were found in serum by the rapid slide agglutination test. Tuberculous lesions in the head and intestinal lymph nodes of market pigs were found by meat inspection in an abattoir. After a complete reconstruction of the pigsty, all the manifestations of mycobacterial infection in pigs disappeared

within several months (I. Pavlik and A. Horvathova, unpublished data).

5.8.4 The Economic Impact of Mycobacterial Infection on Pig Husbandry

Different legislative and different requirements for meat inspection after slaughter in abattoirs have been accepted in various countries. The occurrence of tuberculous lesions in pig lymph nodes is thought to be "underestimated" (Windsor et al., 1984). According to the currently accepted legislature of the EC, it is not necessary to cut lymph nodes from pigs when tuberculous lesions are not prominent above the lymph node surface (Council Directive 91/497/EEC of 29 July 1991 amending and consolidating Directive 64/433/EEC on health problems affecting intra-community trade in fresh meat to extend it to the production and marketing of fresh meat). Therefore, mycobacterial infection of intestinal and head lymph nodes cannot be diagnosed in many cases.

The consequences of the presence of PPM in sawdust and wood products have been analysed in the USA (Songer et al., 1980; Carpenter and Hird, 1986) and Canada (Charette et al., 1989). Despite this, a similar technology was introduced as bedding for piglets and subsequently fattened pigs in the Czech Republic. It involves deep bedding with wood products continuously decomposed by special preparations such as "ENVISTIM" or "EKOSTYM", which contain a complex of cellulose enzymes. Micro-organisms that constitute a natural component of animal faeces and the stable environment make compost from the bedding material and faeces. The bedding material is heated up by the activity of micro-organisms and its temperature is maintained within the thermophilous zone, i.e. in the range of temperatures favouring the activity of micro-organisms and agreeable for animals kept in the stables. The released thermal energy causes water evaporation from the faeces; bedding remains dry and the dry matter of the substrate comprises about 50%.

Concurrently, the amounts of ammonia and other poisonous gasses released to the stable environment markedly decrease. Within several cycles of fattening, a dark brown to black non-odorous material of friable consistency is gradually formed from bedding mixed with faeces. Its volume is substantially smaller in comparison with the initial bedding volume and the end product can be used as a high-quality fertiliser. The additive "ENVISTIM" can reduce the amount of liquid manure by 80–90% and decreases the released ammonia by more than 70%. The benefits of this new ecological pig fattening biotechnology are many:

i. There is no problem with liquid manure disposal because faeces are composted directly in the pens and after several fattening cycles the entire bedding material becomes a powdery substrate that can be used for fertilisation.

ii. The intensity of stable malodour decreases substantially.

iii. Straw or wood waste including sawdust and shavings can be used for bedding.

iv. Keeping the animals in groups in the pens with sufficient room and dry, soft and warm bedding ensures they are in good health.

v. Pigs can be housed at low costs in buildings originally intended for other purposes.

vi. The process of fermentation maintains a temperature of about 30–40 °C.

It was this last mentioned advantage for keeping "pigs in paradise" as described by some farmers that, in fact, became the primary risk factor. PPM, especially *MAC* members (particularly *M. a. hominissuis*), were not killed by this temperature and the ongoing enzymatic processes. The opposite effect was observed on farms using this technology, because *MAC* members rapidly propagated in the initially barely contaminated sawdust and pigs of all age categories became seriously infected. The findings of tuberculous lesions in the intestinal and head lymph nodes of slaughtered pigs were the cause of serious economic losses (Matlova et al., 2004b).

Despite the fact that this technology was more ecologically acceptable, less expensive and more natural for pigs, economic losses resulting from the confiscation of parts of slaughtered animals were high. When lesions were detected in the intestinal lymph nodes (caseous lesions of various sizes), meat was assessed as conditionally consumable after heat treatment. In cases where tuberculous lesions were also detected in parenchymatous organs, whole carcasses were con-

demned. The farmer's profit was decreased by more than 20% (Pavlik et al., 2003).

5.8.5 Tracing of Mycobacterial Infection in Pig Husbandry

Different approaches have been used for the investigation of sources of mycobacteria for pigs from sawdust, wood shavings and other wood products. Previously, all mycobacterial species isolated from wood products and tissues from infected pigs were compared (Tables 5.24a and 5.24b). Due to the fact that *MAC* members have often been isolated from wood products (particularly sawdust), biological experiments on pullets were later applied to epidemiological studies of mycobacterial infections from such sources. That allowed the differentiation of *MAC* members, i.e. *M. a. avium* virulent for birds from *M. a. hominissuis* partly virulent for fowl and *M. intracellulare* that is non-virulent for birds (Piening et al., 1972).

Subsequently, serotyping was introduced for differentiation between *MAC* isolates, which revealed that wood products were most commonly contaminated by various serotypes of *M. a. hominissuis*: serotype 6, serotype 8, or simultaneously serotypes 4, 6, 8 and 9. However, other serotypes have also been described that were not isolated from tuberculous lesions of pigs (Table 5.25).

Recently, IS*1245* restriction fragment length polymorphism analysis has been applied for the evaluation of the effect of sawdust/wood shavings used as bedding on the formation of tuberculous lesions in lymph nodes of pigs. This method confirmed that contaminated sawdust was the source of the *M. a. hominissuis* infection for pigs (Matlova et al., 2004b). The isolates from sawdust and from lymph nodes of pigs were of the same IS*1245* RFLP type. This finding in the Czech Republic led to a relatively early cessation of the use of deep bedding with sawdust on many pig farms.

5.8.6 Health and Food Safety Issues

As mentioned above, particularly *MAC* members of serotypes 4, 6, 8 and 9 (*M. a. hominissuis*) and a number of species of PPM are found in wood products

Table 5.25 Serotype detection of *M. avium* complex members from wood material and pig tissues

Member/Serotype		Japan [6] Sawdust	Pigs	Germany [5] Sawdust	Pigs	Germany [1] Sawdust	Pigs	Czech Republic Sawdust [4]	Pigs [7]	Cattle [2]	Denmark [3] Sawdust	Pigs
M. a. avium	1	0	0	1	0	0	0	0	0	0	1	3
	2	0	0	3	69	0	0	6	421	17	0	327
	3	0	0	0	2	0	0	0	16	0	0	0
M. a. hominissuis	4	1	0	2	1	0	0	1	1	1	3	7
	5	1	0	0	0	0	0	0	0	0	0	0
	6	1	1	6	0	0	0	0	0	0	0	3
	8	1	10	28	10	11	20	15	643	1	1	22
	9	2	2	10	4	4	4	3	74	0	1	7
	10	0	0	0	0	0	0	0	0	0	0	0
	11	0	0	0	0	0	0	0	0	0	0	0
	21	0	0	0	0	0	0	0	0	0	0	0
M. intracellulare	7	0	0	0	0	0	0	0	0	0	0	0
	12	1	0	0	0	0	0	0	0	0	0	0
	13	0	0	0	0	0	0	0	0	0	0	0
	14	3	0	0	0	0	0	0	0	0	0	1
	15	1	0	0	0	0	0	0	0	0	0	0
	16	3	0	0	0	0	0	0	0	0	0	0
	17	0	0	0	0	0	0	0	0	0	0	0
	18	2	0	0	0	0	0	0	0	0	0	0
	19	0	0	0	0	0	0	0	0	0	0	0
	20	0	0	0	0	0	0	0	0	0	0	0
	22–28	0	0	0	0	0	0	0	0	0	0	0
Mixed/unknown		0	0	17	4	3	12	5	338	0	1	9
Total		16	13	67	90	18	36	30	1493	19	7	379

[1] Dalchow W (1988) Tietärztliche Umschau 43:62–74. [2] Dvorska L, Matlova L, Bartos M, Parmova I, Bartl J, Svastova P, Bull TJ, Pavlik I (2004) Veterinary Microbiology 99:239–250. [3] Jorgensen JP (1978) Nord. Vet. Med. 30:155–162. [4] Pavlik I, Matlova L, Dvorska L (2008) Unpublished data. [5] Piening C, Anz W, Meissner G (1972) Dtsch. Tierarztl. Wochenschr. 79:316–321. [6] Sato A, Anada S, Matsuo H, Takebe H, Sato R, Matsuda M (1987) Kekkaku 62:61–66. [7] Shitaye JE, Parmova I, Matlova L, Dvorska L, Horvathova A, Vrbas V, Pavlik I (2006) Veterinarni Medicina 51:497–511.

(Tables 5.24a and 5.24b). After penetration of the host organism through the oral route, these mycobacteria commonly present in other components of the environment are usually stopped by the immune system barrier represented by respective lymph nodes. These mycobacterial species are rarely isolated from tissues of parenchymatous organs or from musculature (Pavlas and Patlokova, 1985). Therefore, food safety (pork or meat products such as sausages, salami, etc.) for consumers is guaranteed by adequate heat treatment.

However, human health may be threatened by the direct handling of wood products contaminated with mycobacteria. People working with sawdust or bark used for decorative purposes in gardens or as substrates for growing plants can occasionally be infected. In botanical gardens, zoological gardens and in greenhouses, sawdust is often used as a decorative mate-

rial on the paths or even on playgrounds for children (Photo 5.52). These pose a risk to visitors, as they can be infected through a break in their skin, or they can breathe them in while playing (Photo 5.35).

It is necessary to consider sawdust stored under unsuitable conditions (often stored in stacks in an open area) as a source of PPM; sawdust may often be wet or even mouldy in deeper layers. The handling of such materials by staff should also be considered risky (Photo 5.53).

Several injuries to people associated with wood, foliage or work in gardens have been described in the available literature (Table 5.26). Various mycobacterial species, including isolates from wood products, have been detected in the skin wounds (Tables 5.24a and 5.24b). The detection of *M. marinum* in such patients may be associated with primary skin injuries and

Table 5.26 Primary human diseases caused by conditionally pathogenic mycobacteria contaminating wood or wooden products

Mode of transmission	Patients[1]	Infected tissue	Isolates[2]	Ref.
Cut while gardening	76/F	Papules, nodules, blister, plaques on periorbital and nasal bridge	*M. chelonae*	[4]
Cut while gardening	62/M	Nodule on dorsum of the right wrist	*M. fortuitum*	[4]
Employed in timber industry	62/M	Plaque on knee, no known injury	*M. marinum*	[1]
Thorn puncture in woods	65/F	Boggy papule on the right fifth finger	*M. marinum*	[4]
Thorn stab to the thumb and regular cleaning of the fish tank filters	56/F	Erythema over a thumb progressed to the form of three cutaneous abscesses over the forearm	*M. marinum*	[3]
Greenstick radius fracture of the wrist	10.7/M	Ulcer on left wrist	*M. ulcerans*	[2]
Abrasion to the hip from foliage	8.8/M	Abscess on right iliac crest	*M. ulcerans*	[2]
Thorn puncture in woods in Nigeria	5/M	Ulcer	*M. ulcerans*	[4]

[1] Age in years/gender (sex): **M** (male), **F** (female).
[2] Mycobacterial species are listed in alphabetical order.
[1] Blacklock ZM, Dawson DJ (1979) Pathology 11:283–287. [2] Goutzamanis JJ, Gilbert GL (1995) Clin. Infect. Dis. 21:1186–1192. [3] Laing RB, Flegg PJ, Watt B, Leen CL (1997) J. Hand Surg.[Br.] 22:135–137. [4] Street ML, Umbert-Millet IJ, Roberts GD, Su WP (1991) J. Am. Acad. Dermatol. 24:208–215.

subsequent wound infections via water contaminated with this species which is frequently found there (Sections 5.2, 6.5 and 7.2). The detection of *M. ulcerans* in the skin of these patients may also be associated with additional wound infection through contaminated water in subtropical and tropical climate zones, where *M. ulcerans* is common (Section 3.1).

5.8.7 Preventive Measures

Even though there was a wide awareness (e.g. in the Czech Republic) of the hazard posed by mycobacteria present in contaminated sawdust used as bedding for pigs (Pavlas and Patlokova, 1985), farmers started to use it again in the mid-1990s. They believed that sawdust used as deep bedding would be more ecologically friendly after enzymatic digestion with "ENVISTIM". However, economic losses for pig breeders were considerable due to a high contamination of deep bedding with PPM (Pavlik et al., 2003; Matlova et al., 2004b).

Farmers from different countries have made efforts to find an alternative to sawdust. In Arizona, USA,

they have started to successfully use "chopped straw" (Songer et al., 1980). Farmers in Germany, meanwhile, ceased to use bedding for pigs (Uhlemann et al., 1975) and farmers in the Czech Republic started to use cheap oat bran. Later, however, they found this to be also contaminated with PPM after long storage in mills (Section 5.6.1.3; Fischer et al., 2004). Farmers searching for different alternatives for bedding in pig boxes (especially for piglets after weaning) must be well apprised of all risks. It is advisable to adhere to the following recommendations:

i. The use of wood shavings from furniture factories that do not process trunks with bark and bast fibre.
ii. Due to the fact that the storage of wood shavings in factories was found to be a critical factor, it is recommended to purchase wood shavings from factories where they are stored in silos or other rooms providing protection against contamination with PPM from the environment. The storage of wood shavings in closed towers immediately after delivery is considered best.

iii. Wood shavings should be transported under dry weather conditions (to protect them from rain and snow) or in bags or closed trucks.

iv. Wood shavings should be stored in shallow layers in dry rooms on farms; that will prevent their soaking with water and the potential propagation of PPM. Wood shavings can also be stored in dry bags in places with no moisture and their use within several months is recommended.

v. The use of "old", wet, mouldy or otherwise devalued wood shavings should be avoided at all costs. This rule should also be respected in cases where wood shavings are not used; the use of sawdust suspected of mycobacterial contamination should not be practiced.

5.8.8 Key Research Issues

Further research should be focused on the following issues:

- Study into the virulence factors of some mycobacterial species (and *MAC* serotypes) that participate in the production of tuberculous lesions.
- Understanding the principles of metabolic processes underlying the propagation of mycobacteria in wood products and their possible application for the acceleration of the mineralisation process.
- Thorough study of the risk posed by PPM to humans if present in different wood products used for decorative purposes in landscape architecture.
- Development of novel methods that will allow the rapid differentiation of isolates within one mycobacterial species for epizootiological and epidemiological studies (and for legal processes).

5.9 Air as a Habitat and Vector for the Distribution of Mycobacteria

I. Pavlik, J.O. Falkinham III and L. Dvorska

Mycobacteria can be readily isolated from air, aerosols or dust and for this reason aerosol transmission is a major vector of human and animal infection. The exposure can occur in natural environments (aerosols generated over lakes and rivers and dust generated by construction, farming or winds), at work (framing, construction or manufacturing) or in the home (showering, cleaning or gardening).

Because of their hydrophobic cell surface, mycobacteria are concentrated at air–water interfaces (Wendt et al., 1980), where they can be transmitted to air in ejected water droplets (Parker et al., 1983). In fact, mycobacteria are highly concentrated in water droplets ejected from water caused by air bubbles rising through the water column (Parker et al., 1983). Further, dust generated by winds and physical disturbances (e.g. plowing or harvesting) also carry mycobacterial cells that adhere to particles. They are present in the air in the following arrangements:

i. Separate cells or in small clusters of cells (5–10 CFU; Parker et al., 1983).
ii. Separate cells or cell clusters in ejected water droplets (Wendt et al., 1980; Parker et al., 1983).
iii. Separate cells or cell clusters complexed with dust particles.

Mycobacteria can easily spread very far through the air and wind. Even though the initial number of mycobacteria in the air is small, they can play a significant role there. In any case, transmission through the air is quite fast and easy on the Earth's surface; mycobacteria can thus be spread to different locations within a continent. Under certain conditions of opportune air circulation, dust particles can even be transferred from the Libyan Sahara in Africa to the Czech Republic in Central Europe within a few days (Photo 5.54).

Few measurements on the aerosol survival of mycobacteria have been made, but, if mycobacteria are like other micro-organisms, light (namely UV), humidity and airborne toxic compounds would be expected to directly influence survival. The relatively slow growth of mycobacteria may be another factor increasing survival. Pigmented PPM (e.g. *M. kansasii* and *M. avium*) might have higher survival as a consequence of protection provided by carotenoid pigments. The ultraviolet light (UV) susceptibility of ESM and PPM is no different from that of other bacteria (David et al., 1971), but the role of carotenoid pigmentation has not been compared between isogenic strains. Organic or inorganic particles can protect ESM and PPM from lethal UV radiation, to which they are exposed while being carried aloft.

After being carried aloft and after the sedimentation of the dust or aerosol particles containing mycobacteria, the fate of these pathogens depends on various factors, among which sufficient humidity and the presence of a suitable substrate are important for their survival. In dark and damp places (such as the floor of houses) ESM and PPM may survive for up to several months. Inflammatory reactions including eye irritation, respiratory infections, wheezing, bronchitis and asthma in workers in water-damaged or "mouldy" buildings have been associated with the presence of high numbers of micro-organisms (Andersson et al., 1997; Torvinen et al., 2006).

Mycobacteria have been recovered from materials collected from water-damaged buildings, in addition to micro-organisms normally associated with building materials (Andersson et al., 1997; Torvinen et al., 2006). During reconstruction, those mycobacteria could be aerosolised in the dust. Although other micro-organisms could be responsible for the respiratory problems, both saprophytic (e.g. *M. terrae*) and pathogenic (e.g. *M. avium*) strains isolated from "mouldy" buildings were capable of inducing inflammatory responses in a mouse macrophage cell line (Huttunen et al., 2000).

The mycobacteria elicited a dose-dependent production of cytokines IL-6 and TNF-α and of nitric oxide and reactive oxygen species from the murine macrophage. Because whole mycobacterial cells were employed in the assays, it is not known whether cell metabolites, that are likely easily aerosolised, were responsible for the induction of inflammatory reactions (Huttunen et al., 2000). Heat-shock proteins from a number of mycobacterial species have been shown to generate Th1-type responses and airway inflammation and hyper responsiveness (Robert et al., 2002). This evidence suggests that mycobacteria or their metabolites may be possible agents causing respiratory disease in individuals exposed to water-damaged buildings.

Under favourable conditions, mycobacteria can get into previously non-infected places, where they do not naturally occur (such as dust and spider's webs in the residences of people), or contaminate material, which was originally sterile, e.g. underground "black" peat (Photos 3.19 and 3.20; Kazda, 2000; Matlova et al.,

2005; Trckova et al., 2005; Trckova et al., 2006a,b). This is a primary risk factor for ESM and PPM transmission when a contaminated substrate becomes the main infection source; mycobacteria can then directly affect the host organisms through:

- Direct inhalation; aerosolised water droplets, less than 10 μm in diameter that reach the bronchioles and alveoli.
- Direct inhalation; particles of floating dust that are up to 10 μm in size (most mycobacteria are 2–5 μm long and 0.2–1.0 μm thick) are not trapped in the ciliary epithelium of the respiratory tract mucosa.
- Mycobacterial contamination of the host skin surface; these usually penetrate into the host organism through an injury or complicate another existing skin disease (such as eczema).
- Mycobacterial contamination of water and foodstuffs; followed by ingestion by the host organisms without heat treatment or some other treatment that can cause devitalisation of mycobacteria.

The inhalation of mycobacterial aerosols can occur from natural bodies of water and in the home, namely showers, as proposed by Collins et al. (1984). The matching identity of DNA fingerprint patterns of *M. avium* isolates from a patient with mycobacterial pulmonary disease and from hot and cold water samples collected in the patient's shower document the transmission of ESM and PPM from water to air (Falkinham et al., 2008).

The inhalation of mycobacteria complexed with dust particles can be viewed as another route of infection, perhaps just as important. Dust generated by the handling of potting soil during gardening activities is another mycobacterial infection source. Patients with mycobacterial pulmonary disease caused by *M. avium*, *M. intracellulare* or *M. kansasii* have been shown to be infected with the same species recovered from the patient's own potting soil. Proof that the infection source was the potting soil was provided by evidence that the patient's and potting soil isolates were clonally related (De Groote et al., 2006). These individuals were elderly and gaunt and lacked the classic predisposing conditions for PPM disease (Prince et al., 1989). However, a significant proportion of the patients had mutations in either the CFTR or α-1-antitrypsin genes (Wang et al., 2005).

During common respiration through the human nostrils, particles larger than 10 μm are trapped within the nasal mucosa and ciliary epithelium along the respiratory tract. The deeper the breath, the greater number of dust particles which are swept with the air stream into deeper parts of the respiratory tract. The dust particles, including mycobacteria, can be deposited in the finest bronchioli and can separately or together with dust particles adversely affect the host organism in the following modes:

Mode 1. Mycobacteria directly infect the respiratory tract of the host.

Mode 2. Dust particles mechanically irritate the airways including pulmonary tissue.

Mode 3. Dust particles are toxic to tissues of the respiratory tract.

Mode 4. Dust particles formed by organic matrices (particularly by whole mites or remnants of their bodies) cause allergic responses accompanied by damage to the airways or pulmonary tissue.

The pathobiological potential latent in the joined adverse effects of mycobacteria (**Mode 1**) and dust particles (**Modes 2 and 3**) is most dangerous. Pneumoconiosis and silicosis are viewed as the main predispositional factors for respiratory tract infections with PPM in humans (Policard et al., 1967; Solomon, 2001). If these diseases occur together with immunosuppression, a clinical disease, i.e. mycobacteriosis of lungs or lymph nodes, develops in patients. At particular risk of this disease are gold miners in South Africa (Corbett et al., 1999, 2000). At the same time, the dust itself may facilitate infection with the causal agent of tuberculosis. Ghio et al. (1990) observed an increased incidence of tuberculosis among silicate workers, which could be explained though the accumulation of iron-complexed dust particles in the lung. Thus, iron is available to dormant mycobacteria as a virulence factor.

No references to related studies, i.e. concurrent effect of pathobiological **Modes 1 and 2** in farm animal herds are available in the literature. We can suppose, however, that dust, in particular from the stable environment, can cause mechanical irritation of the airways and pulmonary tissue of animals and can also be the source of PPM for animals (Table 5.27; Photo 5.55). The presence of ESM and PPM in sta-ble dust can also be a cause of a non-specific allergic tuberculin response during the intravital diagnosis of bovine tuberculosis. Nevertheless, a similar study focused on the effect of pathobiological **Modes 1 and 2** has been undertaken in free-living animals. The presence of "environmental silica" in the UK is viewed as one of the factors increasing the susceptibility of the pulmonary tract of European badgers (*Meles meles*) to *M. bovis* infection (Higgins et al., 1985).

Pathobiological **Mode 4** in combination with **Mode 1** (with present mycobacteria) has been analysed by a number of previous studies. Those were mainly focused on the pathobiological potential of a "bioaerosol". What is a "bioaerosol"?

First, we should define the term "aerosol", which is a dispersion system of solid or liquid particles suspended in a gas. Bioaerosols contain particles or consist of micro-organisms (i.e. bacteria, viruses, fungi, protozoa), their by-products (i.e. endotoxins or mycotoxins) or consist of cell fragments, single- and multi-celled particles, droplets and airborne dust and debris. Although the most common sources for bioaerosols are natural (humans, animals, plants, water and soil), micro-organisms are also used increasingly in biotechnology (Dutkiewicz et al., 1994; Dutkiewicz, 1994). Ellertsen et al. (2005) found that patients with mycobacterial infections exhibit allergic sensitisation more frequently compared with healthy controls. It is conceivable that those predisposed to an allergy are less resistant to mycobacterial infections.

5.9.1 Aerosolisation of Mycobacteria

The first direct evidence of mycobacteria in aerosols was obtained using an Andersen 6-stage cascade sampler (Andersen, 1958; Wendt et al., 1980). Aerosols were collected along the James River in Richmond, Virginia, in an area where 60–70% of individuals had shown skin test sensitivity to mycobacterial antigens (Edwards et al., 1969). In addition to aerosols, water droplets ejected from water were collected (Wendt et al., 1980). The droplets are formed when air bubbles reach the water surface and burst; closure of the resulting crater leads to the ejection of droplets (10–100 μm diameter) that reach heights of 10 cm. Based on that initial data, laboratory experiments were performed to measure the possible concentration of mycobacterial

Table 5.27 Mycobacteria detection rate in dust

Examined dusty material				Culture			Microscopy		
Origin	Country	Specification of dust source	No.	Pos.	%		Pos.	%	Ref.
Houses	Korea	Not specified	111	8	7.2		nt		[2]
Houses	Japan	Not specified	22	17	77.3		nt		[1]
Villages	India	Floor, roof beams, kitchen	150	33	22.0		nt		[3]
Subtotal			283	58	20.5		nt		
Hospital	Japan	Dust from the floors from patient rooms	48	48	100		nt		[7]
Hospital	Romania	Not specified	500	21	4.2		nt		[6]
Subtotal			548	69	12.6		nt		
Pig farms	Japan	Not specified	11	1	9.1		6	54.6	[5]
Pig farms	Czech Republic	Dust and spider webs	117	9	7.7		2	1.7	[4]
Subtotal			128	10	7.8		8	6.3	
Total			959	137	14.3				

[1] Ichiyama S, Shimokata K, Tsukamura M (1988) Microbiol. Immunol. 32:733–739. [2] Jin BW, Saito H, Yoshii Z (1984) Microbiol. Immunol. 28:667–677. [3] Kamala T, Paramasivan CN, Herbert D, Venkatesan P, Prabhakar R (1994) Appl. Environ. Microbiol. 60:2180–2183. [4] Matlova L, Dvorska L, Bartl J, Bartos M, Ayele WY, Alexa M, Pavlik I (2003) Veterinarni Medicina 48:343–357. [5] Saitanu K, Holmgaard P (1977) Nord. Vet. Med. 29:221–226. [6] Toma F (1998) Bacteriol. Virusol. Parazitol. Epidemiol. 43:229–235. [7] Tsukamura M, Mizuno S, Murata H, Nemoto H, Yugi H (1974) Jpn. J. Microbiol. 18:271–277.

cells in droplets ejected from water surfaces. Cells of *M. avium* and *M. intracellulare* strains were suspended in water and air bubbles were passed through the suspension.

Ejected droplets collected at a height of 10 cm above the suspension surface were enriched in mycobacterial numbers that were 10 000 times higher than concentrations in the suspension (Parker et al., 1983). It was proposed that the hydrophobic mycobacterial cells were collected on the surface of the rising air bubbles, which broke upon reaching the surface, creating a crater whose walls were enriched with mycobacterial cells (Parker et al., 1983).

Repeated isolations of *M. kansasii* from water sources for industrial purposes caused diseases in workplaces with the presence of this species. According to the results it is supposed that the infection spread after inhalation of the infectious aerosol (Marks, 1975).

The matching identity of a patient's *M. avium* isolate and those from their shower water (Falkinham et al., 2008) not only proved that aerosols were a source of mycobacterial infection, but further enabled the development of a risk assessment of infection. Both hot and cold water yielded 2 CFU/ml *M. avium* and sediment from the shower head suspended in 40 ml

yielded 240 CFU/ml. Assuming that the flow rate in the shower was 1 gal/min and the infected individual took a 5 min shower, they would be exposed to a total of 38 000 CFU of mycobacteria during the shower. If only 10% of the mycobacteria were aerosolised (i.e. 3800 CFU) and the volume of the bathroom was 800 ft^3, the concentration of mycobacteria in the room would be 4.75 CFU/ft^3. If the person remained in the bathroom another 5 min (i.e. 10 min total duration) and has normal respiration (1 ft^3/min) they would inhale 47.5 CFU of mycobacteria. It is also possible that the entry of water containing the *M. avium* via the mouth and throat also contributed to the infection. Furthermore, mycobacteria in the shower head biofilm would also be expected to be aerosolised and contribute to the aerosol.

Based on the prevailing view that PPM are opportunists of relatively low virulence, the calculated value is surprisingly low. However, the patient had been treated with a fluoroquine for bronchitis immediately prior to the onset of pulmonary mycobacteriosis and the *M. avium* infection could have been a consequence of the absence of any competing microbial biofilm in the airways as well as the bronchitis.

Household hot tubs and spas are generators of aerosols. Unfortunately, if the water in the hot tub or spa contains ESM and PPM, they are likely to be aerosolised for transmission to household members. Hypersensitivity pneumonitis and mycobacterial pulmonary disease have been reported following exposure to hot tubs (Kahana et al., 1997; Embil and Warren, 1997; Mangione et al., 2001). The mycobacteria isolated from hot tubs (e.g. *M. avium*) were likely responsible for the infections based upon the matching identity of patient and hot tub mycobacterial isolates by either RFLP analysis (Mangione et al., 2001) or multilocus enzyme electrophoresis (Kahana et al., 1997; Embil and Warren, 1997). Furthermore, exposure was followed closely by the onset of symptoms and the extent of symptoms was related to the length of exposure (Embil and Warren, 1997). Although these reports do not document the use of disinfectants in the hot tubs, the water had been heated. Mycobacteria are relatively resistant to high temperatures (Schulze-Robbecke and Buchholtz, 1992) and concentrate in hospital hot water systems (du Moulin et al., 1988).

5.9.2 The Formation of Aerosols in Occupational Situations: Indoor Pools and Metalworking Fluids

Hypersensitivity pneumonitis is an occupational hazard of workers in two different industries: automobile manufacturing (i.e. metalwork) and recreation (indoor swimming pools). Pulmonary illness and infection have also been a consequence of exposure to aerosols generated by hot tubs, spas and coolant baths. Respiratory problems have also been associated with exposure to water-damaged buildings during reconstruction and mycobacteria isolated from water-damaged building materials have been shown to provoke inflammatory reactions. The outbreaks share the common feature of aerosol exposure and respiratory illness. It is likely that exposure to aerosols containing mycobacteria is a common feature of the outbreaks and that these mycobacteria or their products could be responsible for the respiratory symptoms.

Epidemiological studies have established that affected individuals shared the following characteristics: the workers were exposed to aerosols generated in the workplace from water that was an integral part of the workplace, namely, metalwork fluid used in grinding operations (Bernstein et al., 1995), indoor swimming pools (Rose et al., 1998) or in the household, e.g.

spas and hot tubs (Kahana et al., 1997; Embil and Warren, 1997; Mangione et al., 2001). Outbreaks of respiratory disease occurred in spite of disinfectant treatment of the water or fluids to reduce the number of micro-organisms.

Metalworking fluids are widely used in a variety of common industrial metal-grinding operations to lubricate and cool both the tool and the working surface. Metalworking fluids are oil–water emulsions that contain paraffin, pine oils, polycyclic aromatic hydrocarbons and heavy metals. Exposure to metalworking fluid aerosols can lead to hypersensitivity pneumonitis and chronic obstructive pulmonary disease. Mycobacteria have been recovered significantly more frequently from metalworking fluid samples collected from facilities where hypersensitivity pneumonitis has been found, compared to facilities that did not have hypersensitivity pneumonitis. In one study it was shown that exposure to metalworking fluid mist resulted in the development of hypersensitivity pneumonitis in 10 workers. Acid-fast micro-organisms identified as mycobacteria were present in the reservoir at a concentration of 10^7 CFU/ml. It was concluded that the presence of mycobacteria in the reservoir was a likely cause of the hypersensitivity pneumonitis, because one patient was infected by a *Mycobacterium* sp. and had antibodies against the reservoir fluid.

Hypersensitivity pneumonitis developed in spite of the disinfection of the metalworking fluid with morpholine, formaldehyde or quaternary ammonium-based disinfectants and mycobacteria were recovered from the metalworking fluid. Mycobacteria are resistant to formaldehyde and quaternary ammonium disinfectants and the heavy metals in metalworking fluids. Furthermore, mycobacteria can grow on the organic compounds in metalworking fluid including paraffin, pine oils and polycyclic aromatic hydrocarbons and can degrade the disinfectant morpholine. It is likely that the mycobacteria present in the water can grow on the organic compounds in metalworking fluids in the absence of competitors following disinfection. Because of the ability of mycobacteria to form biofilms, cleaning would not be expected to eradicate mycobacteria. Disinfectant addition and the cleaning of the reservoir in one facility did not prevent the reappearance of mycobacteria (7×10^5 CFU/ml after 2 weeks). Furthermore, disinfectant treatment would likely result in selection for the mycobacteria remaining after cleaning.

Granulomatous pneumonitis has been reported in lifeguards who worked at an indoor swimming pool that featured waterfalls and sprays. Affected lifeguards with symptoms worked longer hours than unaffected lifeguards demonstrating a dose–response effect. The waterfalls and sprays increased the number of respirable particles 5-fold and the levels of endotoxin 8-fold. Based upon the presence of endotoxin in the aerosol samples, it was suggested that the pneumonitis in the lifeguards was due to endotoxin exposure. However, subsequent data provided evidence of a possible second factor resulting in hypersensitivity pneumonitis. Aerosols containing mycobacteria were shown to cause granulomatous lung disease. High numbers of mycobacteria have, on more than one occasion, been reported in swimming pools, whirlpools and hot tubs. The presence of ESM and PPM in pools, hot tubs and spas is likely due to their resistance to chlorine. The route of infection or exposure is via aerosols.

Water aerosol is rarely encountered on farms. Despite this, various air conditioning systems have recently been introduced into stables to cool the ambient temperature. However, the costs of such systems and their maintenance are high. Accordingly, some companies have started to recommend cooling with water aerosol in summer; these aerosols are produced by a special water pump, which forces water into a thin pipeline equipped with nozzles in stables. Several metal strainers are placed in front of the pump to prevent incrustation and obstruction of the nozzles. Those, however, evidently cannot prevent the penetration of mycobacteria. On one pig farm we detected ESM and PPM in drinking water. Subsequently, 4–8 months later, tuberculous lesions in pulmonary lymph nodes were detected by meat inspection in sows culled for various reasons. *M. a. hominissuis* of serotype 8 and other PPM species were found in the water and tuberculous lesions (J. O. Falkinham III, unpublished data). It follows that novel technologies may pose new risks to animals with regard to PPM infection (Photo 5.11).

5.9.3 The Formation of Dust-Carrying Mycobacteria

Two factors govern the presence of dust-carrying mycobacteria. First, the source material (e.g. soil) must contain environmental mycobacteria and second, there must be some means to generate dust from the material. Solid particles lifted aloft by currents are designated as floating dust; these differ in size, chemical composition and origin. Dust is a general name for solid particles with diameters less than 0.5 mm. Dust is present everywhere and arises mainly from natural processes (Photo 5.13) and human activities (Photo 5.32):

- Farming activities.
- Gardening activities.
- Excavating activities.
- Traffic.
- Restoration of water-damaged structures.

Excessive concentrations of dust particles in the air can harm the respiratory tract. Dust is responsible for the lung disease known as pneumoconiosis, including black lung disease, which occurs among coal miners and is a known risk factor for pulmonary disease caused by PPM (Wolinsky, 1979). In addition to carrying mycobacterial cells, dust particles may also be composed of toxic substances such as heavy metals and organic substances that may have independent effects on the respiratory system.

The appearance of mycobacterial disease in an individual or in individuals not only involves exposure to a mycobacterial source but also the presence of some conditions that predispose the individual to infection. Historically those predisposing factors include pneumoconiosis, silicosis, chronic obstructive pulmonary disease, smoking and alcoholism (Wolinsky, 1979). Recently, other factors including mutations affecting chloride transport (CFTR gene) and α-1-antitrypsin have also been implicated. Under particular conditions, these dust particles can serve as the best "transport" medium for ESM and PPM. Due to contamination with ESM and PPM, dust in human residences, holiday resorts, hospitals, nursing homes, schools and other places can participate in an increased exposure of the "immunopredisposed" population to ESM and PPM. These mycobacteria complexed with dust particles can likewise cause a number of health problems in animal herds.

5.9.3.1 Dust in Potting Activities

Recently it has been shown that potting soil is a source of pulmonary infection by ESM and PPM

(De Groote et al., 2006). The investigations were triggered by reports of high numbers of mycobacteria in peat soil in Finland (Iivanainen et al., 1997) and that soil from potted plants contained high numbers of mycobacteria (Yajko et al., 1995). A large group of elderly, gaunt men and women with pulmonary mycobacteriosis caused by different PPM, including *M. avium* and *M. intracellulare*, provided samples of potting soil they had used. Dusts were generated by dropping patient potting soil samples 30 cm in a hood and particulates collected using an Andersen 6-stage sampler.

PPM were recovered from almost every potting soil sample tested and many were on particles small enough to enter the bronchioles and alveoli. The majority of isolates were *M. avium*, *M. intracellulare* and *M. kansasii*. Fresh potting soil samples purchased from garden supply stores had high numbers of mycobacteria and yielded mycobacterial-laden aerosols. For a number of patients, the species from potting soil aerosols was the same as the patient's isolate and for four such pairs, two showed close matches of DNA fingerprints using pulsed field gel electrophoresis (De Groote et al., 2006).

5.9.3.2 Dust in Households, Hospitals and Other Indoor Environments

ESM and PPM in dust can serve as the aetiological agent of human diseases, especially at present when their immunological profile is changing due to immunosuppressive diseases. According to recent data, more than 50% of patients affected by HIV/AIDS die of mycobacterial infections, the source of which cannot be revealed in most cases. The source of ESM and PPM in such cases is most likely the environment (Falkinham, 1996).

Dust in homes, offices, hospitals and other human environments are mainly generated by the inhabitants, including domestic pets such as dogs, cats and birds. Dust originating mainly from their hair and skin cells that fall out/off is mixed with some atmospheric dust from outdoors. On average, approximately 6 mg/m²/day of house dust is formed in private households, depending primarily on the amount of time spent at home. "Dust bunnies" are little clumps of fluff that form when sufficient dust accumulates. Much dust

is commonly accumulated in spider's webs, under the furniture in households and under the beds in hospitals.

When dust and dust bunnies potentially containing mycobacteria are sucked up with a vacuum cleaner, mycobacteria can be easily detected by culture in the content of the vacuum cleaner bag (Tsukamura et al., 1974; Dvorska et al., 2002). The RFLP method confirmed that dust from one household contained *M. a. hominissuis* isolates of the same serotypes and similar IS*1245* RFLP types, which were detected in the lungs of the landlady suffering from a mycobacteriosis (Photo 5.56; Dvorska et al., 2002).

Particles contaminated with mycobacteria that make up house dust can easily become airborne. Due to this fact people must exercise caution when removing dust, as activity intended to sanitise or remove dust may make it airborne. House dust can be removed by many methods, such as wiping, or sweeping by hand, or with a dust cloth, sponge, duster, broom or by suction with a vacuum cleaner or air filter. The device being used traps the dust; however, some may become airborne and come to settle in the cleaner's lungs, making the activity somewhat hazardous.

With regard to epidemiological aspects, the isolation of the human tuberculosis pathogen *M. tuberculosis* from dust is particularly serious (Table 5.28; Tsukamura et al., 1974). Despite the fact that this finding is rare, *M. tuberculosis* presence in dust may have serious consequences. In Japan an outbreak of pulmonary tuberculosis in four patients who habitually drank and smoked was studied by molecular epidemiology (IS*6110* RFLP typing). It was discovered that they were regular visitors of the same bar, although there was little, if any, contact between them while in or out of the bar (Nakamura et al., 2004). It can, therefore, be supposed that contaminated dust in the bar, which served as a vehicle for these predisposed individuals, was the source of *M. tuberculosis*.

Various mycobacterial species have also been isolated from household dust despite the fact that such an environment does not offer the best conditions for their propagation (Table 5.28). Under certain conditions (especially in damp rooms) dust can be highly contaminated with mycobacteria. Tsukamura et al. (1984) detected 414 mycobacterial colonies in 7 dust samples from hospitals, from which 107 isolates were identified. These included many slow- and fast-growing mycobacterial species (Table 5.28; Photo 5.57).

Table 5.28 Mycobacterial species isolated from dust

Dust origin	Houses						Hospital		Pig herds	
References	Ichiyama et al. [1]		Tsukamura et al. [4]		Tsukamura et al. [5]		Toma [3][2]		Matlova et al. [2]	
Mycobacterial species[1]	No.	%	No.	%	No.	%	No.	%	No.	%
M. agri	0	0	1	0.6	0	0	0	0	0	0
M. avium complex	119	57.8	0	0	3	2.8	2	9.5	0	0
M. a. hominissuis (8)	0	0	0	0	0	0	0	0	3	33.3
M. aurum	0	0	7	3.9	0	0	0	0	0	0
M. flavescens	0	0	1	0.6	4	3.7	0	0	0	0
M. fortuitum	0	0	76	42.7	0	0	0	0	2	22.2
M. gordonae	0	0	32	18.0	2	1.9	3	14.3	0	0
M. chelonae	0	0	2	1.1	0	0	0	0	0	0
M. intracellulare	0	0	1	0.6	0	0	0	0	0	0
M. kansasii	0	0	0	0	0	0	2	9.5	0	0
M. marinum	0	0	0	0	0	0	4	19.1	0	0
M. nonchromogenicum	0	0	45	25.3	2	1.9	0	0	0	0
M. parafortuitum	0	0	5	2.8	0	0	0	0	0	0
M. phlei	0	0	0	0	0	0	2	9.5	0	0
M. pulveris	0	0	0	0	20	18.7	0	0	0	0
M. scrofulaceum	19	9.2	1	0.6	0	0	2	9.5	0	0
M. smegmatis	0	0	0	0	0	0	3	14.3	0	0
M. tuberculosis	0	0	3	1.7	0	0	0	0	0	0
M. terrae complex	0	0	0	0	1	0.9	1	4.8	0	0
M. thermoresistibile	0	0	0	0	44	41.1	0	0	0	0
M. xenopi	0	0	0	0	0	0	2	9.5	1	11.1
Other species or M. sp.	68	33.0	4	2.3	31	29.0	0	0	3	33.3
Total	206		178		107		21		9	

[1] Mycobacterial species are in alphabetical order.
[2] Isolation of some mycobacteria could be connected with water sample examinations from the hospital. (8) Serotype 8.
[1] Ichiyama S, Shimokata K, Tsukamura M (1988) Microbiol. Immunol. 32:733–739. [2] Matlova L, Dvorska L, Bartl J, Bartos M, Ayele WY, Alexa M, Pavlik I (2003) Veterinarni Medicina 48:343–357. [3] Toma F (1998) Bacteriol. Virusol. Parazitol. Epidemiol. 43:229–235. [4] Tsukamura M, Mizuno S, Murata H, Nemoto H, Yugi H (1974) Jpn. J. Microbiol. 18:271–277. [5] Tsukamura M, Mizuno S, Toyama H (1984) Kekkaku 59:625–631.

Accordingly, the use of a central vacuum cleaner system is especially recommended in hospitals because all the dirt is collected to a central unit outside the hospital rooms. The system is usually located in the hospital basement, where the vacuumed mycobacteria do not pose a risk to the patients. Inter-human transmission of ESM and PPM infections is almost impossible (Tsukamura et al., 1974); therefore, close attention should be paid to this risk factor.

Another risk of PPM infections is posed by mites and other invertebrates. They have their own subtle interactions with dust that may have an adverse impact on the health of people. Thus, in many climates it is wise to keep a modicum of airflow going through a house, by keeping doors and windows open or at least slightly ajar. This can reduce the exposition of humans to dust particles. In colder climates, it is essential to manage dust and airflow, since the climate encourages occupants to seal even the smallest air gaps and thus eliminate any possibility of fresh air entering.

An extensive study was performed in 15 randomly selected villages in South India where a high prevalence of infections caused by PPM was recorded in one area. Various mycobacterial species including *MAC* members were detected in the dust from these households. The increased finding of *MAC* in dust correlated with the species of isolates from the sputum of village inhabitants and water from their wells (Kamala et al., 1994).

All three *MAC* members should be viewed as serious pathogens that can spread through dust. Less serotypes are found in dust, in contrast to the infected patients, according to the scarce literature data; however, their frequency in dust and patients is comparable (Table 5.29; Tsang et al., 1992). It is well evident, above all from extensive studies from the USA

Table 5.29 Serotypes of *M. avium* complex detected in house dust and humans

| Member/Serotype No. | Australia (%) [5] | | | USA and Canada [6] | | Germany | | | Japan [4] |
	House dust [5]	Sputum [5]	Extrapulmonary a[2]	Non-AIDS	AIDS	Sputum [1]	Lnn b[3]	All forms [3]	All forms
M. a. avium									
1	0	0	0	147	110	4	1	8	7
2	0	0	0	34	39	13	3	29	1
3	0	0	0	7	5	0	0	0	0
M. a. hominissuis									
4	0	1	0	179	332	1	2	5	2
5	0	2	0	8	1	0	0	0	0
6	2	11	0	50	32	0	2	2	2
8	2	4	0	222	216	7	2	25	3
9	2	6	0	94	45	3	2	9	6
10	0	1	1	9	13	0	0	0	0
11	0	0	0	5	0	0	0	0	0
21	0	0	0	3	2	2	0	0	0
M. intracellulare									
7	3	9	2	14	2	2	1	0	2
12	1	1	1	78	11	0	0	0	1
13	0	2	1	10	3	0	0	0	0
14	4	18	1	63	5	1	1	3	3
15	0	0	2	3	0	0	0	0	9
16	3	4	1	43	1	1	0	3	16
17	1	10	0	57	3	0	0	0	0
18	0	8	0	39	2	0	1	1	6
19	1	4	1	51	4	2	1	5	1
20	3	4	0	11	0	2	1	2	3
22	nt	nt	0	3	0	nt	nt	nt	nt
23	nt	nt	0	8	2	nt	nt	nt	nt
24	nt	nt	0	5	0	nt	nt	nt	nt
25	nt	nt	0	26	0	nt	nt	nt	nt
26	nt	nt	1	4	0	nt	nt	nt	nt
27	nt	nt	0	2	0	nt	nt	nt	nt
28	nt	nt	1	4	0	nt	nt	nt	nt
Mixed/unknown	0	0	2	57	204	0	0	6	45
Total	22	85	14	1236	1032	36	17	98	107

a Extrapulmonal isolates from lymph nodes and infected tissues.
b Head lymph nodes of infected children.
Lnn Lymph nodes. **nt** Not tested.

[1] Anz W, Meissner G (1969) Prax. Pneumol. 23:221–230. [2] Blacklock ZM, Dawson DJ (1979) Pathology 11:283–287. [3] Meissner G, Anz W (1977) Am. Rev. Respir. Dis. 116:1057–1064. [4] Miyachi T, Shimokata K, Dawson DJ, Tsukamura M (1988) Tubercle. 69:133–137. [5] Reznikov M, Dawson DJ (1971) Med. J. Aust. 1:682–683. [6] Tsang AY, Denner JC, Brennan PJ, McClatchy JK (1992) J. Clin. Microbiol. 30:479–484. [7] Uhlemann J, Held R, Müller K, Jahn H, Dürrling H (1975) Monatshefte für Veterinärmedizin 30:175–180.

and Canada, that the range of detected serotypes of *MAC* members in non-AIDS patients is wide (Table 5.29). This may be due to different sources of infection, including *MAC*-contaminated dust.

An analysis of case reports and review articles focused on the analysis of human infections caused by various PPM reveals that the source of infection remains unknown in many cases. After the identification of the species of mycobacteria, we can assume that the infection source could be one of the components of the environment including dust. Some of these cases of immunocompetent patients and immunocompromised patients are presented in Tables 5.30a and 5.30b, respectively.

5.9.3.3 Dust in Animal Stables

Floating dust can be found almost everywhere on farms and in agricultural enterprises. High concentrations of dust can be present in all places where it is generated (depots for straw, hay, corn, feed mixtures, feed supplements, etc.). The main sources of dust in stables with animals are the following:

- Powder feed mixtures.
- Forage (straw and hay) and bedding.
- Passing vehicles (mainly tractors powered by diesel engines).
- Hair and skin cells falling off the animals.
- Dust from the vicinity of farms (dusty roads, soft pens for animals and fields).

The majority of dust particles in the vicinity of farms originate from dusty roads and animal pens with muddy paths (soil, clay, sand and other loose materials). On many farms, various farm outbuildings are situated near animal stables (grain driers, mixing rooms, grain cleaning equipments, etc.) that are substantial sources of dust particles of various sizes and compositions. The animals and the staff are exposed to dust in stables every day. In some agricultural enterprises, staff have started to wear protective masks (e.g. on pig farms; Photo 5.58).

Dust on different surfaces (especially on window sills and different technological elements) in farm outbuildings is one of the most important sources of mycobacteria that are transmitted by drafts into the stables. Spider's webs are also a significant source of dust contaminated with mycobacteria. For example, in herds of pigs affected by mycobacteriosis of the lymph nodes, mycobacteria were detected in 8.5% of examined samples from spider's webs (Matlova et al., 2003). The isolated PPM were of different species, which indicates that they originated from various components of the environment (Table 5.28).

5.9.4 Prevention of Dust Exposure

The prevention of exposure to ESM and PPM complexed with dust particles can be looked at with respect to their places of occurrence:

i.   In an open landscape.
ii. In closed rooms (houses, hospitals, offices, etc.).
iii. On animal farms.

5.9.4.1 Open Landscape

An extensive study in Australia investigated the contamination of drinking water reservoirs with *MAC* members and *M. scrofulaceum* (Tuffley and Holbeche, 1980). After rainfall in the investigated area the water which collects on the roofs of houses is used (Photo 5.10). The outcome of the study was that the most likely source of the above-mentioned PPM was dust, originating from agricultural activities in the farm vicinity. PPM can get into the "roof water" either together with rain or accumulate complexed with dust on the uncovered surfaces of water reservoirs. Subsequently, they are deposited together with the dust particles on the bottom where they form sediment, which often contains either separate mycobacteria or big clusters.

The soil in urban agglomerations contaminated due to environmental pollution can also participate in the increased exposure of the general population not only to toxic substances but also to various micro-organisms including PPM. According to current knowledge, children and adults consume the contaminants from soil and dust unintentionally, although the situation with children is more complex as their occasional geophagia has been documented. Little is known about the actual PPM contamination of dust (Tables 5.27–5.29).

Table 5.30a Dust as a contaminated vehicle infecting immunocompetent human patients with mycobacteria

Mode of infection	Patients[1]	Infected tissue	Isolates[2]	Ref.
Not known	63/M	Pulmonary infection; past history of left upper lobectomy for pulmonary tuberculosis	*M. abscessus*	[12]
Not known	60/M	Sputum	*M. arupense*	[7]
Corneal graft	65/F	Chronic ulceration and stromal infiltration	*M. chelonae*	[1]
Not known	79/M	Sputum isolate from a man (rural Belgium; any contact with elephant)	*M. elephantis*	[10]
Contact lens	35/F	Hard contact lenses worn for 15 years without problems; developed pain in both eyes after receiving new lenses	*M. fortuitum*	[3]
Surgery, contact lens	44/F	Penetrating keratoplasty	*M. fortuitum*	[3]
Surgery, contact lens	80/M	Penetrating keratoplasty	*M. fortuitum*	[3]
Foreign body	26/M	Auto mechanic complaining of a painful red left eye after drilling on metal	*M. fortuitum*	[3]
Not known	65/F	No calcified nodule in the right upper lobe	*M. fortuitum*	[8]
Not known (environmental culture examinations neg.)	5 patients	Keratitis in 36 patients who underwent surgical correction of myopia in one hospital	*M. immunogenum*	[11]
Not known – sputum isolate	?/M	Chronic lung disease without evidence of progression	*M. interjectum*	[6]
No contact with fish	51/F	Pulmonary infection in immunocompetent home worker	*M. marinum*	[5]
Not known	82/F	Tumour-like lesion in the lateral segment of the middle lung lobe	*M. heidelbergense*	[9]
Not known	57/F	Sputum	*M. kumamotonense*	[7]
Wound caused by stapling machine	42/F	Tendonitis of finger following a puncture wound	*M. nonchromogenicum*	[2]
Not known	41/F	Sputum isolates; small peripheral cavitary lesion in the right mid-lung region	*M. parascrofulaceum*	[13]
Contact lens	?/?	Keratitis caused by other mycobacterial species	*M. triviale*	[4]

[1] Age in years/gender (sex): **M** (male), **F** (female), **?** (not known).
[2] Mycobacterial species are in alphabetical order.

[1] Aylward GW, Stacey AR, Marsh RJ (1987) Br. J. Ophthalmol. 71:690–693. [2] Blacklock ZM, Dawson DJ (1979) Pathology 11:283–287. [3] Dugel PU, Holland GN, Brown HH, Pettit TH, Hofbauer JD, Simons KB, Ullman H, Bath PE, Foos RY (1988) Am. J. Ophthalmol. 105:661–669. [4] Ford JG, Huang AJ, Pflugfelder SC, Alfonso EC, Forster RK, Miller D (1998) Ophthalmology 105:1652–1658. [5] Lai CC, Lee LN, Chang YL, Lee YC, Ding LW, Hsueh PR (2005) Clin. Infect. Dis. 40:206–208. [6] Lumb R, Goodwin A, Ratcliff R, Stapledon R, Holland A, Bastian I (1997) J. Clin. Microbiol. 35:2782–2785. [7] Masaki T, Ohkusu K, Hata H, Fujiwara N, Iihara H, Yamada-Noda M, Nhung PH, Hayashi M, Asano Y, Kawamura Y, Ezaki T (2006) Microbiol. Immunol. 50:889–897. [8] Pesce RR, Fejka S, Colodny SM (1991) Am. J. Med. 91:310–312. [9] Pfyffer GE, Weder W, Strassle A, Russi EW (1998) Clin. Infect. Dis. 27:649–650. [10] Potters D, Seghers M, Muyldermans G, Pierard D, Naessens A, Lauwers S (2003) J. Clin. Microbiol. 41:1344. [11] Sampaio JL, Junior DN, de Freitas D, Hofling-Lima AL, Miyashiro K, Alberto FL, Leao SC (2006) J. Clin. Microbiol. 44:3201–3207. [12] Tanaka, Kimoto T, Tsuyuguchi K, Suzuki K, Amitani R (2002) J. Infect. Chemother. 8:252–255. [13] Turenne CY, Cook VJ, Burdz TV, Pauls RJ, Thibert L, Wolfe JN, Kabani A (2004) Int. J. Syst. Evol. Microbiol. 54:1543–1551.

Table 5.30b Dust as a contaminated vehicle infecting immunocompromised human patients with mycobacteria

Patients[1]	Infected tissue	Isolates[2]	Ref.
7 patients	Interstitial infiltrations in lungs	*M. celatum*	[3]
2 patients	Sputum samples without clinical evidence	*M. elephantis*	[6]
?/?	Bronchial aspirate – lung cancer	*M. elephantis*	[6]
12 patients	Fever and respiratory disorders – immunocompromised cancer patients	*M. lentiflavum*	[4]
42/M	Bronchial wash	*M. mageritense*	[7]
55/F	Hickman catheter contamination – immunocompromised patient	*M. neoaurum*	[2]
50/M	Chest radiograph: bilateral infiltration (*HIV+*)	*M. xenopi*	[5]
36/M	Chest radiograph: bilateral nodular abnormality (*HIV+*)	*M. xenopi*	[5]
29/M	Chest radiograph: bilateral diffuse infiltration and bone marrow infection (*HIV+*)	*M. xenopi*	[5]
82/F	Multinodular lung disease	*M. simiae*	[1]

[1] Age in years/gender (sex): **M** (male), **F** (female), **?** (not known).
[2] Mycobacterial species are in alphabetical order.
[1] Braun-Saro B, Esteban J, Jimenez S, Castrillo JM, Fernandez-Guerrero ML (2002) Clin. Infect. Dis. 34:E26-E27. [2] Davison MB, McCormack JG, Blacklock ZM, Dawson DJ, Tilse MH, Crimmins FB (1988) J. Clin. Microbiol. 26:762–764. [3] Piersimoni C, Tortoli E, de Lalla F, Nista D, Donato D, Bornigia S, De Sio G (1997) Clin. Infect. Dis. 24:144–147. [4] Safdar A, Han XY (2005) Eur. J. Clin. Microbiol. Infect. Dis. 24:554–558. [5] Szlavik J, Sarvari C (2003) Eur. J. Clin. Microbiol. Infect. Dis. 22:701–703. [6] Tortoli E, Rindi L, Bartoloni A, Garzelli C, Mantella A, Mazzarelli G, Piccoli P, Scarparo C (2003) Eur. J. Clin. Microbiol. Infect. Dis. 22:427–430. [7] Wallace RJ, Jr., Brown-Elliott BA, Hall L, Roberts G, Wilson RW, Mann LB, Crist CJ, Chiu SH, Dunlap R, Garcia MJ, Bagwell JT, Jost KC, Jr. (2002) J. Clin. Microbiol. 40:2930–2935.

Immunocompromised patients of all age categories should be advised not to go out in strong winds to avoid exposure to dust containing PPM. Pneumonia with concurrent pleuritis was diagnosed in a 17-year-old boy with leukaemia. *M. neoaurum* was isolated from his sweat when monitoring the household where he lived. Although the house was very tidy, we discovered that the boy visited an old clubroom with his friends. There we observed high levels of dust, in which PPM including *M. neoaurum* can be present. Due to that fact, we recommended a change in the boy's behaviour so as to preclude further possible complications caused by PPM (I. Pavlik, unpublished data).

5.9.4.2 Indoors (Houses, Hospital, Offices, etc.)

The protection of people from dust- or aerosol-borne mycobacterial infections can be either active or passive. The former is regular ventilation in rooms and dust removal from contaminated surfaces, e.g. in households or in hospitals using vacuum cleaners as described above.

Passive protection denotes the use of different filter types to ensure non-contaminated air circulation in rooms and that people are protected by the use of personal respirators. Due to the small size of mycobacteria and the physical properties of their surface, HEPA filters have been shown to provide the best protection (Chen et al., 1994; McCullough et al., 1997). Because

of the expense and power requirements of HEPA filters alternatives are needed. Relatively large pore size filters with filter material (e.g. cellulose) coated with paraffin might effectively remove PPM from aerosols.

Mycobacteria are hydrophobic and any aerosolised particles coated with mycobacteria would be expected to adhere to paraffin-coated filters. However, the resistance of mycobacteria in the environment and their ability to survive in respirators remains a problem. Experiments have documented that *M. smegmatis* survived in a respirator for 3 days (Reponen et al., 1999). Preventive measures for the protection of health service workers from human tuberculosis infection should not cease until three successive culture examinations of sputum are negative (Rebmann, 2005).

It is possible to prevent dust sedimentation indoors by frequent ventilation, which ensures the removal of dust particles smaller than 10 μm with mild air movement. The use of "upper-room air ultraviolet germicidal irradiation" for inactivating airborne mycobacteria is necessary to ensure mild and continuous ventilation in rooms. That will ensure the movement of all the dust particles which tend to settle; deposited dust particles cannot be lethally irradiated directly by UV. Because mycobacteria are capable of photoreactivation repair of UV damage (David et al., 1971), UV illumination should be carried out in the dark. The humidity in the patient's room should be maintained at about 50%, because a higher humidity diminishes the effect of UV radiation (Xu et al., 2005).

5.9.4.3 Animal Farms

The amount of dust in the air of an animal house should also be gradually reduced due to the possible transmission of other airborne bacterial, viral and fungal diseases. However, dust represents an unfavourable environment for the survival of mycobacteria, particularly due to its gradual drying in well-ventilated stables. Therefore, it is also advisable to examine this material by microscopy after Ziehl–Neelsen staining. By this examination we can find mycobacteria in dust and webs and ascertain whether the stable environment is massively contaminated, even if subsequent culture examination is negative (Table 5.27). A failure of the cultural detection of mycobacteria in dust and in webs was described by Matlova et al. (2003). It is possible that mycobacteria were killed by the effect of daylight (i.e. by UV radiation) and dehydrated in the dust and spider webs; dehydration is very intense in some stables for pigs without a slurry collecting channel.

Bedding containing sawdust is viewed as a significant source of MAC members for stable dust (Saitanu and Holmgaard, 1977). These authors not only detected the same species of mycobacteria in both pig lymph nodes and sawdust but also in the dust released from sawdust. When farmers ceased to use sawdust for bedding, an increased occurrence of tuberculous lesions in pig lymph nodes was observed for at least the next 9 months. They supposed that PPM contaminating the environment, including dust, could participate in this infection.

5.9.5 Views and Perspectives on the Research

With regard to the fluctuations in nature associated with climatic changes and the increasing occurrence of dust from natural and anthropogenic sources, further research should be aimed at the following:

- Continuing studies of the species profile and quantification of PPM in dust by "culture-independent" methods.
- Due to unfavourable conditions in dust, it is necessary to continue studies of the ability of PPM to survive and propagate here.

- Continuing studies of the clinical significance of PPM most commonly found in dust for people and animals, focusing on pathogenicity and virulence changes in the most significant PPM during their survival in dust.
- Introduction of a system of regular PPM monitoring in hospitals with oncological and otherwise immunocompromised patients. Assessment of the degree of risk according to the contamination of various devices and surfaces with PPM (air holes, ventilators, stocks of medicinal drugs, etc.) and minimisation of the highest risks of exposure to PPM in the population under treatment, based on the conclusions.

5.10 Mycobacteria in Excreta of Birds, Animals and Humans

I. Pavlik and V. Mrlik

Mycobacteria are not only natural constituents of the ecosystem, they are also distributed to the environment through human activities. They enter animal and human bodies from the environment by various routes (mainly by aspiration or ingestion). Mycobacteria are often abundant in different components of the environment such as dust (Section 5.9), water (Section 5.2), foodstuffs (Section 5.7), etc. Mycobacteria are subsequently shed from the bodies of the hosts by various routes. The most common are faeces of animals and stools of people (Tables 5.31 and 5.32). Only a few cases of other excreta such as regurgitation by birds and mammals, urine and vomit have been studied. Their contamination with mycobacteria can be viewed as unusual from the point of view of the ecology of mycobacteria.

5.10.1 Mycobacteria in the Faeces of Birds, Animals and Humans

Under certain conditions (especially in fertilisation), mycobacteria shed by various hosts can be an important factor for their spread to the niches that have not

Table 5.31 Mycobacteria detection in faeces (intestinal content) of birds and animals

Examined faecal samples		No.	Pos.	%	No. of isolates	Ref.
Host	Details					
Birds	Free-living song- and water birds (collected during the summer)	207	62	30.0	76	[2]
	Free-living song- and water birds (on feeding places during the winter)	106	85	80.2	140	[2]
	Zoological garden (intestinal content of dissected birds)	108	43	39.8	52	[2]
	Water birds in a zoological garden (individually collected faeces)	790	79	10.0	79	[5]
	Pheasants kept in an aviary	150	78	52.0	95	[8]
Poultry	Intensive breeding	410	58	14.1	60	[4]
	Extensive breeding	419	174	41.5	219	[4]
	Extensive breeding (faeces collected at euthanasia)	21	9	42.9	9	[10]
	Extensive breeding (cloacae swabs collected at euthanasia)	21	4	19.1	4	[10]
	Extensive breeding (collected 1 week before euthanasia)	168	59	35.1	69	[9]
Pigs	Farm with mycobacteriosis	165	48	29.1	61	[3]
	Farm without mycobacteriosis	110	93	84.5	138	[3]
	Nk	512	41	8.0	42	[1]
	Nk	179	28	15.6	28	[7]
Cattle	Nk	401	345	86.0	525	[1]
	Paratuberculosis-infected herds	155	87	56.1	Nk	[6]
Horses	Nk	150	84	56.0	99	[1]
Sheep	Nk	52	16	30.8	21	[1]
Mixed	Cattle, sheep and horses	603	445	73.8	645	[4]

Nk – Not known.

[1] Beerwerth W (1967) Prax. Pneumol. 21:189–202. [2] Beerwerth W, Kessel U (1976) Prax. Pneumol. 30:374–377. [3] Beerwerth W, Popp K (1971) Zentralbl. Veterinarmed. B 18:634–645. [4] Beerwerth W, Schurmann J (1969) Zentralbl. Bakteriol. [Orig.] 211:58–69. [5] Dvorska L, Matlova L, Ayele WY, Fischer OA, Amemori T, Weston RT, Alvarez J, Beran V, Moravkova M, Pavlik I (2007) Vet. Microbiol. 119:366–374. [6] Glanemann B, Hoelzle LE, Wittenbrink MM (2002) Dtsch. Tierarztl. Wochenschr. 109:528–529. [7] Matlova L, Dvorska L, Bartl J, Bartos M, Ayele WY, Alexa M, Pavlik I (2003) Veterinarni Medicina 48:343–357. [8] Pavlik I, Trcka I, Lamka J, Matlova L (2008) Unpublished data. [9] Shitaye JE, Matlova L, Horvathova A, Moravkova M, Dvorska-Bartosova L, Trcka I, Lamka J, Treml F, Vrbas V, Pavlik I (2008) Veterinarni Medicina 53:101–110. [10] Shitaye JE, Matlova L, Horvathova A, Moravkova M, Dvorska-Bartosova L, Treml F, Lamka J, Pavlik I (2008) Vet. Microbiol. 127:155–164.

been inhabited by them previously. Animal and human excreta primarily composed of urine and faeces/stools can be important sources of mycobacteria. They gain access to the environment in large part through fertilisation or from sewage disposal plants. The subsequent circulation of mycobacteria in the environment depends upon many factors, which affect their penetration into the deeper layers of soil (Section 5.5) or subterranean waters (Section 5.2). Mycobacteria can get into the air by the unprofessional fertilisation of fields with liquid manure, e.g. during windy weather when the arising aerosol can be contaminated by various bacterial species including mycobacteria (Section 5.9).

5.10.1.1 Wild Birds

Birds can play an important role in the ecology of mycobacteria. They can fly long distances, and some of them share a habitat with farm animals and often have

an access to feed stores (Photo 5.59). Studies on the occurrence of mycobacteria in the faeces of wild birds are scarce; the detection of *M. a. paratuberculosis* in faeces from house sparrows (*Passer domesticus*) and tree sparrows (*P. montanus*) living with herds of cattle infected with paratuberculosis has been described in Section 6.4. In a herd of pigs infected with *M. fortuitum,* the same species was detected in the faeces of barn swallows (*Hirundo rustica*; Photo 5.60).

The majority of examinations have been performed on wild birds kept in captivity in zoological gardens or in pheasantries. The frequency of the occurrence of mycobacteria in faeces was associated with both the order of the examined birds (singers, grallatorial birds, Anseriformes and gallinaceous birds, etc.) and the season when samples were collected. In the winter season, the detection of mycobacteria in faeces was more than 2-fold higher than that in the summer. The drying of faeces and temperatures which accelerate their death in the summer seem to have a great effect on the sur-

Table 5.32 Mycobacteria detection in stool samples of humans

No.	Pos.	%	Isolates	Anamnesis (notes)	Ref.
				Examined samples[1]	
				Description of sample origin	
520	0	0	0	Nk (solid conventional culture system)	[4]
15	0	0	0	Sputum-positive patients with pulmonary tuberculosis	[1]
55	4	7.3	4	*HIV+patients*	[1]
31	3	9.7	3	Patients with chronic diarrhoea (>months)	[1]
19	3	15.8	3	Control patients	[1]
368	59	16.0	59[2]	Nk (solid conventional culture system)	[6]
2 176	393	18.1	396	Nk (culture and smear examinations compared[3])	[3]
552	107	19.4	Nk	Nk (solid conventional culture system)	[2]
34	7	20.6	7	*HIV−* patients	[1]
24	5	20.8	5	Patients with suspected pulmonary tuberculosis	[1]
552	170	30.8	Nk	Nk (MGIT liquid culture system)	[2]
50	26	52.0	32	Healthy patients (solid conventional culture system)	[5]
4 396	777	17.7			

[1] Listed according to the mycobacterial frequency in examined stool samples.
[2] Only *Mycobacterium avium* complex isolates were published.
[3] A total of 148 smears were positive after the Ziehl–Neelsen staining (acid-fast bacilli were detected).
Nk Not known.
[1] Conlon CP, Banda HM, Luo NP, Namaambo MK, Perera CU, Sikweze J (1989) AIDS 3:539–541. [2] Hillemann D, Richter E, Rusch-Gerdes S (2006) J. Clin. Microbiol. 44:4014–4017. [3] Morris A, Reller LB, Salfinger M, Jackson K, Sievers A, Dwyer B (1993) J. Clin. Microbiol. 31:1385–1387. [4] Nassal J, Werner-Schieck B (1970) Prax. Pneumol. 24:473–478. [5] Portaels F, Larsson L, Smeets P (1988) Int. J. Lepr. Other Mycobact. Dis. 56:468–471. [6] Yajko DM, Nassos PS, Sanders CA, Gonzalez PC, Reingold AL, Horsburgh CR, Jr., Hopewell PC, Chin DP, Hadley WK (1993) J. Clin. Microbiol. 31:302–306.

vival of mycobacteria there (Table 5.31; Photos 5.60 and 5.61).

The species composition of mycobacteria depends to a large extent on the health status of the birds; e.g. if a flock of grallatorial birds (herons, ibises, spoonbills and bitterns) is affected with avian tuberculosis, the causative agent of the disease, *M. a. avium*, can be detected in their faeces. *M. a. avium* in faeces has been identified in only 26.7% of all mycobacterial isolates. Besides the causative agent of avian tuberculosis, mostly ESM and PPM have been found in bird faeces. Tens of mycobacterial species detected in bird faeces are also found in the environment and are most likely only passively moved through their digestive tract (Table 5.33).

Accordingly, wild birds may play an important role in the ecology of mycobacteria due to their spatial activities and abilities to migrate over long distances. Thus, they interact with different parts of the environment, unreachable for other animals. These are, for instance, newly formed islands due to volcanic activity or abandoned islands, "tepuis" in tropical forests, or caves inhabited by birds, e.g. oilbirds and swiftlets (Salanganes). Mycobacteria could have also been transmitted between continents by birds of passage. These played an important role in the spread

of mycobacterial species before the industrial revolution. The expansion of transport allowed the rapid and easy movement of settlers into almost all parts of the world (early years of the second half of the 19th century), which resulted in the easier spread of different pathogens including mycobacteria (OPM, ESM and PPM).

In captivity, wild birds can respond in unexpected ways to infection with the causative agent of avian tuberculosis. In a flock of water birds infected with *M. a. avium*, higher immunity was detected in the species of the family Threskiornithidae than in the water birds of the family Ardeidae (Photo 3.17; Dvorska et al., 2007). Varying susceptibilities of various orders (or families) of birds to *M. a. avium* infection have been reported in Section 6.6.

5.10.1.2 Domestic Birds

The occurrence of mycobacteria in the faeces of domestic birds and their importance for the ecology of mycobacteria can be evaluated according to the breeding type. The breeding of domestic poultry can basically be categorised into four systems:

Table 5.33 Mycobacterial species detected in faeces or the intestinal content of birds and animals

Host	Anamnestic data	Isolates	Ref.
Wild birds	Faeces of free-living birds	MAA (2,3), *M. diernhoferi*, *M. fortuitum*, *M. gordonae*, *M. novum*, *M. phlei*, *M. terrae*, *M. triviale*, *M.* sp.	[1]
	Intestinal content from dissected birds in a zoo	MAA (2,3), *M. fortuitum*, *M. gordonae*, *M. novum*, *M. terrae*, *M. triviale*, *M.* sp.	[1]
	Faeces from water birds from a zoo (79 isolates)	MAA/21 = 26.7%, MAH/26 = 32.9%, *M.* sp./32 = 40.5%	[2]
Pheasants[1]	Kept in aviary	*M. abscessus*, MAA, MAH, *M. chelonae*, *M. diernhoferi*, *M. flavescens*, *M. fortuitum*, *M. peregrinum*, *M. scrofulaceum*, *M. smegmatis*, *M. szulgai*, *M. terrae*, *M. triviale*	[5]
Hens[2]	Faeces	MAA (2)	[6]
	Faeces from farm surroundings (12 isolates)	MAA/2 = 16.7%, *M. chelonae*/2 = 16.7%, *M. diernhoferi*/1 = 8.3%, *M. fortuitum*/2 = 16.7%, *M. peregrinum*/1 = 8.3%, *M.* sp./4 = 33.3%	[7]
Pigs	Faeces	*M. aquae* (*M. gordonae*), *M. diernhoferi*, *M. fortuitum*, *M. phlei*, *M. nonchromogenicum*, *M.* sp., *M. triviale*, *M. terrae*, *M. vaccae*	[1]
	Faeces	MAH, *M. chelonae*, *M. diernhoferi*, *M. fortuitum*, *M. gordonae*, *M. scrofulaceum*, *M.* sp.	[4]
	Faeces	MAA (2), MAH (8)	[6]
Cattle	Faeces	MAP, MAA, *M. hassiacum*/*M. buckleii*, *M. thermoresistibile*	[3]

[1] *Phasianus colchicus.*
[2] *Gallus gallus f. domesticus.*
MAA M. a. avium. **MAH** M. a. hominissuis. **(2, 3)** Serotypes 2 and 3./**No.** of isolates = %.

[1] Beerwerth W, Popp K (1971) Zentralbl. Veterinarmed. B 18:634–645. [2] Dvorska L, Matlova L, Ayele WY, Fischer OA, Amemori T, Weston RT, Alvarez J, Beran V, Moravkova M, Pavlik I (2007) Vet. Microbiol. 119:366–374. [3] Glanemann B, Hoelzle LE, Wittenbrink MM (2002) Dtsch. Tierarztl. Wochenschr. 109:528–529. [4] Matlova L, Dvorska L, Bartl J, Bartos M, Ayele WY, Alexa M, Pavlik I (2003) Veterinarni Medicina 48:343–357. [5] Pavlik I, Trcka I, Lamka J, Matlova L (2008) Unpublished data. [6] Piening C, Anz W, Meissner G (1972) Dtsch. Tierarztl. Wochenschr. 79:316–321. [7] Shitaye JE, Matlova L, Horvathova A, Moravkova M, Dvorska-Bartosova L, Treml F, Lamka J, Pavlik I (2008) Vet. Microbiol. 127:155–164.

i. Large-scale production of gallinaceous fowl kept in cage systems.

ii. Large-scale production of gallinaceous fowl on ecological farms.

iii. Large-scale production of waterfowl (above all ducks and geese).

iv. Small-scale production of all species of domestic gallinaceous birds and waterfowl.

Large-Scale Production of Gallinaceous Fowl Kept in Cage Systems

This first large-scale production system for the intensive breeding of domestic fowl (*Gallus gallus f.* *domesticus*) and turkeys (*Meleagris gallopavo*) largely restricts the contact of birds with the natural environment. The life of this fowl is quite short for the development of, e.g., avian tuberculosis. Hens (layers) are kept in closed halls for the duration of their lifetime where they can live up to the age of about 1.5 years (hens lay eggs for 9 months on average). Mycobacteria can only gain access to these closed, fully air-conditioned premises with water and granulated diets. Notwithstanding all this, mycobacteria have also been detected in the faeces of such poultry (14.1% in Germany; Table 5.31).

However, ESM and PPM can also penetrate such flocks from other sources. The serological testing of young layers in one flock revealed 85% had antibodies against *M. a. hominissuis* (serotype 8); subsequently, this *Mycobacterium* was also detected in the faeces of the layers by culture. When we visited the farm, we discovered that young layers (pullets) were kept on sawdust, which can often be contaminated with this serotype of *M. a. hominissuis* (Section 5.8).

As far as food safety is concerned, products from such farms are safe and there is no mention in the available literature regarding avian tuberculosis in poultry kept under such conditions. The access of free-living birds (especially house and tree sparrows) to these types of farms is fully prevented. The risk of contaminated feed (especially corn) is also restricted; corn and feeding mixtures are stored in closed silos. Mycobacterial killing would be increased upon exposure to high temperature (65 °C). Considerably unfavourable conditions are caused by such temperatures (including accelerated drying of the feeding mixtures) for the survival of the causative agent of avian tuberculosis and other species of mycobacteria (Section 5.6).

Large-Scale Production of Gallinaceous Fowl on Ecological Farms

However, in the second large-scale production system for gallinaceous fowl on ecological farms, risk factors are present. The kept gallinaceous fowl are in contact with both ESM and PPM. These have been detected in both the faeces of poultry kept on such farms and in wild-living birds (Tables 5.31 and 5.33). The prevention of contact of the kept poultry with wild birds (especially sparrows, pigeons, doves, etc.) is almost impossible. Therefore, in these types of farms it is necessary to take into account the risk of occurrence of avian tuberculosis or the possible occurrence of non-specific responses caused by PPM when the testing with avian tuberculin is performed. Such a situation was described on a similar farm more than 40 years ago after poultry were supplied with surface water contaminated with *M. a. hominissuis* (Kazda, 1966, 1967).

Large-Scale Production of Waterfowl

In the third large-scale production system for water poultry breeding, the contact of kept birds with mycobacteria in the environment cannot effectively be restricted in any way. They are abundant espe-cially in the organically contaminated surface water where waterfowl are kept. ESM and PPM found in water and other components of the environment cause numerous non-specific responses in waterfowl during tuberculin testing with avian tuberculin. The frequent detection of antibodies by serological testing using fast agglutination with antigens prepared from all three *MAC* members could also be explained in this way. It was not possible to confirm avian tuberculosis by both gross and culture examinations of several tens of breeding geese (I. Pavlik, unpublished data).

As far as tuberculin testing is concerned, it is necessary to consider the possible occurrence of avian tuberculosis in both geese and ducks. Despite the fact that waterfowl have long been considered as being quite resistant to avian tuberculosis, this resistance is not absolute. Avian tuberculosis caused by all three serotypes, 1, 2 and 3, of *M. a. avium* was diagnosed in geese and ducks (Section 6.6.2). Little is known about the participation of *M. a. avium* in the ecology of surface waters. Only occasional detection of *M. a. avium* in water has been described (Section 5.2). However, we can suppose that *M. a. avium* growing outside a host organism (in waters, biofilm or soil) loses its virulence for kept birds. Such a loss has been observed, e.g., in isolates growing in soil for 2.5 years (M. Pavlas, personal communication).

Nevertheless, it will also be necessary to evaluate this sphere of effect of the environment on *M. a. avium* virulence in different ecological niches not only by biological experiments on birds but also by genomics, proteomics, lipidomics, etc.

Small-Scale Production of All Species of Domestic Gallinaceous Birds and Waterfowl

These conditions are comparable with the previous system for keeping poultry; however, there exist a difference in that gallinaceous fowl are in contact with mycobacteria from soil that is often infested with them (Section 5.5). We know from extensive studies that besides ESM and PPM, *M. a. avium* of serotypes 2 and 3 have been detected in the faeces of gallinaceous poultry (Tables 5.31 and 5.33). Little is known about the participation of this *MAC* member in the ecology of bacteria as well as in waterfowl. It is only known that the shedding of *M. a. avium* by flocks of infected birds is irregular and its intensity greatly depends on the degree to which parenchymatous organs or intestines have proceeded with the tuberculous process. It was found in a flock of extensively kept 21 hens that only about 20% shed mycobacterial isolates of *M. a. avium*, about one third shed *M. a. hominissuis*, whereas the remaining third were others, mostly ESM species (Table 5.33).

This knowledge is important, especially in conditions of small-scale production of poultry in developing countries and in countries with conventional agricultural small-scale production (Photo 3.12). Bird faeces may not only be a vector of mycobacteria but also a source of infection. The direct contact of hosts with infected faeces has been described, but the transmission of mycobacteria through faeces used for fertilisation, e.g. for vegetables in greenhouses, has not yet been investigated. Results indicate that even clinically healthy birds lacking gross lesions in their organs may shed not only *M. a. avium* but also *M. a. hominissuis* and other PPM species (Tables 5.31 and 5.33). Accordingly, it is necessary to protect animals kept in captivity from contamination with the faeces of free-living and domestic birds (Photo 5.62).

5.10.1.3 The Human Gastrointestinal Tract Including Stools

In contrast to animals (especially ruminants), mycobacteria are only occasionally detected in human stools (Damsker and Bottone, 1985). Nassal and Werner-Schieck (1970) and Conlon et al. (1989) failed to find mycobacteria in any of the hundreds of human faecal samples they examined. Other literature sources describe varying frequencies of mycobacteria in the stools of people ranging between 7.3 and 52.0% (Table 5.32). These considerable differences can be explained by the different exposure of people to mycobacteria, both from food and water or other sources. Food composition for vegetarians (vegans only eat non-heat-treated food) and "almost pure carnivores" (people preferring steaks without garnish) differs considerably. Without regard to water, which is often contaminated with mycobacteria (Section 5.2), humans intake most mycobacteria with raw food of vegetable origin, non-heat-treated milk and meat products (Section 5.7).

The exposure of the human population to mycobacteria in vegetable food is relatively low in comparison with herbivores, because humans only intake vegetables after washing. The parts of vegetables which are in direct contact with soil are often cut off (e.g. vegetable roots) and peeled, and those parts in contact with the surroundings are removed from fruit (bananas, oranges, tangerines and melons). Animals ingest relatively more vegetable feed, which is often contaminated with mycobacteria, or their surface is contaminated with soil often containing mycobacteria (Section 5.5).

The profile of mycobacterial species detected in animals (Table 5.32) and humans (Table 5.33) is comparable. Like animals people are steadily exposed to mycobacteria from the environment. The most important difference between humans and animals is the fact that the causative agent of human tuberculosis *M. tuberculosis* is detected in people. This most often affects the pulmonary tract although extrapulmonary localisations have also been described. With respect to the occurrence of mycobacteria in the digestive tract, their localisation is above all noteworthy. Mycobacteria have been detected in the oral cavity and along almost the entire intestinal tract of humans (Table 5.34).

Considering the importance of humans in the ecology of mycobacteria, we have discovered that humans can be contaminated (in better cases) or infected (in worse cases; particularly immunocompromised people) with mycobacteria. However, very little is known about the ecology of *M. tuberculosis*. The causative agent of human tuberculosis is shed into the environment through various excreta and secretions such as sputum, saliva, urine, stools, etc. (Section 2.2.1). It is assumed that *M. tuberculosis* cannot multiply in the environment. However, due to the fact that like other OPM (e.g. *M. bovis*, *M. caprae*) it is quite resistant to the environment, this issue still remains unanswered (Photo 5.63).

5.10.2 The Transmission of Mycobacteria by Other Excreta of Animals and Humans

Mycobacteria can be also found in other less-known excreta from animals and humans, which can become part of the ecology of mycobacteria under certain conditions. These are mostly urine, regurgitations from vertebrates (cold blooded and warm blooded), damaged tissues of the host (above all pus) and vomit. Even though the occurrence of mycobacteria in these excreta has not been properly investigated yet, it is mentioned here for the sake of completeness.

5.10.2.1 Saliva

Tissues in the oral cavity of humans as well as other tissues of their body can be affected with miliary tuberculosis caused by *M. tuberculosis* (Heigis et al., 2005). Besides saliva, some sputum containing the causative agent of human tuberculosis and various species of ESM and PPM may be present in the oral cavity of patients infected by *M. tuberculosis* (Table 5.34). Mycobacteria can get into the environment through this route from the oral cavity, even though the importance of this transmission route of mycobacteria is very low. Nevertheless, under certain conditions, the oral cavity can be important for the transmission of mycobacterial infections; e.g. in Africa, *M. ulcerans* was transmitted by a playmate's bite (Debacker et al., 2003).

Such transmission routes have not been described in animals yet. However, an interesting case of two dogs has been reported (9- and 10-year-old German shepherds); pulmonary tuberculosis caused by *M. tuberculosis* was detected several days after the owner's hospitalisation, just before his death. Due to the detection of IS*6110* specific for all members of *M. tuberculosis* in the faeces of one of the dogs, the animals were euthanised. IS*6110* was also detected afterwards by the PCR method in the submaxillary lymph nodes and in the intestinal mucosa of both the dogs. The repeated finding of IS*6110* in dog faeces collected in the garden where the dogs lived all their lives and in the balls the dogs played with is most noteworthy from the point of view of ecology (Table 5.35). These balls probably became contaminated in their oral cavity.

Little is known about the ability of *MTC* members to survive or grow in different components of the envi-

Table 5.34 Mycobacterial species detected in the gastrointestinal tract including the mouth of humans

Compartment	Mycobacterial species listed alphabetically	Ref.
Mouthwash[1]	*MI, M. kansasii, M. phlei, M. scrofulaceum, M. smegmatis, M. terrae, M. tuberculosis*	[3]
Mouthwash and sputum[1]	*M. phlei, M. smegmatis, M. tuberculosis*	[3]
Gastric aspirate[1]	*MAC, M. gordonae, M. kansasii, M. malmoense, M. neoaurum, MTC, M. xenopi*	[1]
Gastric content of neonate[2]	*M.* sp. (acquired from baby formula contaminated with *M. xenopi*)	[4]
Intestine	*MAC*	[2]
Stool	*MAC*	[2]
Stool[1]	*MAC, M. chelonae* complex, *M. terrae, MTC*	[1]
Stool (396 isolates from patients with and without tuberculosis)	*MAC* (358/90.4%), *M. chelonae* (1/0.3%), *M. fortuitum* (2/0.5%), *M. gordonae* (10/2.5%), *M. kansasii* (4/1.0%), *M. simiae* (2/0.5%), *M. terrae* (1/0.3%), *M. tuberculosis* (10/2.5%), *M. xenopi* (1/0.3%)	[6]
Stool (26 isolates from healthy patients)	*MAC* (5/19.2%), *M. gordonae* (5/19.2%), *M. malmoense* (2/7.7%), *M. simiae* (14/53.9%)	[5]

[1] Patients with human tuberculosis.
[2] One-day old neonate boy born to a father infected with *M. tuberculosis*.
MAC M. avium complex. *MI M. intracellulare. MTC M. tuberculosis* complex. (No. of isolates/%).
[1] Hillemann D, Richter E, Rusch-Gerdes S (2006) J. Clin. Microbiol. 44:4014–4017. [2] Kiehn TE, Edwards FF, Brannon P, Tsang AY, Maio M, Gold JW, Whimbey E, Wong B, McClatchy JK, Armstrong D (1985) J. Clin. Microbiol. 21:168–173. [3] Mills CC (1972) Appl. Microbiol. 24:307–310. [4] Morita Y, Kimura H, Minakami H, Saitoh M, Kato M, Nagai A, Kozawa K (2002) Pediatr. Infect. Dis. J. 21:987–988. [5] Portaels F, Larsson L, Smeets P (1988) Int. J. Lepr. Other Mycobact. Dis. 56:468–471. [6] Yajko DM, Nassos PS, Sanders CA, Gonzalez PC, Reingold AL, Horsburgh CR, Jr., Hopewell PC, Chin DP, Hadley WK (1993) J. Clin. Microbiol. 31:302–306.

Table 5.35 IS*6110* detected in the environment and in the tissue samples of two male dogs exposed by infected owner with *M. tuberculosis* [1]

Tissue samples	Bojar	Black	Environmental samples	
	IS*6110*		Origin	IS*6110*
Faeces No. 1[1]	–	+	Garden dogs faeces[2]	+
Faeces No. 2[2]	–	–	Garden dogs faeces[3]	+
Blood[2]	–	–	Aquarium sediment[3]	–
Submandibular ln.[2]	+	–	Biofilm in aquarium[3]	–
Spleen[2]	–	–	Butt of cigarettes (workroom)[3]	–
Liver[2]	–	–	Butt of cigarettes (corridor)[3]	–
Retropharyngeal ln.[2]	–	–	Butt of cigarettes (under lathe)[3]	–
Tracheobronchial ln.[2]	–	inh.	Snails from the garden (small)[3]	–
Lungs[2]	–	–	Snails from garden (large)[3]	–
Jejunal mucosa[2]	+	+	Tennis ball of dogs[3]	+
Jejunal ln.[2]	–	–	Plastic foam ball of dogs[3]	+
Ileum[2]	–	+	Garden dogs faeces (day of euthanasia)[3]	+
Ileal ln.[2]	–	–	Faeces of new young dog "Zheryk"[3]	–
Inguinal ln.[2]	–	–	Bone in the garden (toy of dogs)[4]	+
			Butt of cigarettes (main door from workroom)[4]	+
			Butt of cigarettes (corridor before workroom)[4]	–
			Butt of cigarettes (under drilling machine)[4]	–

[1] Collection 42 days after the last contact with the infected owner.
[2] Collection 48 days after the last contact with the infected owner.
[3] Collection 56 days after the last contact with the infected owner.
[4] Collection 99 days after the last contact with the infected owner.
inh. Inhibition of PCR reaction. **ln.** Lymph node.
[1] Pavlik I, Mrlik V, Skoric M, Svobodova J, Svobodova D, Slezakova E, Kralik P, Moravkova M, Tesinska I (2008) Unpublished data.

ronment. The detection of IS*6110* in the faeces of these dogs and in their balls can, however, reveal new and as yet unrecognised transmission routes of this causative agent (Photo 5.63).

5.10.2.2 Regurgitation, Vomit and Leftovers of Feed and Food

Regurgitation is composed of non-digested balls of hair, feathers and bones of prey (often small terrestrial mammals and birds, wing cases of beetles, etc.), which constitutes the food of predators. Regurgitations are expelled from the crop and can often be found in the roosting places where various birds spend the night. Tens of them may be found below a tree, where a predator is resting. The regurgitations of the Eurasian Eagle Owl (*Bubo bubo*) may be 6–15 cm long and 2–3 cm thick.

Regurgitations are not very important from the point of view of the ecology of mycobacteria due to the fact that they are of a hard and rather dry consistency. Mycobacteria most likely do not survive long in them. However, regurgitation from animals infected, e.g., by the causative agent of avian tuberculosis can pose a risk. In one aviary with infected water herons and ibis birds, *M. a. avium* of identical IS*901* RFLP was detected in their regurgitations (Dvorska et al., 2007). These originated from different species of herons, ibises, spoonbills and bitterns. No regurgitations are found, e.g., below white-tailed eagle (*Haliaeetus albicilla*) nests because they are ingested, for instance, by hunting foxes, hedgehogs and badgers (V. Mrlik, personal communication); that fact can play a significant role in the spread of mycobacteria.

Feed leftovers from other predators can take part in the circulation of mycobacteria (e.g. fish leftovers in white-tailed eagle nests, residues of small terrestrial mammals in owl nests). These are left in the place they were caught, either on the ground, on trees or in water; they can also be found in or below nests of birds of prey (Photo 5.64). If they are contaminated with mycobacteria, they can become reservoirs of infection for other hosts, through the same routes as mentioned above for regurgitation. These organic residues potentially infected with mycobacteria can be found hundreds of metres to several kilometres away from the places they were caught (I. Pavlik, unpublished data).

5.10.2.3 Urine

Urine contains a high level of urea that is a very good source of nitrogen for plants. Accordingly, animal urine is often used for the acceleration of the compost fermentation process. However, urine also contains other components, above all various inorganic salts. It is almost sterile compared to faeces/stools, and bacteria (including mycobacteria) are rarely found in it. Nevertheless, alongside the *MTC* members (above all *M. tuberculosis, M. africanum* and *M. bovis*), various slow- and fast-growing mycobacterial species have been found in the urine of people (Table 5.36) and different animal species (Table 5.37). Mycobacteria can be excreted into urine both from kidneys affected by bovine tuberculosis and from infected urinary bladders and tracts. In females, mycobacteria can be found in urine when the reproductive tract is affected. They are rarely found in males, but can potentially be transmitted from the affected accessory glands or testes.

It is evident from Tables 5.36 and 5.37 that as well as OPM, PPM and ESM can be found in urine. The importance of urine in the ecology of mycobacteria has not been sufficiently clarified yet, despite the fact that important causative agents of mycobacterial infections of animals are shed through urine (Table 5.37). Urine is inhibitory and kills bacteria including mycobacteria. For example, *M. a. paratuberculosis* survives in the urine of cattle for only several days. When urine enters the environment (e.g. soil, manure, compost or water) its devitalising effects on mycobacteria lessen or can completely cease. In these cases, urine infected with mycobacteria becomes an important factor for both the spread of mycobacteria and their ecology. The urine of badgers and possums can participate in the interspecies transmission of the causative agent of bovine tuberculosis to cattle (Jackson et al., 1995; Hutchings and Harris, 1999; Hutchings et al., 2001; Jackson, 2002).

5.10.2.4 Non-vertebrates

Even though mycobacteria are acid-fast and resistant to the digestive enzymes of different animals they can also be shed through the secretions of invertebrates and so become sources of mycobacteria for other animals. It is well known that the passage of mycobacteria through the digestive tract of some fly larvae takes just a few tens of minutes. In earthworms (Fischer et al., 2003a) and cockroaches (Fischer et al., 2003b), the passive passage through the intestinal tract after their infection with one dose of, e.g., *M. a. paratuberculosis*, *M. a. avium* and *M. a. hominissuis* lasts up to 72 h. The respective RFLP types of isolates used for infection and the isolates from the faeces of earthworms and cockroaches were always identical. These experiments indicated that the faeces of invertebrates can become part of the environment (Photo 5.65) and subsequently become food for other invertebrates. These can constitute prey for other predators that after death become food for invertebrate animals.

5.10.3 Mycobacteria in Slurry, Liquid Dung and Compost

There are two classes of agricultural fertilisers according to the origin: artificial fertilisers and homestead manure; homestead manure (of farm animal origin) is either solid (manure and compost) or liquid (dung water and slurry). Due to the fact that mycobacteria are an inevitable part of nature (including organic waste), they are also most likely a significant part of the above-mentioned organic fertilisers. The presence of mycobacteria in animal and human faeces/stools (Tables 5.31 and 5.32) and urine (Tables 5.36 and 5.37) is indicative of mycobacterial contamination of organic manure. Accordingly, their detection in manure, compost and slurry can be assumed.

The occurrence of mycobacteria in soil (Section 5.5) and their symbiosis with the roots of definite plant species (Section 6.6) have been described. Consequently, mycobacteria are also found in green matter, i.e. green fodder. Accordingly, the following four items dealing with important factors of ecology and the health of animals and humans present the difficulties arising from the elimination of mycobacteria-containing biomass:

Table 5.36 Mycobacterial species detected in human urine

Isolates	Ref.
M. africanum	[11]
M. avium complex	[10, 13]
M. bovis	[11, 1, 7, 17]
M. fortuitum	[22, 26, 19, 13]
M. gordonae	[15,16, 13]
M. hassiacum	[24]
M. kansasii	[18, 13]
M. malmoense	[12]
M. marinum	[25]
M. neoaurum	[29]
M. scrofulaceum	[14]
M. simiae	[28]
M. terrae	[8], [6]
M. tuberculosis	[11, 9, 2, 21, 20, 5, 3, 27, 23, 4]
M. xenopi	[19]

[1] Albrecht H, Stellbrink HJ, Eggers C, Rusch-Gerdes S, Greten H (1995) Eur. J. Clin. Microbiol. Infect. Dis. 14:226–229. [2] Alsoub H, Al Alousi FS (2001) Ann. Saudi. Med. 21:16–20. [3] Altintepe L, Tonbul HZ, Ozbey I, Guney I, Odabas AR, Cetinkaya R, Piskin MM, Selcuk Y (2005) Ren Fail. 27:657–661. [4] Aslan G, Doruk E, Emekdas G, Serin MS, Direkel S, Bayram G, Durmaz R (2007) Mikrobiyol. Bul. 41:185–192. [5] Boukthir S, Mrad SM, Becher SB, Khaldi F, Barsaoui S (2004) Acta Gastroenterol. Belg. 67:245–249. [6] Carbonara S, Tortoli E, Costa D, Monno L, Fiorentino G, Grimaldi A, Boscia D, Rollo MA, Pastore G, Angarano G (2000) Clin. Infect. Dis. 30:831–835. [7] Chambers MA, Pressling WA, Cheeseman CL, Clifton-Hadley RS, Hewinson RG (2002) Vet. Microbiol. 86:183–189. [8] Chan TH, Ng KC, Ho A, Scheel O, Lai CK, Leung R (1996) Tuber. Lung Dis. 77:555–557. [9] Chaves F, Dronda F, Alonso-Sanz M, Noriega AR (1999) AIDS 13:615–620. [10] Cuervo LM, Martinez FR, Guillot FB, Martinez JC, Font ME (2003) Rev. Cubana Med. Trop. 55:58–60. [11] Grange JM, Yates MD (1992) Br. J. Urol. 69:640–646. [12] Henriques B, Hoffner SE, Petrini B, Juhlin I, Wahlen P, Kallenius G (1994) Clin. Infect. Dis. 18:596–600. [13] Hillemann D, Richter E, Rusch-Gerdes S (2006) J. Clin. Microbiol. 44:4014–4017. [14] Hsueh PR, Hsiue TR, Jarn JJ, Ho SW, Hsieh WC (1996) Clin. Infect. Dis. 22:159–161. [15] Jarikre LN (1991) Am. J. Med. Sci. 302:382–384. [16] Jarikre LN (1992) Genitourin. Med. 68:45–46. [17] Lewis KE, Lucas MG, Smith R, Harrison NK (2003) J. Infect. 46:246–248. [18] Listwan WJ, Roth DA, Tsung SH, Rose HD (1975) Ann. Intern. Med. 83:70–73. [19] Lovodic-Sivcev B, Vukelic A (1999) Med. Pregl. 52:334–342. [20] Mirovic V, Lepsanovic Z (2002) Clin. Microbiol. Infect. 8:709–714. [21] Murcia-Aranguren MI, Gomez-Marin JE, Alvarado FS, Bustillo JG, de Mendivelson E, Gomez B, Leon CI, Triana WA, Vargas EA, Rodriguez E (2001) BMC. Infect. Dis. 1:21. [22] Oren B, Raz R, Hass H (1990) Infection 18:105–106. [23] Rebollo MJ, San Juan GR, Folgueira D, Palenque E, Diaz-Pedroche C, Lumbreras C, Aguado JM (2006) Diagn. Microbiol. Infect. Dis. 56:141–146. [24] Schroder KH, Naumann L, Kroppenstedt RM, Reischl U (1997) Int. J. Syst. Bacteriol. 47:86–91. [25] Streit M, Bohlen LM, Hunziker T, Zimmerli S, Tscharner GG, Nievergelt H, Bodmer T, Braathen LR (2006) Eur. J. Dermatol. 16:79–83. [26] Svahn A, Hoffner SE, Petrini B, Kallenius G (1997) Scand. J. Infect. Dis. 29:573–577. [27] Torrea G, Van de PP, Ouedraogo M, Zougba A, Sawadogo A, Dingtoumda B, Diallo B, Defer MC, Sombie I, Zanetti S, Sechi LA (2005) J. Med. Microbiol. 54:39–44. [28] Valero G, Peters J, Jorgensen JH, Graybill JR (1995) Am. J. Respir. Crit. Care Med. 152:1555–1557. [29] Zanetti S, Faedda R, Fadda G, Dupre I, Molicotti P, Ortu S, Delogu G, Sanguinetti M, Ardito F, Sechi LA (2001) New Microbiol. 24:189–192.

i. A very high content of nitrogen compounds provides favourable conditions for mycobacteria in organic waste and subsequently in the environment: mycobacteria can survive, multiply (e.g. during eutrophisation of water) and spread. Invertebrates (Sections 6.3 and 6.4) and/or vertebrates (Sections 6.6 and 6.7) can become vectors for mycobacteria from slurry tanks, farms and dunghills piled up in the fields.

ii. High densities of captive domestic and wild animals of one or more species (farms, game parks, zoological gardens, etc.) produce rela-

Table 5.37 Mycobacterial species detected in animal urine

Host	Isolates	Ref.
Badgers (*Meles meles*)	*M. bovis*	[2]
Chacma baboons (*Papio ursinus*)	*M. bovis*	[3]
Ferrets (*Mustela putorius* f. *furo*)	*M. bovis*	[4]
Red deer (*Cervus elaphus*)	*M. bovis*	[5]
Rabbits (*Oryctolagus cuniculus*)	*M. a. paratuberculosis*	[1]

[1] Daniels MJ, Henderson D, Greig A, Stevenson K, Sharp JM, Hutchings MR (2003) Epidemiol. Infect. 130:553–559. [2] Gavier-Widen D, Chambers MA, Palmer N, Newell DG, Hewinson RG (2001) Vet. Rec. 148:299–304. [3] Keet DF, Kriek NP, Bengis RG, Grobler DG, Michel A (2000) Onderstepoort J. Vet. Res. 67:115–122. [4] Lugton IW, Wobeser G, Morris RS, Caley P (1997) N. Z. Vet. J. 45:151–157. [5] Palmer MV, Whipple DL, Waters WR (2001) Am. J. Vet. Res. 62:692–696.

tively high amounts of organic waste over a limited area. Thus, high numbers of animals can become "concentrated" sources of different mycobacteria including the causative agents of bovine tuberculosis, *M. bovis*, or paratuberculosis, *M. a. paratuberculosis*.

iii. The products of organic waste fermentation may contain OPM and PPM and thus pose a threat to the health of animal and human populations.

iv. Mycobacteria present in these organic products can be transmitted directly (handling, rising aerosol, contamination of the surface or groundwater, etc.) or indirectly (fertilised agricultural products that are often consumed raw, particularly fruit and vegetables).

Organic matter produced by the fermentation of these organic compounds is an essential part of nature and therefore organic fertilisers must be applied in agriculturally exploited areas. The purpose of this chapter is to describe the occurrence of mycobacteria in organic fertilisers, their resistance against fermentation processes and their subsequent fate in nature.

5.10.3.1 Manure, Slurry and Dung Water

Manure is comprised of animal faeces, urine and bedding (often straw). Urine is mostly absorbed by the bedding material depending on the amount and quality of straw. The production of fertilisers from manure is a complicated biochemical process (Photo 5.66). Various groups of micro-organisms (above all bacteria, fungi and Actinomycetes) participate in this process; they decompose different components of manure and simpler substances are formed. With respect to the ecology of mycobacteria, manure mixed with soil provides ESM and PPM not only with sufficient amounts of nutrients but can also become a source of "massive" amounts of pathogenic mycobacteria for animals and people. Different causative agents of mycobacterial infections are shed through the faeces of the infected animals; the most important are *M. bovis* (Section 2.1.1), *M. a. avium* and *M. a. paratuberculosis* (Section 3.1.3).

Time is the first important factor that causes a reduction in the original amount of OPM in faeces. Due to the fact that there is no evidence of intensive multiplication of OPM outside a host organism (the majority of them do not multiply at all outside a host), this factor is important (see sections mentioned above). The *in vitro* study of the growth capability of different species of mycobacteria revealed that neither *M. bovis* nor *M. tuberculosis* could grow in cattle manure. On the other hand, it was found by microscopy that *M. a. avium* serotype 2 multiplied slightly at 37 °C and that *M. nonchromogenicum*, *M. terrae*, *M. triviale* and other PPM species were able to intensively multiply even at 22 °C (Table 5.38).

The second significant reducing factor for mycobacteria is the intensity of manure fermentation. Aeration, humidity of the environment and temperature are the key factors. The manure ripens for a period of about 2–3 months depending on the season, which significantly affects the humidity and temperature of the manure. However, if fermentation is insufficient, mycobacteria can survive owing to their resistance and can subsequently contaminate the environment for a long time (especially in the vicinity of a dunghill and fertilised soil). A high straw content can contribute to this. *In vitro* experiments have revealed the growth capability of *M. marinum*, *M. gordonae*, *M. scrofulaceum* and other PPM (Table 5.38).

Table 5.38 The growth of mycobacterial species at 22 and 37 °C in media prepared from different substances of autoclaved organic matter "biomass" ([1]; modified)

Runyon's group	Mycobacterial species	Cattle manure	Straw	Peat	Leaves	Sawdust [2]
	M. tuberculosis	−	−	−	−	−
	M. bovis	−	−	−	−	−
I	*M. marinum*	−	+++	−	−	−
II	*M. gordonae*	−	++	−	−	−
	M. scrofulaceum	−	++	++	−	−
III	*M. a. avium* (2)[3]	+	−	++	−	−
	M. a. avium (3)[3]	−	−	++	−	−
	M. a. hominissuis (4)	−	++	++	++	−
	M. a. hominissuis (8)	−	−	++	++	−
	M. gastri	−	−	−	−	−
	M. nonchromogenicum	+++	++	+	++	−
	M. terrae	+++	−	++	++	−
	M. triviale	+++	−	−	−	−
	M. xenopi	−	−	−	−	−
IV	*M. smegmatis*	+++	+++	+	−	−
	M. phlei	+++	−	−	+	−
	M. fortuitum	+++	+++	+	+	−
	M. vaccae	+++	−	+	−	−
	M. diernhoferi	+++	+++	++	++	−
	M. flavescens	+++	+++	−	−	−
pH of substrate used		6.5	5.4	3.8	4.3	4.4

[1] Divided according to Runyon's groups (**I** Photochromogenic species, **II** Scotochromogenic species, **III** Non-photochromogenic species and **IV** Rapid growers).

[2] Sterile sawdust without organic contamination (e.g. soil).

[3] Growth only at 37 °C.

Intensity of growth within 3 months: **−** no growth, **+** separated acid-fast rods and microscopically detected few colony forming units (CFU), **++** microscopic detection of many CFU, **+++** CFU strongly detected by the naked eye. **(2)** Serotype 2.

[1] Beerwerth W (1973) Ann. Soc. Belg. Med. Trop. 53:355–360.

In contrast, PPM contamination of poultry bedding is low. These are likely to be effectively devitalised by fermentation of the bedding that is relatively high in pH (Falkinham et al., 1989). Urine is usually sterile and no bacteria are isolated from healthy individuals. Urine also inhibits *M. a. paratuberculosis* growth, which can be stopped by the addition of 20% urine into the culture medium (Larsen et al., 1956). Direct application of farm animal urine as a fertiliser may be risky because it can burn the vegetation. Urine (usually diluted with drinking, flushing and also rain and surface water) is therefore left standing in tanks. Slurry is formed there, which is considered as a very effective nitrogen–potassium fertiliser depending on its quality. The nutrients present in slurry are available for plant uptake and are fully utilisable immediately after appli-

cation. Owing to the low resistance of mycobacteria to urine and subsequently in slurry (I. Pavlik, unpublished data), this organic manure is safe.

5.10.3.2 Liquid Dung

At present, liquid dung technologies predominate on pig and cattle farms, above all for technical (liquid dung handling is easier) and economic reasons. Liquid dung is usually stored in tanks, which differ in construction and shape: underground or overground concrete tanks, circular enamel-coated steel slurry storage tanks, etc. (Photo 5.67). The solid content separates spontaneously from the liquid part with different contents of organic matter and nutrients. Therefore, regular homogenisation (agitation, usually by submersible propellers) of liquid dung is essential for successful fermentation.

M. a. paratuberculosis was observed to survive in cattle slurry under anaerobic conditions at 5 °C for 252 days (approximately 8 months) and at 15 °C for only 98 days (Jorgensen, 1977).

For example, *M. a. paratuberculosis* was cultivated from samples taken from a liquid dung tank at 2-week intervals for a period of 8 months. No liquid dung was added for 12 months and the content did not undergo any agitation. Several weeks after the last supply of liquid dung, a layer of about 50–70 cm consisting of floating material had formed (straw, parts of non-digested fodder, etc.) on the surface of liquid dung in the tank; that is called the "hat" by farmers. A column of dark brown liquid dung, which was gradually increasing in depth to up to 2–3 m, was present below the hat. The solid particles gradually sedimented to the bottom and a deposit was formed reaching a height of 50–80 cm. Comparable concentrations of *M. a. paratuberculosis* were detected in all of these components during the 8 months (I. Pavlik, unpublished data).

A significant reduction in the number of *M. a. paratuberculosis* was found during anaerobic fermentation of liquid cattle manure under laboratory conditions. *M. a. paratuberculosis* was devitalised by the mesophilic process at 35 °C within 21–28 days; the thermophilic process at 53–55 °C caused killing within 3 h of the initiation of the reaction (Olsen et al., 1985). Therefore, it is necessary to support the introduction of anaerobic fermentation technologies of liquid manure to herds infected with the causative agent of paratuberculosis.

Good quality liquid manure (from cattle, pigs and poultry) is a highly valuable organo-mineral fertiliser, combining the characteristics of manure and mineral fertiliser. Tanks with a sufficient capacity (for at least 6-month storage) is the precondition for obtaining good quality and safe liquid manure.

During the storage (fermentation), nitrogenous organic acids (hippuric and uric acids) and other compounds are decomposed and thereby lose their toxicity for plants. However, mycobacteria are not completely devitalised during the common aerobic fermentation of liquid manure; this was confirmed by their isolation from liquid manure on pig and cattle farms (Table 5.39). Therefore, it is not suitable to apply liquid manure on pastures and crops directly fed to ani-

mals without it first being processed, e.g. by drying or ensilaging (alfalfa/lucerne). It can be used for fertilisation of cereals, root crops, rape, etc.

5.10.3.3 Compost

Composting is basically the acceleration of organic mass decay. It allows utilisation of residual biomass from agricultural premises, remnants from wood-processing plants and municipal waste. It is the most effective way of manure fermentation. Production of compost is best achieved by mixing the biomass with ripening compost, liquid dung or slurry. With respect to mycobacteria, the fact that these are present in the biomass often used for the production of compost is highly risky. According to the composition and production technology, the following categories of compost can be distinguished:

i. *Farm compost*: recycling of organic matter within a farm, e.g. plant remnants (potato haulm, chaff, poppy residues, straw, tree leaves, etc.), devalued feeds, manure, dung water, liquid dung and slurry.
ii. *Commercial compost*: recycling of biodegradable communal and industrial waste (food or paper industry).
iii. *Special compost*: recycling of different cultivating substrates with their subsequent use in horticulture (vegetable leaves, turf, peat or compost obtained with the help of Californian earthworms, i.e. "vermicompost").

Little is known about the occurrence of mycobacteria in compost. In 1978 one of the first publications dealing with this referred to the occurrence of *MAC* members in compost fed to piglets as a supplement. *M. a. avium* of serotypes 2 and 3 and *M. a. hominissuis* serotypes 4, 8 and 9 were isolated from compost. *MAC* members of identical serotypes were isolated from tuberculous lesions of slaughtered pigs. This compost was produced from household waste mixed with peat by a local Dutch company. It was noteworthy that no *MAC* members were isolated from peat; this indicated that compost was the source of infection (Engel et al., 1978). However, the obtained *in vitro* results did not exclude peat as a source of infection for the pigs. Mycobacteria are not only present in

Table 5.39 Mycobacterial species detected in different wastes on cattle and pig farms

Type of waste	Animals	Isolates	Ref.
Manure	Cattle	*MAP*	[5]
Liquid dung	Cattle	*M. fortuitum, MAP, MAH* (8, 9)	[2, 3]
Slurry	Pig	RGM	[6]
Compost	Pig	*MAA* (2, 3), *MAH* (4, 8, 9)	[1]
Liquid dung	Pig	*M. fortuitum, MAH* (8, 9)	[2, 4]

MAP *M. a. paratuberculosis.* **MAH** *M. a. hominissuis.* **MAA** *M. a. avium.* **RGM** Rapidly growing mycobacterial species. **(8, 9)** Serotypes 8 and 9.
[1] Engel HW, Groothuis DG, Wouda W, Konig CD, Lendfers LH (1978) Zentralbl. Veterinarmed. B 25:373–382. [2] Fischer OA, Matlova L, Bartl J, Dvorska L, Svastova P, du Maine R, Melicharek I, Bartos M, Pavlik I (2003) Veterinary Microbiology 91:325–338. [3] Fischer OA, Matlova L, Dvorska L, Svastova P, Bartos M, Weston RT, Kopecna M, Trcka I, Pavlik I (2005) Med. Vet. Entomol. 19:360–366. [4] Fischer OA, Matlova L, Dvorska L, Svastova P, Bartos M, Weston RT, Pavlik I (2006) Folia Microbiol.(Praha) 51:147–153. [5] Jaravata CV, Smith WL, Rensen GJ, Ruzante JM, Cullor JS (2006) Foodborne. Pathog. Dis. 3:212–215. [6] Jones PW, Bew J, Burrows MR, Matthews PR, Collins P (1976) J. Hyg. (Lond) 77:43–50.

peat (Section 5.6.3), but they can also multiply there (Table 5.38).

The growth capability of mycobacteria in straw was mentioned above. However, *in vitro* experiments have shown that mycobacteria can also grow on tree leaves, which are the material for compost production, above all in cities. Both *M. a. hominissuis* serotypes 4 and 8 and other PPM grew on tree leaves (Table 5.38). Accordingly, it is necessary to monitor the entire process and test its effectiveness, i.e. the decomposition of organic matter under temperatures causing the devitalisation of mycobacteria. The process of composting continues for several months (i.e. fast "hot" composting) or up to 3–4 years (slow composting). As far as the ecology of mycobacteria is concerned, the conditions under which compost is produced (in three stages) should be considered.

microflora. It is not suitable to add manure or slurry because they contain only enteric bacteria, which cause putrescence. *M. phlei, M. a. avium* and other thermoresistant mycobacterial species most likely survive the process of mineralisation (Wayne and Kubica, 1986; Pavlas, 1990).

Mycobacterial skin infection caused by *M. thermoresistibile* after a pitchfork injury (6-month history of a violaceous indurated plaque that developed on the left ankle after stepping on gardening shears while spreading mulch containing an unknown type of manure on the lawn) is unusual (Cummings et al., 2000). The name of the *Mycobacterium* indicates that this species is classified among the thermophilic species of mycobacteria, requires incubation at 52 °C and can thus survive the composting fermentation processes.

Stages of Decomposition (The Key Stage of the Fermentation Process with a Concurrent Rise in Temperature)

During the first "aerobic" stage, which lasts for 3–4 weeks, the compost must be appropriately aerated. According to the material, a temperature increase of between 50 and 70 °C is typical. Organic compounds are broken down (mineralised) into nitrates, carbon dioxide, ammonia, amino acids and polysaccharides by the activities of, in particular, aerobic bacteria. A good process of composting can be initiated by the addition of finished compost or earth, which contain soil

Stage of Conversion

The second stage lasts another 4–6 weeks. The temperature gradually decreases and mineralised nutrients become components of the "humus complex". The compost is of a uniform brown colour, has a friable texture and a slight odour resembling forest soil. Its fertilising effect is highest in this stage and mycobacterial contamination is probably at its lowest. These can survive the first stage (mostly thermoresistant species) and start to multiply again in the following stage if the temperature conditions are favourable (Kazda, 2000).

Stage of Synthesis

In the third stage lasting 3–4 weeks, the "live humus" slowly changes to more mineralised "permanent humus". Its fertilising effect is lower in comparison with the second stage compost, but its effect in the soil is higher. Compost at this stage can be secondarily contaminated with mycobacteria from the faeces of infected birds, etc. Due to the fact that the fermentation processes are less intensive, devitalisation of mycobacteria does not occur.

Risk Factors Affecting the Presence of Mycobacteria

The majority of micro-organisms in organic materials are mesophilic; the optimum temperature for their multiplication ranges between 20 and 30 °C. The group of thermophilic aerobic micro-organisms – indispensable to the effectiveness of the composting process – starts to prevail at higher temperatures. The temperatures range between 45 and 65 °C. The pH value of fresh compost ranges between 6 and 8.

First, the relatively low temperatures, which do not kill mycobacteria, should be viewed as a high risk. If manure or liquid dung is composted in small compost piles (of a height of up to 1.5 m), it is difficult to maintain the required temperature for a sufficient time; temperatures below 55 °C for a period of less than 21 days are common in such cases. Various pathogenic micro-organisms including PPM or OPM may be present in compost made this way.

Second, with respect to mycobacteria, potential secondary contamination of compost made in this way (see above) may pose a risk. Mycobacteria can be transmitted into the compost through the faeces of insectivorous species of birds and mammals. They may come into contact with the surface of compost piles or dig corridors through them when pursuing invertebrate pray. ESM and PPM can secondarily contaminate compost with rainwater, dust and other vectors (Photo 5.68).

5.10.3.4 Exposure of Animals and Humans to Waste-Derived Mycobacteria

Animal excreta are also potential sources of micro-organisms (bacteria, viruses and parasites) pathogenic for people and animals. Their amount in excrement varies considerably and largely depends on the health of captive animals. In contrast to manure, the self-heating of liquid dung is not possible (high amounts of technological water, diluted mixtures of faeces, urine, remnants of fodder and other substances). The duration of pathogen survival largely depends on the physical and chemical state of liquid dung and hence can exceed the usual storage time. Some micro-organisms such as salmonellae can multiply in liquid dung (Pell, 1997).

When non-homogenised liquid dung is applied on the field, the soil is unevenly fertilised depending on the proportion of dry matter and nutrients in the used fraction. If the homogenisation is not perfect, the sediment on the bottom can be thicker than 1 m; accordingly, liquid dung in the tanks must be homogenised by means of submersible propellers. They facilitate its transport to the fields, permit complete emptying of the tanks and prevent aerosol formation when liquid dung is drawn (Section 5.9). The hose connector applicator system, which evenly applies the liquid dung, prevents aerosol formation in the fields.

When adopting preventive measures against the spread of mycobacteria from a farm, it is necessary to keep in mind that mycobacteria shed from animal bodies are present in liquid dung and manure. Disinfectants can be used for their reduction (Scanlon and Quinn, 2000), which, however, also kill the fermentation bacteria, Actinomycetes and fungi. Consequently, the fermentation process slows down. With this in mind, it is better to prevent the spread of mycobacterial infection outside the farm by thorough fermentation and a time reserve (e.g. leaving manure in a compost pile for more than a year). If liquid dung and manure are wrongly handled, the causative agent of mycobacterial infection can be transmitted beyond the farm, in particular to wild-living animals. These can then participate in further spread of OPM (Section 2.1.1).

Clothes contaminated with liquid dung, manure or slurry should be washed at temperatures above 45 °C. Laboratory experiments with *M. terrae* performed in Slovenia showed that this mycobacterial species could survive the temperature of 45 °C. Accordingly, temperatures higher than 45 °C are also recommended in laundry procedures for hospital clothing (Fijan et al., 2007).

5.10.4 Views and Perspectives on the Research

With regard to slowly changing feeding habits and the increasing human consumption of vegetables and fruit, it is necessary to take into account the increasing exposure of the human population to various infectious agents. These are pathogens of bacterial, viral and parasitic diseases (Bryan, 1977). Research should therefore be focused on the risks associated with the occurrence of ESM and PPM in different types of fertilisers and their transmission through non-heat-treated vegetables and fruit to animals and people.

It will also be necessary to assess the extent of risk for people working in agriculture exposed to different components in the stable and pasture environments. Ecotourism (holidays on agricultural farms) has not yet been analysed with regard to the occurrence of ESM and PPM in different components of the environment. The localisation of causative agents of various mycobacterial infections (e.g. paratuberculosis or avian mycobacteriosis) in that environment may be unexpected and often unpredictable.

In some cases, high amounts of mycobacteria (ESM and PPM) have been isolated from the faeces of different ruminant species. Therefore, we assume that they can intensively multiply in some parts of the intestinal tract. Their increased concentrations can be supposed in forestomachs (Section 6.2). Further research should therefore be focused on investigation of the species profile and localisation of ESM, PPM and pathogenic mycobacteria (e.g. *M. a. paratuberculosis*) in different parts of the intestinal tract.

Due to the fact that the application of composting technologies is increasingly recommended mainly with regard to the ecological aspects, more intensive studies are necessary to investigate the presence of mycobacteria in composts and their ability to survive there.

Acknowledgements Partially supported by the European Commission PathogenCombat FOOD-CT-2005-007081 (Sections 5.6 and 5.7) and ParaTBTools FP6-2004-FOOD-3B-023106 (Section 5.10). Grants from the Ministry of Agriculture of the Czech Republic NPV 1B53009 partially supported the Sections 5.2, 5.5, 5.6, 5.7, 5.8, 5.9 and 5.10; NAZV QH71054 Section 5.6.3; and NAZV QH81065 Sections 5.5, 5.6, 5.7 and 5.10.

References

Abbes C, Parent LE, Karam A (1993) Ammonia Sorption by Peat and N-Fractionation in Some Peat-Ammonia Systems. Fertilizer Res. 36:249–257

Abdullin K, Morozovskii KK, Kiriliuk DA, Pavlova IP (1969) [Survival of avian tuberculosis mycobacteria in feed grain]. Veterinariia. 46:102–105

Ancusa M, Terbancea W (1970) [Occurrence of tuberculosis bacteria in streams]. Z. Gesamte Hyg. 16:913–916

Andersen AA (1958) New sampler for the collection, sizing, and enumeration of viable airborne particles. J. Bacteriol. 76:471–484

Andersson MA, Nikulin M, Koljalg U, Andersson MC, Rainey F, Reijula K, Hintikka EL, SalkinojaSalonen M (1997) Bacteria, molds, and toxins in water-damaged building materials. Appl. Environ. Microbiol. 63:387–393

Argueta C, Yoder S, Holtzman AE, Aronson TW, Glover N, Berlin OG, Stelma GN, Jr., Froman S, Tomasek P (2000) Isolation and identification of nontuberculous mycobacteria from foods as possible exposure sources. J. Food Prot. 63:930–933

Asadi-Pooya AA, Pnjehshahin MR, Beheshti S (2003) The antimycobacterial effect of honey: an *in vitro* study. Riv. Biol. 96:491–495

Aubuchon C, Hill JJ, Jr., Graham DR (1986) Atypical mycobacterial infection of soft tissue associated with use of a hot tub. A case report. J. Bone Joint Surg. Am. 68:766–768

Ayele WY, Neill SD, Zinsstag J, Weiss MG, Pavlik I (2004) Bovine tuberculosis: an old disease but a new threat to Africa. Int. J. Tuberc. Lung Dis. 8:924–937

Baker MG, Lopez LD, Cannon MC, de Lisle GW, Collins DM (2006) Continuing *Mycobacterium bovis* transmission from animals to humans in New Zealand. Epidemiol. Infect. 134:1068–1073

Bartos M, Hlozek P, Svastova P, Dvorska L, Bull T, Matlova L, Parmova I, Kuhn I, Stubbs J, Moravkova M, Kintr J, Beran V, Melicharek I, Ocepek M, Pavlik I (2006) Identification of members of *Mycobacterium avium* species by Accu-Probes, serotyping, and single IS*900*, IS*901*, IS*1245* and IS*901*-flanking region PCR with internal standards. J. Microbiol. Methods. 64:333–345

Beattie VE, Walker N, Sneddon IA (1998) Preference testing of substrates by growing pigs. Anim. Welfare. 7:27–34

Bedrynska-Dobek M (1966) Investigations of sewage sediment and water from the pond Starorzecze-Naramowice for the presence of tubercle bacilli. Pol. Med. J. 5:1058–1064

Beer AM, Grozeva A, Sagorchev P, Lukanov J (2003a) Comparative study of the thermal properties of mud and peat solutions applied in clinical practice. Biomed. Tech. (Berl). 48:301–305

Beer AM, Junginger HE, Lukanov J, Sagorchev P (2003b) Evaluation of the permeation of peat substances through human skin in vitro. Int. J. Pharm. 253:169–175

Beer AM, Lukanov J, Sagorchev P (2000) The influence of fulvic and ulmic acids from peat, on the spontaneous contractile activity of smooth muscles. Phytomedicine. 7:407–415

Beer AM, Sagorchev P, Lukanov J (2002) Isolation of biologically active fractions from the water soluble components of fulvic and ulmic acids from peat. Phytomedicine. 9: 659–666

Beerwerth W (1971) [Mycobacterial soil flora in the course of the seasons]. Prax. Pneumol. 25:661–668

Beerwerth W (1973) [The use of natural substrates as culture media for mycobacteria]. Ann. Soc. Belg. Med. Trop. 53:355–360

Beerwerth W, Kessel U (1976) [Mycobacteria in the environment of man and animal (proceedings)]. Zentralbl. Bakteriol. [Orig. A]. 235:177–183

Beerwerth W, Schurmann J (1969) [Contribution to the ecology of mycobacteria]. Zentralbl. Bakteriol. [Orig.]. 211: 58–69

Bejsovec J (1962) Spreading of helminths' germs by passage through the intestinal tract of adequate carriers (in Czech). Cs. Parasitol. 9:95–109

Bellometti S, Giannini S, Sartori L, Crepaldi G (1997) Cytokine levels in osteoarthrosis patients undergoing mud bath therapy. Int. J. Clin. Pharmacol. Res. 17:149–153

Bennett SN, Peterson DE, Johnson DR, Hall WN, Robinson-Dunn B, Dietrich S (1994) Bronchoscopy-associated *Mycobacterium xenopi* pseudoinfections. Am. J. Respir. Crit Care Med. 150:245–250

Beran V, Havelkova M, Kaustova J, Dvorska L, Pavlik I (2006) Cell wall deficient forms of mycobacteria: a review. Veterinarni Medicina 51:365–389

Bercovier H, Vincent V (2001) Mycobacterial infections in domestic and wild animals due to *Mycobacterium marinum, M. fortuitum, M. chelonae, M. porcinum, M. farcinogenes, M. smegmatis, M. scrofulaceum, M. xenopi, M. kansasii, M. simiae* and *M. genavense.* Rev. Sci. Tech. 20: 265–290

Bernstein DI, Lummus ZL, Santilli G, Siskosky J, Bernstein IL (1995) Machine Operators Lung - A Hypersensitivity Pneumonitis Disorder Associated with Exposure to Metalworking Fluid Aerosols. Chest. 108:636–641

Besser RE, Pakiz B, Schulte JM, Alvarado S, Zell ER, Kenyon TA, Onorato IM (2001) Risk factors for positive mantoux tuberculin skin tests in children in San Diego, California: evidence for boosting and possible foodborne transmission. Pediatrics. 108:305–310

Blagodarnyi I, Vaksov VM (1972) [Epidemiological and epizootiological significance of effluents coming from antituberculous establishments]. Probl. Tuberk. 50:8–12

Body BA, Boyd JC (1988) Acid-fast staining of urine and gastric contents is an excellent indicator of mycobacterial disease. Am. Rev. Respir. Dis. 137:1514–1515

Bohrerova Z, Linden KG (2006) Ultraviolet and chlorine disinfection of *Mycobacterium* in wastewater: effect of aggregation. Water Environ. Res. 78:565–571

Bonsu OA, Laing E, Akanmori BD (2000) Prevalence of tuberculosis in cattle in the Dangme-West district of Ghana, public health implications. Acta Trop. 76:9–14

Brooks OH (1971) Observations on outbreaks of Battey type mycobacteriosis in pigs raised on deep litter. Aust. Vet. J. 47:424–427

Brooks RW, George KL, Parker BC, Falkinham JO, III, Gruff H (1984) Recovery and survival of nontuberculous mycobacteria under various growth and decontamination conditions. Can. J. Microbiol. 30:1112–1117

Brown J, Tollison JW (1979) Influence of pork consumption on human infection with *Mycobacterium avian-intracellulare*. Appl. Environ. Microbiol. 38:1144–1146

Bryan FL (1977) Diseases Transmitted by Foods Contaminated by Wastewater. J. Food Prot. 40:45–56

Buczowska Z (1965) Tubercle bacilli in the Sewage and in Sewage-receiving waters. Biul. Inst. Med. Morsk. Gdansk. 16:49–56

Buraczewski O, Osinski J (1966) Acid-fast bacilli in sewage. Pol. Med. J. 5:1065–1072

Burchfield SR, Elich MS, Woods SC (1977) Geophagia in Response to Stress and Arthritis. Physiol. Behav. 19:265–267

Carpenter TE, Hird DW (1986) Time-Series Analysis of Mycobacteriosis in California Slaughter Swine. Prev. Vet. Med.. 3:559–572

Carson LA, Bland LA, Cusick LB, Favero MS, Bolan GA, Reingold AL, Good RC (1988) Prevalence of nontuberculous mycobacteria in water supplies of hemodialysis centers. Appl. Environ. Microbiol. 54:3122–3125

Carter G, Wu M, Drummond DC, Bermudez LE (2003) Characterization of biofilm formation by clinical isolates of *Mycobacterium avium.* J. Med. Microbiol. 52:747–752

Cavanagh HM, Hipwell M, Wilkinson JM (2003) Antibacterial activity of berry fruits used for culinary purposes. J. Med. Food. 6:57–61

Chan DKO, Chaw D, Lo CYY (1994) Management of the Sawdust Litter in the Pig-On-Litter System of Pig Waste Treatment. Resour. Conserv. Recy.. 11:51–72

Chanarin I, Stephenson E (1988) Vegetarian diet and cobalamin deficiency: their association with tuberculosis. J. Clin. Pathol. 41:759–762

Chapman JS (1971) The ecology of the a typical mycobacteria Arch. Environ. Health 22:41–46

Chapman JS, Bernard JS, Speight M (1965) Isolation of mycobacteria from raw milk. Am. Rev. Respir. Dis. 91:351–355

Charette R, Martineau GP, Pigeon C, Turcotte C, Higgins R (1989) An outbreak of granulomatous lymphadenitis due to *Mycobacterium avium* in swine. Can. Vet. J. 30:675–678

Chen SK, Vesley D, Brosseau LM, Vincent JH (1994) Evaluation of single-use masks and respirators for protection of health care workers against mycobacterial aerosols. Am. J. Infect. Control. 22:65–74

Cheung PY, Kinkle BK (2001) *Mycobacterium* diversity and pyrene mineralization in petroleum-contaminated soils. Appl. Environ. Microbiol. 67:2222–2229

Chiodini RJ (1989) Crohn Disease and the Mycobacterioses - A Review and Comparison of 2 Disease Entities. Clin. Microbiol. Rev. 2:90–117

Chiodini RJ, Van Kruiningen HJ, Merkal RS (1984) Ruminant paratuberculosis (Johne's disease): the current status and future prospects. Cornell Vet. 74:218–262

Choi JH, Kim YH, Joo DJ, Choi SJ, Ha TW, Lee DH, Park IH, Jeong YS (2003) Removal of ammonia by biofilters: a study with flow-modified system and kinetics. J. Air Waste Manag. Assoc. 53:92–101

Chumkaew P, Karalai C, Ponglimanont C, Chantrapromma K (2003) Antimycobacterial activity of phorbol esters from the fruits of *Sapium indicum.* J. Nat. Prod. 66:540–543

Clark DL, Jr., Anderson JL, Koziczkowski JJ, Ellingson JL (2006) Detection of *Mycobacterium avium* subspecies *paratuberculosis genetic* components in retail cheese curds purchased in Wisconsin and Minnesota by PCR. Mol. Cell Probes. 20:197–202

Codias EK, Reinhardt DJ (1979) Distribution of serotypes of the *Mycobacterium avium-intracellulare-scrofulaceum* complex in Georgia. Am. Rev. Respir. Dis. 119:965–970

Collins CH, Grange JM, Yates MD (1984) Mycobacteria in water. J. Appl. Bacteriol. 57:193–211

Collins CH, Yates MD (1984) Infection and colonisation by *Mycobacterium kansasii* and *Mycobacterium xenopi*: aerosols as a possible source? J. Infect. 8:178–179

Combourieu B, Besse P, Sancelme M, Veschambre H, Delort AM, Poupin P, Truffaut N (1998a) Morpholine degradation pathway of *Mycobacterium aurum* MO1: direct evidence of intermediates by in situ 1H nuclear magnetic resonance. Appl. Environ. Microbiol. 64:153–158

Combourieu B, Poupin P, Besse P, Sancelme M, Veschambre H, Truffaut N, Delort AM (1998b) Thiomorpholine and morpholine oxidation by a cytochrome P450 in *Mycobacterium aurum* MO1. Evidence of the intermediates by in situ 1H NMR. Biodegradation. 9:433–442

Conlon CP, Banda HM, Luo NP, Namaambo MK, Perera CU, Sikweze J (1989) Faecal mycobacteria and their relationship to HIV-related enteritis in Lusaka, Zambia. AIDS. 3:539–541

Conn VM, Franco CM (2004) Analysis of the endophytic actinobacterial population in the roots of wheat (*Triticum aestivum* L.) by terminal restriction fragment length polymorphism and sequencing of 16S rRNA clones. Appl. Environ. Microbiol. 70:1787–1794

Cooney R, Kazda J, Quinn J, Cook B, Muller K, Monaghan M (1997) Environmental mycobacteria in Ireland as a source of non-specific sensitization to tuberculin. Irish Vet. J. 50:370–373

Corbett EL, Churchyard GJ, Clayton T, Herselman P, Williams B, Hayes R, Mulder D, De Cock KM (1999) Risk factors for pulmonary mycobacterial disease in South African gold miners. A case-control study. Am. J. Respir. Crit. Care Med. 159:94–99

Corbett EL, Churchyard GJ, Clayton TC, Williams BG, Mulder D, Hayes RJ, De Cock KM (2000) *HIV* infection and silicosis: the impact of two potent risk factors on the incidence of mycobacterial disease in South African miners. AIDS. 14:2759–2768

Corner LA, Pearson CW (1979) Response of cattle to inoculation with atypical mycobacteria isolated from soil. Aust. Vet. J. 55:6–9

Cosivi O, Grange JM, Daborn CJ, Raviglione MC, Fujikura T, Cousins D, Robinson RA, Huchzermeyer HFAK, de Kantor I, Meslin FX (1998) Zoonotic tuberculosis due to *Mycobacterium bovis* in developing countries. Emerg. Infect. Dis. 4:59–70

Cosivi O, Meslin FX, Daborn CJ, Grange JM (1995) Epidemiology of *Mycobacterium-Bovis* Infection in Animals and Humans, with Particular Reference to Africa. Revue Scientifique et Technique de l Office International des Epizooties. 14:733–746

Costallat LF, Pestana de Castro AF, Rodrigues AC, Rodrigues FM (1977) Examination of soils in the Campinas rural area for microorganisms of the *Mycobacterium avium-intracellulare-scrofulaceum* complex. Aust. Vet. J. 53:349–350

Covert TC, Rodgers MR, Reyes AL, Stelma GN, Jr. (1999) Occurrence of nontuberculous mycobacteria in environmental samples. Appl. Environ. Microbiol. 65:2492–2496

Cowan HE, Falkinham JO, III (2001) A luciferase-based method for assessing chlorine-susceptibility of *Mycobacterium avium*. J. Microbiol. Methods. 46:209–215

Cronin GM, Schirmer BN, Mccallum TH, Smith JA, Butler KL (1993) The Effects of Providing Sawdust to Pre-Parturient Sows in Farrowing Crates on Sow Behavior, the Duration of Parturition and the Occurrence of *Intra-Partum* Stillborn Piglets. Appl. Anim. Behav. Sci. 36:301–315

Cummings GH, Natarajan S, Dewitt CC, Gardner TL, Garces MC (2000) *Mycobacterium thermoresistible* recovered from a cutaneous lesion in an otherwise healthy individual. Clin. Infect. Dis. 31:816–817

Dailloux M, Morlot M, Sirbat C (1980) [Study of factors affecting presence of atypical Mycobacteria in water of a swimming pool (author's transl)]. Rev. Epidemiol. Sante Publique. 28:299–306

Dalchow W (1988) Mycobacteriosis in pigs fed cereal waters (in German). Tietärztliche Umschau. 43:62–74

Damsker B, Bottone EJ (1985) *Mycobacterium avium-Mycobacterium intracellulare* from the intestinal tracts of patients with the acquired immunodeficiency syndrome: concepts regarding acquisition and pathogenesis. J. Infect. Dis. 151:179–181

Danilova AK, Naidenskii MS, Shpits IS, Plotinskii I (1968) [Zoohygienic assessment of various types of peat used for litter]. Veterinariia. 45:88–91

David HL, Jones WD, Jr., Newman CM (1971) Ultraviolet light inactivation and photoreactivation in the mycobacteria. Infect. Immun. 4:318–319

Dawson DJ, Armstrong JG, Blacklock ZM (1982) Mycobacterial cross-contamination of bronchoscopy specimens. Am. Rev. Respir. Dis. 126:1095–1097

De Groote MA, Pace NR, Fulton K, Falkinham JO, III (2006) Relationships between *Mycobacterium* isolates from patients with pulmonary mycobacterial infection and potting soils. Appl. Environ. Microbiol. 72:7602–7606

Debacker M, Zinsou C, Aguiar J, Meyers WM, Portaels F (2003) First case of *Mycobacterium ulcerans* disease (Buruli ulcer) following a human bite. Clin. Infect. Dis. 36:e67-e68

Delaha EC, Garagusi VF (1985) Inhibition of mycobacteria by garlic extract (*Allium sativum*). Antimicrob. Agents Chemother. 27:485–486

Diamond J, Bishop KD, Gilardi JD (1999) Geophagy in New Guinea birds. Ibis. 141:181–193

Dominy NJ, Davoust E, Minekus M (2004) Adaptive function of soil consumption: an in vitro study modeling the human stomach and small intestine. J. Exp. Biol. 207:319–324

Donoghue HD, Overend E, Stanford JL (1997) A longitudinal study of environmental mycobacteria on a farm in south-west England. J. Appl. Microbiol. 82:57–67

Dudley DJ, Guentzel MN, Ibarra MJ, Moore BE, Sagik BP (1980) Enumeration of potentially pathogenic bacteria from sewage sludges. Appl. Environ. Microbiol. 39:118–126

du Moulin GC, Stottmeier KD, Pelletier PA, Tsang AY, Hedley-Whyte J (1988) Concentration of *Mycobacterium avium* by hospital hot water systems. JAMA. 260:1599–1601

du Moulin GC, Stottmeier KD (1986) Waterborne mycobacteria: an increasing threat to health. ASM News. 52:525–529

Durborow RM (1999) Health and safety concerns in fisheries and aquaculture. Occup. Med.-State of the Art Reviews. 14:373–406

Durrling H, Ludewig F, Uhlemann J, Gericke R (1998) Peat as a source of *Mycobacterium avium* infection for pigs. Tierarztliche Umschau. 53:259–261

Dutkiewicz J (1994) Bacteria, Fungi, and Endotoxin As Potential Agents of Occupational Hazard in A Potato Processing Plant. Am. J. Ind. Med.. 25:43–46

Dutkiewicz J, Pomorski ZJH, Sitkowska J, Krysinskatraczyk E, Skorska C, Prazmo Z, Cholewa G, Wojtowicz H (1994) Airborne Microorganisms and Endotoxin in Animal Houses. Grana. 33:85–90

Dvorska L, Bartos M, Ostadal O, Kaustova J, Matlova L, Pavlik I (2002) IS*1311* and IS*1245* restriction fragment length polymorphism analyses, serotypes, and drug susceptibilities of *Mycobacterium avium* complex isolates obtained from a human immunodeficiency virus-negative patient. J. Clin. Microbiol. 40:3712–3719

Dvorska L, Matlova L, Ayele WY, Fischer OA, Amemori T, Weston RT, Alvarez J, Beran V, Moravkova M, Pavlik I (2007) Avian tuberculosis in naturally infected captive water birds of the Ardeidae and Threskiornithidae families studied by serotyping, IS*901* RFLP typing, and virulence for poultry. Vet. Microbiol. 119:366–374

Dvorska L, Parmova I, Lavickova M, Bartl J, Vrbas V, Pavlik I (1999) Isolation of *Rhodococcus equi* and atypical mycobacteria from lymph nodes of pigs and cattle in herds with the occurrence of tuberculoid gross changes in the Czech Republic over the period of 1996–1998. Veterinarni Medicina 44: 321–330

Edwards LB, Acquaviva FA, Livesay VT, Cross FW, Palmer CE (1969) An atlas of sensitivity to tuberculin, PPD-B, and histoplasmin in the United States. Am. Rev. Respir. Dis. 99:Suppl-132

Egamberdiyeva D, Hoflich G (2004) Effect of plant growth-promoting bacteria on growth and nutrient uptake of cotton and pea in a semi-arid region of Uzbekistan. J. Arid Environ. 56:293–301

Eichelsdorfer D (1992) [Examination and evaluation of the hygiene status of natural peloids for human medical use]. Gesundheitswesen. 54:400–405

Eilertsen E (1969) Atypical mycobacteria and reservoir in water. Scand. J. Respir. Dis. Suppl. 69:85–88

Ellertsen LK, Wiker HG, Egeberg NT, Hetland G (2005) Allergic sensitisation in tuberculosis and leprosy patients. Int. Arch. Allergy Immunol. 138:217–224

Embil JM, Warren CPW (1997) Pneumonitis due to *Mycobacterium avium* complex in hot tub water - Infection or hypersensitivity? Chest. 112:1713–1714

Emde KME, Chomyc SA, Finch GR (1992) Initial Investigation on the Occurrence of *Mycobacterium* Species in Swimming Pools. J. Environ. Health. 54:34–36

Engel HW, Berwald LG, Havelaar AH (1980) The occurrence of *Mycobacterium kansasii* in tapwater. Tubercle. 61: 21–26

Engel HW, Groothuis DG, Wouda W, Konig CD, Lendfers LH (1978) "Pig-compost" as a source of *Mycobacterium avium* infection in swine. Zentralbl. Veterinarmed. B. 25:373–382

Ermolenko ZM, Kholodenko VP, Chugunov VA, Zhirkova NA, Rasulova GE (1997) A mycobacterial strain isolated from the oil of the Ukhtinskoe oil field: Identification and degradative properties. Microbiology. 66:542–545

Falkinham JO, III (1996) Epidemiology of infection by nontuberculous mycobacteria. Clin. Microbiol. Rev. 9:177–215

Falkinham JO (2003) The changing pattern of nontuberculous mycobacterial disease. Can. J. Infect. Dis. 14:281–286

Falkinham JO, III (2007) Growth in catheter biofilms and antibiotic resistance of *Mycobacterium avium*. J. Med. Microbiol. 56:250–254

Falkinham JO, III, George KL, Parker BC (1989) Epidemiology of infection by nontuberculous mycobacteria. VIII. Absence of mycobacteria in chicken litter. Am. Rev. Respir. Dis. 139:1347–1349

Falkinham JO, Iseman MD, de Haas P, van Soolingen D (2008) *Mycobacterium avium* in a shower linked to pulmonary disease. J. Water Health. 6:209–213

Falkinham JO, III, Norton CD, LeChevallier MW (2001) Factors influencing numbers of *Mycobacterium avium, Mycobacterium intracellulare*, and other Mycobacteria in drinking water distribution systems. Appl. Environ. Microbiol. 67:1225–1231

Falkinham JO, III, Parker BC, Gruft H (1980) Epidemiology of infection by nontuberculous mycobacteria. I. Geographic distribution in the eastern United States. Am. Rev. Respir. Dis. 121:931–937

Ferreira R, Fonseca LS, Afonso AM, da Silva MG, Saad MH, Lilenbaum W (2006) A report of mycobacteriosis caused by *Mycobacterium marinum* in bullfrogs (*Rana catesbeiana*). Vet. J. 171:177–180

Fertig S (1961) [Area irrigated with sewage. Its hygienic and sanitary evaluation. II. Examination of rodents from fields irrigated with sewage for the presence of tubercle bacilli.]. Acta Microbiol. Pol. 10:429–432

Fijan S, Koren S, Cencic A, Sostar-Turk S (2007) Antimicrobial disinfection effect of a laundering procedure for hospital textiles against various indicator bacteria and fungi using different substrates for simulating human excrements. Diagn. Microbiol. Infect. Dis. 57:251–257

Fine PE (1995) Variation in protection by BCG: implications of and for heterologous immunity. Lancet. 346: 1339–1345

Fischer O, Matlova L, Bartl J, Dvorska L, Melicharek I, Pavlik I (2000) Findings of mycobacteria in insectivores and small rodents. Folia Microbiol. (Praha). 45:147–152

Fischer OA, Matlova L, Bartl J, Dvorska L, Svastova P, du MR, Melicharek I, Bartos M, Pavlik I (2003a) Earthworms (Oligochaeta, Lumbricidae) and mycobacteria. Vet. Microbiol. 91:325–338

Fischer OA, Matlova L, Dvorska L, Svastova P, Pavlik I (2003b) Nymphs of the Oriental cockroach (*Blatta orientalis*) as passive vectors of causal agents of avian tuberculosis and paratuberculosis. Med. Vet. Entomol. 17:145–150

Fischer OA, Matlova L, Dvorska L, Svastova P, Peral DL, Weston RT, Bartos M, Pavlik I (2004) Beetles as possible vectors of infections caused by *Mycobacterium avium* species. Vet. Microbiol. 102:247–255

Fodstad FH (1977) Tuberculin reactions in bulls and boars sensitized with atypical Mycobacteria from sawdust. Acta Vet. Scand. 18:374–383

Frahm H (1959) Die Lebensdauer pathogener Mikroben in Milch und Milcherzeugnissen insbesondere von

Tuberkulosebakterien im Kase. Kieler Milchw. ForschBer. 11:333–339

Framstad T, Hed-Opp G, Rein KA (2001) The use of peat to prevent diarrhoea after weaning. Int. Pig Topics. 16 (3):7–9

Francis J (1973) Letter: Very small public health risk from flesh of tuberculous cattle. Aust. Vet. J. 49:496–497

Fuchs B, Orda J, Pres J, Muchowicz M (1995) The effect of feeding piglets up to the 100th day of their life with peat preparation on their growth and physiological and biochemical indices. Arch. Vet. Pol. 35:97–107

Fuchs V, Kuhnert M, Golbs S, Dedek W (1990) [The enteral absorption of iron (II) from humic acid-iron complexes in suckling piglets using radiolabeled iron (59Fe)]. Dtsch. Tierarztl. Wochenschr. 97:208–209

Garcia MM, Brooks BW, Stewart RB, Dion W, Trudel JR, Ouwerkerk T (1987) Evaluation of gamma radiation levels for reducing pathogenic bacteria and fungi in animal sewage and laboratory effluents. Can. J. Vet. Res. 51:285–289

Garcia MM, McKay KA (1970) Pathogenic microorganisms in soil: an old problem in a new perspective. Can. J. Comp. Med. 34:105–110

Gawaad AAA, Heins B, Stein W (1997) Untersuchung uber die Attraktivwierkung von Lebensmitteln fur Fliegen (Diptera, Calliphoridae, Muscidae). Alimenta. 16:49–52

Gaynor WT, Cousins DV, Friend JA (1990) Mycobacterial Infection in Numbats (*Myrmecobius-Fasciatus*). J. Zoo Wildlife Med. 21:476–479

Gebesh VV, Ianchenko VI, Sukhov IuA (1999) Kaopectate in the combined treatment of patients with intestinal infection (In Russian). Lik. Sprava. 3:140–142

George KL, Parker BC, Gruft H, Falkinham JO, III (1980) Epidemiology of infection by nontuberculous mycobacteria. II. Growth and survival in natural waters. Am. Rev. Respir. Dis. 122:89–94

George RH (1997) Killing activity of microwaves in milk. J. Hosp. Infect. 35:319–320

Ghaemi E, Ghazisaidi K, Koohsari H, Khodabakhshi B, Mansoorian A (2006) Environmental mycobacteria in areas of high and low tuberculosis prevalence in the Islamic Republic of Iran. East Mediterr. Health J. 12:280–285

Ghio AJ, Kennedy TP, Schapira RM, Crumbliss AL, Hoidal JR (1990) Hypothesis: is lung disease after silicate inhalation caused by oxidant generation? Lancet. 336:967–969

Ghittino C, Latini M, Agnetti F, Panzieri C, Lauro L, Ciappelloni R, Petracca G (2003) Emerging pathologies in aquaculture: effects on production and food safety. Vet. Res. Commun. 27 Suppl 1:471–479

Gilardi JD, Duffey SS, Munn CA, Tell LA (1999) Biochemical functions of geophagy in parrots: Detoxification of dietary toxins and cytoprotective effects. J. Chem. Ecol. 25:897–922

Gira AK, Reisenauer AH, Hammock L, Nadiminti U, Macy JT, Reeves A, Burnett C, Yakrus MA, Toney S, Jensen BJ, Blumberg HM, Caughman SW, Nolte FS (2004) Furunculosis due to *Mycobacterium mageritense* associated with footbaths at a nail salon. J. Clin. Microbiol. 42:1813–1817

Glazer C, Martyny J, Rose C (2008) Hot Tub Associated Granulomatous Lung Disease From Mycobacterial Bioaerosols [Interstitial, Inflammatory, and Occupational Lung Disease]. Clini. Pulm. Med. 15:138–144

Gonzalez M, Rodriguez-Bertos A, Gimeno I, Flores JM, Pizarro M (2002) Outbreak of avian tuberculosis in 48-week-old commercial layer hen flock. Avian Dis. 46:1055–1061

Gotze U (1967) [The problem of meat hygiene in pig tuberculosis with special reference to the incidence of mycobacteria in meat from pigs with "isolated" lymph node tuberculosis]. Berl Munch. Tierarztl. Wochenschr. 80:47–49

Graham L, Jr., Warren NG, Tsang AY, Dalton HP (1988) *Mycobacterium avium* complex pseudobacteriuria from a hospital water supply. J. Clin. Microbiol. 26:1034–1036

Grange JM (1996) Mycobacteria and human disease. 2nd ed. London, Arnold, 230 pp

Grange JM, Yates MD (1994) Zoonotic Aspects of *Mycobacterium-Bovis* Infection. Vet. Microbiol. 40:137–151

Grant IR (2005) Zoonotic potential of *Mycobacterium avium* ssp *paratuberculosis*: the current position. J. Appl. Microbiol. 98:1282–1293

Grant IR, Ball HJ, Rowe MT (1996) Thermal inactivation of several *Mycobacterium* spp. in milk by pasteurization. Lett. Appl. Microbiol. 22:253–256

Greenlees KJ, Machado J, Bell T, Sundlof SF (1998) Food borne microbial pathogens of cultured aquatic species. Vet. Clin. North Am. Food Anim Pract. 14:101–112

Groothuis DG (1985) [Swine tuberculosis and public health]. Tijdschr. Diergeneeskd. 110:716–717

Gubler JGH, Salfinger M, Vongraevenitz A (1992) Pseudoepidemic of Nontuberculous Mycobacteria Due to A Contaminated Bronchoscope Cleaning Machine - Report of An Outbreak and Review of the Literature. Chest. 101:1245–1249

Guerin WF, Jones GE (1988) Mineralization of Phenanthrene by A *Mycobacterium* Sp. Appl. Environ. Microbiol. 54:937–944

Gutierrez Garcia JM (2006) Meat as a vector of transmission of bovine tuberculosis to humans in Spain: a historical perspective. Vet. Herit. 29:25–27

Hahn H (1959) Ist die Herstellung von Emmentaler Markenkase aus Rohmilch vom Standpunkt des Lebensmittelhygienikers vertretbar? Tierartl. Umsch. 14:254–256

Hall-Stoodley L, Keevil CW, Lappin-Scott HM (1999) Mycobacterium fortuitum and *Mycobacterium chelonae* biofilm formation under high and low nutrient conditions. J. Appl. Microbiol. 85:60S–69S

Hancox M (2002) Bovine tuberculosis: milk and meat safety. Lancet. 359:706–707

Harrington R, Jr., Karlson AG (1965) Destruction of various kinds of mycobacteria in milk by pasteurization. Appl. Microbiol. 13:494–495

Hartikainen T, Ruuskanen J, Martikainen PJ (2001) Carbon disulfide and hydrogen sulfide removal with a peat biofilter. J. Air Waste Manag. Assoc. 51:387–392

Havelaar AH, Berwald LG, Groothuis DG, Baas JG (1985) Mycobacteria in semi-public swimming-pools and whirlpools. Zentralbl. Bakteriol. Mikrobiol. Hyg. [B]. 180:505–514

Heavey M (2003) Low-cost treatment of landfill leachate using peat. Waste Manag. 23:447–454

Heigis G, Krimmel M, Hoffmann J, Kaiserling E, Reinert S (2005) [Oral manifestation of miliary tuberculosis]. Mund Kiefer Gesichtschir. 9:180–183

Heimann G (1984) Pharmacotherapy of acute infant enteritis (in German). Montsschr. Kinderheilkd. 132:303–305

Hietala SK, Ardans AA (1987) Neutrophil Phagocytic and Serum Opsonic Response of the Foal to *Corynebacterium-Equi*. Vet. Immunol. Immunopathol. 14:279–294

Higgins DA, Kung IT, Or RS (1985) Environmental silica in badger lungs: a possible association with susceptibility to *Mycobacterium bovis* infection. Infect. Immun. 48:252–256

Hilborn ED, Covert TC, Yakrus MA, Harris SI, Donnelly SF, Rice EW, Toney S, Bailey SA, Stelma GN, Jr. (2006) Persistence of nontuberculous mycobacteria in a drinking water system after addition of filtration treatment. Appl. Environ. Microbiol. 72:5864–5869

Holland J, Smith C, Childs PA, Holland AJ (1997) Surgical management of cutaneous infection caused by atypical mycobacteria after penetrating injury: the hidden dangers of horticulture. J. Trauma. 42:337–340

Horsburgh CR, Jr., Chin DP, Yajko DM, Hopewell PC, Nassos PS, Elkin EP, Hadley WK, Stone EN, Simon EM, Gonzalez P (1994) Environmental risk factors for acquisition of *Mycobacterium avium* complex in persons with human immunodeficiency virus infection. J. Infect. Dis. 170:362–367

Hsueh PR, Luh KT (1998) Catheter sepsis due to *Mycobacterium chelonae* - Reply. J. Clin. Microbiol. 36:3444–3445

Hsueh PR, Teng LJ, Yang PC, Chen YC, Ho SW, Luh KT (1998) Recurrent catheter-related infection caused by a single clone of *Mycobacterium chelonae* with two colonial morphotypes. J. Clin. Microbiol. 36:1422–1424

Huang GF, Wong JWC, Wu QT, Nagar BB (2004) Effect of C/N on composting of pig manure with sawdust. Waste Manage. 24:805–813

Hutchings MR, Harris S (1999) Quantifying the risks of TB infection to cattle posed by badger excreta. Epidemiol. Infect. 122:167–173

Hutchings MR, Service KM, Harris S (2001) Defecation and urination patterns of badgers *Meles meles* at low density in south west England. Acta Theriol. 46:87–96

Huttunen K, Ruotsalainen M, Iivanainen E, Torkko P, Katila ML, Hirvonen MR (2000) Inflammatory responses in RAW264.7 macrophages caused by mycobacteria isolated from moldy houses. Environ. Toxicol. Pharmacol. 8:237–244

Iannuzzi L (1968) [On the resistance of pathogenic tubercular mycobacteria in sausages]. Acta Med. Vet. (Napoli). 14:255–261

Ichiriu ET, Bushnell OA (1950) The survival time of *Mycobacterium tuberculosis* in poi; studies in the bacteriology of poi. Hawaii Med. J. 9:163–165

Iivanainen E (1995) Isolation of Mycobacteria from Acidic Forest Soil Samples - Comparison of Culture Methods. J. Appl. Bacteriol. 78:663–668

Iivanainen E, Martikainen P, Vaananen P, Katila ML (1999a) Environmental factors affecting the occurrence of mycobacteria in brook sediments. J. Appl. Microbiol. 86:673–681

Iivanainen E, Sallantaus T, Katila ML, Martikainen PJ (1999b) Mycobacteria in runoff-waters from natural and drained peatlands. J. Environ. Qual. 28:1226–1234

Iivanainen EK, Martikainen PJ, Raisanen ML, Katila ML (1997) Mycobacteria in boreal coniferous forest soils. Fems Microbiol. Ecol. 23:325–332

Iivanainen EK, Martikainen PJ, Vaananen PK, Katila ML (1993) Environmental-Factors Affecting the Occurrence of Mycobacteria in Brook Waters. Appl. Environ. Microbiol. 59:398–404

Ikonomopoulos J, Pavlik I, Bartos M, Svastova P, Ayele WY, Roubal P, Lukas J, Cook N, Gazouli M (2005) Detection of *Mycobacterium avium* subsp. *paratuberculosis* in retail cheeses from Greece and the Czech republic. Appl. Environ. Microbiol. 71:8934–8936

Jackson R (2002) The role of wildlife in *Mycobacterium bovis* infection of livestock in New Zealand. N.Z. Vet. J. 50: 49–52

Jackson R, de Lisle GW, Morris RS (1995) A study of the environmental survival of *Mycobacterium bovis* on a farm in New Zealand. N.Z. Vet. J. 43:346–352

Jamieson W, Madri P, Claus G (1976) Survival of certain pathogenic microorganisms in sea water. Hydrobiologica 50:117–121

Jankowski A, Nienartowicz B, Polanska B, Lewandowicz-Uszynska A (1993) A randomised, double-blind study on the efficacy of Tolpa Torf Preparation (TTP) in the treatment of recurrent respiratory tract infections. Arch. Immunol. Ther. Exp. (Warsz.). 41:95–97

Jeppsson KH (1998) Ammonia emission from different deep-litter materials for growing-finishing pigs. Swed. J. Agric. Res. 28:197–206

Jeppsson KH (1999) Volatilization of ammonia in deep-litter systems with different bedding materials for young cattle. J. Agric. Eng. Res. 73:49–57

Johns T, Duquette M (1991a) Detoxification and Mineral Supplementation As Functions of Geophagy. Am. J. Clin. Nutr.. 53:448–456

Johns T, Duquette M (1991b) Traditional Detoxification of Acorn Bread with Clay. Ecol. Food Nutr. 25:221–228

Jones JJ, Falkinham JO, III (2003) Decolorization of malachite green and crystal violet by waterborne pathogenic mycobacteria. Antimicrob. Agents Chemother. 47:2323–2326

Jones PW, Rennison LM, Matthews PR, Collins P, Brown A (1981) The occurrence and significance to animal health of *Leptospira, Mycobacterium, Escherichia coli, Brucella abortus* and *Bacillus anthracis* in sewage and sewage sludges. J. Hyg. (Lond). 86:129–137

Jones WD, Kubica GP (1965) Differential Grouping of Slowly Growing Mycobacteria Based on Their Susceptibility to Various Dyes. Am. Rev. Respir. Dis. 91:613–615

Jorgensen JB (1977) Survival of *Mycobacterium paratuberculosis* in slurry. Nord. Vet. Med. 29:267–270

Jorgensen JB, Engbaek HC, Dam A (1972) An enzootic of pulmonary tuberculosis in pigs caused by *M. avium*. 2. Bacteriological studies. Acta Vet. Scand. 13:68–86

Kaffka A, Thiele H, Schroder KH (1979) Atypical Mycobacteria in Swimming-Pool Water. Offentliche Gesundheitswesen. 41:405–409

Kahana LM, Kay JM, Yakrus MA, Waserman S (1997) *Mycobacterium avium* complex infection in an immunocompetent young adult related to hot tub exposure. Chest. 111:242–245

Kamala T, Paramasivan CN, Herbert D, Venkatesan P, Prabhakar R (1994) Isolation and Identification of Environmental Mycobacteria in the *Mycobacterium bovis* BCG Trial Area of South India. Appl. Environ. Microbiol. 60: 2180–2183

Kao MM (1993) The Evaluation of Sawdust Swine Waste Compost on the Soil Ecosystem, Pollution and Vegetable Production. Water Sci. Technol. 27:123–131

Kaplan JE, Masur H, Holmes KK (2002) Guidelines for preventing opportunistic infections among *HIV*-infected persons – 2002. Recommendations of the U.S. Public Health Service and the Infectious Diseases Society of America. MMWR Recomm. Rep. 51:1–52

Kasi M, Kausar P, Naz R, Miller LC (1995) Treatment of diarrhoea in infants by medical doctors in Balochistan, Pakistan. J. Diarrhoeal. Dis. Res. 13:238–241

Katayama N, Tanaka Ch, Fujita T, Saitou Y, Suzuki S, Onouchi E (2000) Effect of ensilage on inactivation of *M. avium* subsp. *paratuberculosis*. Grassland Sci. 46:282–288

Katila ML, Iivanainen E, Torkko P, Kauppinen J, Martikainen P, Vaananen P (1995) Isolation of potentially pathogenic mycobacteria in the Finnish environment. Scand. J. Infect. Dis. Suppl. 98:9–11

Kauker E, Rheinwald W (1972) [Studies on the occurrence of atypical mycobacteria, group 3 Runyon, in the bedding material (sawdust) and feed of swine in North Hesse]. Berl Munch. Tierarztl. Wochenschr. 85:384–387

Kawamura Y, Li Y, Liu H, Huang X, Li Z, Ezaki T (2001) Bacterial population in Russian space station "Mir". Microbiol. Immunol. 45:819–828

Kazda J (1966) [Isolation and description of a *Mycobacterium* species, the cause of a para-allergy against tuberculin in poultry]. Zentralbl. Bakteriol. [Orig.]. 199:529–532

Kazda J (1967) [Atypical mycobacteria in drinking water–the cause of para-allergies against tuberculin in animals]. Z. Tuberk. Erkr. Thoraxorg. 127:111–113

Kazda J (1973a) Importance of Water for Spread of Potentially Pathogenic Mycobacteria. 1. Possibilities for Multiplication of Mycobacteria. Zentralblatt fur Bakteriologie Mikrobiologie und Hygiene Serie B-Umwelthygiene Krankenhaushygiene Arbeitshygiene Praventive Medizin. 158:161–169

Kazda J (1973b) [The importance of water for the distribution of potentially pathogenic Mycobacteria. II. Growth of Mycobacteria in water models (author's transl)]. Zentralbl. Bakteriol. [Orig. B]. 158:170–176

Kazda J (1977) [The importance of sphagnum bogs in the ecology of Mycobacteria (author's transl)]. Zentralbl. Bakteriol. [Orig. B]. 165:323–334

Kazda J (1978a) [Multiplication of mycobacteria in the gray layer of sphagnum vegetation (author's transl)]. Zentralbl. Bakteriol. [Orig. B]. 166:463–469

Kazda J (1978b) [The behaviour of *Mycobacterium intracellulare* serotyp Davis and *Mycobacterium avium* in the head region of sphagnum moss vegetation after experimental inoculation (author's transl)]. Zentralbl. Bakteriol. [Orig. B]. 166:454–462

Kazda J (2000) The ecology of mycobacteria. Kluwer Academic Publishers, Dordrecht, Boston, London, 72 pp

Kazda J, Cooney RP, Quinn JP, Cook BR, Muller K, Monaghan M, Keane M (1997) High density of mycobacteria in the bryophyte vegetation (*Musci*) of moorland. Int. Peat J. 7:14–19

Kazda J, Muller K, Irgens LM (1979) Cultivable mycobacteria in sphagnum vegetation of moors in South Sweden and coastal Norway. Acta Pathol. Microbiol. Scand. [B]. 87B:97–101

Keogh BP (1971) Reviews of the progress of dairy science. Section B. The survival of pathogens in cheese and milk powder. J. Dairy Res. 38:91–111

Khoor A, Leslie KO, Tazelaar HD, Helmers RA, Colby TV (2001) Diffuse pulmonary disease caused by nontuberculous mycobacteria in immunocompetent people (hot tub lung). Am. J. Clin. Pathol. 115:755–762

Kim BC, Park JH, Gu MB (2004) Development of a DNA microarray chip for the identification of sludge bacteria using an unsequenced random genomic DNA hybridization method. Environ. Sci. Technol. 38:6767–6774

Kindle G, Busse A, Kampa D, Meyer-Konig U, Daschner FD (1996) Killing activity of microwaves in milk. J. Hosp. Infect. 33:273–278

Kirschner RA, Jr., Parker BC, Falkinham JO, III (1992) Epidemiology of infection by nontuberculous mycobacteria. *Mycobacterium avium*, *Mycobacterium intracellulare*, and *Mycobacterium scrofulaceum* in acid, brown-water swamps of the southeastern United States and their association with environmental variables. Am. Rev. Respir. Dis. 145:271–275

Kirschner RA, Parker BC, Falkinham JO (1999) Humic and fulvic acids stimulate the growth of *Mycobacterium avium*. Fems Microbiol. Ecol. 30:327–332

Kleeberg HH, Nel EE (1973) Occurrence of environmental atypical mycobacteria in South Africa. Ann. Soc. Belg. Med. Trop. 53:405–418

Knezevich M (1998) Geophagy as a therapeutic mediator of endoparasitism in a free-ranging group of rhesus macaques (*Macaca mulatta*). Am. J. Primatol. 44:71–82

Kohlmeier S, Smits TH, Ford RM, Keel C, Harms H, Wick LY (2005) Taking the fungal highway: mobilization of pollutant-degrading bacteria by fungi. Environ. Sci. Technol. 39:4640–4646

Kozhemiakin NG, Sidorenko BV, Ivanov NE (1967) [M. tuberculosis resistance to thermal conditions adopted in the production of cooked sausages]. Vopr. Pitan. 26:61–64

Kreiss K, CoxGanser J (1997) Metalworking fluid-associated hypersensitivity pneumonitis: A workshop summary. Am. J. Ind. Med. 32:423–432

Krulwich TA, Pelliccione NJ (1979) Catabolic Pathways of Coryneforms, Nocardias, and Mycobacteria. Annu. Rev. Microbiol. 33:95–111

Kubalek I, Komenda S (1995) Seasonal variations in the occurrence of environmental mycobacteria in potable water. APMIS. 103:327–330

Kusnetsov J, Torvinen E, Perola O, Nousiainen T, Katila ML (2003) Colonization of hospital water systems by legionellae, mycobacteria and other heterotrophic bacteria potentially hazardous to risk group patients. APMIS. 111:546–556

Larsen AB, Merkal RS, Vardaman TH (1956) Survival time of *Mycobacterium paratuberculosis*. Am. J. Vet. Res. 17:549–551

Larsson LO, Skoogh BE, Bentzon MW, Magnusson M, Olofson J, Taranger J, Lind A (1991) Sensitivity to sensitins and tuberculin in Swedish children. II. A study of preschool children. Tubercle. 72:37–42

Laukkanen H, Soini H, Kontunen-Soppela S, Hohtola A, Viljanen M (2000) A *Mycobacterium* isolated from tissue cultures of mature *Pinus sylvestris* interferes with growth of Scots pine seedlings. Tree Physiol. 20:915–920

Lavoie J, Marchand G, Drolet JY, Gingras G (1995) Biological and Chemical Contamination of the Air in A Grower-

Finisher Pig Building Using Deep-Litter Systems. Can. Agric. Eng. 37:195–203

Le Dantec C, Duguet JP, Montiel A, Dumoutier N, Dubrou S, Vincent V (2002) Occurrence of mycobacteria in water treatment lines and in water distribution systems. Appl. Environ. Microbiol. 68:5318–5325

Lee WJ, Kim TW, Shur KB, Kim BJ, Kook YH, Lee JH, Park JK (2000) Sporotrichoid dermatosis caused by *Mycobacterium abscessus* from a public bath. J. Dermatol. 27:264–268

Lenk T, Benda A (1989) Peat paste - humic acid containing animal health agent for prophylaxis and treatment of calves for diarrhoea (in German). Mh. Vet. Med. 44:563–565

Leoni E, Legnani P, Mucci MT, Pirani R (1999) Prevalence of mycobacteria in a swimming pool environment. J. Appl. Microbiol. 87:683–688

Lepper AWD (1977) Use of Bovine Ppd Tuberculin in Caudal Fold Tests - Reply. Aust. Vet. J. 53:451–452

Lumb R, Stapledon R, Scroop A, Bond P, Cunliffe D, Goodwin A, Doyle R, Bastian I (2004) Investigation of spa pools associated with lung disorders caused by *Mycobacterium avium* complex in immunocompetent adults. Appl. Environ. Microbiol. 70:4906–4910

Lysons RE (1996) Pigs, peat and avian tuberculosis. Proceedings of the 14th IPVS Congress, Bologna, Italy, 7–10 July. 323–323

Mahaney WC, Bezada M, Hancock RGV, Aufreiter S, Perez FL (1996a) Geophagy of Holstein hybrid cattle in the northern Andes, Venezuela. Mt. Res. Dev. 16:177–180

Mahaney WC, Hancock RGV, Aufreiter S, Huffman MA (1996b) Geochemistry and clay mineralogy of termite mound soil and the role of geophagy in chimpanzees of the Mahale Mountains, Tanzania. Primates. 37:121–134

Mallmann WL (1972) Should tuberculosis be eradicated from all species? Health Lab. Sci. 9:133–138

Mangione EJ, Huitt G, Lenaway D, Beebe J, Bailey A, Figoski M, Rau MP, Albrecht KD, Yakrus MA (2001) Nontuberculous mycobacterial disease following hot tub exposure. Emerg. Infect. Dis. 7:1039–1042

Marks J (1975) Occupation and Kansasii infection in Cardiff residents. Tubercle. 56:311–313

Marlier D, Nicks B, Canart B, Shehi R (1994) Comparison of the Evolution of the Bedding of 2 Deep Litter Systems, Deep Sawdust and Deep Straw, for Fattening Pigs. Ann. Med. Vet. 138:43–53

Marsollier L, Stinear T, Aubry J, Saint Andre JP, Robert R, Legras P, Manceau AL, Audrain C, Bourdon S, Kouakou H, Carbonnelle B (2004) Aquatic plants stimulate the growth of and biofilm formation by *Mycobacterium ulcerans* in axenic culture and harbor these bacteria in the environment. Appl. Environ. Microbiol. 70:1097–1103

Martyny JW, Rose CS (1999) Nontuberculous mycobacterial bioaerosols from indoor warm water sources cause granulomatous lung disease. Indoor Air. 9:1–6

Masaki S, Konishi T, Sugimori G, Okamoto A, Hayashi Y, Kuze F (1989) Plasmid profiles of *Mycobacterium avium* complex isolated from swine. Microbiol. Immunol. 33:429–433

Matlova L, Dvorska L, Ayele WY, Bartos M, Amemori T, Pavlik I (2005) Distribution of *Mycobacterium avium* complex isolates in tissue samples of pigs fed peat naturally contaminated with mycobacteria as a supplement. J. Clin. Microbiol. 43:1261–1268

Matlova L, Dvorska L, Bartl J, Bartos M, Ayele WY, Alexa M, Pavlik I (2003) Mycobacteria isolated from the environment of pig farms in the Czech Republic during the years 1996 to 2002. Veterinarni Medicina 48:343–357

Matlova L, Dvorska L, Bartos M, Docekal J, Trckova M, Pavlik I (2004a) Tuberculous lesions in pig lymph nodes caused by kaolin fed as a supplement. Veterinarni Medicina 49:379–388

Matlova L, Dvorska L, Palecek K, Maurenc L, Bartos M, Pavlik I (2004b) Impact of sawdust and wood shavings in bedding on pig tuberculous lesions in lymph nodes, and IS*1245* RFLP analysis of *Mycobacterium avium* subsp. *hominissuis* of serotypes 6 and 8 isolated from pigs and environment. Vet. Microbiol. 102:227–236

Matthews PR, Collins P, Jones PW (1976) Isolation of mycobacteria from dairy creamery effluent sludge. J. Hyg. (Lond). 76:407–413

McClatchy JK (1981) The seroagglutination test in the study of nontuberculous mycobacteria. Rev. Infect. Dis. 3:867–870

McCullough NV, Brosseau LM, Vesley D (1997) Collection of three bacterial aerosols by respirator and surgical mask filters under varying conditions of flow and relative humidity. Ann. Occup. Hyg. 41:677–690

Mediel MJ, Rodriguez V, Codina G, Martin-Casabona N (2000) Isolation of mycobacteria from frozen fish destined for human consumption. Appl. Environ. Microbiol. 66:3637–3638

Mendum TA, Chilima BZ, Hirsch PR (2000) The PCR amplification of non-tuberculous mycobacterial 16S rRNA sequences from soil. FEMS Microbiol. Lett. 185: 189–192

Merkal RS, Crawford JA, Whipple DL (1979) Heat inactivation of *Mycobacterium avium-Mycobacterium intracellulare* complex organisms in meat products. Appl. Environ. Microbiol. 38:831–835

Merkal RS, Lyle PS, Whipple DL (1981) Heat inactivation of *in vivo*- and *in vitro*-grown mycobacteria in meat products. Appl. Environ. Microbiol. 41:1484–1485

Merkal RS, Whipple DL (1980) Inactivation of *Mycobacterium bovis* in meat products. Appl. Environ. Microbiol. 40:282–284

Michel AL (2002) Implications of tuberculosis in African wildlife and livestock. Ann. N.Y. Acad. Sci. 969:251–255

Mijs W, de Hass P, Rossau R, van der Laan T, Rigouts L, Portaels F, van Soolingen D (2002) Molecular evidence to support a proposal to reserve the designation *Mycobacterium avium subsp.* avium for bird-type isolates and '*M. avium subsp. hominissuis*' for the human/porcine type of *M. avium*. Int. J Syst. Evol. Microbiol. 52:1505–1518

Moore JS, Christensen M, Wilson RW, Wallace RJ, Jr., Zhang Y, Nash DR, Shelton B (2000) Mycobacterial contamination of metalworking fluids: involvement of a possible new taxon of rapidly growing mycobacteria. AIHAJ. 61: 205–213

Moravkova M, Bartos M, Dvorska-Bartosova L, Beran V, Parmova I, Ocepek M, Pate M, Pavlik I (2007) Genetic variability of *Mycobacterium avium* subsp. *avium* of pig isolates. Veterinarni Medicina 52:430–436

Muilenberg ML, Burge HT, Sweet T (1998) Hypersensitivity pneumonitis and exposure to acid-fast bacilli in coolant aerosols. J. Allergy Clin. Immunol. 91:311-

Mutimer MD, Woolcock JB (1980) *Corynebacterium equi* in cattle and pigs. Tijdschr. Diergeneeskd. 105:25–27

Naganawa R, Iwata N, Ishikawa K, Fukuda H, Fujino T, Suzuki A (1996) Inhibition of microbial growth by ajoene, a sulfur-containing compound derived from garlic. Appl. Environ. Microbiol. 62:4238–4242

Nakamura Y, Obase Y, Suyama N, Miyazaki Y, Ohno H, Oka M, Takahashi M, Kohno S (2004) A small outbreak of pulmonary tuberculosis in non-close contact patrons of a bar. Internal Medicine 43:263–267

Nakazawa M, Kubo M, Sugimoto C, Isayama Y (1983) Serogrouping of *Rhodococcus equi*. Microbiol. Immunol. 27:837–846

Naser SA, Schwartz D, Shafran I (2000) Isolation of *Mycobacterium avium* subsp. *paratuberculosis* from breast milk of Crohn's disease patients. Am. J. Gastroenterol. 95:1094–1095

Nassal J, Breunig W, Schnedelbach U (1974) [Atypical mycobacteria in fruit, vegetables, and cereals]. Prax. Pneumol. 28:667–674

Nassal J, Werner-Schieck B (1970) [Presence of mycobacteria in human feces]. Prax. Pneumol. 24:473–478

Nel EE (1981) *Mycobacterium avium-intracellulare* complex serovars isolated in South Africa from humans, swine, and the environment. Rev. Infect. Dis. 3:1013–1020

Nicks B (2004) Technical characteristics and environmental aspects of breeding fattening pigs and weaned piglets on accumulated litters. Ann. Med. Vet. 148:31–38

Nicks B, Desiron A, Canart B (1998) Comparison of two litter materials, sawdust and a straw-sawdust mixture, for fattening pigs on deep litter. Annales de Zootechnie. 47:107–116

Nicks B, Laitat M, Farnir F, Vandenheede M, Desiron A, Verhaeghe C, Canart B (2004) Gaseous emissions from deep-litter pens with straw or sawdust for fattening pigs. Anim. Sci. 78:99–107

Nicks B, Laitat M, Vandenheede M, Desiron A, Verhaeghe C, Canart B (2003) Emissions of ammonia, nitrous oxide, methane, carbon dioxide and water vapor in the raising of weaned pigs on straw-based and sawdust-based deep litters. Anim. Res. 52:299–308

Nieminen T, Pakarinen J, Tsitko I, Salkinoja-Salonen M, Breitenstein A, Ali-Vehmas T, Neubauer P (2006) 16S rRNA targeted sandwich hybridization method for direct quantification of mycobacteria in soils. J. Microbiol. Methods. 67:44–55

Niva M, Hernesmaa A, Haahtela K, Salkinoja-Salonen M, Sivonen K, Haukka K (2006) Actinobacterial communities of boreal forest soil and lake water are rich in mycobacteria. Boreal Environ. Res. 11:45–53

Nix DE, Adam RD, Auclair B, Krueger TS, Godo PG, Peloquin CA (2004) Pharmacokinetics and relative bioavailability of clofazimine in relation to food, orange juice and antacid. Tuberculosis. (Edinb). 84:365–373

Norby B, Fosgate GT, Manning EJ, Collins MT, Roussel AJ (2007) Environmental mycobacteria in soil and water on beef ranches: Association between presence of cultivable mycobacteria and soil and water physicochemical characteristics. Vet. Microbiol. 124:153–159

Norton CD, LeChevallier MW, Falkinham JO, III (2004) Survival of *Mycobacterium avium* in a model distribution system. Water Res. 38:1457–1466

Norton JH, Duffield BJ, Coward AJ, Hielscher RW, Nicholls RF (1984) A necropsy technique for cattle to eliminate contamination of lymph nodes by mycobacteria. Aust. Vet. J. 61:75–76

Novotny L, Dvorska L, Lorencova A, Beran V, Pavlik I (2004) Fish: a potential source of bacterial pathogens for human beings. Veterinarni Medicina 49:343–358

O'Donovan G (1987) An ecosystem study of grasslands in the Burren National Park, Co. Clare. Ph.D. Thesis, University of Dublin.

Ogielski L, Zawadzki Z (1961) [Area irrigated with sewage. Its hygienic and sanitary evaluation. III. Studies on the presence of tubercle bacilli in sewage utilized for agricultural purposes.]. Acta Microbiol. Pol. 10:433–437

Olsen JE, Jorgensen JB, Nansen P (1985) On the Reduction of *Mycobacterium-Paratuberculosis* in Bovine Slurry Subjected to Batch Mesophilic Or Thermophilic Anaerobic-Digestion. Agric. Wastes. 13:273–280

Ong HK, Choo PY, Soo SP (1993) Application of Bacterial Product for Zero-Liquid-Discharge Pig Waste Management Under Tropical Conditions. Water Sci. Technol. 27:133–140

Palasek J, Pavlas M, Kubu I (1991) Survival of Salmonellae and mycobacteria in salted and unsalted swine guts used as sausage casing and sausage emulsion of a hard salami. Acta Vet. Brno. 60:375–381

Pande TK, Hiran S, Rao VV, Pani S, Vishwanathan KA (1995) Primary lingual tuberculosis caused by *M. bovis* infection. Oral Surg. Oral Med. Oral Pathol. Oral Radiol. Endod. 80:172–174

Parashar D, Chauhan DS, Sharma VD, Chauhan A, Chauhan SV, Katoch VM (2004) Optimization of procedures for isolation of mycobacteria from soil and water samples obtained in northern India. Appl. Environ. Microbiol. 70:3751–3753

Parker BC, Ford MA, Gruft H, Falkinham JO, III (1983) Epidemiology of infection by nontuberculous mycobacteria. IV. Preferential aerosolization of *Mycobacterium intracellulare* from natural waters. Am. Rev. Respir. Dis. 128:652–656

Parsek MR, Fuqua C (2004) Biofilms 2003: emerging themes and challenges in studies of surface-associated microbial life. J. Bacteriol. 186:4427–4440

Pavicic Z, Balenovic T, Balenovic M, Popovic M, Vlahovic K, Valpotic H, Rudan-Biuk N (2006) The effect of accommodation type and microclimatic conditions in farrowing units on the postural changes of sows and piglet mortality by crushing. Tieraerztliche Umschau. 61:68-+

Pavlas M, Patlokova V (1985) Occurrence of mycobacteria in sawdust, straw, hay and their epizootiological significance. Acta Vet. Brno. 54:85–90

Pavlas M (1990) Thermoresistance of Mycobacteria. Acta Vet. Brno. 59:65–71

Pavlas M, Patlokova V (1977) [Occurrence of *M. avium* and *M. intracellulare* in the organs and muscles of slaughterhouse pigs]. Veterinarni Medicina 22:1–8

Pavlik I, Horvathova A, Dvorska L, Bartl J, Svastova P, du Maine R, Rychlik I (1999) Standardisation of restriction fragment length polymorphism analysis for *Mycobacterium avium* subspecies *paratuberculosis*. J. Microbiol. Methods. 38:155–167

Pavlik I, Matlova L, Dvorska L, Bartl J, Oktabcova L, Docekal J, Parmova I (2003) Tuberculous lesions in pigs in the Czech Republic during 1990–1999: occurrence, causal factors and economic losses. Veterinarni Medicina 48: 113–125

Pavlik I, Matlova L, Dvorska L, Shitaye JE, Parmova I (2005) Mycobacterial infections in cattle and pigs caused by *Mycobacterium avium* complex members and atypical mycobacteria in the Czech Republic during 2000–2004. Veterinarni Medicina 50:281–290

Pavlik I, Matlova L, Gilar M, Bartl J, Parmova I, Lysak F, Alexa M, Dvorska-Bartosova L, Svec V, Vrbas V, Horvathova A (2007) Isolation of conditionally pathogenic mycobacteria from the environment of one pig farm and the effectiveness of preventive measures between 1997 and 2003. Veterinarni Medicina 52:392–404

Pavlik I, Svastova P, Bartl J, Dvorska L, Rychlik I (2000) Relationship between IS*901* in the *Mycobacterium avium* complex strains isolated from birds, animals, humans, and the environment and virulence for poultry. Clin. Diagn. Lab. Immunol. 7:212–217

Peccia J, Hernandez M (2006) Incorporating polymerase chain reaction-based identification, population characterization, and quantification of microorganisms into aerosol science: A review. Atmos. Environ. 40:3941–3961

Pedley JC (1967) The presence of *M. leprae* in human milk. Lepr. Rev. 38:239–242

Pedley S, Bartram J, Rees G, Dufou A, Cotruvo JA (2004) Pathogenic mycobacteria in water. WHO, TJ International (Ltd.), Padstow, Cornwall, UK. 237 pp

Pell AN (1997) Manure and microbes: public and animal health problem? J. Dairy Sci. 80:2673–2681

Pereira MR, Benjaminson MA (1975) Broadcast of microbial aerosols by stacks of sewage treatment plants and effects of ozonation on bacteria in the gaseous effluent. Public Health Rep. 90:208–212

Perez C, Falero A, Hung BR, Tirado S, Balcinde Y (2005) Bioconversion of phytosterols to androstanes by mycobacteria growing on sugar cane mud. J. Ind. Microbiol. Biotechnol. 32:83–86

Pickup RW, Rhodes G, Bull TJ, Arnott S, Sidi-Boumedine K, Hurley M, Hermon-Taylor J (2006) *Mycobacterium avium* subsp. *paratuberculosis* in lake catchments, in river water abstracted for domestic use, and in effluent from domestic sewage treatment works: diverse opportunities for environmental cycling and human exposure. Appl. Environ. Microbiol. 72:4067–4077

Picot B, Paing J, Toffoletto L, Sambuco JP, Costa RH (2001) Odor control of an anaerobic lagoon with a biological cover: floating peat beds. Water Sci. Technol. 44: 309–316

Piening C, Anz W, Meissner G (1972) [Serotyping and its significance for epidemiological studies of porcine tuberculosis in Schleswig Holstein]. Dtsch. Tierarztl. Wochenschr. 79:316–321

Policard A, Gernez-Rieux C, Tacquet A, Martin JC, Devulder B, Le Bouffant L (1967) Influence of pulmonary dust load on the development of experimental infection by *Mycobacterium kansasii*. Nature. 216:177–178

Poptsova NV (1974) [Contamination with *Mycobacterium tuberculosis* of certain environmental objects within the foci of tuberculosis]. Probl. Tuberk. 17–20

Prescott JF (1991) *Rhodococcus equi*: an animal and human pathogen. Clin. Microbiol. Rev. 4:20–34

Prescott JF, Travers M, Yagerjohnson JA (1984) Epidemiological Survey of *Corynebacterium-Equi* Infections on 5 Ontario Horse Farms. Can. J. Comp. Med. Revue Canadienne de Medecine Comparee. 48:10–13

Prince DS, Peterson DD, Steiner RM, Gottlieb JE, Scott R, Israel HL, Figueroa WG, Fish JE (1989) Infection with *Mycobacterium avium* complex in patients without predisposing conditions. N. Engl. J. Med. 321:863–868

Prince KA, Costa AR, Malaspina AC, Luis AF, Leite CQ (2005) Isolation of *Mycobacterium gordonae* from poultry slaughterhouse water in Sao Paulo State, Brazil. Rev. Argent Microbiol. 37:106–108

Pryor M, Springthorpe S, Riffard S, Brooks T, Huo Y, Davis G, Satter SA (2004) Investigation of opportunistic pathogens in municipal drinking water under different supply and treatment regimes. Water Sci. Technol. 50:83–90

Rao SP, Hayashi T, Catanzaro A (2002) Identification of a chemotactic, MCP-1-like protein from *Mycobacterium avium*. Fems Immunol. Med. Microbiol. 33:115–124

Rebmann T (2005) Management of patients infected with airborne-spread diseases: an algorithm for infection control professionals. Am. J. Infect. Control. 33:571–579

Reponen TA, Wang Z, Willeke K, Grinshpun SA (1999) Survival of mycobacteria on N95 personal respirators. Infect. Control Hosp. Epidemiol. 20:237–241

Reuther G (1957) [Interactions between bacteria and fungi in swamp turf.]. Arch. Mikrobiol. 26:93–131

Reznikov M, Dawson DJ (1980) Mycobacteria of the intracellulare-scrofulaceum group in soils from the Adelaide area. Pathology. 12:525–528

Rha YH, Taube C, Haczku A, Joetham A, Takeda K, Duez C, Siegel M, Aydintug MK, Born WK, Dakhama A, Gelfand E (2002) Effect of microbial heat shock proteins on airway inflammation and hyperresponsiveness. J. Immunol. 169:5300–5307

Rheinwald W (1972) Untersuchungen über das Vorkommen atypischer Mykobakterien in Eistreu (Sägemehl) und Futter von Schweinen in Nordhessen. Vet. Med. Diss. Giessen. 142 pp

Rice G, Wright JM, Boutin B, Swartout J, Rodgers P, Niemuth N, Broder M (2005) Estimating the frequency of tap-water exposures to *Mycobacterium avium* complex in the U.S. population with advanced AIDS. J. Toxicol. Environ. Health A. 68:1033–1047

Ristola MA, von Reyn CF, Arbeit RD, Soini H, Lumio J, Ranki A, Buhler S, Waddell R, Tosteson ANA, Falkinham JO, Sox CH (1999) High rates of disseminated infection due to nontuberculous mycobacteria among AIDS patients in Finland. J. Infect. 39:61–67

Rizzuti AM, Cohen AD, Hunt PG, Vanotti MB (1999) Evaluating peats for their capacities to remove odorous compounds from liquid swine manure using headspace "solid-phase microextraction". J. Environ. Sci. Health B. 34:709–748

Robert J, Gantress J, Rau L, Bell A, Cohen N (2002) Minor histocompatibility antigen-specific MHC-restricted CD8 T cell responses elicited by heat shock proteins. J. Immunol. 168:1697–1703

Robinson P, Morris D, Antic R (1988) *Mycobacterium bovis* as an occupational hazard in abattoir workers. Aust. N.Z. J. Med. 18:701–703

Ronn R, Grunert J, Ekelund F (2001) Protozoan response to addition of the bacteria *Mycobacterium chlorophenolicum* and *Pseudomonas chlororaphis* to soil microcosms. Biology and Fertility of Soils 33:126–131

Roost H, Dobberstein I, Kuntsch G, Berber H, Tardel H, Benda A, Helms E (1990) Results and experience obtained from use of peat paste in industrialized piglet raising (in German). Mh.Vet. Med. 45:239–243

Rose CS (1999) Diagnosis and prevention of lung diseases associated with microbial bioaerosols. Semi. Respir. Crit. Care Med. 20:511–520

Rose CS, Martyny JW, Newman LS, Milton DK, King TE, Jr., Beebe JL, McCammon JB, Hoffman RE, Kreiss K (1998) "Lifeguard lung": endemic granulomatous pneumonitis in an indoor swimming pool. Am. J. Public Health. 88:1795–1800

Safranek TJ, Jarvis WR, Carson LA, Cusick LB, Bland LA, Swenson JM, Silcox VA (1987) *Mycobacterium-Chelonae* Wound Infections After Plastic-Surgery Employing Contaminated Gentian-Violet Skin-Marking Solution. N. Engl. J. Med. 317:197–201

Saitanu K, Holmgaard P (1977) An epizootic of *Mycobacterium intracellulare*, serotype 8 infection in swine. Nord. Vet. Med. 29:221–226

Saito H, Tsukamura M (1976) *Mycobacterium intracellulare* from public bath water. Jpn. J. Microbiol. 20:561–563

Saldanha FL, Sayyid SN, Kulkarn I, Sr. (1964) Viability of *M. tuberculosis* in the sanatorium sewage. Indian J. Med. Res. 52:1051–1056

Savov D (1975) [Resistance of tuberculosis mycobacteria in raw-dried and raw-fumigated sausages]. Vet. Med. Nauki. 12:39–43

Scanlon MP, Quinn PJ (2000) Inactivation of *Mycobacterium bovis* in cattle slurry by five volatile chemicals. J. Appl. Microbiol. 89:854–861

Schelonka RL, Ascher DP, McMahon DP, Drehner DM, Kuskie MR (1994) Catheter-related sepsis caused by *Mycobacterium avium* complex. Pediatr. Infect. Dis. J. 13:236–238

Schliesser T, Claus U, Weber A (1972) [Resistance of fast-growing atypical mycobacteria to temperature within the range of short-term pasteurization]. Prax. Pneumol. 26:485–490

Schliesser T, Weber A (1973) [Studies on the tenacity of Mycobacteria of the Runyon Group III in sawdust litter]. Zentralbl. Veterinarmed. B. 20:710–714

Schroder KH, Kazda J, Muller K, Muller HJ (1992) Isolation of *Mycobacterium simiae* from the environment. Zentralbl. Bakteriol. 277:561–564

Schultze WD, Brasso WB (1987) Characterization and identification of *Mycobacterium smegmatis* in bovine mastitis. Am. J. Vet. Res. 48:739–742

Schultze WD, Stroud BH, Brasso WB (1985) Dairy herd problem with mastitis caused by a rapidly growing *Mycobacterium* species. Am. J. Vet. Res. 46:42–47

Schulze-Robbecke R, Buchholtz K (1992) Heat susceptibility of aquatic mycobacteria. Appl. Environ. Microbiol. 58:1869–1873

Schulze-Robbecke R, Fischeder R (1989) Mycobacteria in biofilms. Zentralbl. Hyg. Umweltmed. 188:385–390

September SM, Brozel VS, Venter SN (2004) Diversity of non-tuberculoid *Mycobacterium* species in biofilms of urban and semiurban drinking water distribution systems. Appl. Environ. Microbiol. 70:7571–7573

Shelton BG, Flanders WD, Morris GK (1999) *Mycobacterium* sp. as a possible cause of hypersensitivity pneumonitis in machine workers. Emerg. Infect. Dis. 5:270–273

Shitaye JE, Matlova L, Horvathova A, Moravkova M, Dvorska-Bartosova L, Trcka I, Lamka J, Treml F, Vrbas V, Pavlik I (2008a) Diagnostic testing of different stages of avian tuberculosis in naturally infected hens (*Gallus domesticus*) by the tuberculin skin and rapid agglutination tests, faecal and egg examinations. Veterinarni Medicina 53:101–110

Shitaye JE, Matlova L, Horvathova A, Moravkova M, Dvorska-Bartosova L, Treml F, Lamka J, Pavlik I (2008b) *Mycobacterium avium* subsp. *avium* distribution studied in a naturally infected hen flock and in the environment by culture, serotyping and IS*901* RFLP methods. Vet. Microbiol. 127:155–164

Shitaye JE, Parmova I, Matlova L, Dvorska L, Horvathova A, Vrbas V, Pavlik I (2006) Mycobacterial and *Rhodococcus equi* infections in pigs in the Czech Republic between the years 1996 and 2004: the causal factors and distribution of infections in the tissues. Veterinarni Medicina 51:497–511

Shitaye JE, Tsegaye W, Pavlik I (2007) Bovine tuberculosis infection in animal and human populations in Ethiopia: a review. Veterinarni Medicina 52:317–332

Skoric M, Shitaye EJ, Halouzka R, Fictum P, Trcka I, Heroldova M, Tkadlec E, Pavlik I (2007) Tuberculous and tuberculoid lesions in free living small terrestrial mammals and the risk of infection to humans and animals: a review. Veterinarni Medicina 52:144–161

Skurski A, Szulga T, Wachnik Z, Madra J, Kowalczyk H (1965) Classification of acid-fast bacilli isolated from the milk of cows and from sewage used for fertilizing pastures. I. Pathogenic and saprophytic bacilli. Arch. Immunol. Ther. Exp. (Warsz.). 13:189–196

Slosarek M, Kubin M, Pokorny J (1994) Water as a possible factor of transmission in mycobacterial infections. Cent. Eur. J. Public Health. 2:103–105

Sniezek PJ, Graham BS, Busch HB, Lederman ER, Lim ML, Poggemyer K, Kao A, Mizrahi M, Washabaugh G, Yakrus M, Winthrop K (2003) Rapidly growing mycobacterial infections after pedicures. Arch. Dermatol. 139:629–634

Sobiech T, WACHNIK Z (1966) [Allergic and serologic studies of cattle from areas supplied with city sewage by means of the use of tuberculins from atypical mycobacteria]. Arch. Exp. Veterinarmed. 20:901–908

Solomon A (2001) Silicosis and tuberculosis: Part 2–a radiographic presentation of nodular tuberculosis and silicosis. Int. J. Occup. Environ. Health. 7:54–57

Songer JG, Bicknell EJ, Thoen CO (1980) Epidemiological investigation of swine tuberculosis in Arizona. Can. J. Comp. Med. 44:115–120

Spahr U, Schafroth K (2001) Fate of *Mycobacterium avium* subsp. *paratuberculosis* in Swiss hard and semihard cheese manufactured from raw milk. Appl. Environ. Microbiol. 67:4199–4205

Spiess LD, Lippincott JA (1981) Bacteria isolated from moss and their effect on moss development. Bot. Gaz. 142:512–518

Steed KA, Falkinham JO, III (2006) Effect of growth in biofilms on chlorine susceptibility of *Mycobacterium avium* and *Mycobacterium intracellulare*. Appl. Environ. Microbiol. 72:4007–4011

Stephan R, Schumacher S, Tasara T, Grant IR (2007) Prevalence of *Mycobacterium avium* subspecies *paratuberculosis* in swiss raw milk cheeses collected at the retail level. J. Dairy Sci. 90:3590–3595

Sung N, Collins MT (2000) Effect of three factors in cheese production (pH, salt, and heat) on *Mycobacterium avium* subsp. *paratuberculosis* viability. Appl. Environ. Microbiol. 66:1334–1339

Suppin D, Rippel-Rachle B, Smulders FJM (2007) Screening the microbiological condition of Sushi from Viennese restaurants. Wiener Tierarztliche Monatsschrift. 94:40–47

Sutton AL, Kephart KB, Verstegen MW, Canh TT, Hobbs PJ (1999) Potential for reduction of odorous compounds in swine manure through diet modification. J. Anim. Sci. 77:430–439

Szabo I, Kiss KK, Varnai I (1982) Epidemic pulmonary infection associated with *Mycobacterium xenopi* indigenous in sewage-sludge. Acta Microbiol. Acad. Sci. Hung. 29:263–266

Szabo I, Tuboly S, Szeky A (1975) Swine lymphadenitis due to *Mycobacterium avium* and atypical Mycobacteria. I. Pathological studies. Acta Vet. Acad. Sci. Hung. 25: 67–76

Szulga T, Wieczorek Z, Madra J, Kowalczyk H (1965) Classification of acid-fast bacilli isolated from the milk of cows and from sewage used for fertilizing pastures. 3. Identification of atypical bacilli (2nd and 3rd groups). Arch. Immunol. Ther. Exp. (Warsz.). 13:336–343

Tacquet A, Tison F, Devulder B (1961) [Bactericidal action of yoghurt on mycobacteria.]. Ann. Inst. Pasteur (Paris). 100:581–587

Takai S, Ohbushi S, Koike K, Tsubaki S, Oishi H, Kamada M (1991) Prevalence of Virulent *Rhodococcus-Equi* in Isolates from Soil and Feces of Horses from Horse-Breeding Farms with and Without Endemic Infections. J. Clin. Microbiol. 29:2887–2889

Takai S, Takeuchi T, Tsubaki S (1986) Isolation of *Rhodococcus (Corynebacterium) equi* and atypical mycobacteria from the lymph nodes of healthy pigs. Nippon Juigaku. Zasshi. 48:445–448

Takai S, Tsubaki S (1985) The incidence of *Rhodococcus (Corynebacterium) equi* in domestic animals and soil. Nippon Juigaku. Zasshi. 47:493–496

Takigawa K, Fujita J, Negayama K, Terada S, Yahaji Y, Kawanishi K, Takahara J (1995) Eradication of Contaminating *Mycobacterium-Chelonae* from Bronchofibrescopes and An Automated Bronchoscope Disinfection Machine. Respir. Med. 89:423–427

Taylor RH, Falkinham JO, III, Norton CD, LeChevallier MW (2000) Chlorine, chloramine, chlorine dioxide, and ozone susceptibility of *Mycobacterium avium*. Appl. Environ. Microbiol. 66:1702–1705

Thoen CO, Steele JH (1995) *Mycobacterium bovis* infection in animals and humans. Iowa State University Press, 1st ed., 355 pp

Thoen CO, Steele JH, Gilsdorf MJ (2006) *Mycobacterium bovis* infection in animals and humans. 2nd ed., Blackwell Publishing Professional, Ames, Iowa, USA, 317 pp

Thorel MF, Falkinham JO, Moreau RG (2004) Environmental mycobacteria from alpine and subalpine habitats. Fems Microbiol. Ecol. 49:343–347

Tiquia SM (2002) Evolution of extracellular enzyme activities during manure composting. J. Appl. Microbiol. 92: 764–775

Tobin-D'Angelo MJ, Blass MA, del Rio C, Halvosa JS, Blumberg HM, Horsburgh CR, Jr. (2004) Hospital water as a source of *Mycobacterium avium* complex isolates in respiratory specimens. J. Infect. Dis. 189:98–104

Torvinen E, Meklin T, Torkko P, Suomalainen S, Reiman M, Katila ML, Paulin L, Nevalainen A (2006) Mycobacteria and fungi in moisture-damaged building materials. Appl. Environ. Microbiol.. 72:6822–6824

Trckova M, Matlova L, Dvorska L, Pavlik I (2004) Kaolin, bentonite, and zeolites as feed supplements for animals: health advantages and risks. Veterinarni Medicina 49: 389–399

Trckova M, Matlova L, Hudcova H, Faldyna M, Zraly Z, Dvorska L, Beran V, Pavlik I (2005) Peat as a feed supplement for animals: a review. Veterinarni Medicina 50: 361–377

Trckova M, Zraly Z, Bejcek P, Matlova L, Beran V, Horvathova A, Faldyna M, Moravkova M, Shitaye JE, Svobodova J, Pavlik I (2006a) Effect of feeding treated peat as a supplement to newborn piglets on the growth, health status and occurrence of conditionally pathogenic mycobacteria. Veterinarni Medicina 51:544–554

Trckova M, Zraly Z, Matlova L, Beran V, Moravkova M, Svobodova J, Pavlik I (2006b) Effects of peat feeding on the performance and health status of fattening pigs and environmentally derived mycobacteria. Veterinarni Medicina 51:533–543

Trunova ON (1971) [The effect of various reservoir self-cleaning factors on *Mycobacterium tuberculosis*]. Probl. Tuberk. 49:60–63

Tsang AY, Denner JC, Brennan PJ, McClatchy JK (1992) Clinical and epidemiological importance of typing of *Mycobacterium avium* complex isolates. J. Clin. Microbiol. 30:479–484

Tsavkelova EA, Cherdyntseva TA, Lobakova ES, Kolomeitseva GL, Netrusov AI (2001) [Microbiota of the Orchid rhizoplane]. Mikrobiologiia. 70:567–573

Tsintzou A, Vantarakis A, Pagonopoulou O, Athanassiadou A, Papapetropoulou M (2000) Environmental mycobacteria in drinking water before and after replacement of the water distribution network. Water Air Soil Pollut. 120: 273–282

Tsukamura M, Mizuno S, Murata H, Nemoto H, Yugi H (1974) A comparative study of mycobacteria from patients' room dusts and from sputa of tuberculous patients. Source of pathogenic mycobacteria occurring in the sputa of tuberculous patients as casual isolates. Jpn. J. Microbiol. 18:271–277

Tsukamura M, Mizuno S, Toyama H (1984) [Mycobacteria from dusts of Japanese houses]. Kekkaku. 59:625–631

Tuffley RE, Holbeche JD (1980) Isolation of the *Mycobacterium avium-M. intracellulare-M. scrofulaceum* complex from tank water in Queensland, Australia. Appl. Environ. Microbiol. 39:48–53

Uhlemann J, Held R, Müller K, Jahn H, Dürrling H (1975) Schweinetuberkulose im einem Mastkombinat nach Einstreu von Hobel- und Sägespänen. Monatshefte für Veterinärmedizin. 30:175–180

Ulker N (1967) [Antibacterial action of honey toward different types of *Mycobacterium*]. Turk. Tip. Cemiy. Mecm. 33:282–287

Vaerewijck MJ, Huys G, Palomino JC, Swings J, Portaels F (2005) Mycobacteria in drinking water distribution systems: ecology and significance for human health. FEMS Microbiol. Rev. 29:911–934

Vantarakis A, Tsintzou A, Diamandopoulos A, Papapetropoulou M (1998) Non-tuberculosis mycobacteria in hospital water supplies. Water Air Soil Pollut. 104:331–337

VanTiem JS (1997) The public health risks of cervid production in the United States of America. Rev. Sci. Tech. 16:564–570

Vincke E, Boon N, Verstraete W (2001) Analysis of the microbial communities on corroded concrete sewer pipes – a case study. Appl. Microbiol. Biotechnol. 57:776–785

von Hertzen L, Haahtela T (2006) Disconnection of man and the soil: reason for the asthma and atopy epidemic? J. Allergy Clin. Immunol. 117:334–344

von Reyn CF, Arbeit RD, Horsburgh CR, Ristola MA, Waddell RD, Tvaroha SM, Samore M, Hirschhorn LR, Lumio J, Lein AD, Grove MR, Tosteson AN (2002) Sources of disseminated *Mycobacterium avium* infection in AIDS. J. Infect. 44:166–170

von Reyn CF, Maslow JN, Barber TW, Falkinham JO, III, Arbeit RD (1994) Persistent colonisation of potable water as a source of *Mycobacterium avium* infection in AIDS. Lancet. 343:1137–1141

von Reyn CF, Waddell RD, Eaton T, Arbeit RD, Maslow JN, Barber TW, Brindle RJ, Gilks CF, Lumio J, Lahdevirta J (1993) Isolation of *Mycobacterium avium* complex from water in the United States, Finland, Zaire, and Kenya. J. Clin. Microbiol. 31:3227–3230

Vugia DJ, Jang Y, Zizek C, Ely J, Winthrop KL, Desmond E (2005) Mycobacteria in nail salon whirlpool footbaths, California. Emerg. Infect. Dis. 11:616–618

Wang HC, Liaw YS, Yang PC, Kuo SH, Luh KT (1995) A Pseudoepidemic of *Mycobacterium-Chelonae* Infection Caused by Contamination of A Fiberoptic Bronchoscope Suction Channel. Eur. Respir. J. 8:1259–1262

Wang XJ, Kim J, McWilliams R, Cutting GR (2005) Increased prevalence of chronic rhinosinusitis in carriers of a cystic fibrosis mutation. Arch. Otolaryngol. Head Neck Surg. 131:237–240

Wang Y, Ogawa M, Fukuda K, Miyamoto H, Taniguchi H (2006) Isolation and identification of mycobacteria from soils at an illegal dumping site and landfills in Japan. Microbiol. Immunol. 50:513–524

Wayne LG, Kubica GP (1986) Genus *Mycobacterium* Lehmann and Neumann 1896, 363AL. In: Sneath, P.H.A., Mair, N.S.,

Sharpe, M.E., Holt, J.G. (Eds.), Bergey's manual of systematic bacteriology, 2. The Williams & Wilkins Co., Baltimore. 1436–1457

Wendt SL, George KL, Parker BC, Gruft H, Falkinham JO, III (1980) Epidemiology of infection by nontuberculous Mycobacteria. III. Isolation of potentially pathogenic mycobacteria from aerosols. Am. Rev. Respir. Dis. 122:259–263

Weyer K, Fourie PB, Durrheim D, Lancaster J, Haslov K, Bryden H (1999) *Mycobacterium bovis* as a zoonosis in the Kruger National Park, South Africa. Int. J. Tuberc. Lung Dis. 3:1113–1119

Whipple MJ, Rohovec JS (1994) The Effect of Heat and Low Ph on Selected Viral and Bacterial Fish Pathogens. Aquaculture. 123:179–189

Wilson MJ (2003) Clay mineralogical and related characteristics of geophagic materials. J. Chem. Ecol. 29:1525–1547

Wilson RW, Steingrube VA, Bottger EC, Springer B, Brown-Elliott BA, Vincent V, Jost KC, Jr., Zhang Y, Garcia MJ, Chiu SH, Onyi GO, Rossmoore H, Nash DR, Wallace RJ, Jr. (2001) *Mycobacterium immunogenum* sp. nov., a novel species related to *Mycobacterium abscessus* and associated with clinical disease, pseudo-outbreaks and contaminated metalworking fluids: an international cooperative study on mycobacterial taxonomy. Int. J. Syst. Evol. Microbiol. 51:1751–1764

Windsor RS, Durrant DS, Burn KJ, Blackburn JT, Duncan W (1984) Avian tuberculosis in pigs: miliary lesions in bacon pigs. J. Hyg. (Lond). 92:129–138

Winthrop KL, Abrams M, Yakrus M, Schwartz I, Ely J, Gillies D, Vugia DJ (2002) An outbreak of mycobacterial furunculosis associated with footbaths at a nail salon. N.Engl. J. Med. 346:1366–1371

Wolinsky E (1979) Nontuberculous mycobacteria and associated diseases. Am. Rev. Respir. Dis. 119:107–159

Wolinsky E, Schaefer WB (1973) Proposed numbering scheme for mycobacterial serotypes by agglutination. Int. J. Syst. Bacteriol. 23:182–183

Woolcock JB, Mutimer MD, Farmer AMT (1980) Epidemiology of *Corynebacterium-Equi* in Horses. Res. Vet. Sci. 28:87–90

Xu P, Kujundzic E, Peccia J, Schafer MP, Moss G, Hernandez M, Miller SL (2005) Impact of environmental factors on efficacy of upper-room air ultraviolet germicidal irradiation for inactivating airborne mycobacteria. Environ. Sci. Technol. 39:9656–9664

Yajko DM, Chin DP, Gonzalez PC, Nassos PS, Hopewell PC, Reingold AL, Horsburgh CR, Jr., Yakrus MA, Ostroff SM, Hadley WK (1995) *Mycobacterium avium* complex in water, food, and soil samples collected from the environment of HIV-infected individuals. J. Acquir. Immune. Defic. Syndr. Hum. Retrovirol. 9:176–182

Yoder S, Argueta C, Holtzman A, Aronson T, Berlin OGW, Tomasek P, Glover N, Froman S, Stelma G (1999) PCR comparison of *Mycobacterium avium* isolates obtained from patients and foods. Appl. Environ. Microbiol. 65:2650–2653

Chapter 6

The Occurrence of Pathogenic and Potentially Pathogenic Mycobacteria in Animals and the Role of the Environment in the Spread of Infection

I. Pavlik and J.O. Falkinham III (Eds.)

Introduction

Hosts of ESM, PPM and OPM are protozoa (including amoebae), insects and worms. For the most part, the presence of mycobacteria in these species is due to the widespread distribution of mycobacteria in habitats that are occupied by the animals. In as much as the same overlap of habitats is responsible for human and animal diseases, the protozoa and insects can serve as vectors for human and animal diseases.

Transmission from lower organisms to humans and animals can occur in a variety of ways:

- Lower organisms with mycobacteria on their surface can spread the micro-organisms throughout a niche where they live, but also into other niches or into other compartments in the environment.
- Mycobacteria-infected organisms shed mycobacteria to the environment through their faeces (particularly ruminants), urine, excretions and vomit (above all invertebrate animals), or by other routes (Section 5.10).
- Mycobacteria-infected organisms serve as vectors of mycobacterial transmission through ingestion as food (e.g. *Mycobacterium marinum*-contaminated *Tubifex*, cladocerans, copepods and brine shrimps for aquarium fish; Section 6.3).

- Mycobacteria can also be transmitted passively by parasites that are either ectoparasites or endoparasites (Section 6.3).

6.1 Protozoa – Introduction

J.O. Falkinham III and I. Pavlik

Former members of the kingdom Protozoa are now classified into several new "kingdoms": Amoebozoa, Rhizaria, Chromalveolata and Excavata. Protozoa are, with notable exceptions, single-celled micro-organisms (0.1 to few millimetres; e.g. some dinoflagellates and amoebas) that can ingest smaller micro-organisms by phagocytosis. A number of protists (unicellular eukaryotes) aggregate and form colonies (organisms can again live separately after division) or syncytium (multinuclear stages). Some are highly motile. It has been shown for a number of protists that the size of the prey micro-organisms is a determinant of phagocytosis; if they are large or too small, then the microbe is not ingested. Protozoa inhabit damp or wet habitats (Photo 6.1), not only water but also soil and the surface or the inside of different animal and plant species. Transmission of viable protists can occur via water, aerosols and dust, as well as via higher animal and plant hosts. Protists can survive unfavourable conditions (e.g. starvation) by forming thick-walled, desiccation- and heat-resistant cysts. Transmission of cysts can occur via flowing water, aerosolisation and the generation of dust.

There are more than 92 000 free-living or parasitic species of protozoa that are found throughout the

I. Pavlik (✉)
Head of OIE Reference Laboratories for Paratuberculosis and Avian Tuberculosis, Department of Food and Feed Safety, Veterinary Research Institute, Brno, Czech Republic
e-mail: pavlik@vri.cz

J. Kazda et al. (eds.), *The Ecology of Mycobacteria: Impact on Animal's and Human's Health*, DOI 10.1007/978-1-4020-9413-2_6, © Springer Science+Business Media B.V. 2009

world, including human-engineered habitats (drinking water systems). They are found in either fresh or saline habitats, like the ESM (e.g. Chesapeake Bay of North America). Protists are major contributors to the cycling of organic matter and influence microbial numbers through predation. Protists grow at temperatures of between 10 and 40 °C and can survive low and high temperatures; some are thermophilic and some psychrophilic.

Unlike many other micro-organisms, mycobacteria are not killed by protists following phagocytosis. Laboratory experiments have demonstrated that a variety of ESM and PPM survive and grow in protists. Furthermore, there is evidence that both the mycobacterial-infected host protists and the intracellular mycobacteria grow faster.

6.2 Mycobacteria as Endosymbionts in Protozoa

J.O. Falkinham and I. Pavlik

6.2.1 Presence of Mycobacteria in Protozoa

To test the hypothesis that mycobacteria can serve as food for free-living amoebae, different species of mycobacteria were co-cultured with a strain of *Acanthamoeba castellanii*. A number of species, including *M. kansasii*, *M. gastri*, *M. xenopi* and *M. microti*, were shown to stimulate the growth of *A. castellanii* on water agar medium (Prasad and Gupta, 1977). Further work by those authors demonstrated that the number of acid-fast bacilli in cells of *A. castellanii* increased following inoculation with individual strains of *M. avium*, *M. marinum*, *M. ulcerans*, *M. simiae* or *M. habana* (Prasad and Gupta, 1978).

A similar increase in acid-fast bacilli was observed in a co-culture of a strain of *M. avium* and *A. polyphaga* (Grange et al., 1987). In contrast, inoculation of *A. castellanii* with cells of individual strains of *M. smegmatis*, *M. fortuitum* or *M. phlei* led to rapid increases in numbers of intracellular acid-fast bacilli and death of the amoebae within 5 days (Prasad

and Gupta, 1978). In a survey of 26 species of ESM and PPM, all were shown to survive (non-quantitative colony formation) for up to 5 days in trophozoites and cysts of *A. polyphaga* (Adekambi et al., 2006).

Unfortunately, intracellular growth was not measured. Interestingly, no killing or cytopathic effects were observed in co-cultures of *A. polyphaga*, and for neither *M. smegmatis* nor *M. fortuitum* (Adekambi et al., 2006), in contrast to the results of Prasad and Gupta (1977). This difference could have been due to strain, culture or methodological differences. In all these studies, rapid phagocytosis of mycobacteria, whether they supported amoebae growth or not, was observed in keeping with the higher rates of phagocytosis for more hydrophobic micro-organisms (Monger et al., 1999).

More recent investigations have been initiated to determine whether a relationship exists between the number of protozoa in the environment and the number of mycobacteria. The basis of this hypothesis was that the ESM and PPM are poor competitors in environmental habitats due to their slow growth and impermeability to hydrophilic nutrients. Mycobacterial persistence could be due to their survival and growth in protozoa consistent with their intracellular growth in human and animal macrophages. Based on the examination of eight raw drinking water sources in the USA, there is a correlation (r = 0.559) between total protist numbers and mycobacterial numbers (J. O. Falkinham, unpublished data).

A number of factors undoubtedly contributed to this low correlation coefficient. First, total protozoa were enumerated, not phagocytic protists. Second, the protist species composition was not determined. Based on the feeding preferences of protozoa (Hansen et al., 1997), the failure of some *Mycobacterium* species to support *A. castellanii* growth or the killing of *A. castellanii* by rapidly growing mycobacteria (Prasad and Gupta, 1977), protist species composition would be expected to influence mycobacterial numbers. In spite of the limitations of the survey, it is our working hypothesis that protozoa are important determinants of both the number and distribution of ESM and PPM in the environment. Furthermore, because mycobacteria and protists occupy many of the same habitats, grazing by protozoa is a strong selective force operating upon mycobacterial populations.

6.2.2 Growth in Protozoa

A number of independent laboratory studies have shown that following phagocytosis, mycobacteria, principally *M. avium*, are capable of growth in protozoa. There have been no assessments of mycobacterial growth in protists in natural habitats to date. The high rate of phagocytosis of mycobacteria by protozoa (Strahl et al., 2001) is due, in large part, to their high cell surface hydrophobicity (Van Oss et al., 1975; Monger et al., 1999).

Consequently, there has been strong selection for intracellular survival in protozoa. Intracellular growth of *M. bovis*, *M. bovis* BCG, *M. avium*, *M. paratuberculosis*, *M. intracellulare*, *M. scrofulaceum* and *M. marinum* in *Tetrahymena pyriformis*, *A. polyphaga*, *A. castellanii* and *Dictyostelium discoideum* has been measured by a number of different techniques (Cirillo et al., 1997; Steinert et al., 1998; Miltner and Bermudez, 2000; Strahl et al., 2001; Solomon et al., 2003; Taylor et al., 2003).

Although different strains and methods were employed in the studies, the data collected suggest a relationship between the virulence in animals and growth rate in protists. For example, *M. scrofulaceum*, a poor pathogen in animal models, grew poorly in *T. pyriformis* (Strahl et al., 2001). *M. bovis* BCG strains did not survive well in *A. castellanii*, in contrast to *M. bovis* (Taylor et al., 2003). An *M. marinum* strain unable to grow in macrophages also failed to grow in *D. discoideum* (Solomon et al., 2003). Clearly, a variety of protists offer useful and inexpensive models for the study of growth of OPM in phagocytic cells.

6.2.3 Consequences of Mycobacterial Growth in Protozoa

The ability of *M. avium* strain MAC101 to invade *A. castellanii* was increased in cells following prior growth in that phagocytic amoeba. The increased invasiveness was due to an increase in phagocytosis, which was an adaptation to growth in the intracellular environment because cells grown first in amoebae and then in suspension lost the increased invasiveness. In keeping with the hypothesis that mycobacterial intracellular growth in protozoa mimics growth in animal phagocytic cells, amoebae-grown *M. avium* cells demonstrated increased invasiveness and growth in a human epithelial cell line and human macrophages (Cirillo et al., 1997).

As would be expected from the foregoing, amoebae-grown *M. avium* MAC101 cells introduced via the oral route in C57BL/6 beige mice were better able to colonise the intestinal tract and replicate in the spleen and liver (Cirillo et al., 1997). *M. avium* strain A5 grown in *T. pyriformis* was more virulent in chickens (J. O. Falkinham and W. B. Gross, unpublished data). Thus, like a bacteriophage infection in prokaryotic populations, growth of mycobacteria in protists appears to be subject to self-sustaining (possibly autocatalytic) amplification of mycobacterial numbers. In environmental habitats populated by protozoa, introduction of mycobacteria would be followed by an explosive increase in their numbers.

Measurements of the growth of both *M. avium* and *T. pyriformis* led to the surprising discovery that growth rates of both the *Mycobacterium* and protozoan were increased in infected cells, relative to uninfected cells in the same medium (J. O. Falkinham, unpublished data). Although other studies of mycobacterial growth in protists did not see differences, the differences are small and require large sample numbers. Increased growth of both the intracellular mycobacteria and protozoa supports the notion that mycobacteria should be considered endosymbionts of protozoa. Earlier work has shown that intracellular mycobacterial numbers fell after phagocytosis by *T. pyriformis*, reflecting the higher growth rate of the protozoan (Strahl et al., 2001).

Rather than lead to a dilution and eventual loss, intracellular mycobacterial numbers increased, reflecting an increased rate of growth following phagocytosis. Highest mycobacterial numbers were reached after the protozoan reached its stationary phase and ceased increasing in cell number (Strahl et al., 2001). Thus, intracellular mycobacterial numbers reflect equilibrium between the host's and endosymbiont's physiologic status. This increased growth rate was higher than that reached by suspended cells in laboratory medium. Furthermore, careful measurements of *T. pyriformis* growth rates demonstrated that they were higher as well. A search of the literature revealed that some species of *Tetrahymena* fail to grow in fresh medium if inoculum numbers are low (e.g. <1000) and that a number of different compounds, including fatty acids, could overcome this inoculum effect (Schousboe and Rasmussen, 1994).

Cells of *M. avium*-infected *T. pyriformis*, unlike uninfected cells, were capable of growing from a low inoculum (J. O. Falkinham, unpublished data). Present experiments are testing the hypothesis that the ability of the *M. avium*-infected cells to overcome the growth limitation of low inoculum density is due to the presence of fatty acids originating from the *Mycobacterium*. This data further underscore the complexity of the linkage between mycobacteria and protists in the environment and the selection for mycobacterial intracellular parasitism.

One other consequence of the intracellular growth of mycobacteria in protists is antibiotic resistance. Cells of *M. avium* strain MAC101 grown in cells of *A. castellanii,* in monolayers, were shown to be more resistant to antibiotics, compared to cells grown in laboratory culture medium. As was shown for increased virulence, antibiotic resistance is due to a physiologic adaptation. Cells of *M. avium* MAC101 grown in amoebae and then recovered and grown in laboratory medium lost their antibiotic resistance (Miltner and Bermudez, 2000). Thus, like biofilm-grown cells of *M. avium*, there was an apparent physiologic adaptation to antibiotic resistance, brought about by intracellular growth.

One compelling hypothesis seeking to explain the growth of micro-organisms in protozoa is that they are protected from anti-microbial compounds. Clearly, mycobacteria in cysts would be capable of higher rates of survival because of the presence of the impermeable cyst wall. But what about intracellular mycobacteria in growing protists, are they resistant? This is certainly the case for *Legionella pneumophila* growing in *A. castellanii*; however this is a Gram-negative micro-organism, lacking the thick cell envelope of mycobacteria and is rather sensitive to anti-microbial compounds. In contrast, the environmental mycobacteria are extraordinarily resistant to disinfectants and heavy metals. That intracellular mycobacteria are more resistant to anti-microbial compounds could be measured by isolating cells from protists after growth and exposing them in suspension. However, proving this hypothesis is fraught with difficulty because of the extremely high resistance to disinfectants (Taylor et al., 2000). Increases in disinfectant resistance, due to growth in protists (e.g. two- to fivefold), might not be of high enough magnitude to be recognised statistically.

6.2.4 Spread of Mycobacteria by Soil Protozoa

In soil, protozoa are part of the microedaphon. This is largely constituted by plant-like organisms, whereas from animal-like organisms, above all protozoa are present there. Protozoa participate in the degradation of disintegrating dead biomass here as well as in water. Thus, they play a key role in humification and mineralisation of organic matter, which are the essential factors of soil fertility. Moreover, they are also found in environments rich in both soil bacteria and mycobacteria (Section 5.5).

Food intake can be realised in two or three ways:

Option 1. Diffusion allows the intake of low-molecular-weight compounds through the cytoplasmic membrane.

Option 2. Endocytosis-pinocytosis allows the occlusion of macromolecules, fat droplets and minute particles up to the size of 10 μm.

Option 3. Endocytosis-phagocytosis allows the occlusion of larger food particles, including bacteria and mycobacteria.

6.2.4.1 Soil Protozoa and "Grazing" of Bacteria

Endocytosis (options 2 and 3) is a process whereby cells absorb material from the outside by engulfing it with their cell membrane, which folds around the object leaving it sealed off in a large vacuole. Phagocytised particles inside the phagosome can be digested by lysosome globules. The digestive vacuole is formed by the fusion of a lysosome and a phagosome = phagolysosome. Non-digested remnants are excreted from the cell by exocytosis thorough the cytopyge – cell "anus" that is visible only at the time of defecation. The relationship between bacteria and protozoa can be characterised in three ways (Marciano-Cabral, 2004):

i. Bacteria serve as food for protozoa.
ii. After phagocytosis, bacteria become endosymbionts; this co-existence often ensures survival of both bacteria (including mycobacteria) and protozoa (Table 6.1).

Table 6.1 Isolation of mycobacteria by co-culture with protozoa

Country	Protozoa species	Isolates	Origin of examined samples	References
Switzerland	*Acanthamoeba castellanii*	*M. kansasii*	Hospital water network	[1]
		M. gordonae	Hospital water network	[1]
		M. xenopi	Hospital water network	[1]
Japan	*A. culbertsoni*	*M. vanbaalenii*	Soil at an illegal dumping site and landfills	[2]
		M. mageritense	Soil at an illegal dumping site and landfills	[2]
		M. frederiksbergense	Soil at an illegal dumping site and landfills	[2]
		M. vanbaalenii or *M. austroafricanum*	Soil at an illegal dumping site and landfills	[2]
		M. chubuense or *M. chlorophenolicum*	Soil at an illegal dumping site and landfills	[2]

[1] Thomas V, Herrera-Rimann K, Blanc DS, Greub G (2006) Appl. Environ. Microbiol. 72:2428–2438. [2] Wang Y, Ogawa M, Fukuda K, Miyamoto H, Taniguchi H (2006) Microbiol. Immunol. 50:513–524.

iii. Protozoa become reservoirs of bacteria, which are often pathogenic for other hosts. Among mycobacterial species, in particular the PPM are able to survive and sometimes also to multiply (Table 6.2).

Bacteria, which constitute food for protozoa, are present in soil and their numbers are quite rapidly reduced by protozoa. Their ingestion by protozoa has been described as "grazing"; this mostly reduces the concentration of bacteria transmitted into or present in soil (Ekelund and Ronn, 1994). The bacterial species *Pseudomonas chlororaphis* is used for detection of the intensity of "protozoa grazing" on bacteria in soil. These bacteria are considered as indicator bacteria. On the other hand, mycobacteria (e.g. *M. chlorophenolicum*) are considerably resistant to grazing by soil protozoa. It is assumed that this is mainly due to the hydrophobicity of the mycobacterial cell wall (Ronn et al., 2001).

Many species of soil mycobacteria can also cause degradation of various petroleum products, which can contaminate soil, especially in the mining regions. Therefore, the potential utilisation of these soil mycobacterial species for "cleaning" soil with petroleum products has been investigated intensively (Section 8.3). The ability of these "useful" mycobacteria to survive in soil for a long time, i.e. their ability to survive grazing by protozoa, is of considerable concern. This ability was confirmed for *M. chlorophenolicum* by laboratory experiments (Ronn et al., 2001). The capacity of bacteria including mycobacteria to survive "grazing" by protozoa has metaphorically been designated as "Trojan Horse" tactic and is viewed as a "pre-adaptation" stage of pathogenic intracellular life in higher hosts (Barker and Brown, 1994; Ronn et al., 2001).

6.2.4.2 Views and Perspectives on the Research

Various mutants have been prepared for the study of mechanisms that allow different mycobacterial species to survive the "grazing" phase, which is completed in the protozoan body by the production of the "phagolysosome". With their use, it will be possible to gradually uncover the defence mechanisms against PPM and OPM. From an ecological point of view, the observation that some species survive by encystation is especially significant (Table 6.2). This mode of mycobacterial transmission within the whole biosphere has not been studied thoroughly and remains to be elucidated.

At present, the relationship between protozoa living in the first stomach of adult ruminants is a field which has been insufficiently investigated. These, together with other symbiotic micro-organisms (bacteria and fungi), allow the decomposition of cellulose from vegetable food. Even though Entodiniid ciliates are the primary source of vitamin B_{12} for ruminants, their role as a "Trojan Horse" for mycobacterial infection (e.g. paratuberculosis) has not been studied yet.

Table 6.2 Mycobacteria detected and surviving/dying in protozoa

Country	Protozoa	Isolates	Notes	References
USA	*T. pyriformis*	MAC	Survived, also encystment	[13]
India	*A. castellanii*	MAC	Survived	[6]
Germany	*D. discoideum*	MAC	Mycobacteria killed protozoa	[10]
USA	*A. castellanii*	MAA (1)	Survived	[2]
USA	*A. castellanii*	MAA (1)	Survived and increased resistance[1]	[7]
USA	*T. pyriformis*	MAA (1)	Survived, also encystment	[13]
USA	*A. polyphaga*	MAH (4)	Survived, also encystment	[12]
UK	*A. polyphaga*	MAP	Survived	[8]
UK	*A. polyphaga*	MAP	Survived and grew[2]	[16]
UK	*A. castellanii*	MAP	Survived and grew[2]	[16]
UK	*A. castellanii*	*M. bovis*	Survived	[14]
UK	*A. castellanii*	*M. bovis* BCG	Mycobacteria killed by protozoa	[14]
India	*A. castellanii*	*M. fortuitum*	Mycobacteria killed protozoa in 5 days	[6]
USA	*T. pyriformis*	*M. intracellulare*	Survived, also encystment	[13]
Switzerland	*A. castellanii*	*M. kansasii*	Survived	[4]
India	*A.* spp.	*M. leprae*	Survived, not growing	[5]
Germany	*D. discoideum*	*M. marinum* mutant	Killed protozoa	[11]
India	*A. castellanii*	*M. marinum*	Survived	[6]
Switzerland	*A. polyphaga*	*M. massiliense*	Survived	[1]
India	*A. castellanii*	*M. phlei*	Killed protozoa within 5 days	[6]
USA	*T. pyriformis*	*M. scrofulaceum*	Survived, also encystment	[13]
India	*A. castellanii*	*M. simiae (M. habana)*	Survived	[6]
India	*A. castellanii*	*M. smegmatis*	Killed protozoa within 5 days	[6]
Japan	*A. culbertsoni*	*M. smegmatis*	Killed by protozoa	[15]
USA	*A. castellanii*	*M. smegmatis*	Killed by protozoa within few days	[2]
Germany, USA	*A. castellanii*	*M. smegmatis* mutants	Survived	[9]
India	*A. castellanii*	*M. ulcerans*	Survived	[6]
Japan	*A. culbertsoni*	*M. vanbaalenii*	Survived	[15]
France	*A. polyphaga*	*M. xenopi*	Survived	[3]

[1] Increased resistance to antibiotics and biocides.

[2] Ingested *M. a. paratuberculosis* were protected against chlorine exposure.

***T.** Tetrahymena.* ***A.** Acanthamoeba.* ***D.** Dictyostelium.* **MAA (1)** *M a. avium* (serotype 1). **MAH (4)** *M. a. hominissuis* (serotype 4). **MAP** *M. a. paratuberculosis.* **MAC** *M. avium* complex.

[1] Adekambi T, Reynaud-Gaubert M, Greub G, Gevaudan MJ, La Scola B, Raoult D, Drancourt M (2004) J. Clin. Microbiol. 42:5493–5501. [2] Cirillo JD, Falkow S, Tompkins LS, Bermudez LE (1997) Infect. Immun. 65:3759–3767. [3] Drancourt M, Adekambi T, Raoult D (2007) J.Hosp. Infect. 65:138–142. [4] Goy G, Thomas V, Rimann K, Jaton K, Prod'hom G, Greub G (2007) Res. Microbiol. 158:393–397. [5] Jadin J (2008) Acta Leprologica 59:57–67. [6] Krishna-Prasad B, Gupta S (2008) Current Science 47:245–247. [7] Miltner EC, Bermudez LE (2000) Antimicrob. Agents Chemother. 44:1990–1994. [8] Mura M, Bull TJ, Evans H, Sidi-Boumedine K, McMinn L, Rhodes G, Pickup R, Hermon-Taylor J (2006) Appl. Environ. Microbiol. 72:854–859. [9] Sharbati-Tehrani S, Stephan J, Holland G, Appel B, Niederweis M, Lewin A (2005) Microbiology 151:2403–2410. [10] Skriwan C, Fajardo M, Hagele S, Horn M, Wagner M, Michel R, Krohne G, Schleicher M, Hacker J, Steinert M (2002) Int. J. Med. Microbiol. 291:615–624. [11] Solomon JM, Leung GS, Isberg RR (2003) Infect. Immun. 71:3578–3586. [12] Steinert M, Birkness K, White E, Fields B, Quinn F (1998) Appl. Environ. Microbiol. 64:2256–2261. [13] Strahl ED, Gillaspy GE, Falkinham JO, III (2001) Appl. Environ. Microbiol. 67:4432–4439. [14] Taylor SJ, Ahonen LJ, de Leij FA, Dale JW (2003) Appl. Environ. Microbiol. 69:4316–4319. [15] Wang Y, Ogawa M, Fukuda K, Miyamoto H, Taniguchi H (2006) Microbiol. Immunol. 50:513–524. [16] Whan L, Grant IR, Rowe MT (2006) BMC. Microbiol. 6:63.

6.3 Metazoan Invertebrates

I. Pavlik and J. Klimes

Multicellular invertebrates have adapted to living in various components of the environment; among these are water, soil and air. They have adapted to parasitic life on surfaces or inside the bodies of different hosts. Invertebrates can be a part of the food chain for poikilothermic and homoiothermic animals; therefore, they can play a significant role in the transmission of mycobacteria. Invertebrate animals can become infected by mycobacteria in three different ways:

i. Mycobacteria infect the body surface of protozoa and invertebrates (Diptera, etc.).
ii. Invertebrate animals ingest mycobacteria that are able to survive in their viscera for a long time.
iii. Bloodsucking invertebrate animals can be infected by mycobacteria when they suck blood from the host and can then subsequently infect the next host.

Invertebrate animals also pose the risk of transmission of mycobacteria. Their bodies can protect mycobacteria from desiccation, UV radiation and the effects of different disinfectants. They are protected from these effects by both the proteins that constitute the bodies of invertebrates and the ability of invertebrates to move and escape from the effects of the disinfectants. With respect to the transmission of these mycobacteria, the ability of invertebrates to be carried over great distances by river streams, sea currents (Section 5.2) or air (Section 5.9) poses a great risk.

Besides some mammals and scavengers, in particular some insect species and their larvae that feed on the decomposing carcasses and thus contribute to their disintegration act as saprophages. Fly larvae excrete proteolytic enzymes that cause rapid meat decomposition. The sequence of the appearance of particular species on cadavers is fairly predictable and can be used in estimating the stage of their rotting. ESM and PPM have also been isolated from different invertebrate organisms that feed on faeces and cadavers of infected animals. These invertebrates either migrate or become the prey of predators. Thus, they are dangerous as they can spread various pathogenic agents including mycobacteria.

6.3.1 Worms

The former phylum worms (Vermes) represented a polyphyletic taxon (the artificial assemblage of unrelated groups) and was split many decades ago into several phyla (flatworms, spiny-headed worms, nematodes, annelids, etc.). However, for practical reasons, the historical term "worms" is retained in the following text. Worms belong to the group of organisms which can be termed ubiquitous. They are adapted to different environments and therefore lots of worm species exist. Worms are classified in several phyla; of those, mycobacteria have only been detected in the phylum of segmented worms (Annelida and Nematoda).

6.3.1.1 Earthworms

Earthworms are a significant part of the edaphon, which denotes the live mass of soil, i.e. microscopic and higher organisms living in soil (soil bacteria, fungi, algae, protozoa, worms, adult and larval stages of insects and small terrestrial mammals – moles, mice, hamsters, ground squirrels, etc.).

Earthworms are not found in acidic soils and in very shallow soil horizons. On the other hand, under highly favourable conditions, about 400 earthworms can live in $1 \, m^2$ of good garden soil to a depth of $50 \, cm$, 290 earthworms in meadows and 100 earthworms in fields. Owing to their way of life, they constitute a significant component of soil, because they ingest the soil and carry it up to the surface through their faeces. In this way, the earthworms steadily mix the soil layers and thus increase the depth of the humus layer. By digging out tunnels, the earthworms aerate the soil; this is beneficial for the microbial activity and decomposition of organic residues.

The earthworms predominantly feed on vegetable mass and swallow numerous micro-organisms including mycobacteria together with food. These have been detected in various genera and species of earthworms both from free nature and cattle farms infected with paratuberculosis, from herds of pigs affected by mycobacteriosis and from aviaries with water birds infected with the causative agent of avian tuberculosis (Table 6.3). Due to the fact that these causative agents were also detected in the earthworm bodies, it is necessary to view the transmission of these micro-organisms into the soil and their subsequent transportation from soil on the surface as highly risky. Based on the results obtained by examination of free-living earthworms, we can assume that they may play a significant role in the spread of these pathogens both to small terrestrial mammals (Section 6.7.1) and to birds (Section 6.6) or their predators (Section 6.7).

In dry periods when earthworms live under a damp layer of faeces, the risk of spreading causative agents of mycobacterial diseases is higher. When the spread of the causative agents of paratuberculosis was investigated in herds of cattle and goats, and avian tuberculosis was studied in water birds kept in a zoological garden, earthworms were often found in the surroundings of the stables, in the soil of paddocks, pastures and aviaries. These earthworms had access to animal faeces, feed leftovers and other organic substrates and could ingest them. Accordingly, *in vitro* experiments were performed to analyse the ability of experimentally infected earthworms to spread the causative agent of paratuberculosis in soil. The earthworms could ingest faeces from naturally infected ruminants and shed mycobacteria afterwards through their faeces 2 days after the one-shot infection. No long-term deposition of the causative agent of paratuberculosis was observed in their bodies (Fischer et al., 2003a).

It was found that the same CFU counts of *M. a. paratuberculosis* survived the passage through the digestive tract of earthworms as were present in the ruminant faeces used for the infection. *M. a. paratuberculosis* did not multiply in the digestive tract of earthworms due to the fact that multiplication requires a temperature of between 37 and 40 °C. It is possible that the CFU counts of *M. a. paratuberculosis* were reduced by the defence mechanisms of the earthworms in the intestinal tract (Bilej et al., 2000) or by dilution

of the causative agent of paratuberculosis with the ingested soil. Two *M. a. paratuberculosis* isolates of two different RFLP types were used for the infection. No association between the duration of *M. a. paratuberculosis* shedding through faeces and RFLP type of strain used for the infection of ruminant faeces was found (Fischer et al., 2003a).

The earthworms possess a developed defence system against ingested bacteria based on the lytic activity of coelomic fluid and phagocytosis (Bilej et al., 2000). This system is effective against bacteria, which could be pathogenic for earthworms. The effectiveness of this system against mycobacteria has not yet been studied and remains to be determined. However, due to the fact that the ESM, PPM and OPM were isolated from samples of earthworms originating from the field and also from bodies and faeces of earthworms after experimental infection, it is necessary to view earthworms as a risk factor for the spread of mycobacterial diseases. It is necessary to take these vectors of mycobacteria into consideration, above all when imposing control measures against mycobacterial infections of animals:

i. Disinfection of soil in paddocks and aviaries with an aqueous solution of formaldehyde; it is therefore necessary to safely remove earthworms so that they do not return to the soil and contaminate it with mycobacteria.
ii. With regard to epidemiology and epizootiology, worms can protect the ingested bacteria including mycobacteria from the effects of disinfectants.
iii. During rainfall earthworm tunnels are flooded with water and earthworms leave them and go to the soil surface. During intensive rainfall, the earthworms can be swept together with water from one pasture

Table 6.3 Isolation of mycobacteria from dung worms [1]

Worm species	Isolates	Worms were caught in soil originating from	Disease status in animals
Lumbricus terrestris	*M. terrae*	Nature	Not known (wild animals)
Lumbricus sp.	MAP	Cattle farm DS and surroundings	Paratuberculosis present
Lumbricus sp.	*M. abscessus*	Cattle farm PE and surroundings	Paratuberculosis present
Allolobophora sp.	*M. abscessus*	Cattle farm PE and surroundings	Paratuberculosis present
Unidentified	MAA	The aviary in zoological garden A with water birds	Avian tuberculosis present
Lumbricus terrestris	MAH	The aviary in zoological garden A with water birds	Avian tuberculosis present
Lumbricus rubellus	*M. scrofulaceum*	The fenced run on a pig farm	Mycobacteriosis

MAA M a. avium. **MAH** *M. a. hominissuis.* **MAP** *M. a. paratuberculosis.*

[1] Fischer OA, Matlova L, Bartl J, Dvorska L, Svastova P, du Maine R, Melicharek I, Bartos M, Pavlik I (2003) Vet. Microbiol. 91:325–338

to another, or from mountain pastures into a valley. The ingested mycobacteria released from the bodies of earthworms can thus contaminate other locations.

6.3.1.2 Sludge Worms (*Tubifex tubifex*)

Saprobia are organisms living in sewage, polluted largely with organic substances. The effect of this environment on the presence of different organisms is so pronounced that the degree of contamination can be estimated from their profile. Due to the decomposition of organic matter, oxygen levels decrease considerably and high levels of hydrogen sulphide, methane and other inorganic matters are formed; these are harmful to the organisms. Accordingly, only some organisms have adapted to such an environment and these clean the water by their activity. These include lower organisms, protozoa, worms, rotifers, some molluscs, crustaceans, insect larvae and some bacteria.

Organisms living in mud are sometimes called "sapropelic organisms". According to the contamination level of this water, only defined species can live here; these form characteristic biocenoses:

  i. *Organisms of the polysaprobe zone* live in the most contaminated waters largely inhabited by protozoa (mostly flagellates), annelids (*Tubifex*) and larvae of drone flies (*Eristalis*).
 ii. *Organisms of the mesosaprobe zone* live in less polluted waters where protein matter is dissociated into amino acids and where more oxygen is present due to plant activity. Numerous protozoa live there, worms and rotifers, the mollusc *Sphaerium*, from the crustaceans water louse and numerous insect larvae.
iii. *Organisms of the oligosaprobe zone* live in slightly contaminated waters. In particular, sand fleas and lower crustaceans live there.
 iv. *Organisms of katarrhobia (akeratobia)* live in clean waters in contrast to the above-mentioned saprobiotic organisms.

According to the available literature, the presence and function of mycobacteria during the large-scale degradation of organic matter in contaminated waters have not been systematically studied. Information is very scarce and is mostly concerned with the isolation of mycobacteria from the *Tubifex tubifex*, also called the sludge worm. They inhabit rivers and lakes on several continents. These worms ingest organic detritus and bacteria, which are digested in their intestinal tract. They can absorb smaller molecules through the body wall. They can survive in water without oxygen for months and owing to that can survive in areas heavily polluted with organic matter that almost no other species can endure. By forming a protective cyst, *Tubifex* can survive unfavourable conditions (drought, frost, etc.).

Due to their plentiful occurrence, *Tubifex* have been used as food for tropical fish in both live and freeze-dried forms for more than a century. They can be infected with anaerobic bacteria often found in mud where *Tubifex* live and also enter their intestinal tract. The presence of mycobacteria in *Tubifex* has been studied mainly in association with the occurrence of mycobacterial infections in tropical fish. More than 30 years ago Pattyn et al. (1971) recovered a non-serotypeable *MAC* isolate from *Tubifex* intended as food for fish at fish tanks in a zoological garden in Antwerp.

So far the most extensive study was performed in Germany, where sources of *M. marinum* for infected mangrove killifish (*Rivulus magdalenae*) were investigated. These fish were kept at 24 °C and were fed with only live *Tubifex* harvested in a river (Zwickauer Mulde, Saxonia, Germany). The *Tubifex* were washed in distilled water so as to prevent their additional contamination with mycobacteria. After staining according to Ziehl–Neelsen, numerous AFB were observed by compression microscopy; *M. marinum* was detected by culture. However, examination of the mud where the *Tubifex* was caught only occasionally revealed AFB, and by culture only *M. gordonae* was detected. Oral infection of fish of this species was confirmed by the 8-week experiment with subsequent isolation of *M. marinum* from the clinically ill fish. In the control group fed with *Daphnia pulex* only (their culture examination for mycobacteria was negative), *M. marinum* was not detected in the fish organs (Nenoff and Uhlemann, 2006).

This finding is serious from the ecological and epizootiological point of view, in particular due to the different species composition of mycobacteria detected in *Tubifex* bodies (*M. marinum*) and in mud (*M. gordonae*). It can be explained, e.g., by the fact that *M. marinum* is more virulent for animals and can likely survive the digestive process in the intestinal tract of *Tubifex*. *M. marinum*, whose occurrence in mud and freshwater of rivers in the temperate climatic zone is only sporadic, can subsequently be concentrated in their intestinal tracts. Accordingly, the research should be focused on the study of the ecology of mycobacteria in mud and river sediments and their ability to survive the encystation process in *Tubifex* or other worms.

6.3.1.3 Nematodes

Roundworms (nematodes) are parasitic worms that spend most of their developmental stage and subsequent life in the environment, which is also favourable for mycobacteria: intestinal content and faeces of the host (Section 5.10), humid environment of soil (Section 5.5), the shores of lakes (Section 5.2), shady places of pastures (Section 5.7), etc. Nematodes are well adapted to parasitic life:

  i. They can only move by undulation (e.g. in the intestinal content).
 ii. They ingest food through the mouth opening where they have well-developed cuticular teeth, which disturb tissues of the affected host. They suck food into the digestive tract where it is processed.
iii. The multiplication process is also adapted to parasitic life and is based on the overproduction of eggs. They are highly resistant to unfavourable conditions such as temperature changes and the effect of various chemicals and different humidities (from the aquatic environment up to entire dryness).
 iv. Larvae in the first and second stages usually develop outside the host organism in faeces and the humid environment of pastures and water sources. They mainly feed on bacteria (microbivorous) including mycobacteria (e.g. the causative agent of paratuberculosis). The third stage is negatively geotropic and positively phototrophic. This quality makes them migrate to the surface biofilm in water and to the surface of plants on pastures.
  v. After being ingested by the host, exuviations of larvae occur resulting in the fourth stage. These larvae can transmit various bacteria including

mycobacteria in their intestinal tract and on the surface of their bodies.

How can mycobacteria be transmitted by these nematodes not only in the environment but also among respective hosts? Researchers in Brazil and Australia tried to answer this question. The first study to investigate the possible transmission of PPM through nematodes was performed in South Brazil in the late 1980s (Mota and Sleigh, 1987). They found that PPM (*M. scrofulaceum*, *M. terrae* and *M. intracellulare*) were most often isolated from the mesenteric lymph nodes of cattle with *Oesophagostomum radiatum* present in the intestinal mucus. Mechanical damage to this significant barrier most likely facilitated the transmission of these PPM species (Photo 6.3). Further studies dealing with the significance of nematodes in the spread of mycobacterial infections (paratuberculosis) were published more than two decades later in Australia (Lloyd et al., 2001).

During investigation of sheep faeces for the presence of the causative agent of paratuberculosis, a variety of nematodes were identified. Due to the fact that some of these larvae were used as an effective vaccine after irradiation, the question arose whether *M. a. paratuberculosis* can be transmitted after irradiation with 400 Gy to the vaccinated animals. *In vitro* experiments showed that *M. a. paratuberculosis* infected larvae of *Haemonchus contortus*, *Ostertagia circumcincta* and *Trichostrongylus colubriformis* species. *M. a. paratuberculosis* survived irradiation with 400 Gy even in larvae of *H. contortus*. It was also found that a longer incubation time (7 weeks or more) was necessary for the isolation of *M. a. paratuberculosis* from irradiated larvae. *M. a. paratuberculosis* could be isolated after 4 weeks of culture from non-irradiated larvae (Lloyd et al., 2001).

With respect to the fact that these experiments only highlighted the risks of transmission of *M. a. paratuberculosis* by nematodes (Lloyd et al., 2001), larvae from the faeces of seven naturally infected sheep were examined in subsequent experiments. Of five multibacillary sheep, *M. a. paratuberculosis* was recovered by culture from four of them from larvae of *Trichostrongylus* sp., *Ostertagia* sp. and *Oesophagostomum* sp.; from the remaining sheep, *M. a. paratuberculosis* was obtained by culture "only" from rinse water from larvae. In two sheep with paucibacillary disease, *M. a. paratuberculosis* was not detected in larvae (Whittington et al., 2001).

6.3.1.4 Views and Perspectives on the Research

It follows from the above results that nematodes can spread ESM and PPM both on their surface and also most likely in their intestinal tract. Therefore, the research in the area of ecology of mycobacteria should be focused on the following concerns:

i. Investigation of the significance of other nematode species which are parasites of humans and which cause, e.g., enterobiosis (pinworms), ascariosis (roundworms) and trichinellosis (trichina worms) in the spread of mycobacteria.
ii. Exact localisation of surviving mycobacteria in the bodies or on the surface of larvae.
iii. Evaluation of participation of all developmental stages of nematodes (including resistant eggs) on the transmission of mycobacteria both horizontally and vertically during their development.
iv. Assessment of the significance of larvae penetrating the intestinal wall into the host organism in the transmission of mycobacterial infection and the significance of adults causing damage to tissues by the "toothed mouth".
v. The specification of the significance of nematodes both in the spread of PPM and above all in the control of infections in animals and humans caused by OPM.

6.3.2 Molluscs

Molluscs are the second most numerous phylum of animals, into which more than 150 000 species are classified at present. In the past, molluscs colonised all the main environments existing on the Earth. Therefore, this group of invertebrate animals includes the second highest number of animal species after arthropods. Molluscs live mainly in water (saltwater and freshwater), even though many species are also found ashore. They feed on vegetables and *animals* that can be highly contaminated (infected) with mycobacteria. With respect to their significance for man, molluscs can be divided into the following groups:

i. Economically significant organisms (gastropods, bivalves, cuttlefish, octopuses, squid, etc.) which are used as food by humans.
ii. Agricultural pests (e.g. slugs).
iii. Intermediate hosts of helminths (e.g. great pond snails or great ramshorn snails).

The size of mollusc bodies ranges from several millimetres up to several metres in cephalopods. With respect to the spread of mycobacteria in the environment, the high number of actively or passively moving molluscs is especially important.

6.3.2.1 The Possible Role of Molluscs in the Ecology of Mycobacteria

In an ecosystem, molluscs are significant decomposers of organic matter, both in the aquatic environment and dry land. The major part of mollusc development takes place in shallow salt waters, where most mollusc species have been described so far. In the brackish environment, i.e. in tidal sea zones and freshwater branches, plenty of molluscs are found. This is due to their ability to tolerate changes in salinity and occasional desiccation; however, this has also resulted in lower species diversity compensated for by high number of individuals. Molluscs have also adapted to freshwater and the terrestrial environment. They can survive climatic conditions that are unfavourable for them, i.e. especially desiccation and high temperatures. On land, molluscs have reached the highest species variety in humid warm climates with lush vegetation, in particular in the tropics and subtropics. However, they also survive in arid regions of semi-deserts.

In the sea, molluscs are part of both movable and fast sessile benthos or are buried under the sea bottom. Molluscs can utilise various food sources. Some of them select organic detritus when creeping along the bottom of seas, lakes and rivers. However, other mollusc species consume dead plant leaves or carcasses of different animals. Numerous mollusc species are "grazers" that are grazing on plant vegetation and also protozoa (Section 6.2). Many mollusc species feed on algae on the soil surface or on biofilms of the aquatic environment. They also feed on settling plankton algae, the "plankton rain". With regard to the frequent occurrence of mycobacteria in this environment, we can suppose that molluscs also play a significant role in their ecology.

However, mycobacteria can also be concentrated in mollusc bodies in a different way. The majority of molluscs (above all clams) are classified as microphages, i.e. animals that filter small plankton and organic debris containing mycobacteria.

Molluscs have a feeding structure, the radula; this is a roughened area at the bottom of the mouth mostly composed of chitin and forming a scraping radula which is supported by a cartilaginous odontophore. The radula is a structure used by molluscs to scrape Cyanobacteria and algae off rocks in the aquatic environment (Photo 6.2); on land, they graze plant vegetation. Thus, they make minerals accessible for the growth of other plants. By nibbling at water weeds, water molluscs make plant tissues accessible for bacteria. The plants consequently die and become significant sources of food for bacteria including mycobacteria, which become food for various insect larvae such as dragonflies (Section 6.4.1).

6.3.2.2 Mycobacteria Detected in Molluscs and Their Health Impact

With regard to the possible transmission of mycobacteria in their ecology, it is necessary to mention the following factors concerning molluscs:

i. The majority of molluscs move actively or passively in the environment, whereby they can transmit mycobacteria on their surface or in their viscera.

ii. Molluscs number among those animals with an unstable body temperature (cold-blooded, poikilo-thermic or ectothermic) and thus, the multiplication of mycobacteria in their bodies depends on an ambient temperature.

iii. The skin of a mollusc produces calcium carbonate that forms a hard external shell (sometimes reduced in size, e.g. to cuttlebone, or missing, e.g. in the octopus). This shell protects molluscs from predators; human cases of injury by these shells, which opened an entrance site for a mycobacterial skin infection, have been described (Table 6.4).

So far, mycobacteria have only been detected in different gastropod species (Gastropoda) from the freshwater environment and in bivalves (Bivalvia; Table 6.5).

With respect to the fact that various gastropods are kept, in aquaria infected with *M. marinum*, these may be significant in the process of transmission of this mycobacterial species. Due to the fact that they ingest different organic residues on the bottom such as plants, algae and biofilm on surfaces, they can easily be infected with high numbers of mycobacteria. Various gastropods participate in eating dead fish bodies or laid spawn.

When considering epidemiology, the detection of *M. ulcerans* in tropical gastropods can be significant; these gastropods can constitute not only a vector but also a reservoir. It was found that two aquatic gastropod species cannot cause killing of the ingested *M. ulcerans* in their intestinal tract (Table 6.5). In further experiments, they discovered that *M. ulcerans* was transmitted to biting water bugs (*Naucoris cimicoides*), which fed on these two species of small infected gastropods (Marsollier et al., 2004).

Various undemanding snails such as mimic *Limnaea* (*Pseudosuccinea columella*) are kept in aquaria.

Table 6.4 Isolation of mycobacteria from skin lesions from patients after contact with molluscs (Mollusca; oysters) and echinoderms (Echinodermata; sea urchin)

Mode of transmission	Patients[1]	Infected tissue	Isolates	References
One gallon of oysters harvested and "shucked" from river inlet while on holiday	66/M	Swollen left hand with six non-draining nodular lesions along the ulnar palm	*M. marinum*	[2]
Oyster contact	39/F	Ulcer on the right hand	*M. marinum*	[1]
Injury with sea urchin spines	3 patients	Sea urchin granuloma	*M. marinum*	[3]
Injury with sea urchin spines	4 patients	Sea urchin granuloma	*M.* sp.	[3]

[1] Age in years/gender (sex): **M** (male), **F** (female).
[1] Aubry A, Chosidow O, Caumes E, Robert J, Cambau E (2002) Arch. Intern. Med. 162:1746–1752. [2] Beecham HJ, III, Oldfield EC, III, Lewis DE, Buker JL (1991) Lancet 337:1487. [3] De la TC, Vega A, Carracedo A, Toribio J (2001) Br. J. Dermatol. 145:114–116.

Table 6.5 Isolation of mycobacteria from different molluscs (Mollusca)

Species	Isolates	Notes	References
Snails (*Planorbarius corneus*)	*M. chelonae, M. fortuitum*	Aquarium with healthy ornamental freshwater fish (Czech Republic)	[1]
Oysters (*Crassostrea virginica*)	*M. scrofulaceum, M. gordonae, MAH* (9, 16), *M. terrae* complex, *M. parafortuitum* complex	Raw oysters (Alabama, USA)	[2]
Gastropods (*Bulinus senegalensis*)	*M. ulcerans*	Buruli ulcer endemic regions in the Republic of Benin	[3]
Tropical aquatic snails (*Pomacea canaliculata* and *Planorbis planorbis*)	*M. ulcerans*	Experimental infection with *M. ulcerans* in the aquarium environment; snails were collected in the Daloa region (Ivory Coast)	[4]

MAH (**9, 16**) *M. a. hominissuis* (serotypes 9 and 16).

[1] Beran V, Matlova L, Dvorska L, Svastova P, Pavlik I (2006) J.Fish. Dis. 29:383–393. [2] Hosty TS, McDurmont CI (1975) Health Lab Sci. 12:16–19. [3] Kotlowski R, Martin A, Ablordey A, Chemlal K, Fonteyne PA, Portaels F (2004) J. Med. Microbiol. 53:927–933. [4] Marsollier L, Severin T, Aubry J, Merritt RW, Saint Andre JP, Legras P, Manceau AL, Chauty A, Carbonnelle B, Cole ST (2004) Appl. Environ. Microbiol. 70:6296–6298.

Their eggs are highly resistant and owing to this feature they can also be found in dried feed and thus they can be transmitted through such feed to other aquaria. However, such a mode of transmission of mycobacteria has not yet been reported. *M. marinum* has not been detected in the eggs of crustaceans, such as branchiopod crustaceans used for the feeding of hatched aquarium fish fry (Section 6.5; Photo 6.4).

6.3.2.3 Views and Perspectives on the Research

Further research dealing with the significance of molluscs in the ecology of mycobacteria should be focused on the following:

i. In molluscs ingested food including mycobacteria is digested by salivary enzyme glands from the oesophagus, sacciform stomach and hepatopancreas producing amylase. Even though molluscs can successfully digest cellulose, it is necessary to assess the degree of killing of ingested mycobacteria after their passage through the intestines in the visceral mass.
ii. Information is missing concerning mycobacteria in molluscs that are active predators (highly specialised cephalopods) and catch other invertebrates or fish. With respect to feeding on various prey (also infected by mycobacteria) we can suppose

that mycobacteria will also be detected in the cephalopods.
iii. It is also noteworthy that little is known about the detection of mycobacteria in molluscs that are considerable components of food for people. Many mollusc species (such as bivalves) are eaten raw or only cold smoked. Information on the risk of the occurrence of mycobacteria in these molluscs is missing and remains to be determined.
iv. Accordingly, it is evident from the above-mentioned data that despite the sporadic occurrence of ESM and PPM, it will be necessary to focus on the significance of molluscs in the transmission of OPM (e.g. the increasing incidence of the pathogenic agent of paratuberculosis in surface water) in the near future.

6.3.3 Echinoderms

Echinoderms are marine animals characterised by calcified plates in their subcutis or calcareous spines. This phylum includes, e.g., sea cucumbers (Holothuroidea), starfish (Asteroidea) and sea urchins (Echinoidea). Many inhabitants of seaside regions and especially tourists come into contact with the members of the two latter classes and are injured by the sharp spines of these animals.

Even though it is not known whether mycobacteria are found in these animals, numerous skin injuries, which heal with difficulty, have been noted. Much remains to be discovered regarding the source of mycobacterial skin infections, especially after injury by sea urchins (Table 6.5). Further research should therefore seek to elucidate surface contamination and the occurrence of mycobacteria in the bodies of sea starfish and sea urchins.

6.3.4 Arthropods

Another large group of animals, in which mycobacteria have been detected, are members of the phylum Arthropoda (arthropods). About 80% of described animal species are arthropods. Their body is basically divided into the head (caput), thorax and abdomen. The body of arthropods is covered with a cuticle which is rigid and composed mainly of chitin (sometimes it contains calcium carbonate or wax). The cuticle of arthropods strengthens them against environmental effects and therefore they are able to inhabit all parts of the biosphere. Mycobacteria have so far been detected in three of the four Arthropoda subphyla described. The exception is subphylum Trilobites.

  i. *Subphylum* Crustacea (crustaceans).
 ii. *Subphylum* Chelicerata (chelicerates).
iii. *Subphylum* Tracheata (tracheates).

6.3.4.1 Crustaceans

The fact that the majority of crustacean species live in an environment with an abundant occurrence of mycobacteria should play a significant role during further investigation of crustaceans (Section 5.2). Mycobacteria have been detected in four of the currently accepted classes of crustaceans. The isolations of mycobacteria in crustaceans are shown in Table 6.6. To better analyse the described cases, they can be classified into three groups according to the ecology of mycobacteria:

Group 1: Mycobacteria were detected both in the bodies and eggs of the members of only one class (Branchiopoda – branchiopods); mycobacteria were

detected in the bodies of the examined organisms of members of the class Copepoda (copepods); these crustaceans were used for feeding aquarium fish (Photo 6.4).

Group 2: Mycobacteria were detected in the bodies of woodlice (*Porcellio* sp.; class Malacostraca), which were trapped both in nature and in the agricultural landscape and sawmills that process the tree trunks (*MAC* was detected in both sawmills and sawdust).

Group 3: Among other members of Malacostraca and former class Maxillopoda, *M. marinum* was isolated from patients' skin injured by sharp shells of different crustacean species.

In the first group of cases, the finding of especially *M. marinum* in the eggs of brine shrimp (*Artemia salina*; Table 6.6) is noteworthy. This species lives in salt lakes in semi-desert and desert regions on different continents. As soon as the lakes fill with water after rainfall, brine shrimp begin to multiply hastily even though the eggs can survive for months to years in a dry state under these highly extreme conditions. This phenomenon is designated as anabiosis and allows the lower animals, including crustaceans, to "die" and return to life when conditions are once again favourable.

Because water fleas (copepods; Copepoda) feed on small organic particles including different bacterial species, three mycobacterial species (*M. chelonae*, *M. fortuitum* and *M. flavescens*) have been isolated from water in ombrogenic peat bogs and small lakes where the surface was overgrown with sphagnum; subsequently, these mycobacteria were also detected in 10 bodies of water fleas of *Ceriodaphnia reticulata* species. The concentration of mycobacteria in the summer months of the year 1987 reached a median of 5.1×10^4 CFU with a maximum of 1.2×10^6 CFU at the end of the warm summer (in August and September).

Experiments with radioactively labelled *M. chelonae* under laboratory conditions showed that within 60 min, *Daphnia magna* and *Ceriodaphnia reticulata* could swallow 5.0×10^5–1.0×10^6 CFU and 1.2–2.6×10^5 CFU, respectively. Due to their constant motion the energy expended by these crustaceans is great. Therefore, they must steadily ingest food. In cases where the only source of food was mycobacteria, the water fleas of both species survived for up to 60 hours; if they were kept without food, both of the species survived for only 12 hours (Soeffing, 1990).

Table 6.6 Isolation of mycobacteria from different crustaceans

Class	Species	Isolates	Comments	References
Branchiopoda	Brine shrimp (*Artemia salina*)	*M.* species	Isolation from brine shrimp by aquarist with clinically healthy fish	[3]
		M. marinum	Isolation from sea monkeys eggs by aquarist with skin, *M. marinum* infection on his finger	[2]
	Water flea (*Daphnia* sp.)	*M. marinum*	*M. marinum* detected in mud from the pond from which daphniae were obtained	[6]
	Ceriodaphnia reticulata	*M. chelonae, M. fortuitum, M. flavescens*	Ombrogenic sphagna with the same species of mycobacteria	[7]
Maxillopoda	Barnacles	*M. marinum*	Scraped hands on barnacles while cleaning the underside of the boat – solitary nontender nodules on hands and forearm of 67-year-old man	[5]
Malacostraca	Crab	*M. marinum*	Pinched by the crab while fishing on the South Carolina coast – tender erythematous nodule on the dorsal aspect of the right fourth finger of 42-year-old man	[4]
	Porcellio (*Porcellio* sp.)	*MAC, M.* sp.	Sawmills in Germany (contact with bark and sawdust)	[1]
		M. sp.	Agriculture area and free nature	[1]

MAC M. avium complex.

[1] Beerwerth W, Eysing B, Kessel U (1979) Zentralbl. Bakteriol. [Orig. A] 244:50–57. [2] Beran V, Svobodova J, Pavlik I (2008) Unpublished data. [3] Beran V, Matlova L, Dvorska L, Svastova P, Pavlik I (2006) J.Fish. Dis. 29:383–393. [4] Jernigan JA, Farr BM (2000) Clin. Infect. Dis. 31:439–443. [5] Johnson RP, Xia Y, Cho S, Burroughs RF, Krivda SJ (2007) Cutis 79:33–36. [6] Mansson T (1970) Br. Med. J. 3:46. [7] Soeffing K (2008) Diss. Univ. Hamburg.148 pp.

In the second group of cases, the detection of *MAC* members in woodlice (*Porcellio* sp.) trapped in sawmills that process the tree trunks (Table 6.6) is worthy of special mention. Due to the fact that *MAC* members often contaminate sawdust (Section 5.8), it is possible to explain the contamination of woodlice from these sources. Their contamination with mycobacteria was considerable: 8 of 10 samples from the agricultural landscape and from nature and 39 of 41 samples from sawmills (*MAC* members were detected in 15 samples) were positive. However, detailed data on the vertical transmission of the *MAC* members by woodlice and their significance in the spread of the *MAC* in the environment are not available (Beerwerth et al., 1979).

The third group of cases is, however, characterised by the skin injury of people by crustaceans that effectively fight against a "potential enemy", including man. That is possible owing to a hard exoskeleton covering their bodies, designated as a crust (on which the designation of Crustacea is based), formed by calcium salts. It remains unclear whether the skin of the injured patients was infected immediately after their injury with the crust surface of the contaminated crustaceans or by an additional infection caused by *M. marinum*, which is also found loose in water (Table 6.6).

With regard to the possible spread of PPM (such as *M. marinum* and *MAC* members), it will be useful to focus further research on the following fields:

i. Investigation of the presence of PPM in other crustacean species that often become a significant constituent of food for people (lobsters, prawns, shrimps, etc.).
ii. Verification of the possible survival of PPM in crustaceans under different storage conditions (frozen, salted, cured, etc.) and after various types of culinary processing.
iii. Analysis of the possible survival and transmission of PPM through crustacean eggs that can survive complete desiccation in a state of anabiosis and that can subsequently be a source of mycobacterial infection for aquarium fish.

iv. Assessment of crustacean significance for the spread of PPM in different components of the terrestrial and water environments.

6.3.4.2 Chelicerates

In Chelicerata the first pair of appendages have developed to chelicerae and the second pair of appendages have developed to pedipalps. Several tens of thousands of Chelicerata have been described; it is most likely that their progenitors were the now extinct trilobites. The majority of their members feed on other invertebrates or small vertebrates; besides these, important ectoparasites are classified in this subphylum. The described cases of the detection of mycobacteria in Chelicerata are rare in the literature, despite the fact that they are in close contact with people, especially in developing countries (Table 6.7).

Chelicerata can get infected by mycobacteria through:

i. Sucking blood from hosts (e.g. birds) affected by mycobacteria, e.g. ticks (Blagodarnyi et al., 1971).
ii. Ingestion of tissues infected with mycobacteria (above all the skin of leprosy patients) by skin parasites (e.g. causative agent of scabies).
iii. Preying on other invertebrates (e.g. spiders) that live in environments contaminated with mycobacteria (e.g. sawdust).

How can, e.g., soil mites participate in the ecology of mycobacteria? Their dead bodies become part of the dust where mycobacteria have also been detected (Section 5.8). Different Chelicerata members have been studied in connection with leprosy (Blake et al., 1987; Sreevatsa, 1993). It is however evident that much knowledge in this field remains to be generated.

6.3.4.3 Tracheates

Mycobacteria have been detected in the species belonging to three of five to six (different according to different systems) classes:

i. Millipedes (Diplopoda).
ii. Centipedes (Chilopoda).
iii. Insects (Insecta).

In this chapter, only data concerning the isolation of mycobacteria from the first two mentioned classes of animals (Diplopoda and Chilopoda) will be reported. The isolation of mycobacteria from the large class Insecta is described in a separate Section (Section 6.3.5).

Mycobacteria were only detected in unspecified species of two classes (millipedes – Diplopoda and centipedes – Chilopoda) in Germany, above all in relation to the occurrence of *M. avium* (*MAC*) members in various biotopes (Beerwerth et al., 1979):

- Agricultural landscapes; aside from this, mycobacteria were detected in 14 (46.7%) of 30 samples of millipedes and centipedes; *MAC* members were isolated from 4 (13.3%) samples.
- Sawmills that process tree trunks; mycobacteria were detected in 16 (53.3%) of 30 samples of millipedes and centipedes; *MAC* members were only isolated from 1 (3.3%) sample.

The isolation of *MAC* members from samples of millipedes and centipedes in the biotope of sawmills for processing the trunks of felled trees is not surprising. *MAC* members were often detected both in the bark of the felled trees contaminated with soil during their storage and transport and in sawdust (Section 5.8).

Table 6.7 Isolation of mycobacteria from members of "subphylum" chelicerates (Chelicerata)

Order	Species	Isolates	Comments	References
Spiders (*Araneae*)	Not known	*M.* sp.	Agricultural land, nature and sawmills	[1]
Ticks (Ixodia)	Not known	*M.* sp.	Agricultural land and nature	[1]
	Argas persicus	*M.* sp.	Spontaneously infected bird ticks	[2]
Mites (*Acari*)	*Sarcoptes* sp.	*M. leprae*	Acid-fast rods caused mouse footpad infections (endemic areas with leprosy in India)	[3]

[1] Beerwerth W, Eysing B, Kessel U (1979) Zentralbl. Bakteriol. [Orig. A] 244:50–57. [2] Blagodarnyi I, Makarevich NM, Blekhman IM (1971) Probl. Tuberk. 49:74–76. [3] Narayanan E, Manja KS, Kirchheimer WF, Balasubrahmanyan M (1972) Lepr. Rev. 43:194–198.

Information on the significance of the members of millipedes and centipedes in the ecology of mycobacteria is missing. Accordingly, further research should be focused on this field, not only generally but also on their significance in the spread of PPM (especially *MAC* members).

6.3.5 Insects

Almost one million of the described animal species are at present classified in the class Insecta and new species are constantly being discovered. Insects populate nearly all the components of the biosphere on our planet and can influence the ecology of mycobacteria in many different ways. Apart from the transmission of mycobacteria on their surfaces, insects have a complete digestive system in contrast to all lower animals.

The digestive system can be divided into the following parts:

- Mouthparts (with chewing mandibles) with salivary glands that function to secrete saliva into this area.
- Pharynx.
- Oesophagus.
- Foregut (crop, gizzard) and midgut (stomach), in which proper enzymatic digestion of food occurs.
- Intestine.
- Rectum into which run Malpighian tubules that clean an insect's blood from waste.

Due to the fact that waste is converted into almost waterless urine, which is deposited into the hindgut, this mode of excretion allows extremely efficient water preservation. This ability increases the protection of insects against hot and dry environments. Hence, provided the insects are infected with bacteria, they can spread them throughout the surroundings not only on their surfaces while crawling or flying (mycobacteria can be killed by reduced water activity and solar radiation) but also in their digestive tract. If mycobacteria are resistant to the digestive mechanisms of insects, they can survive dry and unfavourable conditions in the intestinal tract, from which they can later be shed into the environment.

Insects are classified according to the presence or primary absence of wings into two subclasses, Apterygota and Pterygota.

With regard to the ecology of mycobacteria, it is not surprising that since their discovery in the late 1990s, their presence has been noted in a number of insect orders of the subclass Pterygota. An extensive study analysing serotypes of *MAC* in sawdust, insects trapped in the sawdust environment and on/under the bark of trees in forests and other localities revealed a close relationship between insects and the ecology of mycobacteria. The prevailing isolation of *M. a. hominissuis* serotypes 8 and 9 both from sawdust and insects is unambiguous and shows the circulation of mycobacteria in various components of the environment (Table 6.8).

Accordingly, in the following chapters, we will focus on the members of different classes in which mycobacteria were detected, with special emphasis on the ecology of mycobacteria and the risk of their spread into the environment and transmission to susceptible hosts.

6.3.5.1 Dragonflies

At present, up to 6500 dragonfly species are known. The metamorphosis in dragonflies is imperfect; the pupae stage is missing. Development is characterised by several larval stages; nymphae are mainly found in

Table 6.8 Isolation of mycobacteria from sawdust and insects from different environmental niches [1]

Members of *MAC*	Serotypes	Sawdust No.	%	Insects from sawdust No.	%	Insects from woods and other places No.	%
M. a. hominissuis	4	2	7.7	2	3.4	0	0
	8	12	46.2	20	33.9	16	64.0
	9	7	26.9	25	42.4	3	12.0
M. intracellulare	16, 18, 19	0	0	1	1.7	5	20.0
	Mixed	5	19.2	11	18.6	1	4.0
Total		26	100	59	100	25	100

[1] Meissner G, Anz W (1977) Am. Rev. Respir. Dis. 116:1057–1064.

stationary and streaming waters (rivers, puddles, bogs, lakes, ponds, periodic water in forests and meadows, etc.). Larvae of the broad-bodied chaser (*Libellula depressa*) can survive complete dehydration for periods of a few weeks up to months buried in dry mud in anabiotic condition (stiffness). The speed of development of larvae depends upon the food available, water temperature and other factors; the longevity of larvae is also species specific and ranges from between 6 and 7 weeks up to a few years. Larvae of the majority of species are highly predaceous (cannibalism is common among them); they feed on various water invertebrate animals including tadpoles, spawn or fish fingerlings.

Unspecified mycobacteria were detected in 3 of 12 samples of dragonfly larvae (Beerwerth et al., 1979). *M. chelonae* and *M. flavescens* were detected by culture in dragonfly larvae of the following three species: *Leucorrhinia rubicunda*, *Leucorrhinia quadrimaculata* and *Sympetrum danae* (Soeffing, 1990). Laboratory experiments with dragonfly larvae showed that these larvae ingested mycobacteria indirectly through infected water fleas (Section 6.3.4.1). Using radioactively labelled *M. chelonae*, the distribution of radioactivity was observed in different parts of larvae 60 days after the ingestion of infected water fleas (i.e. radioactively labelled sulphur compounds). The highest occurrence was detected in the intestines (30.2%) and cuticle (36.1%); lower radioactivity levels were detected in the head (17.3%) and in fat (16.4%). Hence, these experiments explained the circulation of metabolites of mycobacteria among crustaceans that first filter mycobacteria from water and transmit them through dragonfly larvae, which ingested them (Soeffing, 1990).

On the other hand, imagos often prey on other invertebrates far from water (Photo 8.1). Large chewing mandibles and big eyes are used by the majority of species to skilfully prey on other organisms in the daytime. They usually fly at temperatures above 17 °C; smaller species can fly at relatively low ambient air temperatures (in rare cases exceptionally at 12 °C). The male dragonflies search for their prey mostly near waters, whereas the females prefer preying in places distant from water (meadows, forests, clearings, etc.). They mostly feed on flying insects (gnats, flies and small butterflies) and can swallow high amounts of insects within the space of several minutes. These amounts can weigh as much as the dragonfly itself. However, it is noteworthy that notwithstanding this,

they can fast for weeks. Unspecified mycobacteria were detected in 3 of 28 samples of the bodies of adult dragonflies (Beerwerth et al., 1979).

The finding of *M. ulcerans* in different invertebrate animals (Section 3.1.2), including 6% positivity in dragonfly larvae of the families Gomphidae and Libellulidae in Benin near the Oueme River, should be considered as most consequential for human health (Marsollier et al., 2004). We assume that further details of the ecology of mycobacteria and the role of dragonflies and other invertebrates will soon be revealed.

6.3.5.2 Cockroaches

At present there are 3500–4000 cockroach species found almost all over the world. Many cockroach species have now been spread to regions where they were not originally present, above all through business (ship transport of foodstuffs). However, mycobacteria have only been isolated from some species (see below). Cockroaches are found mainly in locations with sufficient amounts of food composed primarily of organic residues of plants. The majority of cockroach species are, however, omnivorous; they play a significant role in the decomposition of organic matter and are an important part of the food chain. In nature, they live in forests, in residues of leaves and needles (forest litter), on meadows and on vegetation in other parts of the natural environment. Those which lived in close proximity to people (synanthropic species), are mostly found in blocks of flats, in dormitories, rural bakeries, in barracks, hotels and stores of foodstuffs and feeds (Photo 6.5). They most often ingest various organic residues such as bread crumbs, sugar, vegetable and fruit residues, etc.

Evolutionarily speaking, cockroaches belong to the oldest orders of flying insects and are very well adapted to life under unfavourable conditions. This has also enabled them to survive on the Earth for more than 350 million years.

When evaluating the risks associated with the transmission of mycobacteria by cockroaches in the environment, it is necessary to consider many factors that are fundamentally different from factors in other multicellular invertebrate animals. Through a combination of many of these factors, cockroaches can successfully transmit mycobacteria in the following modes

i. On the contaminated body surface; even though cockroaches are not good flyers and only fly short distances, they have very well-developed legs with a strong coxa and shin. There are five segments on each foot. This leg arrangement allows cockroaches to move very fast. Owing to that, they can escape their pursuers and quite quickly spread mycobacteria in the environment.

ii. Due to the fact that cockroaches are herbivores and occasional omnivores, they can become infected with mycobacteria by the ingestion of contaminated plant residues, and in rare cases by the ingestion of small organisms (protozoa, small invertebrates, tissues from dead animals, etc.). The vast majority of cockroaches living in nature are active during the daytime; synanthropic species are, on the other hand, usually active at night when they can move about unnoticed.

iii. With respect to the fact that cockroaches are numbered among long-living insects (up to several years), in the case of their infection with mycobacteria, both insects in different developmental stages (almost 20 are known in some species) and adults can spread them for several months to several years (species *Periplaneta americana* can live up to 4 years).

iv. The risk of the spread of mycobacteria is associated with the ability of cockroaches to perceive very small vibrations and pressure waves, which allow them to successfully escape the enemy. Due to the fact that some cockroaches can survive without water for up to a month and without food for up to 3 months, they may constitute a significant source of mycobacteria under unusual conditions.

In households, in public buildings and, e.g., also in hospitals, the synanthropic cockroach *Blatta orientalis* is frequently found under certain favourable conditions; they are active largely at night and ingest all kinds of waste. In hospitals, cockroaches have been observed to ingest sputum containing the causative agent of human tuberculosis, *M. tuberculosis*. The viability of *M. tuberculosis* in cockroach faeces kept at room temperature, meanwhile, was confirmed after 8 weeks. It follows from these results that *M. tuberculosis* passed through the digestive tract of cockroaches and was shed through faeces (Allen, 1987). Several non-tuberculous mycobacteria species have been found in hospital cockroaches and they may potentially be implicated as being a cause of hospital-acquired infections due to non-tuberculous mycobacteria (Pai et al., 2003).

In feed storehouses on farms keeping cattle, pigs, gallinaceous poultry and other domestic animals, cockroaches can be seen very often (O.A. Fischer, nonpublished observation). Therefore, the question of whether cockroaches can participate in the spread of other species of health-consequential mycobacteria was posed. Experiments performed on oriental cockroaches (*B. orientalis*) showed that *M. a. paratuberculosis* and *M. a. avium* were still shed through faeces 72 hours after the last contact with the source of experimental infection, a suspension of mycobacteria in sucrose. *M. a. avium* was also isolated from the bodies of nymphs 10 days after the last contact with the infectious sugar solution (Fischer et al., 2003b).

The ability of these causative agents of mycobacterial diseases to propagate in their intestinal tract and thereby to be actively shed was another matter of concern. Hence, not only the shedding of the causative agent but also CFU counts in cockroach faeces was investigated. With respect to the fact that both these causative agents (of paratuberculosis and avian tuberculosis) are highly specialised to hosts, i.e. to warm-blooded animals, the finding was not surprising: the CFU counts in cockroach faeces were decreasing in all experiments. No change of RFLP type in either *M. a. paratuberculosis* (IS*900* RFLP analysis) or *M. a. avium* (IS*901* RFLP analysis), which had passed through the intestinal tract, was observed. No propagation was seen in *M. a. hominissuis* which can multiply at temperatures of between 18 and 20 °C (Kazda, 2000; Fischer et al., 2003b).

The following points list information gained and future research concerns based on these fractional published results dealing with the occurrence of mycobacteria in cockroaches in:

i. Storehouses of foodstuffs and feeds: It is necessary to systematically ensure that cockroaches are not present in these locations.

ii. Hospitals (above all in the subtropics and tropics): It is necessary to properly close containers with waste, especially those contaminated with *M. tuberculosis*.

iii. Laboratories for the diagnosis of mycobacterial infections (especially under "field" conditions, not only in the subtropics and tropics): It is necessary to control the contact of cockroaches with all waste and infectious materials.

iv. The ecology of mycobacteria: The role of cockroaches living in a number of niches where mycobacteria are found is not clear.

6.3.5.3 Orthoptera and Homoptera

Orthoptera is an order of insects comprising around 20 000 species, and in nature they are probably best known for their conspicuous acoustic signals (the predominant sound in meadows) and movement (some species have long hind legs which are adapted to jumping – they can jump over 2 m). Their wings are also well developed; these allow them to migrate over long distances. Their strong biting mouthparts are used for the ingestion of food of both vegetable and animal origin (some species feed preferentially on other invertebrate animals). For example, a variety of grasshopper species are omnivorous and locusts are herbivorous.

Homoptera is a group of more than 32 000 species of homopterous insects that have mouthparts of the biting–sucking type.

Even though different members of both these orders live in environments rich in mycobacteria, there are a lack of literature data concerning the occurrence of mycobacteria in these species. Beerwerth et al. (1979) were the only authors to detect one species of unspecified mycobacteria that were not members of the *M. avium* complex among nine investigated samples of grasshoppers and cicadae. An understanding of the significance of the members of both these orders in the ecology of mycobacteria is also missing.

6.3.5.4 Diptera

Diptera are cosmopolitan insects and include an estimated 100 000 species. They are important decomposers of organic substrates in nature. The majority of Diptera members feed on various micro-organisms or fluids (Photo 6.6). During their life cycle (all developmental stages) they also often encounter different disease-causing agents, including mycobacteria. These can be spread by Diptera in any of the developmental stages, i.e. ova, larvae, pupae and imagos. With regard to the ecology of mycobacteria, the participation of Diptera in their spread (or multiplication) can be classified as follows:

1. Transmission on the body surface (horizontal transmission).
2. Transmission through the infected contents of the intestinal tract (horizontal transmission).
3. Transmission through different developmental stages (vertical transmission).

An awareness of these three basic routes for the spread of mycobacteria can be useful in clarifying and solving particular ecological, epidemiological and epizootiological situations, but only together with knowledge of other facts (Photo 6.7). The following are the characteristics of mycobacteria, which can be spread by Diptera and can be crucial factors for their successful transmission or their multiplication in different developmental stages of Diptera:

i. Characteristics of a respective mycobacterial disease-causing agent (above all the ability to multiply outside a host organism).

ii. For ESM, the places of occurrence and requirements for their multiplication are important (temperature, source of food, pH, presence of specific elements, etc.).

iii. For all mycobacterial species, whether or not they are resistant to the environmental conditions where they are present is important (decomposing organs of dead hosts, various components of the environment and different tissues – above all the intestinal tract of Diptera).

With regard to the specific characteristics of Diptera, the following factors should be considered, especially the differences between Diptera and other invertebrates that can play a significant role in the spread of different mycobacteria. Accordingly, it is necessary to synthesise the following pieces of knowledge for every species of Diptera to be able to assess their importance in the ecology of mycobacteria and their health impact:

i. Morphology (segmentation of the body surface, character of mouthparts, etc.), physiological and developmental distinctiveness (minimum and maximum time required to complete different developmental stages) and ethology of respective species of Diptera.

ii. Associations between different species of Diptera and the environment where they occur and their ability to migrate (creeping, swimming, flying, passive movement by being swept by wind, etc.).

iii. The role and involvement of different species of Diptera in the food chains of other invertebrates or vertebrates.

iv. Different feeding characteristics in different stages of respective species: subsisting on many types of food (polyphagy), feeding on blood (hematophagy), faeces (coprophagy), secretions (secretophagy), corpuses (necrophagy), decaying organic matter (saprophagy), etc. It is well known that different developmental stages can feed on different food.

v. The role of different Diptera in the breakdown of organic matter with respect to their digestion (possible spread of mycobacteria through salivary glands, crop contents, faeces or bodily fluids by different stages of respective species of Diptera).

It has been found that different species predominate among the imagos of Diptera trapped in stables than among those in the surrounding environment. In stables, the imagos of Diptera were not diffused proportionally, but gathered on the bodies of animals, on the windows, under the ceiling, in the surroundings of dairy rooms, lighting fixtures, radiators and on the feed. When the weather was cold, the majority of imagos stayed inside the stables. As soon as the rays of sun warmed the outside walls and it became warmer outside than in the stables, different species of Diptera left the stables. They often sat on the gates that were dark in colour (and warmed quicker by sun), stable walls, surrounding flora, boxes for calves and manure which had been disposed of.

Saprophagous and necrophagous species of Diptera have been observed on communal waste and animal cadavers before their transport to rendering plants. Polyphagous and coprophagous species of Diptera were seen, in particular, on animal faeces. Raptorial species feeding on larvae and imagos of other insect species were detected among them. Necrophagous imagos of Diptera were also observed on ripening cherries in the grounds of the farms (O. A. Fischer, personal communication).

Performing a synthesis of all the above-mentioned factors, we can conclude that the importance of Diptera in the ecology of mycobacteria (Photo 6.8) consists above all of the following factors:

i. Females need animal proteins and cholesterol, which they obtain by sucking the blood of vertebrates (mosquitoes) or licking the bodily fluids from animal cadavers (e.g. members of genera *Lucilia*, *Scathophaga* and *Calliphora*) for successful development of ova. They can thus become vectors and spread the causative agents of mycobacterial infections to people and other animals (Section 6.3.5.9; Photos 6.9 and 6.10).

ii. The surface of ova laid in a breeding site that will provide food for larvae (animal cadavers, etc.) may be contaminated with mycobacteria. When the larvae are preyed upon by other invertebrates and vertebrates, further transmission of mycobacteria in the environment or among birds of prey can occur (Photos 6.11 and 6.12).

iii. Larvae usually develop in water (Section 5.2) or putrid (Section 5.5) environments, e.g. on animal cadavers, animal or organic remnants on garbage, animal and human excrements (Section 5.10). The causative agents of mycobacterial infections can therefore be spread, e.g. when larvae are developing in tissues of animals that died of bovine tuberculosis or in the intestinal tract of ruminants that died of paratuberculosis.

iv. Before pupation, larvae often wander from the breeding site over distances of up to several tens of metres. Their bodies then become prey for invertebrates and vertebrates (above all insectivorous mammals and birds), whereby they can participate in the spread of mycobacteria that they carry on or inside their bodies. If they are not preyed upon by predators, they can also transmit mycobacteria to the place where they pupate.

v. In the pupa – the resting stage – larval organs are disintegrated (histolysis) and adult insect organs are formed (histogenesis). It seems that during histolysis, killing of many bacterial species occurs, including mycobacteria. Hence, not the freshly hatched imagos but empty pupae can become sources of mycobacteria due to the fact that the intestinal content adheres to their inner wall so that intestinal bacteria may not destroy the developing larvae.

vi. Imagos of the majority of coprophagous species of Diptera live on various juices and nectars and usually feed from flowers (a number of members of the order Diptera are important pollinators). Thus, the males and females can spread various causative agents of diseases, including different species of mycobacteria. Before egg laying in excrement and cadavers, females need saccharides and vitamins, which they collect from overripe fruit. They prefer pears, peaches and bananas; less frequently they feed on cherries, rapes, raspberries and redcurrants. Among vegetables, they prefer tomatoes and less frequently peppers (Photos 6.13).

According to the relationship between Diptera and human residences (and the resulting risk associated with the transmission of mycobacteria), the following classification was suggested in the 1950s (Gregor and Povolny, 1958):

i. Eusynanthropic (always affiliate with man).
ii. Hemisynanthropic (occasionally affiliate with man).
iii. Asynanthropic (they are found in nature and do not affiliate with man).
iv. Symbovilous (they are found in close proximity to cattle, both on pastures and stables).
v. Causative agents of myiasis (dipterous larvae feeding on live human and vertebrate animal hosts where they develop). Myiasis can be categorised by location as cutaneous, affecting the body orifices (ear, eye, nasal, vaginal and anal) and internal (intestinal and urogenital tract).

In the following sections, the available literature sources dealing with different important mycobacterial species in Diptera will be analysed from the point of view of ecology. Attention will be particularly focused on the causative agents of important mycobacterial diseases and some species of PPM and ESM.

Importance of Different Species of Diptera in the Transmission of the Causative Agent of Bovine Tuberculosis

Transmission of the causative agent of bovine tuberculosis by Diptera was first studied as far back as the end of the 19th century (Celli, 1888, cited in Sylwester, 1960). In simple experiments, flies were left sitting on the entrails of infected cattle. Subsequently, the causative agent of bovine tuberculosis was detected both in the investigated flies and in their faeces. Mechanical transmission of the causative agent

of bovine tuberculosis was detected in the house-fly (*Musca domestica*) twenty years later (Buchanan, 1908, cited in Sylwester, 1960). Before the First World War, the causative agent of tuberculosis was found to survive in the bowels of houseflies for up to 13 days (Graham Smith, 1913, cited in Hawley et al., 1951).

Possible transmission of the causative agent of bovine tuberculosis by three species of Diptera: *Musca domestica*, *Calliphora erythrocephala* and *Lucilia caesar* was not studied until the 1950s. Schlee (1957) left flies of these species in a closed room together with parenchymatous organs which had tuberculous lesions; these organs originating from cows infected by bovine tuberculosis were cut into pieces. Necropsy of the flies and isolation of the causative agent of bovine tuberculosis (among others, by experiments on guinea pigs) from their digestive tract showed that the flies can swallow the causative agent of bovine tuberculosis together with bodily liquids. The infection of a guinea pig with bovine tuberculosis was caused by parenteral application of the homogenate of only one fly after its contact with the organs of tuberculous cows. In systematic experiments, the causative agent of bovine tuberculosis was detected in the digestive tract of flies half an hour after the last contact with the above-mentioned organs showing tuberculous lesions and survived there for the next 20 hours. It was also found that flies can carry the causative agent of bovine tuberculosis on their bodies; it survived there and in the digestive tract of dead flies for 15 and 12 days, respectively.

Importance of Different Species of Diptera in the Transmission of the Causative Agent of Paratuberculosis

The causative agent of paratuberculosis has been isolated from the family of syrphids (Syrphidae), from the second and third larval stages of the species of syrphid fly (*Eristalis tenax*; Table 6.9). Imagos are usually found on the meadow and muckheap flora and other places where they suck nectar from flowers, lick sweet plant juices and excreta from worms and aphids. Females sit on liquid components of manure (mostly contaminated with *M. a. paratuberculosis* on farms with infected ruminants) for several tens of seconds or minutes during egg laying (I. Pavlik, personal observation). This time is evidently not sufficient for con-

Table 6.9 Isolation of mycobacteria from different naturally infected/contaminated developmental stages of Diptera of Syrphidae and Drosophilidae families

Family	Species	Developmental stage	Isolates	Collection site	References
Syrphidae	*Eristalis tenax*	Eggs	Not examined		
		Larvae (first stage)	Not found	Dung storage pits (cattle)	[3]
		Larvae (second stage)	*M. a. paratuberculosis*	Dung storage pits (cattle)	[3]
		Larvae (third stage)	*M. a. paratuberculosis*	Dung storage pits (cattle)	[3]
			M. a. hominissuis (8)	Dung storage pits (pig)	[4]
			M. a. avium (2)	Dung storage pits (pig)	[4]
			M. fortuitum	Dung storage pits (pig)	[4]
			M. chelonae	Dung storage pits (pig)	[4]
		Larvae (all stages)	*M. fortuitum*	Dung storage pits (cattle)	[3]
			M. a. paratuberculosis	Dung storage pits (cattle)	[5]
			M. a. hominissuis (NT)	Pig farm	[6]
			M. fortuitum	Pig farm	[6]
			M. chelonae	Pig farm	[6]
		Puparia	Not found	Farm (cattle)	[3]
			M. a. avium (2)	Dung storage pits (pig)	[4]
			M. a. hominissuis (8)	Dung storage pits (pig)	[4]
			M. a. hominissuis (NT)	Pig farm	[6]
		Exuviae	Not found	Farm (cattle)	[3], [4]
		Imagos	Not found	Farm (cattle)	[3]
			M. avium complex	Stable (pigs)	[2]
			M. a. hominissuis (8)	Dung storage pits (pig)	[4]
			M. a. hominissuis (NT)	Pig farm	[6]
Drosophilidae	*Drosophila* sp.	Imagos	*M. a. hominissuis* (8)	Stable (pigs)	[1]
			M. a. hominissuis (8)	Stable (pigs)	[2]
			M. a. hominissuis (NT)	Pig farm	[6]

(2),(8): serotypes, **(NT)** serotype not tested
[1] Fischer O, Matlova L, Bartl J, Dvorska L, Melicharek I, Pavlik I (2000) Folia Microbiol.(Praha) 45:147–152. [2] Fischer O, Matlova L, Dvorska L, Svastova P, Bartl J, Melicharek I, Weston RT, Pavlik I (2001) Med. Vet. Entomol. 15:208–211. [3] Fischer OA, Matlova L, Dvorska L, Svastova P, Bartos M, Weston RT, Kopecna M, Trcka I, Pavlik I (2005) Med. Vet. Entomol. 19:360–366. [4] Fischer OA, Matlova L, Dvorska L, Svastova P, Bartos M, Weston RT, Pavlik I (2006) Folia Microbiol.(Praha) 51:147–153. [5] Machackova M, Svastova P, Lamka J, Parmova I, Liska V, Smolik J, Fischer OA, Pavlik I (2004) Vet. Microbiol. 101:225–234. [6] Matlova L, Dvorska L, Bartl J, Bartos M, Ayele WY, Alexa M, Pavlik I (2003) Veterinarni Medicina 48:343–357.

tamination of their body surface and that is why they do not play an important role in the ecology (spread) of the causative agent of paratuberculosis in the environment (Photo 6.14). On the other hand, their larvae develop in reservoirs containing liquid manure and *M. a. paratuberculosis* has quite often been isolated from them; 32.2% of infected larvae were found on a heavily infected farm (Fischer et al., 2005).

Therefore, larvae of syrphid flies (*E. tenax*) play an important role in the spread of *M. a. paratuberculosis* on the farms with infected ruminants. They feed on small organic detritus both in highly contaminated stationary water and above all in the liquid component of manure on animal farms. Several hours before pupation, they leave the liquid environment and wander in search of a dry hiding place that may be several tens of metres away from the liquid environment where it had developed (I. Pavlik, unpublished data; Photo 6.15). When crawling, they constitute easy prey for birds, small terrestrial mammals and other predators (Section 6.4.2). If their development into an imago is unsuccessful, *M. a. paratuberculosis* can also be spread in the environment through the dead bodies of the larvae (Photo 6.16).

In a laboratory experiment using high-infection doses (up to 1000 CFU of *M. a. paratuberculosis* in 1 ml of liquid manure), this causative agent was recovered from the intestinal tract of the dissected third stage larvae even 14 days after the last contact with liquid manure. *M. a. paratuberculosis* was also isolated from puparia and exuvia (Table 6.10). This result confirmed that during dramatic changes in internal morphology (processes of histolysis and subsequent histogenesis) the number of bacteria in the body of a developing larva declines. When new tissues are formed, a part of the intestine of ectodermal origin separates and adheres to the inner wall of the puparium. Therefore, when imagos hatch from puparia, their intestinal tract is practically free from bacteria (including mycobacteria). It seems that *M. a. paratuberculosis* likely survives on the inner wall of the puparium, outside the changing body of the imago (it has been described in different bacterial species such as *Escherichia coli* or *Salmonella* sp.). However, the importance of this route of spread of the causative agent of paratuberculosis in the environment is very low in comparison with the migration of the contaminated larvae (Fischer et al., 2001; Fischer et al., 2004; Fischer et al., 2005).

Among other members of Diptera, *M. a. paratuberculosis* has been isolated from non-determined imagos of family Scathophagidae on pastures with grazing cattle infected with paratuberculosis (Fischer et al., 2001; Machackova et al., 2004). The species of this family are raptorial; imagos usually sit on plants and feed on small insects; their larvae live in faeces and prey on the developing larvae of other invertebrates. The imagos examined in the study could become contaminated with *M. a. paratuberculosis* when sitting or laying eggs in fresh faeces (I. Pavlik, personal observation). Also, imagos were usually found in the manure and on the muckheap flora around the manure; they were never observed to enter the stables. Accordingly, the importance of the spread of *M. a. paratuberculosis* through this route (occasional contamination of imagos) does not seem to be important in the ecology of the causative agent of paratuberculosis.

Table 6.10 Isolation of mycobacteria from experimentally infected Diptera

Mycobacterial species	Species	Developmental stage	DPI	Total	Positive	Whole	Pouch	Intestine	References
M. a. paratuberculosis	*Eristalis tenax*	Larvae (third stage)	3	10	6	nt	230	1 358	[2]
			7	10	10	nt	461	1 461	[2]
			14	1	1	nt	2	10	[2]
		Puparium	3	1	1	nt	5	100	[2]
			14	1	0	nt	0	0	[2]
		Imago	14	1	0	nt	0	0	[2]
		Exuviae	14	1	1	1	nt	nt	[2]
M. a. avium (serotype 2)	*Calliphora vicina/Lucilia caesar*	Larvae	4	3/3	2/1	9/20	nt	nt	[1]
			11	3/3	1/0	4/0	nt	nt	[1]
		Puparium	11	3/3	0/0	0/0	nt	nt	[1]
			15	3/3	0/0	0/0	nt	nt	[1]
			18	3/3	0/0	0/0	nt	nt	[1]
		Imago	18	3/nt	0/nt	0/nt	nt	nt	[1]
			29	3/6	0/0	0/0	nt	nt	[1]
		Exuviae	18	3/nt	0/nt	0/nt	nt	nt	[1]
			29	3/6	0/0	0/0	nt	nt	[1]
M. a. paratuberculosis	*Calliphora vicina* (15 imagos)	Head	2	1	1	17	nt	nt	[1]
		Thorax	2	1	1	6	nt	nt	[1]
		Abdomen	2	1	1	25	nt	nt	[1]
		Legs	2	1	0	0	nt	nt	[1]

DPI Days post infection. **CFU** Colony forming units. **Whole** The whole sample was examined. **Pouch** Pouch obtained after intestinal tract removal. **Intestine** Intestinal tract from dissected larvae or puparium.
[1] Fischer OA, Matlova L, Dvorska L, Svastova P, Bartl J, Weston RT, Pavlik I (2004) Med. Vet. Entomol. 18:116–122. [2] Fischer OA, Matlova L, Dvorska L, Svastova P, Bartos M, Weston RT, Kopecna M, Trcka I, Pavlik I (2005) Med. Vet. Entomol. 19:360–366.

When considering epizootiological and epidemiological aspects, it is necessary to view the detection of *M. a. paratuberculosis* in species of the family Calliphoridae (imagos of *Calliphora vicina* and imagos of *L. caesar*) as very serious (Table 6.11; Section 9.7). On farms keeping ruminants, imagos of *C. vicina* have most often been observed on the meadow flora, on ripening fruit (above all cherries) and on the sun-warmed walls of farm buildings. Dead animals on farms (especially calves) were observed to attract imagos of *L. caesar* (O. A. Fischer, personal communication). Several tens of metres away from the emergency slaughterhouse where containers with the intestines from cattle affected by paratuberculosis were present, *M. a. paratuberculosis* was detected in imagos of both these species. The matching identity of

these isolates of *M. a. paratuberculosis* with the isolates from infected cattle was confirmed by the IS*900* RFLP analysis (Fischer et al., 2004).

Importance of the Syrphid Fly (*E. tenax*) in the Transmission of the Causative Agent of Avian Tuberculosis

M. a. avium of serotype 2 was detected on a farm of pigs with tuberculous lesions in the lymph nodes of larvae in the third developmental stage and in puparia of syrphid flies (*E. tenax*; Table 6.11). Faeces from the infected vertebrates (including pigs) or decomposing bodies of dead infected birds or small terrestrial mammals, in which this causative agent was also detected,

Table 6.11 Isolation of mycobacteria from different naturally infected/contaminated developmental stages of other families of Diptera

Family	Species	Developmental stage	Isolates	Collection site	References
Scathophagidae	*Scathophaga* sp.	Imagoes	*M. fortuitum*	Meadow	[1]
			M. fortuitum	Plants – weeds (pig farm)	[2]
			M. fortuitum	Pig farm	[4]
			MAP	Pasture (cattle)	[2]
			MAP	Pasture (cattle)	[3]
Muscidae	*Musca* sp.	Larvae	*M. fortuitum*	Under trough (pig farm)	[1]
			M. chelonae	Pig farm	[4]
			M. sp.	Pig farm	[4]
			MAH	Pig farm	[4]
		Puparia	*MAH*	Pig farm	[4]
			M. fortuitum	Pig farm	[4]
		Imagoes	*M. fortuitum*	Stable (pigs)	[2]
			M. fortuitum	Pig farm	[4]
			M. fortuitum	Delivery room (pig farm)	[1]
			MAH (8)	Stable (pigs)	[2]
			MAH (8)	Delivery room (pig farm)	[1]
			MAH	Pig farm	[4]
			M. sp.	Stable (cattle)	[2]
	Stomoxys calcitrans	Larvae	*M. fortuitum*	Under trough (pig farm)	[1]
		Imagoes	*M. fortuitum*	Stable (pigs)	[2]
			M. fortuitum	Stable (pigs)	[4]
			M. scrofulaceum	Stable (pigs and cattle)	[2]
Calliphoridae	*Lucilia* sp.	Imagoes	*M. phlei*	Dustbin with wastes	[2]
	Lucilia caesar	Imagoes	*MAP*	Emergency slaughterhouse	[2]
	Calliphora vicina	Imagoes	*MAP*	Emergency slaughterhouse	[2]

MAP M. a. paratuberculosis. **MAH** *M. a. hominissuis.* **(8)** Serotype 8.
[1] Fischer O, Matlova L, Bartl J, Dvorska L, Melicharek I, Pavlik I (2000) Folia Microbiol.(Praha) 45:147–152. [2] Fischer O, Matlova L, Dvorska L, Svastova P, Bartl J, Melicharek I, Weston RT, Pavlik I (2001) Med. Vet. Entomol. 15:208–211. [3] Machackova M, Svastova P, Lamka J, Parmova I, Liska V, Smolik J, Fischer OA, Pavlik I (2004) Vet. Microbiol. 101:225–234. [4] Matlova L, Dvorska L, Bartl J, Bartos M, Ayele WY, Alexa M, Pavlik I (2003) Veterinarni Medicina 48:343–357.

may have been the source of infection in the liquid manure (Fischer et al., 2000; Matlova et al., 2003). However, concentrated attention should be paid to the importance of members of the family Syrphidae in the spread of this causative agent.

Possibilities of transmission of *M. a. avium* of serotype 2 were studied experimentally in another two species of Diptera, family Calliphoridae: *C. vicina* and *L. caesar* (Table 6.11; Fischer et al., 2004). When all different developmental stages (larvae, puparia, imagos and exuviae) were cultured, *M. a. avium* of the same serotype 2 (as well as IS*901* RFLP type) was detected in only larvae of both species 4 (*L. caesar*) to 11 (*C. vicina*) days after the last contact with the infected liver of a hen (Table 6.10). Hence, in contrast to the causative agent of paratuberculosis, it is evident that an important role in the spread of the causative agent of avian tuberculosis is played by, in particular, the larvae of Diptera and their possible participation in the food chain when swallowed by a predator (Section 6.4.2).

Importance of Different Species of Diptera in the Transmission of the Causative Agent of Mycobacteriosis

The most frequently detected *Mycobacterium* in different developmental stages of Diptera in the pig herds is the causative agent of avian mycobacteriosis, *M. a. hominissuis*. It has been detected in larvae, puparia and imagos of syrphid flies (*E. tenax*) and in imagos of *Drosophila* sp. (Table 6.9). Among the members of the family Muscidae, this member of the *M. avium* complex was isolated from larvae, puparia and imagos (Table 6.10). However, the body surface was most likely contaminated in all cases. In any case, in laboratory experiments, *M. a. hominissuis* was not detected in any of the developmental stages of two species of Diptera from the family Calliphoridae: *C. vicina* and *L. caesar* after 4 days (Fischer et al., 2004).

With regard to ecological aspects, it is important that besides Drone flies (*E. tenax*), above all coprophagous species *E. arbustorum*, *E. nemorum* and *Syritta* sp. were found in liquid manure and farrowing houses on pig farms. On the other hand, aphidophagous and phytophagous species dominated on muckheap and meadow flora; these were not seen to enter even the pig stables. In the stables, imagos of coprophagous syrphid flies were usually found around the light sources, windows and lighting fixtures. Members of the family Drosophilidae were usually present in the farrowing houses for pigs where they were found on fermenting forage leftovers. Muscidae were found in some stables (above all in the farrowing houses for pigs) throughout the year where they could sustain the spread of mycobacteria in the forage leftovers under certain conditions (Fischer et al., 2001; Fischer et al., 2004; Fischer et al., 2005; Fischer et al., 2006).

Importance of Different Species of Diptera in the Transmission of *M. fortuitum*

M. fortuitum is one of the important fast-growing PPM with respect to human health (Section 3.2.1). Among animals kept in captivity, it is found mainly in cold-blooded vertebrates (Section 6.5) and among domestic animals, especially in the lymph nodes of pigs (Section 6.7) and in cow milk (Section 5.7.4). *M. fortuitum* has been isolated quite often from members of Diptera in almost all stages of development both in pig and cattle herds (Tables 6.9 and 6.11). In one locality, this species was detected not only in different species of Diptera but also in different components of the stables and the surrounding nature (Fischer et al., 2000). With reference to the fact that imagos not only can move between different halls (stables) within a farm but can also fly to meadows in the close surroundings, the risk of spread of this mycobacterial species is quite high.

The isolation of *M. fortuitum* and *M. scrofulaceum* from stable flies (*Stomoxys calcitrans*) on cattle and pig farms is noteworthy and exceptional with regard to the spread of mycobacteria through blood (see more in Section 6.3.5.9; Photo 6.10). Little is known about the participation of different species of Diptera in the transmission of other PPM species. *M. chelonae* was isolated from different developmental stages of the members of two families of Diptera: *E. tenax* and *Musca* sp. (Tables 6.9 and 6.11); *M. phlei* was detected only in the imagos of *Lucilia* sp. caught on a garbage bin (Table 6.11). The risk posed by the stable flies of genus *Lucilia* is constituted by their flying from one source of food to another, which allows transmission of pathogenic micro-organisms. Imagos sit not only on flowers, foodstuffs, particularly fresh meat, milk, ripe fruit but also on excrements, vomit and garbage

(O. A. Fischer, personal communication). *M. phlei,* however, has been isolated from hay and grass and is viewed as non-pathogenic (Wayne and Kubica, 1986). In the ecology of mycobacteria, Diptera seem to be able to spread these ESM; but this ability has not yet been investigated thoroughly.

What Is the Relationship Between the Use of the Fruit Fly (*Drosophila*) as a Model Organism and the Ecology of Mycobacteria?

Drosophila is one of the most investigated invertebrates with respect to physiology and genetics; after man, it is one of the most studied organisms. Therefore, all the studies performed on *Drosophila* as an organism and its tissue culture help the elucidation of many as yet unexplained mysteries of physiology, immunology, genetics, proteomics, lipidomics and other specialisations (Wang et al., 2001; Agaisse et al., 2005; Philips et al., 2005; Dionne et al., 2006). *Drosophila* is also significant as a model for the study of pathogenesis of mycobacterial infections and responses of the host organisms. Some of the other possible non-mammalian models include amoeba *D. discoideum*, nematode *Caenorhabditis elegans* (Kurz and Ewbank, 2007) and aquarium fish (zebrafish; *Brachydanio rerio*; Pozos and Ramakrishnan, 2004).

From the point of view of the ecology of mycobacteria, the studies performed in recent years are noteworthy; these described the wasting and death of *Drosophila* after its contact with *M. marinum.* The penetration of *M. marinum* to haemocytes 48 hours after infection and subsequent multiplication has been described. It led not only to the destruction of cells of the immune system but also to the subsequent death of the entire host organism (Dionne et al., 2003; Dionne et al., 2006). This explains to a considerable degree why *M. marinum* has not been detected in Diptera or other insects in numerous studies of the presence of mycobacteria. *M. marinum* prefers to inhabit aquatic environments where it can infect invertebrate and poikilothermic animals. It can also sometimes infect warm-blooded animals including man (Sections 3.1.7 and 6.5).

On the other hand, *M. smegmatis,* regarded as a typical ESM, was not able to cause any damage to the immune system cells in the bodies of Drosophilae and could not survive in them for a long time

(Dionne et al., 2003). These results explain the relatively low importance of *Drosophila* in the ecology of these mycobacteria. In contrast to *M. marinum, M. smegmatis* is incapable of long-term survival inside the organism of Drosophilae. Detection of *M. a. hominissuis* (serotype 8) in *Drosophila* from a pig herd is noteworthy (Fischer et al., 2001). *M. a. hominissuis* of the same serotype was also detected in pig faeces, in scrapings from the stable surfaces, from *Musca* sp. imagos and the dust in webs on the windows (Fischer et al., 2000).

Imitation of a Tuberculous Process by Diptera

The species of Diptera causing myiasis can complicate the diagnosis of mycobacterial infections. Not only can they spread mycobacteria in wounds or tissues where they parasitise, but their presence in tissues can also imitate infection caused by, e.g. *M. tuberculosis.* In Nigeria, for example, furuncular myiasis of the breast caused by the Tumbu fly (*Cordylobia anthropophaga*) was diagnosed in a woman (furuncular myiasis of the breast). Before the final diagnosis, the skin form of tuberculosis was also considered (Adisa and Mbanaso, 2004). Another case was described in Turkey: the oral cavity of a 15-year-old boy with tuberculous meningitis was affected by myiasis. The clinical symptoms included an opened mouth which was paralysed for a long time. After the 27-day hospitalisation three larvae were found in the aspirate from the buccal cavity; other third stage larvae were collected later. Poor hygiene of the oral cavity with decomposing tissue evidently attracted Diptera of genus *Sarcophaga*. At the same time, the right mandibular lymph node was swollen (Yazar et al., 2005).

Imagos of some species of Diptera are attracted by volatile substances produced by various micro-organisms such as *R. fascians* when different tissues such as the skin of injured vertebrates (e.g. sheep) are affected. They insert eggs into the damaged tissues and the hatched larvae undergo development there before pupation. In sheep it was found that among micro-organisms infecting wounds, *M. aurum* produced volatiles (volatile sulphur compounds including mercaptans and alkyl sulphides) attractive for *Wohlfahrtia magnifica* flies (Diptera: Sarcophagidae) as well as the above-mentioned species *R. fascians* (Khoga et al., 2002).

With regard to the fact that mainly the larvae from the family Calliphoridae cause myiasis, they significantly participate in "cleaning" the damaged tissues in live hosts. Also, they are sometimes used in medicine to effect soft tissue wound healing, in most cases to repair extensively damaged skin or muscles. Due to the fact that imagos of different species of Diptera that insert ova into such wounds live in an environment usually contaminated with mycobacteria, it will be necessary to evaluate their importance in the following processes:

i. In the ecology of mycobacteria, during their spread in nature.
ii. In the transmission of mycobacteria to damaged tissues of hosts.
iii. In the spread of mycobacteria in and around the affected tissue of infected hosts.

The Aims of Other Research Spheres

When all of the above-mentioned analyses are performed, it is necessary to take into account the seasonality of the occurrence of Diptera, above all in climatic zones other than the tropics and subtropics. In warm weather (which can last for only 6 months in the temperate zone and only a few weeks inside the polar circle), Diptera usually complete their development in a few weeks. Under unfavourable conditions (especially temperature, but there may be a lack of food or humidity), development into different stages takes a longer time or is even paused for some time (diapause). Little is known at present about the importance of this noteworthy phase in the development of Diptera and other invertebrates relative to the ecology or epidemiology of mycobacteria. Other questions concerning the survival of mycobacteria in dormant stages of different species of Diptera (ova, larvae, nymphs/puparia or imagos) remain unanswered.

With respect to food and feed safety, it is necessary to keep in mind the fact that a variety of Diptera are attracted to various foodstuffs, which can be contaminated. Imagos of numerous families (such as Calliphoridae, Sarcophagidae, etc.) frequent foodstuffs of animal origin (meat, liver, fish, milk and dairy products). These necrophagous Diptera are attracted by the high content of protein, which is also present in vomit, human stools, faeces and cadavers of animals. On the other hand, canned foodstuffs which are smoked, pickled, adequately salted and spiced (meat, fish, etc.) are less attractive for Diptera. They are evidently discouraged by the presence of a high concentration of salt and the aroma of spices and odours arising from smoked products.

On the other hand, necrophagous and saprophagous species of Diptera prefer overripe fruit and garbage, which contains putrescent meat cuts, bones, animal and vegetable fat, decomposing leftovers. Information on the importance of necrophagous and saprophagous species of Diptera in the ecology of mycobacteria is scarce. These Diptera are likewise strongly attracted by many of the above-mentioned substrates in human settlements of economically less prosperous agglomerations and villages in developing countries. Stores, shops, public transport, households, etc., are not sufficiently protected against the penetration of invertebrates and small vertebrates (Photo 6.7). We can assume that they are important in the spread of mycobacteria.

Because imagos of Diptera can transmit mycobacteria on their bodies or in their digestive tract and Diptera are abundant on cattle and pig farms, it is necessary to investigate the migratory habits of Diptera away from the breeding site. Despite the fact that imagos are usually found at the breeding site because it provides sufficient food and an appropriate breeding substrate for them, they can fly very far. Some Diptera can migrate tens of kilometres, either on the hunt for food or in search of members of the opposite sex before mating. In addition, together with dust particles (Section 5.9), Diptera can be swept with the air to a distance of several tens of kilometres. Accordingly, within the ecology of mycobacteria, these possibilities for the spread of mycobacteria should also be investigated.

The possible transmission of the above-mentioned mycobacterial species by other members of this order has been described in the Section dealing with the bloodsucking members of Diptera (Section 6.3.5.9) but many questions remain to be answered.

6.3.5.5 Beetles

About 450 000 beetle species have been described; they are classified into 166 families. These represent about 40% of all described insects and about 20% of all described animal species. They differ from the other insect orders by virtue of their particularly hard forewings that are the metamorphosed upper pair of wings. The forewings protect the membranous wings (which can be secondarily absent in some species) and allow the easy travel of beetles to distant niches. Beetles have mouthparts that allow the crushing or cutting of food of different types (Photo 5.33). Beetles are found all over the world with the exception of Polar regions and marine environments. With regard of dietary specialisation, beetles can be divided into several groups that can become contaminated with mycobacteria in the following modes:

i. Predators (can become infected with mycobacteria by the ingestion of other infected invertebrate animals).
ii. Parasites (can become infected with mycobacteria from their hosts).
iii. Saprophages (can become infected with mycobacteria by the ingestion of organic residues of vegetable and animal origin).
iv. Scavengers (can become infected with mycobacteria by the ingestion of dead infected animals).
v. Herbivores (can become infected with mycobacteria by the ingestion of contaminated plants).

Beetles in various developmental stages can participate in the transmission of mycobacteria. They propagate by perfect metamorphosis, i.e. in four stages:

i. Eggs – are deposited by the females in different places, e.g. loose on the leaves of nutritive plants, on the soil surface, stones and branches of trees and shrubs, in different hiding places (soil, under bark, in the slots in wood, on flowers and fruits of plants, subterraneous corridors below cadavers that become their food sources, etc.); these places are highly contaminated with mycobacteria.
ii. Larvae – hatch from eggs relatively soon (in 1 or 2 weeks; in some cases, the eggs lie dormant and larvae hatch in spring). Larvae grow rapidly, usually passing through three developmental stages. The duration of the larval stage differs considerably; the larvae of some beetle species can very rapidly become pupae and several generations can exist within a year; in some other species, the larval stage can last for several years.
iii. After a definite time, adult larvae undergo pupation, often in soil or in different secluded places where before pupation the larva spins a cocoon and

is enclosed in a free cocoon (*pupa libera*). Larvae do not ingest food and are in an inactive stage when their bodies undergo the metamorphosis from larva to imago.
iv. After emerging from larvae, adults (imagos) soon copulate. The lifespan of imagos is quite short, males die soon after copulation, females after laying eggs; however, some species survive winter and live until the spring of the subsequent year; other species can live for several years.

Beetles are classified in four suborders:

i. Archostemata is an ancient lineage of beetles consisting of 30 known species; they develop in dead wood.
ii. Myxophaga include about 100 species of small beetles that reach up to 2.7 mm in length.
iii. Carnivora (Adephaga) covers 36 000 species of largely predatory beetles that live and develop both on dry land and in water. The coxae of the last pair of legs are stationary. Coxae divide the first visible segment of the abdomen into two sclerites. The species comprises nine families, of which six live in water.
iv. Polyphaga is the largest suborder of beetles with 325 000 described species of terrestrial, freshwater, carnivorous, herbivorous, saprophagic and saprophytic beetles.

An understanding of how beetles become infected with mycobacteria and their transmission among organisms in respective developmental stages is incomplete. Due to the fact that numerous beetle species constitute food for a variety of invertebrates and vertebrates such as birds, they can become vectors of causative agents of mycobacterial infections. Mycobacteria were described for the first time in ligniperdous beetles trapped in sawdust contaminated with *M. a. hominissuis*. This mycobacterial species was also detected in the larvae of ligniperdous beetles developing in tree bark, which was contaminated with soil (Beerwerth et al., 1979).

When investigating the shedding of live mycobacteria in the faeces of darkling beetles in various developmental stages, we found results comparable to previous experiments on earthworms (Fischer et al., 2003a) and nymphs of the oriental cockroach (Fischer et al., 2003b). If larvae and the next developmental stages of

Tenebrio molitor were in contact with food (bran and peat) naturally infected with mycobacteria for a longer time, they could still be detected by culture from larvae after 14 days and from imagos even after 35 days.

Larvae of *Tenebrio molitor* and *Zophobas atratus* are usually kept as food for birds and small insectivores (Weischner, 1989). If larvae are kept in an environment where OPM are present, they can ingest these mycobacteria and become sources of infection for birds and animals, which are fed on them. The breeders do not consider this risk when they are feeding larvae of darkling beetles to birds and animals or during their handling. Due to the fact that larvae of darkling beetles are potential mechanical vectors of OPM in animal herds, this fact should be kept in mind when measures for control of mycobacterial infections are adopted in animal herds (Photo 6.17).

6.3.5.6 Butterflies

Butterflies are a significant constituent of the environment; their larvae, called caterpillars, are very important for many predators. The vast majority of larvae have biting mouthparts adapted to the ingestion of plant food. Despite a high species variety (the butterfly order includes at least 120 families), systematic research into their role in the ecology of mycobacteria has not yet been performed by any laboratory.

A notable exception are Beerwerth et al. (1979) who examined 19 samples of imagos and 13 samples of larvae originating from various localities in the agricultural landscape and nature. However, they failed to isolate mycobacteria from any of the samples; this provides evidence of the ability of different members of butterflies to enzymatically digest various plant components and mycobacterial walls in their guts. Further knowledge of the interaction of mycobacteria and butterflies remains to be discovered.

6.3.5.7 Hymenopterans

The order of hymenopteran insects (Hymenoptera) includes bumblebees, bees, wasps and ants. These are the most developed insect members and are significant components of the environment. With regard to the ecology of mycobacteria, it is odd that very little is known about the occurrence of mycobacteria in these

invertebrates. PPM that were not members of the *M. avium scrofulaceum* complex were found in 4 (22.2%) of 18 examined bee and bumblebee imagos. Mycobacteria were isolated from 7 (33.3%) of 21 samples of ant imagos; one of these isolates was a member of the *M. avium* complex (Beerwerth et al., 1979).

In the near future, attention will have to be concentrated on the significance of hymenopteran insects in the spread of mycobacteria (e.g. PPM-contaminated honey), with special emphasis on food safety.

6.3.5.8 Hemipterans

The present members of the order Hemiptera were historically classed as belonging to two orders, Homoptera and Heteroptera/Hemiptera (true bugs). These two orders were then combined into the single order Hemiptera. Hemipterans (Hemiptera) make up a group of approximately about 80 000 species of which the majority are herbivores. They have biting–sucking mouthparts, which enable them to suck liquids from terrestrial and aquatic plants and liquids from various animals. Some big species of water bugs not only feed on liquids from the body cavities of other insects, but can also prey on small vertebrates such as frogs, fish and small birds. Some species of bugs (e.g. bedbugs) also live on the blood of different hosts, including man.

The forewings of bugs are hardened (hemelytron is formed in some groups) and the hindwings are membranous and allow the bugs to fly great distances. When the bugs are threatened or when they prey, they release a very pungent liquid that is familiar to most visitors of nature. According to the environment where they live bugs are divided into two groups:

i. Geocorisae: a group of water bugs living beneath the water surface (suborder Nepomorpha) and on the water surface (suborder Gerromorpha). Water bug species breathe in air by means of tracheal tubes on the respiration opening in the abdomen. They carry a bubble of air on their bodies that is held by specialised hairs; this allows them to emerge on the water surface and acquire atmospheric oxygen.

ii. Hydrocorisae: a group of bugs living on dry land (terrestrial organisms); these are the majority of the existing bug species. They mostly live on animal or vegetable juices and are classified in two suborders (Pentatomorpha and Cimicomorpha).

Terrestrial Bugs (Bedbugs) as Pestilent Ectoparasites

M. fortuitum has been isolated from the intestinal tract of bugs of the species *Rhodnius prolixus* (Cavanagh and Marsden, 1969). This species of bug is classified in the family Reduviidae; the other species of this family also feed on caught insects. In the daytime, they actively pursue the insects or lurk in a tangled web of plants. After catching and piercing them, they suck out the content of their inside. These bugs can occasionally bite people.

Among others, the family Cimicidae is classified in the order of bugs (Heteroptera); this is the only evolutionarily young family in this order that includes specialised parasites (both imagos and all five developmental stages of larvae live on blood). They are to such a large extent adapted to parasitic life (suction of blood) that they virtually do not have wings. Originally, bedbugs were narrowly specialised to ectoparasitic life in bat colonies in caves. When people started to use caves as shelters, they were also parasitised by the bugs.

Bedbugs hide in slots and other suitable dark sheltered places close to food sources, and in human residences very often in beds and their surroundings (various slits in the wooden structure of the bed, in the seams of a mattress, etc.). However, they also hide behind picture frames, under wallpaper, behind skirting boards, on the edge of a carpet, below parquetry edging, etc.). Bugs belong to the "gregarious" insects, i.e. they associate with one another in suitable hiding places, where they congregate in groups. In all these niches, mycobacteria are found, especially in dust (Section 5.9.3), which is another possible source of their infection.

In economically developed countries, bedbugs are not present in great numbers due to hygienic measures associated with good housing standards and the introduction of new insecticides. The bugs, however, are not extinct and they can spread locally in accommodation facilities. For example, in the Czech Republic, the last registered increased incidence of bedbugs was associated with the occupying troops coming from Eastern Europe in 1968. In all the remaining countries of Europe, bedbugs spread in accommodation facilities for refugees or in the dwellings of socially deprived inhabitants of different nationalities. In developed countries, both professionals and the non-professional public are not aware of the existence of bugs and the threat posed by them. Accordingly, it is necessary to even today consider the risks that bugs pose in the spread of different infectious agents including mycobacterial infections.

The bodies of imagos and nymphs are about 5 and 1 mm long, respectively. They are typical nest ectoparasites, perfectly adapted to long-time starvation: they can live without blood for up to 1 year and females can use the sperm ejaculated into the body cavity by males as extra-oral nutrition in times of starvation. If a host is available, they prefer sucked blood about once a week, especially at night from sleeping people or animals. Suction by bugs is mostly painless and lasts about 5 minutes; repeated suction leads to skin hypersensitivity, called "cimicosis". This rash which itches does not appear after the first suction, but only after the second and subsequent contacts with bugs and after sensitisation to their toxins, which prevent blood coagulation. Therefore, such dermatitis appears within 2–9 days after the first suction (sensitisation).

Imagos live for 12–18 months. Before laying several tens to hundreds of eggs, females double their weight. The amount of sucked blood containing mycobacteria can therefore be high and constitute a high risk. The significance of bugs is mostly considered in leprosy (Kirchheimer, 1976b; McDougall and Cologlu, 1983). Many of these bug species are specialised to more than one host: they can affect both people and domestic and wild animals. Accordingly, they can also transmit mycobacteria between species. Based on this knowledge, the main routes of transmission of mycobacteria, both OPM and PPM, can be guessed at:

i. OPM can be transmitted through infected blood as bedbugs suck blood from different hosts.
ii. The contaminated body surface of bugs can become a vector for especially PPM from dust and other components of the environment where bugs spend the duration of their life when they do not just suck the blood from a host.
iii. Skin with secondary lesions caused by cimicosis (skin reaction based on hypersensitivity to the toxin against blood coagulation introduced by the bugs) can constitute an open gate to mycobacterial

skin infections caused by both PPM and OPM (e.g. the causative agent of leprosy).

iv. Mycobacteria can also be transmitted through bug faeces; these leave hillocks of faeces dark brown in colour (after digestion of the sucked blood) or light in colour and with a characteristic odour (after ingestion of a different food type).

Accordingly, further research should be focused not only on the direct significance of bugs in the spread of mycobacterial infections in one species of hosts but also on potential interspecies transmission. With respect to the ecology of mycobacteria, it will be necessary to determine the time of survival of mycobacteria on the surface, in the intestinal tract and in the faeces of bugs.

6.3.5.9 Water Bugs

Water bugs live mainly in the aquatic environment and constitute about 10% of all Hemiptera species living in water or on its surface. They pierce their prey with their proboscis and suck out its content. At present, water bugs are classified in four families: Naucoridae, Nepidae, Notonectidae and Corixidae. Some members of the Naucoridae family, e.g. the water bug (*Ilyocoris cimicoides*), are specialised in feeding on the blood of warm-blooded vertebrates including man. This species is found in stationary or slowly flowing waters in the northern temperate climate zone, where they can bite people during bathing (they most often bite the legs). This water bug is colloquially referred to as the "water bee". Mycobacteria have been detected in the bugs because they live in the environments rich in mycobacteria (Section 5.2). However, it remains to be determined whether some bug species live on mycobacteria as, e.g. dragonfly larvae do (Section 6.4).

Water Bugs and Their Significance in the Spread of Mycobacteria

With regard to the ecology of mycobacteria, the significance of water bugs in the spread of *M. ulcerans*, the causative agent of Buruli ulcers, was recently studied in the subtropical and tropical climate zones. The presence of *M. ulcerans* was first described in water bugs in 1999 (Portaels et al., 1999). *M. ulcerans* occurrence in water weeds and other components of the aquatic environment was investigated in Benin and Ghana in Africa. It is noteworthy that *M. ulcerans* was not isolated from mud, roots, stems or leaves of water weeds, but only from water bugs. In addition, the members of genera *Naucoris* and *Diplonychus* very often bite villagers. Such a transmission of *M. ulcerans* to people (mainly through the skin of their lower extremities) was later considered as most probable (Portaels et al., 2001).

Accordingly, these results led to the following studies, which revealed the presence of *M. ulcerans* in salivary glands of water bugs (*N. cimicoides*) fed with artificially infected larvae of *Phormia terrae novae* flies (Marsollier et al., 2002). Similar experiments were also performed with four PPM species that are found in the aquatic environment and can also be transmitted by water bugs (*M. chelonae*, *M. fortuitum*, *M. kansasii* and *M. marinum*). It is also noteworthy that in the experiments with these mycobacterial species, none of them were observed in the tissues from water bugs after staining according to Ziehl–Neelsen. *M. ulcerans* was only found in salivary glands, but not in silks of the raptorial legs or in the ducts of salivary glands leading to the rostrum. It is certainly worthy of note that *M. ulcerans* was present in the salivary glands even after the 105-day experiment. As the quantity of the detected AFB was great, multiplication of *M. ulcerans* inside the water bugs was inferred (Marsollier et al., 2002).

Various water insects from a region with the occurrence of the Buruli ulcer in humans were examined in the Ivory Coast. Out of 80 investigated organisms of the family Naucoridae, *M. ulcerans* was detected by the PCR method in 5 (6.3%) and by culture in 2 (2.5%) cases. The virulence of these two isolates was confirmed by experiments on mice; tails were swollen and inflamed at the site of infection (Marsollier et al., 2002). In the following study, possible routes of *M. ulcerans* transmission to water bugs (*N. cimicoides*) were investigated. Marsollier et al. (2003) described the growth of *M. ulcerans* as being stimulated by the water weed extracts as well as by biofilm on the surface of water weeds. Due to the fact that these water bugs are raptorial, the authors suppose that some "vector" organisms exist that transmit *M. ulcerans* as well as, e.g., cladocerans, which become the food of dragonflies (Sections 6.3.5 and 6.4.1).

With respect to the physiology of the water bugs (*N. cimicoides*) and *M. ulcerans* (possible transmission inside the bodies of water bugs and the participation of *M. ulcerans* in the spread throughout the environment), the following laboratory experiments are also noteworthy (Marsollier et al., 2005). In these experiments, *M. ulcerans* was "microinjected" into the coelom cavities of water bugs. *M. ulcerans* was found in haematocytes after 2 hours, accessory salivary glands after 4 hours and in the rostrum after 14 hours.

These experiments demonstrated both the transport of *M. ulcerans* through salivary ducts in the rostrum and "raptorial legs" and the multiplication capability of *M. ulcerans*. The quantity of *M. ulcerans* in the accessory salivary glands was increasing between 14 days and 11 weeks. After infection of the water bugs with *M. ulcerans* (strain 1615 from Malaysia), capable of producing "mycolactone", the above-mentioned results were also obtained. However, if a mutant was used that did not produce "mycolactone", no such results were obtained. Unfortunately, information on such virulence factors that allow survival and subsequent multiplication in the components of the environment is missing for other PPM. The most recent published results also refer to *M. ulcerans* transmission from "pre-digested" food composed of artificially infected larvae of *P. terrae novae* flies by water bugs of the above-mentioned species (Marsollier et al., 2007). *M. ulcerans* was subsequently detected in different parts of the intestinal tract and in faeces.

Biochemical mechanisms described by these authors will surely initiate further similar research on invertebrates and their significance in the ecology of mycobacteria (Marsollier et al., 2007). Such experiments could also explain the transmission of *M. ulcerans* from the rostrum of water bugs to the skin of injured people during various agricultural activities (work in rice fields), fishing, children playing in water, etc. However, with respect to the ecology of PPM, it is also necessary to consider the results of comparable experiments with *M. marinum*, which however was not detected in any of the examined tissues from these water bugs after 1 week.

6.3.5.10 Bloodsucking Insects

Besides the above-mentioned bugs (Section 6.3.5.8), other insects include fleas, lice, mosquitoes and some species of dipterans. These are classified as parasites that suck blood from humans and other warm-blooded animals (Section 6.3.5.4). The bloodsucking insects (in particular lice, fleas and mosquitoes) are found practically all over the world; most species are significant vectors of many infectious diseases. Mosquitoes only come to a host to suck blood, whereas all developmental stages and the imagos of, e.g., lice and some fleas live their whole life on the surface of a host. They only leave their host for several hours in the case that the host has a high fever or dies. The transmission of mycobacteria through bloodsucking parasites should be taken into consideration in mycobacterial diseases when mycobacteria are present in the peripheral bloodstream (mycobacteraemia) or are plentiful in the affected tissue (most often in the skin).

During blood suction, the insect spreads mycobacteria mostly mechanically through the piercing/sucking or biting mouthparts. If insects are infected by mycobacteria after previous blood suction from an infected host, their transmission occurs in two ways: to the skin and directly to the blood.

The participation of mosquitoes in the transmission of *M. leprae* has been investigated in endemic regions of leprosy for several decades. Golyshevskaya (1991) observed that rod-shaped AFB can change into small coccoid forms under natural conditions. This quality of the causative agent of leprosy is considered as a "trick", which makes transmission by the bloodsucking insect easier. This author also found that mosquitoes of the genus *Aëdes* could spread AFB not only in their imago stage but also in larval stages, in which AFB were also found. Mostly coccoid forms of these AFB were observed in imagos; these mycobacterial forms changed into elongated rod-shaped AFB in the late stages of larval development.

Narayanan et al. (1978) discovered that after sucking blood from humans infected with *M. leprae*, AFB survived in the proboscis of the mosquito *Aëdes aegypti* and in the proboscis of the mosquito *Culex fatigans* for 156 and 144 h, respectively. In the intestines of both mosquito species infected with *M. leprae*, AFB survived for 96 h. The experiments of Banerjee et al. (1991) demonstrated the possible transmission of *M. leprae* from people to young mice after suction by the mosquito *Aëdes aegypti*. Similar capabilities are thought to be possessed by, e.g., sand flies (Sreevatsa et al., 1992). Banerjee et al. (1990, 1991) detected by microscopy after staining according to Ziehl–Neelsen AFB similar to *M. leprae* in the skin of mice infected by the mosquito *Aëdes aegypti* that sucked blood from people suffering from leprosy. Fleas (Narayanan et al., 1978) and sandflies (Sreevatsa et al., 1992) have also been considered as potential vectors of *M. leprae.*

Studies dealing with the detection of *M. leprae* in the intestinal tract of insects, its significance and the risk of its spread were mostly published between the 1970s and 1990s (Narayanan et al., 1972; Geater, 1975; Kirchheimer, 1976a, b; Narayanan et al., 1977; Narayanan et al., 1978; Saha et al., 1985; Sreevatsa et al., 1992). Since that time, the research of mycobacteria in insects has been focused on their significance in the spread of the causative agent *M. ulcerans* (Sections 3.1.2 and 6.3.5.8).

The presence of other mycobacterial species has also been studied in the bloodsucking insects. For instance, unspecified PPM have been detected in mosquitoes in Germany; however, only 1 (2.4%) of 42 examined samples was positive (Beerwerth et al., 1979). *M. fortuitum* and *M. scrofulaceum* have been detected in stable flies (*S. calcitrans*) from pig and cattle herds (Fischer et al., 2001). On the other hand, when Eurasian badgers (*Meles meles*) infected with the causative agent of bovine tuberculosis in UK were examined, *M. bovis* was not detected in any of the caught bloodsucking arthropods: badger fleas, badger lice (*Trichodectes melis*), badger ticks (*Ixodes ricinus*), etc. The results of this study indicate that factors other than the bloodsucking insects participate in the spread of *M. bovis*, most often direct contact (Barrow and Gallagher, 1981). It most probably depends on the character of infection caused by *M. bovis* present in the blood of badgers (mycobacteraemia) for a very short time. The bloodsucking insects can theoretically become infected during this mycobacteraemia, but it is of low significance in the spread of *M. bovis* to other animals and subsequently to the environment.

In view of these results, it will be useful to focus research on the following areas to clarify the spread of mycobacteria through the bloodsucking insects:

- To describe the exact localisation of mycobacteria in the bodies or on the body surface of the bloodsucking insects.
- To describe the capability of different mycobacterial species to survive inside and on the body surface of live or dead bloodsucking insects.
- To clarify the significance of different developmental stages of various species of the bloodsucking

insects by further ecological studies and experiments on hosts.

- To assess the significance of imagos and other developmental stages of the bloodsucking insects in the ecology of mycobacteria, especially of OPM (in particular *M. leprae* and *M. tuberculosis*) in developing countries where people still live under poor hygienic conditions.

6.4 Transmission of Mycobacteria by Vertebrate Predators of Invertebrates

I. Pavlik and J. Klimes

Mycobacteria can often be spread through little expected and even incredible routes. Different developmental stages of invertebrate animals can be contaminated/infected with mycobacteria either on or inside their bodies.

Mycobacteria are acid-fast organisms that resist the effects of the digestive juices of invertebrates (Sections 6.1, 6.2, 6.3 and 6.4.1). Accordingly, after being swallowed by vertebrates, they can most likely pass through their digestive system intact and are shed live in their faeces. Invertebrate animals often become prey for different predatory cold-blooded and warm-blooded vertebrates. The prey can be categorised in the following manner:

i. The main prey: predators prey largely on this animal.
ii. Favourite prey: predators prefer to prey on this animal, even if it is not their main prey.
iii. Opportune prey: predators prey on it only sometimes or in the case of starvation.

According to the above-mentioned simple categorisation, vertebrates can become massively contaminated/infected when mycobacteria are present in their main or favoured prey. Little is known about the fate of mycobacteria inside the intestinal tract of vertebrate predators; the exposure/infection of a host organism to mycobacteria can be confirmed by the examination of faeces or organs. ESM most likely only pass through the intestinal tract or are killed and digested in the intestinal tract, e.g. ESM detected in the intestinal tract of small terrestrial mammals (e.g. *M. phlei*) without gross lesions (Fischer et al., 2000). In contrast, PPM, which are usually encountered by insectivores and small rodents through their prey on invertebrate animals, can survive longer in their tissues and organs. Under certain conditions, PPM can cause gross lesions in their parenchymatous organs (Skoric et al., 2007).

Predators can be categorised into three main groups:

  i. Predatory invertebrates.
 ii. Predatory cold-blooded vertebrates.
iii. Predatory warm-blooded vertebrates.

In the first group, a number of invertebrates are included. The fate of mycobacteria in the bodies of dragonfly larvae has been investigated (Section 8.5). Dragonfly larvae are a constituent of the food chain. They prey on other invertebrate animals or small vertebrates (fish fry) and are also preyed on by many predators. They become prey for different species of predatory water insects, fish, water mammals or birds living around bodies of water. Adult dragonflies also become prey; they are often trapped in the webs constructed by spiders in inshore areas or are preyed on by various dexterous bird species, e.g. the European bee-eater (*Merops apiaster*). The drowned adult dragonflies either become food for other water insect species and fish or small terrestrial mammals.

M. ulcerans is another example of the circulation of mycobacteria in invertebrates. The results of current research highlight the importance of both the water bugs (Portaels et al., 2001) and fish that prey on them (Eddyani et al., 2004; Section 6.3.5.8) in the spread of *M. ulcerans*.

We can in passing mention the fact that non-vigilant damselflies sometimes become the food of insectivorous herbs, e.g. sundews. The several above-mentioned examples show that various mycobacteria contaminating dragonflies in different developmental stages can be spread through various routes of the food chain.

Of the second group of predators, cold-blooded vertebrates, fish that are predators of numerous invertebrates are the most significant members. As described in the previous chapters, besides dragonflies, numerous invertebrate animal species (such as insect larvae and imagos, earthworms and water bugs) can be heavily contaminated/infected with mycobacteria (Sections 6.1, 6.2 and 6.3). All of these can become the prey of numerous predatory species of cold-blooded vertebrates. For instance, the larvae of different fly species

that develop in ruminant faeces are preyed on by other predatory species of ravenous insects. These can also become food for vertebrate predators, particularly cold-blooded animals (toads, salamanders, lizards).

In the third group of predators, warm-blooded vertebrates, in particular insectivorous birds (Section 6.6) and mammalian insectivores such as moles, hedgehogs and shrews are included (Section 6.7.2). *M. a. paratuberculosis* has been detected in the migrating larvae of syrphid flies (*E. tenax*) from a cattle herd infected with paratuberculosis (Fischer et al., 2005). House sparrows (*Passer domesticus*) and tree sparrows (*Passer montanus*) often preyed on them when they were migrating before pupation. *M. a. paratuberculosis* was also detected in the faeces collected from their roosting places (in a corn storehouse) where they spent the nights (I. Pavlik, unpublished data). Thus, the causative agent of paratuberculosis can be spread by an unexpected vector as both sparrow species are herbivores. During nesting time, however, they bring the invertebrates which they catch to their young; these invertebrates provide them with high-quality animal protein (Photo 6.16).

Insectivorous birds, on the other hand, feed almost exclusively on invertebrates. The same mycobacterial species were detected in their faeces (Section 5.10) as in the invertebrate animals (Sections 6.1, 6.2, and 6.3). In one pig herd, *M. fortuitum* was detected in the lymph nodes of pigs that gave positive responses to avian tuberculin. In an extensive study of the ecology of mycobacteria, *M. fortuitum* was detected in different species of Diptera, in the faeces of swallows (*Hirundo rustica*) present under their nests and in the intestinal tract of domestic mice (*Mus musculus*) in this herd (Pavlik et al., 2007). Similarly to other small terrestrial rodents, domestic mice occasionally feed on food of animal origin, in particular on invertebrates in different developmental stages (Section 6.7.3; Photo 6.18).

With regard to the ecology of mycobacteria, it is necessary to emphasise that these predators (birds and small terrestrial mammals) can subsequently be preyed upon by, e.g., birds of prey (goshawk, sparrowhawk, common kestrel, etc.) and carnivores (fox, badger, marten, etc.). After passage through the intestinal tract, mycobacteria can be transmitted over great distances through their faeces into absolutely unexpected niches in the environment (Photo 6.19).

In conclusion, mycobacteria can sometimes gain access to different components of the food chain through little known routes. With respect to future research, many unexpected discoveries concerning mycobacteria as part of food chain of different animals including man can be made.

6.5 Poikilothermic Vertebrates

I. Pavlik and J. L. Khol

The frequency of the occurrence of mycobacterial diseases varies among poikilothermic vertebrate species. Numerous review publications have associated it with the following factors (Thoen and Schliesser, 1984; Wayne and Kubica, 1986; Haas and Fattal, 1990; Matlova et al., 1998; Burge et al., 2004; Novotny et al., 2004):

- The individual difference in the susceptibility of the animals to mycobacterial infections.
- The adverse effect of external factors that contribute to stress in animals.
- Long-term survival abilities of PPM in different environmental components.
- The size of the infective dose and time of exposure to a source of virulent mycobacterial species.

For the above reasons, the exposure of animals kept in captivity to mycobacterial infections is higher than that of free-living animals. The captive animals can be exposed to the effects of the environment such as high temperatures and low humidity in the terrarium, inconsiderate handling, parasitic diseases and other stressful conditions (Cooper, 1961; Ippen, 1964; Thoen and Schliesser, 1984; Conroy and Conroy, 1999). The spectrum of detected mycobacterial species in water and terrarium environments is quite extensive (Table 6.12).

Diagnosis of mycobacterial infections take weeks or months depending on the mycobacterial species. Culture-independent techniques, like mycobacteria species-specific PCR, may help in the rapid identification of infected fish, if it is directly used on infected tissues (Talaat et al., 1997). However, their detection is complicated by the fact that many different mycobacterial species have been isolated from infected fish (Table 6.13).

Table 6.12 Mycobacterial species detected in the environment of aquaria and fish ponds

Species	Fresh water aquaria	Sea water aquaria/terrariums
M. abscessus	[2]	nk/nk
MAC	[3], [5]	nk/nk
M. a. hominissuis (8)	[5]	nk/nk
M. celatum	[3]	nk/nk
M. chelonae	[3], [2], [4]*, [5]	nk/[6]
M. diernhoferi	[3]	nk/nk
M. flavescens	[3], [4]*	nk/[6]
M. fortuitum	[3], [2], [4]*, [5]	[5]/nk
M. gastri	[5]	nk/nk
M. gordonae	[3], [2], [4]*	[5]/[6]
M. kansasii	[3], [2]	[5]/nk
M. marinum	[2]	nk/nk
M. nonchromogenicum	[4]*	nk/[6]
M. peregrinum	[5]	nk/[6]
M. scrofulaceum	[4]*	nk/nk
M. szulgai	[1], [2]	nk/nk
M. triviale	[3]	nk/nk
M. terrae	[3], [2], [4]*, [5]	[5]/[6]
M. vaccae	nk	nk/[6]

* Isolated from the fishpond. **nk** Not known. **(8)** Serotype 8. [1] Abalain-Colloc ML, Guillerm D, Salaun M, Gouriou S, Vincent V, Picard B (2003) Eur. J. Clin. Microbiol. Infect. Dis. 22:768–769. [2] Beran V, Matlova L, Horvathova A, Bartos M, Moravkova M, Pavlik I (2006) Veterinarski Arhiv 76 (Supplement):S33–S39. [3] Beran V, Matlova L, Dvorska L, Svastova P, Pavlik I (2006) J. Fish. Dis. 29:383–393. [4] Haas H, Fattal B (1990) Water Research 24:1233–1235. [5] Pattyn SR, Portaels F, Boivin A, Van den BL (1971) Acta Zool. Pathol. Antverp. 52:65–72. [6] Sussland Z, Prochazkova V (1975) Veterinarstvi 25:320–322.

6.5.1 Fish

Stagnant water and water in tanks and aquaria that is not regularly changed can serve as a suitable environment for the multiplication of mycobacteria and a source of mycobacterial infection for fish (Photo 6.1). Inadequate diets, low concentrations of oxygen in the water, a lack of vitamin C and overfeeding have been identified as the major causes of some outbreaks in captive fish. Due to stressful conditions and the intake of great amounts of food rich in protein, the fish can soon become overweight; the natural resistance of fish consequently decreases and various diseases develop, including mycobacterial infections (Parisot, 1958; Cooper, 1961; Ippen, 1964; Demartinez and Richards, 1991).

Bataillon et al. (Bataillon et al., 1897; Bataillon and Terre, 1897) were the first to document mycobacterial

Table 6.13   Mycobacterial species detected in healthy or spontaneously infected fish during the last two decades

Species	Fresh water fish	Sea fish
M. abscessus	[31], [5], [15], [30]	[22]
M. a. hominissuis (6, 8, 9)	[17]	
M. chelonae	[10], [19], [15], [24], [21], [6], [27], [30]	[3], [4], [8], [22]
M. flavescens	[6]	
M. fortuitum	[7], [5], [5], [29], [25], [17], [24], [21], [6], [27]	[9], [22]
M. gordonae	[24], [21], [6]	
M. haemophilum	[15]	
M. interjectum		[28]
M. kansasii	[27]	
M. marinum	[20], [35], [25], [34], [2], [17], [24], [21], [33], [6]	[14], [12], [9], [11], [28], [33], [34], [26]
M. montefiorense		[13], [18]
M. peregrinum	[21]	
M. peregrinum/septicum	[15]	
M. poriferae	[32]	
M. pseudoshottsii		[26]
M. scrofulaceum		[22], [28]
M. shottsii		[28]
M. simiae	[27]	
M. szulgai	[1]	[28]
M. triplex	[23]	
M. triviale	[17]	
M. terrae	[24], [6]	
M. ulcerans	[16]	

(6, 8, 9) Serotypes 6, 8 and 9.

[1] Abalain-Colloc ML, Guillerm D, Salaun M, Gouriou S, Vincent V, Picard B (2003) Eur. J. Clin. Microbiol. Infect. Dis. 22:768–769. [2] Antychowicz J, Lipiec M, Matusiewicz J (2003) Bulletin of the European Association of Fish Pathologists 23:60–66. [3] Arakawa CK, Fryer JL (1984) Helgolander Meeresuntersuchungen 37:329–342. [4] Arakawa CK, Fryer JL, Sanders JE (1986) J Fish Dis. 9:269–271. [5] Astrofsky KM, Schrenzel MD, Bullis RA, Smolowitz RM, Fox JG (2000) Comp Med. 50:666–672. [6] Beran V, Matlova L, Dvorska L, Svastova P, Pavlik I (2006) J. Fish. Dis. 29:383–393. [7] Bragg RR, Huchzermeyer HF, Hanisch MA (1990) Onderstepoort J. Vet. Res. 57:101–102. [8] Bruno DW, Griffiths J, Mitchell CG, Wood BP, Fletcher ZJ, Drobniewski FA, Hastings TS (1998) Dis. Aquat. Organ 33:101–109. [9] Dalsgaard I, Mellergaard S, Larsen JL (1992) Aquaculture 107:211–219. [10] Daoust PY, Larson BE, Johnson GR (1989) J. Wildl. Dis. 25:31–37. [11] Diamant A, Banet A, Ucko M, Colorni A, Knibb W, Kvitt H (2000) Dis. Aquat. Organ 39:211–219. [12] Hedrick RP, McDowell T, Groff J (1987) J. Wildl. Dis. 23:391–395. [13] Herbst LH, Costa SF, Weiss LM, Johnson LK, Bartell J, Davis R, Walsh M, Levi M (2001) Infect. Immun. 69:4639–4646. [14] Huminer D, Pitlik SD, Block C, Kaufman L, Amit S, Rosenfeld JB (1986) Arch. Dermatol. 122:698–703. [15] Kent ML, Whipps CM, Matthews JL, Florio D, Watral V, Bishop-Stewart JK, Poort M, Bermudez L (2004) Comp Biochem. Physiol C. Toxicol. Pharmacol. 138:383–390. [16] Kotlowski R, Martin A, Ablordey A, Chemlal K, Fonteyne PA, Portaels F (2004) J. Med. Microbiol. 53:927–933. [17] Lescenko P, Matlova L, Dvorska L, Bartos M, Vavra O, Navratil S, Novotny L, Pavlik I (2003) Veterinarni Medicina 48:71–78. [18] Levi MH, Bartell J, Gandolfo L, Smole SC, Costa SF, Weiss LM, Johnson LK, Osterhout G, Herbst LH (2003) J. Clin. Microbiol. 41:2147–2152. [19] Mccormick JI, Hughes MS, Mcloughlin MF (1995) J Fish Dis. 18:459–461. [20] Noga EJ, Wright JF, Pasarell L (1990) J. Comp Pathol. 102:335–344. [21] Pate M, Jencic V, Zolnir-Dovc M, Ocepek M (2005) Dis. Aquat. Organ 64:29–35. [22] Perez AT, Conroy DA, Quinones L (2001) Interciencia 26:252–256. [23] Poort MJ, Whipps CM, Watral VG, Font WF, Kent ML (2006) J Fish Dis. 29:181–185. [24] Prearo M, Zanoni RG, Campo DB, Pavoletti E, Florio D, Penati V, Ghittino C (2004) Vet. Res. Commun. 28 Suppl 1:315–317. [25] Puttinaowarat S, Thompson KD, Kolk A, Adams A (2002) J. Fish Dis. 25:235–243. [26] Ranger BS, Mahrous EA, Mosi L, Adusumilli S, Lee RE, Colorni A, Rhodes M, Small PL (2006) Infect. Immun. 74:6037–6045. [27] Rehulka J, Kaustova J, Rehulkova E (2006) Acta Veterinaria Brno 75:251–258. [28] Rhodes MW, Kator H, Kaattari I, Gauthier D, Vogelbein W, Ottinger CA (2004) Diseases of Aquatic Organisms 61:41–51. [29] Sanders GE, Swaim LE (2001) Comp. Med. 51:171–175. [30] Seok SH, Koo HC, Kasuga A, Kim Y, Lee EG, Lee H, Park JH, Baek MW, Lee HY, Kim DJ, Lee BH, Lee YS, Cho SN, Park JH (2006) Vet. Microbiol. 114:292–297. [31] Teska JD, Twerdok LE, Beaman J, Curry M, Finch RA (1997) J. Aquatic Animal Health 9:234–238. [32] Tortoli E, Bartoloni A, Bozzetta E, Burrini C, Lacchini C, Mantella A, Penati V, Tullia SM, Ghittino C (1996) Comp Immunol. Microbiol. Infect. Dis. 19:25–29. [33] Ucko M, Colorni A (2005) J. Clin. Microbiol. 43:892–895. [34] Ucko M, Colorni A, Kvitt H, Diamant A, Zlotkin A, Knibb WR (2002) Appl. Environ. Microbiol. 68:5281–5287. [35] Wolf JC, Smith SA (1999) Dis. Aquat. Organ 38:191–200.

infections in fish and observed the occurrence of a tuberculous-like disease in carp (*Cyprinus carpio*) that lived in water contaminated with sputum and excretions from tuberculosis sufferers. The authors assumed they had isolated *M. tuberculosis* from the diseased fish. However, further experiments did not confirm the mutual transmission of *M. tuberculosis* between men and fish (Bataillon and Terre, 1897; Terre, 1902). AFB isolated from the fish lesions were later designated as "non-tuberculous" or "atypical" mycobacteria (Thoen and Schliesser, 1984).

Bataillon et al. (1897) observed large distinct lesions with leukocyte aggregates in the abdominal wall and ovary of common carp (*Cyprinus carpio*). Cells with mammal lymphocyte-like functions were not found in these fish even though it was assumed that they were present. After 3–4 days of incubation, microscopic examination of granulomas confirmed the presence of numerous AFB grown on medium at 23 °C. The mycobacterial isolate was designated as *M. piscium*. A spontaneous mycobacterial infection in a sea fish species was detected for the first time in codfish a few years later (Johnstone, 1913).

Mycobacteriosis is one of the most common chronic diseases of freshwater and sea fish in the temperate and tropical climate zones. Based on numerous literature references, we can say that mycobacterial infections of fish have a worldwide distribution (Parisot, 1958; Nigrelli and Vogel, 1963; Ippen, 1964; Van Duijn, 1967; Brownstein, 1978; Thoen and Schliesser, 1984). The mycobacterial species (designated according to the currently accepted taxonomy) isolated most frequently from fish are listed in Table 6.13.

6.5.1.1 Prevalence of Mycobacteria in Fish in the Wild and in Aquacultures

The importance of mycobacterial infections in fish will be demonstrated by several examples. Ippen (1964) reported in a monograph that before the year 1963, mycobacterial infection was described in 10 orders, 85 genera and 123 species of freshwater and sea fish. In the river Yakima in Washington, an 8% prevalence of infection was noted in caught fish of the *Prosopium williamsoni* species, no differences being observed in the occurrence between males and females; nevertheless, the infection was more frequently diagnosed in older fish (Abernethy, 1978).

Mortality caused by mycobacterial infections has often been reported for fish from large water establishments (especially water reservoirs) and from commercial aquaria. However, the occurrence of mycobacterial diseases in fish, amphibians and reptiles kept in zoological gardens has not been adequately investigated and described yet. Perhaps it is due to a lack of adequate gross and bacteriological methods for the examination of cadavers (Aronson, 1926; Cooper, 1961; Ippen, 1964; Thoen and Schliesser, 1984).

The occurrence of mycobacterial infections in salmon incubators, where young fish were fed with raw entrails from adult fish infected with mycobacteria, was a great problem. Due to the infection of fingerlings with mycobacteria, less adult salmon originating from these incubators later returned (Wood and Ordal, 1958).

Mycobacterial infections in fish are, however, a much more common health problem in aquarium fish than in fish from free nature. A 10–15% prevalence of infection has been noted for some species kept in an aquarium (Wolke and Stroud, 1978).

6.5.1.2 Spontaneous and Experimentally Induced Infection

Disease caused by mycobacterial infection can spread widely and become an outbreak with economic consequences. Van Duijn (1967) studied a mycobacterial epizootic outbreak in fish of the *Gymnocorymbus ternetzi* species, during which about 800 fish died. Noga et al. (1990) described natural infection with *M. marinum,* characterised by a persisting skin infection, organ lesions, etc., in the fish species Mozambique tilapia (*Oreochromis mossambicus*) kept in an aquarium. In freshwater carp (*Cyprinus carpio*) in England extensive skin lesions were described (Majeed and Gopinath, 1983). The discovery of mycobacteria in many species of aquarium fish (*Danio malabaricus, G. ternetzi, Haplochromis multicolor, Aphyocharax rubripinnis, Brachydanio albolineatus, Barbus phutunio, Xiphophorus* sp. and *Colisa labiosa*) was reported by Jahnel (1940a, b).

Fish are susceptible to PPM and to OPM (*M. tuberculosis, M. bovis* and *M. avium* complex) that cause tuberculosis in homoiothermic animals. After experimental infection with a pathogenic species of mycobacteria, the condition of the animal usually has a chronic course and clinical signs are not specific for the mycobacterial species administered. Anatomical changes in these fish organisms and in natural hosts of these mycobacteria are comparable. However, in contrast to higher vertebrates, the course of such diseases in poikilothermic animals is much longer and the generalized infection does not occur (Ippen, 1964).

6.5.1.3 Spread of Mycobacterial Infections

The exact routes for the spread of mycobacterial infections in fish are poorly understood. The gills, intestines, hepatopancreas and spleen are affected most. Primary mycobacteriosis only affecting the skin is found occasionally, due to the fact that a general infection is manifested by skin lesions (Jahnel, 1940a). Some authors believe that a vertical transmission takes place from the parents to offspring. According to Nigrelli and Vogel (1963), transovarian spread is strongly probable in ovoviviparous fish. Nevertheless, this route of transmission was not confirmed by the experimental passage of mycobacteria through the spawn and milt of some trout species (Ross, 1970).

Acute forms of the disease are occasionally seen in naturally infected fish, which die without showing clinical signs and hence advanced stages of the disease cannot develop. A systemic disease caused by mycobacteria can affect any organ system, in particular the spleen and hepatopancreas, occasionally kidneys, ovaries, pericardium, gills and other organs. These changes are manifested in the late stage of the disease (Ippen, 1964; Van Duijn, 1967; Wolke, 1975; Wolke and Stroud, 1978).

6.5.1.4 Clinical and Histopathological Findings

Clinical Symptoms

Fish affected by mycobacterial infections exhibit nonspecific signs such as inappetence, progressive wasting, cachexia and changes in behaviour. These symptoms are compatible with those caused by various bacterial diseases. The following clinical signs are especially characteristic of generalised mycobacterial infections in fish: secondary skin lesions including inflammatory injuries, changes in skin pigmentation, ulceration and scale loss (Photos 6.21 and 6.22).

Changes in the skeleton are often diagnosed as lordosis, scoliosis, etc.

Chinabut et al. (1990) described mycobacterial lesions in the *Channa striatus* fish species on both the body surface (loss of pigmentation in the head area, keratitis) and internal organs (small white foci in the gill epithelium). Strikingly enlarged kidneys with small off-white nodules on the surface were found in the affected fish and similar alterations were observed in the liver.

Microscopic Examination

Microscopic examination of lesions revealed classical, focal granulomas. They were comprised of epithelioid lymph nodes (hypertrophied histiocytes with cytoplasm containing hyaloid inclusions, resembling epithelioid cells) and histiocytes filling the central area of necrosis surrounded by a wall of fibroblast cells. The histological findings in fish mycobacteriosis are similar to those in humans but with one difference, i.e. soft and hard tubercles are not present in fish. A soft tubercle is a tubercle showing caseous necrosis at the centre, which is absent in hard tubercles. Fish mycobacteriosis can be distinguished from human tuberculosis by two significant features: (1) the absence of Langhans' cells and caseous necrosis and (2) relatively plentiful AFB in the central area in contrast to higher vertebrates (Amlacher, 1961).

Localisation of Granulomas

Chinabut et al. (1990) detected miliary granulomas in all internal organs, including the gills, of infected fish of the *Channa striatus* species. However, no lesions were found in lymphoid tissue. The central area of necrosis was surrounded by layers of macrophages and epithelioid cells. Melanin-containing cells were found in some lesions, whereas no giant cells were present there. Microscopic examination of the centres of caseous necrosis showed abundant amounts of AFB. The spleen and kidneys were strongly affected and their surface area was largely covered by granulomas. Haematopoietic tissue was considerably disturbed. Focal granulomas were also found on the lamellar tissue of gills, brain, skeleton musculature and in haemorrhagic skin lesions (Parisot, 1958; Ippen, 1964; Thoen and Schliesser, 1984).

Changes in Pigmentation

Melanotic lesions were often found in the skin and spleen by Noga et al. (1990) due to the fact that pigment cells surrounded the skin and spleen lesions. All three types of pigment cells found in fish were detected in inflammatory lesions: melanomacrophages, melanocytes and melanophores.

Melanomacrophages were typical ovoid-shaped cells capable of phagocytosis; their melanosomes were enclosed within membrane vacuoles. Melanocytes and melanophores were usually cells with numerous projections and had an almost star-shaped appearance; their cytoplasm did not contain melanosomes and they were incapable of phagocytosis. Melanomacrophages outnumbered the other pigment cells in the internal organs, in contrast to skin lesions where equal numbers of all three pigment cell types were present (Noga et al., 1990).

Mycobacterial infection is accompanied by the presence of numerous melanomacrophages in the spleen and hepatopancreas and reduced numbers of pancreatic acines; they are replaced here by melanomacrophages. The biological role of melanocytomas during the response to mycobacterial infection has not yet been well characterised. Based on a number of studies focused on infection in lower classes of animals, it is assumed that polyphenol-derived compounds including melanin play an important role in the protection of organisms against invasive pathogens (Roberts, 1975).

6.5.2 Amphibians

Spontaneous mycobacterial diseases are diagnosed occasionally in amphibians. From 234 Amazonian toads and frogs *M. chelonae* was isolated from 6 (2.6%) animals, which all were clinically healthy and without pathological lesions (Mok and Carvalho, 1984). The most common route of infection is alimentary; inadequate keeping conditions in captivity contribute to the development of the disease. The use of some substrates (sawdust, moss, peat, etc.) poses a considerable risk because these can serve as a significant reservoir of mycobacteria (Kazda, 1973b; Pavlas and Patlokova, 1985; Trckova et al., 2005).

Information on mycobacterial diseases in the early 1960s was available for 11 amphibian species (Ippen, 1964). In contrast to fish, spontaneous mycobacterial infections among amphibians are only found in frogs. Rupprecht (1904) was the first to describe AFB and detected them in a greatly enlarged frog liver with prominent nodules on the surface. The most frequently isolated mycobacterial species have been *M. marinum*, *M. fortuitum*, the *M. avium* complex and *M. xenopi* (Kazda and Hoyte, 1972; Kazda, 1973a, b; Thoen and Schliesser, 1984).

6.5.2.1 Spontaneous and Experimental Infections

The onset of disease in amphibians is usually connected with the keeping of animals in sub-optimal conditions. Critical factors include a lack of natural light, low atmospheric moisture and poor quality substrates. However, the highest risk is posed by drinking water of poor quality. Infection often develops in injured and emaciated animals.

Besides the infection of frogs with PPM, a lot of researchers have focused on the issue of the adaptation of homoiothermic animal mycobacteria to amphibian organisms. They infected them, above all, with *M. tuberculosis, M. bovis* and *M. avium* and occasionally found gross lesions in the infected subjects. Most cases were characterised by small nodules containing a vitreous, transparent exudate at the site of inoculation (Baker and Hagan, 1942).

Experimental infection of the African clawed frog (*Xenopus laevis*) did not confirm the participation of *M. marinum* in the initiation and development of lymphoreticular tumours (Clothier and Balls, 1973a, b). A possible use of mycobacterial species isolated from poikilothermic animals for the vaccination of people or animals was investigated, yielding, however, ambiguous results (Schliesser, 1965).

6.5.2.2 Clinical and Histopathological Findings

The affected animals showed non-specific signs, such as a loss of skin pigmentation with subsequent loss of colouration in a small area or over the entire body surface. Open wounds, total apathy, inappetence and a loss of body weight were also observed. Skin forms of the disease were manifested as either granulomatous ulcerations or diffuse mycobacterial dermatitis. In later stages of the disease, internal organs were affected. Miliary or large granulomas were located in the livers, intestines, lungs, kidneys and heart. Granulomas were also found in the brain of one frog infected with *M. marinum* (Shively et al., 1981).

The histological examination of organs from infected animals showed nodules of various sizes. The smallest ones only comprised several cells with abundant cytoplasm and distinct vesicular nuclei. Larger lesions contained epithelioid cells and occasional small lymphocytes. Giant cells were absent and the nodules were not surrounded by connective tissue. Numerous pigment cells surrounding the nodules were observed in some frogs (Rupprecht, 1904; Machicao and Laplaca, 1954; Ippen, 1964).

Shively et al. (1981) achieved noteworthy results in toads infected with *M. marinum*. Microscopic examination of tissues revealed numerous granulomas. Some lesions contained only several macrophages, whereas others were well-developed granulomas containing mycobacteria.

Ramakrishnan et al. (1997) described the course of a granulomatous disease in the leopard frog (*Rana pipiens*) after experimental infection with a *M. marinum* suspension. Colonisation of the liver, spleen, kidneys and lungs occurred in frogs maintained at 25 °C after 2 weeks and granulomatous lesions appeared after 6 weeks. However, mycobacteriosis did not cause any mortality during the 58-week course of the experiment. If, however, *M. marinum* was administered to frogs together with corticoids (hydrocortisone) to produce immunosuppression, the course of the disease was much shorter. Out of seven infected animals, five died within 19 weeks post infection. These experiments unambiguously showed a strong association between the immunocompetence of animals and the course of the disease. The relatively frequent occurrence of *M. marinum* in water can lead to the extinction of an entire population of immunosuppressed frogs.

6.5.3 Reptiles

Reptiles exhibit a relatively high degree of natural resistance to mycobacteria, which is indicated by a low number of described cases of mycobacterial infections (Reichenbach-Klinke and Elkan, 1965). The most significant preconditions for the development of mycobacteriosis are poor microclimatic conditions in the terrarium and stress caused by trapping and transport. Increased susceptibility to PPM may be a result of the "maladaptation syndrome" (Cooper, 1961; Thoen and Schliesser, 1984).

Spontaneous infections are mostly caused by *M. marinum* and *M. chelonae* (Table 6.14) but also by other mycobacterial species (Table 6.15). Mycobacterial infections in several species of snakes, turtles, lizards and crocodiles have been described. However, *M. smegmatis* and *M. phlei* have also been isolated from clinically normal animals (Cooper, 1961; Ippen, 1964; Thoen and Schliesser, 1984). A spontaneous mixed infection caused by *M. a. hominissuis* (serotype 8) and *M. intracellulare* (serotype 19) was detected in a water monitor (*Varanus semiremex*) in Australia (Friend and Russell, 1979). *M. kansasii* has been diagnosed in the Chinese soft shell turtle, *Pelodiscus sinensis* (Oros et al., 2003).

6.5.3.1 The Causes and Spread of Mycobacterial Infections

A progressive mycobacterial disease with tuberculoid granulomas was described for the first time in a naturally infected grass snake (*Natrix natrix*). Miliary nodules were observed in the liver, spleen, kidneys, reproductive system and subcutis, whereas no lesions were found in the lungs (Sibley, 1889). We can learn about the frequency of the occurrence of mycobacteriosis from the study of Snyder (1978), who diagnosed this disease, on the necropsy of animals from captivity. Four (8.0%) of 50 turtles and 31 (14.8%) of 209 snakes and lizards were affected. Meanwhile, the disease appeared in only 0.1–0.5% of reptiles kept in adequate conditions (Brownstein, 1978).

Despite a high natural resistance to mycobacterial infections, reptiles are often used as an experimental model in the study of human infections with PPM. For instance, the effects of *M. ulcerans* species that cause skin lesions in men have been investigated in the lizard *Anolis carolinensis* (Marcus, 1981).

6.5.3.2 Gross and Histopathological Lesions

A disseminated granulomatous form with caseous necrosis is most often observed in reptiles affected

Table 6.14 *M. marinum* and *M. chelonae* detection in captive reptiles and frogs

Mycobacterial species/animal species	Symptoms and infected organs	References
M. marinum		
Royal python (*Python regius*)	Chronic dyspnoe for 18 months, isolation from lung[1]	[4]
Bullsnake (*Pituophis melanoleucus sayi*)	Skin lesions, granulomas in lung, liver, spleen and kidney	[5]
Foxsnake (*Elaphe vulpina vulpina*)	Skin lesions, granulomas in lung, liver, spleen and kidney	[5]
Garter snake (*Thamnophis sirtalis*)	Skin lesions, granulomas in lung, liver, spleen and kidney	[5]
Massassauga snake (*Sistrurus catenatus catenatus*)	Skin lesions, granulomas in lung, liver, spleen and kidney	[5]
Milk snake (*Lampropeltis triangulum triangulum*)	Skin lesions, granulomas in lung, liver, spleen and kidney	[5]
Rat snake (*Elaphe obsoleta obsoleta*)	Skin lesions, granulomas in lung, liver, spleen and kidney	[5]
Rattlesnake (*Crotalus horridus horridus*)	Skin lesions, granulomas in lung, liver, spleen and kidney	[5]
African clawed frog (*Xenopus laevis*)	Swelling with coalescing ecchymoses of the lower mandible	[1]
Leopard frog (*Rana pipiens*)	Chronic granulomatous, non-lethal disease	[9]
African bullfrog (*Pyxicephalus adspersus*)	Infected hind limb	[7]
Bullfrog (*Rana catesbeiana*)	Skin lesions on the head and extremities	[2], [5]
Hawksbill sea turtle (*Eretmochelys imbricata*)	Nk	[11]
M. chelonae		
Kemp's Ridley sea turtle (*Lepidochelys kempii*)	Infected elbow joint swelling and nodules on the skin, isolation from liver, lung, spleen, kidney and pericardium	[3]
Boa constrictor (*Boa constrictor*)	Subcutaneous mass, ulcers on the skin and in the mouth and forced breathing, isolation from skin, subcutis, lung and liver	[8]
Snake	Infected lungs, intestine, heart, peritoneum and CNS	[10]
Granular toad (*Bufo granulosus*)	Without gross lesions	[6]
Giant toad (*Bufo marinus*)	Without gross lesions	[6]

[1] Mixed infection caused by *M. marinum* and *M. haemophilum*.
CNS Central nervous system. **Nk** Not known.

[1] Cannon CZ, Linder K, Brizuela BJ, Harvey SB (2006) Lab Animal 35:19–22. [2] Ferreira R, Fonseca LS, Afonso AM, da Silva MG, Saad MH, Lilenbaum W (2006) Vet. J. 171:177–180. [3] Greer LL, Strandberg JD, Whitaker BR (2003) J. Wildl. Dis. 39:736–741. [4] Hernandez-Divers SJ, Shearer D (2002) J. Am. Vet. Med. Assoc. 220:1661. [5] Maslow JN, Wallace R, Michaels M, Foskett H, Maslow EA, Kiehlbauch JA (2002) Zoo Biology 21:233–241. [6] Mok WY, Carvalho CM (1984) J. Med. Microbiol. 18:327–333. [7] Pizzi R, Miller J (2005) Vet. Rec. 156:747–748. [8] Quesenberry KE, Jacobson ER, Allen JL, Cooley AJ (1986) J. Am. Vet. Med. Assoc. 189:1131–1132. [9] Ramakrishnan L, Valdivia RH, McKerrow JH, Falkow S (1997) Infect. Immun. 65:767–773. [10] Soldati G, Lu ZH, Vaughan L, Polkinghorne A, Zimmermann DR, Huder JB, Pospischil A (2004) Vet. Pathol. 41:388–397. [11] Ucko M, Colorni A (2005) J. Clin. Microbiol. 43:892–895.

by chronic mycobacterial infections transmitted via the alimentary route. High numbers of tubercles are, in particular, found in the spleen and liver. "Secondary tubercles" can develop in bones, lungs, CNS, gonads, subcutis and other organs by a gradual spread of the disease. It is well known from literature data that pulmonary tubercles are a characteristic finding of mycobacterial infections in lizards, snakes and crocodiles (Rhodin and Anver, 1977). White miliary lesions have been observed in the liver and lungs of reticulated pythons (*Python reticulatus;* Olson and Woodard, 1974).

6.5.3.3 Course of Infection

Mycobacterial infection in reptiles has a peracute, acute or chronic form. The peracute form of the disease has only been observed in the case of experimentally induced infection followed by high mortality, associated with a massive intravascular proliferation of AFB and a moderate inflammatory response in the host (Aronson, 1926; Olson and Woodard, 1974).

After spontaneous infection with mycobacteria, the course of mycobacteriosis is not usually acute. The response of an organism to mycobacteria results in a

Table 6.15 Mycobacterial infections in captive reptiles and amphibians caused by mycobacterial species other than *M. marinum*

Mycobacterium species[1]	Origin of examined samples		References
	Reptile group species	Symptoms and infected organs	
M. agri	Lizard	Nk	[9]
MAC	Turtle	Nk	[11]
MAC	Crocodile	Granulomatous lesions	[10]
MAH (8)	Toad	Nk	[4]
MAH (8) and *MI* (18)	Water monitor (*Varanus salvator*)	Anorexia, emaciation, abnormal motion and positions, infected mesenterium, kidney, liver, lung, muscle fascia and spine	[2]
M. chelonae	Kemp's Ridley sea turtle (*Lepidochelys kempii*)	Osteoarthritis	[3]
M. confluentis	Snake	Nk	[9]
M. haemophilum	Snake and turtle	Nk	[9]
M. hiberniae	Snake	Nk	[9]
M. kansasii	Chinese soft shell turtle (*Pelodiscus sinensis*)	Anorexia, respiratory distress, infected lungs and carapace	[6]
M. liflandii	African clawed frog (*Xenopus laevis*)	Nk	[7]
M. neoaurum	Snake	Nk	[9]
M. nonchromogenicum	Turtle	Nk	[9]
M. ulcerans	African tropical clawed frog (*Xenopus tropicalis*)	Lethargy, excess buoyancy, coelomic effusion, cutaneous ulcers and granulomas, visceral granulomas, ulcerative and granulomatous dermatitis, coelomitis and septicaemia	[12]
M sp.	Australian freshwater crocodile (*Crocodylus johnstoni*)	Anorexia, infected liver, lung, spleen and kidney	[1]
	Australian saltwater crocodile (*Crocodylus porosus*)	Cuts and emphysema, infected subcutis, spleen, liver, lung	[1]
	Australian freshwater crocodile (*Crocodylus johnstoni*)	Cuts and emphysema, infected subcutis, spleen, liver, lung	[1]
	Spotted-bellied side-necked turtle (*Phrynops hilarii*)	Ulcer on the mandibles and foot, oedema, anorexia, deterioration followed by death, infected skin, liver and spleen	[8]
	Reticulated python (*Python reticularis*)	Infected liver, lung, kidney, spleen and heart	[5]

[1] Mycobacterial species are in alphabetical order.

CNS Central nervous system. **Nk** Not known. **MAC** *M. avium* complex. **MAH** *M. a. hominissuis*. **MI** *M. intracellulare*. **(8)** Serotype 8.

[1] Ariel E, Ladds PW, Roberts BL (1997) Aust. Vet. J. 75:831–833. [2] Friend SC, Russell EG (1979) J. Wildl. Dis. 15:229–233. [3] Greer LL, Strandberg JD, Whitaker BR (2003) J. Wildl. Dis. 39:736–741. [4] Kazda J, Hoyte R (1972) Zentralbl. Bakteriol. [Orig. A] 222:506–509. [5] Olson GA, Woodard JC (1974) J. Am. Vet. Med. Assoc. 164:733–735. [6] Oros J, Acosta B, Gaskin JM, Deniz S, Jensen HE (2003) Vet. Rec. 152:474–476. [7] Ranger BS, Mahrous EA, Mosi L, Adusumilli S, Lee RE, Colorni A, Rhodes M, Small PL (2006) Infect. Immun. 74:6037–6045. [8] Rhodin AGJ, Anver MR (1977) J. Wildlife Dis. 13:180–183. [9] Soldati G, Lu ZH, Vaughan L, Polkinghorne A, Zimmermann DR, Huder JB, Pospischil A (2004) Vet. Pathol. 41:388–397. [10] Thoen CO, KARLSON AG, Himes EM (1981) Rev. Infect. Dis. 3:960–972. [11] Thoen CO, Richards WD, Jarnagin JL (1977) J. Am. Vet. Med. Assoc. 170:987–990. [12] Trott KA, Stacy BA, Lifland BD, Diggs HE, Harland RM, Khokha MK, Grammer TC, Parker JM (2004) Comp Med. 54:309–317.

suppurative inflammation with signs of anorexia, dyspnoea or pneumonia. Necrotic areas are filled with exudation containing high numbers of extracellular AFB and heterophilic granulocytes (Olson and Woodard, 1974; Brownstein, 1978).

The spontaneous disease usually has a chronic form. The disease with a protracted course is marked by granulomatous lesions with tubercles consisting of a central region containing macrophages and high numbers of intracellularly located AFB. The adjacent tissue is comprised of fusiform histiocytes and reticulocytes. An advanced stage of tubercle development is characterised by central caseous necrosis and the destruction of tubercles. Epithelioid cells predominate in the inflammatory response to mycobacteria in cold-blooded animals. Langhans' cells have not been observed in snakes; however, they have been found in turtles and lizards (Thoen and Schliesser, 1984).

6.5.4 Food Safety Issues

Two risks to food safety posed by infected fish should be considered. The first is the potential infection through the handling and processing of fish before cooking. Skin injuries caused by a sharp knife and fish bones or sharp fins are viewed as dangerous (Photo 6.23). The second is the consumption of raw fish and other seafood by immunocompromised patients (above all with HIV/AIDS). This habit most likely increased the frequency of mycobacterial infections with PPM in patients from Finland (Von Reyn et al., 1996).

6.6 Birds

I. Pavlik, V. Mrlik and J. Klimes

Close attention will be paid to those species of mycobacteria that cause fatal diseases in birds (Photos 3.9 and 6.24). The causative agent of avian tuberculosis, *M. a. avium,* is one of the most significant pathogens (Table 6.16); this species usually causes miliary tuberculosis in fully susceptible birds (especially gallinaceous poultry). However, in more resistant bird species (e.g. predators and pigeons) *M. a. avium* only affects some organs, such as the eyes, skin, bones and limbs. In any case, avian tuberculosis has been discussed in detail in Section 3.1.3.1.

The purpose of this chapter is to give a general overview of the significance of birds in the ecology of mycobacteria. The lack of attention so far paid to this issue is surprising. In many publications dealing with epizootiology and epidemiology, birds are referred to as potential sources of causative agents of mycobacterial infections, but there exist no detailed data in the published literature.

6.6.1 Birds

Birds are one of the most numerous vertebrate classes; almost 10 000 bird species are known at present. Because of their adaptability to different environments and their ability to travel large distances they constitute a significant factor in the spread of mycobacteria in the environment (Photo 6.20). Similarly to other animals, they spread mycobacteria through excrements, body surfaces (largely feathers), infected skin and bird carcasses.

Excrements (Photo 6.25): Mycobacteria are usually spread through faeces mixed with urine that are shed from the cloaca. Numerous mycobacterial species have been detected in bird excrements and the detection rate of mycobacteria has been high (Section 5.10). The faeces of free-living and domestic birds infected with the causative agent of avian tuberculosis pose a great risk to herds of farm animals. This has been discussed from the point of view of the prevention of the spread of the causative agent of avian tuberculosis in Section 7.5. Bird faeces also contain high amounts of organic matter. In countries which have large bird colonies, faeces (designated as guano) which accumulate under bird nests are used as conventional fertiliser in agriculture and as an important economic product (often for export). There has been a paucity of reports on the occurrence of mycobacteria in guano and on the significance of guano in the ecology of mycobacteria.

Body surface, especially feathers (Photos 5.32 and 6.26): Birds can also carry mycobacteria on the parts of their body that are covered with feathers. They keep these areas clean by frequent bathing in different types of surface water, rolling in the dust and by cleaning with the use of their beaks or legs. Some bird species that live in deserts and semi-desert regions (sandgrouse, Pterocliformes) carry drinking water in their breast feathers to their young over distances of several tens of kilometres.

The coccygeal gland, which is used for the lubrication of feathers and thereby for their effective protection, is well developed in the Anseriformes because their life is closely linked with water. On the other hand, ostriches (Struthioniformes) do not have coccygeal gland. However, they care for their feathers by intensive rolling in the dust in different parts of their life environment. These materials, i.e. water and dust, are often contaminated with mycobacteria

Table 6.16   Serotype detection of *Mycobacterium avium* complex members in birds, other animals and humans

Reference		Meissner and Anz [2]				Thoen et al. [3]		Anz and Meissner [1]	
Country: period in years		Germany: 1965–1975 (%)				USA: 1971–1976 (%)		Not known (%)	
Member	Serotype No.	Birds	Cattle	Pigs	Human	Primates	Birds	Birds	Human
M. avium subsp. *avium*									
	1	6 (3.8)	15 (5.7)	21 (2.6)	8 (8.2)	3 (6.8)	60 (89.6)	3 (11.1)	4 (11.1)
	2	117 (74.1)	210 (80.2)	591 (72.8)	29 (29.6)	4 (9.1)	5 (7.5)	23 (85.2)	13 (36.1)
	3	24 (15.2)	32 (12.2)	36 (4.4)	0 (0)	0 (0)	1 (1.5)[2]	0 (0)	0 (0)
	Mixed/unknown	0 (0)	0 (0)	0 (0)	0 (0)	1 (2.3)	0 (0)	0 (0)	0 (0)
Subtotal		147 (93.0)	257 (98.1)	648 (79.8)	37 (37.8)	8 (18.2)	66 (98.5)	26 (96.3)	17 (47.2)
M. avium subsp. *hominissuis*	4	0 (0)	0 (0)	11 (1.4)	5 (5.1)	5 (11.4)	0 (0)	0 (0)	1 (2.8)
	5	0 (0)	0 (0)	0 (0)	0 (0)	0 (0)	0 (0)	0 (0)	0 (0)
	6	0 (0)	0 (0)	0 (0)	2 (2.0)	0 (0)	0 (0)	0 (0)	0 (0)
	8	10 (6.3)	3 (1.2)	113 (13.9)	25 (25.5)	27 (61.4)[1]	0 (0)	1 (3.7)	7 (19.4)
	9	0 (0)	2 (0.8)	11 (1.4)	9 (9.2)	1 (2.3)	0 (0)	0 (0)	3 (8.3)
	10	0 (0)	0 (0)	0 (0)	0 (0)	0 (0)	0 (0)	0 (0)	0 (0)
	11	0 (0)	0 (0)	0 (0)	0 (0)	0 (0)	0 (0)	0 (0)	0 (0)
	21	0 (0)	0 (0)	0 (0)	0 (0)	0 (0)	0 (0)	0 (0)	0 (0)
	Mixed/unknown	1 (0.6)	0 (0)	29 (3.6)	1 (1.0)	0 (0)	1 (1.5)	1 (3.7)	0 (0)
Subtotal		11 (7.0)	5 (1.9)	164 (20.2)	42 (42.9)	33 (75.0)	1 (1.5)	1 (3.7)	11 (30.6)
M. intracellulare,	12–20, 21–22	0 (0)	0 (0)	0 (0)	14 (14.3)	3 (6.8)	0 (0)	0 (0)	8 (22.2)
	Mixed/unknown	0 (0)	0 (0)	0 (0)	5 (5.1)	0 (0)	0 (0)	0 (0)	0 (0)
Subtotal		0 (0)	0 (0)	0 (0)	19 (19.4)	3 (6.8)	0 (0)	0 (0)	8 (22.2)
Mixed *MAC* members/unknown		0 (0)	0 (0)	0 (0)	0 (0)	0 (0)	0 (0)	0 (0)	0 (0)
Total		158 (100)	262 (100)	812 (100)	98 (100)	44 (100)	67 (100)	27 (100)	36 (100)

[1] Cross-reaction detected with serotype 1.
[2] Diagnosed in an imported tree duck kept in quarantine for import, described previously (Thoen and Himes, 1976).
MAC M. avium complex.
[1] Anz W, Meissner G (1969) Prax. Pneumol. 23:221–230. [2] Meissner G, Anz W (1977) Am. Rev. Respir. Dis. 116:1057–1064. [3] Thoen CO, Richards WD, Jarnagin JL (1977) J. Am. Vet. Med. Assoc. 170:987–990.

(Sections 5.2.5 and 5.9.3). Mycobacteria can be spread very quickly among different components of the environment in this way.

The coccygeal gland in parrots (Psittaciformes) is greatly reduced in size. Accordingly, they use powder for cleaning their feathers, which is formed by the disintegration of special feathers present on their back and coccyx. This dust is also likely to be a source of mycobacteria.

Bird skin (Photo 6.27): Besides contaminated feathers, mycobacteria can also be spread through infected and uninfected skin. For example, in a common buzzard (*Buteo buteo*) avian tuberculosis was diagnosed on the unfeathered parts of both limbs, the "bumble-foot" (van Nie, 1981). In gallinaceous poultry infected with avian tuberculosis, lesions on the skin and leg bones are often observed (I. Pavlik. unpublished data). However, the significance of this infected bird skin for the transmission of mycobacteria is not high. The skin surface of clinically healthy birds can also become contaminated with mud, water, blood and the tissues of prey, etc., in the environment. In such cases, the transmission of mycobacteria between various parts of the environment and animals is highly probable.

Bird carcasses and their constituent parts (Photos 3.13, 3.14 and 5.59): Last but not least, bird carcasses also participate in the spread of mycobacteria in the environment. In the event of infection with the causative agent of avian tuberculosis, the tissues of carcasses become sources of mycobacteria for predators and other animals. These are represented by both diurnal and nocturnal birds of prey and other birds, or by other predatory or scavenging animals (Fischer et al., 2000; Sections 6.7 and 7.5).

The remnants of bird carcasses can also be involved in the spread of mycobacteria (e.g. emboweled entrails) due to the fact that predators usually consume only the muscles and/or drink the blood. The intestinal tract of bird carcasses together with the content usually remains at the place where the bird was caught; it can also be delivered to another place if the predator transferred the prey to an undisturbed place for consumption.

In the following section, the occurrence of mycobacteria in birds and their documented or likely significance in the ecology of mycobacteria will be described. Previously, birds have been classified according to their ability of movement as ratites (those

that have lost the ability to fly and can only move by running), diving birds (that have lost the ability to fly and live on water, where they swim very well and dive) and flying birds. Even though the taxonomy of birds has been changed considerably, we will use the above-mentioned classification for the purposes of clarity.

6.6.1.1 Ratites

The ratites, i.e. flightless birds contain the members of four orders: rheas (*Rheiformes*), ostriches (Struthioniformes), cassowaries (*Casuariiformes*) and kiwis (Apterygiformes). It is surprising that there is a paucity of published papers dealing with mycobacteria in these bird orders. The available information concerns in large part avian tuberculosis. Information about ostriches kept on farms has been mostly published in the form of case reports (Sevcikova et al., 1999; Doneley et al., 1999; Garcia et al., 2001). One *M. a. avium* isolate from an ostrich was also studied by IS*901* PCR (Pavlik et al., 2000b). Avian tuberculosis has been diagnosed in farmed emus (*Dromaius novaehollandiae*) in the USA (Shane et al., 1993) and in New Zealand (Davis et al., 1984), and in a brown kiwi (*Apteryx mantelli*) in New Zealand (Davis et al., 1984).

Due to the fact that ostriches and emus are mostly kept on farms, the occurrence of avian tuberculosis in these birds poses significant economic and health problems. In a review article, Tully and Shane (1996) highlighted the problems associated with avian tuberculosis control with regard to the "long-term" contamination of the environment with *M. a. avium* (Section 3.1.3.1). The detection of the causative agent of bovine tuberculosis in a Southern Cassowary (*Casuarius casuarius*), kept in a zoological garden in the Czech Republic may be viewed as a rare event (Pavlik et al., 1998).

Information regarding the occurrence of mycobacteria in the faeces of the ratites (they pass urine separately from faeces in contrast to other birds) or feathers (they often roll in dust that can be contaminated with mycobacteria) is lacking (Photo 6.28).

6.6.1.2 Diving Birds

Penguins (Sphenisciformes) divide their time between aquatic environments and dry land. They have lost their ability to fly and their habitus shapes and other physiological adaptations allow them to be good swimmers, divers and predators of various sea animals. The morphology of penguins' bodies and their physiology and behaviour are highly adapted to life in an aquatic environment. Mycobacteria are found in the environments of the seas and oceans (Section 5.2.5), in fish (Section 6.5), and in different invertebrate animals (Section 6.3) eaten by penguins.

Penguins can play a significant role in ecology of mycobacteria. We cannot expect the frequent occurrence of mycobacteria in Antarctica, but the seasonal migration of these birds and especially the travel of some species to warmer waters (the northernmost place where they can be found are the Galápagos Islands) can play a role in the spread of mycobacteria. Some penguin species nest in underground holes, e.g. African penguins (*Spheniscus demersus*; Photo 6.29), Humboldt's penguins (*Spheniscus humboldti*) and Little Penguins (*Eudyptula minor*); this is also noteworthy with regard to the ecology of mycobacteria.

6.6.1.3 Flying Birds

A majority of bird species can fly; that often allows them to traverse great distances, including the oceans (Photo 6.30). The type of movement is variable; however, they can fly within a limited space or they can migrate over long distances (transcontinental migration). For example, the Arctic tern (*Sterna paradisaea*) flies 35 000 km annually only taking into account migration. Nevertheless, birds can also fly considerable distances within the area of their home. For example, the genuine sovereigns of the sky, albatrosses (order Procellariiformes), bring food for their young from a distance of up to thousands of kilometres from their nest colonies.

With respect to the fact that birds live in all the components of the Earth's surface, they become the important vectors of mycobacteria in the environment. Flying birds are numbered among the most significant sources of the causative agent of avian tuberculosis *M. a. avium* (Section 3.1.3.1). However, they can also participate in the spread of other mycobacterial species. Besides the causative agent of avian tuberculosis, various PPM have been isolated from bird organs such as:

M. fortuitum (Hoop et al., 1994; Hoop et al., 1996; Matlova et al., 2003).

M. terrae (Dvorska et al., 2007).

M. gordonae (Hoop et al., 1996; Butler et al., 2001; Matlova et al., 2003).

M. nonchromogenicum (Hoop et al., 1996; V. Mrlik and I. Pavlik, unpublished data).

M. simiae (Travis et al., 2007).

M. gastri (Matlova et al., 2003; Shitaye et al., 2008).

M. intracellulare and *M. diernhoferi* (Shitaye et al., 2008).

M. ulcerans, *M. scrofulaceum*, *M. chelonae* and *M. phlei* (Matlova et al., 2003).

Seventy days after experimental infection with *M. intracellulare* and *M. fortuitum,* no clinical symptoms or gross lesions were observed in budgerigars (*Melopsittacus undulatus*; Ledwon et al., 2008). Comparable results have been obtained after infection with *M. a. hominissuis*; that also did not induce any gross lesions in the birds (Morita et al., 1999; Pavlik et al., 2000b; Matlova et al., 2003; Dvorska et al., 2007). In gallinaceous poultry it was found that its presence in drinking water caused positive reactions during avian tuberculin testing (Kazda, 1967a, b, c).

6.6.2 The Impact of Birds Through Their Transmission of Selected Mycobacterial Species

There is a scarcity of information regarding the significance of birds in the ecology of other mycobacterial species. Accordingly, the following chapters will be focused primarily on the *MTC* members, specifically on *M. tuberculosis* and *M. bovis* and from other mycobacterial species, *M. genavense*.

6.6.2.1 *M. tuberculosis*

Little is known about the significance of birds in the spread of *MTC* members, even though some parrots are kept as pets and gallinaceous fowl come into close contact with other domestic birds, animals, and people on farms. The contact of pets with people infected by *M. tuberculosis* may be sometimes very close; in some cases, such pets share the owner's plate (Photos 6.31–6.36). Two of the seven currently accepted pathogenic species *M. tuberculosis* and *M. bovis* belonging to the *MTC* will be the focus of interest in the following Section (besides *M. bovis* BCG).

M. tuberculosis has occasionally been diagnosed in birds. Among bird infections published recently, we can mention the detection of *M. tuberculosis* in a common canary (*Serinus canaria*) and a blue-fronted Amazon parrot (*Amazona aestiva*) in Switzerland (Hoop, 2002) and in Germany (Peters et al., 2007). *M. tuberculosis* was likewise diagnosed in green-winged macaws (*Ara chloroptera*) in Switzerland (Steinmetz et al., 2006) and in the USA (Washko et al., 1998).

After experimental infection of budgerigars (*Melopsittacus undulatus*) with *M. tuberculosis,* no gross lesions or clinical signs were detected after 70 days (Ledwon et al., 2008). It seemed to be due to the high natural immunity of these birds against this infection; the key factors of infection development are direct contact with people infected by *M. tuberculosis* and the existence of some predisposition factors, such as the age of the birds, another disease, etc. (Thoen et al., 2006b). The significance of birds in the transmission of the causative agent of human tuberculosis to people (including family members) with whom they are in close contact is unknown.

6.6.2.2 *M. bovis*

There are only a small number of cases which report the diagnosis of *M. bovis* in birds in the available literature. However, in Russia, this causative agent was isolated from gallinaceous poultry kept together with cattle affected by bovine tuberculosis (Kurmanbaev and Blagodarnyi, 1981).

Granulomatous processes diagnosed by histology with the concurrent detection of *M. bovis* (by culture for 2 months) developed under experimental conditions in European starlings (*Sturnus vulgaris*) and American crows (*Corvus brachyrhynchos*) after intraperitoneal infection. However, oral or intraperitoneal infection failed to produce clinical disease in these birds within 2 months and *M. bovis* could not be detected in the faeces of the infected birds (Butler et al., 2001).

Experimental intratracheal infection of pigeons (*Columba livia*) has failed to produce gross lesions or induce disease. After the administration of *M. bovis* through both routes, occasional shedding of the pathogen has been observed (Fitzgerald et al., 2003). Comparable results were obtained using the same routes of infection in mallard ducks (*Anas platyrhynchos*; Fitzgerald et al., 2005) and wild turkeys (*Meleagris gallopavo*; Clarke et al., 2006). After experimental infection of budgerigars (*Melopsittacus undulatus*) with *M. bovis,* lesions were detected after 70 days by radiological examination; these were subsequently confirmed by histology and culture examination (Ledwon et al., 2008).

The above-mentioned results strongly suggest that there exists a considerable natural immunity in birds and explain the negative results of the examination of 262 birds found in the surroundings of cattle herds infected with *M. bovis* in the USA (Pillai et al., 2000).

6.6.2.3 *M. genavense*

M. genavense is a mycobacterial species (often diagnosed in recent times) which causes fatal disease in birds. It was described for the first time in 1993 (Bottger et al., 1993). The *in vitro* isolation of this species is difficult; it often takes several months, which is the main problem when diagnosing this infection. More than 100 studies describing the occurrence of *M. genavense* in both birds and humans have been published. It is noteworthy that besides birds, *M. genavense* has also been diagnosed in other animals: a monkey in Germany (Steiger et al., 2003), ferrets in Australia (Lucas et al., 2000), a cat in Ireland (Hughes et al., 1999) and a dog in the USA (Kiehn et al., 1996). Bringing together these studies and also a review article of Bercovier and Vincent (2001) it can be said that *M. genavense* has been detected in Europe, Australia and the USA.

With respect to the fact that *M. genavense* has quite frequently been diagnosed in exotic birds (Bercovier and Vincent, 2001), we can suppose that this causative agent is also found in tropical regions of other countries that are natural homeland for parrots. In the Czech Republic, *M. genavense* was detected in a blue-headed parrot (*Pionus menstruus*); the bird was imported from Surinam (J. E. Shitaye, I. Grymova, J. Svobodova, M. Moravkova, I. Pavlik, unpublished data; Photo 6.37).

Even though Bercovier and Vincent (2001) supposed that *M. genavense* could cause infections in up to 10% of patients with HIV/AIDS, the epizootiology, epidemiology and ecology of this causative agent remain unclear. From the environment, *M. genavense* has, for example, been isolated from water in Finland (Hillebrand-Haverkort et al., 1999) and effluent from a hospital in China (Lin et al., 2007). However, a more extensive study dealing with the ecology of this causative agent has not yet been carried out. Infection from the environment was thought to be responsible based on the anamnestic data of three Finnish patients infected with *M. genavense* who all lived at a distance of about 1 km from the source of infection (Ristola et al., 1999). However, a complex study that would shed light on the occurrence of *M. genavense* in the environment (most likely in water) has not yet been conducted. Nevertheless, there is no doubt that birds play a significant role in the spread of this causative agent (Hoop et al., 1993; Portaels et al., 1996; Kiehn et al., 1996; Ramis et al., 1996; Tell et al., 2003; Manarolla et al., 2007; Foldenauer et al., 2007).

6.7 Mammals

I. Pavlik, V. Mrlik and J. Klimes

The species diversity of mammals, with only just over 4000 described species, is low compared to other animal groups (e.g. insects). Nevertheless, mammals play an important role in the ecology of mycobacteria because they are capable of traversing long distances in the environment and their ways of life are varied. Accordingly, mammals can participate in the spread of mycobacteria to almost all components of the environment. With regard to the ecology of mycobacteria, it is also important that the majority of these mammalian species are not "specialised" in only one of the above-mentioned components. Mammals can spread mycobacteria in the two following ways: actively or passively.

i. *Active spread of mycobacteria* (especially OPM and PPM) has been described in Chapters 2 and 3. In the event of organ or tissue infection of mammals including man (intestinal tract, pulmonary tract, skin, etc.), mycobacteria causing the infection are subsequently actively shed. For instance, when pulmonary tissue is affected, mycobacteria can also be shed via the intestinal tract; the swallowed mycobacteria survive the passive transport through the body of a host (they are resistant to the acidic environment and numerous digestive enzymes).

ii. *Passive spread of mycobacteria* is (in this book) the transmission of the same/lower amount of mycobacteria than the host organism was originally contaminated /infected with.

Passive spread of mycobacteria can occur in the following ways:

- On the body surface: dust released from, in particular, hair, skin and wool (Section 5.9).
- On the surface of mucous membranes: above all the mucosa of the oral cavity (Section 5.10.2).
- Through the intestinal tract: ruminants, especially, are important vectors (Section 5.10).
- Excretions and secretions: mainly milk (Section 5.7.4).

Mycobacteria can be spread directly (by predation) between respective animals or groups of animals of one species or between different mammalian species. Mycobacteria can also be spread through indirect contact with the environment, which is either only temporarily contaminated with mycobacteria or colonised for a long time, e.g. sphagnum (Section 5.1), water (Section 5.2) and soil (Section 5.5). Inside a population, the horizontal spread of mycobacteria is possible, mainly through the contact between animals that live in herds, packs or small families. The spread of *M. microti* is intensive, for example, and occurs mainly between small terrestrial mammals that live in underground burrows and bring up their offspring in nests.

The transmission of mycobacteria between animals may be horizontal or vertical. Horizontal transmission from the mother to the young can be considerably influenced by the way in which offspring are reared and can be classified as:

- *Precocial:* The newborn animals leave their nests shortly after birth; they are well covered with hair and have developed senses, which enable survival in the absence of the mother; they are exposed to the risk of transmission of mycobacteria especially when they suck milk.

- *Altricial:* The newborn animals stay in the nests where they were born for a longer time and are exposed to a high risk of transmission of mycobacteria from the mother and the ambient environment.

These two types of young animal rearing do not, however, rule out the intrauterine vertical transmission of mycobacteria. This type of transmission has been relatively well recognised for the causative agent of paratuberculosis (Ayele et al., 2001). These types of young animal rearing cannot likewise decrease their infection rates if they are fed by their parents with mycobacteria containing feed (e.g. infected animals or animals that died of avian tuberculosis). In such cases, the risk of shedding the mycobacteria through faeces should be considered first and then the infection of the young animals, which will sooner or later die of the disease.

With regard to the spread of mycobacteria (especially causative agents of the diseases leading to a weakening or death of mammals), it is necessary to keep in mind that migrating animals are source of feed for carnivores and birds of prey (Section 6.6). These concentrate on the weak animals in a herd. When they are caught the causative agent of bovine tuberculosis, for instance, may spread from infected buffalos and antelopes to lions in South Africa (Thoen et al., 2006b).

Mammals inhabit almost all components of the environment together with many ESM and PPM; the environment can also be contaminated with OPM, e.g. *M. bovis*. According to the environment inhabited by mammals, they can be classified as mammals living a substantial part of their lives in water, underground and on the land, in trees or in the air.

Accordingly, in the following sections of this chapter we will describe the importance of different mammalian species for the ecology of mycobacteria in the environment, which they permanently inhabit or just visit temporarily.

6.7.1 Mammals Living in Water for a Substantial Part of Their Lives

The sea mammals spend a substantial part of their lives in water; some species such as the common seal or sea lions generally only emerge from the water to breed in the sun. The females deliver and nurse the young on the land. Little is known about their importance in the ecology of mycobacteria. Only a small number of publications deal with mycobacteria in different water mammals: generalised mycobacteriosis caused by *M. smegmatis* in a California sea lion (*Zalophus californicus*; Gutter et al., 1987), cutaneous mycobacteriosis caused by *M. fortuitum* in the Southern sea lion *Otaria flavescens* (Lewis, 1987), systemic *M. marinum* infection in the Amazon manatee (*Trichechus inunguis*; Morales et al., 1985), mycobacteriosis caused by *M. marinum*, *M. fortuitum* and *M. kansasii* in two captive Florida manatees, *Trichechus manatus latirostris* (Sato et al., 2003) and cutaneous mycobacteriosis caused by *M. fortuitum* in the harbour seal (*Phoca vitulina concolor*; Wells et al., 1990).

Detection of serum antibodies against *M. marinum*, *M. fortuitum* and *M. chelonae* in Atlantic bottlenose dolphins (*Tursiops truncatus*) in the USA attests to the contact of water mammals with PPM (Beck and Rice, 2003). However, whether or not mycobacteria are present in their intestinal tract and how important the spread of mycobacteria in aquatic environments is are questions that remain to be answered.

The detection of *MTC* members in different water mammal species over the last few decades can be considered as important from the point of view of the potential infection of other animals including man (Photo 6.38). One of the first instances of detection published in the available literature is that of *MTC* from the lung or lymph node lesions of six captive seals in Australia between January and September 1986 (Cousins et al., 1990). Since that time, *MTC* members have been detected on other continents, both from animals in captivity (Forshaw and Phelps, 1991; Thorel et al., 1998; Thoen et al., 2006b) and in the wild (Cousins et al., 1993; Bernardelli et al., 1996). Based on molecular analysis, some isolates were designated as a new *MTC* member *M. pinnipedii* (Cousins etal., 2003).

Little is known at present about the ecology of *MTC* members in the water mammals. It is likely that these water mammals become infected in coastal waters where sewage from cities, especially from slaughterhouses, is discharged. If animals affected by tuberculosis are slaughtered there, the water animals become infected when ingesting the organic remnants infected with *MTC* members. The horizontal or vertical transmission of *MTC* members between free-living animals remains to be explored (Photo 6.39).

6.7.2 Mammals Living a Substantial Part of Their Lives Underground and on the Land

Most mammals spend a substantial part of their lives on the land or underground. Many mammalian species inhabit both these environments, where they are often in contact with ESM, PPM and possibly also with OPM.

6.7.2.1 Insectivores

Insectivores (Insectivora) live both underground and on the land. They largely feed on invertebrate animals and small vertebrates; this is reflected in the range of mycobacterial species detected in them. Even though over 370 species of insectivores that live all over the world have been described, the detection of mycobacteria in these animals has been reported only very rarely (Table 6.17).

The isolation of the causative agent of avian tuberculosis *M. a. avium* from a hedgehog (western European hedgehog; *Erinaceus europaeus*) and from a European shrew (*Sorex araneus*; Table 6.17) is worthy of special mention. Due to the fact that these insectivores occasionally feed on small vertebrates (e.g. dead birds) as well as invertebrates, this route of infection is most likely. They can also become infected through this route when eating dead animals infected with *M. bovis* (Lugton et al., 1995; Ragg et al., 2000). Infection caused by *M. marinum* in a western European hedgehog (*Erinaceus europaeus*) kept in captivity has also been described. This animal was kept in an empty aquarium where fish were previously kept. The source of infection was probably the contaminated walls of the aquarium because *M. marinum* is often found in fish (Section 6.5.1).

The detections of *M. chelonae* and *M. vaccae* in two species of the family Erinaceidae (gymnures and hedgehogs; Table 6.17) can be explained by the occurrence of these mycobacterial species in soil (Section 5.5), earthworms (Section 6.3.1), arthropods (Section 6.3.4) and insects (Section 6.3.5). Due to the fact that these invertebrates often constitute food for the insectivores, mycobacteria could be most likely spread from contaminated soil through this route.

The *MTC* member *M. microti*, which was detected in the European shrew (*S. araneus*) in 1946 in Great Britain (Lapage, 1947; cited by Stunkard et al., 1975), can also be found in the insectivores. Invertebrate animals infected with the causative agent of bovine tuberculosis were likely the sources of *M. bovis* for two (1.2%) of 162 moles in the UK (Table 6.17).

Even though the literature data concerning the occurrence of mycobacteria in the insectivores are scarce, these can be considered as potential vectors of mycobacterial species with consequences for the health of domestic animals under specific conditions. Viewed in this light, the migration of insectivores at times of food shortage (in the Northern hemisphere during the autumn months) to pig and cattle stables can help to clarify some unexplained cases of avian tuberculosis in domestic animals (I. Pavlik, unpublished data). Accordingly, when preventive measures against the causative agents of mycobacterial infections are imposed on domestic animal farms, it is necessary to consider these risks, which can be reduced by regular rodent control.

6.7.2.2 Lagomorphs

A variety of lagomorphs (especially rabbits) are often used in different biological experiments, e.g. with the causative agent of paratuberculosis (Hirch, 1956; Rankin, 1958; Mokresh et al., 1989; Mokresh and Butler, 1990; Mondal and Sinha, 1992; Mondal et al., 1994; Manabe et al., 2003; Vaughan et al., 2005), *MAC* (Collins et al., 1983; Emori et al., 1993) or *MTC* (Dorman et al., 2004) members. Rabbits have also been experimentally infected with *M. lepraemurium* (Collins et al., 1983). Despite this, very little is known about the ecology of ESM and PPM in other lagomorphs.

The results concerning free-living animals mainly deal with pathogenic species, i.e. *M. a. paratuberculosis*, *M. bovis* and *M. a. avium* (Table 6.18); data concerning other mycobacterial species found in the available literature are scarce. With respect to the invasive capability of some species of Lagomorpha (e.g. wild rabbits), knowledge of their role in the spread of mycobacteria is important.

Free-Living Rabbits and Mycobacteria

A number of recent publications have focused on the transmission of primarily the causative agent of

Table 6.17 Isolation of mycobacteria from different insectivores

Family	Species	Isolates	Comments	References
Gymnures and hedgehogs (*Erinaceidae*)	Western European hedgehogs (*Erinaceus europaeus*)	*MAA* (2)	Isolation from mesenteric lymph nodes of one of five free-living, apparently healthy animals in the Berkshire Downs (UK)	[4]
	Western European hedgehogs (*Erinaceus europaeus*)	*M. marinum*, many CFU were isolated	Otodermatitis was diagnosed in one animal purchased 3.5 months earlier, which had been kept in a large fish tank for 7 weeks (its diet consisted of moist dog food and vegetables). After 18 months, exhausted animal died with confirmed *M. marinum* infection in cervical lesions and lungs (New York, USA)	[6]
	Western European hedgehogs (*Erinaceus europaeus*)	*M. bovis*	Four animals with gross pulmonary lesions	[3]
	Western European hedgehogs (*Erinaceus europaeus*)	*M.* sp.	One of three examined animals trapped in the vicinity of the sawdust stack in one pig breeding herd	[7]
Shrews (*Soricidae*)	Lesser shrew (*Crocidura suaveolens*)	*M. chelonae*, 1 CFU was isolated	One of 22 apparently healthy animals from animal-free house (shed, yard) was infected in liver without gross lesions (Brno, Czech Republic)	[2]
	European shrew (*Sorex araneus*)	*M. vaccae*, 1 CFU was isolated	One of four apparently healthy animals from animal-free house (shed, yard) was infected (1 CFU isolated) in organs from abdominal cavity besides the liver (Brno, Czech Republic)	[2]
	European shrew (*Sorex araneus*)	*MAA* (1), 1 CFU was isolated	One of two apparently healthy animals from a road at the edge of a forest further than 1 km from houses and animal barn was infected in organs from abdominal cavity besides the liver (Brno, Czech Republic)	[2]
	European shrew (*Sorex araneus*)	*M. microti*	By *M. microti* caused tuberculous lesions in lungs, spleen, liver, kidney and lymph nodes in 8 (1.5%) of 550 animals	[5][1]
Moles (*Talpidae*)	Mole (*Talpa europaea*)	*M. bovis*	Two (1.2%) of 162 examined animals in the UK during the years 1971–1996	[1]
	Mole (*Talpa europaea*)	*MAA* (2), *MAH* (4), *M.* sp.	Three of 10 examined animals trapped in the vicinity of the sawdust stack in one pig breeding herd	[7]

[1] Lapage 1947, as quoted by [5].

MAA M. avium subsp. *avium*. *MAH M. avium* subsp. *hominissuis*. (2) Serotype 2. **CFU** Colony Forming Units.

[1] Delahay RJ, Cheeseman CL, Clifton-Hadley RS (2001) Tuberculosis.(Edinb.) 81:43–49. [2] Fischer O, Matlova L, Bartl J, Dvorska L, Melicharek I, Pavlik I (2000) Folia Microbiol.(Praha) 45:147–152. [3] Lugton IW, Johnstone AC, Morris RS (1995) N. Z. Vet. J. 43:342–345. [4] Matthews PR, McDiarmid A (1977) Res. Vet. Sci. 22:388. [5] Stunkard JA, Migaki G, Robinson FR, Christian J (1975) Lab Anim Sci. 25:723–734. [6] Tappe JP, Weitzman I, Liu S, Dolensek EP, Karp D (1983) J. Am. Vet. Med. Assoc. 183:1280–1281. [7] Windsor RS, Durrant DS, Burn KJ (1984) Vet. Rec. 114:497–500.

Table 6.18 Isolation of mycobacteria from spontaneously infected lagomorphs

Species	Country	Isolates	Comments	References
Rabbit (*Oryctolagus*	Scotland	*MAP*	Diagnosed histologically only	[1]
cuniculus)	Scotland	*MAP*	Present also in faeces and urine	[4]
	Scotland	*MAP*	With histopathological lesions	[8]
	Scotland	*MAP*	Present in testes, uterus, placenta, faeces and milk	[7]
	Scotland	*MAP*	With histopathological lesions also	[6]
	New Zealand	*M. bovis*	Tuberculous lesions in organs	[5]
Rabbits[1]	USA	*MAP*	Faeces were not identified according to animal species	[11]
European hare (*Lepus europaeus*)	Scotland	*MAP*	Without histopathological lesions	[2]
	Czech Republic	*MAP*	Without gross lesions	[9]
	England	*MAA* (2) *M.* sp.	Four animals infected One animal infected	[10]
European hare (*Lepus europaeus* f. *occidentalis*)	New Zealand	*M. bovis*	With histopathological lesions	[3]

[1] Faecal samples (pellets) were collected from eastern cottontail rabbit (*Sylvilagus floridanus*) and white-tailed jackrabbit (*Lepus townsendii*).
MAP *M. a. paratuberculosis*. **MAA** *M.* a. *avium*. (**2**) Serotype 2.
[1] Angus KW (1990) J.Comp Pathol. 103:101–105. [2] Beard PM, Daniels MJ, Henderson D, Pirie A, Rudge K, Buxton D, Rhind S, Greig A, Hutchings MR, McKendrick I, Stevenson K, Sharp JM (2001) J. Clin. Microbiol. 39:1517–1521. [3] Cooke MM, Jackson R, Coleman JD (1993) N.Z. Vet. J. 41:144–146. [4] Daniels MJ, Henderson D, Greig A, Stevenson K, Sharp JM, Hutchings MR (2003) Epidemiol. Infect. 130:553–559. [5] Gill JW, Jackson R (1993) N.Z. Vet. J. 41:147. [6] Greig A, Stevenson K, Henderson D, Perez V, Hughes V, Pavlik I, Hines ME, McKendrick I, Sharp JM (1999) J. Clin. Microbiol. 37:1746–1751. [7] Judge J, Kyriazakis I, Greig A, Allcroft DJ, Hutchings MR (2005) Appl. Environ. Microbiol. 71:6033–6038. [8] Judge J, Kyriazakis I, Greig A, Allcroft DJ, Hutchings MR (2005) Appl. Environ. Microbiol. 71:6033–6038. [9] Machackova M, Svastova P, Lamka J, Parmova I, Liska V, Smolik J, Fischer OA, Pavlik I (2004) Vet. Microbiol. 101:225–234. [10] Matthews PR, Sargent A (1977) Br. Vet. J. 133:399–404. [11] Raizman EA, Wells SJ, Jordan PA, DelGiudice GD, Bey RR (2005) Can. J. Vet. Res. 69:32–38.

paratuberculosis from free-living rabbits to ruminants (especially grazing cattle). Rabbit faeces ingested by cattle together with feed may become a dangerous source of, in particular, *M. a. paratuberculosis* (Daniels et al., 2001; Daniels and Hutchings, 2001; Daniels et al., 2003; Judge et al., 2005; Judge et al., 2006). However, little is known about the importance of rabbit faeces for the survival of *M. a. paratuberculosis* in the environment. The occurrence of *M. a. paratuberculosis* poses a problem because their *in vitro* culture detection either fails or their incubation time necessary for their detection may be several months longer than that of isolotes detected in cattle. The matching identity of RFLP types of *in vitro* growing isolates found both in rabbits and domestic ruminants has also been confirmed by the IS*900* RFLP method (Greig et al., 1999).

No ecological study dealing with *M. a. paratuberculosis* survival in faeces (pellets), urine and other excretions or secretions of rabbits has been performed yet. It has not been determined how long *M. a. paratuberculosis* present in the above-mentioned biological materials is virulent for a variety of host (above all ruminants). Information on the ecology of different phyla of *M. a. paratuberculosis*, which do not grow *in vitro* and are found in rabbits, is also missing.

Free-Living Hares and Mycobacteria

Even though a variety of leporid species (family Leporidae) are widely spread all over the world, the published data concerning the occurrence of mycobacteria in these animals are scarce (Table 6.18). The detection of *M. bovis* in the European hare (*Lepus europaeus* f. *occidentalis*) should be also viewed as noteworthy (Cooke et al., 1993).

In the Czech Republic, molecular epidemiology was investigated in a herd of beef cattle infected with

M. a. paratuberculosis. A different RFLP type B-C9 was detected in an isolate from the large intestine of one hare, which was not detected in a previous study performed on rabbits and domestic ruminants in Scotland. The RFLP type B-C1, meanwhile, was identified in cattle, in the larvae of syrphid flies (*E. tenax*) in a slurry reservoir, in soil on pastures and in a caught roe deer (*Capreolus capreolus*; Machackova et al., 2004).

It was thought that *M. a. paratuberculosis* of RFLP type B-C9 could circulate among the free-living hares, ruminants or wild boar living in this mentioned area. In the study, this RFLP type B-C9 was also detected in the free-living red deer despite the fact that the RFLP type B-C1 was detected in an infected heifer living free for more than 8 months. This RFLP type was also found in the free-living red deer and roe deer which shared the pastures and most likely met at the feed-troughs at times of pasture shortage. It was later found that several red deer (*Cervus elaphus*), had been inhabiting this area for several months; these escaped from a farm of cervids in the area. This herd came from an area where this very RFLP type B-C9 was found in grazing cattle (Pavlik et al., 2000a). However, the source of infection for this hare infected with *M. a. paratuberculosis* of RFLP type B-C9 could not be revealed even by subsequent monitoring of paratuberculosis in the free-living ruminants across the entire territory of the Czech Republic including this area (M. Kopecna, J. Lamka, I. Pavlik, unpublished data).

Views and Perspectives on the Research

After introduction of culture-independent detection techniques for various mycobacterial species, it will be possible to perform various ecological studies in Lagomorpha. These will surely help to explain the circulation of *M. a. paratuberculosis* in, e.g., rabbit burrows, in the surrounding soil, groundwater, roots of plants in the neighbourhood, etc.

From the point of view of hosts, i.e. organisms of various Lagomorpha species, the role of coprophagy has not yet been explained despite the fact that it is an important part of their physiology (Hirakawa, 2001). It is possible that coprophagy and repeated "re-infection" of the hosts with *M. a. paratuberculosis* alone may lead to the changes in its virulence, capability of *in vitro* growth, etc.

Therefore, to clarify the spread of *M. a. paratuberculosis* to various hosts, above all predators including man, it will be suitable to focus on the investigation of the distribution of the causative agent of paratuberculosis in the organisms of different species of Lagomorpha. Such studies have already been performed in wild rabbits (Table 6.18), but have not yet been conducted in other species of Lagomorpha that are present in abundance on pastures with grazing cattle. Their meat is considered as a culinary delicacy in many countries.

6.7.2.3 Rodents

Even though rodents are the most species-rich order of mammals and various species of rodents are used as laboratory animals, knowledge of the occurrence of mycobacteria in free-living animals and their importance in the ecology of mycobacteria is scarce. Attention has been concentrated only on the study of the causative agent of tuberculosis in free-living small terrestrial mammals. It was described for the first time in Great Britain (Wells and Oxon, 1937) and the causative agent was later designated as *M. tuberculosis* subsp. *muris* (Reed, 1957). Subsequently, the causative agent was renamed *M. microti* (Wayne and Kubica, 1986) and classified as an *MTC* member (Frota et al., 2004).

M. microti was isolated successively from different species of rodents (Table 6.19). It has been supposed since the 1990s that small terrestrial mammals (especially rodents) come into contact with *M. microti* through the consumption of contaminated food of vegetable and animal origin (Chitty, 1954). Cannibalism is definitely an important source of infection for other animals in the population as well. However, with respect to the findings of tuberculous lesions in skin and subcutis, their transmission via injured skin is highly probable (Cavanagh et al., 2002; I. Pavlik, M. Heroldova, M. Skoric, unpublished data).

Due to the difficult *in vitro* isolation of the causative agent of tuberculosis *M. microti*, little is known about its occurrence in small terrestrial mammals. However, it may be more widely spread within the animal and human populations than has been assumed so far. The detection of *M. microti* in animal species other than rodents attests to that (Table 6.20). The sources of their infection may be both the infected rodents and insectivores, and perhaps some components of the

Table 6.19   Isolation of mycobacteria from spontaneously infected free-living rodents

Species	Country	Isolates	Comments	References
Field vole (*Microtus agrestis*)	UK	*M. microti*	Analysis of data from Dr. A. Q. Wells (1933–1938) concerning 952 animals	[3]
Field vole (*Microtus agrestis*) and bank vole (*Clethrionomys glareolus*)	UK	*M. microti*	About 13 (7.2%) of 180 voles had clinical signs of tuberculosis; dissection revealed further 25 (13.9%) voles with caseous abscesses under the skin or in the abdomen	[2]
Vole	NL	*M. microti*	Five isolates identified as genotype "Vole"	[6]
Common vole (*Microtus arvalis*)	CR	*M. chelonae* 1 CFU, *M. fortuitum* 1 CFU	Two of eight apparently healthy animals were infected; one in abdominal organs* and one in liver without gross lesions	[5]
Common vole (*Microtus arvalis*)	CR	*MAP* (3 CFU)	One of 35 animals was positive in intestine on infected cattle farm with *MAP*	[8]
Yellow-necked mouse (*Apodemus flavicollis*)	CR	*M.* sp. 1 and 4 CFU, *MAA* (1) 10 CFU	Three of 15 apparently healthy animals were infected in abdominal organs* without gross lesions	[5]
Woodmouse (*Apodemus sylvaticus*)	SC	*MAP*	Trapped in areas with cattle herds infected with *MAP*	[1]
Lesser shrew (*Crocidura suaveolens*)	CR	*MAP*	One of four animals was positive in intestine on infected cattle farm with *MAP*	[8]
House mouse (*Mus musculus*)	CR	*M.* sp. 1 CFU	One of 33 apparently healthy animals was infected in abdominal organs* without gross lesions	[5]
House mouse (*Mus musculus*)	CR	*M. fortuitum*	One infected animal on pig farm with mycobacteriosis caused by *M. fortuitum*	[9]
Braun rat (*Rattus norvegicus*)	CR	*MAP*	Trapped in infected cattle herd with *MAP*	[7]
Braun rat (*Rattus norvegicus*)	SC	*MAP*	Trapped in areas with infected cattle herds with *MAP*	[1]
Braun rat (*Rattus norvegicus*)	UK	*M. bovis*	Five (1.2%) of 412 trapped animals in cattle herds infected with *M. bovis*	[4]

* Without liver. **CFU** Colony Forming Units. **MAA (1)** *M. a. avium* of serotype 1. **MAP** *M. a. paratuberculosis*. **NL** The Netherlands. **CR** Czech Republic. **SC** Scotland.

[1] Beard PM, Daniels MJ, Henderson D, Pirie A, Rudge K, Buxton D, Rhind S, Greig A, Hutchings MR, McKendrick I, Stevenson K, Sharp JM (2001) J. Clin. Microbiol. 39:1517–1521. [2] Cavanagh HM, Hipwell M, Wilkinson JM (2003) J. Med. Food 6:57–61. [3] Chitty D (1954) Ecology 35:227–237. [4] Delahay RJ, Cheeseman CL, Clifton–Hadley RS (2001) Tuberculosis.(Edinb.) 81:43–49. [5] Fischer O, Matlova L, Bartl J, Dvorska L, Melicharek I, Pavlik I (2000) Folia Microbiol.(Praha) 45:147–152. [6] Kremer K, van Soolingen D, van Embden J, Hughes S, Inwald J, Hewinson G (1998) J. Clin. Microbiol. 36:2793–2794. [7] Pavlik I, Alexa M, Ayele W (2008) Unpublished data. [8] Pavlik I, Horvathova A (2008) Unpublished data. [9] Pavlik I, Matlova L, Gilar M, Bartl J, Parmova I, Lysak F, Alexa M, Dvorska-Bartosova L, Svec V, Vrbas V, Horvathova A (2007) Veterinarni Medicina 52:392–404.

environment that are unknown at present. With regard to the potential infection of people, *M. microti* was long viewed as a causative agent with low virulence for man. Due to that fact, along with *M. bovis* BCG some attenuated strains of *M. microti* (M-vaccine or Wells vaccine) were used for the vaccination of children (Sula and Radkovsky, 1976; Manabe et al., 2002). The further development of molecular methods have meant that the identification of the *MTC* including *M. microti* has been more successful during the last decade. It has revealed that various forms of human infection caused by *M. microti* are found in both immunosuppressed and in immunocompetent patients (Table 6.21).

Among other mycobacterial species, *M. bovis* and various ESM and PPM have been isolated from different naturally infected free-living rodent species. The detection of the causative agent of avian tuberculosis *M. a. avium* and the causative agent of paratuberculosis *M. a. paratuberculosis* (Table 6.19) is noteworthy. However, their occurrence will most likely be of little importance as regards epizootiology in contrast to the occurrence of, e.g., *M. a. paratuberculosis* in rabbits in Scotland (Section 6.7.2.2).

Mycobacterial infections have been detected only sporadically in rodents kept in captivity. One of the rare publications describes disseminated mycobacteriosis

Table 6.20 Isolation of *M. microti* from spontaneously infected animals other than small terrestrial mammals

Animals species	Country	Comments	References
Charolais-Hereford crossbred bull	UK	Small miliary lesions observed in lungs in slaughterhouse	[5]
Cow	UK	One isolate identified as genotype "Vole"	[6]
Domestic cat	NL	None	[4]
	UK	Eight isolates identified: two as genotype "Vole" and six as genotype "Llama"	[6]
Dog (4-year-old Beauceron)	F	Acute peritonitis with abscesses in spleen, liver and mesenterium surfaces caused by genotype "Llama"	[2]
Badger (*Meles meles*)	UK	One isolate identified as genotype "Vole"	[6]
Ferret (*Mustela putorius* f. *furo*)	NL	Isolated in 1993 from a ferret kept as a household pet in the western part of the Netherlands (genotype "Vole")	[6], [10]
Domestic pig	NL	Isolated in 1965 in the southern part of the Netherlands (genotype "Vole")	[10]
	UK	One isolate identified as genotype "Vole"	[6]
	NL	None	[3]
	UK	Calcified granuloma lesions in submandibular lymph nodes	[9]
Llama (*Lama glama*)	SW	In two captive animals multiple yellowish, confluent, caseous nodules were found in lungs, liver, spleen and hepatic, mesenteric and bronchial lymph nodes	[7]
Llama	NL	One isolate identified as genotype "Llama"	[6]
		Generalised caseous tuberculosis in different organs	[8]
	BE	Isolated in 1968 from llama born in the Antwerp zoo (genotype "Llama")	[10]
Cape rock hyrax (*Procavia capensis*)	SA	Isolated in 1958 (dassie bacillus, genotype "Llama")	[10]
	SA	Lung infection in imported animal to Australia	[1]

UK United Kingdom. **NL** The Netherlands. **F** France. **SW** Switzerland. **BE** Belgium. **SA** South Africa.

[1] Cousins DV, Peet RL, Gaynor WT, Williams SN, Gow BL (1994) Vet. Microbiol. 42:135–145. [2] Deforges L, Boulouis HJ, Thibaud JL, Boulouha L, Sougakoff W, Blot S, Hewinson G, Truffot-Pernot C, Haddad N (2004) Vet. Microbiol. 103:249–253. [3] Huitema H, Jaartsve FH (1967) Antonie Van Leeuwenhoek J. Microbiol. 33:209–&. [4] Huitema H, van VLOT (1960) Antonie Van Leeuwenhoek 26:235–240 [5] Jahans K, Palmer S, Inwald J, Brown J, Abayakoon S (2004) Vet. Rec. 155:373–374. [6] Kremer K, van Soolingen D, van Embden J, Hughes S, Inwald J, Hewinson G (1998) J. Clin. Microbiol. 36:2793–2794. [7] Oevermann A, Pfyffer GE, Zanolari P, Meylan M, Robert N (2004) J. Clin. Microbiol. 42:1818–1821. [8] Pattyn SR, Antoine-Portaels F, Kageruka P, Gigase P (1970) Acta Zool. Pathol. Antverp. 51:17–24. [9] Taylor C, Jahans K, Palmer S, Okker M, Brown J, Steer K (2006) Vet. Rec. 159:59–60. [10] van Soolingen D, van der Zanden AG, de Haas PE, Noordhoek GT, Kiers A, Foudraine NA, Portaels F, Kolk AH, Kremer K, van Embden JD (1998) J. Clin. Microbiol. 36:1840–1845.

caused by *M. chelonae*. In one breed of golden hamsters, the paws were primarily affected, followed by the dissemination of infection into the liver, spleen and lungs (Karbe, 1987). In one breed, the tails of immunosuppressed laboratory mice were affected (Mahler and Jelinek, 2000).

Views and Perspectives on the Research

Regarding the extremely fecund invasive species of rodents, it will be useful to perform ecological studies that would be able to reveal the circulation of *M. microti* in different components of the environment and to clarify the infections of other animals including people. *M. microti* can also be spread by domestic cats and ferrets that catch wild rodents (Baxby et al., 1994).

Above all else the introduction of culture-independent methods can help elucidate the circulation of ESM and PPM in the environment and their potential transport through the intestinal tract of rodents. We can suppose that mycobacteria swallowed by small terrestrial mammals pass through their digestive tract undamaged and are shed via faeces. It is highly probable that PPM can survive for a long time in the intestinal tract or other tissues and thus they can be transmitted across great distances into different niches

Table 6.21 *M. microti* infection in humans

Country	Genotype	Patients[1]	Comments	References
G	Vole	69/M	*Sepsis tuberculosa acutissima* (immunocompetence)	[2]
	Llama	48/M	Pulmonary tuberculosis (*HIV*+ white man)	[3]
	Vole	53/M	Pulmonary tuberculosis (*HIV*–)	[5]
	Llama	58/M	Pulmonary tuberculosis (*HIV*–)	[5]
NL	Llama	1 patient	None	[4]
	Vole	4 patients	None	[4]
	Vole	12/?	Alveolar lavage fluid of kidney transplantation patient	[6]
	Vole	41/?	Perfusion fluid of dialysis patient	[6]
	Vole	34/M	Lung biopsy (a persistent cough and undefined abnormalities on the chest X ray – immunocompetence)	[6]
	Vole	39/M	Sputum (*HIV*+)	[6]
USA	Nk	39/M	Pulmonary tuberculosis (*HIV*+, homosexual patient)	[1]

[1] Age in years/gender (sex): **M** (male), ? (not known).
G Germany. **NL** The Netherlands. **USA** United States of America. **Nk** Not known.
[1] Foudraine NA, van Soolingen D, Noordhoek GT, Reiss P (1998) Clin. Infect. Dis. 27:1543–1544. [2] Geiss HK, Feldhues R, Niemann S, Nolte O, Rieker R (2005) Infection 33:393–396. [3] Horstkotte MA, Sobottka I, Schewe CK, Schafer P, Laufs R, Rusch-Gerdes S, Niemann S (2001) J. Clin. Microbiol. 39:406–407. [4] Kremer K, van Soolingen D, van Embden J, Hughes S, Inwald J, Hewinson G (1998) J. Clin. Microbiol. 36:2793–2794. [5] Niemann S, Richter E, Dalugge-Tamm H, Schlesinger H, Graupner D, Konigstein B, Gurath G, Greinert U, Rusch-Gerdes S (2000) Emerg. Infect. Dis. 6:539–542. [6] van Soolingen D, van der Zanden AGM, de Haas PEW, Noordhoek GT, Kiers A, Foudraine NA, Portaels F, Kolk AHJ, Kremer K, van Embden JDA (1998) J. Clin. Microbiol. 36:1840–1845.

within the environment. Due to the fact that rodents also become important source of food for different predators (mammals, reptiles, birds, etc.) mycobacteria can spread in this way quite quickly. Other causative agents of different serious diseases with zoonotic potential can also be spread by small terrestrial mammals (Skoric et al., 2007).

Free-living small terrestrial mammals including rodents may also be affected by leprosy caused by *M. lepraemurium* (Krakower and Gonzalez, 1937; Shepard, 1960; Johnstone, 1987; Rojas-Espinosa and Lovik, 2001). The ecology of the causative agent of this disease remains to be explored.

6.7.2.4 Omnivores

Omnivores are species of animals that feed on both plants and animals. Examples of omnivores are organisms from the order of even-toed ungulates (Artiodactyla) – wild boar (*Sus scrofa*), from the order of carnivores (Carnivora) – almost all species of bears and from the order of primates (Primates), e.g. chimpanzees (*Pan troglodytes*) and man (*Homo sapiens*).

Wild boar can be considered at present as a "successful" animal species owing to their adaptability to the rapidly changing life environment. Information on the occurrence of OPM, PPM and ESM in wild boars in a number of countries on different continents can be obtained from published review articles and books (Little et al., 1982; de Lisle et al., 2001; Coleman and Cooke, 2001; Machackova et al., 2003; Thoen et al., 2006b).

Particular attention is drawn to the wild boar primarily due to their significance in the spread of causative agents of bovine tuberculosis (Woodford, 1982; Tweddle and Livingstone, 1994; McInerney et al., 1995; Aranaz et al., 1996; Serraino et al., 1999; Bollo et al., 2000; Coleman and Cooke, 2001; Nugent et al., 2001; Prodinger et al., 2002; Nugent et al., 2002; Parra et al., 2003; Aranaz et al., 2004; Prodinger et al., 2005; Parra et al., 2006), paratuberculosis (Machackova et al., 2003; Alvarez et al., 2005) and avian tuberculosis (Trcka et al., 2006; Photo 5.34). Moreover, genetic resistance against bovine tuberculosis has been investigated in wild boar (Acevedo-Whitehouse et al., 2005). However, there has been a lack of studies on other mycobacterial species (Corner et al., 1981; Trcka et al., 2006).

With respect to the ecology of mycobacteria, it is necessary to note that wild boar are omnivorous animals and, hence, they do not eat only plants, but also attack bird nests, catch small vertebrates or weak animals and the young of large animal species (hares, roe deer, etc.). They also feed on carcasses of free-living animals and sometimes those of dead animals held in captivity and inadequately disposed of. They will not refuse animal remnants that are not suitable for human consumption (e.g. entrails and intestines) and can be found on the waste dumps outside human residences. All these materials can be infected or contaminated with OPM, PPM and ESM. Wild boars also like to lie in mud pools and morasses in different parts of the environment (Photo 6.40). It follows that they will spread mycobacteria throughout the environment not only via their excrements (above all faeces and urine) but also on their body surface. This has been confirmed by the detection of different PPM and ESM species in their gut contents and faeces, intestinal mucosa and mesenteric lymph nodes (Corner et al., 1981; Trcka et al., 2006).

The considerable migratory activities of wild boar are significant with regard to epizootiology and epidemiology. They can also overcome different natural barriers including broad rivers (Machackova et al., 2003). However, wild boars migrate if there is a lack of food or they are disturbed (especially at times when they are hunted, in the mushroom-picking season, when forest fruits ripen and are collected by man, etc.), as documented in New Zealand (Nugent et al., 2002) and the Czech Republic (R. Urbanec, personal communication).

The faeces of wild boar, rich in remnants of organic materials, also represent a significant source of nutrients for mycobacteria. There is little information available regarding the survival of mycobacteria in sumps contaminated with wild boar faeces as well as on the survival and propagation of causative agents of bovine tuberculosis, paratuberculosis and avian tuberculosis in the environment. The significance of omnivores has not yet been adequately investigated with regard to ecology and deserves attention in the future.

6.7.2.5 Herbivores

Herbivores, especially all ruminants (cattle, sheep, goats, red deer and antelopes; suborder ruminants

(Ruminantia), order Artiodactyla, are the most economically significant animals and the most important vector of mycobacteria. Particular attention has been paid to the occurrence of the causative agents of bovine tuberculosis or paratuberculosis in different publications and books (de Lisle et al., 2001; Cousins and Roberts, 2001; Manning and Collins, 2001; Kennedy and Benedictus, 2001; Pavlik, 2006; Thoen et al., 2006a). Less attention, however, has been paid to infections of ruminants caused by *M. a. avium* or PPM (Quigley et al., 1997; O'Grady et al., 2000; Miller et al., 2002; Pavlik et al., 2002; Machackova et al., 2004; Dvorska et al., 2004; Pavlik et al., 2005; Machackova-Kopecna et al., 2005). The latter mycobacterial species have been investigated primarily in relation to their occurrence in faeces and lymph nodes in, e.g., cattle giving non-specific responses to bovine tuberculin testing (Sections 5.10 and 7.4).

The occurrence of mycobacterial species other than *MTC* has on rare occasions been investigated in other ruminants (primarily in Cervidae from game preserves). *M. kansasii* and *M. a. silvaticum* have been diagnosed in, e.g., llama (Johnson et al., 1993) and in a roe deer (Jorgensen and Clausen, 1976). As far as the ecology of mycobacteria is concerned, publications dealing with PPM occurrence in fallow deer and red deer faeces can be considered as noteworthy (Weber, 1982; Weber and Gurke, 1992).

The purpose of the present chapter is to review the data documenting the significance of herbivores in the ecology of mycobacteria. Herbivores live in environments with rich vegetation, which form the base of their diet. This environment, including plants, may be contaminated with mycobacteria. Mycobacteria usually come into contact with plant surfaces together with rain water (Section 5.2), dust (Section 5.9), soil (Section 5.5) or organic fertilisers (Section 5.10.3). Moreover, root systems are abundant in mycobacteria that can also penetrate root tissues (Sections 2.1 and 5.6).

Why are herbivores considered to be the most significant animals in the ecology of mycobacteria (Photo 6.41)? This question can be answered after consideration of the following factors:

i. Herbivores live in an ecosystem where plants that are often contaminated with mycobacteria grow (see above).
ii. Mycobacteria can survive the passage through the intestinal tract.

iii. Some mycobacterial species can also propagate in the intestinal tract of herbivores.

iv. The faeces of herbivores are a significant source of organic and inorganic matter, which contribute to the occurrence of soil and water microflora including mycobacteria in the environment.

v. Due to the seasonal vegetative cycles and shortages of food, up to 1.5–2.0 million animals of different herbivorous species migrate in search of food, e.g. in East Africa.

All the above-mentioned factors consequently increase the occurrence and the range of mycobacterial species present in the faeces of herbivores (Section 5.10). From the point of view of nutrition, herbivores can be classified as species that live:

i. Permanently in an aquatic environment, i.e. in the oceans and the seas. The members of order Sirenia, in which mycobacteria have been detected, are the Amazon manatee (*Trichechus inunguis*) and the Florida manatee (*Trichechus manatus latirostris;* Section 6.7.1).

ii. Temporarily in an aquatic environment, especially the hippopotamus (*Hippopotamus amphibius*), classified as a non-ruminant mammal in order Artiodactyla.

iii. In treetops, eating foliage; mostly various species of the order Primates, e.g. langurs and colobuses (family Cercopithecidae).

iv. On the land's surface; these herbivores can be classified according to their nutrition as species that eat:

- Plants together with roots (mostly sheep and goats).
- Lichen and mosses in the areas of tundra (especially reindeer).
- The vegetation in mountain areas (in particular sheep, goats and free-living yaks).
- Grassland and pasture grass (above all cattle, the majority of antelope species and cervids).
- The foliage of shrubs and trees (some antelope species, cervids, goats and giraffes).

Plants Together with Roots

There is a lack of understanding of the significance of domestic sheep and goats in the spread of mycobacte-

ria. As they graze on vegetation on pastures, these animals, due to the shape of their lips, can pull out plants together with their roots from the ground. The roots can be colonised by various mycobacterial species (Section 5.6) that can be subsequently spread over great distances through the faeces of sheep that migrate in search of new pastures.

Lichen and Mosses in the Areas of Tundra

There are also very little data on the occurrence of PPM and ESM in reindeer (*Rangifer tarandus*) and other herbivores that live on lichens and mosses in tundra areas. Moreover, in these areas, favourable conditions exist for mycobacteria, which can live in peat bogs and the water sediment, thus becoming sources of mycobacteriosis for humans, e.g. in Finland (Section 5.4).

The Vegetation in Mountainous Areas

In mountains, mycobacteria spread mainly in the direction of gravitational pull. Mycobacteria are gradually shifted by glaciers and flushed with water from melting snow and ice into valleys. They can only be moved in the opposite direction by wind that sweeps dust particles containing mycobacteria or by animals. Domestic sheep and goats commonly graze in all parts of mountains, even in places where the quality of vegetation is poor. Free-living ruminants can move to almost any area on a mountain on which vegetation is found.

Grassland and Pasture Grass

Due to the gradual deforestation of landscapes and the formation of grassland, herbivores have become the most important factors in the spread of mycobacteria. At present we can see the increasing significance of ruminants. Before man-made changes of the landscape, herds of, e.g., bison (*Bison bison*) in North America and various species of antelopes in Africa were the most numerous among the herds of large mammals existing on the Earth. Accordingly, it is not necessary to describe the significance of these vertebrates in the spread of mycobacteria in detail. On the other hand, the significance of the odd-toed ungulates is gradually decreasing and thus their significance for the spread of mycobacteria is also diminishing.

The Foliage of Shrubs and Trees

Mycobacteria are also found on the foliage of trees and shrubs; they gain access to these organisms in the same way as they do to plants. Owing to the hydrophobic quality of their cell wall, mycobacteria can adhere to the leaf surface. Herbivores that eat them (e.g. giraffes – the tallest living terrestrial vertebrates) can significantly participate in the spread of these mycobacterial species.

The routes of the spread of mycobacteria associated with the nutrition and way of life of herbivores can be summarised as follows:

  i. The most common route of the spread of mycobacteria is shedding through faeces and less frequently through urine (Section 5.10; Photos 6.41 and 6.42).
 ii. Spitting of the contents of their three-part stomachs by camels (suborder Tylopoda) when they defend themselves can be considered an unusual route for the spread of mycobacteria.
iii. Mycobacteria can be spread by herbivores through the contaminated surface of their bodies. In particular, members of species that temporarily inhabit aquatic environments (e.g. hippopotamuses and elephants) can spread mycobacteria from their skin to various water reservoirs such as watering places, lakes and rivers. The skin and hair of herbivores provide generally hospitable conditions for the survival and spread of mycobacteria to the surroundings (Photo 6.43).

Herbivores play the leading role in the circulation of mycobacteria in nature due to their high agility and their ability to migrate. These vertebrates can even traverse water reservoirs. Also, taking into consideration the above-mentioned routes of the spread of mycobacteria, the movement of herbivores can be categorised as follows:

- **Migration.** The significance of herbivores in the spread of mycobacteria consists in their ability to migrate over great distances. An example of such long-distance movement is the seasonal migration of antelopes in some parts of Africa to pastures found tens or hundreds of kilometres away. An example of migration over a short distance is the daily movement of different species within their home range.

- **Transport of herbivores by humans.** Mycobacteria are spread by herbivores to regions that they did not formerly inhabit mostly due to transport organised by humans, e.g. cattle, sheep, goats and horses imported to America, Australia, New Zealand and the Hawaiian Islands.

- **Sale and exchange of herbivores.** At present, the main risk factors for the transmission of mycobacteria are the transfer of breeding animals between zoological gardens (e.g. elephants, giraffes and rhinoceroses) and the sale of especially highly bred ruminants (above all various breeds of cattle and sheep are sold). These herbivores subsequently become involved in the ecology of mycobacteria through manure and dung water commonly used for fertilisation of agricultural soil.

Herbivores can also be involved in the spread and propagation of mycobacteria through their body remnants – cadavers. After being trapped by predators (or by human hunters), or after death from other causes, their bodies decompose or are eaten by carnivores and scavengers. In these cases, OPM or PPM play a significant role from the point of view of epizootiology and epidemiology owing to the fact that they can be transmitted to other hosts. These factors and risks will be discussed in Section 6.7.2.6, dealing with carnivores.

The spread of mycobacteria can be facilitated by high concentrations of animals, e.g. keeping herbivores at a high density in a small area (farmed cervids or herds kept in zoological gardens). This is a precondition for the survival and spread of mycobacteria. These animals heavily contaminate paddocks and pastures with their urine and faeces. Further study should be focused on the occurrence of OPM (e.g. *M. tuberculosis* in elephants; *M. bovis* in bison) under such conditions because it is unknown at present whether they can propagate outside the host organism.

Large herbivores (elephants, rhinoceroses and giraffes) held in captivity for labour, in circuses, and zoological gardens (Photo 6.44) are an example of the involvement of herbivores in the ecology of different mycobacterial species under new epizootiological and epidemiological conditions. There is a lack of published results on the occurrence of mycobacteria in these animals. No study has been undertaken to investigate the mycobacterial species profile in their faeces or urine.

Tuberculosis in elephants was first described more than 2000 years ago according to a review article by Montali et al. (2001). It follows, therefore, that *M. bovis* can be detected in these animals living in captivity (Payeur et al., 2002; Lyashchenko et al., 2006). However, attention has primarily been drawn to these animals because they are susceptible to *M. tuberculosis* as has been documented on different continents, e.g. Russia (Cherniak et al., 1963), UK (Garrod, 1875), USA (Thoen et al., 1977; Greenberg et al., 1981; Mann et al., 1981; Saunders, 1983; Furley, 1997; Mikota et al., 2001; Payeur et al., 2002; Lyashchenko et al., 2006), Sweden (Lewerin et al., 2005), Sri Lanka (Pinto et al., 1973) and Germany (von Benten et al., 1975).

In these epizootiological and epidemiological studies, *M. tuberculosis* was diagnosed both in the infected breeders, veterinarians, businessmen, etc. and in large herbivores, primarily elephants, rhinoceroses and giraffes in captivity (Cherniak et al., 1963; Greenberg et al., 1981; Michalak et al., 1998; Davis, 2001; Lewerin et al., 2005; Lyashchenko et al., 2006). The most thorough study of the spread of *M. tuberculosis* among large herbivores was performed in one zoological garden in Sweden using I6110 RFLP and spoligotyping methods. The infected animals (five elephants and one giraffe) were either in direct or indirect contact with one another. The transmission of *M. tuberculosis* from an elephant to a giraffe documented by Lewerin et al. (2005) can be viewed as one of the study's most noteworthy findings.

There are a scarcity of available literature data concerning other mycobacterial species isolated from elephants held in captivity. In one zoological garden in the USA, *M. szulgai* was isolated from two of three African elephants (*Loxodonta africana*): from the joint of a dead 34-year-old elephant and from the joints and lungs of one 35-year-old elephant after euthanasia. From another 55-year-old euthanised elephant, *M. smegmatis* was isolated from a lung granuloma only (Lacasse et al., 2007). In Canada, non-specified scotochromogenous isolates were obtained from two elephants (Lutze-Wallace and Turcotte, 2006) and in the USA, *M. intracellulare* was isolated from the lungs lacking gross lesions of one euthanised 41-year-old Indian elephant (*Elephas maximus*) affected by chronic arthritis (Lyashchenko et al., 2006).

In 2000, a new fast-growing *Mycobacterium* species *M. elephantis* was described; it was originally isolated from the pulmonary abscess of an elephant (Shojaei et al., 2000). In a biennial study performed in Canada (province of Ontario) 11 isolates were identified as *M. elephantis*: 10 isolates originated from the sputum of 10 patients and 1 isolate was from the axillary lymph node, affected with caseating granulomas, of a 27-year-old male. The patients suffered from three different chronic diseases (Turenne et al., 2002). A year later, another three isolates from three patients were identified in Italy: two isolates from sputum and one isolate from bronchial aspirate (Tortoli et al., 2003). The studies indicated that these species were most likely PPM. Their primary source was apparently the environment and not the elephants, in which this species had been detected for the first time (none of the patients mentioned contact with an elephant in their past history). The ecology of this newly described species is unknown.

6.7.2.6 Carnivores

Carnivores (Carnivora) are usually raptorial animals designated as predators, as they mostly feed on other animals, both live and dead. Animals that consume in large part already dead animals (carrions) are called scavengers. Carnivores, especially scavengers, are considered to be "naturally highly resistant" to mycobacterial infections. Nevertheless, they can become infected with mycobacteria (see below). There are no great differences in the transmission of mycobacteria from ingested prey to carnivores irrespective of whether it is caught prey or an infected cadaver. However, when hunting live prey, predators can become injured and subsequently infected with mycobacteria (most often the skin and subcutis are damaged). Scavengers are exposed to an increased risk of infection from cadaver tissues if the animal died of a serious mycobacterial infection.

Among OPM, it is the *MTC* members that, in particular, pose a great risk to carnivores. Of the eight species recognised at present and classified as *MTC*, the following four pathogenic species have been isolated from spontaneously infected terrestrial carnivores; they are listed from the most to the least frequently detected ones: *M. bovis*, *M. caprae*, *M. microti* and *M. tuberculosis*. *M. pinnipedii* has not been isolated from spontaneously infected terrestrial carnivores yet; it has been mostly detected in pinnipedians (Section 6.7.1). The two remaining virulent *MTC* species *M. canettii* and *M. africanum* have not been detected in carnivores either. It may be due to the fact that *M. canettii* is found in the human population only very rarely. For example, in 2004, *M. canettii* was isolated in Australia from only 1 (0.1%) of 787 human patients suffering from tuberculosis (Lumb et al., 2006).

M. africanum. It is noteworthy that *M. africanum*, which is mostly found in the inhabitants of Africa, has not been isolated from carnivores yet (Haas et al., 1997; Bonard et al., 2000; Viana-Niero et al., 2001; Niemann et al., 2002; Niobe-Eyangoh et al., 2003; Niemann et al., 2003; Niemann et al., 2004; Cadmus et al., 2006; Addo et al., 2007; de Jong et al., 2007). When Africans migrated, they spread *M. africanum* to their countries of destination and it has been detected in, e.g., Great Britain (Wilkins et al., 1986; Grange and Yates, 1989), Germany (Richter et al., 2003), Italy (Lari et al., 2005; Lari et al., 2007), USA (Brudey et al., 2004; Desmond et al., 2004) and Portugal (David et al., 2005). In Germany, *M. africanum* was diagnosed in an immunosuppressant-treated administrative assistant in an admittance centre for refugees. She was in frequent contact with migrating Africans who most likely infected her with *M. africanum* (Schmid et al., 2003).

Besides sputum, *M. africanum* has been detected in the skin (Baril et al., 1995; Iborra et al., 1997), urine and genitourinary tract of infected patients (Grange and Yates, 1992). In an ecological study, *M. africanum* was isolated from sewage, along with *M. tuberculosis* (Nguematcha and Le Noc, 1978). *M. africanum* has also been isolated from different species of domestic and wild animals kept in captivity: monkeys (Thorel, 1980) and hyrax (*Procavia habessinica*; Pavlik et al., 2005). Among domestic animals, *M. africanum* has been diagnosed in cattle (Alfredsen and Saxegaard, 1992; Weber et al., 1998; Erler et al., 2003; Cadmus et al., 2006; Rahim et al., 2007) and domestic pigs (Alfredsen and Saxegaard, 1992). Accordingly, it seems almost unbelievable that *M. africanum* has not yet been isolated from the organs or faeces of a carnivore. Very little is known about the transmission of the causative agent of this disease or its importance in the ecology of mycobacteria and potential reservoirs of infection outside a host: these areas remain to be explored.

M. bovis BCG. On the other hand, it is not surprising that *M. bovis* BCG has not yet been isolated from any spontaneously infected carnivore. That may be due to the low residual virulence of these vaccination strains. For example, in cats subcutaneously infected with *M. bovis* BCG, no clinical signs of disease were observed (Legendre et al., 1979). *M. bovis* BCG could potentially be transmitted to carnivores (especially household pets) and to the environment by children with post-vaccination complications or through the waste from medical facilities where the vaccination of children was performed. The numerous experiments dealing with *M. bovis* BCG vaccination of free-living animals, i.e. European badgers (*Meles meles*) in UK (Delahay et al., 2001; Gormley and Costello, 2003; Lesellier et al., 2006), ferrets (*Mustela putorius* f. *furo*) in New Zealand (Qureshi et al., 1999; Cross et al., 2000) and white-tailed deer (*Odocoileus virginianus*) in USA (Palmer et al., 2007) can be considered significant from the point of view of ecology.

With regard to other mycobacterial species, all *MAC* members, *M. fortuitum*, *M. smegmatis* and other PPM species have been isolated from spontaneously infected carnivores (see below).

Cats and Dogs as Examples of the Ecology of Mycobacteria in Carnivores

Keeping conditions of cats and dogs significantly affect the range of mycobacterial species that become causative agents of their infections and which can be spread by these animals in the environment (Photos 6.45, 6.46 and 6.47):

i. Household pets almost never leave the houses or flats of their owners. ESM and PPM can participate in the development of skin and respiratory infections in these animals due to unfavourable breeding conditions (dry and hot air, etc.).

ii. In households with gardens where the kept cats and dogs can run free or go out with their owners/breeders for a walk. These come into contact with ESM and PPM species through soil dust and especially through surface water (the risk increases after rain).

iii. In the mating season, cats, in particular, may leave the houses or flats of their owners/breeders for several days or weeks. After their return, they are often emaciated, because it is difficult for them to obtain food on their own and are injured after fighting with other members of their species or other animals.

iv. In towns and villages, the animals may live free as "street cats and street dogs". Their numbers can reach hundreds of thousands and even millions, especially in developing countries. Little is known about the importance of cats and dogs in the ecology of mycobacteria under these conditions.

v. Hunting dogs constitute an independent group with respect to the risk of mycobacterial infection. The most common site of pathogen penetration into their organism is wounded skin. The highest risk and resulting infection can be expected during the hunting of foxes and other animals. When wild boars are hunted, the skin of the dogs is often injured and contaminated with soil, water and mud. Therefore, it is necessary to always clean and disinfect any injury.

Even though cat and dog breeding has become massively popular, research into mycobacterial infections is quite difficult, in large part, due to the following three reasons:

i. The first reason is a considerable array of clinical manifestations, depending on the type of affected tissue or organ; some of the clinical symptoms develop secondarily owing to disseminated infection.

ii. The second reason is a long incubation period, which often hinders identification of the infection sources. For example, analysis of the clinical course of disease in 19 cats infected with various *MTC* members in the UK demonstrated that the disease may last from 1 to 48 months. In the case where the disease had a duration of 48 months, short-term remissions of the disease were observed in the cat (Gunn-Moore et al., 1996).

iii. The third reason is the low frequency of the testing of biological samples from cats and dogs in mycobacteriology laboratories.

Accordingly, rather limited information from the point of view of the ecology of mycobacteria is available; insufficient investigation of the occurrence of mycobacteria in the faeces of cats and dogs living in different conditions described above is one reason

(Section 5.10). Due to great differences in ecology among various mycobacterial species, this Section will be focused on infections in cats and dogs concentrating on different groups of mycobacterial species that have something in common. This approach will allow insight into potential ecological relationships between the occurrence of mycobacteria and the environment of other carnivores.

M. tuberculosis Complex and *M. tuberculosis* in Cats and Dogs

Infections caused by pathogenic species of *MTC* can occur in both cats (Table 6.22) and dogs (Table 6.23). Infections with *M. tuberculosis* have shown different clinical manifestations. According to the anamneses, contact with a person infected with *M. tuberculosis* was usually identified as the source of infection.

Considering the ecology of all *MTC* members and the potential infection of people, their shedding through the faeces of cats and dogs is important, even though it has not been studied yet. When handling cat and dog faeces, it is necessary to observe basic hygienic rules including the protection of children from contact with them. Litter boxes containing the faeces and urine of household pets can be contaminated with *MTC* members. Accordingly, it is necessary to dispose of the content of these litter boxes, ideally by incineration, and not by composting, which does not necessarily safely devitalise mycobacteria including *MTC* members (Section 5.10.3).

M. bovis in Cats and Dogs

M. bovis infections have also been detected in cats and dogs (Tables 6.22 and 6.23). In the anamneses of these animals, contact with other infected animals (especially cattle) has usually been described; the most frequent transmission route of *M. bovis* infection is the drinking of raw milk from infected cows. However, the sources of infection could not be established in many infected cats and dogs. The localisation of infection in submandibular lymph nodes indicated an oral route of infection (de Bolla, 1994; Table 6.22). The infection pressure on cats can be relatively high during bovine tuberculosis outbreaks on cattle farms; in the UK, for example, *M. bovis* was detected in 6 (16.7%) of 36 examined cats between 1971 and 1996 (Delahay et al., 2001).

Table 6.22 Infections in cats caused by members of *M. tuberculosis* complex

Breed (age in years)	Symptoms	Infection localization	Isolates	References
Domestic short hair (adult)	Fever, lethargy, anorexia, dehydration, polypnoe, dyspnoe, enforced bronchial lung sounds	Lungs and pulmonary ln	*MTC*	[4]
Short-haired cat (7.0)	Emaciation, apathy	Submandibular, mesenteric, bronchial ln, lungs and gut	*MTB* and/or *M. bovis*	[2]
Chinchilla cat (3.0)	Enlarged and firm submandibular ln	Submandibular ln	*MTB* and/or *M. bovis*	[3]
Chinchilla cat (3.0)	Enlarged submandibular ln	Submandibular ln	*MTB* and/or *M. bovis*	[7]
Tabby cat (adult)	Limp caused by a diseased joint	Joint tissue	*MTB* and/or *M. bovis*	[7]
Domestic cat (adult)	Nk	Disseminated infection in lungs, liver	*M. bovis*	[8]
Domestic cat (adult)	Weight loss, respiratory distress, swelling in the upper neck region	Submandibular, retropharyngeal prescapular, subscapular ln	*M. bovis*	[10]
Siamese cat (7.0)	Anorexia, dysphagia, lethargy, swelling on submandibular and left ear areas	Liver, spleen, lungs and ln	*M. bovis*	[11]
Cat (0.4)	Discharging submandibular lesion of 5 cm in diameter	Submandibular ln	*M. bovis*	[6]
Cat (0.5)	Chronic progressive respiratory embarrassment and dry rales on auscultation without cough and discharges	Miliary lesions throughout the lungs and few nodules in the spleen	*M. bovis*	[6]
Cat (adult)	Euthanised cat (no sign of illness) after the contact with infected possum with *M. bovis*	Single 2 cm in diameter granulomatous lesion in one lung	*M. bovis*	[6]
Two urban domestic cats (Nk)	Nk	Nk	*M. bovis*	[1]
Cat (adult)	Euthanized animal, clinically symptoms Nk	Lung tuberculosis with similar focus in the mesenteric ln	*M. microti*	[5]
Five cats (Nk)	Nk	Nk	*M. microti*, genotype Llama	[9]
One cat (Nk)	Nk	Nk	*M. microti*, genotype Vole	[9]

MTC M. tuberculosis complex. *MTB M. tuberculosis*. **Nk** Not known. **ln** lymph nodes.

[1] Aranaz A, Liebana E, Mateos A, Dominguez L, Vidal D, Domingo M, Gonzolez O, Rodriguez-Ferri EF, Bunschoten AE, van Embden JD, Cousins D (1996) J. Clin. Microbiol. 34:2734–2740. [2] Blunden AS, Smith KC (1996) Vet. Rec. 138:87–88. [3] de Bolla GJ (1994) Vet. Rec. 134:336. [4] Hartmann K, Gerle K, Hirschberger J, Reischl U, Hermanns W (2000) Tierarztliche Praxis Ausgabe Kleintiere Heimtiere 28:197–202. [5] HUITEMA H, van VLOT (1960) Antonie Van Leeuwenhoek 26:235–240. [6] Isaac J, Whitehead J, Adams JW, Barton MD, Coloe P (1983) Aust. Vet. J. 60:243–245. [7] Jones JW, Jenkins PA (1995) Lancet 346:442–443. [8] Kaneene JB, Bruning-Fann CS, Dunn J, Mullaney TP, Berry D, Massey JP, Thoen CO, Halstead S, Schwartz K (2002) Am. J. Vet. Res. 63:1507–1511. [9] Kremer K, van Soolingen D, van Embden J, Hughes S, Inwald J, Hewinson G (1998) J. Clin. Microbiol. 36:2793–2794. [10] Monies B, Jahans K, de la RR (2006) Vet. Rec. 158:245–246. [11] Orr CM, Kelly DF, Lucke VM (1980) J. Small Anim Pract. 21:247–253.

In New Zealand, clinical signs and the distribution of tuberculous lesions were analysed in 57, mostly free-living, cats infected with *M. bovis* in an area inhabited by *M. bovis*-infected marsupials (possums). Skin lesions predominated among the clinical symptoms; these were detected in 32 (56.1%) cats. Lymphadenopathy, wasting, respiratory problems and depressions were detected in eight (14.0%), seven (12.3%), four (7.0%) and two (3.5%) cats, respectively. *M. bovis* was most often found in the skin lesions; it was isolated from all 33 cats with skin lesions in the study of De Lisle et al. (1990). This fact pointed to a skin infection most often contracted through injuries, which the cats sustained during hunting and especially the fighting associated with mating. Little is known about the importance of *M. bovis* contamination

Table 6.23 Infections in dogs caused by different members of *M. tuberculosis* complex

Breed (age in years)	Symptoms	Infection localization	Isolates	References
Mixed setter/cocker (6.0)	Three-year history of lethargy, inappetence, regurgitation, lameness and severe respiratory distress associated with pleural effusion	Generalised chronic granulomatosis	*M. tuberculosis*	[2]
Mixed breed (0.4)	Right submandibular region (bite wounds)	Granulomatous lymphadenitis	*M. tuberculosis*	[5]
Yorkshire Terrier (3.5)	Anorexia, vomiting and productive coughing for several months to a year	Granulomatous inflammation in lungs, liver, kidneys and omentum	*M. tuberculosis*	[7]
Yorkshire (5.0)	Weight loss	Lungs tuberculosis	*M. tuberculosis*	[8]
Wire-haired fox terrier (5.0)	Anaemia, vomiting and bronchitis, weight loss	Generalised tuberculosis in lungs, liver and kidneys	*M. tuberculosis*	[1]
Irish setter (5.0)	Weight loss, blood-stained diarrhoea	Generalised tuberculosis	*M. tuberculosis*	[1]
Miniature pincher (0.8)	Purulent discharge from nose and eyes, inappetence	Local abdominal tuberculosis (liver and mesenteric lymph nodes)	*M. tuberculosis*	[1]
Airedale, terrier (5.0)	Weight loss, apathy, anaemia, uraemia, ascites, purulent metritis	Generalised tuberculosis	*M. tuberculosis*	[1]
Boxer (6.0 to 7.0)	Weight loss, enlarged bulk of belly	Abdominal tuberculosis with exudative peritonitis	*M. tuberculosis*	[1]
German Shepherd (7.0)	Lean and inert, dyspnoe and uraemia	Generalised tuberculosis	*M. tuberculosis*	[1]
Airedale, terrier (5.0)	Coughing, inert	Lung tuberculous	*M. tuberculosis*	[1]
Border collie (6.0)	Inappetence, weight loss, increased thirst, vomiting, fever	Granulomatous inflammation in liver, spleen, pancreas, lungs, etc.	*M. bovis*	[4]
German Shepherd (8.0)	Cough, inappetence and weight loss	Lungs	*M. bovis*	[6]
Beauceron dog (4.0)	Anorexia, ptyalism, depression, lameness of the right hind leg with painful and crepitant joints, polypnoe, tachycardia, fever	Acute peritonitis	*M. microti,* genotype Llama	[3]

[1] Berg OA (1956) Acta Tuberc. Scand. 32:351–361. [2] Clercx C, Coignoul F, Jakovljevic S, Balligand M, Mainil J, Henroteaux M, Kaeckenbeeck A (1992) J. Am. Animal Hospital Assoc. 28:207–211. [3] Deforges L, Boulouis HJ, Thibaud JL, Boulouha L, Sougakoff W, Blot S, Hewinson G, Truffot-Pernot C, Haddad N (2004) Vet. Microbiol. 103:249–253. [4] Ellis MD, Davies S, McCandlish IA, Monies R, Jahans K, Rua-Domenech R (2006) Vet. Rec. 159:46–48. [5] Foster ES, Scavelli TD, Greenlee PG, Gilbertson SR (1986) J. Am. Vet. Med. Assoc. 188:1188–1190. [6] Gay G, Burbidge HM, Bennett P, Fenwick SG, Dupont C, Murray A, Alley MR (2000) N.Z. Vet. J. 48:78–81. [7] Hackendahl NC, Mawby DI, Bemis DA, Beazley SL (2004) J. Am. Vet. Med. Assoc. 225:1573–7, 1548. [8] Struckmann B (1991) Kleintierpraxis 36:587.

of different components of the environment for the development of these skin forms of the disease; the risk level of *M. bovis* transmission to humans from these components of the environment remains obscure (Section 2.1).

The second most frequent form of cat infection in New Zealand is generalised tuberculosis, which was diagnosed in 12 (21.1%) of 57 cats. Among other organs, the intestinal, pulmonary, cephalic and axillary lymph nodes of 11 (19.3%) cats were affected (de Lisle et al., 1990). When night vision cameras were used in the area, the free-living cats were seen to feed on the cadavers of different animal species (especially possums) infected with *M. bovis* (Ragg et al., 2000). That identified the source of *M. bovis* for free-living cats; on the other hand, cats living in proximity to man most often become infected by drinking infected raw cow's milk. As documented in Spain, however, the source of infection for cats is not always only cattle affected by *M. bovis*. The genotypes of the *M. bovis* isolates from cats differed from the genotypes of the isolates from cattle and other wild animals in the same area (Aranaz et al., 1996). Accordingly, other at present unrecognised sources of *M. bovis* may also be associated with the ecology of *M. bovis* in different insufficiently investigated components of the environment (Section 2.1).

In France, the presence of mycobacteria in the lymph node biopsy specimens of 245 patients with cat scratch disease was investigated. The following mycobacteria were isolated from 10 (4.1%) patients: *M. tuberculosis* (5 patients), *MAC* (3 patients), *M. fuerthenensis* and *M. gordonae,* both in one patient (Rolain et al., 2006). According to the species profile, other sources of infection may be to blame other than contaminated cat claws. Nevertheless, these may become sources of *M. bovis*, as their shedding through urine and saliva has been documented (Monies et al., 2000). Identical I6*110* RFLP types of isolates in urban cats and inhabitants of Buenos Aires in Argentina have provided evidence of the transmission of *M. bovis* from cats to people. However, because the identical RFLP type was also detected in cattle, it is obvious that cats and people can also contract infection from drinking raw milk (van Soolingen et al., 1994). Comparable results were obtained by a molecular epidemiology study using spoligotyping, which revealed that the isolates from cats and cattle were of the most common spoligotype; this confirms the previously obtained results, i.e. the role of cattle as the main source of infection for cats (Zumarraga et al., 1999).

M. microti in Cats and Dogs

The infection of cats and dogs by *M. microti* (Tables 6.22 and 6.23) can be explained by the occurrence of this mycobacterial species in small terrestrial mammals. Nevertheless, it is necessary to mention that *M. microti* has not yet been detected in free-living small terrestrial mammals on continental Europe. The detection of *M. microti* in dogs and cats then gives indi-

rect evidence of its occurrence in free-living animals, or in the environment of other parts of the world. The problem of the flow of *M. microti* within the population of wild animals and also within the environment remains to be explored.

M. avium Complex in Cats and Dogs

With regard to ecology, the occurrence of other mycobacterial species, in particular the three most important *MAC* members which differ in pathogenicity, is particularly noteworthy. These were detected in both cats (Table 6.24) and dogs (Table 6.25). Based on available information we can guess what their sources are. We should not underestimate the importance of free-living animals infected with the causative agent of avian tuberculosis, *M. a. avium*.

However, the sources of other *MAC* members, i.e. *M. a. hominissuis* and *M. intracellulare*, remain obscure. We can assume that the infection source is not always the most important factor for the development of a clinical disease, which is often fatal for an infected cat or dog, but may be a weakened host organism as is the case for immunosuppressed humans.

PPM in Cats and Dogs

However, infections caused by various PPM species, especially *M. scrofulaceum*, *M. kansasii* and *M. fortuitum*, have also been detected in cats and dogs. It is clear that these PPM species can, besides skin infections, cause lethal organ or disseminated infections in many individuals (Tables 6.26 and 6.27). On the one hand, little is known about the effect of the immunocompetence of cats and dogs on the development and course of the diseases caused by PPM. On the other hand, little is known of the importance of these infected cats and dogs for the further spread of PPM in the environment.

In the catskin infection cases, a variety of PPM species have been isolated from the infected skin and subcutis (Table 6.26). Accordingly, we can assume that the infectious agent penetrates the injured skin of infected animals in a number of ways: for example, injuries that a cat sustains during fighting at a time

Table 6.24   Infections in cats caused by different members of *M. avium* complex

Breed (age in years)	Symptoms	Infection localization	Isolates	References
Shorthair cat (7.0)	Moderate weight loss and difficulty in walking	Lungs, liver, ln	MAC	[1]
Siamese cat (5.0)	Emaciation, inappetence, fever, large ln in the left submandibular area	Lungs, heart, lungs, liver, submandibular ln	MAC	[4]
Domestic short hair (11.0)	After the kidney transplantation mild depression and unkempt coat	Disseminated infection	MAC	[5]
Siamese cat (1.5)	Anorexia, lethargy and anaemia	Liver, lungs, spleen, gut	MAC	[6]
Siamese cat (2.2)	Anorexia, fever and abdominal tenderness	Lungs, ln, gut	MAC	[6]
Domestic long hair (6.5)	Left-sided *otitis externa* and peripheral vestibular disease	Mass from the ear canal and salivary gland	MAC	[7]
Domestic short hair (4.0)	Progressive weight loss, anaemia, fever, sporadic vomiting after eating	Liver, spleen, lung, gut, mesenteric ln	MAC	[8]
Domestic short hair (11.0)	Localised subcutaneous swelling on the nasal bridge	Nasal cavity	MAC	[9]
Domestic short hair (5.0)	Non-healing ulcerated mass with fistulous tracts on the left hind paw	Skin of the hind limb	MAC	[10]
Abyssinian cat (5.0)	Weight loss and anorexia, abdominal mass (removed surgically)	Lungs and ln, omentum	MAC	[12]
Domestic short hair (6.0)	Inappetence, emaciation, subcutaneous lump behind the jaw	Lymphoid tissue behind the jaw	MAA (IS*901*+)	[13]
Calico cat (13.0)	Emaciation, dehydration, multiple spherical nodules in subcutaneous tissue	Skin, lungs, jejunum, liver	MAA (1)	[2]
Domestic short hair (0.6)	Anorexia and unspecified chronic respiratory disease	Lungs and ln	MAA (1)	[11]
Domestic cat (0.6)	Skin on paw of the right hind leg	Skin granulomas	MAA (2)	[14]
Siamese cat (2.0)	Anorexia, fever, vomiting and weight loss	Lung ln	MAH (4)	[6]
Domestic short hair (8.0)	Corneal opacity	Eye (corneal granuloma)	MI	[3]

MAC M. avium complex (serotypes 1 to 28). *MAA M. a. avium* (serotypes 1, 2, 3, genotype IS*901*+ and IS*1245*+). *MAH M. a. hominissuis* (serotypes 4–6, 8–11, 21, genotype IS*901*- and IS*1245*+). *MI M. intracellulare* (serotypes 7, 12–20, 22–28, genotype IS*901*- and IS*1245*-). **(1)** Serotype 1. **ln** lymph nodes.

[1] Barry M, Taylor J, Woods JP (2002) Can. Vet. J. 43:369–371. [2] Buergelt C, Fowler J, Wright P (1982) California Veterinarian 10:13–15. [3] Deykin AR, Wigney DI, Smith JS, Young BD (1996) Australian Veterinary Practitioner 26:23. [4] Drolet R (1986) J. Am. Vet. Med. Assoc. 189:1336–1337. [5] Griffin A, Newton AL, Aronson LR, Brown DC, Hess RS (2003) J. Am. Vet. Med. Assoc. 222:1097–1098. [6] Jordan HL, Cohn LA, Armstrong PJ (1994) J. Am. Vet. Med. Assoc. 204:90–93. [7] Kaufman AC, Greene CE, Rakich PM, Weigner DD (1995) J. Am. Vet. Med. Assoc. 207:457–459. [8] Latimer KS, Jameson PH, Crowell WA, Duncan JR, Currin KP (1997) Vet. Clin. Pathol. 26:85–89. [9] Malik R, Gabor L, Martin P, Mitchell DH, Dawson DJ (1998) Aust. Vet. J. 76:604–607. [10] Miller MA, Fales WH, McCracken WS, O'Bryan MA, Jarnagin JJ, Payeur JB (1999) Vet. Pathol. 36:161–163. [11] Morfitt DC, Matthews JA, Thoen CO, Kluge JP (1989) J. Vet. Diagn. Invest 1:354–356. [12] Sieber-Ruckstuhl NS, Sessions JK, Sanchez S, Latimer KS, Greene CE (2007) Veterinary Record 160:131–132. [13] Stevenson K, Howie FE, Low JC, Cameron ME, Porter J, Sharp JM (1998) Vet. Rec. 143:109–110. [14] Suter MM, von Rotz A, Weiss R, Mettler C (1984) Zentralbl. Veterinarmed. A 31:712–718.

Table 6.25 Infections in dogs caused by different members of *M. avium* complex

Breed (age in years)	Symptoms	Infection localization	Isolates	References
Maltese cross (3.0)	Diarrhoea, inappetence, anaemia, vomitus	Lungs, gut, liver, kidneys and ln	*MAC*	[3]
Mixed-breed hound dog (young adult)	Posterior limb paralysis and proprioceptive deficits	Spinal cord	*MAC*	[4]
Basset Hound (5.0)	Progressive right forelimb lameness, cervical pain and generalised, large, firm ln	Osteomyelitis	*MAC*	[1]
Shih Tzu Poodle cross (2.0)	Anaemia, abdominal pain, lethargy, splenomegaly	Mesenteric ln	*MAC*	[5]
Basset Hound (3.0)	Anorexia, shivering, hyperthermia, generalized lymphadenopathy	Tracheobronchial ln, spleen, liver, small intestine, bone marrow, lungs	*MAC*	[6]
Basset Hound (4.0)	Weight loss, diarrhoea, lymphadenopathy	Peripheral ln, liver, spleen, bone marrow	*MAA* (1)	[7]
Labrador Retriever (7.0)	Shifting leg lameness and reduced exercise tolerance	Enlarged liver with pink-yellow plaques	*MAA* (2)	[2]
Basset Hound (2.0)	Weight loss, vomiting, hepatosplenomegaly	Lungs, abdominal cavity, ln	*MAH* (4)	[1]
Basset Hound (3.0)	Polyarthritis, diarrhoea, weight loss, enlargement of ln, liver and spleen; non-regenerative anaemia	Miliary tuberculosis	*MAH* (4)	[1]
Basset Hound (2.0)	Vomiting, weight loss and lethargy (jaundice, anaemia, cirrhosis, splenomegaly, thickened small and large intestines)	Miliary tuberculosis	*MAH* (8)	[1]

MAC M. a. complex. *MAA M. a. avium. MAH M. a. hominissuis.* (**1**) Serotype 1. **ln** lymph nodes.
[1] Carpenter JL, Myers AM, Conner MW, Schelling SH, Kennedy FA, Reimann KA (1988) J. Am. Vet. Med. Assoc. 192:1563–1568. [2] Friend SC, Russell EG, Hartley WJ, Everist P (1979) Vet. Pathol. 16:381–384. [3] Horn B, Forshaw D, Cousins D, Irwin PJ (2000) Aust. Vet. J. 78:320–325. [4] Kim DY, Cho DY, Newton JC, Gerdes J, Richter E (1994) Vet. Pathol. 31:491–493. [5] O'Toole D, Tharp S, Thomsen BV, Tan E, Payeur JB (2005) J. Vet. Diagn. Invest 17:200–204. [6] Shackelford CC, Reed WM (1989) J. Vet. Diagn. Invest 1:273–275. [7] Walsh KM, Losco PE (1984) J. Am. Animal Hospital Assoc. 20:295–299.

of increased sexual activity, during hunting and subsequent fighting with the prey. Due to the fact that PPM have also been isolated from damaged skin exudates (White et al., 1983), it is necessary to consider cats as a potential source of PPM for other animals, for the owner and last but not least for the environment they live in. Accordingly, careful handling of the infected cats and dogs is necessary, basic hygienic principles should be observed and the contact of the infected animals with children prevented. We assume that the handling of their dead bodies need not be mentioned. Nevertheless, in Japan, a strong positive reaction to human tuberculin and positive response in the QuantiFERON test detecting gamma-interferon was found in a prosector who performed a necropsy on dog infected with *M. tuberculosis*.

Feline Leprosy and Canine Leproid Granuloma Syndrome

At present, four main mycobacterial infection syndromes can be distinguished in cats (GunnMoore and Shaw, 1997):

i. Classical tuberculosis.
ii. Feline tuberculosis syndrome.
iii. PPM mycobacteriosis.
iv. Feline leprosy syndrome.

The final syndrome mentioned above, feline leprosy syndrome, has been described by various authors (Suter et al., 1984; Barrs et al., 1999; Davies et al., 2006). However, histopathological lesions resembling

Table 6.26 Infections in cats caused by different potentially pathogenic mycobacteria

Breed (age in years)	Symptoms	Infection localization	Isolates	References
Feral – two animals (Nk)	Nk	Gastrointestinal tract	*MAP*	[2]
Short haired (4.0)	Increased respiratory effort and dyspnoea	Lungs	*M. fortuitum*	[4]
Domestic (6.0)	Recurring cutaneous and movable abscesses	Several subcutaneous nodules	*M. fortuitum*	[5]
Domestic short hair (4.0)	Cutaneous nodules in inguinal area	Skin	*M. fortuitum*	[15]
Domestic short hair (2.0)	Cutaneous nodules, fistulous tracts in flank and lumbar area	Skin and exudate	*M. fortuitum*	[15]
Domestic short hair (2.0)	Skin lesions	Ventral abdominal wall	*M. fortuitum*	[11]
Domestic short hair (7.0)	Draining cutaneous tracts on the abdomen and thorax	Chronic suppurative cellulitis	*M. fortuitum*	[12]
Domestic short hair (6.0)	Subcutaneous nodules on the dorsum originally diagnosed as a bite wounds	Pyogranulomatous panniculitis	*M. fortuitum/peregrinum* group	[17]
Persian (11.0)	Emaciation, skin nodules	Skin (feline leprosy syndrome)	*M. haemophilum**	[1]
Short hair (2.0)	Ulcerated subcutaneous nodules	Skin	*M. lepraemurium*	[3]
Domestic short hair (10.0)	Cutaneous nodules in paralumbar area	Skin	*M. phlei*	[15]
Domestic short hair (8.0)	Emaciation, skin nodules	Disseminated infection in skin, lungs, lymph nodes and one eye	*M. simiae*	[6]
Domestic short hair (8.0)	Skin lesions	Inguinal and proximal hind leg	*M. smegmatis*	[10]
Domestic short hair (1.5)	Skin lesions	Right axilla and ventral chest	*M. smegmatis*	[10]
Domestic short hair (4.0)	Skin lesions	Inguinal and ventral abdomen	*M. smegmatis*	[10]
Domestic short hair (7.0)	Skin lesions	Inguinal region	*M. smegmatis*	[10]
Domestic short hair (1.5)	Painful subcutaneous granuloma on the ventral region of the abdomen	Skin	*M. smegmatis*	[14]
Domestic short hair (1.5)	Nodular dermal thickening on a digit	Pyogranuloma with AFB	*M. terrae* complex	[8]
Domestic short hair (1.5)	Productive cough, fever	Lungs	*M. thermoresistibile*	[7]
Domestic short hair (10.5)	Skin lesion in the lumbosacral area (numerous fluctuating nodules up to 3 cm)	Subcutaneous nodules and aspirated fluid, lymph nodes	*M. thermoresistibile*	[16]
Domestic short hair (4.0)	Rapidly growing firm subcutaneous mass in the left submandibular region	Submandibular lymph node, abdominal fluid	*M. xenopi*	[9]
Domestic (Nk)	Skin lesion	Skin	*M. xenopi*	[13]

* Isolate resembled that of *M. haemophilum*, but differed in 6 of 300 bases. **Nk** Not known. **AFB** Acid-fast bacilli. ***MAP*** *M. a. paratuberculosis.*

[1] Barrs VR, Martin P, James G, Chen S, Love DN, Malik R (1999) Australian Veterinary Practitioner 29:159. [2] Corn JL, Manning EJ, Sreevatsan S, Fischer JR (2005) Appl. Environ. Microbiol. 71:6963–6967. [3] Courtin F, Huerre M, Fyfe J, Dumas P, Boschiroli ML (2007) J. Feline Med. Surg. 9:238–241. [4] Couto SS, Artacho CA (2007) Veterinary Pathology 44:543–546. [5] Dewevre P, McAllister H, Schirmer R, Weinacker A (1977) J. Am. Anim. Hosp. Assoc. 13:68–70. [6] Dietrich U, Arnold P, Guscetti F, Pfyffer GE, Spiess B (2003) J.Small Anim Pract. 44:121–125. [7] Foster SF, Martin P, Davis W, Allan GS, Mitchell DH, Malik R (1999) J.Small Anim Pract. 40:433–438. [8] Henderson SM, Baker J, Williams R, Gunn-Moore DA (2003) J. Feline Med. Surg. 5:37–41. [9] MacWilliams PS, Whitley N, Moore F (1998) Vet. Clin. Pathol. 27:50–53. [10] Malik R, Hunt GB, Goldsmid SE, Martin P, Wigney DI, Love DN (1994) J. Small Animal Pract. 35:524–530. [11] Michaud AJ (1994) Feline Practice 22:7–9. [12] Monroe WE, August JR, Chickering WR, Sriranganathan N (1988) Compendium on Continuing Education for the Practicing Veterinarian 10:1044–1048. [13] Tomasovic AA, Rac R, Purcell DA (1976) Aust. Vet. J. 52:103. [14] Weber A, Wachowitz R, Pfleghaar WS, Heim U, Schaal KP (2000) Tierarztliche Umschau 55:251. [15] White SD, Ihrke PJ, Stannard AA, Cadmus C, Griffin C, Kruth SA, Rosser EJ, Jr., Reinke SI, Jang S (1983) J. Am. Vet. Med. Assoc. 182:1218–1222. [16] Willemse T, Groothuis DG, Koeman JP, Beyer EG (1985) J. Clin. Microbiol. 21:854–856. [17] Youssef S, Archambault M, Parker W, Yager J (2002) Can. Vet. J. 43:285–287.

Table 6.27 Infections in dogs caused by mycobacteria other than members of *M. tuberculosis* and *M. avium* complexes

Breed (age in years)	Symptoms	Infection localization	Isolates	References
Doberman Pinscher (3.0)	Progressive cough, depression, lethargy, fever	Lungs with nodules	*M. fortuitum*	[5]
Doberman Pinscher (1.2)	Multiple subcutaneous abscesses	Disseminated subcutaneous infection	*M. fortuitum*	[2, 5]
Pembrokeshire Corgi (1.0)	Apathy, inappetence, productive cough	Lungs	*M. fortuitum*	[4]
Mixed breed (0.5)	Fever, emaciation, laminitis	Lungs, hypertrophic osteopathy	*M. fortuitum*	[7]
Whippet (3.0)	Persistent dermatosis, cough, respiratory distress, anorexia, lethargy	Heart, lungs, kidney, spleen, liver	*M. kansasii*	[6]
Basset Hound (3.0)	Fever, hind limb laminitis, ocular discharge	Skin, lungs, eye	*M. smegmatis*	[3]
Basset Hound (0.9)	Inappetence, lethargy, vomiting, diarrhoea and fever	Lungs, abdominal cavity	*M.* sp.	[1]

[1] Carpenter JL, Myers AM, Conner MW, Schelling SH, Kennedy FA, Reimann KA (1988) J. Am. Vet. Med. Assoc. 192:1563–1568. [2] Fox LE, Kunkle GA, Homer BL, Manella C, Thompson JP (1995) J. Am. Vet. Med. Assoc. 206:53–55. [3] Grooters AM, Couto CG, Andrews JM, Johnson SE, Kowalski JJ, Esplin RB (1995) J. Am. Vet. Med. Assoc. 206:200–202. [4] Irwin PJ, Whithear K, Lavelle RB, Parry BW (2000) Aust. Vet. J. 78:254–257. [5] Jang SS, Eckhaus MA, Saunders G (1984) J. Am. Vet. Med. Assoc. 184:96–98. [6] Pressler BM, Hardie EM, Pitulle C, Hopwood RM, Sontakke S, Breitschwerdt EB (2002) J. Am. Vet. Med. Assoc. 220:1336–4. [7] Wylie KB, Lewis DD, Pechman RD, Cho DY, Roy A (1993) J. Am. Vet. Med. Assoc. 202:1986–1988.

human leprosy have also been observed on the skin of dogs. The lesions are designated as "canine leproid granuloma syndrome" (Malik et al., 1998; Hughes et al., 2000; Malik et al., 2001). The failure to detect mycobacteria *in vitro* is a common feature of this disease. The ecology and the source of the causative agent(s) of these skin lesions remain obscure. The description of skin lesions of mycobacterial origin in cattle in Ireland is noteworthy. Mycobacteria could not be isolated from them *in vitro* either, even though they were observed in the lesions after staining according to Ziehl–Neelsen (J. Kazda, unpublished data). The questions as to whether the aetiological agent causing these lesions in dogs, cats and cattle is the same and whether it also affects other animal species remain open.

Other Carnivores and the Ecology of Mycobacteria

The importance of other carnivores is due to their way of life (mostly exoanthropic) very low. Mycobacteria are resistant to the environment of the intestinal tract and can pass along the entire tract and are subsequently shed to the environment via faeces. The digestive tract of predators is relatively short, and thus its devitalising effect on mycobacteria may not be very strong. Therefore, a relatively easy transport of mycobacteria

through the intestinal tract can be expected in predators, even though information on this topic is scarce. Most widely researched is the occurrence and pathogenesis of infections caused by *M. bovis* in other carnivores. Numerous recent review articles (Gallagher and Clifton-Hadley, 2000; De Vos et al., 2001; de Lisle et al., 2001; Bengis et al., 2002; Alexander et al., 2002; Michel, 2002) and books (Thoen et al., 2006b) deal with the occurrence of *M. bovis* in different species of free-living carnivores.

Bovine tuberculosis has been detected and studied in detail in the following free-living carnivores in different parts of the world:

Ferret (*Mustela putorius* f. *furo*) in New Zealand (Ragg et al., 2000; de Lisle et al., 2005)

Iberian lynx (*Lynx pardina*) in Spain (Briones et al., 2000; Perez et al., 2001; Aranaz et al., 2004)

Red fox (*Vulpes vulpes*) in Spain (Martin-Atance et al., 2005)

Badger (*Meles meles*) in the UK and Ireland (Tuyttens et al., 2000; Wilkinson et al., 2000; Gavier-Widen et al., 2001; Delahay et al., 2001; Chambers et al., 2002; Hancox, 2004; Scantlebury et al., 2004; Fend et al., 2005; Costello et al., 2006)

Red fox (*V. vulpes*), ferret (*Mustela putorius* f. *furo*), American mink (*Mustela vison*), ermine (*Mustela erminea*) in the UK (Delahay et al., 2001)

Black bear (*Ursus americanus*), coyote (*Canis latrans*), red fox (*V. vulpes*), bobcat (*Lynx rufus*) in the USA (Bruning-Fann et al., 1998; Bruning-Fann et al., 2001)

Coyote (*Canis latrans*) in Canada (Sangster et al., 2007)

Wolf (*Canis lupus*) in the USA (Lutze-Wallace et al., 2005)

Raccoon (*Procyon lotor*) in the USA (Palmer et al., 2002)

Lion (*Panthera leo*) in Africa (Cleaveland et al., 2005; Kirberger et al., 2006)

However, the purpose of the present chapter is not to describe the course and spread of the causative agent of bovine tuberculosis in carnivores, but rather its importance in the ecology of mycobacteria. Except for the studies concerning the occurrence of *M. bovis* in the faeces, excretions and secretions of different carnivores, publications dealing with the isolation of other mycobacterial species are scarce.

Accordingly, information on the importance of predators in the ecology of mycobacteria is also not abundant. Nevertheless, the environment which they inhabit (e.g. setts, dense shrubs, tree nests, tree holes, lofts and cellars in the houses) can become infected through their faeces, secretions or excretions if these contain and shed mycobacteria. When investigating the potential contamination of the environment, the fact that predators can bury prey under the ground or leaves or drag it up trees where they store it during periods of food shortage should be considered. Under particular conditions, various mycobacterial species can be spread through unexpected routes.

6.7.3 Mammals Spending a Substantial Part of Their Lives in Flight and in Trees

Chiropterans (order: Chiroptera) inhabit all continents (except Antarctica) and are the second largest mam-

Table 6.28 Isolation of mycobacteria from naturally and artificially infected chiropterans

Infection	Species	Mycobacterial species	Comments	References
Natural	Florida free-tailed bat (*Tadarida cynocephala*)	III Runyon Group*	Isolated from the livers of 10 animals from Arizona and Texas (USA)	[3]
	Fruit bat (*Pteropus medius*)	*M. bovis*	The right pleural cavity was occupied by adhesions densely infiltrated with caseating tubercles, the pericardium showed sparse milia and the epicardium more – diagnosed in one zoo in dead animal in 1925	[6]
	Fruit bat (*Cercocebus fuliginosus*)	*M. bovis*	The right lung was in condition of caseous pneumonia and the bronchial glands on that side were large and caseous throughout (nothing abnormal in abdomen) – diagnosed in one zoo in dead animal in 1926	[4]
	Fruit bat (*Pteropus giganteus*)	*M. bovis*	Tuberculous lesions were confined to the abdominal lymphatic system only – died after 8 years in the zoo in 1930	[5]
Artificial	Fruit bat (*Eidolon helvum*)	*M. ulcerans*	Isolation from inoculated web muscles 4 weeks after infection	[1]
	Eastern pipistrels (*Pipistrellus subflavus*), Florida free-tailed bat (*Tadarida cynocephala*)	*M. marinum*	Several bats developed footpad lesions (site of infection) similar to those in mice, but none developed significant visceral infections within the period they could be kept alive, usually 5–14 days after the infection	[2]

* Non-photochromogenic species.

[1] Church JC, Griffin ER (1968) J. Pathol. Bacteriol. 96:508–512. [2] Clark HF, Chepard CC (1963) J. Bacteriol. 86:1057–&. [3] Disalvo AF, Ajello L, Palmer JW, Winkler WG (1969) Am. J. Epidemiol. 89:606. [4] Griffith A (1928) J. Hyg. 29:198–218. [5] Hamerton A (1931) Proc. Zool. Soc. London 527–555. [6] Scott H (1926) Proc. Zool. Soc. London 231–244.

malian order after rodents. Various species of chiropterans feed on, e.g. insects, fish or the blood of animals including cattle. Little attention has been paid to the detection of mycobacteria in these animals so far. Only one reference concerning the detection of mycobacteria in livers without gross lesions of 10 Florida free-tailed bats (*Tadarida cynocephala*) from the wild is available (Table 6.28).

M. bovis has been detected in only three animals of different species living in captivity in zoological gardens. The source of infection could not be determined. The anamnestic data of all three cases revealed that the animals were sometimes fed with condensed milk. Other species of mycobacteria were detected in different tissues of bats only after their artificial infection (Table 6.28).

Due to the fact that some chiropterans also feed on blood, the question of the risk associated with the spread of, e.g., *M. bovis* remains open. The occurrence and survival of different mycobacterial species in faeces from chiropterans in the places where they stay overnight (above all in caves) remain obscure.

Acknowledgements Partially supported by the European Commission PathogenCombat FOOD-CT-2005-007081 (Section 6.7) and by the Ministry of Agriculture of the Czech Republic NPV 1B53009 (Chapter 6) and NAZV QH81065 (Section 6.7).

References

Abernethy CSLJE (1978) Mycobacteriosis in mountain whitefish (*Prosopium williamsoni*) from the Yakima River, Washington. Wildlife Dis. 14:333–336

Acevedo-Whitehouse K, Vicente J, Gortazar C, Hofle U, Fernandez-de-Mera IG, Amos W (2005) Genetic resistance to bovine tuberculosis in the Iberian wild boar. Mol. Ecol. 14:3209–3217

Addo K, Owusu-Darko K, Yeboah-Manu D, Caulley P, Minamikawa M, Bonsu F, Leinhardt C, Akpedonu P, Ofori-Adjei D (2007) Mycobacterial species causing pulmonary tuberculosis at the korle bu teaching hospital, accra, ghana. Ghana. Med. J. 41:52–57

Adekambi T, Ben Salah S, Khlif M, Raoult D, Drancourt M (2006) Survival of environmental mycobacteria in *Acanthamoeba polyphaga*. Appl. Environ Microbiol. 72:5974–5981

Adisa CA, Mbanaso A (2004) Furuncular myiasis of the breast caused by the larvae of the Tumbu fly (*Cordylobia anthropophaga*). BMC. Surg. 4:5 (available from: http://www.biomedcentral.com/1471-2482/4/5)

Agaisse H, Burrack LS, Philips JA, Rubin EJ, Perrimon N, Higgins DE (2005) Genome-wide RNAi screen for host factors required for intracellular bacterial infection. Science. 309:1248–1251

Alexander KA, Pleydell E, Williams MC, Lane EP, Nyange JFC, Michel AL (2002) *Mycobacterium tuberculosis*: An emerging disease of free-ranging wildlife. Emerg. Infect. Dis. 8:598–601

Alfredsen S, Saxegaard F (1992) An outbreak of tuberculosis in pigs and cattle caused by *Mycobacterium africanum*. Vet. Rec. 131:51–53

Allen B (1987) Excretion of viable tubercle bacilli by *Blatta orientalis* (the Oriental cockroach) following ingestion of heat-fixed sputum smears: a laboratory investigation. Trans. R. Soc. Tropical Med. Hyg. 81:98–99

Alvarez J, de Juan L, Briones V, Romero B, Aranaz A, Fernandez-Garayzabal JF, Mateos A (2005) *Mycobacterium avium* subspecies *paratuberculosis* in fallow deer and wild boar in Spain. Vet. Rec. 156:212–213

Amlacher RE (1961) Taschenbuch der Fischkrankheiten. Jena, Gustav Fisher Verlag, 286 pp.

Aranaz A, de Juan L, Montero N, Sanchez C, Galka M, Delso C, Alvarez J, Romero B, Bezos J, Vela AI, Briones V, Mateos A, Dominguez L (2004) Bovine tuberculosis (*Mycobacterium bovis*) in wildlife in Spain. J. Clin. Microbiol. 42:2602–2608

Aranaz A, Liebana E, Mateos A, Dominguez L, Vidal D, Domingo M, Gonzolez O, Rodriguez Ferri EF, Bunschoten AE, vanEmbden JDA, Cousins D (1996) Spacer oligonucleotide typing of *Mycobacterium bovis* strains from cattle and other animals: A tool for studying epidemiology of tuberculosis. J. Clin. Microbiol. 34:2734–2740

Aronson JD (1926) Spontaneous tuberculosis in salt water fish. J. Inf. Dis. 39:315–320

Ayele WY, Machackova M, Pavlik I (2001) The transmission and impact of paratuberculosis infection in domestic and wild ruminants. Veterinarni Medicina 46:205–224

Baker J, Hagan W (1942) Tuberculosis of Mexican platyfish (*Platypoecilus maculatus*). J. Infect. Dis. 70:248–252

Banerjee R, Banerjee BD, Chaudhury S, Hati AK (1991) Transmission of viable *Mycobacterium leprae* by *Aedes aegypti* from lepromatous leprosy patients to the skin of mice through interrupted feeding. Lepr. Rev. 62:21–26

Banerjee R, Chaudhury S, Hati AK (1990) Transmission of *Mycobacterium-leprae* from lepromatous leprosy patients to the skin of mice through intermittent feeding. Trop. Geograp. Med. 42:97–99

Baril L, Caumes E, Truffotpernot C, Bricaire F, Grosset J, Gentilini M (1995) Tuberculosis caused by *Mycobacterium-africanum* associated with involvement of the upper and lower respiratory-tract, skin, and mucosa. Clin. Inf. Dis. 21:653–655

Barker J, Brown MRW (1994) Trojan-horses of the microbial world – protozoa and the survival of bacterial pathogens in the environment. Microbiology-UK. 140:1253–1259

Barrow PA, Gallagher J (1981) Aspects of the epidemiology of bovine tuberculosis in badgers and cattle. I. The prevalence of infection in two wild animal populations in south-west England. J. Hyg. (Lond). 86:237–245

Barrs VR, Martin P, James G, Chen S, Love DN, Malik R (1999) Feline leprosy due to infection with novel mycobacterial species. Aust. Vet. Pract. 29:159–164

Bataillon E, Dubard J, Terre L (1897) Un nouveau type de tuberculose. Compte Rendu des Séances de la Société de Biologie. 49:446–449

Bataillon E, Terre L (1897) La form saprophytique de la tuberculose humaine et de la tuberculose aviare. Compte Rendu des Séances de la Société de Biologie. 124:1339–1400

Baxby D, Bennett M, Getty B (1994) Human cowpox 1969–93: a review based on 54 cases. Br. J. Dermatol. 131:598–607

Beck BM, Rice CD (2003) Serum antibody levels against select bacterial pathogens in Atlantic bottlenose dolphins, *Tursiops truncatus*, from Beaufort NCUSA and Charleston Harbor, Charleston, SC, USA. Marine Environ. Res. 55:161–179

Beerwerth W, Eysing B, Kessel U (1979) Mycobacteria in arthropodes of different biotopes (in German). Zentralbl Bakteriol [Orig A]. 244:50–57

Bengis RG, Kock RA, Fischer J (2002) Infectious animal diseases: the wildlife/livestock interface. Rev. Sci. Tech. 21:53–65

Bercovier H, Vincent V (2001) Mycobacterial infections in domestic and wild animals due to *Mycobacterium marinum*, *M. fortuitum*, *M. chelonae*, *M. porcinum*, *M. farcinogenes*, *M. smegmatis*, *M. scrofulaceum*, *M. xenopi*, *M. kansasii*, *M. simiae* and *M. genavense*. Rev. Sci. Tech. 20:265–290

Bernardelli A, Bastida R, Loureiro J, Michelis H, Romano MI, Cataldi A, Costa E (1996) Tuberculosis in sea lions and fur seals from the south-western Atlantic coast. Rev. Sci. Tech. 15:985–1005

Bilej M, De Baetselier P, Beschin A (2000) Antimicrobial defense of the earthworm. Folia Microbiol. (Praha). 45:283–300

Blagodarnyi IA, Makarevich NM, Blekhman IM (1971) Isolation of atypical mycobacteria from spontaneously infected bird ticks *Argas persicus* (in Russian). Probl Tuberk. 49(6):74–6

Blake LA, West BC, Lary CH, Todd JR (1987) Environmental nonhuman sources of leprosy. Rev. Infect. Dis. 9:562–577

Bollo E, Ferroglio E, Dini V, Mignone W, Biolatti B, Rossi L (2000) Detection of *Mycobacterium tuberculosis* complex in lymph nodes of wild boar (*Sus scrofa*) by a target-amplified test system. J. Vet. Med. B Infect. Dis. Vet. Public Health. 47:337–342

Bonard D, Msellati P, Rigouts L, Combe P, Coulibaly D, Coulibaly IM, Portaels F (2000) What is the meaning of repeated isolation of *Mycobacterium africanum*? Int. J. Tuberculosis Lung Dis. 4:1176–1180

Bottger EC, Hirschel B, Coyle MB (1993) *Mycobacterium-genavense* Sp-Nov. Int. J. Syst Bacteriol. 43:841–843

Briones V, de Juan L, Sanchez C, Vela AI, Galka M, Montero, Goyache J, Aranaz A, Dominguez L (2000) Bovine tuberculosis and the endangered Iberian lynx. Emerg. Infect. Dis. 6:189–191

Brownstein DG (1978) Reptilian Mycobacteriosis. In: Proc. Mycobacterial Infections in Zoo Animals, Mountali R.J. (ed.), Smithsonian Institution Press, Washington, D.C. 265–268

Brudey K, Gutierrez MC, Vincent W, Parsons LM, Salfinger M, Rastogi N, Sola C (2004) *Mycobacterium africanum* genotyping using novel spacer oligonucleotides in the direct repeat locus. J. Clin. Microbiol. 42:5053–5057

Bruning-Fann CS, Schmitt SM, Fitzgerald SD, Fierke JS, Friedrich PD, Kaneene JB, Clarke KA, Butler KL, Payeur JB, Whipple DL, Cooley TM, Miller JM, Muzo DP (2001) Bovine tuberculosis in free-ranging carnivores from Michigan. J. Wildl. Dis. 37:58–64

Bruning-Fann CS, Schmitt SM, Fitzgerald SD, Payeur JB, Whipple DL, Cooley TM, Carlson T, Friedrich P (1998) *Mycobacterium bovis* in coyotes from Michigan. J. Wildl. Dis. 34:632–636

Burge EJ, Gauthier DT, Ottinger CA, Van Veld PA (2004) *Mycobacterium*-inducible Nramp in striped bass (*Morone saxatilis*). Infect. Immun. 72:1626–1636

Butler KL, Fitzgerald SD, Berry DE, Church SV, Reed WM, Kaneene JB (2001) Experimental inoculation of European starlings (*Sturnus vulgaris*) and American crows (*Corvus brachyrhynchos*) with *Mycobacterium bovis*. Avian Dis. 45:709–718

Cadmus S, Palmer S, Okker M, Dale J, Gover K, Smith N, Jahans K, Hewinson RG, Gordon SV (2006) Molecular analysis of human and bovine tubercle bacilli from a local setting in Nigeria. J. Clin. Microbiol. 44:29–34

Cavanagh P, Marsden PD (1969) Bacteria isolated from gut of some reduviid bugs. Trans. R. Soc. Tropical Med. Hyg. 63:415–416

Cavanagh R, Begon M, Bennett M, Ergon T, Graham IM, de Haas PE, Hart CA, Koedam M, Kremer K, Lambin X, Roholl P, Soolingen DD (2002) *Mycobacterium microti* infection (vole tuberculosis) in wild rodent populations. J. Clin. Microbiol. 40:3281–3285

Chambers MA, Pressling WA, Cheeseman CL, Clifton-Hadley RS, Hewinson RG (2002) Value of existing serological tests for identifying badgers that shed *Mycobacterium bovis*. Vet. Microbiol. 86:183–189

Cherniak VZ, Kokurichev PI, Dobin MA, Epsthein I (1963) A Case of tuberculosis in an elephant (in Russian). Tr. Leningr. Sanitarnogig. Med. Inst. 83:179–180

Chinabut S, Limsuwan C, Chanratchakool P (1990) Mycobacteriosis in the snakehead, *Channa striatus* (Fowler). J. Fish Dis. 13:531–535

Chitty D (1954) Tuberculosis among wild voles – with a discussion of other pathological conditions among certain mammals and birds. Ecology. 35:227–237

Cirillo JD, Falkow S, Tompkins LS, Bermudez LE (1997) Interaction of *Mycobacterium avium* with environmental amoebae enhances virulence. Infect. Immun. 65:3759–3767

Clarke KR, Firlgerald SD, Hattey JA, Bolin CA, Berry DE, Church SV, Reed WM (2006) Experimental inoculation of wild turkeys (*Meleagris gallopavo*) with *Mycobacterium bovis*. Avian Dis. 50:131–134

Cleaveland S, Mlengeya T, Kazwala RR, Michel A, Kaare MT, Jones SL, Eblate E, Shirima GM, Packer C (2005) Tuberculosis in Tanzanian wildlife. J. Wildl. Dis. 41:446–453

Clothier RH, Balls M (1973a) Mycobacteria and lymphoreticular tumors in *Xenopus-Laevis*, South-African Clawed Toad. 1. Isolation, Characterization and Pathogenicity for *Xenopus* of *M Marinum* Isolated from Lymphoreticular Tumor-Cells. Oncology. 28:445–457

Clothier RH, Balls M (1973b) Mycobacteria and Lymphoreticular Tumors in *Xenopus-Laevis*, South-African Clawed Toad. 2. Have Mycobacteria A Role in Tumor Initiation and Development. Oncology. 28:458–480

Coleman JD, Cooke MM (2001) *Mycobacterium bovis* infection in wildlife in New Zealand. Tuberculosis. (Edinb.). 81:191–202

Collins P, Matthews PR, McDiarmid A, Brown A (1983) The pathogenicity of *Mycobacterium avium* and related mycobacteria for experimental animals. J. Med. Microbiol. 16:27–35

Conroy G, Conroy DA (1999) Acid-fast bacterial infection and its control in guppies (*Lebistes reticulatus*) reared on an ornamental fish farm in Venezuela. Vet. Rec. 144:177–178

Cooke MM, Jackson R, Coleman JD (1993) Tuberculosis in a free-living brown hare (*Lepus europaeus occidentalis*). N. Z. Vet. J. 41:144–146

Cooper JE (1961) Bacteria. In: Diseases of the Reptilia. In: Cooper J.E., Jackson O.F. (eds.). Vol. 1, Academic Press, London-New York-Toronto-Sydney-San Francisco. 197–191

Corner LA, Barrett RH, Lepper AW, Lewis V, Pearson CW (1981) A survey of mycobacteriosis of feral pigs in the Northern Territory. Aust. Vet. J. 57:537–542

Costello E, Flynn O, Quigley F, O'Grady D, Griffin J, Clegg T, McGrath G (2006) Genotyping of *Mycobacterium bovis* isolates from badgers in four areas of the Republic of Ireland by restriction fragment length polymorphism analysis. Vet. Rec. 159:619–623

Cousins DV, Bastida R, Cataldi A, Quse V, Redrobe S, Dow S, Duignan P, Murray A, Dupont C, Ahmed N, Collins DM, Butler WR, Dawson D, Rodriguez D, Loureiro J, Romano MI, Alito A, Zumarraga M, Bernardelli A (2003) Tuberculosis in seals caused by a novel member of the *Mycobacterium tuberculosis* complex: *Mycobacterium pinnipedii* sp nov. Int. J. Syst. Evol. Microbiol. 53:1305–1314

Cousins DV, Francis BR, Gow BL, Collins DM, McGlashan CH, Gregory A, Mackenzie RM (1990) Tuberculosis in captive seals: bacteriological studies on an isolate belonging to the *Mycobacterium tuberculosis* complex. Res. Vet. Sci. 48: 196–200

Cousins DV, Roberts JL (2001) Australia's campaign to eradicate bovine tuberculosis: the battle for freedom and beyond. Tuberculosis.(Edinb.). 81:5–15

Cousins DV, Williams SN, Reuter R, Forshaw D, Chadwick B, Coughran D, Collins P, Gales N (1993) Tuberculosis in wild seals and characterisation of the seal bacillus. Aust. Vet. J. 70:92–97

Cross ML, Labes RE, Mackintosh CG (2000) Oral infection of ferrets with virulent *Mycobacterium bovis* or *Mycobacterium avium*: susceptibility, pathogenesis and immune response. J. Comp. Pathol. 123:15–21

Daniels MJ, Ball N, Hutchings MR, Greig A (2001) The grazing response of cattle to pasture contaminated with rabbit faeces and the implications for the transmission of paratuberculosis. Vet. J. 161:306–313

Daniels MJ, Henderson D, Greig A, Stevenson K, Sharp JM, Hutchings MR (2003) The potential role of wild rabbits *Oryctolagus cuniculus* in the epidemiology of paratuberculosis in domestic ruminants. Epidemiol. Infect. 130:553–559

Daniels MJ, Hutchings MR (2001) The response of cattle and sheep to feed contaminated with rodent faeces. Vet. J. 162:211–218

David S, Barros V, Portugal C, Antunes A, Cardoso A, Calado A, Sancho L, de Sousa JG (2005) Update on the Spoligotypes of *Mycobacterium tuberculosis* complex isolates from the Fernando Fonseca Hospital (Amadora-Sintra, Portugal). Rev. Port. Pneumol. 11:513–531

Davies JL, Sibley JA, Myers S, Clark EG, Appleyard GD (2006) Histological and genotypical characterization of feline cutaneous mycobacteriosis: a retrospective study of formalin-fixed paraffin-embedded tissues. Vet. Dermatol. 17: 155–162

Davis GB, Watson PR, Billing AE (1984) Tuberculosis in A Kiwi (Apteryx-Mantelli). N. Z. Vet. J. 32:30–30

Davis M (2001) *Mycobacterium tuberculosis* risk for elephant handlers and veterinarians. Appl. Occup. Environ. Hyg. 16:350–353

de Bolla GJ (1994) Tuberculosis in a cat. Vet. Rec. 134:336

de Jong BC, Hill RC, Aiken A, Jeffries DJ, Onipede A, Small RM, Adegbola RA, Corrah TR (2007) Clinical presentation and outcome of tuberculosis patients infected by *M-africanum* versus *M-tuberculosis*. Int. J. Tuberc. Lung Dis. 11:450–456

de Lisle GW, Collins DM, Loveday AS, Young WA, Julian AF (1990) A report of tuberculosis in cats in New Zealand, and the examination of strains of *Mycobacterium bovis* by DNA restriction endonuclease analysis. N. Z. Vet. J. 38:10–13

de Lisle GW, Mackintosh CG, Bengis RG (2001) *Mycobacterium bovis* in free-living and captive wildlife, including farmed deer. Rev. Sci. Tech. 20:86–111

de Lisle GW, Yates GF, Caley P, Corboy RJ (2005) Surveillance of wildlife for *Mycobacterium bovis* infection using culture of pooled tissue samples from ferrets (*Mustela furo*). N. Z. Vet. J. 53:14–18

De Vos V, Bengis RG, Kriek NPJ, Michel A, Keet DF, Raath JP, Huchzermeyer HFKA (2001) The epidemiology of tuberculosis in free-ranging African buffalo (*Syncerus caffer*) in the Kruger National Park, South Africa. Onderst. J. Vet. Res. 68:119–130

Delahay RJ, Cheeseman CL, Clifton-Hadley RS (2001) Wildlife disease reservoirs: the epidemiology of *Mycobacterium bovis* infection in the European badger (*Meles meles*) and other British mammals. Tuberculosis. (Edinb.). 81:43–49

Demartinez MCC, Richards RH (1991) Histopathology of Vitamin-C-deficiency in A Cichlid, *Cichlasoma-Urophthalmus* (Gunther). J. Fish Dis. 14:507–519

Desmond E, Ahmed AT, Probert WS, Ely J, Jang Y, Sanders CA, Lin SY, Flood J (2004) *Mycobacterium africanum* cases, California. Emerg. Infect. Dis. 10:921–923

Dionne MS, Ghori N, Schneider DS (2003) *Drosophila melanogaster* is a genetically tractable model host for *Mycobacterium marinum*. Infect. Immun. 71:3540–3550

Dionne MS, Pham LN, Shirasu-Hiza M, Schneider DS (2006) Akt and FOXO dysregulation contribute to infection-induced wasting in Drosophila. Curr. Biol. 16:1977–1985

Doneley RJ, Gibson JA, Thorne D, Cousins DV (1999) Mycobacterial infection in an ostrich. Aust. Vet. J. 77:368–370

Dorman SE, Hatem CL, Tyagi S, Aird K, Lopez-Molina J, Pitt ML, Zook BC, Dannenberg AM, Jr., Bishai WR, Manabe YC (2004) Susceptibility to tuberculosis: clues from studies with inbred and outbred New Zealand White rabbits. Infect. Immun. 72:1700–1705

Dvorska L, Matlova L, Ayele WY, Fischer OA, Amemori T, Weston RT, Alvarez J, Beran V, Moravkova M, Pavlik I (2007) Avian tuberculosis in naturally infected captive water birds of the Ardeidae and Threskiornithidae families studied by serotyping, IS*901* RFLP typing, and virulence for poultry. Vet. Microbiol. 119:366–374

Dvorska L, Matlova L, Bartos M, Parmova I, Bartl J, Svastova P, Bull TJ, Pavlik I (2004) Study of *Mycobacterium avium* complex strains isolated from cattle in the Czech Republic between 1996 and 2000. Vet. Microbiol. 99:239–250

Eddyani M, Ofori-Adjei D, Teugels G, De Weirdt D, Boakye D, Meyers WM, Portaels F (2004) Potential role for fish in transmission of *Mycobacterium ulcerans* disease (*Buruli ulcer*): An environmental study. Appl. Environ. Microbiol. 70: 5679–5681

Ekelund F, Ronn R (1994) Notes on protozoa in agricultural soil with emphasis on heterotrophic flagellates and naked amoebae and their ecology. FEMS Microbiol. Rev. 15: 321–353

Emori M, Saito H, Sato K, Tomioka H, Setogawa T, Hidaka T (1993) Therapeutic efficacy of the benzoxazinorifamycin KRM-1648 against experimental *Mycobacterium avium* infection induced in rabbits. Antimicrob. Agents Chemother. 37:722–728

Erler W, Kahlau D, Martin G, Naumann L, Schimmel D, Weber A (2003) On the epizootiology of tuberculosis of cattle in the Federal Republic of Germany. Berliner und Munchener Tierarztliche Wochenschrift. 116:288–292

Fend R, Geddes R, Lesellier S, Vordermeier HM, Corner LA, Gormley E, Costello E, Hewinson RG, Marlin DJ, Woodman AC, Chambers MA (2005) Use of an electronic nose to diagnose *Mycobacterium bovis* infection in badgers and cattle. J. Clin. Microbiol. 43:1745–1751

Fischer O, Matlova L, Bartl J, Dvorska L, Melicharek I, Pavlik I (2000) Findings of mycobacteria in insectivores and small rodents. Folia Microbiol. (Praha). 45: 147–152

Fischer O, Matlova L, Dvorska L, Svastova P, Bartl J, Melicharek I, Weston RT, Pavlik I (2001) Diptera as vectors of mycobacterial infections in cattle and pigs. Med. Vet. Entomol. 15:208–211

Fischer OA, Matlova L, Bartl J, Dvorska L, Svastova P, du Maine R, Melicharek I, Bartos M, Pavlik I (2003a) Earthworms (Oligochaeta, Lumbricidae) and mycobacteria. Vet. Microbiol. 91:325–338

Fischer OA, Matlova L, Dvorska L, Svastova P, Bartl J, Weston RT, Pavlik I (2004) Blowflies *Calliphora vicina* and *Lucilia sericata* as passive vectors of *Mycobacterium avium* subsp. avium, *M. a. paratuberculosis* and *M. a. hominissuis*. Med. Vet. Entomol. 18:116–122

Fischer OA, Matlova L, Dvorska L, Svastova P, Bartos M, Weston RT, Kopecna M, Trcka I, Pavlik I (2005) Potential risk of *Mycobacterium avium* subspecies *paratuberculosis* spread by syrphid flies in infected cattle farms. Med. Vet. Entomol. 19:360–366

Fischer OA, Matlova L, Dvorska L, Svastova P, Bartos M, Weston RT, Pavlik I (2006) Various stages in the life cycle of syrphid flies (*Eristalis tenax*; Diptera: Syrphidae) as potential mechanical vectors of pathogens causing mycobacterial infections in pig herds. Folia Microbiol. (Praha). 51:147–153

Fischer OA, Matlova L, Dvorska L, Svastova P, Pavlik I (2003b) Nymphs of the Oriental cockroach (*Blatta orientalis*) as passive vectors of causal agents of avian tuberculosis and paratuberculosis. Med. Vet. Entomol. 17:145–150

Fitzgerald SD, Boland KG, Clarke KR, Wismer A, Kaneene JB, Berry DE, Church SV, Hattey JA, Bolin CA (2005) Resistance of mallard ducks (*Anas platyrhynchos*) to experimental inoculation with *Mycobacterium bovis*. Avian Diseases. 49:144–146

Fitzgerald SD, Zwick LS, Berry DE, Church SV, Kaneene JB, Reed WM (2003) Experimental inoculation of pigeons (*Columba livia*) with *Mycobacterium bovis*. Avian Dis. 47:470–475

Foldenauer U, Curd S, Zulauf I, Hatt JM (2007) *Ante mortem* diagnosis of mycobacterial infection by liver biopsy in a budgerigar (*Melopsittacus undulatus*). Schweizer Archiv fur Tierheilkunde. 149:273–276

Forshaw D, Phelps GR (1991) Tuberculosis in a captive colony of pinnipeds. J. Wildl. Dis. 27:288–295

Friend SC, Russell EG (1979) *Mycobacterium intracellulare* infection in a water monitor. J. Wildl. Dis. 15:229–233

Frota CC, Hunt DM, Buxton RS, Rickman L, Hinds J, Kremer K, van Soolingen D, Colston MJ (2004) Genome structure in the vole bacillus, *Mycobacterium microti*, a member of the *Mycobacterium tuberculosis* complex with a low virulence for humans. Microbiology. 150:1519–1527

Furley CW (1997) Tuberculosis in elephants. Lancet. 350: 224–224

Gallagher J, Clifton-Hadley RS (2000) Tuberculosis in badgers; a review of the disease and its significance for other animals. Res. Vet. Sci. 69:203–217

Garcia A, LeClear CT, Gaskin JM (2001) *Mycobacterium avium* infection in an ostrich (*Struthio camelus*). J. Zoo. Wildl. Med. 32:96–100

Garrod AH (1875) Report on the Indian elephant which died in the society's gardens on July 7th, 1875. Proc. Zool. Soc. Lond. 542–543

Gavier-Widen D, Chambers MA, Palmer N, Newell DG, Hewinson RG (2001) Pathology of natural *Mycobacterium bovis* infection in European badgers (*Meles meles*) and its relationship with bacterial excretion. Vet. Rec. 148:299–304

Geater JG (1975) The fly as potential vector in the transmission of leprosy. Lepr. Rev. 46:279–286

Golyshevskaya VI (1991) The role of coccoid ultrafine forms of mycobacteria in the transmission of the mycobacterial infection. Pneumoftiziologia. 40:11–13

Gormley E, Costello E (2003) Tuberculosis and badgers: New approaches to diagnosis and control. J. Appl. Microbiol. 94 Suppl:80S–86S

Grange JM, Dewar CA, Rowbotham TJ (1987) Microbe Dependence of *Mycobacterium-Leprae* – A possible intracellular relationship with protozoa. Int. J. Leprosy Other Mycobacterial Dis. 55:565–566

Grange JM, Yates MD (1989) Incidence and Nature of Human Tuberculosis Due to *Mycobacterium africanum* in Southeast England – 1977–87. Epidemiol. Infect. 103:127–132

Grange JM, Yates MD (1992) Survey of mycobacteria isolated from urine and the genitourinary tract in south-east England from 1980 to 1989. Br. J. Urol. 69:640–646

Greenberg HB, Jung RC, Gutter AE (1981) Hazel elephant is dead (of Tuberculosis). Am. Rev. Respir. Dis. 124:341–341

Gregor F, Povolny D (1958) [Study on a classification of synanthropic flies (Diptera).]. J. Hyg. Epidemiol. Microbiol. Immunol. 2:205–216

Greig A, Stevenson K, Henderson D, Perez V, Hughes V, Pavlik I, Hines ME, McKendrick I, Sharp JM (1999) Epidemiological study of paratuberculosis in wild rabbits in Scotland. J. Clin. Microbiol. 37:1746–1751

Gunn-Moore DA, Jenkins PA, Lucke VM (1996) Feline tuberculosis: A literature review and discussion of 19 cases caused by an unusual mycobacterial variant. Vet. Rec. 138: 53–58

GunnMoore D, Shaw S (1997) Mycobacterial disease in the cat. In Practice. 19:493

Gutter AE, Wells SK, Spraker TR (1987) Generalized Mycobacteriosis in A California Sea Lion (*Zalophus-Californicus*). J. Zoo Animal Med. 18:118–120

Haas H, Fattal B (1990) Distribution of mycobacteria in different types of water in Israel. Wat. Res. 24:1233–1235

Haas WH, Bretzel G, Amthor B, Schilke K, Krommes G, RuschGerdes S, StichtGroh V, Bremer HJ (1997) Comparison of DNA fingerprint patterns of isolates of *Mycobacterium africanum* from east and west Africa. J. Clin. Microbiol. 35:663–666

Hancox M (2004) Does badger culling make TB worse? Lett. Appl. Microbiol. 39:311–312

Hansen PJ, Bjornsen PK, Hansen BW (1997) Zooplankton grazing and growth: Scaling within the 2–2,000-mu m body size range. Limnol. Oceanogr. 42:687–704

Hawley JE, Penner LR, Wedberg SE, Kulp WL (1951) The Role of the House Fly, *Musca-Domestica*, in the Multiplication of Certain Enteric Bacteria. Am. J. Tropical Med. 31: 572–582

Hillebrand-Haverkort ME, Kolk AHJ, Kox LFF, Ten Velden JJAM, Ten Veen JH (1999) Generalized *Mycobacterium genavense* infection in *HIV*-infected patients: Detection of the *Mycobacterium* in hospital tap water. Scand. J. Infect. Dis. 31:63–68

Hirakawa H (2001) Coprophagy in leporids and other mammalian herbivores. Mammal Rev. 31:61–80

Hirch A (1956) Infection of hamsters and rabbits with *Mycobacterium johnei*. J. Comp. Pathol. 66:260–269

Hoop RK (2002) *Mycobacterium tuberculosis* infection in a canary (*Serinus canana L.*) and a blue-fronted Amazon parrot (*Amazona amazona aestiva*). Avian Dis. 46:502–504

Hoop RK, Bottger EC, Ossent P, Salfinger M (1993) Mycobacteriosis due to *Mycobacterium genavense* in 6 pet birds. J. Clin. Microbiol. 31:990–993

Hoop RK, Bottger EC, Pfyffer GE (1996) Etiological agents of mycobacterioses in pet birds between 1986 and 1995. J. Clin. Microbiol. 34:991–992

Hoop RK, Ehrsam H, Ossent P, Salfinger M (1994) [Mycobacteriosis of ornamental birds – frequency, pathologo-anatomic, histologic and microbiologic data]. Berl Munch. Tierarztl. Wochenschr. 107:275–281

Hughes MS, Ball NW, Love DN, Canfield PJ, Wigney DI, Dawson D, Davis PE, Malik R (1999) Disseminated *Mycobacterium genavense* infection in a FIV-positive cat. J. Feline Med. Surg. 1:23–29

Hughes MS, James G, Ball N, Scally M, Malik R, Wigney DI, Martin P, Chen S, Mitchell D, Love DN (2000) Identification by 16S rRNA gene analyses of a potential novel mycobacterial species as an etiological agent of canine leproid granuloma syndrome. J. Clin. Microbiol. 38: 953–959

Iborra C, Cambau E, Lecomte C, Grosset J, Bricaire F, Caumes E (1997) Cutaneous tuberculosis: Study of 4 cases. Annales de Dermatologie et de Venereologie. 124:139–143

Ippen R (1964) Vergleichende pathologische Untersuchungen über die spontane und experimentelle Tuberkulose der Kaltblüter. Akademie Verlag GmbH, Berlin, 90 pp.

Jahnel J (1940a) Die Fischtuberkulose. Wochenschr. f. Aquarien- und Terrarienkd. 37:317–321

Jahnel J (1940b) Spontaninfektionen mit saürefesten Stäbchen bei Fischen. Wien. tierärztl. Mschr. 27:289–302

Johnson CT, Winkler CE, Boughton E, Penfold JW (1993) *Mycobacterium kansasii* infection in a llama. Vet. Rec. 133:243–244

Johnstone J (1913) Disease conditions of fishes (Tubercular lesions in a cod). Rept. Lancashire Sea – Fish Lab. 21: 20–23

Johnstone PAS (1987) The Search for Animal-Models of Leprosy. Int. J. Lepr. Other Mycobact. Dis. 55:535–547

Jorgensen JB, Clausen B (1976) Mycobacteriosis in a Roe-Deer caused by Wood-Pigeon mycobacteria. Nordisk Vet. Med. 28:539–546

Judge J, Kyriazakis I, Greig A, Allcroft DJ, Hutchings MR (2005) Clustering of *Mycobacterium avium* subsp. *paratuberculosis* in rabbits and the environment: How hot is a hot spot? Appl. Environ. Microbiol. 71:6033–6038

Judge J, Kyriazakis I, Greig A, Davidson RS, Hutchings MR (2006) Routes of intraspecies transmission of *Mycobacterium avium* subsp *Paratuberculosis* in rabbits (*Oryctolagus cuniculus*): A field study. Appl. Environ. Microbiol. 72: 398–403

Karbe E (1987) Disseminated mycobacteriosis in the golden hamster. Zentralbl. Veterinarmed. B. 34:391–394

Kazda J (1967a) Mycobacteria in drinking-water as cause of parallergy to tuberculins in animals. I. Parallergic reactions in fowl. Zentralblatt fur Bakteriologie Parasitenkunde Infektionskrankheiten und Hygiene Abteilung 1. 203:92

Kazda J (1967b) Mycobacteria in drinking-water as cause of parallergy to tuberculins in animals. 2. Parallergic reactions to Mammalian and Avian Tuberculins in Cattle. Zentralblatt fur Bakteriologie Parasitenkunde Infektionskrankheiten und Hygiene Abteilung 1. 203:190–198

Kazda J (1967c) Mykobakterien Im Trinkwasser Als Ursache der Parallergie Gegenuber Tuberkulinen Bei Tieren. 3. Taxonomische Studie Einiger Rasch Wachsender Mykobakterien und Beschreibung Einer Neuen Art – *Mycobacterium Brunense* N Sp. Zentralblatt fur Bakteriologie Parasitenkunde Infektionskrankheiten und Hygiene Abteilung 1. 203:199

Kazda J (1973a) Importance of water for distribution of potentially pathogenic mycobacteria. 2. Growth of Mycobacteria in water models. Zentralblatt fur Bakteriologie Mikrobiologie und Hygiene Serie B-Umwelthygiene Krankenhaushygiene Arbeitshygiene Praventive Medizin. 158: 170–176

Kazda J (1973b) Importance of Water for Spread of Potentially Pathogenic Mycobacteria. 1. Possibilities for Multiplication of Mycobacteria. Zentralblatt fur Bakteriologie Mikrobiologie und Hygiene Serie B-Umwelthygiene Krankenhaushygiene Arbeitshygiene Praventive Medizin. 158:161–169

Kazda J (2000) The ecology of mycobacteria. Kluwer Academic Publishers, Dordrecht, Boston, London, 72 pp.

Kazda J, Hoyte R (1972) Concerning ecology of *Mycobacterium intracellular* serotype Davis. Zentralblatt fur Bakteriologie Mikrobiologie und Hygiene Series A-Medical Microbiology Infectious Diseases Virology Parasitology. 222: 506–509

Kennedy DJ, Benedictus G (2001) Control of *Mycobacterium avium* subsp *paratuberculosis* infection in agricultural species. Revue Scientifique et Technique de l Office International des Epizooties. 20:151–179

Khoga JM, Toth E, Marialigeti K, Borossay J (2002) Fly-attracting volatiles produced by *Rhodococcus fascians* and *Mycobacterium aurum* isolated from myiatic lesions of sheep. J. Microbiol. Methods. 48:281–287

Kiehn TE, Hoefer H, Bottger EC, Ross R, Wong M, Edwards F, Antinoff N, Armstrong D (1996) *Mycobacterium genavense* infections in pet animals. J. Clin. Microbiol. 34:1840–1842

Kirberger RM, Keet DF, Wagner WM (2006) Radiologic abnormalities of the appendicular skeleton of the lion (*Panthera leo*): Incidental findings and *Mycobacterium bovis*-induced changes. Vet. Radiol. Ultrasound. 47:145–152

Kirchheimer WF (1976a) Recent advances in experimental leprosy. South. Med. J. 69:993–996

Kirchheimer WF (1976b) The role of arthropods in the transmission of leprosy. Int. J. Lepr. Other Mycobact. Dis. 44: 104–107

Krakower C, Gonzalez LM (1937) Spontaneous leprosy in a mouse. Science. 86:617–618

Kurmanbaev K, Blagodarnyi I (1981) Interrelationship of human and animal tuberculosis (in Russian). Problemy Tuberkuloza. 6–8

Kurz CL, Ewbank JJ (2007) Infection in a dish: High-throughput analyses of bacterial pathogenesis. Curr. Opin. Microbiol. 10:10–16

Lacasse C, Terio K, Kinsel MJ, Farina LL, Travis DA, Greenwald R, Lyashchenko KP, Miller M, Gamble KC (2007) Two cases of atypical mycobacteriosis caused by *Mycobacterium szulgai* associated with mortality in captive African elephants (*Loxodonta africana*). J. Zoo Wildlife Med. 38:101–107

Lapage G (1947) Tuberculosis of voles and shrews. Nature. 160:687–687

Lari N, Rindi L, Bonanni D, Rastogi N, Sola C, Tortoli E, Garzelli C (2007) Three-year longitudinal study of genotypes of *Mycobacterium tuberculosis* isolates in Tuscany, Italy. J. Clin. Microbiol. 45:1851–1857

Lari N, Rindi L, Sola C, Bonanni D, Rastogi N, Tortoli E, Garzelli C (2005) Genetic diversity, determined on the basis of katG463 and gyrA95 polymorphisms spoligotyping and IS*6110* typing, of *Mycobacterium tuberculosis* complex isolates from Italy. J. Clin. Microbiol. 43:1617–1624

Ledwon A, Szeleszczuk P, Zwolska Z, ugustynowicz-Kopec E, Sapierzynski R, Kozak M (2008) Experimental infection of budgerigars (*Melopsittacus undulatus*) with five *Mycobacterium* species. Avian Pathology. 37:59–64

Legendre AM, Easley JR, Becker PU (1979) In vivo and in vitro responses of cats sensitized with viable *Mycobacterium bovis* (BCG). Am. J. Vet. Res. 40:1613–1619

Lesellier S, Palmer S, Dalley DJ, Dave D, Johnson L, Hewinson RG, Chambers MA (2006) The safety and immunogenicity of Bacillus Calmette-Guerin (BCG) vaccine in European badgers (*Meles meles*). Vet. Immunol. Immunopathol. 112:24–37

Lewerin SS, Olsson SL, Eld K, Roken B, Ghebremichael S, Koivula T, Kallenius G, Bolske G (2005) Outbreak of *Mycobacterium tuberculosis* infection among captive Asian elephants in a Swedish zoo. Vet. Rec. 156: 171–175

Lewis J (1987) Cutaneous mycobacteriosis in a Southern sealion. Aquatic Mammals. 13:105–108

Lin SP, Yang YZ, Tan WG, Cheng JQ, Luo DQ (2007) [An epidemiological study on the ecological environment related to Nontuberculous mycobacteria in Shenzhen city of Guangdong province]. Zhonghua Liu Xing. Bing. Xue. Za Zhi. 28:430–432

Little TW, Swan C, Thompson HV, Wilesmith JW (1982) Bovine tuberculosis in domestic and wild mammals in an area of Dorset. III. The prevalence of tuberculosis in mammals other than badgers and cattle. J. Hyg. (Lond). 89: 225–234

Lloyd JB, Whittington RJ, Fitzgibbon C, Dobson R (2001) Presence of *Mycobacterium avium* subspecies *paratuberculosis* in suspensions of ovine trichostrongylid larvae produced in faecal cultures artificially contaminated with the bacterium. Vet. Rec. 148:261–263

Lucas J, Lucas A, Furber H, James G, Hughes MS, Martin P, Chen SC, Mitchell DH, Love DN, Malik R (2000) *Mycobacterium genavense* infection in two aged ferrets with conjunctival lesions. Aust. Vet. J. 78:685–689

Lugton IW, Johnstone AC, Morris RS (1995) *Mycobacterium bovis* infection in New Zealand hedgehogs (*Erinaceus europaeus*). N. Z. Vet. J. 43:342–345

Lumb R, Bastian I, Crighton T, Gilpin C, Haverkort F, Sievers A (2006) Tuberculosis in Australia: Bacteriologically confirmed cases and drug resistance, 2004: A report of the Australian Mycobacterium Reference Laboratory Network. Commun. Dis. Intell. 30:102–108

Lutze-Wallace C, Berlie-Surujballi G, Barbeau Y, Bergeson D (2005) Strain typing of *Mycobacterium bovis* from a 1978 case of tuberculosis in a wolf (*Canis lupis*) from Manitoba. Can. Vet. J. 46:502–502

Lutze-Wallace C, Turcotte C (2006) Laboratory diagnosis of bovine tuberculosis in Canada for calendar year 2005. Can. Vet. J. 47:871–873

Lyashchenko KP, Greenwald R, Esfandiari J, Olsen JH, Ball R, Dumonceaux G, Dunker F, Buckley C, Richard M, Murray S, Payeur JB, Andersen P, Pollock JM, Mikota S, Miller M, Sofranko D, Waters WR (2006) Tuberculosis in elephants: Antibody responses to defined antigens of *Mycobacterium tuberculosis*, potential for early diagnosis, and monitoring of treatment. Clinical and Vaccine Immunology. 13: 722–732

Machackova M, Matlova L, Lamka J, Smolik J, Melicharek I, Hanzlikova M, Docekal J, Cvetnic Z, Nagy G, Lipiec M, Ocepek M, Pavlik I (2003) Wild boar (*Sus scrofa*) as a possible vector of mycobacterial infections: Review of literature and critical analysis of data from Central Europe between 1983 to 2001. Veterinarni Medicina 48: 51–65

Machackova M, Svastova P, Lamka J, Parmova I, Liska V, Smolik J, Fischer OA, Pavlik I (2004) Paratuberculosis in farmed and free-living wild ruminants in the Czech Republic (1999–2001). Vet. Microbiol. 101:225–234

Machackova-Kopecna M, Bartos M, Straka M, Ludvik V, Svastova P, Alvarez J, Lamka J, Trcka I, Treml F, Parmova I, Pavlik I (2005) Paratuberculosis and avian tuberculosis infections in one red deer farm studied by IS*900* and IS*901* RFLP analysis. Vet. Microbiol. 105: 261–268

Machicao N, Laplaca E (1954) Lepra-Like Granulomas in Frogs. Lab. Invest. 3:219–227

Mahler M, Jelinek F (2000) Granulomatous inflammation in the tails of mice associated with *Mycobacterium chelonae* infection. Lab. Anim. 34:212–216

Majeed SK, Gopinath C (1983) Cutaneous tuberculosis in the Carp, Cyprinus-Carpio l. J. Fish Dis. 6:313–316

Malik R, Love DN, Wigney DI, Martin P (1998) Mycobacterial nodular granulomas affecting the subcutis and skin of dogs (canine leproid granuloma syndrome). Aust. Vet. J. 76: 403–7, 398

Malik R, Martin P, Wigney D, Swan D, Slatter PS, Cibilic D, Allen J, Mitchell DH, Chen SC, Hughes MS, Love DN (2001) Treatment of canine leproid granuloma syndrome: Preliminary findings in seven dogs. Aust. Vet. J. 79:30–36

Manabe YC, Dannenberg AM, Jr., Tyagi SK, Hatem CL, Yoder M, Woolwine SC, Zook BC, Pitt ML, Bishai WR (2003) Different strains of *Mycobacterium tuberculosis* cause various spectrums of disease in the rabbit model of tuberculosis. Infect. Immun. 71:6004–6011

Manabe YC, Scott CP, Bishai WR (2002) Naturally attenuated, orally administered *Mycobacterium microti* as a tuberculosis vaccine is better than subcutaneous *Mycobacterium bovis* BCG. Infect. Immun. 70:1566–1570

Manarolla G, Liandris E, Pisoni G, Moroni P, Piccinini R, Rampin T (2007) *Mycobacterium genavense* and avian polyomavirus co-infection in a European Goldfinch (*Carduelis carduelis*). Avian Pathol. 36:423–U108

Mann PC, Bush M, Janssen DL, Frank ES, Montali RJ (1981) Clinicopathologic Correlations of Tuberculosis in Large Zoo Mammals. J. Am. Vet. Med. Assoc. 179:1123–1129

Manning EJB, Collins MT (2001) Introduction – Mycobacterial infections in domestic and wild animals. Revue Scientifique et Technique de l Office International des Epizooties. 20:9–9

Marciano-Cabral F (2004) Introductory remarks: Bacterial endosymbionts or pathogens of free-living amoebae. J. Eukaryot. Microbiol. 51:497–501

Marcus LC (1981) Veterinary Biology and Medicine of Captive Amphibians and Reptiles. Lea and Febiger, Philadelphia, 239 pp.

Marsollier L, Andre JP, Frigui W, Reysset G, Milon G, Carbonnelle B, Aubry J, Cole ST (2007) Early trafficking events of *Mycobacterium ulcerans* within *Naucoris cimicoides*. Cell Microbiol. 9:347–355

Marsollier L, Aubry J, Coutanceau E, Andre JP, Small PL, Milon G, Legras P, Guadagnini S, Carbonnelle B, Cole ST (2005) Colonization of the salivary glands of *Naucoris cimicoides* by *Mycobacterium ulcerans* requires host plasmatocytes and a macrolide toxin, mycolactone. Cell Microbiol. 7:935–943

Marsollier L, Aubry J, Saint-Andre JP, Robert R, Legras P, Manceau AL, Bourdon S, Audrain C, Carbonnelle B (2003) [Ecology and transmission of *Mycobacterium ulcerans*]. Pathol. Biol. (Paris). 51:490–495

Marsollier L, Robert R, Aubry J, Saint Andre JP, Kouakou H, Legras P, Manceau AL, Mahaza C, Carbonnelle B (2002) Aquatic insects as a vector for *Mycobacterium ulcerans*. Appl. Environ. Microbiol. 68:4623–4628

Marsollier L, Severin T, Aubry J, Merritt RW, Saint Andre JP, Legras P, Manceau AL, Chauty A, Carbonnelle B, Cole ST (2004) Aquatic snails, passive hosts of *Mycobacterium ulcerans*. Appl. Environ. Microbiol. 70:6296–6298

Martin-Atance P, Palomares F, Gonzalez-Candela M, Revilla E, Cubero MJ, Calzada J, Leon-izcaino L (2005) Bovine tuberculosis in a free ranging red fox (*Vulpes vulpes*) from Donana National Park (Spain). J. Wildl. Dis. 41:435–436

Matlova L, Dvorska L, Bartl J, Bartos M, Ayele WY, Alexa M, Pavlik I (2003) Mycobacteria isolated from the environment of pig farms in the Czech Republic during the years 1996 to 2002. Veterinarni Medicina. 48:343–357

Matlova L, Fischer O, Kazda J, Kaustova J, Bartl J, Horvathova A, Pavlik I (1998) The occurrence of mycobacteria in invertebrates and poikilothermic animals and their role in the infection of other animals and man. Veterinarni Medicina. 43:115–132

McDougall AC, Cologlu AS (1983) Lepromatous leprosy in man; depth of the cellular infiltrate and bacillary mass in relation to the possibility of transmission of leprosy by biting arthropods. Ann. Trop. Med. Parasitol. 77: 187–193

McInerney J, Small KJ, Caley P (1995) Prevalence of *Mycobacterium bovis* infection in feral pigs in the Northern Territory. Aust. Vet. J. 72:448–451

Michalak K, Austin C, Diesel S, Bacon JM, Zimmerman P, Maslow JN (1998) *Mycobacterium* tuberculosis infection as a zoonotic disease: Transmission between humans and elephants. Emerg. Infect. Dis. 4:283–287

Michel AL (2002) Implications of tuberculosis in African wildlife and livestock. Ann. N. Y. Acad. Sci. 969:251–255

Mikota SK, Peddie L, Peddie J, Isaza R, Dunker F, West G, Lindsay W, Larsen RS, Salman MD, Chatterjee D, Payeur J, Whipple D, Thoen C, Davis DS, Sedgwick C, Montali RJ, Ziccardi M, Maslow J (2001) Epidemiology and diagnosis of *Mycobacterium tuberculosis* in captive Asian elephants (*Elephas maximus*). J. Zoo Wildl Med. 32:1–16

Miller JM, Jenny AL, Payeur JB (2002) Polymerase chain reaction detection of *Mycobacterium tuberculosis* complex and *Mycobacterium avium* organisms in formalin-fixed tissues from culture-negative ruminants. Vet. Microbiol. 87: 15–23

Miltner EC, Bermudez LE (2000) *Mycobacterium avium* grown in *Acanthamoeba castellanii* is protected from the effects of antimicrobials. Antimicrobial Agents and Chemotherapy. 44:1990–1994

Mok WY, Carvalho CM (1984) Occurrence and experimental infection of toads (*Bufo marinus* and *B. granulosus*) with *Mycobacterium chelonei* subsp. *abscessus*. J. Med. Microbiol. 18:327–333

Mokresh AH, Butler DG (1990) *Granulomatous enteritis* following oral inoculation of newborn rabbits with *Mycobacterium paratuberculosis* of bovine origin. Can. J. Vet. Res. 54:313–319

Mokresh AH, Czuprynski CJ, Butler DG (1989) A rabbit model for study of *Mycobacterium paratuberculosis* infection. Infect. Immun. 57:3798–3807

Mondal D, Sinha RP (1992) Pathogenicity of Caprine Strain of *Mycobacterium paratuberculosis* in Rabbit. Ind. J. Anim. Sci. 62:1117–1120

Mondal D, Sinha RP, Gupta MK (1994) Effect of combination therapy in *Mycobacterium paratuberculosis* infected rabbits. Indian J. Exp. Biol. 32:318–323

Monger BC, Landry MR, Brown SL (1999) Feeding selection of heterotrophic marine nanoflagellates based on the surface hydrophobicity of their picoplankton prey. Limnology and Oceanography. 44:1917–1927

Monies RJ, Cranwell MP, Palmer N, Inwald J, Hewinson RG, Rule B (2000) Bovine tuberculosis in domestic cats. Vet. Rec. 146:407–408

Montali RJ, Mikota SK, Cheng LI (2001) *Mycobacterium tuberculosis* in zoo and wildlife species. Revue Scientifique et Technique de l Office International des Epizooties. 20:291–303

Morales P, Madin SH, Hunter A (1985) Systemic *Mycobacterium marinum* infection in an *Amazon manatee*. J Am Vet. Med. Assoc. 187:1230–1231

Morita Y, Maruyama S, Hashizaki F, Katsube Y (1999) Pathogenicity of *Mycobacterium avium* complex serovar 9 isolated from painted quail (*Excalfactoria chinensis*). J. Vet. Med. Sci. 61:1309–1312

Mota E, Sleigh AC (1987) Water-Contact Patterns and *Schistosoma mansoni* Infection in A Rural-Community in Northeast Brazil. Revista do Instituto de Medicina Tropical de Sao Paulo. 29:1–8

Narayanan E, Manja KS, Kirchheimer WF, Balasubrahmanyan M (1972) Occurrence of *Mycobacterium leprae* in arthropods. Lepr. Rev. 43:194–198

Narayanan E, Sreevatsa, Kirchheimer WF, Bedi BM (1977) Transfer of leprosy bacilli from patients to mouse footpads by *Aedes aegypti*. Lepr. India. 49:181–186

Narayanan E, Sreevatsa, Raj AD, Kirchheimer WF, Bedi BM (1978) Persistence and distribution of *Mycobacterium leprae* in *Aedes aegypti* and *Culex fatigans* experimentally fed on leprosy patients. Lepr. India. 50:26–37

Nenoff P, Uhlemann R (2006) Mycobacteriosis in mangrove killifish (*Rivulus magdalenae*) caused by living fish food (*Tubifex tubifex*) infected with *Mycobacterium marinum*. Dtsch. Tierarztl. Wochenschr. 113:230–232

Nguematcha R, Le Noc P (1978) [Detection of pathogenic mycobacteria in the environment of the medical units and of the slaughter-house of an African town (author's transl)]. Med. Trop.(Mars.). 38:59–63

Niemann S, Kubica T, Bange FC, Adjei O, Browne EN, Chinbuah MA, Diel R, Gyapong J, Horstmann RD, Joloba ML, Meyer CG, Mugerwa RD, Okwera A, Osei I, Owusu-Darbo E, Schwander SK, Rusch-Gerdes S (2004) The species *Mycobacterium africanum* in the light of new molecular markers. J. Clin. Microbiol. 42:3958–3962

Niemann S, Rusch-Gerdes S, Joloba ML, Whalen CC, Guwatudde D, Ellner JJ, Eisenach K, Fumokong N, Johnson JL, Aisu T, Mugerwa RD, Okwera A, Schwander SK (2002) *Mycobacterium africanum* subtype II is associated with two distinct genotypes and is a major cause of human tuberculosis in Kampala, Uganda. J. Clin. Microbiol. 40: 3398–3405

Niemann S, Rusch-Gerdes S, Schwander SK (2003) Is *Mycobacterium africanum* subtype II (Uganda I and Uganda II) a genetically well-defined subspecies of the *Mycobacterium tuberculosis* complex? Authors' reply. J. Clin. Microbiol. 41:1345–1348

Nigrelli RF, Vogel H (1963) Spontaneous tuberculosis in fishes and other cold-blooded vertebrates with special reference to *Mycobacterium fortuitum* Cruz from fish and human lesions. Zoologica. 48:131–144

Niobe-Eyangoh SN, Kuaban C, Sorlin P, Cunin P, Thonnon J, Sola C, Rastogi N, Vincent V, Gutierrez MC (2003) Genetic biodiversity of *Mycobacterium tuberculosis* complex strains from patients with pulmonary tuberculosis in Cameroon. J. Clin. Microbiol. 41:2547–2553

Noga EJ, Wright JF, Pasarell L (1990) Some unusual features of mycobacteriosis in the cichlid fish *Oreochromis mossambicus*. J. Comp Pathol. 102:335–344

Novotny L, Dvorska L, Lorencova A, Beran V, Pavlik I (2004) Fish: A potential source of bacterial pathogens for human beings. Veterinarni Medicina. 49:343–358

Nugent G, Fraser KW, Asher GW, Tustin KG (2001) Advances in New Zealand mammalogy 1990–2000: Deer. J. R. Soc. N. Z. 31:263–298

Nugent G, Whitford J, Young N (2002) Use of released pigs as sentinels for *Mycobacterium bovis*. J. Wildl. Dis. 38:665–677

O'Grady D, Flynn O, Costello E, Quigley F, Gogarty A, McGuirk J, O'Rourke J, Gibbons N (2000) Restriction fragment length polymorphism analysis of *Mycobacterium avium* isolates from animal and human sources. Int. J. Tuberc. Lung Dis. 4:278–281

Olson GA, Woodard JC (1974) Miliary tuberculosis in a reticulated python. J. Am. Vet. Med. Assoc. 164: 733–735

Oros J, Acosta B, Gaskin JM, Deniz S, Jensen HE (12-4-2003) *Mycobacterium kansasii* infection in a Chinese soft shell turtle (*Pelodiscus sinensis*). Vet. Rec. 152:474–476

Pai HH, Chen WC, Peng CF (2003) Isolation of non-tuberculous mycobacteria from hospital cockroaches (*Periplaneta americana*). J. Hosp. Infect. 53:224–228

Palmer MV, Thacker TC, Waters WR (2007) Vaccination of white-tailed deer (*Odocoileus virginianus*) with *Mycobacterium bovis* bacillus Calmette Guerin. Vaccine. 25: 6589–6597

Palmer MV, Waters WR, Whipple DL (2002) Susceptibility of raccoons (*Procyon lotor*) to infection with *Mycobacterium bovis*. J. Wildl. Dis. 38:266–274

Parisot TJ (1958) Tuberculosis of fish; a review of the literature with a description of the disease in salmonoid fish. Bacteriol. Rev. 22:240–245

Parra A, Fernandez-Llario P, Tato A, Larrasa J, Garcia A, Alonso JM, Hermoso dM, Hermoso dM (2003) Epidemiology of *Mycobacterium bovis* infections of pigs and wild boars using a molecular approach. Vet. Microbiol. 97: 123–133

Parra A, Garcia A, Inglis NF, Tato A, Alonso JM, Hermoso dM, Hermoso dM, Larrasa J (2006) An epidemiological evaluation of *Mycobacterium bovis* infections in wild game animals of the Spanish Mediterranean ecosystem. Res. Vet. Sci. 80:140–146

Pattyn SR, Portaels F, Boivin A, Van den BL (1971) Mycobacteria isolated from the aquaria of the Antwerp Zoo. Acta Zool. Pathol. Antverp. 52:65–72

Pavlas M, Patlokova V (1985) Occurrence of Mycobacteria in Sawdust, Straw, Hay and Their Epizootiological Significance. Acta Veterinaria Brno. 54:85–90

Pavlik I (2006) The experience of new European Union Member States concerning the control of bovine tuberculosis. Vet. Microbiol. 112:221–230

Pavlik I, Bartl J, Dvorska L, Svastova P, du Maine R, Machackova M, Ayele WY, Horvathova A (2000a) Epidemiology of paratuberculosis in wild ruminants studied by restriction fragment length polymorphism in the Czech Republic during the period 1995–1998. Vet. Microbiol. 77: 231–251

Pavlik I, Bartl J, Parmova I, Havelkova M, Kubin M, Bazant J (1998) Occurrence of bovine tuberculosis in animals and humans in the Czech Republic in the years 1969 to 1996. Veterinarni Medicina. 43:221–231

Pavlik I, Dvorska L, Matlova L, Svastova P, Parmova I, Bazant J, Veleba J (2002) Mycobacterial infections in cattle in the Czech Republic during 1990–1999. Veterinarni Medicina. 47:241–250

Pavlik I, Matlova L, Gilar M, Bartl J, Parmova I, Lysak F, Alexa M, Dvorska-Bartosova L, Svec V, Vrbas V, Horvathova A (2007) Isolation of conditionally pathogenic mycobacteria from the environment of one pig farm and the effectiveness of preventive measures between 1997 and 2003. Veterinarni Medicina 52:392–404

Pavlik I, Svastova P, Bartl J, Dvorska L, Rychlik I (2000b) Relationship between IS*901* in the *Mycobacterium avium* complex strains isolated from birds, animals, humans, and the environment and virulence for poultry. Clin. Diagn. Lab. Immunol. 7:212–217

Pavlik I, Trcka I, Parmova I, Svobodova J, Melicharek I, Nagy G, Cvetnic Z, Ocepek M, Pate M, Lipiec M (2005) Detection of bovine and human tuberculosis in cattle and other animals in six Central European countries during the years 2000–2004. Veterinarni Medicina 50:291–299

Payeur JB, Jarnagin JL, Marquardt JG, Whipple DL (2002) Mycobacterial isolations in captive elephants in the United States. Ann. N. Y Acad. Sci. 969:256–258

Perez J, Calzada J, Leon-Vizcaino L, Cubero MJ, Velarde J, Mozos E (31-3-2001) Tuberculosis in an Iberian lynx (*Lynx pardina*). Vet. Rec. 148:414–415

Peters M, Prodinger WM, Gummer H, Hotzel H, Mobius P, Moser I (2007) *Mycobacterium tuberculosis* infection in a blue-fronted amazon parrot (*Amazona aestiva aestiva*). Vet. Microbiol. 122:381–383

Philips JA, Rubin EJ, Perrimon N (2005) Drosophila RNAi screen reveals CD36 family member required for mycobacterial infection. Science. 309:1251–1253

Pillai SD, Widmer KW, Ivey LJ, Coker KC, Newman E, Lingsweiler S, Baca D, Kelley M, Davis DS, Silvy NJ, Adams LG (2000) Failure to identify non-bovine reservoirs of *Mycobacterium bovis* in a region with a history of infected dairy-cattle herds. Prev. Vet. Med. 43: 53–62

Pinto MR, Jainudeen MR, Panabokke RG (1973) Tuberculosis in a domesticated Asiatic elephant *Elephas maximus*. Vet. Rec. 93:662–664

Portaels F, Chemlal K, Elsen P, Johnson PD, Hayman JA, Hibble J, Kirkwood R, Meyers WM (2001) *Mycobacterium ulcerans* in wild animals. Rev. Sci. Tech. 20:252–264

Portaels F, Elsen P, Guimaraes-Peres A, Fonteyne PA, Meyers WM (1999) Insects in the transmission of *Mycobacterium ulcerans* infection. Lancet. 353:986–986

Portaels F, Realini L, Bauwens L, Hirschel B, Meyers WM, de Meurichy W (1996) Mycobacteriosis caused by *Mycobacterium genavense* in birds kept in a zoo: 11-year survey. J. Clin. Microbiol. 34:319–323

Pozos TC, Ramakrishnan L (2004) New models for the study of Mycobacterium-host interactions. Curr. Opin. Immunol. 16:499–505

Prasad BNK, Gupta SK (1977) Suitability of Mycobacteria As Food for Free-Living Amoebas. Curr. Sci. 46:710–712

Prasad BNK, Gupta SK (1978) Preliminary-report on engulfment and retention of Mycobacteria by trophozoites of Xenically grown *Acanthamoeba castellanii* Douglas, 1930. Curr. Sci. 47:245–246

Prodinger WM, Brandstatter A, Naumann L, Pacciarini M, Kubica T, Boschiroli ML, Aranaz A, Nagy G, Cvetnic Z, Ocepek M, Skrypnyk A, Erler W, Niemann S, Pavlik I, Moser I (2005) Characterization of *Mycobacterium caprae* isolates from Europe by mycobacterial interspersed repetitive unit genotyping. J. Clin. Microbiol. 43:4984–4992

Prodinger WM, Eigentler A, Allerberger F, Schonbauer M, Glawischnig W (2002) Infection of red deer, cattle, and humans with *Mycobacterium bovis* subsp. *caprae* in western Austria. J. Clin. Microbiol. 40:2270–2272

Quigley FC, Costello E, Flynn O, Gogarty A, McGuirk J, Murphy A, Egan J (1997) Isolation of mycobacteria from lymph node lesions in deer. Vet. Rec. 141:516–518

Qureshi T, Labes RE, Cross ML, Griffin JF, Mackintosh CG (1999) Partial protection against oral challenge with *Mycobacterium bovis* in ferrets (*Mustela furo*) following oral vaccination with BCG. Int. J. Tuberc. Lung Dis. 3:1025–1033

Ragg JR, Mackintosh CG, Moller H (2000) The scavenging behaviour of ferrets (*Mustela furo*), feral cats (*Felis domesticus*), possums (*Trichosurus vulpecula*), hedgehogs (*Erinaceus europaeus*) and harrier hawks (*Circus approximans*) on pastoral farmland in New Zealand: Implications for bovine tuberculosis transmission. N. Z. Vet. J. 48: 166–175

Rahim Z, Mollers M, te Koppele-Vije A, de Beer J, Zaman K, Matin MA, Kamal M, Raquib R, van Soolingen D, Baqi MA, Heilmann FG, van der Zanden AG (2007) Characterization of *Mycobacterium africanum* subtype I among cows in a dairy farm in Bangladesh using spoligotyping. Southeast Asian J. Trop. Med. Public Health. 38:706–713

Ramakrishnan L, Valdivia RH, McKerrow JH, Falkow S (1997) *Mycobacterium marinum* causes both long-term subclinical infection and acute disease in the leopard frog (*Rana pipiens*). Infect. Immun. 65:767–773

Ramis A, Ferrer L, Aranaz A, Liebana E, Mateos A, Dominguez L, Pascual C, FdezGarayazabal J, Collins MD (1996) *Mycobacterium genavense* infection in canaries. Avian Dis. 40:246–251

Rankin JD (1958) The experimental production of Johne's disease in laboratory rabbits. J. Pathol. Bacteriol. 75:363–366

Reed GB (1957) Genus *Mycobacterium* (species affecting warm-blooded animals except those causing leprosy). In: Breed R.S., Murray E.G.D., Smith N.R. (Eds.): Bergey's Manual of Determinative Bacteriology.

7th ed. The Williams and Wilkins Co., Baltimore. 703–704

Reichenbach-Klinke H, Elkan E (1965) The Principal Diseases of Lower Vertebrates. Academic Press, New York, 1965.xii+ 600 pp.

Rhodin AGJ, Anver MR (1977) Myobacteriosis in turtles – Cutaneous and hepatosplenic involvement in a *Phrynops-hilari*. J. Wildl. Dis. 13:180–183

Richter E, Weizenegger M, Rusch-Gerdes S, Niemann S (2003) Evaluation of genotype MTBC assay for differentiation of clinical *Mycobacterium tuberculosis* complex isolates. J. Clin. Microbiol. 41:2672–2675

Ristola MA, von Reyn CF, Arbeit RD, Soini H, Lumio J, Ranki A, Buhler S, Waddell R, Tosteson ANA, Falkinham JO, Sox CH (1999) High rates of disseminated infection due to non-tuberculous mycobacteria among AIDS patients in Finland. J. Infect. 39:61–67

Roberts Rj (1975) Melanin-containing cells of teleost fish and their relation to disease. In: The Pathology of Fishes. In: Ribelin, W.E., Migaki, G. (Eds.).University of Wisconsin Press, Madison. 399–428

Rojas-Espinosa O, Lovik M (2001) *Mycobacterium leprae* and *Mycobacterium lepraemurium* infections in domestic and wild animals. Rev. Sci. Tech. 20:219–251

Rolain JM, Lepidi H, Zanaret M, Triglia JM, Michel G, Thomas PA, Texereau M, Stein A, Romaru A, Eb F, Raoult D (2006) Lymph node biopsy specimens and diagnosis of cat-scratch disease. Emerg. Infect. Dis. 12:1338–1344

Ronn R, Grunert J, Ekelund F (2001) Protozoan response to addition of the bacteria *Mycobacterium chlorophenolicum* and *Pseudomonas chlororaphis* to soil microcosms. Biol. Fertil. Soils. 33:126–131

Ross AJ (1970) Mycobacteriosis among Pacific salmonid fishes. In: A Symposium on Diseases of Fishes and Shellfishes. Snieszko, S.F. (Ed.). Amer. Fish. Soc., Spec. Publ., No.5, Washington D.C.

Rupprecht J (1904) Über saürefeste Bazillen nebst Beschreibung eines Falles von spontaner Froschtuberkulose. Inaug. Diss. Freiburg.

Saha K, Jain M, Mukherjee MK, Chawla NM, Chaudhary DS, Prakash N (1985) Viability of *Mycobacterium leprae* within the gut of *Aedes aegypti* after they feed on multibacillary lepromatous patients: A study by fluorescent and electron microscopes. Lepr. Rev. 56:279–290

Sangster C, Bergeson D, Lutze-Wallace C, Crichton V, Wobeser G (2007) Feasibility of using coyotes (*Canis latrans*) as sentinels for bovine mycobacteriosis (*Mycobacterium bovis*) infection in wild cervids in and around Riding Mountain National Park, Manitoba, Canada. J. Wildl. Dis. 43: 432–438

Sato T, Shibuya H, Ohba S, Nojiri T, Shirai W (2003) Mycobacteriosis in two captive Florida manatees (*Trichechus manatus latirostris*). J. Zoo Wildl. Med. 34:184–188

Saunders G (1983) Pulmonary *Mycobacterium-tuberculosis* infection in a circus elephant. J. Am. Vet. Med. Assoc. 183:1311–1312

Scantlebury M, Hutchings MR, Allcroft DJ, Harris S (2004) Risk of disease from wildlife reservoirs: Badgers, cattle, and bovine tuberculosis. J. Dairy Sci. 87:330–339

Schlee K (1957) Über die Verbreitung und Übertragung von *Mycobacterium tuberculosis* durch Fliegen in Schlachtstätten. München, Tierärztliche Fakultät der Ludwig-Maximilians-Universität, 31 pp.

Schliesser TH (1965) Infektionen bei Tieren mit "atypischen" Mykobakterien der Gruppen I, II and IV nach Runyon. Prax. Pneumol. 19:544–550

Schmid K, Schoerner C, Drexler H (2003) [Occupationally acquired tuberculosis in an administrative assistant: Aspects of an expert report]. Dtsch. Med. Wochenschr. 128:432–434

Schousboe P, Rasmussen L (1994) Survival of *Tetrahymena-Thermophila* at low initial cell densities – Effects of lipids and long-chain alcohols. J. Eukaryot. Microbiol. 41:195–199

Serraino A, Marchetti G, Sanguinetti V, Rossi MC, Zanoni RG, Catozzi L, Bandera A, Dini W, Mignone W, Franzetti F, Gori A (1999) Monitoring of transmission of tuberculosis between wild boars and cattle: Genotypical analysis of strains by molecular epidemiology techniques. J. Clin. Microbiol. 37:2766–2771

Sevcikova Z, Ledecky V, Capik I, Levkut M (1999) Unusual manifestation of tuberculosis in an ostrich (*Struthio camelus*). Vet. Rec. 145:708

Shane SM, Camus A, Strain MG, Thoen CO, Tully TN (1993) Tuberculosis in commercial emus (*Dromaius novaehollandiae*). Avian Dis. 37:1172–1176

Shepard CC (1960) The experimental disease that follows the injection of human leprosy bacilli into foot-pads of mice. J. Exp. Med. 112:445–454

Shitaye JE, Matlova L, Horvathova A, Moravkova M, Dvorska-Bartosova L, Treml F, Lamka J, Pavlik I (2008) *Mycobacterium avium* subsp *avium* distribution studied in a naturally infected hen flock and in the environment by culture, serotyping and IS*901* RFLP methods. Vet. Microbiol. 127:155–164

Shively JN, Songer JG, Prchal S, Keasey MS, Thoen CO (1981) *Mycobacterium marinum* Infection in Bufonidae. J. Wildl. Dis. 17:3–7

Shojaei H, Magee JG, Freeman R, Yates M, Horadagoda NU, Goodfellow M (2000) *Mycobacterium elephantis* sp. nov., a rapidly growing non-chromogenic *Mycobacterium* isolated from an elephant. Int. J. Syst. Evol. Microbiol. 50 Pt 5:1817–1820

Sibley. W.K. (1889) Über Tuberkulose bei Wirbeltieren. Virchows Arch. Path. Anat. 116:104–115

Skoric M, Shitaye EJ, Halouzka R, Fictum P, Trcka I, Heroldova M, Tkadlec E, Pavlik I (2007) Tuberculous and tuberculoid lesions in free living small terrestrial mammals and the risk of infection to humans and animals: A review. Veterinarni Medicina. 52:144–161

Snyder RL (1978) Historical aspects of tuberculosis in the Philadelphia Zoo. In: Proc. Symp. Mycobacterial Infections in Zoo Animals, RJ Montali (Ed.), Smithsonian Institution Press, Washington, D.C., 33–44

Soeffing K (1990) Verhaltensökologie der Libelle *Leucorrhinia rubicunda* L. unter besonderer Berücksichtigung nahrungsökologischer Aspekte. Diss. Univ. Hamburg. 148 pp.

Solomon JM, Leung GS, Isberg RR (2003) Intracellular replication of *Mycobacterium marinum* within *Dictyostelium discoideum*: Efficient replication in the absence of host coronin. Infect. Immun. 71:3578–3586

Sreevatsa (1993) Leprosy and arthropods. Ind. J. Lepr. 65: 189–200

Sreevatsa, Girdhar BK, Ipe IM, Desikan KV (1992) Can sand-flies be the vector for leprosy? Int. J. Lepr. Other Mycobact. Dis. 60:94–96

Steiger K, Ellenberger C, Schuppel KF, Richter E, Schmer-bach K, Krautwald-Junghanns ME, Wunnemann K, Eulen-berger K, Schoon HA (2003) Uncommon mycobacterial infections in domestic and zoo animals: Four cases with spe-cial emphasis on pathology. Deutsche Tierarztliche Wochen-schrift. 110:382–388

Steinert M, Birkness K, White E, Fields B, Quinn F (1998) *Mycobacterium avium* bacilli grow saprozoically in cocul-ture with *Acanthamoeba polyphaga* and survive within cyst walls. Appl. Environ. Microbiol. 64:2256–2261

Steinmetz HW, Rutz C, Hoop RK, Grest P, Bley CR, Hatt JM (2006) Possible human-avian transmission of *Mycobacterium tuberculosis* in a green-winged macaw (*Ara chloroptera*). Avian Dis. 50:641–645

Strahl ED, Gillaspy GE, Falkinham JO (2001) Fluores-cent acid-fast microscopy for measuring phagocytosis of *Mycobacterium avium*, *Mycobacterium intracellulare*, and *Mycobacterium scrofulaceum* by *Tetrahymena pyriformis* and their intracellular growth. Appl. Environ. Microbiol. 67: 4432–4439

Stunkard JA, Migaki G, Robinson FR, Christian J (1975) Shrews: A review of the diseases, anomalies, and parasites. Lab Anim Sci. 25:723–734

Sula L, Radkovsky J (1976) Protective Effects of *M-Microti* Vaccine Against Tuberculosis. J. Hyg. Epidemiol. Microbiol. Immunol. 20:1–6

Suter MM, von Rotz A, Weiss R, Mettler C (1984) [Atypical mycobacterial skin granuloma in a cat in Switzerland]. Zen-tralbl. Veterinarmed. A. 31:712–718

Sylwester K (1960) The possibility of infecting the medium with Salmonella *Calliphora erythrocephala* Meig developmental forms (in Polish). Weterynaria. 9:9–31

Talaat AM, Reimschuessel R, Trucksis M (1997) Identification of mycobacteria infecting fish to the species level using poly-merase chain reaction and restriction enzyme analysis. Vet. Microbiol. 58:229–237

Taylor RH, Falkinham JO, III, Norton CD, LeChevallier MW (2000) Chlorine, chloramine, chlorine dioxide, and ozone susceptibility of *Mycobacterium avium*. Appl. Environ. Microbiol. 66:1702–1705

Taylor SJ, Ahonen LJ, De Leij FAAM, Dale JW (2003) Infection of *Acanthamoeba castellanii* with *Mycobacterium bovis* and *M-bovis* BCG and survival of *M-bovis* within the amoebae. Appl. Environ. Microbiol. 69:4316–4319

Tell LA, Leutenegger CM, Larsen RS, Agnew DW, Keener L, Needham ML, Rideout BA (2003) Real-time polymerase chain reaction testing for the detection of *Mycobacterium genavense* and *Mycobacterium avium* complex species in avian samples. Avian Diseases. 47:1406–1415

Terre L (1902) Essai sur la tuberculose des vertébrés a sang froid. Étude de pathologie experimentale et comparée. These Lyon. Zentr. Bakteriol. Parasitenk. Abt. I, Ref. 33:210–

Thoen C, Lobue P, de K, I (2006a) The importance of *Mycobacterium bovis* as a zoonosis. Vet. Microbiol. 112: 339–345

Thoen CO, Richards WD, Jarnagin JL (1977) Mycobacteria iso-lated from exotic animals. J. Am. Vet. Med. Assoc. 170: 987–990

Thoen CO, Schliesser TA (1984) Mycobacterial infections in cold-blooded animals. In: The mycobacteria. Kubica G.P., Wayne l.G.(eds.), Part B, New York and Basel, 1297–1311.

Thoen CO, Steele JH, Gilsdorf MJ (2006b) *Mycobacterium bovis* infection in animals and humans. 2nd ed., Blackwell Publishing Professional, Ames, Iowa, USA, 317 pp.

Thorel MF (1980) Isolation of *Mycobacterium africanum* from Monkeys. Tubercle. 61:101–104

Thorel MF, Karoui C, Varnerot A, Fleury C, Vincent V (1998) Isolation of *Mycobacterium bovis* from baboons, leopards and a sea-lion. Vet. Res. 29:207–212

Tortoli E, Rindi L, Bartoloni A, Garzelli C, Mantella A, Maz-zarelli G, Piccoli P, Scarparo C (2003) *Mycobacterium ele-phantis:* Not an exceptional finding in clinical specimens. Eur. J. Clin. Microbiol. Infect. Dis. 22:427–430

Travis EK, Junge RE, Terrell SP (2007) Infection with *Mycobac-terium simiae* complex in four captive *Micronesian kingfish-ers*. J. Am. Vet. Med. Assoc. 230:1524–1529

Trcka I, Lamka J, Suchy R, Kopecna M, Beran V, Moravkova M, Horvathova A, Bartos M, Parmova I, Pavlik I (2006) Mycobacterial infections in European wild boar (*Sus scrofa*) in the Czech Republic during the years 2002 to 2005. Veteri-narni Medicina. 51:320–332

Trckova M, Matlova L, Hudcova H, Faldyna M, Zraly Z, Dvorska L, Beran V, Pavlik I (2005) Peat as a feed supplement for animals: A review. Veterinarni Medicina. 50:361–377

Tully TN, Shane SM (1996) Husbandry practices as related to infectious and parasitic diseases of farmed ratites. Rev. Sci. Tech. 15:73–89

Turenne C, Chedore P, Wolfe J, Jamieson F, May K, Kabani A (2002) Phenotypic and molecular characterization of clini-cal isolates of *Mycobacterium elephantis* from human speci-mens. J. Clin. Microbiol. 40:1230–1236

Tuyttens FAM, Macdonald DW, Rogers LM, Cheeseman CL, Roddam AW (2000) Comparative study on the consequences of culling badgers (*Meles meles*) on biometrics, population dynamics and movement. J. Animal Ecol. 69:567–580

Tweddle NE, Livingstone P (1994) Bovine tuberculosis control and eradication programs in Australia and New Zealand. Vet. Microbiol. 40:23–39

Van Duijn C (1967) Diseases of fishes. Cox and Wyman Ltd., London, Reading and Fakenham, 309 pp.

van Nie GJ (1981) Avain Tuberculosis in a Free-Living Buz-zard with Bumblefoot. Tijdschrift Voor Diergeneeskunde 106:1033–1036

Van Oss C, Gillman C, Neumann A (1975) Phagocytic Engulf-ment and Cell Adhesiveness. Marcel Dekker, Inc., New York.

van Soolingen D, de Haas PE, Haagsma J, Eger T, Hermans PW, Ritacco V, Alito A, van Embden JD (1994) Use of various genetic markers in differentiation of *Mycobacterium bovis* strains from animals and humans and for studying epi-demiology of bovine tuberculosis. J. Clin. Microbiol. 32: 2425–2433

Vaughan JA, Lenghaus C, Stewart DJ, Tizard ML, Michalski WP (2005) Development of a Johne's disease infection model in laboratory rabbits following oral administration of *Mycobac-terium avium* subspecies *paratuberculosis*. Vet. Microbiol. 105:207–213

Viana-Niero C, Gutierrez C, Sola C, Filliol I, Boulah-bal F, Vincent V, Rastogi N (2001) Genetic diversity of

Mycobacterium africanum clinical isolates based on IS*6110*-restriction fragment length polymorphism analysis, spoligo-typing, and variable number of tandem DNA repeats. J. Clin. Microbiol. 39:57–65

von Benten K, Fiedler HH, Schmidt U, Schultz LC, Hahn G, Dittrich L (1975) [Occurrence of tuberculosis in zoo mammals; a critical evaluation of autopsy material from 1970 to the beginning of 1974]. Dtsch. Tierarztl. Wochenschr. 82: 316–318

von Reyn CF, Arbeit RD, Tosteson ANA, Ristola MA, Barber TW, Waddell R, Sox CH, Brindle RJ, Gilks CF, Ranki A, Bartholomew C, Edwards J, Falkinham JO, O'Connor GT, Jacobs NJ, Maslow J, Lahdevirta J, Buhler S, Ruohonen R, Lumio J, Vuento R, Prabhakar P, Magnusson M (1996) The international epidemiology of disseminated *Mycobacterium avium* complex infection in AIDS. Aids 10:1025–1032

Wang T, Lafuse WP, Zwilling BS (2001) NFkappaB and Sp1 elements are necessary for maximal transcription of toll-like receptor 2 induced by *Mycobacterium avium*. J. Immunol. 167:6924–6932

Washko RM, Hoefer H, Kiehn TE, Armstrong D, Dorsinville G, Frieden TR (1998) *Mycobacterium tuberculosis* infection in a green-winged macaw (*Ara chloroptera*): Report with public health implications. J. Clin. Microbiol. 36:1101–1102

Wayne LG, Kubica GP (1986) Family Mycobacteriaceae Chester 1897, 63 AL. In: Sneath, P.H.A., Mair, N.S., Holt, J.G.(Eds.), Bergey's Manual of Systematic Bacteriology, Part 2, 1st ed. Williams and Wilkins, Baltimore, MD, 1599 pp.

Weber A (1982) Occurrence of Atypical Mycobacteria in Feces Samples of Game Animals. Berliner und Munchener Tierarztliche Wochenschrift. 95:30–32

Weber A, Gurke R (1992) Bacteriological Investigations on the Presence of *Mycobacterium-paratuberculosis* in Fecal Samples of Fallow Deer (*Dama-Dama L*). Zeitschrift fur Jagdwissenschaft. 38:55–59

Weber A, Reischl U, Naumann L (1998) Isolation of *Mycobacterium africanum* from a bull in northern Bavaria. Berliner und Munchener Tierarztliche Wochenschrift. 111:6–8

Weischner M (1989) Der große Schwarzkäfer als Futterinsekt *Zophobas morio*. Gefiederte Welt. 113:89

Wells A, Oxon D (1937) Tuberculosis in wild vole. Lancet. 229:1221–1221

Wells SK, Gutter A, Vanmeter K (1990) Cutaneous Mycobacteriosis in A Harbor Seal – Attempted Treatment with Hyperbaric-Oxygen. J. Zoo Wildl. Med. 21:73–78

White SD, Ihrke PJ, Stannard AA, Cadmus C, Griffin C, Kruth SA, Rosser EJ, Jr., Reinke SI, Jang S (1983) Cutaneous atypical mycobacteriosis in cats. J. Am. Vet. Med. Assoc. 182:1218–1222

Whittington RJ, Lloyd JB, Reddacliff LA (2001) Recovery of *Mycobacterium avium* subsp. *paratuberculosis* from nematode larvae cultured from the faeces of sheep with Johne's disease. Vet. Microbiol. 81:273–279

Wilkins EG, Griffiths RJ, Roberts C (1986) Pulmonary tuberculosis due to *Mycobacterium bovis*. Thorax. 41:685–687

Wilkinson D, Smith GC, Delahay RJ, Rogers LM, Cheeseman CL, Clifton-Hadley RS (2000) The effects of bovine tuberculosis (*Mycobacterium bovis*) on mortality in a badger (*Meles meles*) population in England. J. Zool. 250: 389–395

Wolke RE (1975) Pathology of bacterial and fungal diseases affecting fishes. In: The Pathology of fishes. Ribelin, W.E., Migali, G.(Eds.), The University of Wisconsin Press, Madison Wisconsin. 33–116

Wolke RE, Stroud RK (1978) Piscine mycobacteriosis. In: Proc. Symp. Mycobacterial Infections in Zoo Animals, Montali R.J.(Ed.), Smithsonian Institution Press, Washington D.C. 269–275

Wood JW, Ordal EJ (1958) Tuberculosis in Pacific salmon and steelhead trout. Oreg. Fish Comm. Contrib. 25:1–38

Woodford MH (1982) Tuberculosis in wildlife in the Ruwenzori National Park, Uganda (Part II). Trop. Anim Health Prod. 14:155–160

Yazar S, Dik B, Yalcin S, Demirtas F, Yaman O, Ozturk M, Sahin I (2005) Nosocomial Oral Myiasis by *Sarcophaga* sp. in Turkey. Yonsei Med. J. 46:431–434

Zumarraga MJ, Martin C, Samper S, Alito A, Latini O, Bigi F, Roxo E, Cicuta ME, Errico F, Ramos MC, Cataldi A, van Soolingen D, Romano MI (1999) Usefulness of spoligotyping in molecular epidemiology of *Mycobacterium bovis*-related infections in South America. J. Clin. Microbiol. 37:296–303

Chapter 7

Transmission of Mycobacteria from the Environment to Susceptible Hosts

I. Pavlik, J.O. Falkinham III and J. Kazda (Eds.)

Introduction

As a consequence of the widespread distribution and diversity of sources of ESM and PPM, a large number of possible sources and modes of transmission from the environment to susceptible hosts can be identified. Disease caused by PPM is not restricted to lung infection; disseminated (e.g. bacteraemia), gastro-intestinal (e.g. Crohn's disease) and skin infections (e.g. granulomas) have also been reported. That further increases the possible modes of transmission.

7.1 Modes of Transmission of Mycobacteria

J.O Falkinham III and I. Pavlik

Natural and potable water have been proven to be a source of pulmonary as well as disseminated disease (HIV/AIDS) suggesting that transmission from water can occur via either aerosols (i.e. mycobacteria-containing droplets in a shower) or ingestion (i.e. in HIV/AIDS patients). Also, the close similarity of DNA fingerprints of mycobacterial isolates from patients with pulmonary disease and their potting soil

proves that dust and soil particulates are vectors of transmission.

Aerosolisation of mycobacteria in hot tubs or spas, indoor swimming pools and metalworking fluid (metal removal fluid) has been proposed as a route of transmission leading to both infection and hypersensitivity pneumonitis. It has been proposed that mycobacteria in damp or water-damaged housing materials can lead to respiratory problems and even pulmonary infection. Entrainment of mycobacterial cells in either water or soil requires agitation, any process that leads to the aeration of water (e.g. bubbling, high flow rates, water falls), dust formation in soil (e.g. farming operations, wind) or dust formation during restoration of water-damaged buildings (e.g. Hurricane Katrina in New Orleans, USA) would increase the number of mycobacteria in air. Once aerosolised, the size and weight of droplets or particles would influence the length of time the mycobacteria remained aloft and the distance over which they could be transmitted.

7.2 Demonstrated Modes of Mycobacterial Transmission from Environmental Sources

J.O. Falkinham III

7.2.1 Transmission via Ingestion

The first proof that infection occurred as a consequence of exposure to ESM and PPM showed that water was the likely source of infection in HIV/AIDS

I. Pavlik. (✉)
Head of OIE Reference Laboratories for Paratuberculosis and Avian Tuberculosis, Department of Food and Feed Safety Veterinary Research Institute Brno Czech Republic
e-mail: pavlik@vri.cz

J. Kazda et al. (eds.), *The Ecology of Mycobacteria: Impact on Animal's and Human's Health*,
DOI 10.1007/978-1-4020-9413-2_7, © Springer Science+Business Media B.V. 2009

patients from Boston, USA (von Reyn et al., 1994). DNA fingerprinting involved comparison of large restriction fragments generated by rare-cutting restriction endonucleases and separated by pulsed field gel electrophoresis. The evidence showed an identity of large fragment patterns of isolates from HIV/AIDS patients and from water in a coffee pot in an HIV/AIDS treatment clinic (von Reyn et al., 1994).

In addition to potable water, it was shown that an HIV/AIDS patient, who was exposed to water in the Charles River, was infected with a strain of *Mycobacterium avium* whose large restriction fragment pattern was identical to an *M. avium* isolate from the Charles River (von Reyn et al., 1994). It is possible that this individual was infected via mycobacterial-laden droplets ejected from the river's surface.

7.2.2 Transmission via Aerosolised Water Droplets in Showers in the Home

The IS*1245*/IS*1311* restriction fragment length polymorphism pattern of an isolate of *M. avium* from an immunocompetent patient was shown to be clonally related to those of *M. avium* isolates recovered from the patient's water supply and showerhead sediment (Falkinham et al., 2008). Although there was no perfect match of RFLP patterns with the isolate from the patient, there were only two band differences out of 15–20 bands. Although the *M. avium* isolates from the patient's water and showerhead samples were related, there were band differences in the RFLP patterns, demonstrating that there is clonal variation of *M. avium* within a household water system. It is likely that clonal variation occurs in the patient's mycobacterial population, thus requiring the recovery of multiple isolates from both the patient and their environment to find RFLP matches.

7.2.3 Transmission via Aerosols from Hot Tubes and Spas in the Home

M. avium pulmonary infection has been linked with the presence of *M. avium* in hot tubs and spas. Fingerprint analysis has further demonstrated the matching identity of patient(s) and hot tub or spa isolates, identifying the source of the infection. Although aerosol samples were not collected, it is likely that infection occurred as a consequence of inhalation of mycobacteria-laden aerosols. Mycobacteria are enriched and concentrated in droplets (Parker et al., 1983).

7.2.4 Transmission via Aerosols at Work

Outbreaks of hypersensitivity pneumonitis associated with the work place are possibly due to the aerosolisation of environmental mycobacteria. "Lifeguard Lung" has been linked to the exposure of lifeguards to aerosols from indoor swimming pools (Rose et al., 1998). Unfortunately, neither aerosol samples of the facility nor mycobacterial cultures from the lifeguards were collected, prohibiting definitive proof of mycobacterial aetiology. Hypersensitivity pneumonitis has been reported in mechanics exposed to aerosols of metalworking fluid (metal recovery fluid) used in automotive drilling, cutting, grinding and finishing operations to cool the tool and part and to carry off the metal particles. A novel species, *M. immunogenum*, was isolated from metalworking fluid and is capable of growing in the oil–water emulsions that are found in the metalworking fluid (Wallace et al., 2002).

In both these workplace situations, proof of ESM is lacking. However, the circumstances strongly support the role of mycobacteria. ESM and PPM are present in water used in pools and metalworking fluid and are resistant to disinfection. Biofilms in both systems would further increase disinfectant resistance and lead to persistence in a system in spite of draining and cleaning. In fact, the outbreaks are associated with the use of disinfectants; consistent with the hypothesis that disinfection leads to selection for mycobacteria. Further, mycobacteria are capable of degrading the oils and surfactants in metalworking fluids (Wallace et al., 2002).

7.2.5 Transmission via Aerosolised Dusts

Potting soil contains high levels of peat or sphagnum vegetation in which mycobacteria have been found in concentrations of up to 10^6 CFU/g (Iivanainen et al., 1997). If the potting soil is dry, the particles can be released as dust during handling. To mimic the normal handling of the potting soils, samples were dropped from a height of 30 cm and the aerosol was collected (Andersen, 1958). Aerosols collected from the 79 potting soil provided by the 21 elderly patients with mycobacterial pulmonary disease harboured a large and diverse population of mycobacteria; most frequently *M. avium* and *M. intracellulare*. For 14 (67%) of the 21 patients the potting soil yielded the same *Mycobacterium* species as from the patient. Thus, potting soils are potential sources of infection by environmental mycobacteria (De Groote et al., 2006).

7.2.6 Transmission via Direct Physical Contact

Mycobacterial diseases of the skin (i.e. granulomas) have been typically associated with *M. marinum* infection with water as the possible source. Infections have been reported in both humans and fish. Infection in either food fish (e.g. fish-farming) or aquaria (e.g. tropical fish) is important because of the potential economic loss. Human infection is commonly associated with trauma and skin abrasions that are work (fishing) or hobby (aquaria) related. To date there have been no published fingerprinting studies showing the identity or near identity (clonality) of isolates from the environment and the infected individual.

Recently, an outbreak of mycobacteriosis has been investigated in an aquarium in Florida (J. O. Falkinham, unpublished data). The fish were shown to be infected with *M. marinum* and the colony morphology of all the isolates was unique and identical. Significant numbers of *M. marinum* CFU of the same unique colonial morphology were recovered from biofilm samples which were collected from a pipe feeding the aquarium (260 CFU/4 cm^2) and a 2.5 cm diameter plastic ring from an inline filter (21 000 CFU/ring). The fingerprinting of the *M. marinum* will now follow. Other samples from the aquarium, including filter material, biofilms and water yielded other mycobacteria including, *M. chelonae*, *M. haemophilum*, *M. kansasii* and *M. gordonae*. Thus, that particular aquarium and likely

many other aquaria, harbour a diverse population of mycobacteria, many in high numbers.

7.2.7 Transmission via Insect Bite

One of the recent and significant contributions to understanding mycobacterial ecology has come from studies of outbreaks of "Buruli Ulcer" in Africa and Australia. The ulcer is caused by *M. ulcerans*, one of the few mycobacteria producing an extracellular toxin. Historic observations documented that the majority of cases were associated with the close proximity to river courses and that wearing long sleeved shirts and long trousers offered protection from infection. The current hypothesis is that biting insects (e.g. "water bugs") serve as the vector of infection. They have been shown to carry mycobacteria, including *M. ulcerans* and can transmit *M. ulcerans* via biting a susceptible host.

7.3 Risk Assessment of Transmission of Mycobacterial Disease from Environmental Sources

J.O. Falkinham III and J. Kazda

7.3.1 General Considerations

A major difficulty in assessing the risk posed by environmental mycobacteria in either water or soil is that the dose required for the appearance of disease is unknown. Further, the incubation period between exposure and evidence of infection (i.e. immune reaction) or disease is unknown. As a consequence, the only risk analysis that can be performed is based on retrospective analysis. Such an approach is flawed as the concentration and species distribution of mycobacterial cells in a putative environmental source (e.g. shower water or potting soil) at the time of the initial exposure is unknown. The studies below are based on retrospective analysis where the environmental samples were collected long after the patient's diagnosis. In the following analyses, it has been assumed that the number and species composition in the environmental samples has not changed since the patient's exposure.

7.3.2 Risk Assessment of M. avium in Drinking Water for Immuno-Deficient Individuals

It has been proven that the source of *M. avium* infection in simian immunodeficiency virus (SIV)-infected macaques was their drinking water; no other water source was available (Mansfield et al., 1995; Mansfield and Lackner, 1997). *M. avium* infection developed in 21 (31%) of 67 of the SIV-infected macaques that were not treated with any anti-viral or antibiotic agents. *M. avium* was recovered from 11 (44%) of 25 hot- or cold-water samples. There was no difference in the recovery of *M. avium* from hot and cold water. Eight (53%) of 15 animal isolates shared an identical or nearly identical DNA fingerprint pattern as one of the five water isolates tested. Four (27%) other animal isolates had the same DNA fingerprint patterns as a second water isolate. The three (20%) remaining animal isolates had different profiles that were distinct and not seen in the remaining three water isolates, each of which had a unique profile. The diversity of *M. avium* fingerprint patterns demonstrated in these studies is similar to that shown in other studies. *M. avium* is a very diverse species. The macaques were housed at the New England Primate Centre and the water system served by that Centre was surveyed in the 1980s for *M. avium* (Du Moulin and Stottmeier, 1986).

The concentration of *M. avium* in the water of the New England Primate Centre ranged from 10 to 500 CFU/100 ml. Assuming that the SIV-infected macaques drank 100 ml of water per day, they were exposed to between 10 and 500 *M. avium* cells per day on a continuous basis. The mean survival of the macaques with *M. avium* infection was approximately 1000 days following the SIV infection. Thus, those monkeys were cumulatively exposed to between 10 000 and 500 000 CFU following the SIV infection. However, only 31% of those macaques showed evidence of *M. avium* infection; 69% were free of *M. avium*.

On the assumption that it is appropriate to extend this analysis from SIV-infected macaques to *HIV*-infected humans, the data suggest that one-third of *HIV*-infected humans, exposed to a cumulative dose of between 10 000 and 500 000 CFU of *M. avium* through water, would develop *M. avium* infection. It is probable that this risk is shared by patients with malignancy (i.e. leading to immune deficiency) and being treated with immune-suppressive agents (e.g. transplant patients). Interleukin-12 deficiency is also a risk factor for *M. avium* bacteraemia whose route of infection is water. Five of fourteen (36%) young patients with an IL-12 deficiency (average age 10 years) were shown to have *M. avium* infections (Altare et al., 1998).

Using the assumptions above, one-third of individuals with an IL-12 deficiency exposed to a cumulative dose of 35 000–2 000 000 CFU of *M. avium* through water will develop *M. avium* bacteraemia. The IL-12 deficient individuals appear to have somewhat lower probability of contracting *M. avium* infections than HIV/AIDS patients.

7.3.3 Risk Analysis for M. avium Infection via Natural Aerosols

M. avium and *M. intracellulare* are readily aerosolised from water (Wendt et al., 1980; Parker et al., 1983). In fact, *M. avium* and *M. intracellulare* strains are highly concentrated (i.e. average = 3000-fold) in droplets ejected from water (Parker et al., 1983). Based on field (Wendt et al., 1980) and laboratory (Parker et al., 1983) experiments, approximately 3000 CFU of *M. avium* or *M. intracellulare* would be naturally aerosolised per square metre per day from water containing three CFU of *M. avium* or *M. intracellulare*. That would lead to an average concentration of *M. avium* or *M. intracellulare* cells at ground level of three CFU/m^3. The number aerosolised would rise in direct proportion to the number in the water (Parker et al., 1983).

Aerosols generated naturally from water containing 10–500 CFU of *M. avium* or *M. intracellulare* would contain approximately 10–500 CFU/m^3 of *M. avium* or *M. intracellulare*. If an adult inhales 6 m^3 of air per day, they would be infected with between 60 and 3000 CFU/day of *M. avium* or *M. intracellulare*. It is likely that aerosols in showers have even higher concentrations of *M. avium* (though the length of exposure is shorter). Eight of nineteen (42%) patients with pulmonary alveolar proteinosis (and all smokers) were infected with *M. avium* (Witty et al., 1994).

Aerosol-borne *M. avium* might not be the only source of infection, because cigarettes contain *M. avium* (Eaton et al., 1995). However, if aerosols were the source and we assume that infection follows a long exposure of 10 years, 42% of patients with pulmonary alveolar proteinosis exposed to a cumulative aerosol dose of 220 000 –11 000 000 CFU of *M. avium* will develop *M. avium* pulmonary disease.

7.3.4 Risk Assessment of M. avium Infection via Shower Aerosols

Because of the presence of *M. avium* in drinking water (Du Moulin and Stottmeier, 1986; Du Moulin et al., 1988; Fischeder et al., 1991; von Reyn et al., 1993; von Reyn et al., 1994; Glover et al., 1994; Covert et al., 1999; Falkinham, 2002), it is likely that an individual can be infected by exposure to shower aerosols. Recently, samples of hot and cold water and a sample of biofilm collected from a showerhead were received from a patient with pulmonary *M. avium* infection. *M. avium* was recovered from the cold water (2 CFU/ml), hot water (2 CFU/ml) and showerhead biofilm (240 CFU/cm^2). Fingerprinting by IS*1245*/IS*1311* RFLP demonstrated that the patient, water and biofilm isolates were clonally related (see above).

Assuming that the flow rate in the shower is 1 gal/min and that the person takes a 5 min shower, the individual would be exposed to a total of 38 000 CFU of mycobacteria during the shower. The number of mycobacteria in water is an underestimate, because that calculation does not include any mycobacteria released from the biofilm and entering the water during showering. This could be a considerable number because of the large number of mycobacteria in the showerhead biofilm, as reported by others (Schulze-Robbecke and Fischeder, 1989; Iivanainen et al., 1999).

If only 10% of the mycobacteria were aerosolised (i.e. 3800 CFU) and the volume of the bathroom was 800 ft^3, the concentration of mycobacteria in the room would be 4.75 CFU/ft^3. If the person remained in the bathroom another 5 min (i.e. 10 min total duration) and has normal respiration (1 ft^3/min), they would inhale

47.5 CFU. Mycobacteria in the showerhead biofilm would also be expected to be aerosolised and their contribution to aerosol-borne mycobacteria substantial because of their large numbers in the showerhead.

7.3.5 Risk Assessment of Mycobacteria in Potting Soil

Identity of PFGE fingerprints from isolates of mycobacteria isolated from patients and their potting soil (De Groote et al., 2006) make a calculation of the exposure generated by working with potting soil worthwhile. Aerosol samples obtained by dropping a sample of potting soil 30 cm yielded between 0.04 and 0.4 CFU/liter and as high as 2400/283 liter were gathered within the 10 min collection time (De Groote et al., 2006). Because the Andersen 6-Stage Cascade Sampler collects air at approximately the same rate as an individual breathing (28.3 liter/min), a 10 min exposure would result in the inhalation of between 11 and 113 CFU.

As noted above, even higher numbers (i.e. 2400 CFU) could be inhaled for some samples. Not all particles with associated mycobacteria would be expected to reach the deepest portion of the lung; only those collected on stages 4, 5 and 6 (Andersen, 1958). In one example, representatives of a single *M. intracellulare* clone were recovered from stages 4 (1.87 CFU/liter), 5 (0.24 CFU/liter) and 6 (0.003 CFU/liter). Exposure to that potting soil aerosol would lead to inhalation of 529 CFU (stage 4), 68 CFU (stage 5) and 8.5 CFU (stage 6). Notwithstanding the fact that some of the elderly patients had predisposing conditions for mycobacterial disease (e.g. CFTR mutations), all patients were infected with mycobacteria and were exposed to between 10 and 2400 CFU of mycobacteria if their work with potting soil generated dust and their exposure was limited to 10 min.

7.3.6 The Possible Convergence Towards Pathogenicity in Environmentally Derived Mycobacteria: An Ecological Approach

The phylogeny of mycobacteria indicates that pathogenic species originated from saprophytic ones.

The following hypothesis will attempt to indicate one of the possible ways in which pathogenicity developed in previously saprophytic species.

As mentioned elsewhere, the grey layer of sphagnum vegetation has been found to maintain favourable conditions for the multiplication of environmentally derived mycobacteria, a proportion of which are potentially pathogenic. Intact sphagnum bogs are known to be biotopes that tend to be stable for a very long time. In a sphagnum bog in north Germany, acid-fast rods which morphologically resembled mycobacteria were found in the whole peat profile up to 110 cm beneath the surface of the vegetation. This indicated that acid-fast microorganisms were present more than 1000 years ago and that the colonisation started at the same time as the development of this sphagnum bog (Kazda, 2000).

Similar results obtained in numerous other sphagnum bogs have confirmed that sphagnum biotopes can be considered to be one of the oldest sources of mycobacteria. Furthermore, an examination carried out in three sphagnum bogs in north Germany over a period of 4 years revealed the permanent presence of mycobacteria. These favourable conditions, obtained repeatedly in moderate climate zones from spring to autumn, facilitate the continuing multiplication of mycobacteria. Similar results were obtained from sphagnum bogs in Central and Eastern Europe, Scandinavia, Ireland, New Zealand and North and South America. Thus, intact sphagnum vegetation represents a biotope with a closed microbial population, where mycobacteria as habitat microorganisms thrive continuously, depending on abiotic factors such as temperature and humidity.

The extreme environmental conditions created by acidification exclude the growth of higher plants, thus enabling the development of very large, homogenous sphagnum carpets. Humans and most animals have avoided these hostile environments, but environmental mycobacteria adapted to these conditions, successfully colonising sphagnum bogs from the outset. An increase in the human population resulted in new settlements, followed by the enlargement of arable and pasture fields. Hitherto, intact moorland biotopes were colonised for agriculture. By invading sphagnum bogs, humans and animals entered these biotopes by force, manipulated and finally destroyed them. The niches exclusively used by mycobacteria were now overlapped by humans and animals. This resulted initially in a strong interaction between mycobacteria and macroorganisms.

The natural closed cycle of mycobacterial growth in sphagnum bogs was interrupted. The mycobacteria were compelled to colonise other kinds of environments, because the niches in sphagnum bogs were no longer suitable. As an alternative, they colonised the mucous membranes, particularly the respiratory tract of macroorganisms. Such an invasion of humans by mycobacteria was also observed recently in Finland, where the peatlands were intensively ditched for forestry between 1975 and 1990. In this region, the colonisation of the respiratory tract of humans by mycobacteria increased fivefold (Iivanainen et al., 1993). The authors stressed that such widespread manipulation of sphagnum bogs had influenced the increase in clinical isolation of mycobacteria from humans.

Attempts to colonise sphagnum bogs for agricultural purposes have varied from country to country. In north Germany, one approach to the colonisation of sphagnum bogs was initiated by the King of Denmark more than 250 years ago. Farmers from the South were invited to develop settlements in large sphagnum bogs and introduce agriculture. This experiment failed due to the high acidity of the cultivated arable soil and the settlers fell into poverty and misery. Close contact with the sphagnum bogs and housing in poor peat dwellings (Photo 7.1) enhanced the flooding with mycobacteria. Intercurrent diseases of the farmers and their families, combined with the hostile cold and humid climate of peatlands may have suppressed their immune system.

Furthermore, malnutrition and a diet high in vegetables, especially savoy, could have supported the development of mycobacterioses. In a study using savoy (*Brassica oleracea* var. *sabauda*), its growth-stimulating effect in vitro on *M. tuberculosis*, *M. bovis*, *M. kansasii* and the *MAC* was monitored (Kazda, 1978). Oral application of savoy juice to guinea pigs had an aggravating effect on an infection with *M. kansasii*. This *Mycobacterium*, which is either not at all or only slightly pathogenic for guinea pigs provoked heavy mycobacterioses of the lung, liver and spleen, when savoy juice was added to the ordinary diet.

Thus, the gate for a massive invasion of the respiratory tract by mycobacteria was open. As supposed for other bacteria and viruses, this high level of permanent exposure may have aided the development of parasitic forms of mycobacteria. Once restricted to humans and animals, these mycobacteria specialised in growth within their living niches. The part of the microbial population exhibiting mutants with pathogenic properties benefited and, by invading susceptible, immunocompromised hosts, mycobacterioses could develop. In fact, the number of tuberculosis cases among moorland farmers was much higher than the average. Mycobacteria, originally saprophytic, found other niches for multiplication and the species, which were able to combine parasitic properties with growth in the environment, were more successful. This enabled them to survive under the different conditions created by civilisation.

With regard to mycobacteria as a cause of disease, PPM species create a transitional stage between the saprophytic and the obligate pathogenic group. Their pathogenic character fluctuates, because PPM can also exist as environmental saprophytes. Some of them, such as *M. a. hominissuis*, *M. marinum* or *M. xenopi*, are more often associated with mycobacterioses. Others, like *M. gordonae* or *M. terrae* are generally regarded as contaminants if found in lesions. However, in extensive studies, *M. gordonae* was classified as having a moderate pathogenicity for humans and experimental animals (Fasske and Schroder, 1989). *M. terrae* has been described in connection with synovitis and osteomyelitis (May et al., 1983). Collins et al. concluded that all mycobacterial species may have the potential to cause disease under permissive conditions (Collins et al., 1986). The recent increase of mycobacterioses in immunocompromised HIV/AIDS patients, provoked by mycobacteria until this point regarded as non-pathogenic, can support this hypothesis.

7.4 Allergic Reactions Provoked by Environmental Mycobacteria

J. Kazda

One of the oldest immunological tests, developed by Robert Koch in the late nineteenth century is the tuberculin test. It has been used for more than 100 years for the diagnosis of tuberculosis in cattle. The response to the tuberculin injection is described as delayed-type hypersensitivity, mediated by a population of sensitised T cells and such reactions have specific and non-specific components.

7.4.1 Non-specific Reactions to Tuberculins

The occurrence of non-specific tuberculin reactions in animals greatly supported research into the ecology of mycobacteria. In the advanced phase of tuberculosis eradication, animals which had shown positive reactions failed to display any evidence of tuberculosis during detailed *post mortem* examinations. These non-specific reactions, explained mainly by sensitisation from environmental mycobacteria (Worthington, 1967), were not satisfactorily documented.

One of the first detailed ecological studies was carried out in a large chicken breeding farm in south Bohemia. Over three consecutive years many positive results were observed for avian tuberculin. In the experiment, 600 chickens were randomly chosen, tested with avian tuberculin and then autopsied. Strong positive reactions to avian tuberculin developed in 78.3% of cases, indicating avian tuberculosis. The autopsies revealed no tuberculous lesions, but it was possible to cultivate "*M. brunense*" (Kazda, 1966) from the liver and spleen. This *Mycobacterium* provoked positive reactions to avian tuberculin in experimentally infected chickens. It was later re-identified as *M. intracellulare* serotype 8, now *M. avium* subsp. *hominissuis* (Kubin et al., 1969).

The source of these mycobacteria was found to be moorland water used as an additional water supply during the summer months. The tanks used for storing the drinking water were exposed to sunshine during the summer months. The water temperature in the tanks exceeded 30 °C. It also contained a high concentration of the described mycobacteria. When pure well water was introduced instead, the non-specific reactions to avian tuberculin lowered within a few months to 2.8% (Kazda, 1967b).

In the same moorland region, non-specific tuberculin reactions in cattle occasionally occurred. The same *Mycobacterium* species as above were found in a watering place surrounded by sphagnum moss. This *Mycobacterium* was able to provoke positive reactions to bovine tuberculin in 52.7% of tested cattle and the reaction size was 0.95–3.8 cm (Kazda, 1967a). Further non-specific tuberculin reactions in cattle on pastures occurred in north Germany. They were caused by *M. a. hominissuis* found in a moorland pond used as a watering place. This *Mycobacterium*, as well as *M. a. avium*, was able to grow in samples of this water, following sterile filtration with incubation at 31 °C (Kazda, 1973). The question then arose as to whether moorland pond water could be regarded as a source of these mycobacteria. Over the next 3 years, this pond was examined from spring through to autumn for the presence of mycobacteria. The above-mentioned species were not found, which indicated that this moorland water was not the reservoir of these *Mycobacterium* species.

Deviation of the delayed-type hypersensitivity, caused by environmental mycobacteria is of great importance in veterinary medicine, as the tuberculin test is one of the most suitable methods for programs to eradicate bovine tuberculosis. In the advanced stadium, when tuberculosis infection subsides, non-specific reactions have often been recorded, especially in Finland and the USA (Freerksen, 1960).

Concerning the pathogenicity of *M. a. hominissuis,* similarly to *M. a. avium*, this species provoked the Yersin-type septic tuberculosis in chicken following intravenous application. Using labelled cells, it was evident that pathogenic *M. a. avium* is capable of multiplying in chicken organs, whereas the numbers of *M. a. hominissuis* continually diminished (Hampl et al., 1969).

An ecological approach was applied in New Zealand in order to find the cause of non-specific reactions to tuberculin. Concerning bovine tuberculosis, the possum (*Trichosurus vulpecula*) is known to be its natural reservoir, sharing pastures together with cattle. Furthermore, in some parts of the country, frequent non-specific tuberculin reactions were common. The mild climate of New Zealand, especially on the North Island creates favourable conditions for slowly growing mycobacteria to thrive. Unlike European sphagnum associations where rapidly growing mycobacteria predominated, a total of 89.7% of the isolated mycobacterial strains were slowly growing. Most of them (55.8%) belonged to the new species *M. cookii* (Kazda et al., 1990), isolated from *Sphagnum cristatum* and *Sphagnum falcatulum*, common in New Zealand. *M. cookii*, whose optimum growth is at 31 °C, was also found in *Sphagnum falcatulum*, surrounding the Waimangu crater near the hot fumaroles. These strains were adapted to the higher temperature and

grew optimally at 37 °C. *Sphagnum falcatulum*, floating on the surface of the Crater Lake released *M. cookii* in water acidified by sulphuric gas to pH 2.4. The high CFU concentrations (5.8×10^5 and 6.1×10^6/ml were isolated from these adverse water conditions) document the unusual adaptability of this *Mycobacterium* species.

The favourable temperature conditions of New Zealand aid the growth of the mycobacteria not only in sphagnum but also in ponds, which are often surrounded by this vegetation and used as watering places for cattle. Of 71 water samples collected, 43.9% contained mycobacteria 62.9% of were *M. cookii* (Kazda and Cook, 1988). In an experiment carried out to establish if this *Mycobacterium* is capable of inducing non-specific sensitisation to bovine tuberculin in cattle, it was found that all animals developed a reaction to bovine tuberculin following intradermal sensitisation. Using *M. cookii* sensitin, it was possible to distinguish the different kinds of sensitisation (Monaghan et al., 1991).

The development of non-specific sensitisation to tuberculin has had a decisive influence on the route of oral transmission. In most cases it is the surface water as described above. In special cases, the digestion of sensitising mycobacteria could take place directly from the vegetation. In the relatively mild climate of Ireland it is common for animals to spend the winter in the fields. Early in the spring, the grass is scarce but moss is abundant because of the low vegetation canopy and heavy rainfall. Bryidae such as *Hylocomium splendens, Breutelia chrysocoma* and *Thuidium tamariscinum* are common in the pasture fields and are a rich source of *M. hiberniae* (Kazda et al., 1997). For more detailed information see Section 5.3.3a.

7.4.2 Development of Cutaneous Basophilic Hypersensitivity Provoked by Mycobacteria

This reaction, known also as the Jones-Mote Reaction, first observed following the repeated intracutaneous application of albumin, was originally regarded as a transient reaction of the tuberculin type. A detailed description of this reaction has shown the accumulation of basophilic leukocytes in *dermis superficialis* and the reaction was designated as cutaneous basophilic hypersensitivity. This reaction develops

5 days after the antigen application compared with 3 weeks for the tuberculin reaction (Stadecker and Leskowitz, 1973). Digestion of mycobacteria caused an increase in basophilic leukocytes in the blood, as shown by the application of *M. fortuitum* in drinking water (Kazda, 1977). This type of reaction to mycobacteria is mentioned as one of the possible interactions with environmental mycobacteria because it can be provoked by an extremely small amount of mycobacterial cells.

7.4.3 Allergic Reactions of the Digestive Tract Caused by Mycobacteria

In addition to the above described responses, environmental mycobacteria can cause an allergic reaction in the small intestine. In the author's experiments, *M. a. hominissuis* was used for oral sensitisation of guinea pigs and rabbits. Three weeks later, a suspension of the same mycobacteria was administered orally to the previously sensitised animals and to the control group. The sensitised animals showed weight loss at the start of the second day, later refused food and died within 12 days of application. The autopsies revealed hyperaemia and swelling of the mucosa in the duodenum, jejunum and ileum beginning on the second day of oral application.

Continued administration led to heavy haemorrhagic oedema and extended haemorrhages in all parts of the small intestine. Micro-morphological examinations confirmed heavy oedematous swelling of the mucosa showing an infiltration with plasma cells and numerous haemorrhages. These lesions did not occur in the control group. Furthermore, it could be shown that the intensity of the allergy in the digestive tract was dependent on the concentration of *M. a. hominissuis* applied orally to sensitised animals. The effective dose in guinea pigs and rabbits was 10^6/dose.

This indicates that an allergic reaction of the digestive tract in individuals, previously sensitised by environmental mycobacteria can be provoked by the digestion of PPM. It seems possible that disorders can occur when sufficient concentrations of mycobacteria are ingested via drinking water or contaminated vegetables. The majority of the human population is sensitised to mycobacterial antigens by the BCG vaccination or previously by environmental mycobacteria. The allergic reaction in the small intestine could cause transient types of disorders, the aetiology of which is still not explained. In such cases, mycobacteria should be taken into consideration as possible causative agents.

7.4.4 Deviation of BCG Protection Caused by Environmental Mycobacteria

In the late 1950s, 20 years after the evaluation of the vaccine was begun, varied protective effects against tuberculosis were evident. Compared with the high protective effect of 75% in the UK, vaccinated individuals in Puerto Rico showed only 30% protection. It was found that on the one hand, environmental mycobacteria can offer some protection against tuberculosis and leprosy but on the other hand, they may mask the protection by BCG vaccines. Furthermore, contact with environmental mycobacteria probably plays an important role in building up the infant immune system (Fine et al., 2001).

It was found that the cross-reaction induced by environmental mycobacteria can influence the protection by the BCG and vice versa. The degree of protection against tuberculosis and leprosy varied and the efficacy against leprosy was greater than against tuberculosis: in Malawi 54 vs. 11%, in Kenya 81 vs. 22% and in South India 20 vs. 0%, respectively. These differences (besides other factors) may be hypothetically explained by the occurrence of local mycobacterial flora.

There is evidence that exposure of animals to environmental mycobacteria can protect against the challenge with *M. tuberculosis*. Based on guinea pig experiments, intradermal exposure to the *MAC* imparted 50%, in the case of *M. kansasii* 85% and with *M. fortuitum* 15% as much protection as the BCG (Palmer and Long, 1966). There is a general acceptance that exposure to mycobacteria can interfere with subsequent responses to other mycobacterial species. Contact with environmental mycobacteria could mask the protective effect of the superimposed vaccine. The BCG-vaccine trials against leprosy in Uganda showed a positive effect of the BCG-mediated of the immune response caused by environmental mycobacteria (Stanford and Paul, 1973). *M. vaccae* and *M. nonchromogenicum* widely distributed in soil and water were considered to be responsible for a beneficial effect of the BCG.

Furthermore, it was found that the products of these mycobacteria can not only influence the development of natural immunity, but support the therapy of tuberculosis alone or in combination with antibiotics (Stanford, 1992).

In experiments using mice models the most common environmental mycobacteria from Malawi, *M. scrofulaceum, M. vaccae, M. fortuitum, M. chelonae* and the *MAC* were tested. It was found that environmental mycobacteria raise the immune responses that control the multiplication of BCG, curtailing the vaccine-induced immune response before it is fully developed. The environmental mycobacteria promote a distinct level of protective immunity, also against the inoculated BCG. It is very important that the protective effect was not sufficient to reduce the multiplication of *M. tuberculosis* (Brandt et al., 2002).

On the other hand, environmental mycobacteria can have an aggravating effect on leprosy in nu/nu mice when inoculated together with *M. leprae* (Section 2.2).

The role which environmental mycobacteria play in heterologous immunity is poorly understood. Their ubiquitous incidence in soil, water and dust means they often interact with human and animal macroorganisms. They commonly occur in a mixture of various species, in most experiments however, only single mycobacterial strains have been used separately. The distribution of mycobacteria in the environment has been evaluated indirectly, in most cases using specific sensitins. This delayed type of sensitivity is transient depending on the concentration of mycobacteria and the duration of the exposition. In the future, sophisticated molecular genetic methods should be used to detect the whole spectrum of environmental mycobacteria in different biotopes.

7.5 The Prevention of Mycobacterial Infections Derived from the Environment

I. Pavlik, V. Mrlik and J. Kazda

The presence of environmental mycobacteria on farms can be associated with or without the appearance of mycobacterial infections in animals. How is this possible? One of the main reasons is the purchase of infected animals, e.g. piglets reared on another farm

where they became infected by peat fed as a supplement (Section 5.6.3). Mycobacterial infections in such pigs become "economically" apparent only after slaughter despite the fact that the animals were infected when young (usually between birth and weaning). Financial losses due to the presence of tuberculous lesions in the intestinal and head lymph nodes (submandibular, retropharyngeal and parotid) can reach tens of percent (Pavlik et al., 2003). Another example is the transmission of the causative agent of paratuberculosis to a herd of ruminants by the purchase of a young animal from an infected herd. Clinical manifestations in ruminants can appear after several years without the contamination of the farm environment with *M. a. paratuberculosis* (Ayele et al., 2001).

Accordingly, when a survey is made, it is necessary to first collect a thorough anamnesis in every herd and only then to attempt to explain the results of the study of the environment. Investigations should be focused on the following risk factors in animal herds:

- Bedding.
- The technology of animal housing.
- Water supply for animals in stables, paddocks, pastures and during transport; water used for swimming, bathing or showering animals.
- Forage (green, preserved, dried fermented) and grain-based feed (sources of feed mixture components including diet preparation).
- Feed supplements.
- Liquid and solid manure and urine (the method of handling, their removal from stables and storage).
- Contact with excreta from other animals.

In the case of a higher frequency of tuberculous lesions, e.g. in pigs (tens of percent of affected animals), caused by ESM and PPM, a massive reservoir of infection is likely present in the environment. Occasional detections of mycobacteria, e.g. in pig faeces or in faeces attached to the walls and floors of stables, are not usually very significant. Direct microscopic examination of samples after staining according to Ziehl–Neelsen can give general information on the intensity of infection. The analysis of laboratory results can be of help for the confirmation of infection and may allow the formulation of specific and effective measures:

- The analysis of the species profile of mycobacteria from the infected tissues of animals and from the environment.
- Serotyping of *MAC* members.
- RFLP analyses carried out on all *M. avium* species and the results of other methods used for the study of molecular epizootiology/epidemiology.

7.5.1 Basic Principles of Prevention in Different Types of Herds

During the analysis of all of these data, massive infections (most often in water and feed) and other risk factors (most often bedding) should be primarily considered. Animal farming can be divided into the following five groups from the point of view of the prevention of mycobacterial infections:

- Conventional herds of domestic animals in old buildings.
- Conventional herds of domestic animals in new buildings.
- Ecological herds of domestic animals.
- The keeping on pasture of domestic and wild animals (above all on farms).
- Captive-bred animals (above all in zoos).

7.5.1.1 Conventional Herds of Domestic Animals in Old Buildings

"The exhaustion" of the environment should first be assessed when the occurrence of ESM and PPM is analysed in these herds. The presence of mycobacteria in examined samples from the environment (scrapings of faeces and bedding residues from the walls, floors and troughs, soil from animal paddocks) may often document contamination of the stable environment with mycobacteria. However, a potential source of infection cannot be detected from these results because these samples mostly contain mycobacteria from animal faeces. The examination of solid and liquid manure and dung water for mycobacteria fails to provide useful information relating to the search for a reservoir of ESM and PPM infections.

In old buildings (stables) for keeping animals, it is also necessary to focus attention on the wooden fences and barriers. In one pig herd, *M. a. hominissuis* was detected on the wooden fences, which consisted of old, damp wood. A sow was infected by nibbling on the fences (J. Docekal and M. Pavlas, personal communication; Photos 7.2 and 7.3). In old buildings, small terrestrial mammals are often present and can transmit the causative agent of avian tuberculosis to farm animals (above all pigs). Surprisingly, pigs of all age categories are dexterous predators and can easily catch and swallow small, careless terrestrial mammals (I. Pavlik, unpublished data) which migrate to animal farms due to lack of food (in the northern hemisphere in autumn and winter months; Photos 7.4 and 7.5).

7.5.1.2 Conventional Herds of Domestic Animals in New Buildings

On agricultural farms, ESM and PPM are found less frequently in the environment of new buildings than of old buildings. The first vectors of ESM and PPM are usually animals transported into the buildings; mycobacteria can be present both on their body surface and in their intestinal tract. In pig herds, it is therefore suitable to suggest cleaning the animals with warm water and a brush, e.g. prior to the sows entering the farrowing house. This is routinely performed on farms with a high standard of animal hygiene.

The second frequent route of PPM transmission to a herd is contaminated feed or feed supplements. Such a case was documented on a pig farm in Portugal. In a new building, the farmer began to give peat to piglets from day five of life, not only as a feed supplement but also as bedding for weaned piglets. The numerous findings of tuberculous lesions in lymph nodes after the slaughtering of the first market pigs could thus be explained immediately.

The third frequent transmission route is through the animals' water. New technologies, above all those used for watering the piglets, benefit from naturally heated water in the plastic pipeline systems. The heat-carrying capacity of this material is lower than that of a metal piping system and thus, the heat energy from the environment can be transferred more easily to cold water in the pipelines, ranging between 8 and 10 °C when the water is drawn from a well that taps water sources to depths of at least 5–10 m below ground level. Pipelines situated under the ceiling of stables and corridors can be warmed to 20 or 30 °C quite easily, especially in the summer months. Biofilm formed in the pipeline system can subsequently detach and clog up the drinkers for piglets. Clusters of several tens to hundreds of AFB can be detected by microscopy. Piglets can usually be infected in this way after weaning as they drink a lot of water (they no longer intake milk from their mother, but drink more water due to the stress of weaning).

In many herds, antibiotics are given to piglets through the water supply (provided they suffer from diarrhoea) or vitamin C which increases their immunity after weaning (Photo 7.6). The vitamin carrier material is fructose or sucrose, which attenuates the acidic taste and the piglets readily drink such "flavoured" water. However, the fact that large biofilms are gradually formed in such pipelines during the summer months and subsequently released into the water represents a risk. Mycobacteria growing in biofilm can cause the formation of tuberculous lesions in the lymph nodes of pigs. Nevertheless, they can only be diagnosed after slaughter at the age of 6–8 months.

7.5.1.3 Ecological Herds of Domestic Animals

The first and highest risk in ecological herds is the contact of farm animals with free-living animals (Photos 7.4 and 7.5), in particular the contact with birds and small terrestrial mammals in the feeding and watering places. Protective measures (above all nets) are imposed on some farms; these prevent contact with free-living birds. However, it is difficult to restrict the movement of small terrestrial mammals and their contact with domestic animals, although it is known that small terrestrial vertebrates (both rodents and insectivores) can be infected with the causative agent of avian tuberculosis (Photo 7.7).

The second high risk is the presence of mycobacteria in the soil of paddocks for pigs (Photo 7.8). ESM and PPM were observed to be present in the lymph nodes of up to 70% of these pigs, even though no gross lesions were seen in the lymph nodes (I. Pavlik, unpublished data; I. Moser, personal communication), this due to the fact that tuberculous lesions are formed 10–12 weeks after infection. Thus, the quality and safety of "high-quality meat" produced from ecological herds can be markedly reduced owing to the presence of mycobacteria and can so pose a threat to immunosuppressed consumers.

The third risk factor for the potential mycobacterial infection of pigs is the keeping of animals of different age categories together (e.g. several sows and piglets of different age categories are kept together in paddocks). The infection of piglets in this environment can originate from soil and adult animals (sows) and may be of a high intensity.

7.5.1.4 Animals on Pastures

The exposure of animals on pastures to ESM and PPM becomes evident (in the case of seasonal grazing), mainly after the removal of animals from pastures because of a decreasing quality of grass (in the northern hemisphere before winter begins). Plants enter a period of vegetative inactivity and become overgrazed. The most intensive infection of animals with ESM and PPM from soil occurs during this period. Non-specific responses are frequently found to bovine tuberculin testing in early winter; these should be distinguished by differential diagnosis using simultaneous testing with bovine and avian tuberculin (Photo 7.9).

Accordingly, it is useful to use the simultaneous tuberculin testing at once. Another possible approach is to perform tuberculin testing in spring (March or April), when the sensitisation to ESM and PPM in stabled animals for the most part disappears. Non-specific responses to bovine tuberculin have been detected in this season in cattle and sheep kept on pastures (I. Pavlik and M. Pavlas, unpublished data; Photo 7.9).

Another important source of mycobacteria for grazing animals is water and the water-contaminated surface of pastureland. Mycobacteria are found in the surface water (particularly in rain water and surface water in areas used by people). Accordingly, it is more appropriate to use artificial watering places which use cistern water, or water conducted through pipelines. It is advisable to remove the sediment from the watering places because clusters containing millions of mycobacteria have been detected there (I. Pavlik, unpublished data; Photo 7.10).

7.5.1.5 Animals Bred in Captivity

The first risk factor for animals bred in captivity (mainly in zoological gardens) is "the exhaustion" of the environment due to the long-term stay of animals in one place (one paddock, aviary or cage). Such a place becomes contaminated with organic mass (above all faeces and feed leftovers) over a long period of time and a variety of mycobacteria including *M. a. paratuberculosis* can be detected there (I. Pavlik, unpublished data; Photo 7.11). Therefore, it is often necessary to perform mechanical cleaning of pens, ideally with hot water and steam under pressure.

The second risk factor is posed by decorative wooden subjects (roots, decomposing trunks, bark, sawdust, etc.; Photo 7.12) in which *M. a. hominissuis* is often found and subsequently flushed with rain water into ponds or other water containers (e.g. in paddocks for wild ruminants or in terrariums; I. Pavlik, unpublished data). It is usually more suitable to use shavings (or sawdust from trees without bark) that are almost sterile so as to decrease the risk of contaminating the surrounding environment with ESM and PPM (Matlova et al., 2003).

The third risk factor for the transmission of mycobacteria is the use of feed from non-tested (or not sufficiently tested) animal herds. In one rearing station for birds of prey, *M. a. avium* transmission was documented as having come from infected pheasants from an old peasantry (3- to 4-year-old culled hens were fed to them; Photos 3.7 and 3.8). For example, in several zoos in the Czech Republic between 1969 and 1996, bovine tuberculosis was detected in the following predators: lions (*Panthera leo*), cheetahs (*Acinonyx jubatus*) and wolves (*Canis lupus*) Pavlik et al., 1998). These animals were fed with condemned meat from an abattoir where cattle lacking tuberculous lesions, but yielding a positive response to bovine tuberculin, were slaughtered (M. Pavlas, personal communication).

7.5.2 Bedding in Animal Herds and Risks of Mycobacterial Infection

Bedding is a highly risky factor in all types of herds because PPM can be transmitted through it. Above all, bedding containing wood products is highly risky for all types of herds. The bedding materials are most often contaminated with PPM from sawmills that process entire trunks or crushed bark used for decoration in landscape architecture (Section 5.8; Photo 5.52). Their use can be categorised relating to the risk of infection for kept animals with the present PPM species (mostly

MAC members) according to the type of housing and species of animals as follows:

- Pig herds in large-scale and small-scale production conditions.
- Herds of different categories of cattle.
- Flocks of gallinaceous birds and waterfowl.
- Animal herds in zoos.

7.5.2.1 Pig Herds in Large-Scale and Small-Scale Production Conditions

The risk posed by the use of sawdust as deep bedding has been reported by a number of authors from many countries. Even the enzymatic fermentation of sawdust for deep bedding (e.g. with the use of the preparation ENVISTIM) does not lead to their devitalisation (Section 5.8). Less is known about the use of sawdust in farrowing houses. Many breeders use sawdust to dry piglets after birth, the use of sawdust on the floor of pens for piglets so as to better isolate their bodies from the concrete floor is common in pens. Sawdust is also added to the "nests" of piglets, where it helps keep the environment dry and teaches piglets to use assigned areas in the pens for urination and defecation. If the sawdust is highly contaminated with PPM, piglets can only get infected after the "sugaring" of the floor or the sprinkling of the floor with sawdust (Photos 5.51 and 5.52).

It is necessary to draw farmers' attention to the risks connected with the use of other raw materials. In one herd, sawdust was replaced with bran (Photos 5.42 and 5.51). This is waste that accumulates from corn processing; "stockpiles" of bran are formed in some years. Accordingly, farmers cheaply acquire large amounts of bran and store it in shelters on non-reinforced floors. Damp soil causes the paper bags to become moist and the bran then becomes wet and mouldy. *M. a. hominissuis* was found to be abundant in the bottom part of bags. With regard to the frequent presence of ESM and PPM in soil (Section 5.5), the detections were not surprising. However, it is noteworthy that *M. a. hominissuis* was also isolated from the upper parts of bags containing dry bran.

Accordingly, we examined bran from bags stored in a dry place on wooden pallets, which prevented the bags from absorbing moisture from the ground

(Photo 5.42). Nevertheless, the detection rate of mycobacteria was comparable with the case mentioned above. The source was found to be the "mills" from which the bran originated. It was stored there for several years and only put into the bags when farmers wished to purchase it. The expiry date was set as 6 months from the placing of the bran in a bag and not from the date when it was obtained after flour production. Because this "relatively dry" bran has a high absorption capability, it most likely became contaminated with PPM during storage.

Another risk factor is the use of sawdust after the weaning of piglets on both large-scale and small-scale production farms. Three different kinds of use can be distinguished:

The Use of Sawdust During Transport of Weaned Piglets to New Stables

In some cases, sawdust may be highly contaminated with ESM and PPM (Section 5.8). The weaned piglets can get infected during transport within the course of approximately an hour if sawdust is put on the floor of the vehicles for their transportation (in most cases lorries and tractor trailers). In Holland, the contact of piglets with sawdust during transport which lasted approximately 1 hour was sufficient for their infection (R. Komijn, personal communication). Thus, in the Czech Republic, a 30 km journey necessary for the transport of weaned piglets to another farm appeared to be highly risky (Z. Rihacek, personal communication). In another pig herd, the finding of tuberculous lesions in the lymph nodes of slaughtered pigs decreased from 20% to less than 1% when sawdust was replaced with high-quality dry straw, which was not a source of mycobacteria during their transport to another farm which lasted approximately 1 hour (I. Pavlik, unpublished data).

The Use of Sawdust as Deep Bedding

The use of sawdust as deep bedding for weaned piglets should be viewed as highly risky. The analysis of different layers of deep bedding (surface, depths of 20 cm and 40 cm) showed the same frequency of ESM and PPM and also the same species profile. This was mainly due to the high activity of piglets and later to

that of growing pigs. Their relatively thin legs sink deep into the bedding and they repeatedly "mix" the deep bedding with their snouts (Photo 7.13).

The Use of Sawdust on the Pen Floor to Prevent Slipping and Its Effect on the Ethology of Piglets

Some farmers wrongly assume that they will prevent the infection of their piglets with mycobacteria by limited the use of sawdust. Sawdust making up a layer of 1 cm is usually laid on the floor ("sugaring" or sprinkling with sawdust; Photo 5.42) of pens for weaned piglets with a concrete floor and a warmed bed. Sawdust is spread over the area of the pen where the piglets are supposed to move and later on also lie. In direct contrast to this, sawdust is not put on other parts of the floor in pens (most often not on the floor near the bottom grate, through which urine and faeces flow away) and that area is repeatedly made wet with cold water during the daytime. Piglets thus learn to urinate and defecate through the bottom grate, where the excrements can easily flow out and thereby, do not dirty the pens and their body surfaces. If faeces remain lying in the corner of pens for a long time, different species of Diptera visit it to lay eggs. In the summer and winter seasons, larvae develop there and can be involved in the spread of mycobacteria in a herd of infected pigs (Photo 7.14).

7.5.2.2 Herds of Cattle

Sawdust is used as bedding for cattle herds, especially in cases where there is a lack of straw or other type of conventional bedding (Photo 7.13). This happens especially in extremely dry years, when the amount of straw available for this purpose is insufficient, or after an accident (e.g. the loss in a fire of straw or haystacks intended for bedding). A lack of bedding material is typical for the subtropical zone. In some cow herds, sawdust is used as the main bedding material; this is usually spread over the areas where cows lie. If there is a lack of straw, sawdust mixed with straw can be sprinkled over selected areas of the stables, most often the manure corridors, with the aim of increasing the absorption capability of the bedding. In herds of young cattle, sawdust is only used in the manure corridors.

The use of sawdust contributes to the risk of the occurrence of PPM, especially the presence of *M. a. hominissuis* (Section 5.8). However, other *MAC* members and other species of PPM (above all *M. fortuitum*) can also cause the sensitisation of cattle. After testing with bovine tuberculin, numerous non-specific responses can be observed; these must be distinguished by simultaneous tuberculin testing. Another risk factor is the occurrence of mastitis caused by *M. fortuitum*, commonly diagnosed in subtropical and tropical regions (Section 5.7.4). The slow decomposition of sawdust in the soil is an ecological problem due to the fact that manure containing sawdust is less airy and becomes "heavy". It is, therefore, more suitable not to use sawdust in cattle herds and to maintain litter-free stables (the areas where cows lie are often covered with rubber, which provides thermal insulation and prevents the risk of pressure sores and teat lesions).

7.5.2.3 Breeds of Gallinaceous and Water Fowl

Sawdust and shavings are often used for conventional breeds of gallinaceous birds. They may pose a risk, primarily due to the fact that they cause the sensitisation of poultry, which then exhibit non-specific responses to avian tuberculin. This testing is particularly necessary on small and ecological farms, whose young pullets are sold on to other farms or whose eggs are sold at markets.

On water fowl farms which produce ducklings for reintroduction into the wild, rearing breeds of wild ducks such as *Anas platyrhynchos*, sawdust and shavings are placed in the nests (I. Pavlik, personal observation). Ducks or geese then lay eggs on the dry ground and the egg surface is not contaminated with mud containing bacteria, which can cause the decreased hatchability of ducklings or goslings. Nevertheless, PPM present in sawdust can cause non-specific sensitisation to avian tuberculin.

7.5.2.4 Breeds of Wild Animals in Captivity

The risks associated with the possible infection of wild animals in captivity can be categorised into four groups according to their type of use.

The Transport of Wild Animals

The duration of animal transport and the quality of the environment are important factors from the point of view of mycobacterial infections. A case of aquarium fish transportation from countries in southeastern Asia in fertiliser bags is particularly bad. The water in the bags for fish transportation was highly organically polluted and the percentage of dead fish in one delivery reached several tens of percent. The presence of *M. marinum* was detected in all of the examined dead fish from a group of several thousand aquarium fish of different species. All of the fish from the delivery were euthanised and the water and fish were safely disposed of (I. Pavlik and P. Lescenko, unpublished data). If organic residues, including the carcasses of animals imported in this manner are insufficiently devitalised, they may constitute a risk of new mycobacterial species being introduced into the environment of target countries.

In birds, *M. genavense* was detected in the Czech Republic for the first time in the intestines of a blue headed parrot (*Pionus menstreeus*) imported from Surinam in 2005 (I. Pavlik, V. Grymova, R. Halouzka, J. E. Shitaye, J. Svobodova, M. Moravkova, unpublished data; Photo 6.37). *M. a. paratuberculosis* was spread to the Czech Republic by imported ruminants; it was subsequently detected in different components of the environment, in invertebrate animals, wild ruminants, hares and other animals (Section 5.10.2).

Considering the ecology of mycobacteria, we can conclude that all imported animals may be a potential source. Accordingly, it is necessary to abide by the import quarantine for at least 28 days after the import of the animals. However, if possible, it should be prolonged, in zoological gardens for example, to several months due to the long incubation period of many mycobacterial infections.

Wild Animal Shops

The close contact of animals with the environment in exotic animal shops, which can be highly contaminated with mycobacteria, can pose the highest hazard. Transmissions of, for example, *M. marinum* from an aquarist shop have been described, not only to a variety of poikilothermic animals (Section 6.5) but also to insectivores (Section 6.7.1) and shop staff (Section 3.7.7).

From the ecological point of view, it is also necessary to consider the possible risk of the transmission of different mycobacterial species by plants or decorative objects (e.g. old roots). Accordingly, it is necessary to properly wash aquarium plants (including roots) and to hold the decorative wooden objects in boiling water before their introduction into the aquarium. It is also necessary to wash new sand in boiling water before use. Dried, granulated feed or self-made and similarly treated feed is most suitable for fish. It is well known that drying for a period of a few days to weeks causes the devitalisation of ESM and PPM. Store-bought feed should also be considered as risky: *M. marinum* has been detected in live *Tubifex* (Section 6.3.1.2) and shrimp ova (Section 6.3.4.1).

Store-bought bedding for rabbits and other pets, particularly sawdust and shavings, must look dry and "fresh" at first sight. It is safer to buy shavings, which are less frequently infected with PPM. Other items are more or less safe for animals (pets) with respect to mycobacterial contamination.

Zoological Gardens

Zoological gardens are highly risky when considering the ecology of mycobacteria and mycobacterial infections. It is above all due to the high variety and concentration of animals kept there. The environment in the majority of zoological gardens is affected by the long-term captivity of animals. Tens of different ESM and PPM species are found, mostly in the moist environments of paddocks, cages and aviaries. Mycobacteria are mostly found in drinkers, ponds, wet peat and wet sand around the feeding places. Other "matrices" that are hazardous owing to the potential multiplication of mycobacteria are wooden decorative items and soil used for growing decorative plants (Photo 7.12).

Dry environments free from food leftovers, faeces and urine can be viewed as the surest path to prevention. Other principles of prevention are comparable with those for farm animals (Section 7.5.1).

Special Interest (Small) Animals Kept as Pets

The mutual transmission of the causative agents of mycobacterial infections among animals kept in households as pets is highly risky. Even though their owners care selflessly for such animals, over-caring can in fact constitute a pre-dispositional factor for the development of a mycobacterial infection in many cases. Mouth-to-parrot beak feeding, for example, can result in the transmission of the causative agent of human tuberculosis *M. tuberculosis* (Sections 2.2.1 and 6.6; Photo 6.31). All of the above-mentioned principles are valid for the hygiene of pet breeding. The risk of transmitting the causative agents of different mycobacterial infections should be emphasised because animal owners can also become infected. Children are particularly at risk due to the frequent occurrence of geophagia and the habit of inserting toys shared with the animals into their mouths (Section 5.10.2).

7.5.2.5 Prevention of the Spread of Mycobacteria Through Bedding

The transmission of mycobacteria from bedding to different animal herds can be prevented by the adoption of the following general principles:

i. On the whole, the use of sawdust should be considered as risky. Sawdust or shavings obtained from wood processing plants (e.g. furniture companies) may be subsequently contaminated with mycobacteria if stored in inadequate conditions (humidity causing fermentation, storage of sawdust on a non-reinforced floor, e.g. in a loft, storage for an extended period; Photos 5.24 and 5.51).

ii. If wood products are used, it must be borne in mind that despite sterilisation by heat or chemical treatment, these can get subsequently contaminated with PPM through the faeces of birds, small terrestrial mammals, dust, etc. (Photo 5.52). Accordingly, the best method of prevention is not to use sawdust, shavings, bark or other wood products.

iii. The use of straw stored in a wet place (e.g. outdoor stacks; Photo 7.15) is likewise highly risky. Mycobacteria can be transmitted to the straw through surface water, bird faeces and small terrestrial mammals. The use of dry straw from the lower parts of a stack usually contaminated with faeces and urine from small terrestrial mammals may also pose a risk (Photo 5.43).

iv. The use of peat as bedding is also hazardous. Farmers keeping pigs and turkeys are recommended to use peat as soft bedding, which prevents pressure-induced limb lesions in pigs and limb and pectoral muscle lesions in turkeys. However, peat is often contaminated with mycobacteria including the causative agent of avian mycobacteriosis *M. a. hominissuis* and the causative agent of avian tuberculosis *M. a. avium* (Photo 7.16).

v. The use of adequately stored dry straw as bedding appears to be most convenient (Photo 7.15). The use of shavings has been shown to be suitable for technological reasons in some herds. However, it must originate from companies which produce furniture from wood lacking bark residues which can often be contaminated with *M. a. hominissuis*.

vi. Another important factor is the storage of straw and shavings. Straw should be kept in a dry place, ideally in a roofed shelter. After being air-blown straight from the production room through a special pipeline, shavings should be stored in inaccessible silos in the factory. The storage time is also important. Shavings should be used within several weeks of their delivery to a farm; straw should be used before the following harvest.

7.5.3 The Effect of Animal Housing Technology on the Spread of Mycobacteria

Animal housing built on a reinforced concrete floor is the best (Photo 7.17). A sufficient amount of good quality straw or other raw materials (e.g. hay of lower quality; Photo 5.43) should be used if old technologies are being utilised for the bedding. Paddocks with earthen floors are most dangerous from the standpoint of the transmission of mycobacteria (Photo 5.16). On older farms, these can be highly contaminated with animal faeces and urine and are usually muddy all year round. Such soft paddocks can become sources of causative agents causing gross lesions in lymph nodes, especially of pigs: mycobacteria (Section 5.5.4.3) and *Rhodococcus equi* (Section 5.5.5; Photo 5.36).

The rotting parts of wooden fencing constitute another risk factor. They are often nibbled on by the housed animals, which can then become infected by the PPM present in them. They are most often contaminated with *M. a. hominissuis* which penetrates

damaged mucosa in the oral cavity and reaches the blood and lymph vessels. PPM subsequently cause the formation of tuberculous lesions in the corresponding head (submaxillary, retropharyngeal and parotid) lymph nodes (J. Docekal and M. Pavlas, unpublished data).

A lot of dust is usually present in herds of animals who subsist on dry diets. The dust itself can irritate the respiratory tract of the animals (especially pigs) and it is also usually contaminated with PPM (Photos 7.2 and 7.3). Accordingly, the stables should be effectively ventilated and the accumulating dust should be quickly disposed of. Other preventive measures include the removing of settled dust from the stables at least once every 6 months and whitewashing once a year (Photo 5.55).

7.5.4 The Water Supply for Animals and the Risk of the Spread of Mycobacteria

Water and the water sediment in watering systems are very frequent sources of mycobacteria for infected animal herds (Photos 5.12 and 5.15). Watering systems for animals on pastures are different to those of stables (usually on animal farms or in the halls and pavilions in zoological gardens; Photos 5.14 and 5.15).

7.5.4.1 The Water Supply for Animals on Pastures, in Game Parks and on Deer Farms

The different types of water supply for animals can be categorised as follows:

- Natural watering places (burns, ponds, lakes, morasses, etc.; Photos 5.18, 5.19, 5.20 and 5.21).
- Delivered water available directly from transport cisterns or added to water reservoirs or watering places (Photo 7.10).
- Water from artesian wells pumped by windmills (most common in Australia and North America) and human or animal-power pumped water (above all in desert and semi-desert areas – the Sahara, the Sahel, etc.; Photo 5.13).

- Water conducted under pressure through pipelines into watering places (Photo 7.10).

In all these cases, ESM and PPM colonisation of the formed sediment and biofilms can be expected during hot periods. Therefore, it is necessary to keep the surroundings of watering places dry. When using mobile water reservoirs, the watering place should be changed at least once a month to let the damp area on the pasture dry. As far as ESM and PPM occurrence are concerned, water from artesian and conventional wells is generally the least contaminated; in contrast, slack surface water is the most contaminated (Photos 5.18, 5.19, 5.20 and 5.21).

It follows that the best pastures should be covered with grass and that no pools should be allowed to form during rainy periods. The access of captive animals to natural water sources should be prevented (also due to different parasitic disorders) and they should be given well water (Photo 7.10). Although this way of watering is rather expensive, it is safer in hot months.

7.5.4.2 The Water Supply for Animals on Farms and in Zoological Gardens

Better conditions for the supply of water and for watering animals exist in closed buildings (in stables on farms or in pavilions in zoological gardens). Technologically speaking, water is mostly enclosed in the piping during transport and thus, protected from surface contamination. Nevertheless, it is necessary to follow the guidelines in order to reduce the risk of animals being exposed to ESM and PPM that might affect their health (Section 7.4). Watering systems can be categorised according to the water supply system and water sources as follows.

The Delivery of Cistern Water to Stables

This system is old-fashioned, expensive and laborious. The delivery of water to animals depends upon human factors, which brings about numerous risks. The case of *M. bovis* detection on a cattle farm after the eradication of bovine tuberculosis was an unusual one. After a well dried up during the hot months, heifers housed in the stables were given cistern water tapped from a nearby burn twice a day (in the morning and the evening). Several months later, positive responses to bovine tuberculin were detected in almost one half of them. The causative agent of bovine tuberculosis was not detected in the sediment from the brook. It was found through later investigation of the cistern which was in fact contaminated with the source of infection. Subsequently, the cistern was lent to another farm where bovine tuberculosis was also detected and where it was used for dung water transport. It was used afterwards on the original farm as a drinking water cistern for cattle without prior washing or disinfection and in this way they transmitted *M. bovis* to the herd (I. Pavlik senior and J. Subrt, unpublished data).

Supplying Water Under Overpressure

Overpressure reservoirs are necessary for the supply of water under overpressure; gravity causes the water to flow out to different types of watering places (Photo 7.18). If this system is used, it is necessary to routinely check (once a month) whether the expansion tank is protected with a cover (to protect the water from the penetration of mycobacteria present in dust, bird faeces, invertebrates, etc.). The tanks should be cleaned at least twice a year. Large clusters of AFB can be observed in the sediment (rust coloured), often with a high content of iron (stimulation of mycobacterial growth; Photos 5.12 and 5.15). If the temperature outside is high, mycobacteria present in the reservoir deposits can easily multiply (Photo 7.19). In hot months, the outflow is increased causing their agitation and thus ESM and PPM from the sediment gain access to the intestinal tract of animals (I. Pavlik, unpublished data; Matlova et al., 2003).

Overpressure Water Supply (the Flow of Water to Watering Places Under Pressure Straight from the Water Piping)

In the compression system, water flows under pressure straight from the water piping and thus no expansion reservoirs are necessary. Nevertheless, it is necessary to disinfect and clean the water piping regularly to prevent the formation of biofilms usually colonised with ESM and PPM (Section 4.7; Photo 7.20). If mycobacteria are released spontaneously into water or after disinfection (Section 5.2.8), massive infection of animals can occur.

Water Supply from Surface Water

For the most part, the use of surface water (often contaminated with mycobacteria) is not allowed for the watering of animals (Photos 5.18, 5.19, 5.20 and 5.21). In one pig herd, pond water was used for animal watering during a hot period when water in the farm well dried up. A variety of PPM including *M. fortuitum*, *M. diernhoferi* and *M. a. hominissuis* were detected in that water and later in the lymph nodes of pigs (R. Axman, J. Bartl, I. Pavlik, L. Matlova, unpublished data).

In conclusion we can say that an overpressure system supplied from water piping appears to be suitable for watering animals. The watering of animals with surface water in herds with a high density of animals is inconvenient. Highly polluted water from natural sources (rivers, burns and ponds) is likewise unsuitable. If water reservoirs and expansion tanks are used, the repeated cleaning of mud and sediment (several times a year) is recommended.

7.5.5 The Protection of Feed and Feed Supplements from Mycobacteria

Feeds and feed supplements, e.g. deep mine peat, often become sources of mycobacteria for animals. Peat is a source of ESM or can be contaminated by water and soil. Accordingly, a number of preventive measures should be adopted for the protection of feed and feed supplements from contamination with mycobacteria present in water, soil and dust.

7.5.5.1 The Protection of Fodder

Fodder (often referred to as grange feed) is usually grown and harvested on farms and fed to the animals kept there. The risks of mycobacterial contamination were discussed previously (Section 5.6). When mowing meadows and fields, it is necessary to bear in mind free-living animals, which may be disturbed or killed by this activity. It is necessary to search for dead animals and remove them before mowing.

Harvested forage plants from pastures or fields cannot be effectively protected from contamination because dust, rainwater and other components of the environment in industrialised regions contain high levels of ESM and PPM. The preventative treatment of pastures by adding lime to a concentration of 800–1000 kg/ha can generally adjust the pH of acidic soil, which is a good environment for the survival of mycobacteria. The period of vegetation dormancy is the best time for its application, which should be performed at least once a year (Photo 7.21).

Due to the fact that mycobacteria are particularly sensitive to desiccation (Chapter 4), the rapid sun-drying of hay is an important prevention practice. The sun-dried hay (*M. a. paratuberculosis* is devitalised by the drying process) is suitable for the feeding of the animals. This drying process results in decreased amounts of *M. a. paratuberculosis* in pastures and consequently a low risk of infection for ruminants. Taking into account the age-dependent increased resistance of ruminants to infection with *M. a. paratuberculosis*, animals over the age of 18 months can be fed hay. The same measures can also be adopted in neighbouring pastures, which provide hay for free-living animals.

Green fodder given to the animals in stables should not be exposed to the sun for a long period of time. Green fodder should be stored in a shaded place in thin layers up to several tens of centimetres. Fodder can "self-heat" and if stored for several days, it is not unusual for the propagation of fast-growing mycobacteria (e.g. *M. fortuitum* and *M. chelonae*) to follow.

7.5.5.2 The Protection of Feed Grain

It is best to store feed grain (corn, corn meal, bulk feed mixtures and granulated diets) in dry and sealed containers or in covered tanks (Photo 5.39). The temperature and humidity during storage should be kept at low levels so as to reduce the risk of multiplication of ESM and PPM from the environment.

The bodies of dead birds (possibly containing the causative agent of avian tuberculosis) and of small terrestrial mammals (at risk of being contaminated with different mycobacterial species) can be found in feed grain. If dead bodies of animals are introduced together with the feed or if they are cut, crushed and mixed into the granules (as in the production of granulated diets), a severe threat exists of extensive contamination (Photo 7.5). A variety of mycobacteria from cadavers can be spread throughout feeds by beetle larvae or imagos (Section 6.3.5.5). That route of transmission of the causative agent of avian tuberculosis was revealed in a pig herd kept under poor hygienic conditions where the corn meal was contaminated with both faeces and the bodies of dead birds and small terrestrial mammals. The breeder wrongly assumed that the animals are thus provided with "safe" animal protein (I. Pavlik, J. Bartl, L. Matlova, unpublished data).

Another important risk factor is the use of liquid feeds (Photos 3.20 and 6.12). This feed is prepared several hours before feeding by mixing the feed in pure water or in water containing whey or yeast (soaked in water for several days). Cold drinking water with quite a low level of mycobacteria is usually used. However, the use of warm water is risky. In one pig herd, water from an old electric water heater was used for mixing the diet. *M. xenopi* was later isolated from tuberculous lesions in the intestinal lymph nodes of the pigs, from the water and from the liquid feed. Culture examination of the dry diet was negative for ESM and PPM; it follows that *M. xenopi* most likely originated from the warm water (I. Pavlik and M. Pavlas, unpublished data; Photo 3.27).

For this reason, it is safer to use cold drinking water (water from warm water conduits can be contaminated with high amounts of "thermophilic" mycobacteria, especially *M. xenopi*). This information is also important for the preparation of milk diets for piglets and milk replacements for calves (Photo 6.13).

7.5.5.3 The Protection of Feed for Carnivores and Primates in Captivity and for Domestic Pets

Purchased granulated feed for carnivores (dogs and cats) is not usually contaminated with mycobacteria (I. Pavlik, unpublished data). Nevertheless, feed (meat) from infected animals may pose a problem (Photo 7.22). In the 1960s and 1970s in a few zoos in the Czech Republic, bovine tuberculosis was diagnosed in feline beasts fed with condemned meat from the

slaughter of cattle responsive to bovine tuberculin (M. Pavlas, personal communication). The surface contamination of meat granules or meat with PPM is not typical and represents the least significant hazard for kept carnivores.

The sources and routes of the spread of bovine tuberculosis in zoo-kept primates cannot be explained in many cases. The high density of captive animals and the limited space in many zoos contribute to a high risk of infection. The importation of animals from abroad or the exchange of breeding stallions for the prevention of "inbreeding" are the most frequent causes of transmission of infection. In a few cases the breeders or visitors to the primates were suspected to have spread the infection to them (C. O. Thoen, personal communication). To a large extent, senior citizens coming with grandchildren and great-grandchildren constitute sources of mycobacteria for the exhibited animals. The transmission usually occurs when the animals are given feed brought by the visitors because these can be contaminated with aerosol, saliva and/or sputum containing mycobacteria.

Accordingly, it is recommended to sell feed to the visitors which they can give the animals. The animals can also be kept behind glass barriers, separating them from visitors; this has been as applied in many modern zoos.

7.5.5.4 The Protection of Feed and Feed Supplements

It was found recently that feed supplements (peat, kaolin, bentonite, etc.) used to prevent diarrhoea in young animals and in the prevention of pressure lesions may pose a risk (Section 5.6.3). Due to the fact that these raw materials are almost sterile upon extraction from the mine, they can be protected from mycobacterial contamination. However, their high absorption capability is problematic because they freely absorb mycobacteria from the environment (from dust, surface water, etc. after extraction).

After it underwent irradiation treatment with gamma rays, several companies sold peat as a feed supplement for piglets between 2001 and 2005. They ensured the absolute sterility of the peat (I. Pavlik, unpublished data) by transporting it in sealed bags and it was safe for the health of piglets. However, irradiated foodstuffs and feed supplements have been banned for feeding purposes since 2005. Accordingly, bentonite or kaolin must be certified as being mycobacteria-free before they are given to animals.

7.5.6 The Prevention of the Spread of Mycobacteria Though the Correct Handling and Storage of Excrement

Mycobacteria are present in the faeces and urine of animals and possess the ability to survive fermentation for some period (Section 5.10). Herds can be protected from the adverse effects of mycobacteria in the following sections:

7.5.6.1 Contact with Faeces and Urine of All Types of Domestic and Wild Animal Herds Should Be Very Limited

Provided they have sufficient room in free stall barns, some animals (e.g. llamas or pigs) can defecate and urinate in an assigned area in which they do not lie afterwards (Photo 7.17). However, if the concentration of animals in pens is high, this may not be possible and the repeated contact of animals with their excrement can result in the spread of PPM shed through faeces (for detailed description see item 5).

7.5.6.2 Transport Vehicles Can Contribute to the Spread of Mycobacteria (Faecal Tanks, Trailers Contaminated with Manure, Tractors, etc.)

In one cattle herd infected with *M. a. paratuberculosis* the causative agent with an identical genotype was also detected on the tractor wheels which transported manure during the day and delivered the feed to the feeding table in the mornings and evenings (I. Pavlik, unpublished data). Another risk is posed by these vehicles when they deliver feed, for example, to the silage pits. Undesired manure thus comes into contact with the silage (it interferes with milk fermentation, which is rather putrescent in places contaminated with manure). Manure can contain ESM and PPM as well as mycobacteria causing serious diseases

in animals (e.g. *M. bovis*, *M. a. paratuberculosis*, *M. a. avium* and *M. a. hominissuis*).

7.5.6.3 The Tools (Shovels, Brooms, etc.) and Clothes of Breeders Contaminated with Excrement and Solid or Liquid Manure Can Be Sources of Mycobacteria

The different tools used for the maintenance of animals can also be contaminated with faeces, urine, solid or liquid manure. In one herd, we isolated the causative agent of paratuberculosis from a broom, which was used for sweeping out faeces and urine from calf pens and sweeping out feed leftovers from their troughs (I. Pavlik, unpublished data).

In one cattle herd where paratuberculosis prevention failed, *M. a. paratuberculosis* was detected on the gum boots and the dirty aprons of the farmers (two men) who milked the cows. It was found after several months that these two men wore the same working clothes when they occasionally assisted with the feeding of milk to calves. The wives of the farmers were, in fact, in charge of looking after the calves. This explained the isolation of *M. a. paratuberculosis* from the calves' stable: it was transmitted from the between-pen-corridors to the brooms used for the cleaning of pens and troughs. Thus, the programme for paratuberculosis control failed because the calves likely got infected by licking the infected gum boots and clothes.

Contaminated machines, tools, clothes and shoes of herdsmen can be the means of the transmission of various causative agents of diseases to different parts of a farm, where they would not usually be found (e.g. trough and milk containers for calves, troughs for piglets, etc.). Control measures can fail and the epizootiological situation becomes unfavourable.

7.5.6.4 Herdsmen with Poor Hygienic Habits Can Infect Animals Through Their Stool and Urine

It has been found, especially in herds affected by bovine tuberculosis, that some of them were infected with *M. bovis* from the herdsmen. The infected sputum from herdsmen (pulmonary form of bovine tuberculosis) was the most common route of transmission of bovine tuberculosis. However, in several herds, bovine tuberculosis was detected in the kidneys of herdsmen and herdswomen and in the intestines of one herdsman. These people defecated and urinated in the troughs or in the feed of the animals and that was evidently the main route of transmission of *M. bovis* to these herds (I. Pavlik senior, personal communication). Accordingly, it is necessary to give exact and clear instructions to staff with poor hygienic habits or with diminished mental capacity on how to behave to prevent the spread of the different causative agents of diseases through the above-mentioned routes.

7.5.6.5 The Faeces and Urine of Other Animals as Sources of Mycobacteria

The excrement of other infected animal species can also constitute sources of mycobacteria. The faeces of small terrestrial mammals, domestic animals (dogs and cats) and wild animals, especially birds, can pose a high risk. This is due in large part to the rapid multiplication of these animals. They can gain access to closed rooms and granaries where they contaminate fodder, feed grain and feed supplements (for more details see Section 7.5.7.1; Photo 7.5).

Free-living animals can also become sources of mycobacteria because they infect the pastures. Animals can shed, e.g., *M. bovis* through sputum, faeces or urine. These risks vary between different pastures and different farms and preventative measures must be adopted for each situation.

7.5.7 The Prevention of the Transmission of Mycobacteria Among Animals, People and the Environment

Mycobacteria can be transmitted either by direct or indirect contact. The direct contact of animals and people can be highly risky as far as the transmission of mycobacterial infections is concerned. Indirect contact is contact with various components of the environment (vehicles of mycobacteria) including faeces, urine, manure, water and soil. The multiplication of mycobacteria in some components of the environment can proceed very rapidly. Contact with such

a contaminated environment can sometimes be more risky than direct contact, i.e. with an uninfected host (Photo 2.2).

Peat (Section 5.1) and "peatland run-off water" can typically be vehicles of mycobacteria (Section 5.4). In Finland, increasing numbers of cases of mycobacteriosis in human patients can be explained by the drying of the peat-bogs with the aim of obtaining new land. Dust rising from the peatland and the run-off water is most likely an important source of PPM for people (Section 5.5).

7.5.7.1 The Prevention of the Transmission of Mycobacteria Through Direct Contact

Various animals come into direct or indirect contact with men and other animals due to their way of life. They can be categorised into three groups according to the possible spread of different infection agents to the captive domestic animals or human population. Animals are defined according to their relationship to human dwellings, animal farms and close surroundings.

i. **Eusynanthropic** (permanently synanthropic) species can survive all year round in human dwellings and in animal stables where they find hiding places and food (Photos 6.34 and 6.35). These are, e.g. the common house mouse (*Mus musculus*), black rat (*Rattus rattus*), brown rat (*Rattus norvegicus*), house sparrow (*Passer domesticus*), homing pigeon (*Columba livia* f. *domestica*), etc.

ii. **Hemisynanthropic** (temporarily synanthropic) species enter these places for the same reasons, however in different seasons (e.g. only in winter) and only for transitional periods (Photos 6.19 and 6.20). These are, e.g. the European polecat (*Putorius putorius*), beech marten (*Martes foina*), lesser shrew (*Crocidura suaveolens*), common shrew (*Sorex araneus*), Apodemus mouce (*Apodemus* sp.), barn swallow (*Hirundo rustica*), house martin (*Delichon urbica*), etc.

iii. **Exoanthropic** (non-synanthropic) species live in nature independently of man and do not purposely enter their dwellings. These are, e.g. the European mole (*Talpa europaea*), western European hedgehog (*Erinaceus europaeus*), eastern white-breasted hedgehog (*Erinaceus concolor*), European starling (*Sturnus vulgaris*), great spotted woodpecker (*Dendrocopos major*), etc.

Avian tuberculosis is detected relatively often in wild synanthropic and hemisynanthropic birds living in close proximity to human dwellings. It has been isolated in some of following species: the barn swallow (*Hirundo rustica*), house martin (*Delichon urbica*), house sparrow (*Passer domesticus*) and tree sparrow (*Passer montanus*), turtle dove (*Streptopelia turtur*), Eurasian collared dove (*Streptopelia decaocto*) and rook (*Corvus frugilegus*). Accordingly, the contact of kept gallinaceous birds (outside flocks), water fowl, ring-necked pheasants (*Phasianus colchicus*), grey partridges (*Perdix perdix*), and the common quail (*Coturnix coturnix*), kept in aviaries with birds can be quite dangerous (Section 6.6; Photos 3.8 and 3.9).

Avian tuberculosis can also affect other free-living animals that move among flocks of poultry, pheasants and wild boar such as small terrestrial mammals, sewer-rats, etc. (Section 6.7). The causative agent of bovine tuberculosis in ruminants is spread by badgers (*Meles meles*) in Great Britain and in Ireland and by the Australian marsupial possum (*Trichosurus vulpecula*) in New Zealand. The causative agent of paratuberculosis in domestic and wild ruminants can be spread by wild rabbits.

The migration of mammals and birds plays an important role in the spread of mycobacteria in the environment. During these migrations, the causative agents of important mycobacterial infections can be spread. The migration of mammals can be classified according to distance as migration (longer distances up to several hundreds of kilometres) or dispersion (wandering, the movement of shorter distances from several hundreds of metres up to tens of kilometres). From the point of view of the transmission of mycobacteria, the migration of animals can be characterised as follows.

i. **Two-directional migration (re-migration):** The regular seasonal migration of mammals searching for food (typical above all for terrestrial mammals found in Africa and tundra and for mammals living in the oceans), the migration of mammals during the reproduction period, etc.

ii. **Unidirectional migration:** This is due to the high multiplication rate of some mammalian species; this type of migration ensures the spread of some parts of the population to areas that have not been inhabited by them yet (e.g. the spread of the elk population southward).

iii. **Vertical migration:** The migration of mammalian populations from mountain regions to lowlands and vice versa (e.g. the migration of red deer in search of food).

iv. **Horizontal migration:** The migration of mammalian populations in lowlands due to high rates of multiplication and conversely due to decreased population density.

The mutual contact of animals kept in captivity with other animals takes the following forms:

- Contact on pastures (lifelong, regular and/or occasional; Photo 7.23).
- Contact in houses (farms, pavilions in zoos, etc.; Photo 5.60).
- Contact in aviaries and pheasantries (Photo 6.33).
- Contact at watering and feeding places (feed-troughs, etc.; Photo 6.32).
- Other contact (e.g. birds of prey with the keeper at the falconry; Photo 6.31).

The age of animals can also be an important factor for the spread of mycobacteria among them. In domestic poultry, *M. a. avium* is above all shed by older birds. If they are in contact with young birds, these become massively infected. Older breeders often leave several-year-old hens in the flock, despite the fact that they can constitute the highest risk of infection for the younger birds (Photo 3.9).

7.5.7.2 The Prevention of the Transmission of Mycobacteria Through Indirect Contact

Indirect contact can be categorised into three main types:

Contact Mediated by Vehicles (e.g. Water, Soil, Dust, Feed and Pastures)

Preventative control measures adopted for the use of these vehicles can limit the spread of mycobacteria to the environment and subsequently back to the host, i.e. it is necessary to carry out mechanical cleaning of the mangers and their surroundings after having finished the winter feeding of wild animals. This is carried out for the prevention of the transmission of mycobacteria through contaminated feed. It is necessary to plough up the land about five to 10 m around the mangers to a depth of at least 30 cm and mix lime with the soil. The same preventative measures should also be applied to the pastures in areas not covered with grass or other vegetation and where animals have spent a good deal of time (mobile watering places).

The stables, rooms for feed preparation and stores should be protected against the invasion of free-living birds and mammals. Their faeces can often be a source of microorganisms, not only ESM and PPM but also OPM. With respect to the high susceptibility of pigs to mycobacterial infections, these measures should also be adopted in stables where the pigs are kept. It is also necessary to limit or prevent contact with free-living birds (seasonal presence of swallows and martins and the year-round presence of sparrows and doves; small terrestrial rodents and insectivores). The environment of nests can be highly contaminated with faeces from the infected birds, especially at the time of nesting Photos 6.19 and 7.7). After contamination of the feed or drinking water, these can become the sources of mycobacterial infection. In one herd of pigs, swallow and sparrow nests were present just above the troughs of young breeding boars and the causative agent of avian tuberculosis was later detected in their intestinal lymph nodes (I. Pavlik, unpublished data).

In modern halls with closed ventilation systems, the access of free-living birds and mammals to the stables where domestic animals are kept is limited or completely prevented. If, for instance, avian tuberculosis is diagnosed in pigs, other sources of infection should be investigated in such herds. On one pig farm, we established that migrating small terrestrial mammals were the source of infection. They manage to gain access to feed stores and other parts of the buildings especially just before the onset of cold weather (in autumn in the northern hemisphere; Photo 7.4). Pigs in stables have been observed to catch small terrestrial mammals (I. Pavlik, unpublished data).

The circulation of mycobacteria in the environment of infected herds can be prevented by the elimination of all infected animals. Clinically ill animals that shed large amounts of mycobacteria into the environment are the primary sources of OPM and PPM. Accordingly, it is necessary to cull these animals from the herd

and to only introduce healthy animals from an uncontaminated environment. Healthy animals must be protected from contact with potential sources of diseases (Photo 3.16).

Accidental or Intentional Contact with Animal Cadavers

It is necessary to continuously remove the dead animal cadavers from herds (Photos 5.48, 5.49 and 5.58). Mycobacteria can survive in cadavers for a very long time compared to other bacteria (*Salmonella* sp., *Escherichia coli*, *Proteus* sp.). Therefore it is unsuitable (and to a certain degree risky) to leave these organic residues as "carrion" in the places where they were caught.

The viscera and entrails of caught or slaughtered animals should be transported to rendering plants, incinerated or buried outside the breeding facility at a depth of at least 80 cm (using disinfectants). The compact dead bodies of animals should also be transported to rendering plants for their safe disposal. During their dissection in the rendering plants, samples for laboratory analysis must be collected; these will be of help for the detection of potential mycobacterial infections. The meat of ill animals can also be a source of mycobacterial infection for carnivores and herbivores (especially carrion beetles; Photo 7.22).

Contact with Trapped Free-Living Animals

In particular, zoologists, veterinarians and other members of staff in zoos who care for wild animals are exposed to a considerable risk of mycobacterial infection. The causative agent of bovine tuberculosis identified, e.g., in wild ruminants and carnivores in captivity can spread very easily directly by droplet infection (the aspiration of infected aerosol). Oral infection by swallowing the causative agent of bovine tuberculosis through a contaminated vehicle is less frequent.

The contact of domestic and wild birds and animals with man are varied and sometimes unpredictable. In herds of domestic animals, the close contact of visitors with animals is usually prevented because the breeder is afraid of the transmission of the causative agent of some disease into his herd. Accordingly, this chapter is mainly focused on the contact of people with wild animals.

7.5.7.3 The Prevention of the Transmission of Mycobacteria from Animals to People

Contact with animals in rescue stations may be highly risky, namely if an animal is weak or exhibits unusual behaviour. The case of a young black stork (*Ciconia nigra*) which had lost the ability to fly is a good example. The miliary form of avian tuberculosis was diagnosed in the bird a few weeks later. Before the diagnosis, the bird was kept in an aviary, other birds were kept there afterwards. However, they were placed there without prior mechanical cleaning and disinfection. It was found, through detailed anamnesis, that antibodies against *M. a. avium* were detected in water birds kept in this aviary. Nevertheless, the causative agent of avian tuberculosis was not detected by culture examinations of faeces at regular 3-month intervals (I. Pavlik, unpublished data; Photo 7.24).

In zoological gardens, the main route of transmission of mycobacteria is from the captive animals to the keepers. However, mycobacteria can also be transferred to the visitors. The direct contact of children with various animal species, e.g. in zoo corners, should also be considered as risky. A herd of free-range Vietnamese pigs was kept in a "Tierpark" in Germany. They increased in number to almost 100 animals within 3 years. The Vietnamese pigs became the most popular animals and the visiting rate increased by more than 100%. After the diagnosis of bovine tuberculosis in the parenchymatous organs and joints of six animals, all of the captive pigs were slaughtered, including several tens of guinea pigs. *M. bovis* was also detected in these as well as in one captive Eurasian lynx (*Lynx lynx*) and red deer (*Cervus elaphus*). The fact that hundreds of children were in close contact with the animals over the course of several years is worrying due to the fact that many of the animals could shed large amounts of *M. bovis* to the environment (J. Kazda and I. Pavlik, unpublished data; Photo 7.24).

When **amateurs, laymen or children** selflessly treat infected animals, they can become infected by *M. bovis*. It is important to know that the majority of free-living infected animals in the advanced stages of a disease lose timidity and can be more easily caught and unprofessionally treated by laymen. The hazard of

infection for people may be lowered by the consistent observance of basic principles for prevention and personal hygiene.

Another risk may be the contact of **hunters** of game with caught free-living animals, their bodies and parenchymatous organs. These people can also become infected with PPM and OPM due to the fact that many hunters eat the raw liver of the caught animals. They believe that the power of the animals and many vitamins from the liver will enter their bodies. Some hunters believe that salt and pepper, which they always take with them for this purpose, can devitalise both the causative agents of bovine tuberculosis and other potential infectious zoonotic diseases (I. Pavlik, unpublished data). Eating insufficiently heat-treated specialities made from killed wild animals can also pose a risk (Section 9.9).

7.5.8 Preventive Measures in Hospitals and Other Facilities

In hospitals, homes of the elderly and the long-term care homes, patients are exposed to the risk of mycobacterial infection (above all *M. tuberculosis*) from other people (including the nursing personnel) and also from different animals and the environment. Quite frequent nosocomial infections caused by mycobacteria from the environment testify to these risks (Section 3.3). Both living organisms (invertebrates and vertebrates including man) and different non-living vehicles (water, dust, non-heat-treated food, etc.) can become the sources of mycobacteria present in the environment of these facilities. Living organisms transmitting mycobacteria to the environment of hospitals can be categorised into two groups: humans and animals.

7.5.8.1 Humans (Patients, Staff, Visitors and Other Individuals)

People themselves are the most common source of mycobacteria in hospitals. The transmission of different species of mycobacteria by patients with suspected clinical disease can be assumed to take place. It is not necessary to give detailed information on the transmission of the causative agent of human tuberculosis, *M. tuberculosis*, owing to the fact that it is generally known and also much literature data are available. Inter-human transmission is not suspected in patients infected by other species of mycobacteria (Sections 3.3 and 7.3.). However, the transmission of different PPM species from these patients to the environment of hospital facilities is possible. Nursing personnel, including surgeons can become the sources of ESM and PPM (see below).

In Israel, cases of nosocomial infection caused by *M. jacuzzii* at the surgical site infection were described in 15 patients after plastic reconstruction of the breast. *M. fortuitum*, *M. kansasii* and *M. gordonae* were isolated from the examined drinking water. Air samples from the operating room were sterile. *M. jacuzzii* was isolated from the body surface of only one of three surgeons: from the eyebrows, scalp, face, nose, ears and groin. Subsequent search revealed *M. jacuzzii* in his bed linen, pillows, towels, bathrobe, car air-conditioning system and from water in his personal outdoor whirlpool. *M. jacuzzii* was also isolated from another family member who had used the whirlpool. The effective prevention measures adopted by this surgeon were as follows, no use of the outside whirlpool, careful washing of the body surface with soap and the covering of the whole body surface during operations (Rahav et al., 2006).

As mentioned above, patients themselves can become the source of the contamination of the environment with various PPM species. For example, in the case of skin massively contaminated with *M. chelonae*, it is shed to the environment with pus and other damaged tissues. In facilities with poor hygiene, the environment can soon be contaminated with mycobacteria. After propagation, various PPM species may threaten other patients and/or the nursing personnel.

It follows from the wide variety of above-mentioned possible routes of transmission of mycobacteria that it is necessary to adopt numerous preventive measures. Accordingly, it is necessary to strictly require that visitors use headwear when visiting immunosuppressed patients and observe the ban on bringing "pets" into hospitals, etc.

7.5.8.2 Animals as Vectors of Mycobacteria

The detection of invertebrate animals such as cockroaches living in, for example, hospitals can also indicate an environment highly contaminated with mycobacteria. These most often become infected by the ingestion of organic debris in their surroundings (Section 6.3.5.2). Due to the fact that they can also spread *M. tuberculosis* through their faeces, the adoption of hygiene measures and disinsection is very important. A variety of other arthropod species are also potential sources of mycobacteria in hospitals in tropical and subtropical regions (Sections 6.3.4 and 6.3.5).

The risk of the transmission of mycobacteria in hospitals by blood sucking insects: such as bugs (Section 6.3.5.8), fleas, lice and mosquitoes (Section 6.3.5.4) has long been under discussion. It is highly likely that they play some role in the transmission of mycobacteria and constitute a certain risk in the ecology of mycobacteria. There is no doubt about the presence of mycobacteria in the sucking and licking mouthparts of Diptera and especially in their excreta (Section 6.3.5.9). Accordingly, the installation of mosquito nets against airborne invertebrates is necessary, particularly in less developed countries in the tropics and subtropics, alongside the adoption of essential hygienic measures.

However, reptiles such as geckos are often present in the health-care facilities of the subtropics and tropics; these move rapidly over walls and prey on the invertebrates. Thus, they are welcomed by the personnel and patients because they remove these tiresome insects. Despite this, it is necessary to seriously consider the potential risk of involvement of these animals in the ecology of mycobacteria in these health-care facilities and to instead seek to reduce the numbers through the use of effective nets.

In the often primitive health institutions of developing countries, it is also advisable to invest money into regular rodent control, which at least reduces the number of small terrestrial mammals inside hospitals and the surroundings. These ubiquitous animals can spread not only a variety of ESM and PPM but also species of OPM through their faeces (Sections 6.7.1 and 6.7.3).

7.5.8.3 Water and Dust as Important Vehicles of Mycobacteria in Hospitals

Among other risk factors for the spread of ESM and PPM in the environment of hospitals, water (Section 5.2) and dust (Section 5.9) appear to be most important. It would be naive to believe that the spread of

ESM and PPM through the drinking water and dust of these facilities will be prevented (Photo 5.57). From the point of view of the patients' infection, wards can be categorised as being either "conventional" or "specialised". The second group comprises both operating rooms and immunosuppressed patients' wards (oncology wards, HIV/AIDS patients, etc.). Strict precautions must be applied here.

Various sophisticated devices including under-pressure systems (hydraulic gradient between different rooms), HEPA filters, etc. are used for the protection of these rooms against contamination with bacteria. However, drinking water is often ignored; this is tested for the presence of a number of bacteria, not however, for mycobacteria. Aerosol potentially contaminated with PPM from the water main is created when flushing the toilet, brushing teeth and washing one's face. Accordingly, it would be best if no water main were installed in the wards which house the most susceptible patients (e.g. after bone marrow transplantation or cytostatics treatment).

Patients might well be supplied with, for example, tea made with boiled drinking water or other mycobacteria-free drinks for a certain period of time. They could also use sterile distilled water for washing and basic hygiene. Special devices that do not use flowing drinking water could be installed to remove urine and stools. With regard to the potential contamination of drinking water with mycobacteria, an ozonisation treatment system should be used (Section 7.5.9). The installation of a new water piping system is recommended and the distance between the disinfection ozonisation tank and the appliance (basin, toilet, shower room, etc.) should be as short as possible. This should be disinfected by the flow of water with a high content of ozone before the first use. These measures will serve to prevent the spread of mycobacteria to these special hospital facilities.

In wards where the maintenance of a sterile environment is not necessary, drinking water treated with the "anti-*Legionella*" filter system is best. However, these must be regularly checked and changed according to the manufacturer's instructions (usually once a month). The fact that pulmonary mycobacteriosis may be associated with the inhalation of mycobacteria from rising aerosols from water that can contain ESM and PPM is often disregarded (Sections 3.3 and 5.9). For this reason, only a covered toilet should be flushed (so that the aerosol cannot leak to the surrounding envi-

ronment of the room). It is best to leave the room and not to re-enter until the aerosol is deposited in the toilet bowl, i.e. in 15–20 min. Teeth should not be brushed using flowing tap water because aerosol containing mycobacteria can in such instances be directly inhaled. Therefore, sterile or ozonised water is preferable for cleaning teeth.

While removing the dust and webs from these "super clean wards" in the hospitals, it is suitable to use vacuum cleaners with central suction that removes the dust and deposits it outside these rooms. The removal of the smallest particles to which mycobacteria are most often attached (up to $10 \, \mu m$) is thus ensured (Section 5.9).

7.5.8.4 Non-heat Treated Foodstuffs as an Important Source of Mycobacteria

The foodstuffs, which provide favourable conditions for the growth and subsequent transport of mycobacteria, have been discussed previously (Section 5.7). It is more important to focus primarily on heat-treated food for immunosuppressed patients in hospitals. It is of course possible to thoroughly wash vegetables, fruit or other foodstuffs of vegetable origin (sprouted seeds, roots of some plants, etc).

However, the effort to supply patients with the necessary fibre and vitamins in a natural form can result in the development of nosocomial infections caused by various PPM species (Section 3.3). Owing to the fact that mycobacteria are highly hydrophobic (Section 4.3), they strongly attach to the surface of plants and foodstuffs of vegetable origin and it is not possible to remove them either by rinsing or by washing under a strong stream of water (I. Pavlik, unpublished data; Section 5.7). Therefore, giving non-heat-treated food to these patients is highly risky.

7.5.8.5 Other Risk Factors Resulting from the Ecology of Mycobacteria

Flowers have also been identified as risk factors in hospital facilities for highly immunosuppressed patients. In this way patients are exposed to a risk posed not only by the flowers with attached mycobacteria on their surface (Sections 5.6 and 5.7) but also from tap water (Section 5.2) in a vase in the patient's room, most often for several days. The risk posed by mycobacteria present in the soil of flower pots is evident (Section 5.5) and it is essential that there be a ban, for instance, on the reporting of plants in these hospital facilities (Section 5.1.2).

Cigarettes have also been identified as risk factors for PPM transmission in hospital facilities. These should be considered as risk factors in facilities for highly immunocompromised patients because PPM have also been detected in cigarette tobacco (Eaton et al., 1995).

7.5.9 Experiences with Water Supplies Free of Mycobacteria

As described previously (Section 7.5.8.3), in special hospital departments mycobacterial contamination of drinking water is considered as a risk factor and must be avoided. For this purpose, experiences in the breeding of mycobacteria-free experimental animals are recommended, especially for the patient's hygiene. Chlorination reduced the count of mycobacteria but was not sufficient to keep water outlets mycobacteria-free (Slosarek et al., 1994; Falkinham, 1997). One of the first attempts to obtain mycobacteria-free drinking water was made in a newly constructed facility for experimental animals in Germany (Kazda, 1981; Kazda, 2000). Compact equipment for water treatment with ozone was installed and the disinfection of drinking water took place in a small tank using a high ozone concentration. This water was first used for the rinsing and disinfection of the pipes, and thereafter, the concentration of ozone in drinking water was reduced to a minimum which did not influence the taste of the water, although it was sufficient to prevent the colonisation of pipes and outlets with mycobacteria. The efficacy was monitored weekly by screening for mycobacteria by smears and cultures. The water was free of mycobacteria for the duration of the experiment which lasted 12 years. Furthermore, biological observation of experimental animals (guinea pigs, white and nu/nu mice and nine-banded armadillos; *Dasypus novemcinctus*) did not show any side effects due to the ozone content in drinking water.

The presence of mycobacteria in drinking water was underestimated until heavy infections in HIV/AIDS patients and endemics of *M. kansasii* and *M. xenopi*

in the immunocompetent population were recorded. Furthermore, a high concentration of mycobacteria in drinking water can provoke allergic reactions in the digestive tract, as described in Section 7.4.3. The guidelines for the bacteriological control of public drinking water need to be completed by the inclusion of examination for mycobacteria, especially at the outlets. The elimination of mycobacterial contamination of drinking water should be one of the urgent objectives of the public hygiene sector.

In recent years the technology of ozone treatment has improved and evolved and its potential for successful application has increased. These experiences can be used to prevent mycobacterial contamination of drinking water.

Acknowledgements Partially supported by the Ministry of Agriculture of the Czech Republic NPV 1B53009 and NAZV QH81065 (Section 7.5).

References

Altare F, Durandy A, Lammas D, Emile JF, Lamhamedi S, Le Deist F, Drysdale P, Jouanguy E, Doffinger R, Bernaudin F, Jeppsson O, Gollob JA, Meinl E, Segal AW, Fischer A, Kumararatne D, Casanova JL (1998) Impairment of mycobacterial immunity in human interleukin-12 receptor deficiency. Science. 280:1432–1435

Andersen AA (1958) New sampler for the collection, sizing, and enumeration of viable airborne particles. J. Bacteriol. 76:471–484

Ayele WY, Machackova M, Pavlik I (2001) The transmission and impact of paratuberculosis infection in domestic and wild ruminants. Veterinarni Medicina. 46:205–224

Brandt L, Feino CJ, Weinreich OA, Chilima B, Hirsch P, Appelberg R, Andersen P (2002) Failure of the *Mycobacterium bovis* BCG vaccine: Some species of environmental mycobacteria block multiplication of BCG and induction of protective immunity to tuberculosis. Infect. Immun. 70:672–678

Collins CH, Grange JM, Yates MD (1986) Unusual opportunist mycobacteria. Med. Lab. Sci. 43:262–268

Covert TC, Rodgers MR, Reyes AL, Stelma GN, Jr. (1999) Occurrence of nontuberculous mycobacteria in environmental samples. Appl. Environ. Microbiol. 65:2492–2496

De Groote MA, Pace NR, Fulton K, Falkinham JO, III (2006) Relationships between *Mycobacterium* isolates from patients with pulmonary mycobacterial infection and potting soils. Appl. Environ. Microbiol. 72:7602–7606

Du Moulin GC, Stottmeier KD, Pelletier PA, Tsang AY, Hedley-Whyte J (1988) Concentration of *Mycobacterium avium* by hospital hot water systems. JAMA. 260:1599–1601

Du Moulin GC, Stottmeier KD (1986) Waterborne mycobacteria: an increasing threat to health. ASM News. 52:525–529

Eaton T, Falkinham JO, III, von Reyn CF (1995) Recovery of *Mycobacterium avium* from cigarettes. J. Clin. Microbiol. 33:2757–2758

Falkinham JO (1997) Transmission of mycobacteria. In: Mycobacteria I. Basic aspects. 178–209.Gangadharam PRJ, Jenkins PA, Eds. Chapman & Hall, New York

Falkinham JO, III (2002) Nontuberculous mycobacteria in the environment. Clin. Chest Med. 23:529–551

Falkinham JO, Iseman MD, de Haas P, van Soolingen D (2008) *Mycobacterium avium* in a shower linked to pulmonary disease. J. Wat. Health. 6:209–213

Fasske E, Schroder KH (1989) Granulomatous pulmonary reactions after instillation of *Mycobacterium gordonae*. Light and electron microscopic investigations on the rat model. Med. Microbiol. Immunol. 178:149–161

Fine PEM, Floyd S, Stanford JL, Nkhosa P, Kasunga A, Chaguluka S, Warndorff DK, Jenkins PA, Yates M, Ponninghaus JM (2001) Environmental mycobacteria in northern Malawi: implications for the epidemiology of tuberculosis and leprosy. Epidemiology and Infection. 126:379–387

Fischeder R, Schulze-Robbecke R, Weber A (1991) Occurrence of mycobacteria in drinking water samples. Zentralbl. Hyg. Umweltmed. 192:154–158

Freerksen E (1960) Die Sogenannten Atypischen Mycobakterien. Klinische Wochenschrift. 38:297–309

Glover N, Holtzman A, Aronson T, Froman S, Berlin OGW, Dominguez P, Kunkel KA, Overturf G, Stejma G, Smith C, Yakrus M (1994) The isolation and identification of *Mycobacterium avium* complex (MAC) recovered from Los Angeles potable water, a possible source of infection in AIDS patients. Int. J. Environ. Health Res. 4:63–72

Hampl J, Kazda J, Simon V (1969) Distribution of 35S-labelled *Mycobacterium avium* in organism of chicken after intravenous infection. Zentralblatt fur Bakteriologie Parasitenkunde Infektionskrankheiten und Hygiene Abteilung 1-Originale Medizinisch Hygiensche Bakteriologie Virusforschung und Parasitologie. 210:531–537

Iivanainen E, Northrup J, Arbeit RD, Ristola M, Katila ML, von Reyn CF (1999) Isolation of mycobacteria from indoor swimming pools in Finland. APMIS. 107:193–200

Iivanainen EK, Martikainen PJ, Raisanen ML, Katila ML (1997) Mycobacteria in boreal coniferous forest soils. Fems Microbiol. Ecol. 23:325–332

Iivanainen EK, Martikainen PJ, Vaananen PK, Katila ML (1993) Environmental-factors affecting the occurrence of mycobacteria in brook waters. Appl. Environ. Microbiol. 59:398–404

Kazda J (1966) [Isolation and description of a *Mycobacterium* species, the cause of a para-allergy against tuberculin in poultry]. Zentralbl. Bakteriol. [Orig.]. 199:529–532

Kazda J (1967a) Mycobacteria in drinking-water as cause of parallergy to tuberculins in animals. 2. Parallergic reactions to mammalian and avian tuberculins in cattle. Zentralblatt fur Bakteriologie Parasitenkunde Infektionskrankheiten und Hygiene Abteilung 1. 203:190–198

Kazda J (1967b) [Atypical mycobacteria in drinking water–the cause of para-allergies against tuberculin in animals]. Z. Tuberk. Erkr. Thoraxorg. 127:111–113

Kazda J (1973) Importance of water for spread of potentially pathogenic mycobacteria. 1. Possibilities for multiplication of Mycobacteria. Zentralblatt fur Bakteriologie Mikrobiolo-gie und Hygiene Serie B-Umwelthygiene Krankenhaushygiene Arbeitshygiene Praventive Medizin. 158:161–169

Kazda J (1977) Course of Jones-Mote reaction in guinea-pigs after sensitization with Mycobacteria. Zentralblatt fur Bakteriologie Mikrobiologie und Hygiene Series A-Med. Microbiol. Inf. Dis. Virol. Parasitol. 237:80–89

Kazda J (1978) [Stimulatory action of *Brassica oleracea* var. sabauda (Savoy) on mycobacteria and mycobacterial infection (author's transl)]. Zentralbl Bakteriol [Orig A]. 242:365–374

Kazda J (2000) The ecology of mycobacteria. Kluwer Academic Publishers, Dordrecht, Boston, London, 72 pp.

Kazda J, Cook BR (1988) Mycobacteria in pond waters as a source of non-specific reactions to bovine tuberculin in New Zealand. N.Z. Vet. J. 36:184–188

Kazda J, Cooney R, Quinn PJ, Cook BM, Muller K, Monaghan ML, Keane M (1997) High density of mycobacteria in the bryophyte vegetation (*Musci*) of moorland. Int. Peat J. 7:14–19

Kazda J, Stackebrandt E, Smida J, Minnikin DE, Daffe M, Parlett JH, Pitulle C (1990) *Mycobacterium-Cookii* Sp-Nov. Int. J. Systematic Bacteriol. 40:217–223

Kazda JM (1981) Nine-banded armadillos in captivity: prevention of losses due to parasitic diseases. Some remarks on mycobacteria-free maintenance. Int. J Lepr. Other Mycobact. Dis. 49:345–346

Kubin M, Matuskova E, Kazda J (1969) *Mycobacterium Brunense* N Sp identified as serotype Davis of Group-3 (Runyon) Mycobacteria. Zentralblatt fur Bakteriologie Parasitenkunde Infektionskrankheiten und Hygiene Abteilung 1-Originale Medizinisch Hygiensche Bakteriologie Virusforschung und Parasitologie. 210:207–211

Mansfield KG, Lackner AA (1997) Simian immunodeficiency virus-inoculated macaques acquire *Mycobacterium avium* from potable water during AIDS. J. Inf. Dis. 175:184–187

Mansfield KG, Pauley D, Young HL, Lackner AA (1995) *Mycobacterium-avium* complex in Macaques with AIDS is associated with a specific strain of simian immunodeficiency virus and prolonged survival after primary infection. J. Inf. Dis. 172:1149–1152

Matlova L, Dvorska L, Bartl J, Bartos M, Ayele WY, Alexa M, Pavlik I (2003) Mycobacteria isolated from the environment of pig farms in the Czech Republic during the years 1996 to 2002. Veterinarni Medicina. 48:343–357

May DC, Kutz JE, Howell RS, Raff MJ, Melo JC (1983) *Mycobacterium terrae* tenosynovitis: chronic infection in a previously healthy individual. South. Med. J. 76:1445–1447

Monaghan ML, Kazda JF, Mcgill K, Quinn PJ, Cook BM (1991) Sensitization of cattle to bovine and avian tuberculins with *Mycobacterium Cookii*. Veterinary Record. 129:383–383

Palmer CE, Long MW (1966) Effects of Infection with Atypical Mycobacteria on Bcg Vaccination and Tuberculosis. Am. Rev. Respir Dis. 94:553–568

Parker BC, Ford MA, Gruft H, Falkinham JO, III (1983) Epidemiology of infection by nontuberculous mycobacteria. IV. Preferential aerosolization of *Mycobacterium intracellulare* from natural waters. Am Rev Respir Dis. 128:652–656

Pavlik I, Bartl J, Parmova I, Havelkova M, Kubin M, Bazant J (1998) Occurrence of bovine tuberculosis in animals and humans in the Czech Republic in the years 1969 to 1996. Veterinarni Medicina. 43:221–231

Pavlik I, Matlova L, Dvorska L, Bartl J, Oktabcova L, Docekal J, Parmova I (2003) Tuberculous lesions in pigs in the Czech Republic during 1990–1999: occurrence, causal factors and economic losses. Veterinarni Medicina. 48:113–125

Rahav G, Pitlik S, Amitai Z, Lavy A, Blech M, Keller N, Smollan G, Lewis M, Zlotkin A (2006) An outbreak of *Mycobacterium jacuzzii* infection following insertion of breast implants. Clin. Infect. Dis. 43:823–830

Rose CS, Martyny JW, Newman LS, Milton DK, King TE, Jr., Beebe JL, McCammon JB, Hoffman RE, Kreiss K (1998) "Lifeguard lung": endemic granulomatous pneumonitis in an indoor swimming pool. Am. J. Public Health. 88:1795–1800

Schulze-Robbecke R, Fischeder R (1989) Mycobacteria in biofilms. Zentralbl. Hyg. Umweltmed. 188:385–390

Slosarek M, Kubin M, Pokorny J (1994) Water as a possible factor of transmission in mycobacterial infections. Cent. Eur. J. Public Health. 2:103–105

Stadecker MJ, Leskowitz S (1973) The cutaneous basophil response to particulate antigens. Proc. Soc. Exp. Biol. Med. 142:150–154

Stanford JL (1992) Immunotherapy with *Mycobacterium vaccae* for pulmonary tuberculosis. Phil. J. Microbiol. Infect. Diss. 21:73–75

Stanford JL, Paul RC (1973) A preliminary report on some studies of environmental mycobacteria. Ann. Soc. Belg. Med. Trop. 53:389–393

von Reyn CF, Waddell RD, Eaton T, Arbeit RD, Maslow JN, Barber TW, Brindle RJ, Gilks CF, Lumio J, Lahdevirta J (1993) Isolation of *Mycobacterium avium* complex from water in the United States, Finland, Zaire, and Kenya. J. Clin. Microbiol. 31:3227–3230

von Reyn CF, Maslow JN, Barber TW, Falkinham JO, Arbeit RD (1994) Persistent colonization of potable water as a source of *Mycobacterium avium* infection in Aids. Lancet. 343: 1137–1141

Wallace RJ, Zhang YS, Wilson RW, Mann L, Rossmoore H (2002) Presence of a single genotype of the newly described species *Mycobacterium immunogenum* in industrial metal-working fluids associated with hypersensitivity pneumonitis. Appl. Environ. Microbiol. 68:5580–5584

Wendt SL, George KL, Parker BC, Gruft H, Falkinham JO, III (1980) Epidemiology of infection by nontuberculous Mycobacteria. III. Isolation of potentially pathogenic mycobacteria from aerosols. Am Rev Respir Dis. 122: 259–263

Witty LA, Tapson VF, Piantadosi CA (1994) Isolation of Mycobacteria in patients with pulmonary alveolar proteinosis. Medicine. 73:103–109

Worthington RW (1967) Mycobacterial sensitins and the nonspecific reactor problem. Onderstepoort J. Vet. Res. 34: 345–348

Chapter 8

Biological Role of Mycobacteria in the Environment

J. Kazda and J.O. Falkinham III (Eds.)

Introduction

Mycobacteria are generally regarded as causing diseases. Little is known, however, about the role and behaviour of saprophytic mycobacteria in the environment. In this chapter, their role in the degradation of organic compounds and in the stimulation of the growth of some protozoa is described. Special attention is paid to the mycobacteria as a source of nutrients for water fleas and dragonfly larvae in oligotrophic moorland waters.

8.1 Selection for Intracellular Growth

J.O. Falkinham III

It is compelling to speculate, as others have, that the ability of a microorganism to survive intracellular killing and persist and grow in protozoa and amoebae leads to intracellular pathogenesis in animals. That is certainly the case for mycobacteria. Environmental mycobacteria and protists occupy the same habitats. Mycobacterial growth rates are not high enough to allow them to outgrow the losses due to protozoa and amoebae predation; consequently, persistence in that habitat requires intracellular survival. Not only do

mycobacteria survive in protozoa and amoebae but also grow intracellularly at rates greater than those possible in the surrounding medium or habitat (Cirillo et al., 1997; Steinert et al., 1998; Strahl et al., 2001).

As a consequence, mycobacterial numbers increase in the presence of protists to levels above those of other cells. Furthermore, evidence that protozoa- or amoebae-grown mycobacteria have higher invasive rates than those grown outside the protists (Cirillo et al., 1997) means that the introduction of protozoa and amoebae to a habitat occupied by mycobacteria would lead to acceleration in the rate of increase in mycobacterial numbers.

One final interesting fact is that cells of *Mycobacterium avium*-infected *Tetrahymena pyriformis* grow faster than uninfected cells (J. O. Falkinham, unpublished). Furthermore, *M. avium*-infected *T. pyriformis* cells are capable of growing from a low inoculum density, unlike uninfected cells (Falkinham J, unpublished). This data strongly suggests that the intracellular growth of mycobacteria in protozoa and amoebae is symbiotic; each gains from the infection. Based on the fact that growth from a low inoculum density is promoted by fatty acids (Schousboe and Rasmussen, 1994) and mycobacterial envelopes are rich in fatty acids (Brennan and Nikaido, 1995), it is tempting to propose that the stimulation of growth at low densities is due to the presence of mycobacterial fatty acids.

J. Kazda (✉)
University of Kiel, Kiel, Germany
e-mail: J.Kazda@t-online.de

J. Kazda et al. (eds.), *The Ecology of Mycobacteria: Impact on Animal's and Human's Health*, DOI 10.1007/978-1-4020-9413-2_8, © Springer Science+Business Media B.V. 2009

8.2 Degradation of Organic Compounds by Mycobacteria

A. Jesenska and J. Damborsky

8.2.1 Organic Pollutants in the Environment

Polycyclic aromatic hydrocarbons (PAHs), aromatic hydrocarbons benzene, toluene, ethyl toluene, xylene (BTEX), heterocyclic compounds or organohalides form a very diverse class of environmental pollutants. Anthropogenic and natural sources in combination with global transport result in their worldwide distribution. Interest in biodegradation mechanisms and the environmental fate of these compounds are prompted by their ubiquitous distribution and their potentially deleterious effects on living organisms.

The fate of organic pollutants in the environment depends on their chemical properties and potential chemical and biological transformations. Due to their lipophilic nature, these pollutants have a high potential for biomagnification through trophic transfers. They are also known to exert acutely toxic effects and some of them possess mutagenic, teratogenic or carcinogenic properties. Many synthetic compounds are resistant to biodegradation and persist in the environment. There are several possible reasons for the persistence of a pollutant in the environment:

i. The absence of microorganisms with the capability of degrading a compound.

ii. The presence of unfavourable environmental conditions for biodegradation, i.e. permeability of the subsurface to air and water, temperature, pH, salinity, water content and presence of inhibitory chemicals.

iii. Unfavourable substrate concentration, i.e. the pollutant may be present in too high a concentration, leading to toxicity, or in too low a concentration, failing to induce the degradative enzymes or being below the threshold concentration.

iv. Lack of bioavailability of the pollutant because of incorporation into humic substances or strong adsorption to soil particles (Fetzner and Lingens, 1994).

Bioremediation of polluted environments is based on the capacity of microorganisms to transform pollutants to non-toxic products. These microbial conversions require the expression of enzymes that have sufficient catalytic activity towards the compounds of interest. The use of the microbial metabolic potential for eliminating environmental pollutants provides a safe and economically viable alternative to their disposal in waste dump sites and to commonly used physico-chemical strategies. Microorganisms, including mycobacteria, capable of mineralising a variety of toxic compounds under laboratory conditions have been isolated.

8.2.2 Organic Pollutants Biodegradable by Mycobacteria

PAHs consist of two or more fused aromatic rings and are present as natural constituents in fossil fuels, can be formed during the incomplete combustion of organic material and are present in relatively high concentrations in products of fossil fuel refining. PAHs released into the environment may originate from gasoline and diesel fuel combustion and tobacco smoke (Kanaly and Harayama, 2000). Aromatic hydrocarbons (BTEX) are formed during the refining of petroleum and chemical manufacturing, the manufacture of paints, lacquers, adhesives and solvents. In fuels and solvents, these compounds are often used as mixtures.

Heterocyclic-ring compounds contain carbon atoms and one or more heteroatoms, defined here as nitrogen, oxygen and sulphur. Morpholine, for example, is a common additive used for pH adjustment in both fossil fuel and nuclear power plant steam systems (Wackett and Hershberger, 2001). Halogenated hydrocarbons cover a broad range of substances, which include chlorinated, brominated and iodinated primary and secondary haloalkanes, halocycloalkanes, haloalcohols, halogenated amides, haloethers, halogenated aromatics and others. They are produced industrially in large amounts for use as solvents, cleaning agents, intermediates for further chemical synthesis, pesticides, insecticides, etc. (Janssen et al., 1994). The representative groups of environmental pollutants biodegradable by mycobacteria are listed in Fig. 8.1.

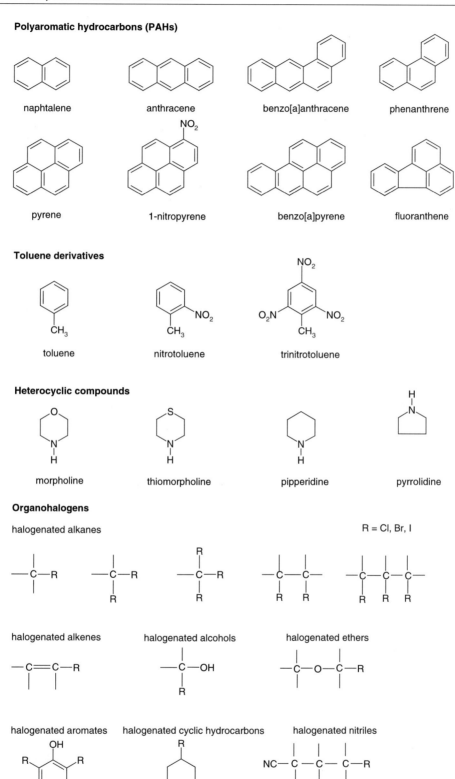

Fig. 8.1 Examples of organic compounds biodegradable by mycobacteria

8.2.3 Microbial Metabolism of Organic Pollutants

Microorganisms have been found to degrade PAHs via either metabolism or cometabolism. Cometabolism is important for the degradation of mixtures and high-molecular-weight PAHs. In contrast, two- to four-ring PAHs have been known for years to be growth substrates for bacteria (Habe and Omori, 2003). Bacteria with an acquired ability to metabolise PAHs obtain a substantial benefit over their competitors by their ability to utilise extra metabolic energy and carbon sources. The fundamental strategy to metabolise PAHs is derived from that of the single-ring benzenoid aromatic hydrocarbons. An aromatic nucleus can undergo dioxygenation by the action of a dioxygenase.

Following dioxygenation, a dihydrodiol with saturated carbon atoms at C1 and C2 is produced, which further undergoes enzymatic dehydrogenation yielding a catechol. The catecholic ring is aromatic but is now activated by the presence of two hydroxylic groups (Fig. 8.2). The activation facilitates an enzymatic ring cleavage reaction. This process can be repeated to metabolically deconstruct the neighbouring aromatic ring (Wackett and Hershberger, 2001).

The microbial catabolism of BTEX compounds has been widely studied, particularly under aerobic conditions (Gibson and Subramanian, 1984). The defining characteristics of the reaction are that BTEX must be activated to overcome stability and subsequently split to open the aromatic ring. Microbes use molecular oxygen in the reaction, which is highly exergonic. BTEX compounds are oxidised by dioxygenases or

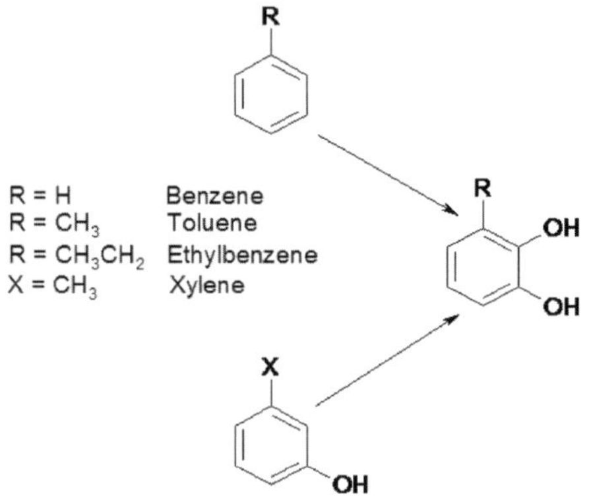

Fig. 8.3 Aerobic metabolism of BTEX compounds and other aromatic hydrocarbons proceeding through catechol intermediates (Wackett and Hershberger, 2001)

monooxygenases in different organisms (Fig. 8.3). For example, toluene is known to be oxidised by dioxygenases to yield *cis*-1,2-dihydroxyhydro-3-methyl-3,5-cyclohexadiene. A dehydrogenase oxidises this intermediate to 3-methylcatechol, which is a common substrate for ring cleavage dioxygenase enzymes (Wackett and Hershberger, 2001).

There are likely hundreds of thousands of heterocyclic-ring structures, if one considers all known alkaloids, plant pigments and porphyrin-like derivatives produced biologically. It is impossible to simply represent various biochemical mechanisms necessary to catabolise all of these compounds and experimental data are available for only a limited number of examples (Kaiser et al., 1996). Therefore, here we illustrate only a basic mechanism by which the heterocyclic ring can be opened. This mechanism

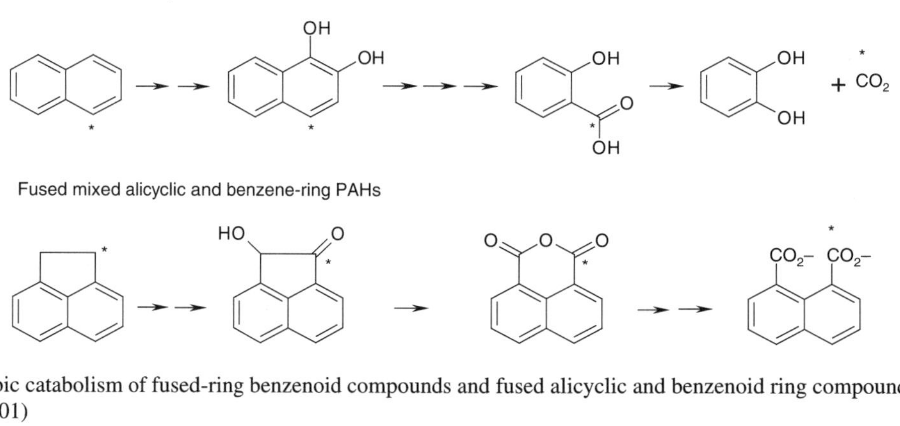

Fig. 8.2 Aerobic catabolism of fused-ring benzenoid compounds and fused alicyclic and benzenoid ring compounds (Wackett and Hershberger, 2001)

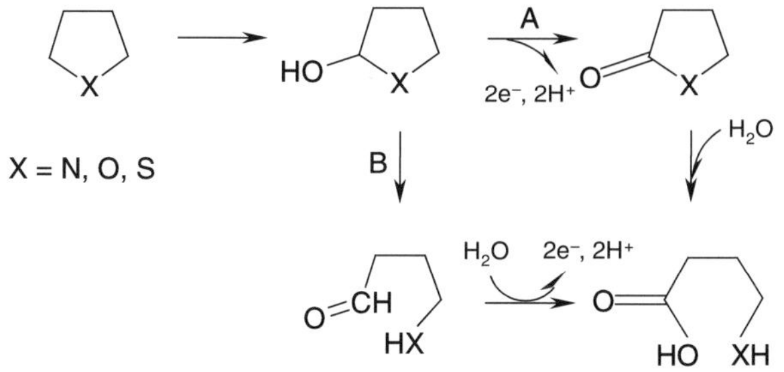

X = N, O, S

Fig. 8.4 Metabolic reactions used by bacteria for opening nitrogen, oxygen and sulphur heterocycles (Wackett and Hershberger, 2001)

is fundamentally the same with nitrogen, oxygen and sulphur heterocyclic rings. The metabolic strategy is to hydroxylate the ring at a carbon atom adjacent to the heteroatom. Thereafter, two different reaction pathways can ensue with the same outcome, i.e. the opening of the ring to generate a carboxylic acid.

In pathway A, dehydrogenation generates a ring ester, amide or thioester, depending on the identity of the heteroatom. All of these intermediates can be cleaved by hydrolysis to open the ring. Pathway B shows an isomerisation of 2-hydroxyheteroatom compounds, with the ring opening to produce an open-ring aldehyde product (Fig. 8.4). The isomerisation can be enzymatically catalysed or occur spontaneously,

depending upon the reactivity of the compound (Wackett and Hershberger, 2001).

Organohalides are compounds containing a carbon bonded to chlorine, bromine and iodine. The key reaction during microbial degradation of halogenated compounds is an actual dehalogenation reaction. During this reaction, the halogen substituent, which is usually responsible for the toxic and xenobiotic character of the compound, is replaced by a hydrogen or hydroxyl group. Halogen removal reduces both recalcitrance to biodegradation and the risk of forming toxic intermediates during subsequent metabolic steps. Microorganisms that can utilise halogenated compounds as a growth substrate often produce enzymes whose function is carbon–halogen bond cleavage (Janssen et al., 2001). There is no single metabolic mechanism that will apply to all organohalides. Fundamentally, four different mechanisms are observed in biodehalogenation (Fig. 8.5).

The nucleophilic displacement mechanism is predominantly represented by displacement with water (hydrolysis). This mechanism often occurs with singly halogenated alkyl halides. A reductive dehalogenation requires the input of two electrons and protons to generate a carbon–hydrogen bond and HCl, HBr or HI, respectively. These reactions are more commonly catalysed by anaerobic bacteria. Oxygenation represents a versatile mechanism for dehalogenation.

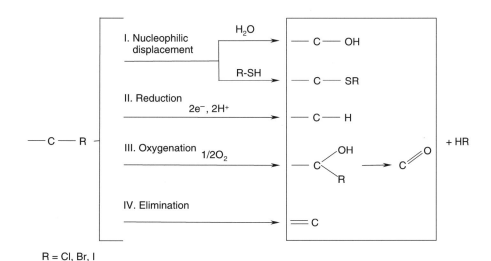

Fig. 8.5 Haloorganic compounds are metabolised via four fundamental mechanisms capable of cleaving the carbon–halogen bond (Wackett and Hershberger, 2001)

Oxygenation of an alkyl halide generates an unstable *gem*-haloalcohol, which undergoes the rapid spontaneous elimination of HCl, HBr or HI, to yield a carbonyl compound. The fourth pathway of dehalogenation is via elimination of a proton and a halide anion from adjacent carbon atoms to yield an additional bond between the carbon atoms (Wackett and Hershberger, 2001).

8.2.4 Mycobacteria Involved in Biodegradation of Organic Pollutants

Mycobacteria possessing the ability to degrade organic pollutants have mostly been isolated from various kinds of environment previously exposed to particular xenobiotic compounds (Table 8.1). Two toluene-degrading *M.* sp. strains T103 and T104 were isolated from a rock surface biomass in a freshwater stream contaminated with toluene (Tay et al., 1998). *Mycobacterium* sp. strain RP1 degrading morpholine was isolated from contaminated sludge collected in the wastewater treatment unit of a chemical plant (Poupin et al., 1998).

M. sp. strain KR20 originates from the polycyclic aromatic hydrocarbon-contaminated soil of a former gasworks plant site (Rehmann et al., 2001). The pure cultures of mycobacteria were obtained by enrichment culture techniques, which allow the selective cultivation of one or more isolates from a complex mixture typically found in soil and groundwater. The method typically relies on using a particular organic compound as the sole carbon source or as the nitrogen, sulphur or phosphorus source (Wackett and Hershberger, 2001). *M.* sp. strain LB501T utilises anthracene as the sole source of carbon and energy (van Herwijnen et al., 2003). The *M.* sp. strains AP1 (Vila et al., 2001), CP1 and CP2 (Lopez et al., 2005) were isolated based upon their ability to use fluoranthene as the sole source of carbon and energy. The endosulfane insecticide is used by the *M.* sp. strain ESD as the only source of sulphur (Sutherland et al., 2002).

Bastiaens et al. (2000) compared two different methods by which PAH-utilising bacteria from PAH-contaminated soil and sludge were obtained: (1) shaken enrichment cultures in liquid mineral medium in which PAHs were supplied as crystals and (2) enrichment and recovery from hydrophobic membranes containing adsorbed PAHs. Both techniques were successful, but different selected bacterial isolates were able to grow on PAHs as the sole source of carbon and energy. The *Sphingomonas* spp. is mainly selected for liquid enrichment, whereas the membrane method led exclusively to selection of *M.* spp. (Bastiaens et al., 2000). Analysis of the 16S rRNA gene sequence of mycobacterial degraders and their growth characteristics reveals these isolates to be fast-growing mycobacteria. Some isolates such as *M. austroafricanum* strain GTI-23 (Bogan et al., 2003), *M. frederiksbergense* strain LB501T (Wick et al., 2003) or *M. hodleri* (Kleespies et al., 1996) were taxonomically classified as new species based on their phylogenetic characteristics such as unique fatty acid pattern, distinctness of their physiological properties and the uniqueness of the sequence of their 16S ribosomal DNA.

Enzymatic pathways enabling biodegradation of organic pollutants evolved as an adaptation to the presence of organic pollutants in the environment. Identification of metabolites formed from organic pollutants indicated the type of enzymes and metabolic pathway(s) leading to the degradation of organic pollutants. Various methods were used to analyse the metabolites and to monitor the kinetics of the biodegradation of organic pollutants. Gas chromatography was used to analyse the metabolites formed during biodegradation of 1,2-dibromoethane by *M.* sp. strain GP1 (Poelarends et al., 1999) or the metabolites of the biodegradation of β-endosulfane by *M.* sp. strain ESD (Sutherland et al., 2002).

In order to acquire more information on the degradation pathways of anthracene by the *M.* sp. strain LB501T, extracts of cultures for metabolites were analysed by gas chromatography–mass spectrometry (van Herwijnen et al., 2003). Gas chromatography–mass spectrometry in combination with high-performance liquid chromatography was used to analyse metabolites in the biodegradation pathway of fluoranthene in the *M.* sp. strain KR20 (Rehmann et al., 2001) and in *M.* sp. strains CP1 and CP2 (Lopez et al., 2005). The use of *in situ* ^{1}H nuclear magnetic resonance spectroscopy allowed the determination of two intermediates in the biodegradative pathway of morpholine in *M.* sp. strain RP1 (Poupin et al., 1998) and *M. aurum* strain MO1 (Combourieu et al., 2000).

Table 8.1 Mycobacterial isolates with degradative capabilities

Organic compounds	Isolates	Compounds	References
PAHs	*M.* sp. CH1	3–4-ring PAHs, pyrene, phenanthrene, fluoranthene	[7, 17]
	M. sp. CH2	3–4-ring PAHs	[17]
	M. sp. BB2	Fluoranthene, phenanthrene, pyrene	[2]
	M. sp. BG1	Phenanthrene	[14]
	M. frederiksbergense LB501T	Anthracene	[42, 46–48]
	M. sp. KR2	Pyrene	[32]
	M. sp. KR20	Fluoranthene, pyrene	[31]
	M. sp. 1B	Fluoranthene	[11, 12]
	M. RJGII-135	Pyrene, benz[a]pyrene, benz[a]anthracene	[3, 34]
	M. austroafricanum GTI-23	Pyrene, fluoranthene, phenanthrene	[1]
	M. hodleri sp. EM12	Fluoranthene	[23]
	M. frederiksbergense	Pyrene, fluoranthene, phenanthrene	[49]
	M. sp. AP1	Pyrene	[26, 44]
	M. sp. CP1, CP2, CFt2, CFt6	Fluoranthene, pyrene	[25]
	M. sp.6Py1	Pyrene, phenanthrene	[24]
	M. vanbaalenii PYR-1	Naphthalene, anthracene, phenanthrene, pyrene	[4, 19]
BTEX	*M.* sp. T103, T104	Toluene	[39]
	M. sp. HL 4-NT-1	1-Nitrotoluene, trinitrotoluene	[36, 45]
Heterocyclic compounds	*M.* sp. RP1	Morpholine, pyrrolidine, piperidine	[30, 35, 40]
	M. aurum MO1	Morpholine, triomorpholine, piperidine	[9, 10]
Organohalides	*M.* sp. GP1	Halogenated aliphatic hydrocarbons, alcohols	[29]
	M. smegmatis and others	Halogenated aliphatic hydrocarbons	[22]
	M. avium N85	Halogenated aliphatic hydrocarbons	[20, 28]
	M. bovis 5033/66	Halogenated aliphatic hydrocarbons	[21]
	M. sp. TA5, TA27	1,1,1-Trichloroethane	[50]
	M. vaccae JOB-5	Trichloroethylene, chlorobutane, chlorinated alcohols, dichlorinated alkenes	[13, 16, 43]
	M. sp. JS60	Vinyl chloride	[8]

Table 8.1 (continued)

Organic compounds	Isolates	Compounds	References
	M. aurum L1	Vinyl chloride	[18]
	M. sp. ESD	Endosulfane	[37, 38]
	M. sp. M156	Chlorinated and brominated styrenes	[6, 33]
	M. fortuitum CG-2	Chlorinated phenols	[15, 41]
	M. chlorophenolicum PCP-I	Pentachlorophenol	[5, 27]

M. Mycobacterium.

[1] Bogan BW, Lahner LM, Sullivan WR, Paterek JR (2003) Journal of Applied Microbiology 94:230–239. [2] Boldrin B, Tiehm A, Fritzsche Ch (1993) Applied and Environmental Microbiology 59:1927–1930. [3] Brezna B, Khan AA, Cerniglia CE (2003) FEMS Microbiology Letters 223:177–183. [4]. Brezna B, Kweon O, Stingley RL, Freeman JP, Khan AA, Polek B, Jones RC, Cerniglia CE (2006) Applied Microbiology and Biotechnology 71:522–532. [5] Briglia M, Eggen RIL, van Elsas DJ, de Vos WM (1994) International Journal of Systematic Bacteriology 44:494–498. [6] Chion ChKChK, Askew SE, Leak DJ (2005) Applied and Environmental Microbiology 71:1909–1914. [7] Churchill SA, Harper JP, Churchill PF (1999) Applied and Environmental Microbiology 65:549–552. [8] Coleman NV, Spain JC (2003) Journal of Bacteriology 185:5536–5545. [9] Combourieu B, Besse P, Sanceleme M, Veschambre H, Godin J-P, Monteil A, Veschambre H, Delort AM (2000) Applied and Environmental Microbiology 66:3187–3193. [10] Combourieu B, Besse P, Sanceleme M, Veschambre H, Delort AM, Poupin P, Truffaut N (1998) Applied and Environmental Microbiology 64:153–158. [11] Dandie CE, Bentham RH, Thomas SM (2006) Applied Microbiology and Biotechnology 71:59–66. [12] Dandie CE, Thomas SM, Bentham RH, McClure NC (2004) Journal of Applied Microbiology 97:246–255. [13] Fairlee JR, Burback BL, Perry JJ (1997) Canadian Journal of Microbiology 43:841–846. [14]. Guerin WF, Jones GE (1988) Applied and Environmental Microbiology 54:937–944. [15] Haggblom MM, Nohynek LJ, Salkinoja-Salonen MS (1988) Applied and Environmental Microbiology 54:3043–3052. [16] Halsey KH, Sayvedra-Soto LA, Bottomley PJ, Arp DJ (2005) Applied Microbiology and Biotechnology 68:794–801. [17] Harper JP, Churchill PF, Lokey-Flippo L, Lalor MM (2005) J Environ Sci Health A Tox Hazard Subst Environ Eng 40:493–507. [18] Hartmans S, de Bont JAM (1992) Applied and Environmental Microbiology 58:1220–1226. [19] Heitkamp MA, Freeman JP, Miller DW, Cerniglia CE (1988) Appl. Environ. Microbiol. 54:2556–2565. [20] Jesenska A, Bartos M, Czernekova V, Rychlik I, Pavlik I, Damborsky J (2002) Applied and Environmental Microbiology 68:3724–3730. [21] Jesenska A, Pavlova M, Strouhal M, R. C, Tesinska I, Monincova M, Prokop Z, Bartos M, Pavlik I, Mobius P, Nagata Y, Damborsky J (2005) Applied and Environmental Microbiology 71:6736–6745. [22] Jesenska A, Sedlacek I, Damborsky J (2000) Applied and Environmental Microbiology 66:219–222. [23] Kleespies M, Kroppenstedt RM, Rainey FA, Webb LE, Stackebrandt E (1996) International Journal of Systematic Bacteriology 46:683–687. [24] Krivobok S, Kuony S, Meyer Ch, Louwagie M, Willison JC, Jouanneau Y (2003) Journal of Bacteriology 185:3828–3841. [25] Lopez Z, Vila J, Grifoll M (2005) Journal of Indian Microbiology and Biotechnology 32:455–464. [26] Lopez Z, Vila J, Minguillon C, Grifoll M (2006) Applied Microbiology and Biotechnology 70:747–756. [27] Mieting R, Karlson U (1996) Applied and Environmental Microbiology 62:4361–4366. [28] Pavlova M, Jesenska A, Kivana M, Prokop Z, Konecna H, Sato Y, Tsuda M, Nagata Y, Damborsky J (2007) Journal of Structural Biology 157:384–392. [29] Poelarends GJ, van Hylckama Vlieg JET, Marchesi JR, Freitas dos Santos LM, Janssen DB (1999) Journal of Bacteriology 181:2050–2058. [30] Poupin P, Truffaut N, Combourieu B, Besse P, Sanceleme M, Veschambre H, Delort AM (1998) Applied and Environmental Microbiology 64:159–165. [31] Rehmann K, Hertkorn N, Kettrup AA (2001) Microbiology 147:2783–2794. [32] Rehmann K, Noll HP, Steinberg CE, Kettrup AA (1998) Chemosphere 36:2977–2992. [33] Rigby SR, Matthews CS, Leak DJ (1994) Bioorganic and Medical Chemistry 2:553–556. [34] Schneider J, Grosser R, Jayasimhulu K, Xue W, Warshawsky D (1996) Applied and Environmental Microbiology 62:13–19. [35] Sielaff B, Andreesen JR (2005) Microbiology 151:2593–603. [36] Spiess T, Desiere F, Fischer P, Spain JC, Knackmuss H-J, Lenke H (1998) Applied and Environmental Microbiology 64:446–452. [37] Sutherland TD, Horne I, Harcourt RL, Russel RJ, Oakeshott JG (2002) Journal of Applied Microbiology 93:380–389. [38] Sutherland TD, Horne I, Russel R, Oakeshott JG (2002) Applied and Environmental Microbiology 68:6237–6245. [39] Tay STL, Hemond HF, Polz MF, M. CC, Dejesus I, Krumholtz LR (1998) Applied and Environmental Microbiology 64:1715–1720. [40] Trigui M, Pulvin S, Truffaut N, Thomas D, Poupin P (2004) Research in Microbiology 155:1–9. [41] Uotila JS, Kitunen VH, Saastamoinen T, Coote T, Haggblom MM, Salkinoja-Salonen MS (1992) Journal of Bacteriology 174:5669–5675. [42] van Herwijnen R, Springael D, Slot P, Govers HAJ, Parsons JR (2003) Applied and Environmental Microbiology 69:186–190. [43] Vanderberg LA, Perry JJ (1994) Canadian Journal of Microbiology 40:169–172. [44] Vila J, Lopez Z, Sabate J, Minguillon C, Solanas AM, Grifoll M (2001) Applied and Environmental Microbiology 67:5497–5505. [45] Vorbeck C, Lenke H, Fischer P, Spain JC, Knackmuss H-J (1998) Applied and Environmental Microbiology 64:246–252. [46] Wick LY, de Munain AR, Springael D, Harms H (2002) Applied Microbiology and Biotechnology 58:378–385. [47] Wick LY, Pasche N, Bernasconi SM, Pelz O, Harms H (2003) Applied and Environmental Microbiology 69:6133–6142. [48] Wick LY, Pelz O, Bernasconi SM, Andersen N, Harms H (2003) Environmental Microbiology 5:672–680. [49] Willumsen P, Karlson U, Stackebrandt E, Kroppenstedt RM (2001) Journal of Systematic and Evolutionary Microbiology 51:1715–1722. [50] Yagi O, Hashimoto A, Iwasaki K, Nakajima M (1999) Applied and Environmental Microbiology 65:4693–4696.

Genetic analyses have significantly contributed to a better understanding of the biodegradation mechanisms of organic pollutants by mycobacteria. The involvement of plasmids in the biodegradation of PAHs has been observed in the *M.* sp. strain BG1. Phenanthrene degradation by this strain BG1 is plasmid mediated since plasmid loss in nutrient-grown cultures accompanied loss of the phenanthrene-degrading phenotype (Guerin and Jones, 1988).

Cytochrome P450 (*morA*), which is involved in the biodegradation of morpholine, piperidine and pyrrolidine in *M.* sp. strain RP1, was cloned and sequenced. Four open reading frames were detected in this DNA fragment. The first encoded a cytochrome P450 designated as *morA*, which was the second member of the *CYP151* family and was named *CYP151A2*. The second open reading frame (*morB*) featured a [3Fe–4S] type of ferredoxin. A third gene (*morC*), exhibiting the sequence identity of known reductases and a fourth truncated gene encoding a putative glutamine reductase (*orf1'*), were found downstream of *morB*. The recombinant *MorA* cytochrome P450 was purified to homogeneity from *Escherichia coli*. The purified enzyme was a monomeric soluble protein and catalysed the ring cleavage of the secondary amines. It was found that pyrrolidine is the preferred substrate for this monooxygenase (Trigui et al., 2004).

Propene monooxygenase has been cloned from *M.* sp. strain M156 capable of degrading chlorinated and brominated styrenes based on hybridisation with *amoABCD* genes from *Rhodococcus corallinus* B276. Site-directed mutagenesis of active site residues led to a systematic alternation of the stereoselectivity of styrene oxidation, presumably by producing different orientations for substrate binding during catalysis (Chion et al., 2005). Massive random sequencing of environmental DNA and microbial genome-sequencing projects has shown a large diversity of sequences that are not employed by known catabolic pathways. The corresponding proteins may have novel functions and selectivity and can be valuable for biotransformation in the future. Identification of genes coding for dehalogenating enzymes, haloalkane dehalogenases in slow-growing *M. tuberculosis* strain H37Rv (Jesenska et al., 2000) and their cloning from *M. bovis* 5033 (Jesenska et al., 2005) represent an example in which the genome-sequencing project has led to the discovery of novel enzymes.

It has been speculated that most degradation in nature occurs via consortial metabolism. This metabolism is characterised by sequential conversions in which part of the pathway is found in one microorganism and part in others. This is a difficult conjecture to prove or disprove and impossible to consider properly without knowing the genetic and enzymatic details of the biodegradation derived from thousands of studies conducted with microbial pure cultures (Wackett and Hershberger, 2001). Fast-growing mycobacteria are considered essential members of the PAH-degrading bacterial community in PAH-contaminated soils. A microbial consortium that rapidly mineralised the environmentally persistent pollutant benzo[a]pyrene has been recovered from soil (Kanaly et al., 2000).

The consortium cometabolically converted benzo[a]pyrene to CO_2 when it was grown on diesel fuel and the extent of benzo[a]pyrene mineralisation was dependent on both diesel fuel and benzo[a]pyrene concentrations. DNA sequence analysis showed that the consortium included members related to the genera *Mycobacterium* and *Sphingobacterium*. The combined culture of *M. vaccae* and *R.* sp. strain R-22 may represent another example of such a system. *M. vaccae* transforms benzene to phenol and subsequently hydroquinone. Hydroquinone is an important dead-end product because it is potentially an active intermediate responsible for the toxicity of benzene (Fairlee et al., 1997).

Bioremediation of trichloroethylene as a model substrate by an aerobic consortium containing *Pseudomonas putida*, *Pseudomonas fluorescens*, *M.* sp. and *Nocardia paraffinae* was studied by Meza et al. (2003). Using isolated enzymes, TCE degradation intermediates appeared to be inhibitory to the oxygenase enzymes, thereby diminishing the overall degradation. To study the natural role and diversity of the mycobacterial community in contaminated soils, a culture-independent fingerprinting method based on PCR combined with denaturing gradient gel electrophoresis (DGGE) was developed by Leys et al. (2005).

New PCR primers were selected specifically targeting the 16S rRNA genes of fast-growing mycobacteria. The PCR products were examined on 1.5% agarose gels and directly used for DGGE analysis on polyacrylamide gels. Single-band DGGE profiles of amplicons were obtained for most of the tested mycobacterial isolates. Strains belonging to the same species revealed identical DGGE fingerprints and in most cases these fingerprints were typical for one species, allowing partial differentiation between species in a mycobacterial community (Leys et al., 2005).

8.2.5 *Mycobacterium vanbaalenii Strain PYR-1 Utilising Polycyclic Aromatic Hydrocarbons*

M. vanbaalenii strain PYR-1 is one of the best-studied mycobacterial strains which is able to metabolise a wide range of low- and high-molecular-weight PAHs. This strain PYR-1 has been isolated from petroleum-contaminated estuarine sediment based on its ability to degrade pyrene. The isolate was identified as *M.* sp. on the basis of its cellular and colony morphology, Gram-positive staining and strong acid-fast reactions, diagnostic biochemical tests, 66.6% G+C content of the DNA and the presence of high-molecular-weight mycolic acids in its cell wall (Heitkamp et al., 1988).

In 2002, Khan et al. (2002) reclassified *M.* sp. strain PYR-1 as *M. vanbaalenii* sp. nov. Sequencing of 16S rDNA, fatty acid methyl esters, DNA–DNA hybridisation and PFGE analysis of the total genomic DNA were used to determine the taxonomic relationship of the strain PYR-1 to other closely related *M.* species: *M. aurum* strain ATCC 23366[T], *M. vaccae* strain ATCC 15438[T] and *M. austroafricanum* strain ATCC 33464[T]. Based on phylogenetic analysis, it was concluded that strain PYR-1 represents a novel species of the genus *Mycobacterium*, for which the name *M. vanbaalenii* sp. nov. was proposed.

This strain PYR-1 is the first reported bacterium able to extensively mineralise pyrene and other PAHs containing four aromatic rings; it is unable to utilise pyrene as the sole source of carbon and energy in comparison to *M.* sp. strains AP1 (Vila et al., 2001) or CP1, CP2, CFt2 and CFt6 (Lopez et al., 2006). Surprisingly, *M. vanbaalenii* strain PYR-1 is able to mineralise four-ringed PAHs faster, compared to two- or three-ringed PAHs displaying lower hydrophobicity and stability. A pure culture is able to mineralise over 60% of pyrene to CO_2 after 96 hours. Analyses by UV, infrared, mass and nuclear magnetic resonance

spectrometry and gas chromatography have identified *cis-* and *trans-*4,5-dihydrodiols of pyrene, which suggests the existence of multiple pathways for the initial oxidative attack on pyrene.

Experiments studying the incorporation of $^{18}O_2$ confirmed that the pyrene *trans-*4,5-dihydrodiol formation was catalysed by monooxygenase and the formation of pyrene *cis-*4,5-dihydrodiol was a dioxygenase-catalysed reaction. The major metabolites 4-phenantroic acid, 4-hydroxyperinaphthenon and cinnamic phthalic acids were identified as ring fission products. The mechanism of oxidation is unique since *M. vanbaalenii* strain PYR-1 has both mono- and dioxygenase to catalyse the initial attack on PAHs (Heitkamp et al., 1988). The identification and isolation of metabolites that are released during the fluoranthene degradation indicate that several pathways (Fig. 8.6) operate simultaneously during degradation by the strain PYR-1 (Kelley et al., 1993).

Besides pyrene and fluroanthene, *M. vanbaalenii* strain PYR-1 is capable of degrading other polyaromatic hydrocarbons, including naphthalene, anthracene, phenanthrene, 1-nitropyrene and benzo[a]pyrene. Biodegradation pathways have been proposed based on the identification of the initial ring oxidation and ring cleavage metabolites (Kelley et al., 1990; Heitkamp et al., 1991; Moody et al., 2001, 2004). As the organism utilises multiple pathways to degrade some PAHs, its genome likely encodes a number of enzymes involved in these pathways.

The proteins involved in PAH degradation have been examined by the proteomic approach. Two-dimensional gel electrophoresis of proteins from cultures of *M. vanbaalenii* strain PYR-1 was used to detect proteins, whose abundance increased after phenanthrene, dibenzothiaphene and pyrene exposure. At least six major proteins of different sizes were expressed. In addition to the expression of catalase peroxidase, a 50-kDa polypeptide with an N-terminal sequence similar to those of other dioxygenases was expressed. The oligonucleotide probe designed from this protein sequence was used to screen dioxygenase-positive clones from the genomic library of *M.* sp. strain PYR-1. As a result of cloning and sequencing of a dioxygenase-positive clone with the probe designed from this protein sequence, the genes encoding the dioxygenase large subunit (*nidA*), dioxygenase small subunit (*nidB*) and dehydrogenase (*nidD*) genes were obtained. This gene arrangement of the α and β

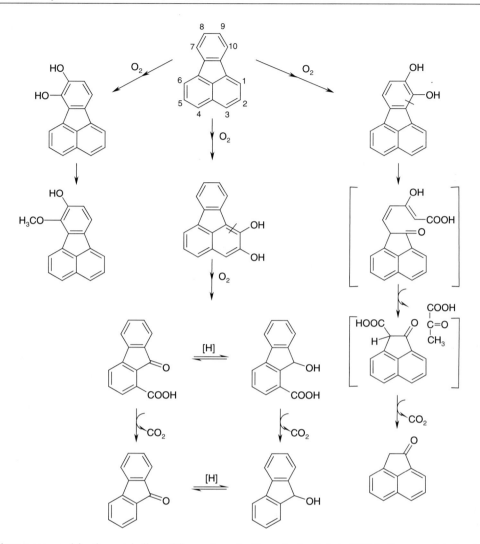

Fig. 8.6   Pathways proposed for the metabolism of fluoranthene by *M. vanbaalenii* strain PYR-1. Compounds in brackets were not identified (Kelley et al., 1993)

subunits of PAH–dioxygenase has rarely been found in other bacterial PAH–dioxygenase systems (Khan et al., 2001).

The genes involved in the degradation of phenanthrene via the phthalic acid pathway were identified in 2004. Southern blot analysis was used to identify the clones containing the *nidA* gene. Sequence analysis revealed a number of additional genes involved in PAH degradation. The presence of genes encoding putative ABC transporters, which may play a role in the uptake of hydrocarbons or in the efflux of metabolites, was confirmed (Stingley et al., 2004). A gene cluster encoding the 20-kDa polypeptide showing upregulated expression in response to exposure of *M. vanbaalenii*

strain PYR-1 to several PAHs was isolated and characterised. A novel dioxygenase gene designated as *nidA3B3* was cloned and the substrate specificities and transformation rates of recombinant proteins were studied.

The dioxygenase showed a wide substrate specificity towards PAHs and its PAH-induced upregulation in the strain PYR-1 was confirmed by reverse transcription-PCR analysis (Kim et al., 2006). In 2006, Brezna et al. (2006) amplified three cytochrome P450 monooxygenases genes (*cyp151*, *cyp150* and *cyp50*) by PCR. The complete sequence of these genes was determined. Genes *cyp151* and *cyp150* were cloned and the proteins were expressed in *E. coli*. Thirteen other mycobacterium strains were screened for the presence of *pipA*, *cyp150* and *cyp51* genes, as well as the initial PAH dioxygenase (*nidA* and *nidB*). The results indicated that many of the *M.* spp. surveyed contain both monooxygenases and dioxygenases to degrade PAHs. The wide substrate specificity of *M. vanbaalenii* strain PYR-1 towards a variety of PAHs makes this bacterium useful for bioremediation processes. Its ability to form sterically and optically pure arene dihydrodiols can be exploited for the biosynthesis of pharmaceutical and industrially important compounds.

8.2.6 Distribution of Haloalkane Dehalogenase Enzymes in Mycobacteria

Mycobacterium sp. strain GP1 was isolated by prolonged batch enrichment from a mixed bacterial culture, which was able to mineralise 1,2-dibromoethane (Poelarends et al., 1999). Strain GP1 is able to grow on 1,2-dibromoethane, both in liquid cultures and on plates. This strain is a member of the subgroup of thermosensitive, fast-growing mycobacteria. Strain GP1 is the first member of the genus *Mycobacterium* known to degrade short-chain halogenated aliphatics. There are two different dehalogenating enzymes present in the strain GP1:

(i) a hydrolytic dehalogenase, which converts haloalkanes to the corresponding alcohols and halide ions and;

(ii) a haloalcohol dehalogenase converting haloalcohols to the corresponding epoxides and halide ions.

Strain GP1 is the first organism found to produce both dehalogenating enzymes. Together, hydrolytic dehalogenase and haloalcohol dehalogenase convert 1,2-dibromoethane to ethylene oxide (Poelarends et al., 1999). The haloalkane dehalogenase gene of GP1 designated as *dhaAf* is very similar to the *dhaA* gene found in the *Rhodococcus rhodochrous* strain NCIMB 13064. The haloalkane dehalogenase encoded by *dhaAf* is identical to DhaA, except for three amino-acid substitutions and a 14-amino-acid extension at the C-terminus. Nucleotide sequence analysis indicated

that the *dhaAf* gene was formed by the fusion of a *dhaA* gene with the last 42 nucleotides of the *hheB* gene, which encodes a haloalcohol dehalogenase (Poelarends et al., 1999).

A very different approach for the isolation of haloalkane dehalogenase genes was used by Jesenska et al. (2000). A search of genetic databases with the sequences of known haloalkane dehalogenases revealed the presence of three different genes encoding the putative haloalkane dehalogenases in the genome of *M. tuberculosis* strain H37Rv (Fig. 8.7).

Two orthologous haloalkane dehalogenase genes designated *dmbA* and *dmbB* were cloned from *M. bovis* 5033/66, a bacterium whose genome is 99.9% identical to *M. tuberculosis* strain H37Rv. DmbA and DmbB proteins were heterologously expressed in *E. coli*, purified to homogeneity and positively tested for enzymatic activity. The DmbB protein had to be expressed in fusion with thioredoxin to obtain a soluble protein. The optimum temperature of DmbA and DmbB proteins determined with 1,2-dibromoethane is 45 °C. The melting temperature assessed by circular dichroism spectroscopy is 47 °C for DmbA and 57 °C for DmbB. The optimum pH of DmbA depends on the composition of the buffer with maximal activity at pH 9.0. DmbB had an optimum pH at pH 6.5. PCR screening of 48 isolates from various hosts revealed that both these genes are widely distributed among species of the *M. tuberculosis* complex, including *M. bovis*, *M. bovis* BCG, *M. africanum*, *M. caprae*, *M. microti* and *M. pinnipedii*. The third haloalkane dehalogenase from *M. bovis* strain 5033/66, designated DmbC, was biochemically characterised only recently (A. Jesenska, unpublished data). The DmbC protein was expressed in *M. smegmatis*. Its optimum temperature is 45 °C and optimum pH 8.3. Its activity on halogenated substrates is about two levels of magnitude lower, compared with the activity of DmbA and DmbB.

Another haloalkane dehalogenase has been identified by sequence comparisons in the genome of *M. avium* strain 104, the bacterium isolated from a swine's mesenteric lymph node. Heterologous expression of the *dhmA* gene in *E. coli* strain GI724 resulted in a dehalogenase hydrolysing a wide range of halogenated compounds (Jesenska et al., 2002). Recently, *dhmA* was recloned to a newly constructed pK4RP rhodococcal expression system, which allowed for the expression of DhmA in a soluble and stable form (Pavlova et al., 2007). The catalytic pentad of DhmA

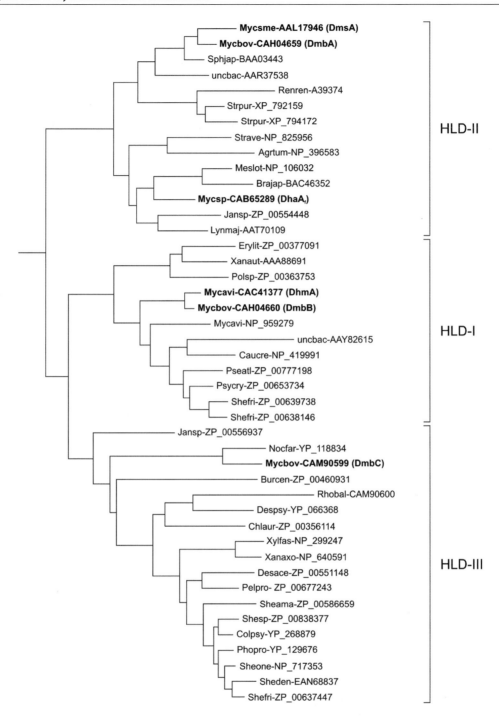

Fig. 8.7 Phylogenetic tree of haloalkane dehalogenases calculated by the maximum likelihood method. The subdivision of the haloalkane dehalogenase family into three subfamilies (HLD-I, HLD-II and HLD-III) is indicated. Names of individual proteins include the genus and species abbreviated to three letters and the NCBI accession number. Experimentally characterised haloalkane dehalogenases from mycobacterial species (DmbA, DmbB, DmbC, DhmA, DmsA and DhaAf) are highlighted (Chovancova et al., 2007)

was deduced from the sequence comparisons and homology model and confirmed by site-directed mutagenesis. A two-step reaction mechanism was proposed for this enzyme.

The biological role of haloalkane dehalogenases in bacteria, which do not grow on halogenated compounds as carbon or energy sources is currently unknown. The presence of the haloalkane dehalogenase genes in the genomes of mycobacteria, including the obligatory pathogenic mycobacterial species *M. tuberculosis* and *M. bovis*, and the potentially pathogenic *M. avium*, indicates that haloalkane dehalogenases may be involved in the protection of mycobacteria against halogenated substances produced by the immune system of host organisms. Alternatively, these enzymes may act on a physiological substrate of unknown chemical structure. More research is needed to test different hypotheses proposed for the role of haloalkane dehalogenases in mycobacteria.

8.2.7 Conclusions

Organic compounds can be degraded by a variety of naturally occurring soil bacteria, including mycobacteria. Until now, about 50 mycobacterial isolates capable of degrading various organic pollutants have been isolated, mainly from contaminated sites. Analysis of their 16S rRNA sequences has suggested an affiliation with the genus *Mycobacterium* and has led to their clustering together with fast-growing mycobacteria. The enzyme-catalysed conversion of halogenated organic compounds, however, has also been confirmed in slow-growing mycobacteria. Whether the primary role of these enzymes is the transformation of halogenated substances or whether these activities originate from enzymatic promiscuity is currently unknown.

8.3 Mycobacterial Volatilisation, Sequestering or Precipitation of Toxic Heavy Metals

J.O. Falkinham III

A mercury-, cadmium- and copper-resistant strain of *M. scrofulaceum* has been shown to be capable of metal volatilisation, metal sequestering and metal

precipitation. Mercury resistance was shown to be due to the presence of plasmid-encoded mercuric reductase, whose reduction of Hg^{2+} to Hg^0 led to mercury loss from the medium (Meissner and Falkinham, 1984). Resistance to cadmium in the strain was due to the accumulation of Cd^{2+} in the particulate, envelope fraction; in the susceptible plasmid-free segregant of the strain, Cd^{2+} accumulation was primarily in the soluble, intracellular fraction (Erardi et al., 1989).

Copper resistance was due to a sulphate-dependent precipitation of the metal as black, insoluble copper sulphide (Erardi et al., 1987). All three mechanisms, volatilisation, sequestering and precipitation, not only conferred metal resistance on the mycobacterial cells but also removed the toxic heavy metal from the environment. Consequently, cells other than the mycobacteria would be protected from the heavy metal because of its removal from the soluble phase.

8.4 The Biological Role of Mycobacteria as Nutrients for Dragonfly Larvae and Cladocera

J. Kazda

The majority of dragonflies are dependent on heterotrophic water ponds rich in organic substances because their larvae require abundant nutrients, especially in the first phase of development. For a long time it could not be satisfactorily explained why some species of dragonflies can colonise oligotrophic ponds in sphagnum bogs, which are very poor in nutrients. What is known, however, is that environmental saprophytic mycobacteria have an important biological role in the colonisation of oligotrophic moorland waters by dragonflies. A variety of saprophytic mycobacteria number among the habitat microorganisms of sphagnum biotopes and multiply in floating sphagnum, covering parts of small water bodies. Mycobacteria are released onto the surface water layer but dragonflies' larvae are not able to utilise microorganisms directly. As found elsewhere, Cladocera are able to filter out bacteria present in water and use them as nutrients (Brandelberger, 1988). These Cladocera, especially *Ceriodaphnia reticulata*, are caught by the young dragonfly larvae in their first phase of development (Thorp and Cothran, 1984). Thus, as floating sphagnum contains a high concentration of mycobacteria, enhanced numbers of mycobacteria have been found in both Cladocera and dragonfly larvae living in these biotopes, near sphagnum (Soeffing, 1988).

To confirm this nutritive chain, a model tritium-labelled mycobacterium was used. It was shown that Cladocera filter out mycobacteria, while dragonfly larvae catch them and incorporate them in their organs. The *Leucorrhinia rubicunda* larvae acquire mycobacterial flora in their nutritive chain by the ingestion of Cladocera (Soeffing and Kazda, 1993). Mycobacteria growing in floating sphagnum enable colonisation of boreal and sub-boreal moorland water ponds by dragonflies (Photo 8.1). They contribute to the nutrition of Cladocera and their predators, young moorland dragonflies larvae. These mycobacteria resemble such saprophytic species as *M. sphagni, M. gordonae* and *M. komossense* (Kazda, 2000).

This is the first evidence that saprophytic mycobacteria play an important role in the ecology of Cladocera and insects such as moorland dragonflies (Photo 8.1). On the other hand, it is possible that the adult dragonflies, whilst laying eggs on the surface of sphagnum floating in different ponds, may contribute to the spread of some mycobacterial species within moorland waters.

References

Bastiaens L, Springael D, Wattiau P, Harms H, de Wachter R, Verachtert H, Diels L (2000) Isolation of adherent polycyclic aromatic hydrocarbon (PAH)-degrading bacteria using PAH-sorbing carriers. Applied and Environmental Microbiology. 66:1834–1843

Bogan BW, Lahner LM, Sullivan WR, Paterek JR (2003) Degradation of straight-chain aliphatic and high-molecular-weight polycyclic aromatic hydrocarbons by a strain *Mycobacterium austroafricanum*. Journal of Applied Microbiology. 94:230–239

Brandelberger H (1988) Untersuchungen zur Funktionsmorphologie des Filterapparates von Cladoceren. Inaug. Diss. University of Kiel.

Brennan PJ, Nikaido H (1995) The envelope of Mycobacteria. Annual Review of Biochemistry. 64:29–63

Brezna B, Kweon O, Stingley RL, Freeman JP, Khan AA, Polek B, Jones RC, Cerniglia CE (2006) Molecular characterization of cytochrome P450 genes in the polycyclic aromatic hydrocarbon degrading *Mycobacterium vanbaalenii* PYR-1. Applied Microbiology and Biotechnology. 71:522–532

Chan Kwo Chion CK, Askew SE, Leak DJ (2005) Cloning, expression and site-directed mutagenesis of the propene monooxygenase genes from *Mycobacterium* sp. strain M156. Applied and Environmental Microbiology. 71: 1909–1914

Cirillo JD, Falkow S, Tompkins LS, Bermudez LE (1997) Interaction of *Mycobacterium avium* with environmental amoebae enhances virulence. Infection and Immunity. 65:3759–3767

Combourieu B, Besse P, Sanceleme M, Godin J-P, Monteil A, Veschambre H, Delort AM (2000) Common degradative pathways of morpholine, thiomorpholine, and piperidine by *Mycobacterium aurum* MO1: evidence from 1H-nuclear magnetic resonance and ionspray mass spectrometry performed directly on the incubation medium. Applied and Environmental Microbiology. 66:3187–3193

Erardi FX, Failla ML, Falkinham, III JO (1987). Plasmid-encoded copper resistance and precipitation by *Mycobacterium scrofulaceum*. Applied and Environmental Microbiology 53: 1951–1954

Erardi FX, Failla ML, Falkinham JO (1989) Accumulation and transport of cadmium by tolerant and susceptible strains of *Mycobacterium-Scrofulaceum*. Antimicrobial Agents and Chemotherapy. 33:350–355

Fairlee JR, Burback BL, Perry JJ (1997) Biodegradation of groundwater pollutants by a combined culture of *Mycobacterium vaccae* and a *Rhodococcus* sp. Canadian Journal of Microbiology. 43:841–846

Fetzner S, Lingens F (1994) Bacterial dehalogenases: biochemistry, genetics, and biotechnological applications. Microbiological Reviews. 58:641–685

Gibson DT, Subramanian V (1984) Microbial degradation of organic compounds. D. T. Gibson (ed.), Microbial degradation of organic compounds. Dekker, New York

Guerin WF, Jones GE (1988) Mineralization of anthracene by a *Mycobacterium* sp. Applied and Environmental Microbiology. 54:937–944

Habe H, Omori T (2003) Genetics of polycyclic aromatic hydrocarbon metabolism in diverse aerobic bacteria. Bioscience, Biotechnology and Biochemistry. 67:225–243

Heitkamp MA, Freeman JP, Miller DW, Cerniglia CE (1988) Pyrene degradation by a *Mycobacterium* sp.: identification of ring oxidation and ring fission products. Applied and Environmental Microbiology. 54:2556–2565

Heitkamp MA, Freeman JP, Miller DW, Cerniglia CE (1991) Biodegradation of 1-nitropyrene. Archives of Microbiology. 156:223–230

Janssen DB, Oppentocht JE, Poelarends GJ (2001) Microbial dehalogenation. Current Opinion in Biotechnology. 2:254–258

Janssen DB, Pries F, van der Ploeg JR (1994) Genetics and biochemistry of dehalogenating enzymes. Annual Review of Microbiology. 48:163–191

Jesenska A, Bartos M, Czernekova V, Rychlik I, Pavlik I, Damborsky J (2002) Cloning and expression of the haloalkane dehalogenase gene *dhmA* from *Mycobacterium avium* N85 and preliminary characterization of DhmA. Applied and Environmental Microbiology. 68:3724–3730

Jesenska A, Pavlova M, Strouhal M, R. C, Tesinska I, Monincova M, Prokop Z, Bartos M, Pavlik I, Rychlik I, Mobius P, Nagata

Y, Damborsky J (2005) Cloning, biochemical characterization, and distribution of mycobacterial haloalkane dehalogenases. Applied and Environmental Microbiology. 71:6736–6745

Jesenska A, Sedlacek I, Damborsky J (2000) Dehalogenation of haloalkanes by *Mycobacterium tuberculosis* H37Rv and other mycobacteria. Applied and Environmental Microbiology. 66:219–222

Kaiser JP, Feng Y, Bollag DM (1996) Microbial metabolism of pyridine, quinoline, acridine, and their derivatives under aerobic and anaerobic conditions. Microbiological Reviews. 60:483–198

Kanaly RA, Bartha R, Watanabe K, Harayama S (2000) Rapid mineralization of benzo[*a*]pyrene by a microbial consortium growing on diesel fuel. Applied and Environmental Microbiology. 66:4205–4211

Kanaly RA, Harayama S (2000) Biodegradation of high-molecular-weight polycyclic aromatic hydrocarbons by bacteria. Journal of Bacteriology. 182:2059–2067

Kazda J (2000) The ecology of mycobacteria. Kluwer Academic Publishers, Dordrecht, Boston, London, 72 pp.

Kelley I, Freeman JP, Cerniglia CE (1990) Identification of metabolites from degradation of naphthalene by a *Mycobacterium* sp. Biodegradation. 1:283–290

Kelley I, Freeman JP, Evans FE, Cerniglia CE (1993) Identification of metabolites from the degradation of fluoranthene by *Mycobacterium* sp. strain PYR-1. Applied and Environmental Microbiology. 59:800–806

Khan AA, Kim S-J, Paine DD, Cerniglia CE (2002) Classification of a polycyclic aromatic hydrocarbon-metabolizing bacterium, *Mycobacterium* sp. strain PYR-1, as *Mycobacterium vanbaalenii* sp. nov. International Journal of Systematic Bacteriology. 52:1997–2002

Khan AA, Wang R-F, Cao W-W, Doerge DR, Wennerstrom D, Cerniglia CE (2001) Molecular cloning, nucleotide sequence, and expression of genes encoding a polycyclic aromatic ring dioxygenase from *Mycobacterium* sp. strain PYR-1. Applied and Environmental Microbiology. 67:3577–3585

Kim S-J, Kweon O, Freeman JP, Jones RC, Adjei MD, Jhoo J-W, Edmondson RD, Cerniglia CE (2006) Molecular cloning and expression of genes encoding a novel dioxygenase involved in low- and high-molecular-weight polycyclic aromatic hydrocarbon degradation in *Mycobacterium vanbaalenii* PYR-1. Applied and Environmental Microbiology. 72:1045–1054

Kleespies M, Kroppenstedt RM, Rainey FA, Webb LE, Stackebrandt E (1996) *Mycobacterium hodleri* sp. nov., a new member of the fast-growing mycobacteria capable of degrading polycyclic aromatic hydrocarbons. International Journal of Systematic Bacteriology. 46:683–687

Leys NM, Ryngaert A, Bastiaens L, Wattiau P, Top EM, Verstraete W, Springael D (2005) Occurrence and community composition of fast-growing *Mycobacterium* on soils contaminated with polycyclic aromatic hydrocarbons. FEMS Microbiology Ecology. 51:375–388

Lopez Z, Vila J, Grifoll M (2005) Metabolism of fluoranthene by mycobacterial strains isolated by their ability to grow in fluoranthene or pyrene. Journal of Indian Microbiology and Biotechnology. 32:455–464

Lopez Z, Vila J, Minguillon C, Grifoll M (2006) Metabolism of fluoranthene by *Mycobacterium* sp. strain AP1. Applied Microbiology and Biotechnology. 70:747–756

Meissner PS, Falkinham JO, III (1984) Plasmid-encoded mercuric reductase in *Mycobacterium scrofulaceum*. Journal of Bacteriology. 157:669–672

Meza L, Cutright TJ, El-Zahab B, Wang P (2003) Aerobic biodegradation of trichloroethylene using a consortium of five bacterial strains. Biotechnology Letters. 25:1925–1932

Moody JD, Freeman JP, Doerge DR, Cerniglia CE (2001) Degradation of phenanthrene and anthracene by cell suspensions of *Mycobacterium* sp. strain PYR-1. Applied and Environmental Microbiology. 67:1476–1483

Moody JD, Freeman JP, Fu PP, Cerniglia CE (2004) Degradation of benzo[a]pyrene by *Mycobacterium vanbaalenii* PYR-1. Applied and Environmental Microbiology. 70:340–345

Pavlova M, Jesenska A, Klvana M, Prokop Z, Konecna H, Sato Y, Tsuda M, Nagata Y, Damborsky J (2007) The identification of catalytic pentad in the haloalkane dehalogenase DhmA from *Mycobacterium avium* N85: reaction mechanism and molecular evolution. Journal of Structural Biology. 157:384–392

Poelarends GJ, van Hylckama Vlieg JET, Marchesi JR, Freitas dos Santos LM, Janssen DB (1999) Degradation of 1,2-dibromoethane by *Mycobacterium* sp. strain GP1. Journal of Bacteriology. 181:2050–2058

Poupin P, Truffaut N, Combourieu B, Besse P, Sanceleme M, Veschambre H, Delort AM (1998) Degradation of morpholine by an environmental *Mycobacterium* strain involves a cytochrome P-450. Applied and Environmental Microbiology. 64:159–165

Rehmann K, Hertkorn N, Kettrup AA (2001) Fluoranthene metabolism in *Mycobacterium* sp. strain KR20: identity of pathway intermediates during degradation and growth. Microbiology. 147:2783–2794

Schousboe P, Rasmussen L (1994) Survival of *Tetrahymena–Thermophila* at low initial cell densities – effects of lipids and long-chain alcohols. Journal of Eukaryotic Microbiology. 41:195–199

Soeffing K (1988) The importance of mycobacteria for the nutrition of larvae of *Leucorrhinia rubicunda* (L.). Odonatologica. 17:227–233

Soeffing K, Kazda J (1993) Die Bedeutung der Mykobakterien in Torfmoosrasen bei der Entwicklung von Libellen in Moorgewässern. Telma. 23:261–269

Steinert M, Birkness K, White E, Fields B, Quinn F (1998) *Mycobacterium avium* bacilli grow saprozoically in coculture with *Acanthamoeba polyphaga* and survive within cyst walls. Applied and Environmental Microbiology. 64:2256–2261

Stingley RL, Khan AA, Cerniglia CE (2004) Molecular characterization of a phenanthrene degradation pathway in *Mycobacterium vanbaalenii* PYR-1. Biochemical and Biophysical Research Communications. 322: 133–146

Strahl ED, Gillaspy GE, Falkinham JO, III (2001) Fluorescent acid-fast microscopy for measuring phagocytosis of *Mycobacterium avium*, *Mycobacterium intracellulare*, and *Mycobacterium scrofulaceum* by *Tetrahymena pyriformis* and their intracellular growth. Applied and Environmental Microbiology. 67:4432–4439

Sutherland TD, Horne I, Harcourt RL, Russel RJ, Oakeshott JG (2002) Isolation and characterization of a *Mycobacterium* strain that metabolizes the insecticide endosulfan. Journal of Applied Microbiology. 93:380–389

Tay STL, Hemond HF, Polz MF, M. CC, Dejesus I, Krumholtz LR (1998) Two new *Mycobacterium* strains and their role in toluene degradation in a contaminated stream. Applied and Environmental Microbiology. 64:1715–1720

Thorp JH, Cothran ML (1984) Regulation of fresh-water community structure at multiple intensities of dragonfly predation. Ecology. 65:1546–1555

Trigui M, Pulvin S, Truffaut N, Thomas D, Poupin P (2004) Molecular cloning, nucleotide sequencing and expression of genes encoding a cytochrome P-450 system involved in secondary amine utilization in *Mycobacterium* sp. strain RP1.Research in Microbiology. 155:1–9

van Herwijnen R, Springael D, Slot P, Govers HAJ, Parsons JR (2003) Degradation of anthracene by *Mycobacterium* sp. strain LB501T proceeds *via* a novel pathway, through *o*-phthalic acid. Applied and Environmental Microbiology. 69:186–190

Vila J, Lopez Z, Sabate J, Minguillon C, Solanas AM, Grifoll M (2001) Identification of a novel metabolite in the degradation of pyrene by *Mycobacterium* sp. strain AP1: action of the isolate on two- and three-ring polycyclic aromatic hydrocarbons. Applied and Environmental Microbiology. 67:5497–5505

Wackett LP, Hershberger CD, Eds. (2001) Biocatalysis and biodegradation: Microbial transformation of organic compounds. ASM Press, 300 pp

Wick LY, Pasche N, Bernasconi SM, Pelz O, Harms H (2003) Characterization of multiple-substrate utilization by anthracene-degrading *Mycobacterium frederiksbergense* LB501T. Applied and Environmental Microbiology. 69:6133–6142

Chapter 9

Key Research Issues

J. Kazda, I. Pavlik, J.O. Falkinham III and K. Hruska (Eds.)

Introduction

We expect that mycobacteria will be responsible for ever greater levels of morbidity and mortality in the future because of aging human populations and increasing numbers of immune-suppressed individuals. Much of the impact will be a consequence of human activities, such as decontamination of drinking water and the accompanying selection of mycobacteria. Furthermore, because of the positive and negative impacts of mycobacteria on the environment, ecologists should consider mycobacteria in their plans and predictions. These considerations lead us to suggest the following areas for mycobacteria research.

9.1 Recruiting of Students for Training in Mycobacteriology

It is our first recommendation that increased emphasis be placed on training mycobacteriologists. Because of the difficulty in mycobacterial culture and the long incubation periods required, few graduate students are sufficiently motivated to study the mycobacteria. With the advent of genomics, mycobacteriologists are no longer constrained by these aspects. That notwithstanding, we need to recruit a new generation of students for training in mycobacteriology.

9.2 Possible Requirements for Monitoring Levels of the *Mycobacterium avium* Complex in Drinking Water

The *M. avium* complex (*MAC*) has already been included in the US Environmental Protection Agency "Candidate Contaminant List" for possible regulation in drinking water. Before a final decision can be made, it is important to provide a risk assessment of *MAC* organisms in drinking water, specifically, how many and what type of *MAC* cells are linked with disease. Furthermore, based on the present data, *MAC* members display significant differences in virulence. Unfortunately, markers of virulence for the *MAC* are unknown. It is logical to ensure that markers of virulence be identified before performing dose–response studies in suitable animal models. These markers must be amenable to widespread use in the drinking water industry, be relatively simple to identify (e.g. culture or PCR-based methods) and inexpensive.

9.3 Mechanisms of Mycobacterial Adaptation

Because of the impact of mycobacterial adaptations on their characteristics (e.g. increased antimicrobial resistance following growth in biofilms or protozoa and amoebae and increased virulence following passage in protozoa and amoebae), a systematic study of mechanisms and consequences of mycobacterial adaptation

J. Kazda (✉)
University of Kiel, Kiel, Germany
e-mail: J.Kazda@t-online.de

J. Kazda et al. (eds.), *The Ecology of Mycobacteria: Impact on Animal's and Human's Health*,
DOI 10.1007/978-1-4020-9413-2_9, © Springer Science+Business Media B.V. 2009

is needed. For example, biofilm growth leads to resistance to antibiotics and disinfectants. This means that cells spontaneously released from biofilms in drinking water systems are transiently more resistant to disinfection.

Likewise, mycobacterial cells released from catheter biofilms are more antibiotic resistant. If treatment is based on the susceptibility of suspension-grown cells, the applied dosage may be insufficient to kill the mycobacteria in either a patient or a drinking water system. Because mycobacteria, protozoa and amoebae occupy the same habitats, it is likely that mycobacteria in the environment are more virulent compared to those cells grown in laboratory media. Thus, our assessments of mycobacterial virulence that are currently based on laboratory-media-grown cells are underestimates of virulence of cells which infect humans and animals. This should be considered as part of any risk assessment.

Finally, a better understanding of the life cycle of mycobacteria is needed, from different growth stages through to dormancy. Such knowledge may contribute to our understanding of the pathogenesis and drug susceptibility of these organisms.

9.4 Sequence and Genomics of Mycobacterial Plasmids

Although the genomes of a number of mycobacterial species have been sequenced and annotated (e.g. *M. tuberculosis*, *M. bovis*, *M. a. hominissuis* and *M. a. paratuberculosis*), few mycobacterial plasmids have been sequenced. Amongst members of the *MAC* and *M. scrofulaceum* species, plasmids share sequence similarity (based on DNA:DNA hybridisation) and individual strains can harbour as much as 30% of the genomic DNA in plasmids. This means that up to 1 megabase (mB) of DNA is outside the chromosome. Clearly, this extranuclear DNA is contributing to the phenotype of the acid-fast bacilli, but to date, only a handful of genes have been identified. These include those for mercury-, cadmium- and copper resistance, a restriction endonuclease and conjugal DNA transmission.

9.5 Mycobacterial Vaccines

Because of the role of environmental mycobacteria in diseases of agriculturally important animals (e.g. fish and pigs), support should be provided for the development of effective vaccines. Clearly, such efforts should go hand-in-hand with the development of measures to prevent the contraction of disease by either ingestion (water treatment or filtration) or aerosol transmission (paraffin-coated filters bind with mycobacteria). Anti-mycobacterial vaccines for use on food animals such as pigs and fish would be expected to reduce the possibility of infection and disease transmission by a food vector. Identification of gene products conferring immunity to infection in animals may provide clues and guidance for the development of anti-tuberculosis vaccines for humans.

9.6 New Targets for Anti-Mycobacterial Therapy

Because of the impact of mycobacterial infection on human morbidity and mortality, mycobacteriologists should always be aware that the results of their investigations may help identify novel targets for drug therapy. For example, the discovery that mycobacterial dormancy can be induced by a gradual decrease in oxygen concentration led to trials demonstrating the susceptibility of such anaerobic-adapted cells to metronidazole.

9.7 Cell Envelope Modifications, Including Cell Wall-Deficient Mycobacteria

A number of investigators have observed the loss of acid-fastness of intracellular mycobacteria. This loss is also exhibited by drug-exposed cells. The loss of acid-fastness of mycobacteria in tissues and cells poses difficulties in the rapid diagnosis of infection and suggests that there are life-cycle stages that lack adequate description. Structural and physiological changes accompanying the loss of acid-fastness (perhaps stages lacking a cell wall) may lead to long-term persistence, resulting from dormancy, and prevent sterilisation. The use of PCR-based methods and *in situ* hybridisation may provide proof for the existence of non-acid-fast or cell wall-deficient mycobacterial cells. Furthermore, they may open new avenues for detecting mycobacterial persistence and identifying agents that could kill mycobacteria in such growth stages.

9.8 Mycobacterial Food-Safety Issues

Mycobacteria are mainly found in non-heat-treated (vegetables, meat, milk, etc.) and inadequately heat-treated (desiccation, cold fermentation, cold smoking, inadequate pasteurisation, etc.) foodstuffs. The eating habits of inhabitants of developed countries are changing (increasing numbers of vegetarians and vegans do not eat heat-treated food) and hygienic conditions in developing countries are often not ideal. Therefore, this research sphere requires special focus. Mycobacteria are present in many different foodstuffs and raw materials of animal and vegetable origin. Therefore, research should be focused on the use of techniques, including genomics, proteomics and lipidomics in the study of the survival of mycobacteria in foodstuffs and the importance of their virulence genes or cell components causing disease in hosts.

9.9 Possible Risk of Autoimmune or Autoinflammatory Diseases Triggered by Mycobacteria

K. Hruska

9.9.1 Autoimmunity and Bacterial Triggers

M. a. paratuberculosis (*MAC*) and other agents (*Clostridium* spp., *Campylobacter jejuni*, *Campylobacter feacalis*, *Listeria monocytogenes*, *Brucella abortus*, *Yersinia pseudotuberculosis*, *Yersinia entero-colica*, *Klebsiella* spp., *Chlamydia* spp., *Eubacterium* spp., *Peptostreptococcus* spp., *Bacteroides fragilis*, *Enterococcus feacalis* and *Escherichia coli*) are suspected to be possible triggers of Crohn's disease, a chronic autoimmune inflammatory bowel condition (Carbone et al., 2005), with similar pathological changes to paratuberculosis, an infectious disease of cattle (Chiodini, 1989). The American Academy of Microbiology Colloquium "Microbial Triggers of Chronic Human Illness" listed a number of chronic diseases for which there is suspicion of infectious aetiology. A number of chronic human illnesses are triggered by microorganisms, among them Crohn's disease, which "does not result from infection alone but from the confluence of infection and genetic susceptibility. Susceptible individuals, who carry the nucleotide-binding oligomerisation domain (NOD) or tumour necrosis factor polymorphisms, may respond to certain commensal intestinal flora, stimulating acute inflammation that leads to chronic inflammation and colitis" (Carbone et al., 2005). The Colloquium also stated that it can be extremely difficult to prove that a pathogen is the cause of chronic disorder when the onset of disease begins some time after exposure. Often, it is not practical or even possible to use Koch's postulates to prove the infectious nature of chronic illnesses. The Colloquium also recommended the consideration of other important aspects, such as if a pathogen or pathogen genes are present in diseased tissues, if the disease is multifocal, involving more than one organ, if similar diseases are known to be infectious and if animal models suggest an infectious origin (Carbone et al., 2005). These requirements are all fully met by Crohn's disease.

Peptidoglycans producing muramyl dipeptide (MDP) are the common inflammatory triggers of mycobacterial cells. Their ability to enhance the immune response to antigens has been used for decades in antibody production with Freund's adjuvant, composed of inactivated and dried mycobacteria. *MAP* cells contain peptidoglycans and heat-shock proteins, which are able to initiate the inflammatory changes in the intestine (Elzaatari et al., 1995; Chamaillard et al., 2003; Kobayashi et al., 2005; MacDonald and Monteleone, 2005; Maeda et al., 2005). The generally accepted characterisation of Crohn's disease as an autoimmune disorder does not assume that inflammation is a result of infection. Recently, it was demonstrated that the sensitisation of mice with complete Freund's adjuvant creates a condition in which dysregulation of a single cytokine leads to arthritis by triggering TNF-α-driven osteoclastogenesis (Geboes et al., 2007). The pathogenesis of some types of arthritis is similar to Crohn's disease and both disorders develop in some patients at the same time, possibly sharing some common genetic control in inflammatory pathways (Ho et al., 2005). The clinical improvements observed for Crohn's disease and rheumatoid arthritis after treatment with anti-IL-6 receptor antibody or infliximab, inhibiting the action of tumour necrosis factor-α (Keating and Perry, 2002; Nahar et al., 2003; Nishimoto, 2005), support the hypothesis of similar pathogenesis and triggers.

McGonagle and McDermott (2006) have suggested the following definition for autoinflammation: "Self-directed inflammation, whereby local factors at sites predisposed to disease lead to activation of innate immune cells, including macrophages and neutrophils, with resultant target tissue damage." The authors define Crohn's disease as a polygenic disease with a genetically defined autoinflammatory component and cite authors who have shown that the disease-associated mutation occurs in a protein involved in innate immune responses (Hugot et al., 1996, 2001; Ogura et al., 2001). The NOD proteins NOD1 and NOD2 have important roles in innate immunity as sensors of microbial components derived from bacterial peptidoglycan and contribute to the maintenance of mucosal homeostasis and the induction of mucosal inflammation (Strober et al., 2006). Muramyl dipeptide activation of NOD2 regulates innate responses to intestinal microflora by downregulating multiple TLR responses and suggests that the absence of such regulation leads to increased susceptibility to Crohn's disease (Watanabe et al., 2008).

The probability of prolonged contact with the risk molecules originating from *MAP* present in milk and beef is higher than that from other pathogens possessing the same triggers mentioned above. Moreover, *MAP* triggers can affect and prime immune systems during the sensitive period of early postnatal development of many newborns. Baby foods produced from *MAP*-contaminated milk are widely used and although mycobacteria should be inactivated, the chemical components of the cells represent a risk (Hruska et al.,

2005). Moreover, the quantity of peptidoglycans consumed by infants can be increased if water, contaminated with mycobacteria, is used. Babies consume daily doses of 400–800 ml of milk for several months. Their sensitisation by the *MAP* cell molecules is thought to act in a way similar to food allergens. Pathological changes can develop many years later. Baby foods, based on cow milk from herds suffering from paratuberculosis and reconstituted with contaminated water, may have contributedto the increase in Crohn's disease incidence reported mainly in prosperous countries. Formula feeding is employed more and more and baby foods are produced and sold the world over without any specific restrictions. A review of 17 studies concluded that breastfeeding is associated with a lower risk of Crohn's disease (Klement et al., 2004). This result should be interpreted not only as an undoubted confirmation of the benefits of breastfeeding, but also alternatively as pointing to a higher risk associated with the use of formula. A lower incidence of Crohn's disease in some ethnic groups, not associated with variation in NOD2 (Cavanaugh, 2006), could be connected to their approach to breastfeeding.

The inhalation of mycobacteria with aerosol or water droplets (Blanchard and Syzdek, 1972; Wendt et. al., 1980) represents another possible mode of sensitisation. In 1994, Gent et al. hypothesised that the incidence of Crohn's disease has increased in developed countries over the past 50 years in connection with the hot tap water and bathrooms available to future Crohn's disease patients early in life. The association of hot water supply in early childhood with Crohn's disease was also identified by Duggan et al. (1998). The possible involvement of surface water, aerosols and tap water in Crohn's disease incidence is also supported by other authors. *MAP* has been detected in the water of a river running down hill pastures grazed by livestock in which paratuberculosis is endemic. The river flows through a city in which a highly significant increase in cases of Crohn's disease was previously reported. The aerosol carrying *MAP* from the river is suspected as being a trigger for Crohn's disease (Pickup et al., 2005). From the study of another catchment the authors suggested a model of transmission of *MAP* to humans via dairy products and hypothetical routes including domestic livestock and wildlife reservoirs, agricultural practices, water from runoff, aerosols from contaminated rivers and the drinking of water from the domestic supply (Pickup et al., 2006).

The apparent lack of *MAP* infectivity in humans, although not ruled out, and the fact that the intestinal environment is not suitable for *MAP* multiplication do not exclude the development of chronic inflammation as a result of the long-term effect of bacterial triggers from dead cells. Clinical changes and severe health problems can develop after time and *MAP* need not be present in Crohn's disease intestinal lesions. A combination of the quantity of cells, duration of consumption, genetic disposition, age and general body conditions may apparently play an important role in the development of Crohn's disease. Application of a classical experimental "infection" is impossible in humans, but studies on animal and cell models should contribute to confirming the risk hypothesis. It is evident, however, that neither mycobacteria in food nor formula feeding alone is responsible for the development of Crohn's disease. Similarly, mycobacterial triggers may play a role not only in Crohn's disease but also in other autoimmune diseases. *MAP* should be considered as an intestinal allergen as well as a zoonotic pathogen. Nevertheless, as Crohn's disease is not a typical infection, direct contact with animals is not necessary for the development of autoinflammatory bowel disease.

9.9.2 Paratuberculosis

Paratuberculosis (Johne's disease) and Crohn's disease are the focus of growing interest, with the number of research projects and published results doubling between 1994 and 2003 (Hruska, 2004). Paratuberculosis is a widely distributed infectious disease of cattle and other domestic and wild ruminants caused by *MAP* (Kennedy and Benedictus, 2001). Up to 70% of dairy herds suffer from this disease in most European countries, the United States and Canada. The most important losses caused by the presence of clinically ill animals have been thoroughly described and include loss of milk production and poor body condition followed by death or culling. In contrast, losses arising from subclinical disease have not been well documented and contradictory results have been published to date (Hasonova and Pavlik, 2006). In 1998, in the United States, they were estimated as amounting to about $1.5

billion per year (Stabel, 1998). Paratuberculosis is a notifiable disease for OIE, but it is not yet classified as an emerging disease or zoonosis. Thus, fully in agreement with the present norms of food and veterinary inspections, dead mycobacteria are present in milk, cheese and beef intended for retail. An OIE Technical Disease Card on paratuberculosis is not yet available. Milk and meat from infected animals are not banned if numbers of bacteria (other than mycobacteria) and cell contamination are under the prescribed limits. Possible contamination of milk and meat by mycobacteria is not considered if animals do not exhibit pathological lesions at slaughter. Sound evidence of paratuberculosis in individual animals is rather difficult as infected cattle do not always shed *MAP* in faeces or milk. Serological methods have low sensitivity and specificity, and cultivation of the causative agent, although considered the gold standard, takes several months, with some *MAP* forms not growing *in vitro* at all (Beran et al., 2006). As a result, no restrictions have been applied to the trade of animals and the use of *MAP*-contaminated milk and beef. Nevertheless, some countries have started national programs of paratuberculosis control or certification of paratuberculosis-free herds.

If the disease is not efficiently controlled, it is guaranteed to spread *MAP* to most animals in the herd, although genetic influence on the susceptibility of cattle to paratuberculosis has been reported (Koets et al., 2000). Subsequently, as a result of different stress factors such as parturition, malnutrition, transportation, some animals suffer from the clinical form of the disease. In such cases the massive shedding of *MAP* in faeces contaminates the environment and is responsible for the transmission of the disease to other animals. Most susceptible are calves during their first weeks of life. Evidence of the pathogen can be found not only in the intestine but also in milk, lymph nodes and different parenchymatous organs, even in clinically normal animals (Pavlik et al., 2000; Ayele et al., 2004; Brady et al., 2008).

MAP is very resistant to high temperatures and chlorine treatment. The organism can remain cultivable in lake water for more than 600 days and may persist for up to 800 days (Pickup et al., 2005). *MAP* needs up to 4 months for cultivation with some forms not growing in vitro at all. However, the concentration of *MAP*, quoted in colony-forming units, does not give any reliable information on the total number of cells present, which is estimated to be in the order of 10^5/ml of milk or potable water (Lehtola et al., 2006). Molecular techniques enable more rapid and very specific detection of *MAP* and its quantification in milk, cheese and meat. It is obvious that milk and beef contaminated by *MAP* can appear in the market. Confirmed *MAP* isolates were cultured from 1.8% and *MAP* DNA was isolated from 11.8% of commercially pasteurised milk samples in the United Kingdom (Grant et al., 2002). Similar data have been reported from the United States (Ellingson et al., 2005). In Switzerland, 19.7% of bulk-tank milk samples were IS*900*PCR-positive (Stephan et al., 2002). Goat's tank-milk and ewe's tank-milk samples were also PCR-positive for the IS*900* fragment (23.0 and 23.8%, respectively), revealing the presence of *MAP* in Switzerland (Muehlherr et al., 2003). *MAP* is also able to survive the ripening of cheese (Mason et al., 1997; Donaghy et al., 2004; Ikonomopoulos et al., 2005). Nevertheless, milk products and beef are not the only possible source of consumption of mycobacterial peptidoglycans.

9.9.3 Crohn's Disease

Crohn's disease, a type of inflammatory bowel disease, was first described in 1932. The number of sufferers has increased in recent decades. Crohn's disease affects men and women equally and seems to run in some families. About 20% of people with Crohn's disease have a blood relative with some form of inflammatory bowel disease, most often a brother or sister and sometimes a parent or child. Crohn's disease can occur in people of all age groups, but it is more often diagnosed in people between the ages of 20 and 30. Although the causative agent is not known, the autoimmune character of the disease is generally accepted. The clinical signs and pathological changes in the intestine are similar to those of paratuberculosis, a chronic devastating enteritis very common in cattle herds, caused by *MAP*, with no doubt regarding its infectious origin. Mycobacteria in the environment, food and water, the chemical components of the bacterial cell walls, unchanged during boiling or cooking, the genetic predisposition of people with a NOD2 mutation and a number of other factors might play a role in this mycobacterial mystery.

The highest reported prevalence of Crohn's disease to date is in north-eastern Scotland, where almost 0.15% of the population suffer from the disease. Based on the latest epidemiological research, there are 8 00 000 people in the United States who suffer from Crohn's disease. In the United States, in 1990, Crohn's disease costs between $1.0 and $1.2 billion. Other countries with a high prevalence of Crohn's disease are Canada, Sweden, Norway, Germany, United Kingdom, Netherlands, Belgium, France, Switzerland, Austria, Spain, Portugal, Greece, Italy, Ireland, Australia, New Zealand and many countries of Eastern and Central Europe. In all these countries, bovine paratuberculosis is a common disease found in 30–70% of cattle herds. The prevalence of paratuberculosis is unknown in sheep, goats and game ruminants in most countries. Some authors have described a parallel increase in paratuberculosis and Crohn's disease prevalence and have discussed the possible links (Hermon-Taylor and Elzaatari, 2005).

The general consensus of experts participating in a recent colloquium on *MAP* is that there are reasons to suspect a role for the bacterium in Crohn's disease. Muramyl dipeptide of the *MAP* cell wall and the frequent contamination of water, beef, milk and dairy products with *MAP* have also been mentioned. Crohn's disease can result from different infectious agents and different underlying genetic or immune factors. The effect of triggers early in life and an onset of clinical disease commonly after sexual maturity are evident similarities between paratuberculosis and Crohn's disease, according to the report (Nacy and Buckley, 2008). However, an increase in Crohn's disease among juveniles (in the United States, 50% of patients are children) could support the hypothesis that *MAP* in infant formula and/or mycobacteria in tap water may play a role in autoinflammatory diseases that manifest clinical symptoms only many years later.

The situation in the Czech Republic is noteworthy in this regard. Paratuberculosis was formerly only a sporadic disease and baby food was a domestic product until the international market for animals and dairy products was opened in 1990. After the import of heifers and dairy products began, an increase in paratuberculosis was found in slaughtered cattle (Vecerek et al., 2003). The number of patients treated for CD increased more than 4-fold from 1995 to 2007 and in young people (age 0–19), 5.2-fold. The possible link between increasing CD incidence and dead

MAP cells in baby food and the consumption of milk contaminated by *MAP* has been therefore suggested (Hruska et al., 2005).

The risk of *MAP* is well documented, but the consequences are not yet fully accepted. However, the risk for consumers, although only hypothetical according to some critics, cannot be neglected. In the present international market, food safety and consumer protection is a priority. National control programmes for eradication of paratuberculosis should be initiated, although the long latent period between infection and the clinical form of the disease, as well as the absence of a reliable diagnostic method for recognition of paratuberculosis-suffering animals, brings with it many difficulties. Certification of paratuberculosis-free herds and *MAP*-free milk or milk with a low contamination of *MAP* is possible using real-time PCR and *MAP*-specific DNA sequences. Such measures should at least be applied to protect people at high risk, e.g. children and Crohn's disease sufferers and their direct relatives. Paratuberculosis is a real threat through mycobacterial trigger dissemination, and although it is an enormous economical and political problem, it must be controlled. Nevertheless, no problem disappears if the known facts are not accepted and even hypothetical risks should be treated as real ones.

Comprehensive information on Crohn's disease is available from the US National Digestive Diseases Information Clearinghouse: http://digestive.niddk.nih.gov/ddiseases/pubs/crohns/index.htm#what

Acknowledgements Section 9.9 was partially supported by the European Commission PathogenCombat FOOD-CT-2005-007081.

References

Ayele WY, Bartos M, Svastova P, Pavlik I (2004) Distribution of *Mycobacterium avium* subsp *paratuberculosis* in organs of naturally infected bull-calves and breeding bulls. Vet. Microbiol. 103:209–217

Beran V, Havelkova M, Kaustova J, Dvorska L, Pavlik I (2006) Cell wall deficient forms of mycobacteria: a review. Veterinarni Medicina 51:365–389

Blanchard DC, Syzdek LD (1972) Concentration of Bacteria in Jet Drops from Bursting Bubbles. J. Geophys. Res. 77:5087–5099

Brady C, O'Grady D, O'Meara F, Egan J, Bassett H (2008) Relationships between clinical signs, pathological changes and tissue distribution of *Mycobacterium avium* subspecies *paratuberculosis* in 21 cows from herds affected by Johne's disease. Vet. Rec. 162:147–152

Carbone KM, Luftig RB, Buckley MR (2005) Microbial triggers of chronic human illness. Am. Acad. Microbiol. Colloq. 1–14

Cavanaugh J (2006) NOD2: Ethnic and geographic differences. World J. Gastroenterol. 12:3673–3677

Chamaillard M, Girardin SE, Viala J, Philpott DJ (2003) Nods, Nalps and Naip: intracellular regulators of bacterial-induced inflammation. Cell. Microbiol. 5:581–592

Chiodini RJ (1989) Crohn's disease and the mycobacterioses – A review and comparison of two disease entities. Clin. Microbiol. Rev. 2:90–117

Donaghy JA, Totton NL, Rowe MT (2004) Persistence of *Mycobacterium paratuberculosis* during manufacture and ripening of cheddar cheese. Appl. Environ. Microbiol. 70:4899–4905

Duggan AE, Usmani I, Neal KR, Logan RFA (1998) Appendicectomy, childhood hygiene, *Helicobacter pylori* status, and risk of inflammatory bowel disease: a case control study. Gut. 43:494–498

Ellingson JLE, Anderson JL, Koziczkowski JJ, Radcliff RP, Sloan SJ, Allen SE, Sullivan NM (2005) Detection of viable *Mycobacterium avium* subsp. *paratuberculosis* in retail pasteurized whole milk by two culture methods and PCR. J. Food Prot. 68:966–972

Elzaatari FAK, Naser SA, Engstrand L, Burch PE, Hachem CY, Whipple DL, Graham DY (1995) Nucleotide-Sequence Analysis and Seroreactivities of the 65 K Heat-Shock Protein from *Mycobacterium-Paratuberculosis*. Clin. Diag. Lab. Immunol. 2:657–664

Geboes L, De KB, Van BM, Kelchtermans H, Mitera T, Boon L, De Wolf-Peeters C, Matthys P (2007) Freund's complete adjuvant induces arthritis in mice lacking a functional interferon-gamma receptor by triggering tumor necrosis factor alpha-driven osteoclastogenesis. Arthritis Rheum. 56:2595–2607

Gent AE, Hellier MD, Grace RH, Swarbrick ET, Coggon D (1994) Inflammatory Bowel-Disease and Domestic Hygiene in Infancy. Lancet. 343:766–767

Grant IR, Ball HJ, Rowe MT (2002) Incidence of *Mycobacterium paratuberculosis* in bulk raw and commercially pasteurized cows' milk from approved dairy processing establishments in the United Kingdom. Appl. Environ. Microbiol. 68:2428–2435

Hasonova L, Pavlik I (2006) Economic impact of paratuberculosis in dairy cattle herds: a review. Veterinarni Medicina 51:193–211

Hermon-Taylor J, Elzaatari FAK (2005) The *Mycobacterium avium* subspecies *paratuberculosis* problem and its relation to the causation of Crohn disease. In: S. Pedley et al., (Eds.), Pathogenic mycobacteria in water: A guide to public health consequences, monitoring and management. WHO 2004. 74–94

Ho P, Bruce IN, Silman A, Symmons D, Newman B, Young H, Griffiths CEM, John S, Worthington J, Barton A (2005) Evidence for common genetic control in pathways of inflammation for Crohn's disease and psoriatic arthritis. Arthritis Rheum. 52:3596–3602

Hruska K (2004) Research on paratuberculosis: Analysis of publications 1994–2004. Veterinarni Medicina 49:271–282

Hruska K, Kralik P, Bartos M, Pavlik I (2005) *Mycobacterium avium* subsp. *paratubeculosis* in powdered infant milk: paratuberculosis in cattle – the public health problem to be solved. Veterinarni Medicina 50:327–335

Hugot JP, Chamaillard M, Zouali H, Lesage S, Cezard JP, Belaiche J, Almer S, Tysk C, O'Morain CA, Gassull M, Binder V, Finkel Y, Cortot A, Modigliani R, Laurent-Puig P, Gower-Rousseau C, Macry J, Colombel JF, Sahbatou M, Thomas G (31-5-2001) Association of NOD2 leucine-rich repeat variants with susceptibility to Crohn's disease. Nature. 411:599–603

Hugot JP, Laurent-Puig P, Gower-Rousseau C, Olson JM, Lee JC, Beaugerie L, Naom I, Dupas JL, Van GA, Orholm M, Bonaiti-Pellie C, Weissenbach J, Mathew CG, Lennard-Jones JE, Cortot A, Colombel JF, Thomas G (29–2–1996) Mapping of a susceptibility locus for Crohn's disease on chromosome 16. Nature. 379:821–823

Ikonomopoulos J, Pavlik I, Bartos M, Svastova P, Ayele WY, Roubal P, Lukas J, Cook N, Gazouli M (2005) Detection of *Mycobacterium avium* subsp. *paratuberculosis* in retail cheeses from Greece and the Czech Republic. Appl. Environ. Microbiol. 71:8934–8936:

Keating GM, Perry CM (2002) Infliximab – An updated review of its use in Crohn's disease and rheumatoid arthritis. Biodrugs. 16:111–148

Kennedy DJ, Benedictus G (2001) Control of *Mycobacterium avium* subsp. *paratuberculosis* infection in agricultural species. Revue Scientifique et Technique de l Office International des Epizooties. 20:151–179

Klement E, Cohen RV, Boxman J, Joseph A, Reif S (2004) Breastfeeding and risk of inflammatory bowel disease: a systematic review with meta-analysis(1–3). Am. J. Clin. Nutr. 80:1342–1352

Kobayashi KS, Chamaillard M, Ogura Y, Henegariu O, Inohara N, Nunez G, Flavell RA (2005) Nod2-dependent regulation of innate and adaptive immunity in the intestinal tract. Science. 307:731–734

Koets AP, Adugna G, Janss LLG, van Weering HJ, Kalis CHJ, Wentink GH, Rutten VPMG, Schukken YH (2000) Genetic variation of susceptibility to *Mycobacterium avium* subsp. *paratuberculosis* infection in dairy cattle. J. Dairy Sci. 83:2702–2708

Lehtola MJ, Torvinen E, Miettinen IT, Keevil CW (2006) Fluorescence in situ hybridization using peptide nucleic acid probes for rapid detection of *Mycobacterium avium* subsp. *avium* and *Mycobacterium avium* subsp. *paratuberculosis* in potable-water biofilms. Appl. Environ. Microbiol. 72:848–853

MacDonald TT, Monteleone G (2005) Immunity, inflammation, and allergy in the gut. Science. 307:1920–1925

Maeda S, Hsu LC, Liu HJ, Bankston LA, Iimura M, Kagnoff MF, Eckmann L, Karin M (2005) Nod2 mutation in Crohn's disease potentiates NF-kappa B activity and IL-10 processing. Science. 307:734–738

Mason O, Rowe MT, Ball HJ (1997) Is *Mycobacterium paratuberculosis* a possible agent in Crohn's disease? Implications for the dairy industry. Milchwissenschaft-Milk Science International. 52:311–316

McGonagle D, McDermott MF (2006) A proposed classification of the immunological diseases. PLoS Med. 3:1242–1248

Muehlherr JE, Zweifel C, Corti S, Blanco JE, Stephan R (2003) Microbiological quality of raw goat's and ewe's bulk-tank milk in Switzerland. J. Dairy Sci. 86:3849–3856

Nacy C, Buckley M (2008) *Mycobacterium avium paratuberculosis*: Infrequent human pathogen or public health thread? Report from the American Academy of Microbiology, 37 pp, http://www.asm.org/ASM/files/ccLibraryFiles/Filename/000 000004169/MAP.pdf

Nahar IK, Shojania K, Marra CA, Alamgir AH, Anis AH (2003) Infliximab treatment of rheumatoid arthritis and Crohn's disease. Ann. Pharmacother. 37:1256–1265

Nishimoto N (2005) Clinical studies in patients with Castleman's disease, Crohn's disease, and rheumatoid arthritis in Japan. Clin. Rev. Allerg. Immunol. 28:221–229

Ogura Y, Bonen DK, Inohara N, Nicolae DL, Chen FF, Ramos R, Britton H, Moran T, Karaliuskas R, Duerr RH, Achkar JP, Brant SR, Bayless TM, Kirschner BS, Hanauer SB, Nunez G, Cho JH (31–5–2001) A frameshift mutation in NOD2 associated with susceptibility to Crohn's disease. Nature. 411:603–606

Pavlik I, Matlova L, Bartl J, Svastova P, Dvorska L, Whitlock R (2000) Parallel faecal and organ *Mycobacterium avium* subsp. *paratuberculosis* culture of different productivity types of cattle. Vet. Microbiol. 77:309–324

Pickup RW, Rhodes G, Arnott S, Sidi-Boumedine K, Bull TJ, Weightman A, Hurley M, Hermon-Taylor J (2005) *Mycobacterium avium* subsp. *paratuberculosis* in the catchment area and water of the River Taff in South Wales, United Kingdom, and its potential relationship to clustering of Crohn's disease cases in the city of Cardiff. Appl. Environ. Microbiol. 71:2130–2139

Pickup RW, Rhodes G, Bull TJ, Arnott S, Sidi-Boumedine K, Hurley M, Hermon-Taylor J (2006) *Mycobacterium avium* subsp. *paratuberculosis* in lake catchments, in river water abstracted for domestic use, and in effluent from domestic sewage treatment works: Diverse opportunities for environmental cycling and human exposure. Appl. Environ. Microbiol. 72:4067–4077

Stabel JR (1998) Johne's disease: a hidden threat. J. Dairy Sci. 81:283–288

Stephan R, Buhler K, Corti S (2002) Incidence of *Mycobacterium avium* subspecies *paratuberculosis* in bulk-tank milk samples from different regions in Switzerland. Vet. Rec. 150:214–215

Strober W, Murray PJ, Kitani A, Watanabe T (2006) Signalling pathways and molecular interactions of NOD1 and NOD2. Nat. Rev. Immunol. 6:9–20

Vecerek V, Kozak A, Malena M, Tremlova B, Chloupek P (2003) Veterinary meat inspection of bovine carcasses in the Czech Republic during the period of 1995–2002. Veterinarni Medicina 48:183–189

Watanabe T, Asano N, Murray PJ, Ozato K, Tailor P, Fuss IJ, Kitani A, Strober W (2008) Muramyl dipeptide activation of nucleotide-binding oligomerization domain 2 protects mice from experimental colitis. J. Clin. Invest. 118:545–559

Wendt SL, George KL, Parker BC, Gruft H, Falkinham JO (1980) Epidemiology of Infection by Nontuberculosis Mycobacteria. 3. Isolation of Potentially Pathogenic Mycobacteria from Aerosols. Am. Rev. Respir. Dis. 122:259–26

Chapter 10

Photographs

I. Pavlik and K. Hruska (Eds.)

The fascinating story of the ubiquitous mycobacteria and their never-ending threat to human and animal health comes to a close with the following set of photographs. Both the breadth and depth of material on mycobacteria are illustrated with the short captions and links to relevant chapters of this book are provided. Readers should be aware that all photos were taken as working documents for laboratory and travel reports over the last three decades. This chapter therefore does not have the aim of representing issues relating to mycobacteria as a set of "artistic photos." Rather, the intention of the authors was to summarize the story by showing at least some of environments where mycobacteria survive or thrive, often posing risks to other inhabitants of the habitat, including invertebrates, amphibians, fish or mammals, and even humans. The history of mycobacteria begins thousands of years ago, long before some of the species were recognized as pathogens and from ancient times they represent a permanent health risk to animals and humans. The scientific description of mycobacteria has a much shorter history. Two papers cited in this book appeared in 1875 and since that time thousands more have been published. The methods used in mycobacterial research have developed from simple observation, microscopy, culture, experiments on animals and case studies to the use of sophisticated instruments, applied in electron and confocal microscopy, liquid culture with automatic mycobacterial growth detection, radioisotope techniques, genetics, immunology, molecular biology, genomics, proteomics, etc.

I. Pavlik (✉)
Head of OIE Reference Laboratories for Paratuberculosis and Avian Tuberculosis, Department of Food and Feed Safety, Veterinary Research Institute, Brno, Czech Republic
e-mail: pavlik@vri.cz

J. Kazda et al. (eds.), *The Ecology of Mycobacteria: Impact on Animal's and Human's Health*,
DOI 10.1007/978-1-4020-9413-2_10, © Springer Science+Business Media B.V. 2009

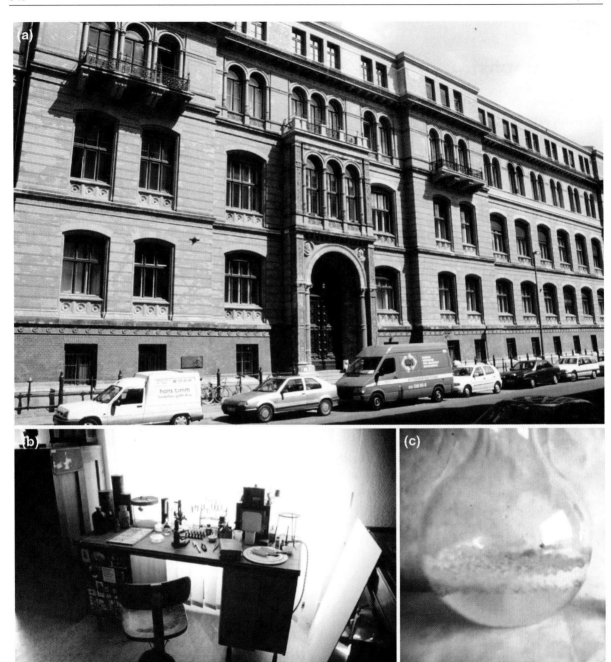

Photo 1.1 The epoch of research on mycobacteria began after the discovery (and successful *in vitro* isolation) of the causative agent of human tuberculosis *Mycobacterium tuberculosis*. It happened in 1882, in Berlin, in the institution shown in the photographs (**a, b**). Mycobacteria, which resembled moulds ("mycos" in Greek), were observed to grow in liquid medium covered with a film. However, the discoverers were surprised because microscopic examination revealed bacteria. Accordingly, the genus of these bacteria was designated by a name which incorporated both fungi and bacteria: *Myco + bacterium* (**c**). Photo I. Pavlik

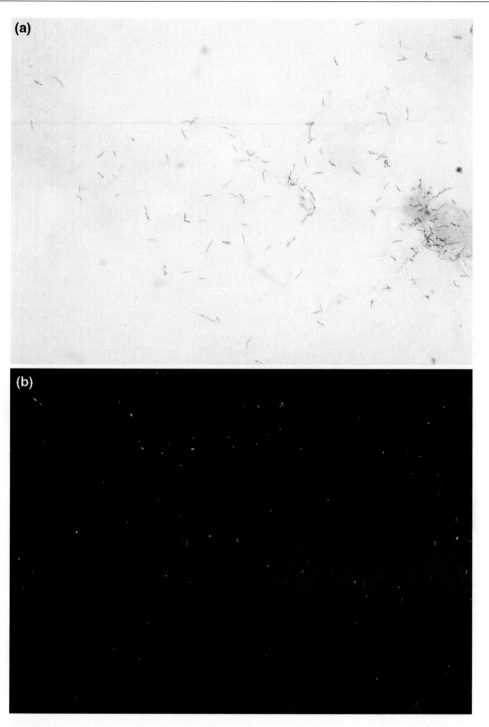

Photo 2.1   People with open pulmonary tuberculosis can shed large amounts of *M. tuberculosis* into the environment through sputum and also through faeces after swallowing sputum. (**a**) *M. tuberculosis* can be detected in sputum by microscopy; staining according to Ziehl–Neelsen, magnification 1000× (Photo J. Svobodova and V. Mrlik). (**b**) *M. tuberculosis* can be also detected in sputum by fluorescent staining at lower magnification (200×). Due to the fact that fluorescent staining is more sensitive, it allows the examination of more fields of view than staining according to Ziehl–Neelsen at magnification 1000×; its reliability is comparable to that of Ziehl–Neelsen. However, fluorescent staining cannot be used for the examination of samples from the environment (e.g. soil, dust, animal faeces), in which fluorescent bacteria and microparticles other than the causative agent of human tuberculosis are often present (Photo L. Mezensky and J. Svobodova)

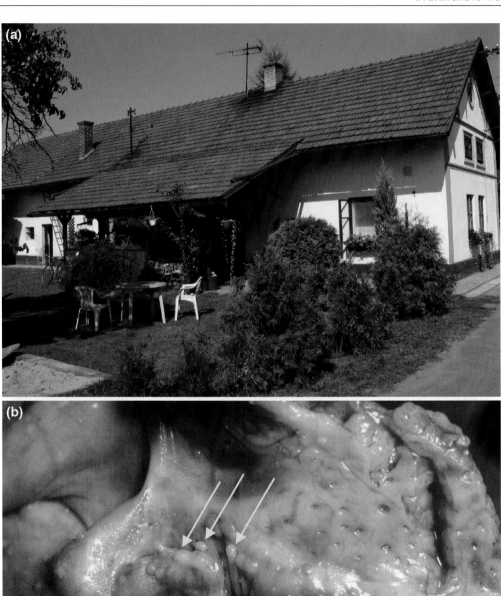

Photo 2.2 A pig keeper working on a family farm (**a**) (the left-most door is the stable entrance) shed *M. tuberculosis* through his sputum over a long period of time. This was detected by culture examination of his sputum. Tuberculous lesions were also detected in the submandibular lymph nodes of two finished pigs from this farm. By subsequent examination of other pigs from the herd, *M. tuberculosis* was also detected by the PCR method in the tonsils of a sow; numerous small lentil-like nodules shown by the arrows are evident on their surface (**b**) (I. Pavlik, unpublished data; Photo I. Pavlik)

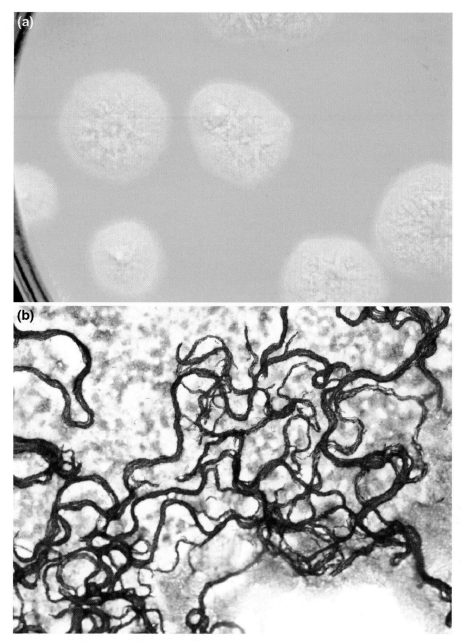

Photo 2.3 *M. tuberculosis* grown on solid media typically forms colonies with an elevated profile (**a**); however, these are also formed by other mycobacterial species found in the environment. Accordingly, it is always necessary to identify the species of mycobacterial isolate. Isolates of *M. tuberculosis* from humans and animals growing in a liquid medium adopt typical cord-like shapes, designated as "cords" (**b**). They are also apparent at small magnifications, e.g. 100×, after Ziehl–Neelsen staining; isolates growing in this form are designated as "cord factor" positive (Photo L. Mezensky and J. Svobodova). However, these cords can, under certain conditions, also be formed by other mycobacterial species (e.g. *M. kansasii* or *M. marinum*) that are causative agents of human diseases and are present in the environment. Nevertheless, these mycobacterial species are usually sensitive to different antituberculotic and antibiotic drugs than *M. tuberculosis*; due to this fact, an exact identification of the causative agent is necessary for the successful treatment of a mycobacterial disease (Sections 3.1.1 and 3.1.5). *M. tuberculosis* colonies of the same morphology and positive for the "cord factor" were isolated from the lungs and mesenteric lymph nodes of a dog (5.5 years old Doberman), who lived in contact with its owner who had open tuberculosis and also from the sputum of the owner [1]

Source: [1] Pavlik I, Trcka I, Parmova I, Svobodova J, Melicharek I, Nagy G, Cvetnic Z, Ocepek M, Pate M, Lipiec M (2005) Veterinarni Medicina 50:291–299.

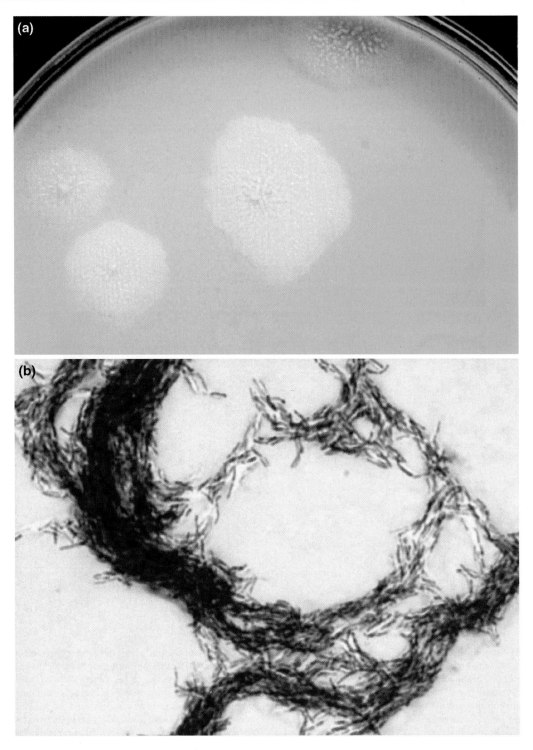

Photo 2.4   In contrast to *M. tuberculosis, M. bovis* BCG form typically flatter and less wrinkled colonies when grown on a solid medium (**a**). Even though no information regarding the detection of this member of the *M. tuberculosis* complex is available in the literature, the detection of *M. bovis* BCG in the environment may be expected primarily in areas with experimental vaccination of animals, particularly cattle and/or badgers. *M. bovis* BCG also forms cords when growing in liquid media (**b**); however, these are usually less intensively expressed than *M. tuberculosis* cords (Photo 2.3) and individual mycobacteria retain less carbolfuchsin after treatment with the Ziehl–Neelsen method; magnification 200× (Photo L. Mezensky and J. Svobodova)

Photo 2.5 *M. caprae* was detected outside of a host organism in environmental samples from an infected family farm: scrapings from the wall behind the first infected cow, standing leftmost at the back, scrapings from a manger for calves positioned opposite on the right-hand side and scrapings from the floor near a pig pen situated in the room behind the stable. The source of infection for all of the 29 heads of cattle (through direct contact in the stable) and five pigs (they were fed pooled raw milk from the cows) was a 14-year-old cow (anergent to bovine tuberculin) diagnosed with tuberculosis of lungs and pulmonary lymph nodes [1]. These results revealed the occurrence of *M. caprae* in an environment comparable with the environment where *M. bovis* is found, i.e. in stables for cattle infected with bovine tuberculosis (Photo I. Pavlik)

Source: [1] Pavlik I, Bures F, Janovsky P, Pecinka P, Bartos A, Dvorska L, Matlova L, Kremer K, van Soolingen D (2002) Veterinarni Medicina 47:251–263.

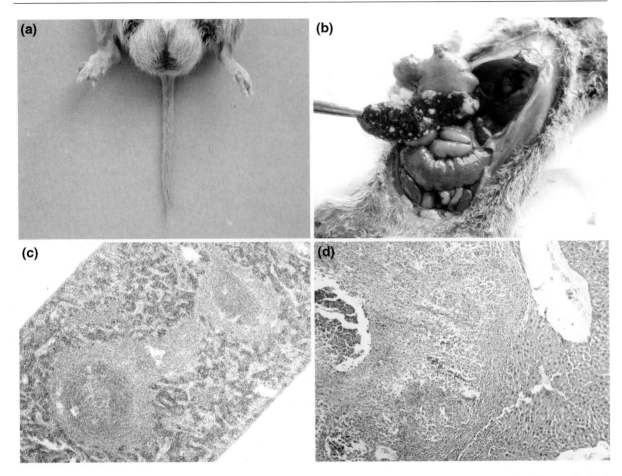

Photo 2.6 While investigating the occurrence of *M. microti* in small terrestrial mammals in the Czech Republic, in one location we encountered a syndrome of the thickening of legs, testicle swelling (**a**), and the occurrence of tuberculoid lesions in the parenchymatous organs (liver) (**b**) of common voles (*Microtus arvalis*) (Photo I. Pavlik). Histopathological examination of the spleen (**c**) and liver (**d**) (haematoxylin–eosin staining, magnification 100×, Photo M. Skoric) revealed pyogranulomas characterized by central suppuration, the presence of cellular detritus, degenerated and intact neutrophilic granulocytes, macrophages and lymphocytes. Larger, often, confluent nodules showed signs of caseation in their centres and partial mineralization, and were surrounded by epithelioid macrophages, sporadic giant polycaryotic cells and lymphocytes with neutrophilic granulocytes. On the periphery it was possible to observe the proliferation of fibroblasts. Later, it was found that brucellosis, subsequently described as a new species of *Brucella microti* [1, 2] was the cause of a relatively common disease in the voles. The following case should be regarded as serious: on 31 August 2000, a female worker was bitten on the finger through thick leather gloves by a trapped male vole (with a relatively large body weight of 50 g)

that had subcutaneous abscesses below the skin on the back abdomen and paws. This male originated from a locality monitored over a long period of time, where up to 30% of small terrestrial mammals had the above-mentioned symptoms and where *B. microti* was subsequently detected. On 22 September 2000, a fever developed in that worker whose finger was bitten (from 39.5 to 40.0 °C for several days) with concurrent lower limb swelling (ankles and feet), symptoms of joint pain, headache and general weakness. The infection was cured by the oral administration of high doses of antibiotics and thus far no complications have been observed (J. Nesvadbova, I. Pavlik, M. Skoric, unpublished data)

Source: [1] Hubalek Z, Scholz HC, Sedlacek I, Melzer F, Sanogo YO, Nesvadbova J (2007) Vector.Borne.Zoonotic.Dis. 7:679–687. [2] Scholz HC, Hubalek Z, Sedlacek I, Vergnaud G, Tomaso H, Al Dahouk S, Melzer F, Kampfer P, Neubauer H, Cloeckaert A, Maquart M, Zygmunt MS, Whatmore AM, Falsen E, Bahn P, Gollner C, Pfeffer M, Huber B, Busse HJ, Nockler K (2008) International Journal of Systematic and Evolutionary Microbiology 58:375–382.

Photo 2.7 The contact of some sea mammals when ashore is very close, e.g. in sea lions (Mar del Plata, Argentina). In the event of the occurrence of infection caused by *M. pinnipedii*, it is most likely transmitted by direct contact. However, there is a lack of information regarding the occurrence of *M. pinnipedii* in the resting places of these mammals and on its ecology in general at present (Photo I. Pavlik)

Photo 2.8   Deserted old farmhouse in detail (**a**). Old reservoir for surface water running down a slope surrounded by sphagnum (**b**). Southern slopes in coastal Norway with an old deserted farmhouse formerly with high rates of leprosy incidence; the water supply was transported by wooden pipes from a water reservoir (**c**) (Photo J. Kazda)

Photo 2.9 Sphagnum bog in the Naustdal region (Norway) (**a**), the origin of the water supply of the farm, which previously had a high prevalence of leprosy. Non-cultivable acid-fast bacilli found in a smear from the homogenised footpad of a white mouse inoc-ulated with sphagnum, collected in a region of Norway where leprosy formerly prevailed (**b**) (Ziehl–Neelsen staining, magnification 1000×, Photo J. Kazda)

Photo 2.10   A typical washing area in the tropics. As it is not possible to fully dry underwear and linen in the humid climate, *M. leprae* can access humans and their dwellings (Photo J. Kazda)

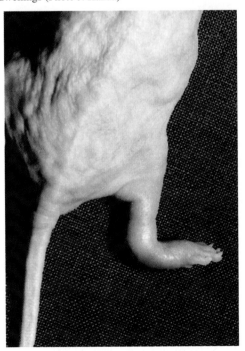

Photo 2.11   (continued)

Photo 2.12 A region of leprosy epidemicity in Bombay where environmentally derived *M. leprae* was detected. Besides the obvious barefoot factor, additional contamination can take place as a result of the unhygienic washing of dishes (Photo J. Kazda)

Photo 2.11 Swelling of the footpad and cutaneous leproma on the lateral body sites in a nude mouse, inoculated 6 months previously with both *M. leprae* and *M. intracellulare* serotype 19. The non-pathogenic serotype 19 of *M. intracellulare* considerably enhanced the pathogenicity of *M. leprae* in the nude mouse (Photo J. Kazda)

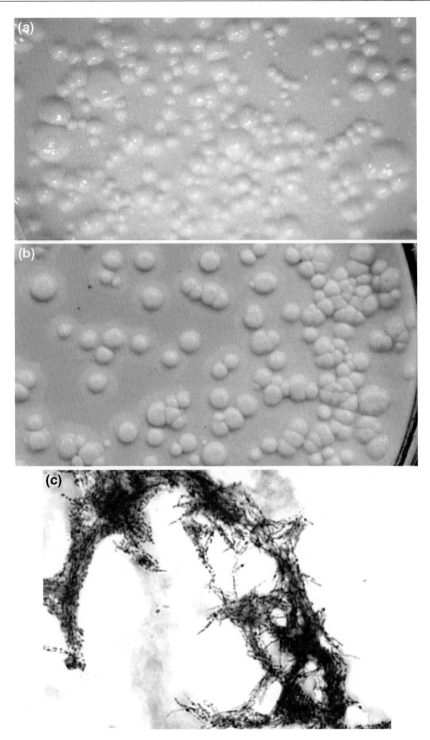

Photo 3.1 *M. kansasii* grows on solid media in the form of prominent colonies. It is classified as a photochromogenic mycobacterial species, occurring above all in the environment and producing pigment upon exposure to light: before exposure to light (**a**) and after (**b**). *M. kansasii* grows *in vitro* in the form of long, slim, highly granulated rods, which are usually arranged in loose filaceous structures (**c**) (Ziehl–Neelsen staining; magnification 1000×, Photo L. Mezensky and J. Svobodova). These apparent cords can result in a mistaken diagnosis of infection with *M. tuberculosis*. Accordingly, it is necessary to precisely identify these mycobacterial isolates, especially in samples from the external environment

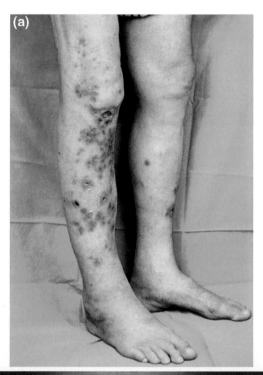

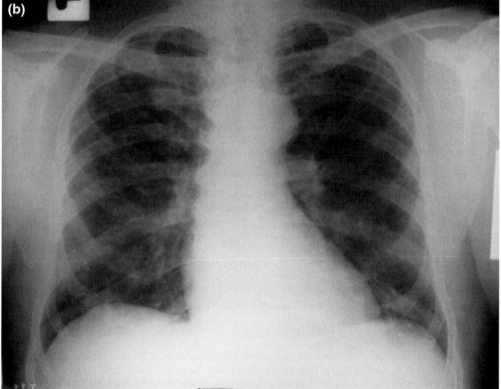

Photo 3.2 *M. kansasii* causes pulmonary and skin infections in predisposed people. In this patient skin efflorescence with resulting necrotizing vasculitis is apparent (**a**) and a diffuse micronodular pattern in pulmonary tissue was detected by RTG of the lungs (**b**) (Photo R. Tyl)

Photo 3.3 Biofilms can increase rapidly in number in pipelines through which highly organically polluted water flows (tanks for fresh water fish aquaculture are supplied with surface water) (**a**). In biofilms, both *M. kansasii* and other mycobacterial species with consequences for health have been detected; these can multiply in this environment if the temperature is convenient (Chapter 4). Water contaminated with released biofilms or freely floating mycobacteria can pose a risk to host organisms and feeds, foods, the aquarium environment, fish present in the aquarium and other animals, etc. It was discovered on a pig farm that plastic water piping was a source of ESM and PPM (**b**). In the summer months, water became warm there relatively quickly due to its better heat conductivity in comparison with metal piping. The surface of the plastic material evidently facilitated the production and subsequent persistence of biofilms on the piping walls. After the occurrence of tuberculous lesions caused by various species, especially PPM including *M. kansasii*, in the head and mesenteric lymph nodes of pigs, the breeder was forced to exchange the plastic for metal piping (Photo I. Pavlik)

Photo 3.4   A professional aquarist stored rain water in these blue barrels (**a**) and used that water, either directly or after dilution with tap water, in tanks for the spawning of some fish species. On the back wall of the room, we can see the water distribution system using plastic piping abundant in biofilms. A temperature of above 25 °C was maintained throughout the year in the rooms with tanks used for breeding (**b**). During warm and sunny weather (i.e. in the summer months in the northern hemisphere) the temperature in this room reached up to between 28 and 35 °C. During the night, when tap water was not in use, water temperature could increase relatively quickly in the piping. A higher temperature of water and in the walls of piping where biofilm is formed supports its growth and results in an increased multiplication of both *M. kansasii*, mainly found in endemic regions, and other PPM species. *M. kansasii* poses a real risk, because it was also detected in aquarium fish. (**c**) In aquaria, biofilms are formed on the walls (left) and on different decorative elements, e.g. the tree roots (in the middle). In particular in aquaria with tropical fish, biofilms become a suitable environment for the survival and multiplication of different PPM species, including *M. kansasii*. Due to the fact that biofilms often constitute food for different animal species (in this case for a catfish of *Baryancistrus* sp. L018), both the fish in the aquarium and the aquarist or people handling the contaminated water or other parts of aquarium, such as sand, plants, fish, filters, etc. can become infected with PPM (Section 6.5.1; [1]; Photo I. Pavlik) Source: [1] Rehulka J, Kaustova J, Rehulkova E (2006) Acta Veterinaria Brno 75:251–258.

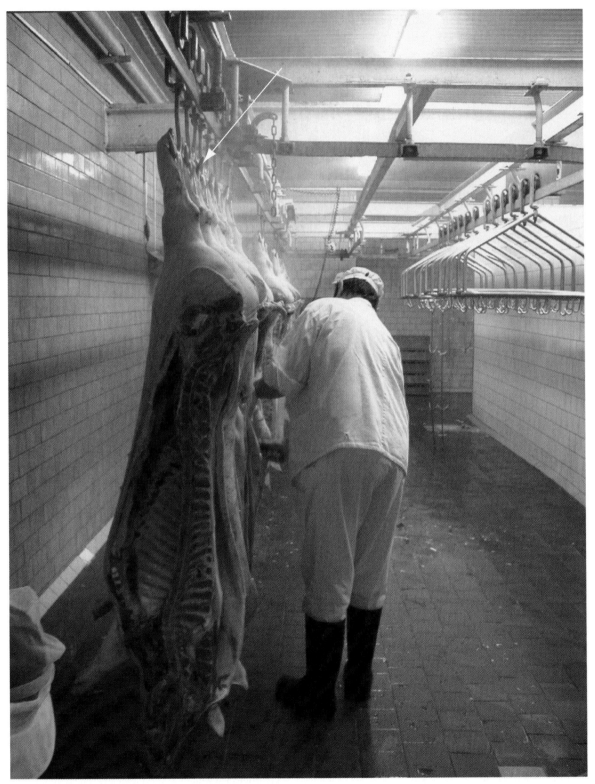

Photo 3.5   The picture shows a slaughterhouse situated in an enzootic region of *M. kansasii* in the Czech Republic; this mycobacterial species was detected in condensed water near the traverse rails (top left indicated by the arrow). The surface of pig carcasses could thus become contaminated with the condensed water that was dripping down (J. E. Shitaye, A. Horvathova, M. Moravkova, I. Pavlik, unpublished data; Photo I. Pavlik)

Photo 3.6   Moss vegetation (Bryidae), as found in the Ivory Coast, is very abundant at the edges of small streams (Photo J. Kazda)

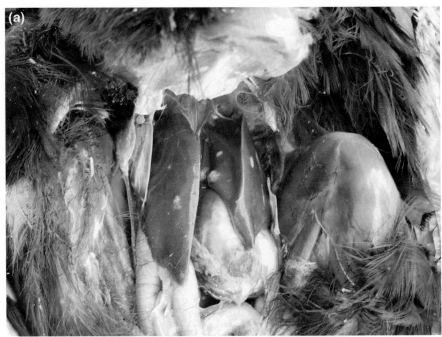

Photo 3.7 Avian tuberculosis particularly affects gallinaceous domestic poultry, in which most cases of the disease have been diagnosed. The sources of infection often remain undetected. However, in many cases, free living birds are considered to be responsible. They are also most likely the sources of *M. a. avium* for common pheasants (*Phasianus colchicus*) kept in aviaries (**a**) (Photo J. Lamka; [2]) and little egrets (*Egretta garzetta*) kept in zoological gardens (**b**) (Photo I. Pavlik; [1])

Source: [1] Dvorska L, Matlova L, Ayele WY, Fischer OA, Amemori T, Weston RT, Alvarez J, Beran V, Moravkova M, Pavlik I (2007) Vet.Microbiol. 119:366–374. [2] Pavlik I, Matlova L, Gilar M, Bartl J, Parmova I, Lysak F, Alexa M, Dvorska-Bartosova L, Svec V, Vrbas V, Horvathova A (2007) Veterinarni Medicina 52:392–404.

Photo 3.8 In captivity, pheasants (*Phasianus colchicus*) are usually kept at high densities in aviaries; this often leads to cannibalism (**a**). Even though the earthy soil becomes muddy after the rain, *M. a. avium* has rarely been detected in the soil, mud or invertebrates (J. Lamka, I. Pavlik, I. Trcka, unpublished data). Although pheasants in captivity are usually kept in aviaries protected by wire mesh, e.g. sparrows (*Passer* sp.) can penetrate the wire (**b**). They enter both the watering places and feeding places of the pheasants. In some breeds, house sparrows (*Passer domesticus*) nest in the area of the pheasantry, this was also the case in this breed, where a nest was found at top left between the tin roof and wire mesh (yellow arrow). The several-day-old young of common pheasants (*Phasianus colchicus*) and the environment of the aviary can thus be exposed to the source of *M. a. avium* from sparrows, a risk which is also real for pig herds ([1]; Photo I. Pavlik)

Source: [1] Matlova L, Dvorska L, Bartl J, Bartos M, Ayele WY, Alexa M, Pavlik I (2003) Veterinarni Medicina 48:343–357

Photo 3.9 Gallinaceous poultry (*Gallus domesticus*) kept extensively in small flocks are considered as the main reservoir of *M. a. avium* for other animals (**a**). Various domestic and wild animals come into contact with them in different situations. While searching for food, poultry often wander into the sur-roundings of farms, where they come into contact with various animals (**b**). In this game preserve, they came into close proximity with a European roe deer (*Capreolus capreolus*) visible in the top right-hand corner of the picture (**c**) (Photo I. Pavlik)

Photo 3.10 (continued)

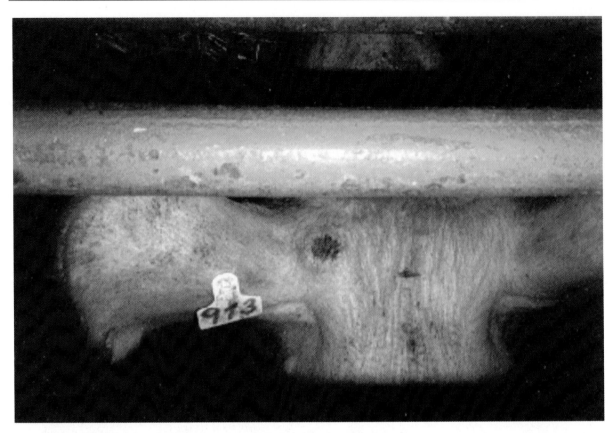

Photo 3.11 When attempting to diagnose avian tuberculosis in domestic pigs, reactions to avian tuberculin are often detected (necrosis is evident on the base of the right ear of the pig within 48 h of testing with avian tuberculin). After slaughter, no tuberculous lesions either in the parenchymatous organs or in the intestinal lymph nodes were found. *M. a. hominissuis* was only isolated from intestinal lymph nodes and from the environment where, moreover, a variety of PPM were detected ([1]; Photo R. Axman)
Source: [1] Pavlik I, Matlova L, Gilar M, Bartl J, Parmova I, Lysak F, Alexa M, Dvorska-Bartosova L, Svec V, Vrbas V, Horvathova A (2007) Veterinarni Medicina 52:392–404.

Photo 3.10 Dead birds become part of different components of the environment. The photo shows the dead body of a barn swallow (*Hirundo rustica*) in a reservoir for liquid manure in a pig herd (**a**). It is likely that the bodies of larvae and puparia (i.e. two brown puparia on the top and on the down left site) of syrphid flies (*Eristalis tenax*) became infected with *M. a. avium* (serotype 2) in a reservoir for liquid manure in a pig herd [1]. Avian tuberculosis was detected in the domestic pigs of one herd. During investigation of sources of infection, *M. a. avium* was identified in 5 out of 17 house sparrows (*P. domesticus*) that were shot to death; these were, therefore, the likely sources of infection. Dead (yellow arrow shows dead sparrow), weakened or non-alert birds infected *ante finam* with *M. a. avium* can transmit the infection directly to domestic pigs, which, in turn are good hunters of, e.g. young birds who are not yet proficient fliers or small terrestrial mammals (**b**) (I. Pavlik and J. Bartl, unpublished data; Photo I. Pavlik)
Source: [1] Fischer OA, Matlova L, Dvorska L, Svastova P, Bartos M, Weston RT, Pavlik I (2006) Folia Microbiol. (Praha) 51:147–153.

Photo 3.12 (continued)

Photo 3.13 Among water birds, in particular the species of ibises and spoonbills belonging to the family Threskiornithidae are resistant to *M. a. avium* [1]. These birds live on fish and a variety of other aquatic animals; occasionally, they also prey on small terrestrial animals. With respect to their potential infection with the causative agent of avian tuberculosis, they have most likely developed natural resistance against this infection. This photo shows the sacred ibis (*Threskiornis aethiopicus*) and the African Spoonbill (*Platalea alba*) near Cape Town, Republic of South Africa (Photo I. Pavlik)
Source: [1] Dvorska L, Matlova L, Ayele WY, Fischer OA, Amemori T, Weston RT, Alvarez J, Beran V, Moravkova M, Pavlik I (2007) Vet.Microbiol. 119:366–374.

Photo 3.12 The intramuscular infection of pullets is often used to investigate virulence for birds. A fully virulent strain of *M. a. avium* causes miliary tuberculosis of the parenchymatous organs, especially in the liver (arrow shows liver with miliary tuberculosis) (**a**). However, some strains lose their virulence; such cases have been described for collection strains and isolates from the environment [1]. After intramuscular infection, gross lesions are not usually detected (pullet on the right). Nevertheless, caseous matter is often observed at the site of inoculation in pectoral muscle as shown by the arrow (**b**) (Photo I. Pavlik)
Source: [1] Pavlik I, Svastova P, Bartl J, Dvorska L, Rychlik I (2000) Clin.Diagn.Lab Immunol. 7:212–217.

Photo 3.14 To prevent the spread of OPM and PPM including *M. a. avium,* it is necessary to adopt the measure of washing footwear which may be contaminated with faeces or soil; these can subsequently become sources of infection for other animals. Breeders use various simple, but effective equipments for the cleaning of and the washing of footwear (in Slovakia, Japan and the Czech Republic (**a–c**), respectively (Photo I. Pavlik)

(a) **(b)**

(c)

Photo 3.15 Peat bogs number among the significant sources of *M. a. hominissuis* in the environment (Section 5.1.1). The almost completely mineralized black peat contains low amounts of mycobacteria. However, after being mined from deposits, peat is exposed to rain, dust, etc. (**a**), which are besides the faeces of birds and other animals, the main sources of *M. a. hominissuis* and to a lesser extent *M. a. avium* [1, 2]. The storage (packing) of peat immediately after extraction failed to prevent contamina-

tion with *M. a. hominissuis* (**b**). Mycobacteria penetrated either through tears in the packaging or through rain and dust during the mining of peat in surface mines (**c**) (I. Pavlik, L. Matlova, P. Bejcek, unpublished data; Photo I. Pavlik)
Source: [1] Matlova L, Dvorska L, Ayele WY, Bartos M, Amemori T, Pavlik I (2005) J. Clin. Microbiol. 43:1261–1268. [2] Matlova L, Dvorska L, Bartl J, Bartos M, Ayele WY, Alexa M, Pavlik I (2003) Veterinarni Medicina 48:343–357.

(a)

(b)

(c)

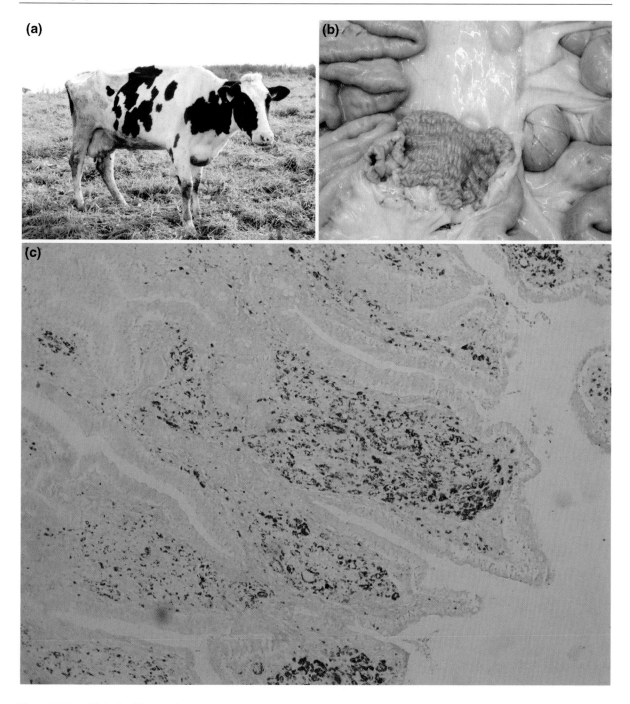

Photo 3.16 Clinically ill cows infected with paratuberculo-sis can shed watery faeces heavily contaminated with *M. a. paratuberculosis* for weeks or months (**a**). The causative agent of paratuberculosis is present here in the intestinal mucosa which has increased several folds in height resembling the gyres of the brain (especially in small intestine: in the end of jejunum and in the whole ileum (**b**); [1]). Therefore, this gross lesion is desig-nated as gyrification. In the upper part of the mesentery, one evi-dently enlarged jejunal lymph node can be seen (Photo I. Pavlik).

Histology (magnification 200×; Ziehl–Neelsen staining, Photo T. Amemori) shows the mucosa of intestinal villi stained green, filled with billions of *M. a. paratuberculosis* rods present in clus-ters, stained red (**c**). The pathogen is irregularly secreted into the intestinal content and is subsequently shed with faeces into the environment

Source: [1] Amemori T, Matlova L, Fischer OA, Ayele WY, Machackova M, Gopfert E, Pavlik I (2004) Veterinarni Medicina 49:225–236.

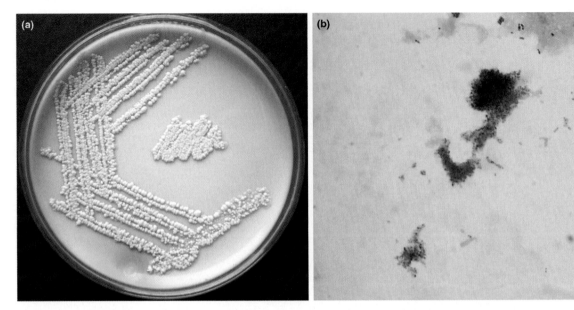

Photo 3.17 *M. malmoense* growth on Middlebrook agar medium is characterized by prominent creamy colonies; another contaminting bacterium/mould colony cab be seen at teh bottom (**a**). *M. malmoense* grows in clusters, usually formed by small rod- or cocci-shaped cells containing numerous dark-staining granules (**b**) (magnification 1000× after staining according to Ziehl-Neelsen, Photo J. Kaustova and V. Ulmann). Due to their morphology, they resemble small rods of *M. a. avium*, which can be isolated from both infected animals and occasionally from environmental samples

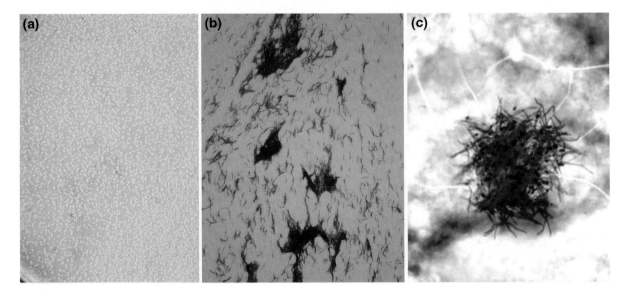

Photo 3.18 The growth of *M. xenopi* on a solid medium is characterized by very small pale yellow colonies (**a**). Over the course of time their colouring gradually changes to a darker yellow, which is more intense in colour than the pigments of other mycobacterial species that produce pigment without access to light (Photo L. Mezensky and J. Svobodova). *M. xenopi* grows either separately or grouped into clusters *in vitro*, as shown by the photo of an isolate from a domestic pig mesenteric lymph node lacking tuberculous lesions (**b**) (magnification 400×; Photo V. Mrlik). The individual mycobacteria are rather long and granulated, as evident from the second isolate from human sputum. Due to the fact that after multiplication, individual mycobacteria remain interconnected, a "bird nest" arrangement can be observed by microscopy (**c**) (magnification 1000×; Ziehl–Neelsen staining; Photo L. Mezensky and J. Svobodova)

Photo 3.19 In a pig farrowing unit, *M. xenopi* was isolated from peat which was fed as a supplement immediately after birth (**a**) (for more details, see Section 5.6.3.1). *M. xenopi* was also isolated from spider webs present on the door in the pig farrowing unit (**b**) ([1]; Photo I. Pavlik). Under certain condi-tions, different components of the environment, and not only as expected water from the warm water main, can become sources of *M. xenopi*

Source: [1] Matlova L, Dvorska L, Bartl J, Bartos M, Ayele WY, Alexa M, Pavlik I (2003) Veterinarni Medicina 48:343–357.

Photo 3.20 In one herd of market pigs, *M. xenopi* was detected in warm water from three old electric water heaters (**a**). This water was always used for the dilution of a bulk feed in the feed preparation room; water was added just before the feed was prepared (in the vessel on the left) (**b**). In this way pre-pared warm liquid feed was fed to pigs in metal bowls several times a day (**c**). In almost 20% of the market pigs, tubercu-lous lesions were detected in the cephalic and mesenteric lymph nodes and *M. xenopi* was identified (M. Pavlas, L. Matlova, I. Pavlik, unpublished data; Photo I. Pavlik)

(a)

(b)

Photo 3.21 On the pig farm presented here, sewage was discharged into a surface lagoon (**a**). Due to the terrible smell of sewage which was seeping through the plastic foils placed on the bottom of the lagoon, its use was banned. As it dried, dust containing a variety of PPM including one isolate of *M. xenopi* was released (**b**). The remainder of the dry sewage can be seen on the bottom of the photo; a pump was installed behind the lagoon, which transported the sewage through a pipeline to the surrounding fields (I. Pavlik and L. Matlova, unpublished data; Photo I. Pavlik). In Hungary, the spread of *M. xenopi* through the dust from a sedimentation tank left to dry was described [1]

Source: [1] Szabo I, Kiss KK, Varnai I (1982) Acta Microbiol.Acad.Sci.Hung. 29:263–266.

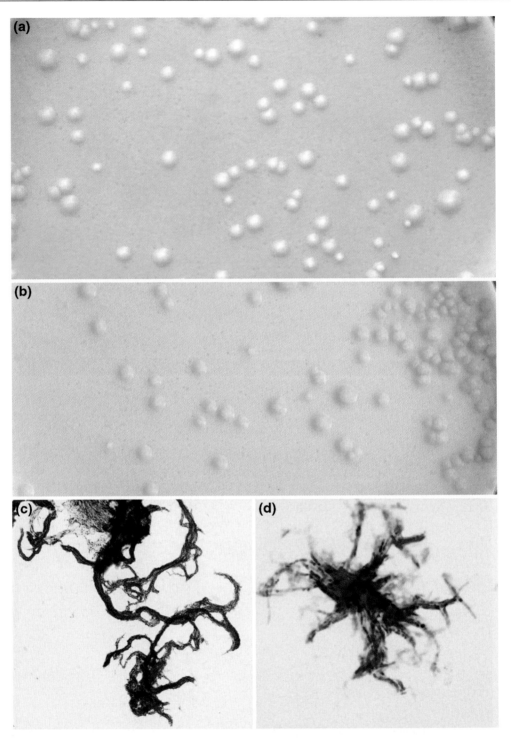

Photo 3.22 *M. marinum* grows on Lowenstein–Jensen medium: before exposure to light (**a**) and after (**b**). After Ziehl–Neelsen staining at magnification 200× *M. marinum* sometimes also appears as cords resembling those of *M. tuberculosis* with the difference that the cords of *M. marinum* form more loose aggregates (**c**). However, at higher magnification (800×) the rods of *M. marinum* are longer; these are usually granulated and more similar to *M. kansasii* (**d**) (Photo L. Mezensky and J. Svobodova)

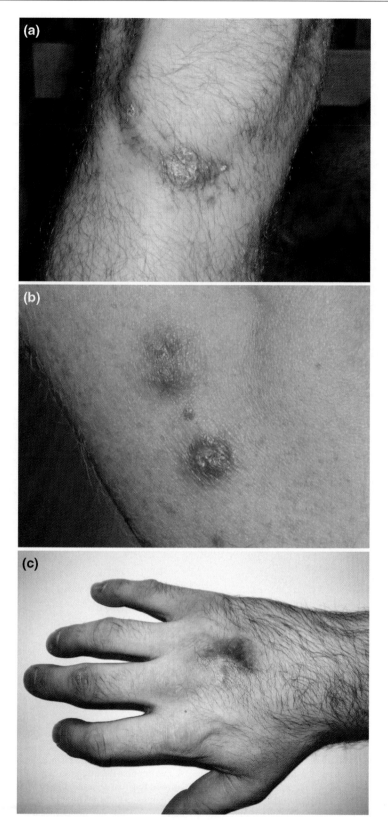

Photo 3.23   Skin diseases caused by *M. marinum* on the right knee (**a**), on the right shoulder (**b**) and on the dorsum of the right hand (**c**) (Photo J. Kaustova)

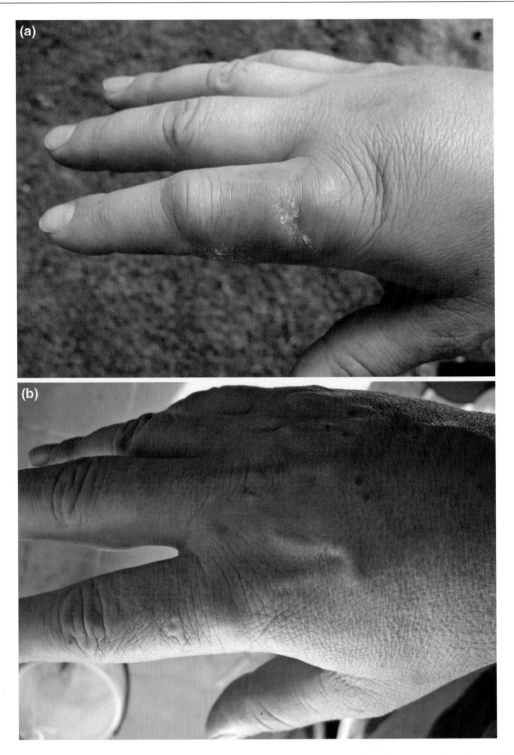

Photo 3.24 An *M. marinum* infection on the skin of the right forefinger of a professional aquarist; the infection occurred at the site of an injury he sustained whilst cleaning aquaria. A "fish tank granuloma" was diagnosed 28 months later by histological and subsequent culture detection of *M. marinum*. However, afraid of the side effects of the drugs to which the isolate *M. marinum* was sensitive, the patient refused targeted treatment. A mild course of infection was still present in the skin of the surrounding tissue after almost 6 years: the course of disease 30 months (**a**) and 59 months (**b**) after injury (I. Pavlik, J. Svobodova, L. Hartman, unpublished data; Photo I. Pavlik)

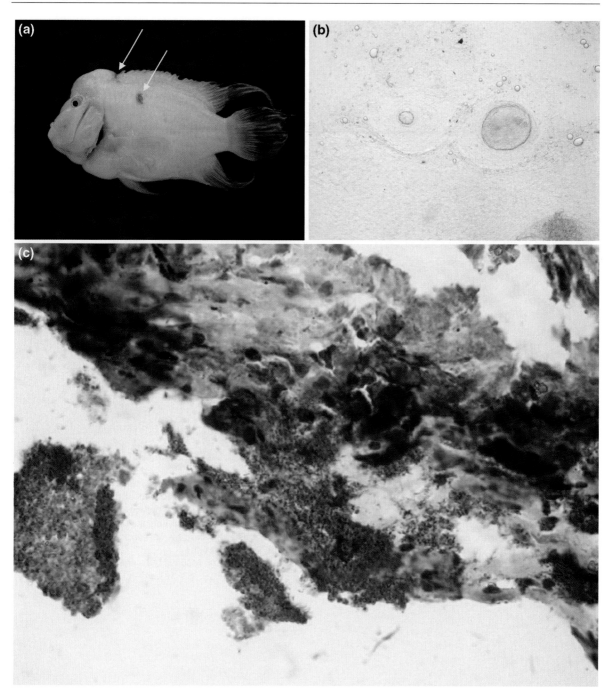

Photo 3.25   The most common source of *M. marinum* for people is the environment of tropical aquaria and/or the infected fish themselves. *M. marinum* often causes infection of their skin (two arrows); in the fish species *Cyphotilapia frontosa* (**a**), it usually occurs at the site of skin injury, sustained while fighting over females or territories in a tank. After the development of skin lesions, which are incurable in most cases, both the breeding and display fish lose their value; they can also become sources of *M. marinum* for the remaining fish present in the tank and for people who are in contact with their bodies or the water environment. By microscopy, granulomas can be detected in a fresh preparation (magnification 40×) (**b**) and after specific staining according to Ziehl–Neelsen numerous mycobacteria can be seen in the skin and surroundings of the injured tissue (**c**) (Photo L. Novotny, magnification 400×). However in tropical fish, different organs can also be infected by mycobacteria as described in detail in Section 6.5.1

Photo 3.26 *M. marinum* was detected in different components of an aquarium with infected tropical fish (mycobacteriosis of skin above the right eye, shown by the arrow, in the fish species *Cyphotilapia frontosa*): in biofilms on the plant leaves and walls of the tanks, in the bottom sediment, in water, in filters, etc. (I. Pavlik, V. Beran., L. Matlova, unpublished data; Photo L. Novotny). We failed to detect *M. marinum* in an aquarium with tropical fish that were clinically healthy over a long period of time. However, other PPM species, which can also cause skin infections in both fish and other animals, and in humans, were isolated from different components of the environment in these aquaria [1]. Nevertheless, the presence of certain predisposition factors such as a weakened immune system, skin injury, etc. is necessary as is the case for *M. marinum* infections. For more details, see Section 6.5.1

Source: [1] Beran V, Matlova L, Dvorska L, Svastova P, Pavlik I (2006) J.Fish.Dis. 29:383–393.

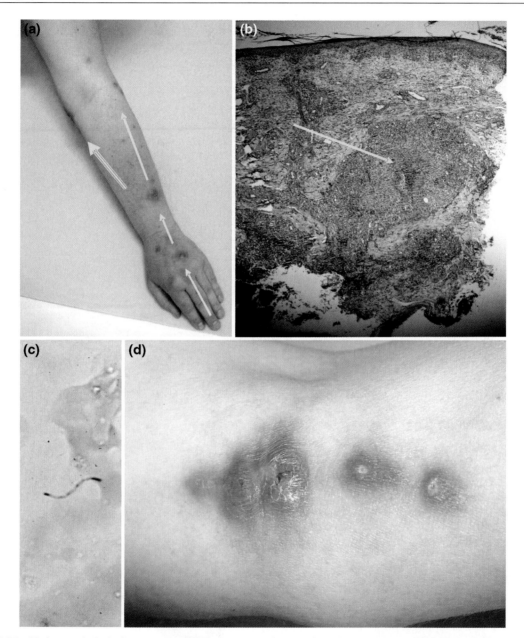

Photo 3.27 Various pathological processes resulting in mortality were observed in fish (scalars, tetras, zebra and neon fish and cichlids) over several months. The finger of a 53-year-old aquarist sustained a slight injury from the dorsal fin of a bullhead, while she was cleaning the aquarium without gloves. Two to three weeks later, pain, swelling and flare of the distal phalanx of the left middle finger developed. Subsequently, a painless inflammatory red bulge appeared on the dorsum of the hand (**a**) (Photo J. Stork). Histological examination provided evidence of lymphocyte exocytosis in the epidermis; perivascular interstitial infiltration was observed in the upper corium and nodular infiltrations or abscesses from neutrophil granulocytes were seen in the middle and lower dermis (yellow arrows) (**b**) Photo J. Stork). Long, thin acid-fast rods with indicated granules were detected sporadically in the tissue of a nodule on the hand dorsum after Ziehl–Neelsen staining (**c**) (magnification 1000×, Photo P. Jezek). After 7 days of culture at 33 °C, *M. marinum* was isolated. This is a case of the sporotrichoid form of the disease where the infection spreads through the lymph vessels indicated by the yellow arrow; a bold arrow shows the course of the palpable fibrotic lymph vessel (**a**). Therefore, nodules were also formed at places distant from the primary site of *M. marinum* penetration (the inside of the elbow in this case) (**d**) (Photo J. Stork). The patient was administered a combined therapy of etambutol and ciprofloxacin for 10 months which was determined by the sensitivity of clinical isolates. Control culture examination of a tissue biopsy from a regressed nodule 7 months later was negative (J. Stork, P. Jezek, M. Havelkova, M. Bodnarova, unpublished data)

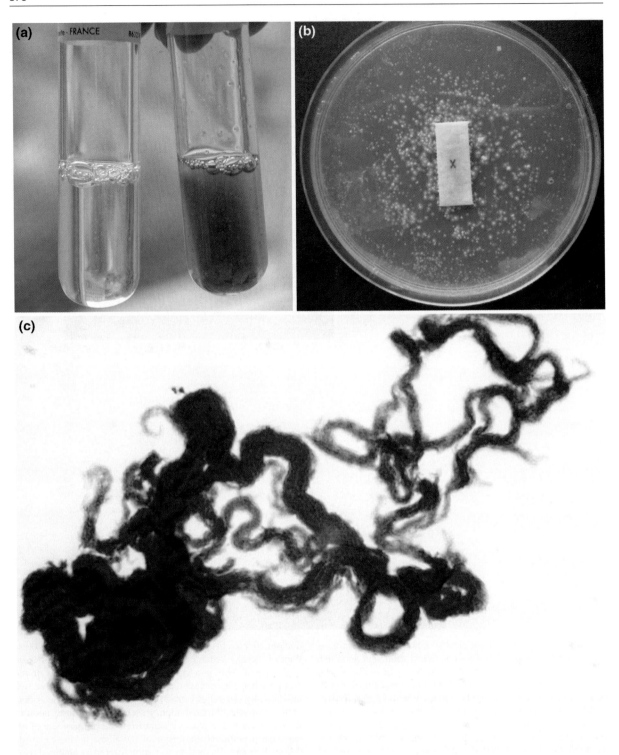

Photo 3.28 *M. haemophilum* grows in both liquid media (left, negative growth; right, positive growth) (**a**) and on solid Middlebrook medium with X factor which stimulates growth (**b**). *M. haemophilum* after Ziehl–Neelsen staining; magnification 200× (**c**). The formed cords may resemble *M. tuberculosis* (Photo J. Kaustova)

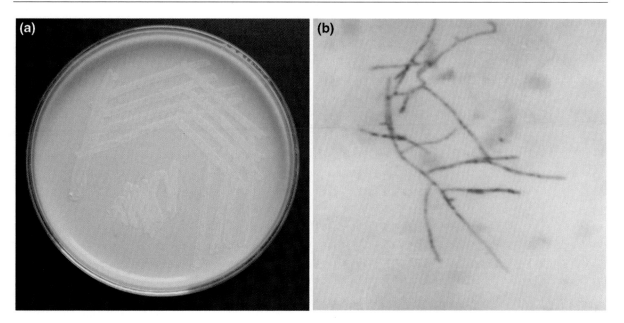

Photo 3.29 *M. celatum* grows on Lowenstein–Jensen medium as creamy colonies (**a**) and after Ziehl–Neelsen staining single branching is usually observed (**b**) (Photo J. Kaustova and V. Ulmann, magnification 1000×)

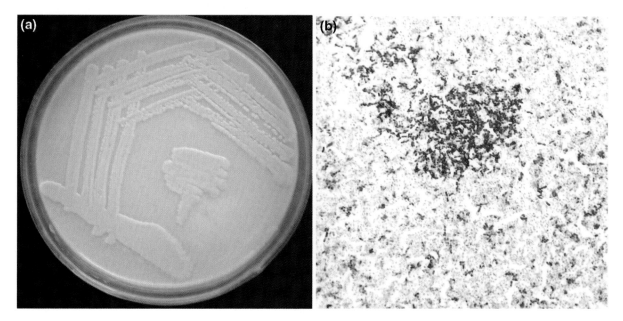

Photo 3.30 *M. mucogenicum* grows on Lowenstein–Jensen medium as creamy colonies (**a**). After Ziehl–Neelsen staining and 1000× magnification, small rods and cocci arranged in clusters of different sizes can be observed (**b**) (Photo J. Kaustova and V. Ulmann)

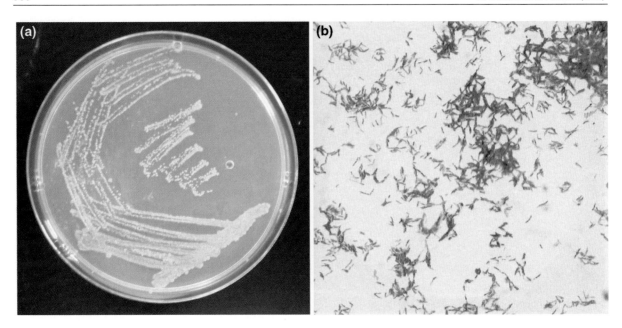

Photo 3.31   *M. lentiflavum* grows on Middlebrook medium as *yellow* colonies (**a**) and after Ziehl–Neelsen staining and 1000×
magnification thin and rather long single rods or small clusters can be observed (**b**) (Photo J. Kaustova and V. Ulmann)

Photo 3.32   *M. fortuitum* forms rosette clustered, waxy colonies, which are sometimes spherical in shape (**a**). Colonies of a relatively "young" isolate are cream coloured, "older" isolates absorb the green dye from Malachite green present in the egg medium, in this case Lowenstein–Jensen medium (**b**) (Photo L. Mezensky and J. Svobodova)

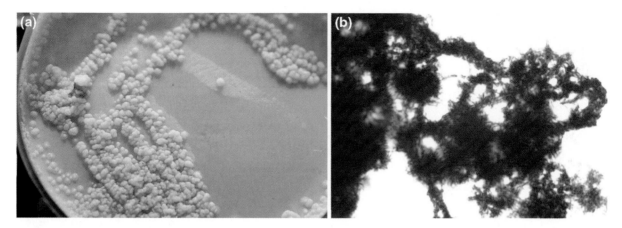

Photo 3.33 *M. terrae* often grows in primary culture as dry and rough colonies, which may resemble *M. tuberculosis* depending on their growth stage (**a**). Accordingly, the identification of this mycobacterial species commonly isolated from the environment must be performed carefully. By microscopic examination of an isolate after Ziehl–Neelsen staining and 200× magnification, cords resembling those of *M. tuberculosis* can often be observed (**b**) (Photo L. Mezensky and J. Svobodova)

Photo 5.1 The vaulted surface of a sphagnum bog shows a relief of relatively dry hummocks created by moderately hygrophilous sphagnum species and a shallow wet part occupied by highly hygrophilous sphagnum species (Photo J. Kazda)

Photo 5.2   A profile view of the moderately hygrophilous sphagnum association: the actively growing green region on the top, the deeper grey layer (the habitat of mycobacteria) and the brown stratum on the bottom (Photo J. Kazda)

Photo 5.3 *Sphagnum squarosum*: The *green* region on the top, showing the growth of the heads and the development of capsules (Photo K. Muller)

Photo 5.4 Hygrophilous sphagnum surrounding small streams come into close contact with surface water. Mycobacteria which thrive in the grey layer of sphagnum are released by rain or flooding into surface water, which serves as a vector for their further spread (Photo J. Kazda)

Photo 5.5 Under favourable conditions Bryidae create hummocks, a life form which is very similar to that of sphagnum (Photo J. Kazda)

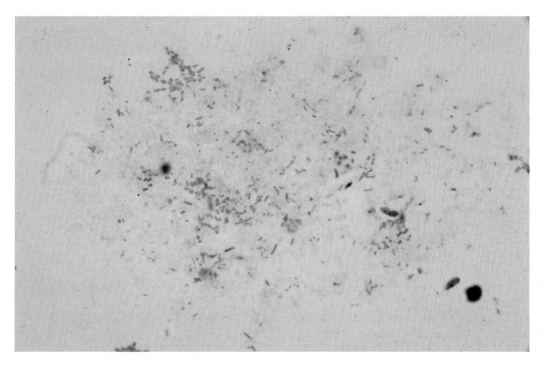

Photo 5.6 The grey layer of *Hylocomium splendens,* collected in fields in Ireland, was shown to contain an unusually high density of acid-fast bacilli in direct smears (Photo J. Kazda)

Photo 5.7 A profile view of *Hylocomium splendens*, the grey layer of which offers favourable conditions for the growth of mycobacteria (Photo J. Kazda)

Photo 5.8 Direct ingestion of moss (Bryidae) containing mycobacteria on a pasture when grazing is scarce (early April) in Ireland. *M. hiberniae*, common in this kind of moss, has provoked a tuberculin reaction similar to that of bovine tuberculosis (Photo J. Kazda)

(a)

(b)

Photo 5.9   Water aerosol is one of the most important sources of mycobacteria in the air. At the ocean shore, aerosol is mostly formed above the surf (**a**); it was found, e.g. in the USA, that it is a cause of non-specific responses to human tuberculin testing in people (Section 5.2.5.2). Inland water aerosol formed by irrigation with surface water (e.g. river or lake water) during agricultural activities can also be a source of mycobacteria (**b**); high concentrations of especially ESM and PPM can be present there in the summer months (Section 5.2.5.4; Photo I. Pavlik)

Photo 5.10 In Australia (e.g. Orange, New South Wales), rain water reservoirs are often situated close to houses in the countryside, most often in land used for agriculture (**a**). This water is subsequently drawn from the reservoirs to be used in the household and given to animals as drinking water. During heavy rain, the ground dust swirls and the dust present in the air is pulled down; both may contain ESM and PPM (**b**). Contaminated dust was most likely the source of *M. scrofulaceum* in these reservoirs of rain water in Australia ([1]; Section 5.9.4.1; Photo I. Pavlik) Source: [1] Tuffley RE, Holbeche JD (1980) Appl. Environ. Microbiol. 39:48–53.

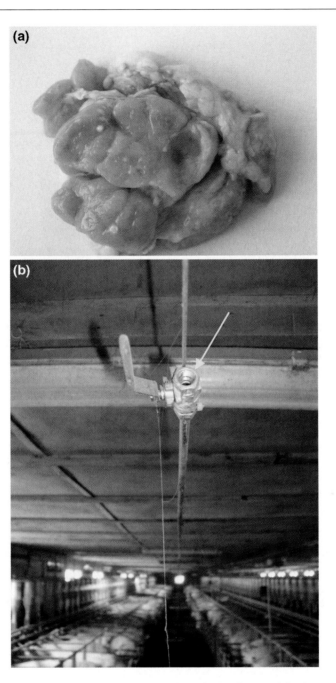

Photo 5.11 Water aerosol, which may contain ESM and PPM, is purposely produced in stables only very rarely if pigs are present. The risk of pulmonary infection for pigs by the aspiration of mycobacteria under these conditions is unclear but can be demonstrated in a case from one herd with the occurrence of tuberculous lesions in the pulmonary lymph nodes caused by *M. a. hominissuis* (serotype 8) (**a**). During the summer months sows were cooled with water aerosol produced in piping nozzles placed under the ceiling in stables for inseminated sows (during cooler months, the water supply of the piping was cut off, as shown in the photograph by the arrow) (**b**). *M. a. hominissuis* of an identical serotype was isolated from drinking water for pigs in the warm months of the previous season. Metal sieves in a pressure appliance which injects water into the piping trapped a major part of the released biofilms and thus prevented the obstruction of small nozzles in the metal piping; however, they most likely failed to prevent the spread of mycobacteria into the aerosol. This was the route by which mycobacteria caused respiratory tract infections in sows (I. Pavlik, J. Volavka, A. Horvathova, J. E. Shitaye, unpublished data; Photo I. Pavlik)

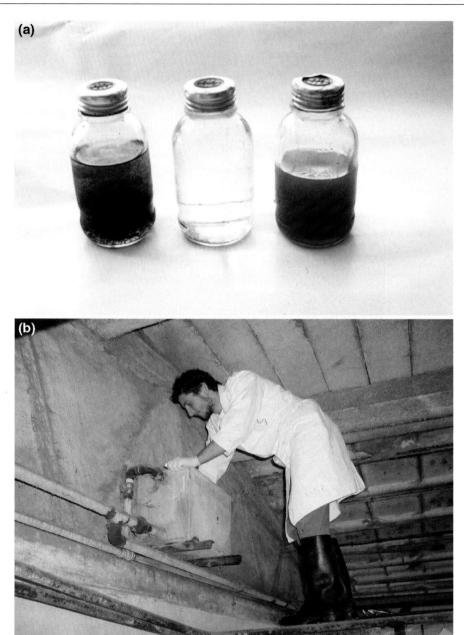

Photo 5.12 The most common source of ESM and PPM in herds of pigs, cattle and other animals kept in captivity is drinking water. On the left is a bottle with sediment collected from a plastic expansion tank; in the middle is a bottle with water supplied from the breeder's well, which is about 5 m deep; that water displayed only slight turbidity. The bottle on the right contains sediment from a metal expansion tank, supplied with drinking water from the town water supply (**a**). It is likely that oxidised iron present in the sediment from the tank (**b**), together with favourable temperature stimulated the growth of ESM and PPM. Different mycobacterial species were detected in the above-mentioned three samples: *M. fortuitum*, *M. gordonae* and *M. chelonae* (along with the causative agent of avian tuberculosis *M. a. avium*). Mycobacteria caused the formation of tuberculous lesions in the submaxillary and mesenteric lymph nodes of market pigs in this herd. These mycobacteria caused nonspecific reactions to avian tuberculin in gilts intended for sale and in the basic herd sows, because culture examination failed to detect *M. a. avium* either in the environment or in their tissues ([1]; Photo I. Pavlik)

Source: [1] Matlova L, Dvorska L, Bartl J, Bartos M, Ayele WY, Alexa M, Pavlik I (2003) Veterinarni Medicina 48:343–357.

Photo 5.13 The arid environments of desert and semi-desert regions do not support the occurrence of mycobacteria, which survive and multiply above all in damp conditions. The quickly drying surface waters present after the short rainy season in the Sahara and the subsequent direct solar radiation and high day-time temperatures quickly devitalize all mycobacteria (**a**) (I. Pavlik, unpublished data). Scanty tree vegetation grows around the watering places for camels (shown by the *arrow*) (**b**). Water is drawn from deep wells, which contain almost no ESM and PPM. This fact may also explain the very low incidence of nonspecific responses to bovine tuberculin testing in the animals of this region (J. Zinsstag, personal communication; Photo I. Pavlik)

Photo 5.14 Mycobacteria (ESM and PPM) have been detected in water, but especially in biofilms, which are formed very easily and quickly in aquatic environments under certain conditions. If the organic pollution of water is high, and light can gain access, algae are also present in biofilms, which are often rich in mycobacteria; the inner walls absorb green dye from the algae present in the above shown swimming pool for seals (**a**). However, biofilms of comparable composition and colour can also be observed in watering places for animals on pastures (**b**). Thus the warmed water becomes a significant source of ESP and PPM over the summer months (I. Pavlik, I. Trcka, M. Kopecna, unpublished data; Photo I. Pavlik)

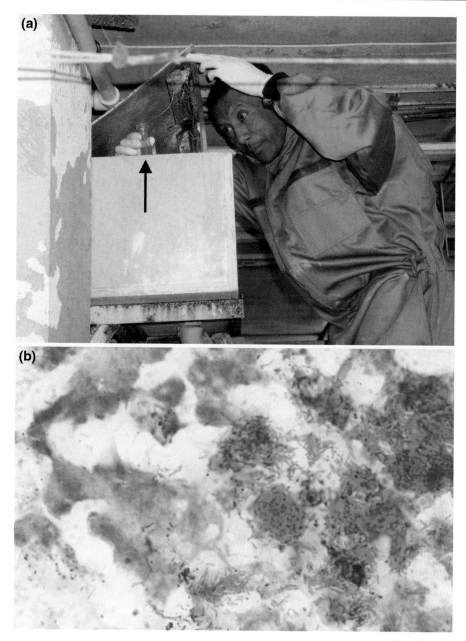

Photo 5.15 When investigating sources of ESM and PPM for pigs with a high occurrence of tuberculous lesions in their cephalic or mesenteric lymph nodes, it was found that the sediments present in plastic expansion tanks often contained iron oxide. In this pig herd with an increased incidence of nonspecific responses to avian tuberculin, ESM and PPM were detected in sediment from these expansion tanks (usually several expansion tanks are placed in every stable). Due to the fact that the source of water supply is identical, the examination of sediment from only several of them is sufficient. As shown in the upper part of the photograph, the water distribution piping was made from plastic material. It seems that some metal elements in the water distribution piping were the sources of iron oxide in the sediment (indicated by the arrow) as evident from the lower part of the photograph (metal elbow of the distribution piping) (**a**). Due to the fact that the piping in stables is usually placed under the ceiling, water in the piping can reach quite high temperatures, going above 30 °C in the warm months. Subsequently, large clusters of mycobacteria can be observed in the sediment by direct microscopy at magnification 1000× after Ziehl–Neelsen staining (**b**). These clusters of mycobacteria spread from the sediment in a tank to animals when they drink the water. The water steadily whirls and the animals become infected with fresh clusters of mycobacteria (I. Pavlik, I. Trcka, J. E. Shitaye, A. Horvathova, unpublished data; Photo I. Pavlik)

Photo 5.16 Surface water is rarely used today as drinking water for domestic pigs. In one herd, ESM and PPM were detected in troughs with rain water (**a**); these were filled over the winter season, when the pig paddocks were not used. The troughs were not cleaned in spring and the pigs most likely became infected with some species of ESM or PPM through this route. In autumn, nonspecific responses to avian tuberculin were detected in gilts reared in these pens. It is possible to occasionally encounter soft paddocks for pigs (paddocks with uncon-solidated covering, most often earthen ground). These paddocks should be considered highly risky from the point of view of the mycobacterial infection of pigs. During prolonged rainfall, the earth turns to mud, which is usually a source of different ESM and PPM present in the soil (**b**). Domestic pigs often dig and bathe in such mud as do free living wild boars (Section 5.5; I. Pavlik, P. Miskovic, A. Horvathova, unpublished data; Photo I. Pavlik)

Photo 5.17 Surface water is the most common source of mycobacteria for pigs kept occasionally in paddocks. Farmers do not usually consider this risk of ESM and PPM species infection. In this herd (**a**), sources of mycobacteria were investigated. Drinking water was supplied from the farm's own deep well and was directly pumped into the watering places. Accordingly, no mycobacteria were detected in this water. However, rain water retained in only a part of the concrete container during rainfall was a source of mycobacteria. Rainwater from the drain on the stable roof ran straight into the paddocks (**b**). It was retained on uneven parts of the concrete paddock only during rainfall and then dried up. In both paddocks, ESM and PPM were detected in water and in wet organic remnants after rainfall (Photo I. Pavlik)

Photo 5.18 On cervids' farms or in game parks, a passing brook is a water source for animals. It may often flow continuously through, e.g. a game park with white fallow deer (*Dama dama*) if it does not dry up over the year (**a**). In New Zealand, however, a farmer solved the problem of the drying up of a small river by building little lakes on his red deer farm (*Cervus* *elaphus*). Water was retained in these lakes over a long period of time (**b**). The risk of the occurrence of mycobacteria in the water and sediment from these water sources for animals is high; it can also become a source of infection for other animals if the causative agent of paratuberculosis or bovine tuberculosis is present in a herd (Photo I. Pavlik)

Photo 5.19 The adoption of preventative measures is necessary for both collected rain water and stagnant water in small lakes, pools and ponds. It is necessary to cover the rainwater containers. Water will then be protected from dust, which often spreads mycobacteria into water (**a**). In game parks with ungulate animals, sources of running water are sometimes missing (neither rivers nor brooks pass through the area). Celestial pools, which retain water for almost the whole of the dry period without rain, are the only sources of drinking water (**b**). In this game park, mouflons (*Ovis aries musimon*) can be seen; they often come and drink this water (Photo I. Pavlik). When water disappears by evaporation, organic pollution increases. The water surface becomes covered by vegetation and the watering place is very muddy around the edges. These watering places are sources of both mycobacteria and parasites for animals present in the environment (**c**) (Photo J. Lamka). The prevention of animal infection by these waterborne pathogens is the focus of Section 7.5

Photo 5.20 In zoological gardens, surface and rain water is often used as an important decorative element (**a**). With regard to the risk of the spread of mycobacteria, this water, often highly polluted with organic material (eutrophic), can be highly detrimental to the health of the animals. The fact that different species of wild birds that may be sources of, for example, the causative agent of avian tuberculosis also frequent this water poses a health risk as well (Section 6.6). In the photograph, we can see several free living and captive bird species together. Lakes are also often used as barriers separating animals of different species and may also function to prevent visitors from entrance (**b**). In addition, for the sake of variety and to increase the attractiveness of the zoo for visitors, various species of other animals are kept in such small lakes (semi-free) such as koi-carps (a red fish below the water surface on the left of the photograph) and European pond terrapins (Photo I. Pavlik). Potential infections of these poikilothermic animals with mycobacteria have been described in Section 6.5

Photo 5.21 Surface water in the paddocks in zoological gardens can pose a risk for some animals, which may not be considered by the breeders. The adult orangutans in one such paddock were observed to be little interested in the water surface (**a**). However, the young animals were curious. One of them searched the water surface of a canal for different items to play with (**b**). The young animal licked these items and subsequently drank contaminated surface water (**c**). Due to the frequent occurrence of ESM and PPM in surface water (often in zoological gardens with high organic pollution – eutrophic), the infection of animals can be expected. This does not manifest itself in disease, but antibodies are produced against these mycobacteria. These antibodies subsequently complicate the diagnosis of human or bovine tuberculosis in primates kept under such conditions. High levels of antibodies against *M. a. hominissuis* serotypes 8 and 9, which are often present in surface water, were detected in primates in a zoology garden. Antibodies against the causative agent of avian tuberculosis were not detected in these animals because their contact with free living birds is only occasional (I. Pavlik, J. Bartl, L. Matlova, unpublished data; Photo I. Pavlik)

Photo 5.22 Liquid waste from domestic animal farms (sewage, liquid manure, dung water, etc.) is often emptied into a field, especially after the harvest of agricultural crops. During its application to the soil in countries with developed agriculture, strict rules are followed. For example, sewage water can be applied to soil with the use of hose pipes resulting in a decrease in ammonium release and aerosol production (**a**). Nevertheless, a highly risky way of liquid waste application can be sometimes seen in these countries as well (**b**). Mycobacteria, other bacteria with detrimental consequences on health, and parasites and viruses can be spread to the environment via several routes. First, aerosols are formed during the emptying of liquid waste; these may contain conditionally pathogenic bacteria or even pathogenic microorganisms. If the soil is frozen, sewage can be flushed away to the surface waters, i.e. to brooks or rivers. These can consequently spread various pathogens, including mycobacteria, over long distances (Photo I. Pavlik)

Photo 5.23 The risk of contamination of different surfaces on animal farms through sewage water may be high in some cases. During rainfall, water mixed with animal urine and faeces and other farm waste including traces of disinfectants, medicaments, etc. often leaks from dunghills. The eutrophication of such waste water is high as is the occurrence of ESM and PPM in the sewage. Sewage water from the above pig stable was discharged during heavy rainfall to a pasture for calves and heifers from another stable. These were exposed to mycobacteria while grazing for more than 8 months every year. Subsequently, nonspecific responses to bovine tuberculin testing were detected (Photo I. Pavlik)

Photo 5.24 The risk of surface water contamination with, in particular, *M. avium* complex members is high in wood-processing plants. After the snow had melted, water was discharged from the above shown old sawdust pile in spring; *M. a. hominissuis* (serotypes 8 and 9) was isolated from that water. Despite that, sawdust from the pile was used as bedding on a heifer farm. Numerous responses were detected to bovine tuberculin testing. Simultaneous testing with avian tuberculin elicited positive responses in all tested animals. The above-mentioned sawdust was the source of *M. avium* complex infection for heifers. This water was discharged from the sawdust pile to a brook outside the farm which several kilometres downstream ran into a big river (I. Pavlik, unpublished data; Photo I. Pavlik). The spread of *M. a. hominissuis* (serotypes 8 and 9) in this way, from sawdust piles in wood-processing plants into the water of adjacent rivers, was observed more than 30 years ago (M. Pavlas, unpublished data)

Photo 5.25 Small ponds for the purification of water were first built as far as back as the middle ages in village centres throughout Europe (Photo I. Pavlik). Waste water from villages was discharged into these ponds and was retained there (organic matter was intensively mineralized in the sediments). The clean surface water, free from the majority of organic matter, ran into a water stream (most often to a brook or a river). Water flowing from such ponds was relatively clean even though horses were taken to that water and various domestic animals drank from it. The fact that trout which require clean and well-aerated water with no organic pollution lived in the brooks flowing out from the ponds attested to its cleanness (Pavlik I. senior, personal communication). Nevertheless, the sediment from such ponds is abundant in ESM and PPM. When water from these sources has been used as drinking water for animals, their infection with ESM and PPM, e.g. *M. a. hominissuis* (serotype 8) has been detected [1, 2]
Source: [1] Kazda J (1966) Zentralbl.Bakteriol.[Orig.] 199:529–532. [2] Kazda J, Cook BR (1988) New Zealand Veterinary Journal 36:184–188.

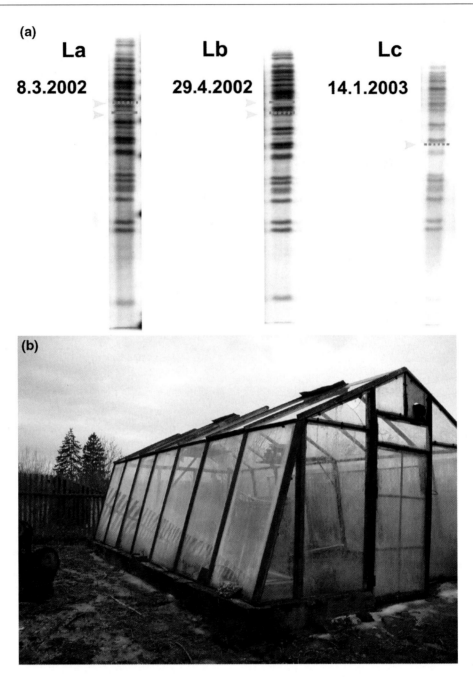

Photo 5.26 The source of *M. a. hominissuis* (serotype 8) for a patient with an avian mycobacteriosis infection of the lungs was investigated; that causative agent was repeatedly detected in his sputum. It was found by IS*1245* RFLP analysis that two isolates, La and Lb, from sputum were very similar (they differed only in the transposition of one band). The examination of samples from the environment of his residence revealed *M. a. hominissuis* of identical serotype 8 and comparable IS*1245* RFLP profile (isolate Lc) (**a**). This isolate originated from a glasshouse; the patient used rain water or water from a nearby brook during dry periods (*blue* barrels for water can be seen inside the glasshouse) for watering plants there (**b**). Waste water (sewage) from the neighbouring house was discharged into the brook. Due to the fact that the patient did not use peat for fertilization (which is often contaminated with *M. avium* complex; Section 5.1), it is highly probable that the source of *M. a. hominissuis* (serotype 8) was present in either contaminated soil in the glasshouse or water used for watering the plants (I. Pavlik, O. Ostadal, L. Dvorska-Bartosova, M. Bartos, unpublished data; Photo I. Pavlik)

Photo 5.27 The occurrence of bovine tuberculosis was suspected on a farm when an old emaciated cow was out of necessity slaughtered and gross examination was performed. While investigating the sources of the causative agent, it was found that sewage, including liquid manure, was being discharged outside the farm (shown by the arrow) (**a**). The sewage outflow ran through an open sewer to a sludge (sedimentation) pool, which was situated several 100 m away from the farm beyond the pastures (**b**). Several times per year during heavy rainfall, the expanded outflow was released over pastureland. In the dry season when there was a lack of feed, various free living ruminant species and lagomorphs were seen in the adjacent pastures as well as at the exposed outflow channel. In this case, the transport of waste through an open sewer can facilitate the spread of different causative agents of mycobacterial infections, including, e.g. the causative agent of bovine tuberculosis (I. Pavlik, L. Dvorska-Bartosova, W. A. Ayele, unpublished data; Photo I. Pavlik)

Photo 5.28 Aerosol produced during the treatment of sewage, liquid manure, etc. is a particularly significant risk factor for the spread of mycobacteria to the environment. Aerosols are formed during the aeration of these effluents; this accelerates degra-dation processes and most likely reduces the concentration of mycobacteria. This risk should be considered when treating the sewage of animal farms where there is an occurrence of bovine tuberculosis and paratuberculosis (Photo I. Pavlik)

Photo 5.29 The layer of arable soil, which is abundant in mycobacteria (Section 5.5) is often very thin. Dry subsoil with a low content of organic matter does not support their growth and only small amounts of mycobacteria are generally isolated from them (**a**). Similarly unfavourable conditions for the survival and multiplication of mycobacteria are most probably found in arid regions where rapid erosion occurs (**b**). However, much remains to be learned about the ecology of mycobacteria in these locations and further investigation is necessary (Photo I. Pavlik)

Photo 5.30 A variety of natural factors and in particular different human agricultural activities are involved in soil erosion. The arable soil layer becomes denuded above all by the deforestation of soil and its subsequent use as pastureland for ruminants; New Zealand, South Island (**a**). Repeated grazing on pastures by sheep and goats that can pull out plants together with their roots devastate an area's vegetation; Mediterranean region (**b**). The present mycobacteria activate both the cell and humoral components of the immune systems of these animals. The mycobacteria cause nonspecific responses, e.g. during the serological diagnosis of paratuberculosis or skin tuberculin testing for the diagnosis of bovine tuberculosis. The ecology of mycobacteria in arid pasturelands is not well described at present; Cyprus (**c**). However, we can suppose that various mycobacterial species adapt to the unfavourable conditions to a certain degree. Accordingly, it remains to be elucidated, which mycobacterial species live in symbiosis, e.g. with plant roots that have survived intensive grazing by small ruminants (Photo I. Pavlik)

Photo 5.31 Mycobacteria from soil are transported to other components of the environment especially by surface water (**a**). The intensity of their washing out from the soil depends on many factors which require further investigation. Dust storms in the arid regions in Asia or Africa are very common events (**b**). However, no information about the sweeping of mycobacteria together with the dust from the soil towards the surface of the Indian Ocean, as shown by this photograph, has been published. There is also a paucity of knowledge regarding the effect of these mycobacteria on the health status of organisms living in coastal regions (Photo I. Pavlik)

Photo 5.32 Mycobacteria from soil are spread in the environment through both the activities of humans and animals. Vehicles that pass over dusty roads or fields cause an intensive whirl of dust, which sticks to their wheels and other surfaces (**a**). We detected mycobacteria in the dust from the wheels of a tractor by microscopy after Ziehl–Neelsen staining (**b**); after isolation, these were identified as *M. a. hominissuis*. However, mycobacteria can also be spread on the body surface of animals (**c**). For example, after rolling in ash, this common kestrel (*Falco tinnunculus*) can on the same day fly to places, which may be tens of kilometres away (Photo I. Pavlik)

Photo 5.33 From the point of view of the ecology of mycobacteria, soil seems to be the most significant component of the environment (Section 5.5). Pathogenic mycobacteria become involved in ecology in many ways; of these, their accessing of soil and water should be considered most significant. They are spread in the soil by both invertebrate and vertebrate animals. Of invertebrate animals, dung beetles in Africa are the most illustrative example; this beetle buries a ball with its "offspring" in the soil (**a**). Thus, it can also be involved in the spread of, e.g. the causative agent of paratuberculosis or bovine tuberculosis in the soil. Vertebrates can also be involved in the spread of mycobacteria in the soil. In an infected cattle herd, *M. a. paratu-* *berculosis* was detected in drying faeces (in the part which was still wet) (**b**). Also significant from the point of view of ecology is the fact that small terrestrial mammals – common voles (*Microtus arvalis*) bred excessively in that pastureland – drove holes through the faeces. The lower parts of faeces were also ingested by, e.g. earthworms that spread it to deeper layers of soil. In an infected cattle herd, *M. a. paratuberculosis* was also isolated from earthworms ([1]; Photo I. Pavlik)

Source: [1] Fischer OA, Matlova L, Bartl J, Dvorska L, Svastova P, du MR, Melicharek I, Bartos M, Pavlik I (2003) Vet.Microbiol. 91:325–338.

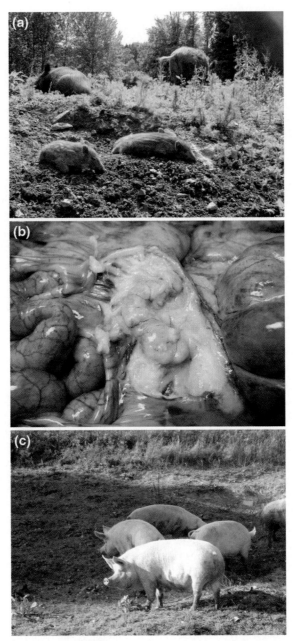

Photo 5.34 Ingestion of soil (geophagy) is a phenomenon seen in many animals [3]. They may ingest soil due to its mineral substances or microelements, which are essential for them. The behaviour of young wild piglets (*Sus scrofa*) may constitute a typical example of geophagy as the soil can supplement their organisms with deficient iron (**a**). Domestic piglets require supplementation with iron, e.g. administered parenterally (Photo I. Pavlik). However, piglets can become infected with PPM from soil. Some PPM species, e.g. *M. a. hominissuis*, can cause tuberculous lesions in the mesenteric lymph nodes of wild boars (**b**) similar to lesions caused by *M. a. avium* [1, 2], (Photo J. Lamka). In domestic pigs kept on farms practicing organic agriculture, the occurrence of tuberculous lesions in cephalic and mesenteric lymph nodes is more frequent than in wild boars living under comparable conditions (**c**). It is probably due to the higher susceptibility of bred domestic pigs to intracellular parasites, which has been noted in domestic pigs kept on farms (Photo I. Pavlik) Source: [1] Machackova M, Matlova L, Lamka J, Smolik J, Melicharek I, Hanzlikova M, Docekal J, Cvetnic Z, Nagy G, Lipiec M, Ocepek M, Pavlik I (2003) Veterinarni Medicina 48:51–65. [2] Trcka I, Lamka J, Suchy R, Kopecna M, Beran V, Moravkova M, Horvathova A, Bartos M, Parmova I, Pavlik I (2006) Veterinarni Medicina 51:320–332. [3] Trckova M, Matlova L, Dvorska L, Pavlik I (2004) Veterinarni Medicina 49: 389–399.

Photo 5.35 Children come into contact with various sources of mycobacteria from soil and other components of the environment. Water, soil and wood products are considered the most significant sources of *M. avium* complex members. In the zoo pre-sented in the picture, *M. a. hominissuis* was isolated from bark and sawdust used mainly as absorptive and decorative material on the ground of the paddock (I. Pavlik, V. Mrlik, M. Moravkova, unpublished data; Photo I. Pavlik)

Photo 5.36 Gross lesions caused by *Rhodococcus equi* in the lymph nodes of pigs resemble tuberculous lesions caused by mycobacteria ([1, 2]; Section 5.5.5). In one pig herd, an increased incidence of gross (tuberculoid) lesions in cephalic and mesenteric lymph nodes was observed; this resulted in considerable economic losses. We failed to isolate the expected mycobacteria from these gross lesions; nevertheless, we succeeded in isolating *R. equi* (**a**) (Photo I. Pavlik and I. Parmova). While investigating the source of infection, it was found that horses were kept near the paddocks for pigs (**b**). Horses are considered the primary source of *R. equi*, an organism which has now adapted to soil and is often found there today. Subsequently, the sources of gross lesions in pigs were investigated in the other four pig herds, from which only *R. equi* was isolated (**c**). Due to the use of earthen paddocks in all four herds for pigs of different age categories, contaminated soil was the most likely source of *R. equi* (I. Pavlik, A. Horvathova, J. E. Shitaye, I. Parmova, Z. Voslajer, unpublished data; Photo I. Pavlik)

Source: [1] Dvorska L, Parmova I, Lavickova M, Bartl J, Vrbas V, Pavlik I (1999) Veterinarni Medicina 44:321–330. [2]Shitaye JE, Parmova I, Matlova L, Dvorska L, Horvathova A, Vrbas V, Pavlik I (2006) Veterinarni Medicina 51:497–511.

Photo 5.37 During the production of silage, there are numerous opportunities for mycobacterial contamination from the environment (Section 5.6.1.2); the most common source of mycobacteria in silage is soil. Harvested vegetables and silage transported to the silo are often contaminated with soil. The muddy, unpaved lane on the right was the source of PPM, which was later isolated from the silage pile at the back on the right (**a**). With a view to the possible prevention of silage contamination with ESM and PPM from the environment, the use of plastic covers – green bags containing silage on the left – has been shown to be highly useful at present. In particular, the use of agricultural machinery with clean wheels can prevent the soil contamination of silage. In an infected cattle herd, *M. a. paratuberculosis* was isolated from the wheels of a tractor (**b**) which was used for forage delivery, manure distribution and silage production at the harvest time of maize (I. Pavlik, L. Matlova, J. Bartl, unpublished data; Photo I. Pavlik)

(a)

(b)

Photo 5.38 Silage can also become contaminated with mycobacteria after its production. The highest risk for the transmission of mycobacteria is posed by birds. Particularly during the cold months of the year when food is scarce, they are attracted by the silage holes where they most commonly find the grains of fermented maize. Due to that fact the silage piles should always be covered (**a**). On a cattle farm, *M. a. avium* was isolated from exposed silage; this caused nonspecific responses during the serological diagnosis of paratuberculosis. Nevertheless, various contamination risks also exist on ruminant farms, where silage is protected by a cover at all times before consumption. Red deer infected with paratuberculosis were kept on a farm near the forest in the background of the photo (**b**). After several years, the transmission of *M. a. paratuberculosis* of identical IS*900* RFLP type to heifers kept in close proximity (in the foreground of the photograph) was documented. We can only speculate as to the sources of infection; however, contaminated forage including silage has been considered as a possibility (I. Pavlik, J. E. Shitaye, M. Moravkova, unpublished data; Photo I. Pavlik)

Photo 5.39 Feed mixture management in large-scale systems of pig production is well organized nowadays. The old and unsuitable single-cover metal containers (container on the right) have recently been replaced by plastic double-cover containers (**a**). The feeding mixture does not attach to the walls of the latter; the previously used containers were sources of mycotoxins produced by growing moulds, and often also ESM and PPM (M. Pavlas and I. Pavlik, unpublished data). Today, feeding mixtures are protected in almost all pig herds from contamination by bird faeces, often infected with *M. a. avium*. Consequently, e.g. pigeons (in the middle) and sparrows (on the right), the most common "visitors" to these containers, can only eat feed remnants on the surface of the containers (**b**). Needless to say, these remnants are not fed to pigs, and bird faeces cannot contaminate the feeding mixtures. The bodies of dead birds are not mixed with the feeds either, although it occurred previously on farms where old open containers for feeding mixtures were used (Photo I. Pavlik)

Photo 5.40 The waste of bakeries which is usually fed to pigs is safe from the point of view of their potential contamination with mycobacteria. High-temperature baking safely destroys mycobacteria. However, their storage on the floor may be risky because they can subsequently become contaminated with ESM and PPM. On this farm, mycobacteria were detected by culture from the floor of a storehouse, which was dirty from mud brought on the wheels of a tractor delivering the old bread and rolls (**a**). Due to the fact that this soil had been wet for a long time, mycobacteria might have multiplied there during the warm months. Moreover in this pig herd, PPM were also isolated from the ground in front of the troughs for pigs, where rolls were thrown. The risk of mycobacterial contamination was also posed by free moving birds (especially sparrows). The cat in the photograph was one of approximately ten, which had free access everywhere and were kept on the farm to catch small terrestrial mammals (**b**) (I. Pavlik, L. Matlova, M. Krpec, unpublished data; Photo I. Pavlik)

Photo 5.41 The waste from bakeries may become contaminated with ESM and PPM present in the dust on the floor. *M. a. hominissuis* was detected in bakery waste on this cattle farm (mainly bread was fed). Sawdust freely stored on the floor on the left might have been the source of the pathogens (**a**). Birds could also be sources of *M. a. hominissuis* (**b**). Bird spoors in the dust among the loafs of bread provided evidence of their presence (I. Pavlik, L. Matlova, I. Trcka, B. Klimes, unpublished data; Photo I. Pavlik)

Photo 5.42 In a large-scale system of pig production, sawdust was used in the pens for neonatal piglets. It was scattered on the floor so that piglets could lie down on a dry surface (**a**). Sawdust can absorb moisture from both the floor and their body surfaces. However, due to the fact that the sawdust was stored under less than ideal conditions (on the floor), it was contaminated with *M. a. hominissuis* (**b**). When fattening was completed, i.e. 6–7 months later, tuberculous lesions in submaxillary and mesenteric lymph nodes caused by *M. a. hominissuis* were detected in 20% of pigs, which were reared "on sawdust". Subsequently, the farmer purchased bran in sacks, which he applied over the sawdust no longer in use at that time. Surprisingly, no decrease in the incidence of tuberculous lesions was observed after 8 months, when pigs kept "on bran" were slaughtered. Subsequent investigations revealed that *M. a. hominissuis* was present in both bran poorly stored on damp sawdust and bran stored in dry sacks in a dry and clean room (**c**). A primary contamination of the bran in the mills and bakeries where it was stored for several years due to a lack of customers was the cause of the problem ([1]; Photo I. Pavlik)

Source: [1] Fischer OA, Matlova L, Dvorska L, Svastova P, Peral DL, Weston RT, Bartos M, Pavlik I (2004) Vet.Microbiol. 102:247–255.

Photo 5.43 Properly stored and dry (usually mycobacteria-free) straw can be used, e.g. in pens for piglets instead of highly risky sawdust (Section 5.8) or bran (Section 5.6.1.3). However, straw must be cut into short pieces (about 5–8 cm long), which do not plug the opening in the grids, and allow faeces to fall into the reservoir present under the grid (**a**). Consequently, e.g. the larvae of invertebrates cannot develop and ESM or PPM cannot multiply in the straw and faecal remnants. The poor storage of straw poses a great risk for the spread of ESM and PPM to pig herds, cattle and other animals (**b**). Straw, stored on a damp floor (wetted by dung water from underneath on this farm), quickly loses its good qualities. The risk is posed primarily by mycotox-ins from moulds that grow through it and from ESM and PPM that multiply here during warmer months. *M. fortuitum* was also isolated from this straw. The storage of straw in the open air, where it is not protected from rainfall is also unsuitable (**c**). Various mycobacterial species have been detected in damp straw stored in this way, through which vegetation was growing ([1, 2]; Photo I. Pavlik)

Source: [1] Matlova L, Dvorska L, Bartl J, Bartos M, Ayele WY, Alexa M, Pavlik I (2003) Veterinarni Medicina 48:343–357. [2] Pavlik I, Matlova L, Gilar M, Bartl J, Parmova I, Lysak F, Alexa M, Dvorska-Bartosova L, Svec V, Vrbas V, Horvathova A (2007) Veterinarni Medicina 52:392–404.

Photo 5.44 Peat is characterized by its very good dietetic qualities, because of which its use as a feed supplement, especially for piglets after birth and after weaning, was initiated [3]. In many large-scale systems of pig production peat has started to be generally used and is fed as a supplement to neonatal piglets. The same parts of peat and feed for suckling piglets can be seen in the feeder (**a**); piglets start to ingest the mixture on the first day of life. Thus, they imitate the behaviour of piglets in the wild, who, in this way, supplement the iron deficiency in their diet. Tuberculous lesions in head and mesenteric lymph nodes were detected in these piglets 6–7 months later after slaughter (**b**). The isolated *M. a. hominissuis* most often of serotypes 8 and 9, which was also detected in peat fed as a supplement, was of comparable IS*1245* RFLP types ([1, 2]; Photo I. Pavlik)

Source: [1] Matlova L, Dvorska L, Ayele WY, Bartos M, Amemori T, Pavlik I (2005) J. Clin. Microbiol. 43:1261–1268. [2] Matlova L, Dvorska L, Bartl J, Bartos M, Ayele WY, Alexa M, Pavlik I (2003) Veterinarni Medicina 48:343–357. [3] Trckova M, Matlova L, Hudcova H, Faldyna M, Zraly Z, Dvorska L, Beran V, Pavlik I (2005) Veterinarni Medicina 50:361–377.

Photo 5.45 Kaolin is used as a dietetic supplement and sometimes even as a medicament due to its good absorption qualities [3]. Kaolin is extracted by surface mining and is usually processed in adjacent factories (see the factory on the horizon in the photo) (**a**). Kaolin is transported to factories in the form of a water suspension. Surface water from the mines, which is often contaminated with ESM and PPM, is usually used for kaolin levigation (**b**) (Section 5.2.5). It was found in this mine and in this kaolin-processing plant that mycobacteria present in the water used for kaolin transport survived the following processing stages: transport, levigation, separation, coagulation and the subsequent drying process right up to the final product (**c**) [1]. In market pigs fed kaolin as well as in piglets fed peat as a supplement after birth, an increased incidence of tuberculous lesions in the lymph nodes was detected after slaughter ([2]; Photo I. Pavlik)

Source: [1] Matlova L, Dvorska L, Bartl J, Bartos M, Ayele WY, Alexa M, Pavlik I (2003) Veterinarni Medicina 48:343–357.
[2] Matlova L, Dvorska L, Bartos M, Docekal J, Trckova M, Pavlik I (2004) Veterinarni Medicina 49:379–388. [3] Trckova M, Matlova L, Dvorska L, Pavlik I (2004) Veterinarni Medicina 49:389–399.

Photo 5.46 A 17-year-old boy suffering from non-Hodgkin's lymphoma was anxiously attended to by his mother: All meals she prepared for him were thoroughly heat treated. However, the mother also gave him non-heat-treated green parsley haulms which she scattered on fried eggs (the egg yolk was almost raw). *M. engbaekii* was detected in the soil, roots and on the leaves of the parsley, although this mycobacterial species was not isolated from the patient. The eggs originated from three small flocks of hens, kept under very primitive conditions with a high risk of the occurrence of avian tuberculosis. After the chemotherapy of this oncology patient was initiated, small parts of mucosa that had separated from his intestines were delivered for testing. *M. a. avium* was detected there by the IS*901* PCR method. The above-mentioned eggs were the source of this causative agent. In addition, *M. neoaurum* was detected in the lung exudates from the patient. However, its source could not be revealed either in the environment of his home or in the hospital where he was admitted (I. Pavlik, J. Sterba, P. Kralik, I. Tesinska, V. Mrlik, unpublished data; Photo I. Pavlik)

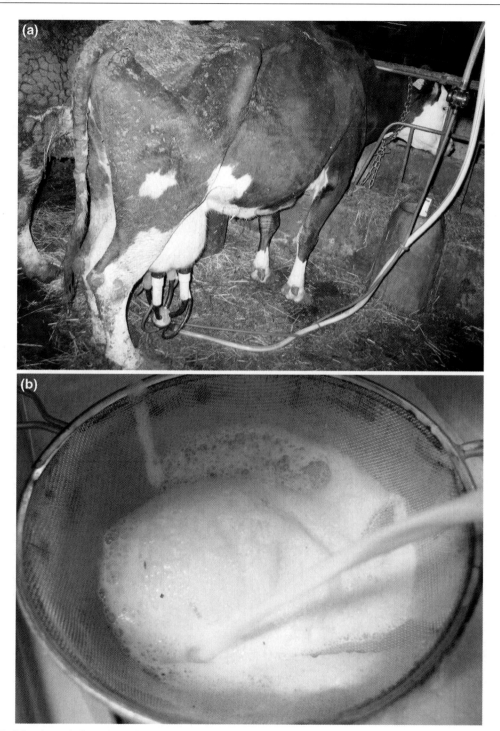

Photo 5.47 Mycobacteria from the environment may contaminate milk, particularly during milking. The milking of cows performed in stables (in stalls) should be considered highly risky in comparison with milking in a parlour because the risk of milk contamination with mycobacteria from the environment is high (**a**). Immediately after milking (shortly before milk filtration), pieces of straw can be seen on the strainer (**b**). Accordingly, the filters used for the mechanical purification of milk must be regularly changed in accordance with the manufacturer's instructions. In the event of "saturation" (exceeding) of their capability to filter the milk, the contaminating materials from the environment (including faeces) can only be washed out. *M. fortuitum*, *M. chelonae*, *M. a. hominissuis*, *M. phlei* and other species were detected in the bedding from this type of stable and subsequently in milk as mentioned in Section 5.7.4 (I. Pavlik and L. Matlova, unpublished data; Photo I. Pavlik)

Photo 5.48 Animal carcasses, which decompose in the environment, can become significant sources of mycobacteria and other pathogens. Organic tissues are decomposed by both physical and biological factors; mycobacteria can be spread in the environment through many routes that have been described in different sections of this book. It often happens that farmers do not arrange for the safe removal of dead animals from the environs of large herds of sheep by veterinary sanitary services. (**a**) Various animals come to such localities and ingest the remnants of carcasses or take them to distant places (e.g. birds). They can thus be involved in the further spread of mycobacteria in the environment. Sometimes, a sheep may die far from the pasture and its body will not be immediately found after death (**b**). The decomposition of organic parts and their ongoing degradation proceed relatively quickly on the pastures, above all in the summer months, and only the remains of bones and wool provide evidence of a dead sheep (**c**). I. Pavlik, O. A. Fischer, L. Matlova, unpublished data; Photo I. Pavlik

Photo 5.49 The bodies of large ruminants are significant sources of different mycobacterial species in the environment, including the causative agents of bovine tuberculosis or paratuberculosis. In areas where red deer are farmed (*Cervus elaphus*), e.g. remnants of the exposed parts of their carcasses or whole bodies, can be seen (**a**). Different carnivores and scavengers are attracted to the environs of farms in such cases. These are sometimes lured by the owners who subsequently hunt them. In the surroundings of pastures with extensively farmed large domestic ruminants (especially cattle), it is sometimes possible to see either separate bones or skeletons of dead cows or cattle of different age categories, which attests to the previous presence of bodies of dead animals. It is necessary to note that this phenomenon can be encountered above all in developing countries, where some animals are not registered (**b**). In game parks, remains of dead ruminants can sometimes be present; the photo shows the skeletal remains of a dead sika deer (*Cervus nippon*) on Hokkaido Island in Japan (**c**). Photo I. Pavlik

Photo 5.50 The dead bodies of animals can also be present in sewage waters. The photo shows the bodies of two cats thrown into a rainwater reservoir on a farm of cattle infected with paratuberculosis (**a**). In this cattle herd, *M. a. paratuberculosis* was detected in, e.g. invertebrate animals (e.g. larvae of drone flies), in liquid manure, in waste milk and in the intestinal lymph nodes of cats, which were fed with waste fresh cow's milk (I. Pavlik, O. A. Fischer, W. Y. Ayele, L. Matlova, unpublished data). On another farm with cattle infected with *M. a. paratuberculosis,* a cow died and was transported to the cafilery 3 days after death. Due to the high day and night temperatures, which were in the region of 25–35 °C, a fluid discharged from the natural body openings; this was washed off by heavy rainfall to a nearby brook (**b**). Photo I. Pavlik

Photo 5.51 Sawdust is one of the most risky materials for pig herds due to the fact that it may be contaminated with different *M. avium* complex members; *M. a. hominissuis* (serotypes 6 and 8) has been isolated most commonly. Poorly stored sawdust can easily become contaminated with mycobacteria present in surface water, in dust and in the faeces of birds and mammals, etc. (**a**) Under humid conditions and higher temperatures, above 20 °C, various ESM and PPM species start to multiply in the sawdust. Sawdust used as bedding ensures that floors remain dry (**b**). Piglets also learn to lie on places without grids where they later urinate and defaecate. Consequently, faeces and urine do not contaminate the floor in pens and the animals are clean during the whole fattening period. The freshly weaned piglets are exposed to high levels of stress in the pens. It was observed at the same time that all piglets intensively ingested sawdust from the ground (**c**). Thus, these animals became heavily infected with *M. a. hominissuis*. After reaching body weights of 115–120 kg, their infection was manifested by an increased incidence of tuberculous lesions in the intestinal and submaxillary lymph nodes ([1, 2]; Photo I. Pavlik)

Source: [1] Matlova L, Dvorska L, Bartl J, Bartos M, Ayele WY, Alexa M, Pavlik I (2003) Veterinarni Medicina 48:343–357. [2] Matlova L, Dvorska L, Palecek K, Maurenc L, Bartos M, Pavlik I (2004) Veterinary Microbiology 102:227–236.

Photo 5.52 Tree bark most often becomes contaminated with mycobacteria on the floor, after it has been peeled. The most common sources of mycobacteria from the environment, which contaminate bark and other wood waste from wood-processing plants, are soil, dust and rainwater (**a**). Bark and other wood products stored in piles may heat up (steam can escape out of the piles of bark) (**b**). If humidity is sufficiently high, mycobacteria can multiply there; *M. avium* complex members are usually present in bark as well as in sawdust. Accordingly, bark used as decoration or bedding, e.g. in zoological or botanical gardens, must be stored under dry conditions, for example, on a ground paved with stone and under a roof (**c**). Bark or whole trunks with bark (**d**) originating from wood stockpiles stored for a long time in an unsuitable environment (**e**) and exposed to rainfall and birds' faeces (**f**), can be a source of ESM and PPM (Photo I. Pavlik)

Photo 5.53 If sawdust is stored under unsuitable conditions, it may become heavily contaminated with various ESM and PPM, usually *M. a. hominissuis* [1, 2]. Mycobacteria can often be detected in sawdust directly after Ziehl–Neelsen staining (**a**), as demonstrated by these two clusters of mycobacteria. In contrast to dust, mycobacteria can nearly always be detected by culture. Accordingly, the unprofessional handling of sawdust without the use of protective tools (especially an aspirator) carries with it certain risks (**b**). Not least when the dust rises, because personnel or animals may be exposed to the inhalation of dust and with it mycobacteria (Photo I. Pavlik)

Source: [1] Matlova L, Dvorska L, Bartl J, Bartos M, Ayele WY, Alexa M, Pavlik I (2003) Veterinarni Medicina 48:343–357. [2] Matlova L, Dvorska L, Palecek K, Maurenc L, Bartos M, Pavlik I (2004) Veterinary Microbiology 102:227–236.

Photo 5.54 On 25 March 2007, cars (**a**) and plants (**b**) in central Europe were covered with a reddish dust after an overnight rainfall. The dust originated from the desert of Libya (Africa). It was swept over the Mediterranean Sea for more than 2000 km during a dust storm. Mycobacteria together with other microorganisms present in the dust particles can survive transport over such long distances (Photo I. Pavlik)

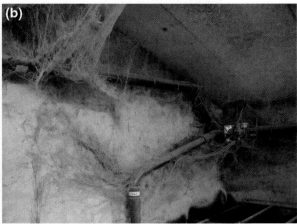

Photo 5.55 A lot of dust is also created by agricultural activities which involve the handling of various dry materials such as feeding mixtures, dry hay, straw, etc. Concentrations of dust particles can reach high levels during certain activities, especially in closed rooms (e.g. in stables for animals). The photograph documents the production of dust during the distribution of straw to cattle boxes (**a**). The dust contains both mycobacteria and other microorganisms (e.g. fungal spores) that are dangerous for animals and people. The dust particles themselves (in particular those, which are smaller than 10 μm) are hazardous for the respiratory tract, because they can cause both mechanical irritation and also elicit various allergic responses with a subsequent increase in sensitivity to air-borne infections. Spider webs covered with thick layers of dust bear witness to the dusty environment of stables and rooms for feed preparation (**b**) (Photo I. Pavlik)

Photo 5.56 Immunosuppressed patients can often be exposed to *M. avium* complex (*MAC*) members from varied and even unexpected sources. During investigations in the residence of a patient suffering from pulmonary mycobacteriosis (infection with *M. a. hominissuis* of serotypes 6 and 9), various *MAC* members were isolated from different components of the environment. *M. a. avium* (serotype 2) was isolated from soil in a paddock for hens, even though avian tuberculosis had never manifested itself in the hens according to the patient's (landlady) account (**a**). *M. a. hominissuis* (serotype 9) was isolated from dust present in the vacuum cleaner used for vacuuming throughout the house (**b**). However, *Mycobacterium* sp. were also isolated from dust present on the ground in the barn, where the patient regularly collected eggs (**c**); these were laid on the seat of an old tractor that had not been used for several years ([1]; Photo I. Pavlik)

Source: [1] Dvorska L, Bartos M, Ostadal O, Kaustova J, Matlova L, Pavlik I (2002) J.Clin.Microbiol. 40:3712–3719.

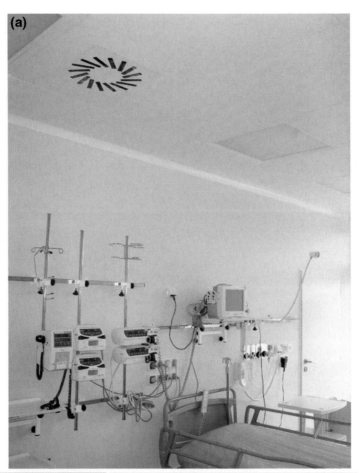

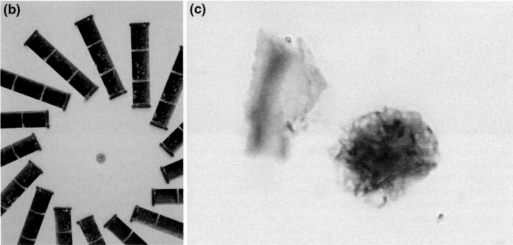

Photo 5.57 The abundant dust found in closed rooms is usually sucked away by air conditioning. Such was the case in a closed specialized hospital ward for immunosuppressed patients; an opening is present on the left of the ceiling, through which the air was sucked away (**a**). Dust settled around the openings of the vent (white clusters of dry dust present on the black surfaces of the vent openings) (**b**). By direct microscopic examination after Ziehl–Neelsen staining, red-stained mycobacteria were observed in a dust cluster (**c**). It seems that they were relatively quickly destroyed in the dry environment, because we failed to detect them by culture in these samples (I. Pavlik and V. Mrlik, unpublished data; Photo I. Pavlik)

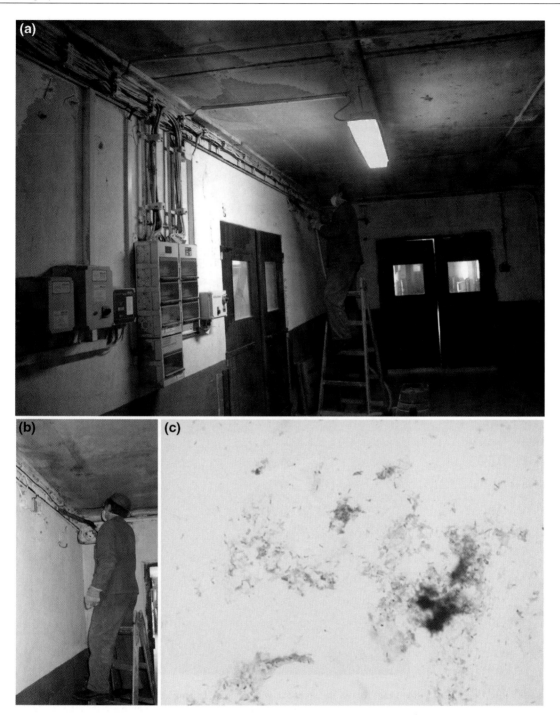

Photo 5.58 Mycobacteria can often be detected in the dust outside animal stables and feed preparation rooms. In a herd of finishing pigs with an increased incidence of tuberculous lesions in the intestinal and pulmonary lymph nodes, dust from passageways was also examined (**a**). Dust from the surroundings of the farm rather than dust from the stables was streaming into these passageways. However, as part of preventive measures, dust was also regularly removed from various parts of equipment present in these passageways by a vacuum cleaner. The personnel were obliged to use protective masks during this activity to prevent a potential infection with mycobacteria (**b**) due to the fact that mycobacteria were detected by microscopy after Ziehl–Neelsen staining in different dust samples from the passageways (**c**) (Photo I. Pavlik)

Photo 5.59 Bird faeces play a significant role in the spread of mycobacteria, in large part due to the fact that the majority of bird species are able to fly and can thus overcome numerous barriers. Birds can produce volumes of faeces which can be infected by various species of mycobacteria. The photograph shows house martin's nests with faeces on the surface below (**a**). They were produced by chicks during the nesting season. In a hall of pens for calves, tens to hundreds of swallows and martins were nesting for many days or weeks under the roof (**b**). Therefore, the exposure of calves to mycobacteria from the faeces of these birds was obvious. Moreover, these birds may also become involved in the ecology of mycobacteria through their whole bodies. In a pig herd, the body of a dead pigeon was found at the bottom of a liquid dung reservoir after its discharge (**c**). This could be a route for the spread of *M. a. avium* to liquid dung and subsequently to the larvae of drone flies in another pig herd, and such a case was documented by Fischer et al. [1]. It is noteworthy that *M. a. avium* was not detected in the lymph nodes of pigs during a 5-year investigation, even though it was isolated from the drone fly larvae (I. Pavlik and L. Matlova, unpublished data; Photo I. Pavlik)

Source: [1] Fischer OA, Matlova L, Dvorska L, Svastova P, Bartos M, Weston RT, Pavlik I (2006) Folia Microbiol.(Praha) 51:147–153.

Photo 5.60   Bird faeces may be involved in the circulation of mycobacteria in the environment of wild animals and ruminants (ungulates) kept in game parks. The environments of game parks are comparable in many aspects to natural outdoor conditions, at least with regard to the free movement of birds. Different species of birds are attracted by the feed supplied to the kept animals, especially in the winter season. These are, e.g. wood pigeons (*Columba palumbus*) (**a**) or Eurasian jays (*Garrulus glandarius*) (**b**). However, these birds are not only attracted by the feeds in the animal troughs, they can also be drawn to grains of undigested corn after their passage through the intestinal tract of, e.g. red deer (*Cervus elaphus*). Some grains can be seen among decomposing droppings (**c**). Their digestibility in birds is enhanced by their absorption of moisture during passage through the intestinal tract. Accordingly, birds search for them and after their ingestion they can become infected with mycobacteria, which may be present in the faeces (Photo J. Lamka)

Photo 5.61 Mycobacteria can also spread throughout the environment of game parks in summer by birds on the lookout for food. They are attracted by the availability of feed provided to the animals kept in zoological gardens. A variety of raven species can, due to their intelligence, overcome different barriers and also their fear of people. These birds (jungle crow; *Corvus macrorhynchos*) (**a**) are often found in the immediate vicinity of sika deer (*Cervus nippon*). Common magpies (*Pica pica*) often search for feed in the paddocks of ruminants and other animals in zoos (**b**). These birds were fearless and entered the glass aviary through an opening in the top. (**c**) Their faeces can become a source of mycobacteria for both animal feed and various components of the environment (Photo I. Pavlik)

Photo 5.62 It is very difficult to protect animals kept in captivity from the faeces of wild birds. It is an almost insoluble problem under small-scale production conditions, because, e.g. free-ranging pigeons may come into contact with many other bird species (**a**). In a seaside zoological garden, an aviary with water fowl was protected by an intact wire mesh, which prevented the penetration of smaller birds and their intake of feed (**b**). However, the seagulls (herring gull; *Larus argentatus*) would sit on the supporting columns and their faeces fell to the ground and on the water surface and feed containers. Clinically ill free-ranging birds (in this case a domestic pigeon that became wild) lose their timidity (**c**). While searching for feed, they come into close contact with domestic birds. Pigs on farms have also been observed to catch weak pigeons (Pavlik I. senior, personal communication; Photo I. Pavlik)

Photo 5.63 The faeces of infected people as well as of dogs may become sources of mycobacteria in the environment with significant consequences for health (e.g. *M. tuberculosis*). An infected secondary school caretaker transmitted *M. tuberculosis* to his two dogs. *M. tuberculosis* was detected by the IS*6110* PCR method in the faeces of the dogs after many weeks (10-week-old faeces in the garden are indicated by a wooden spatula) (**a**). However, *M. tuberculosis* was also detected in the favourite toys of the dogs (a bone and balls). The photograph (**b**) shows tennis balls contaminated with *M. tuberculosis*. (I. Pavlik, V. Mrlik, M. Moravkova, I. Slana, J. Svobodova, P. Kralik, unpublished data; Photo I. Pavlik)

Photo 5.64   Birds may be involved in many ways in the circulation of mycobacteria. Various samples for mycobacterial examination were collected from the nest of a White-tailed Eagle (*Haliaeetus albicilla*; Photo I. Pavlik) in a European larch (*Larix decidua*). One of the adult birds flew continuously above the nest (**a**). Chick faeces and soil from the bottom of the nest were collected (**b**) (Photo J. Vesely). The remains of feed in the nest were also sampled for culture examination of mycobacteria (Photo V. Mrlik). As can be seen in the photo, the leftovers comprised parts of several fish and a piece of skin from a wild boar as shown in the lower right-hand corner (**c**). The recently described *M. arupense* was isolated from the soil present in the nest (V. Mrlik, I. Pavlik, M. Moravkova, unpublished data)

Photo 5.65 Invertebrate animals can also be involved in the circulation of mycobacteria. Different species of Diptera are usually present in pig herds. Mycobacteria were detected in the larvae of Diptera developing in those parts of pig faeces, which did not fall through the grate in pens to the pools underneath (**a**). Various mycobacterial species were also isolated from the faeces of Diptera and from the bodies of imagos (**b**) in pig stables (**c**). Therefore, it is necessary to regularly apply insecticides that can reduce their amount to a certain degree (**d**) (Photo I. Pavlik)

Photo 5.66   Mycobacteria can only be destroyed in manure if an adequate fermentation process occurs there during long-term storage. Water intensively evaporates from fermented manure transported to a field in winter if its temperature is high (**a**). The temperature inside the manure piles is relatively high even in frosty weather when the outside temperature falls to –20 °C (**b**). These temperatures can destroy OPM and other mycobacteria if they are exposed for a long time (**c**). Manure which fermented in this way from cattle suffering an outbreak of paratuberculosis was repeatedly culture negative for *M. a. paratuberculosis*. Nevertheless, the use of this manure is not recommended for pastures and fields with growing green fodder. It is more suitable for use in fields of rapeseed, potatoes, soya, etc. (I. Pavlik, P. Miskovic, L. Matlova, Z. Rozsypalova, unpublished data; Photo I. Pavlik)

Photo 5.67 Mycobacteria can survive longer in liquid manure than in solid manure due to the hospitable conditions of the former. In the event of the insufficient homogenization of liquid manure and the production of a "crust" on the surface, fermentation is insufficient (**a**). The crust formation may be prevented by agitation and aeration of the liquid manure. The temperatures in liquid manure do not reach high values even in the summer months, when the outside temperature exceeds 30 °C (**b**). Fermentation is slow at these temperatures. For example, the causative agent of paratuberculosis can survive in non-homogenized liquid manure for up to 8 months (I. Pavlik, Z. Rozsypalova, P. Miskovic, unpublished data). The discharge of liquid manure from the reservoir to a pasture on a farm with cattle infected with *M. a. paratuberculosis* should therefore be considered a highly risk development (**c**). *M. a. paratuberculosis* was repeatedly detected by culture in liquid manure and in mud from the pasture (Photo I. Pavlik)

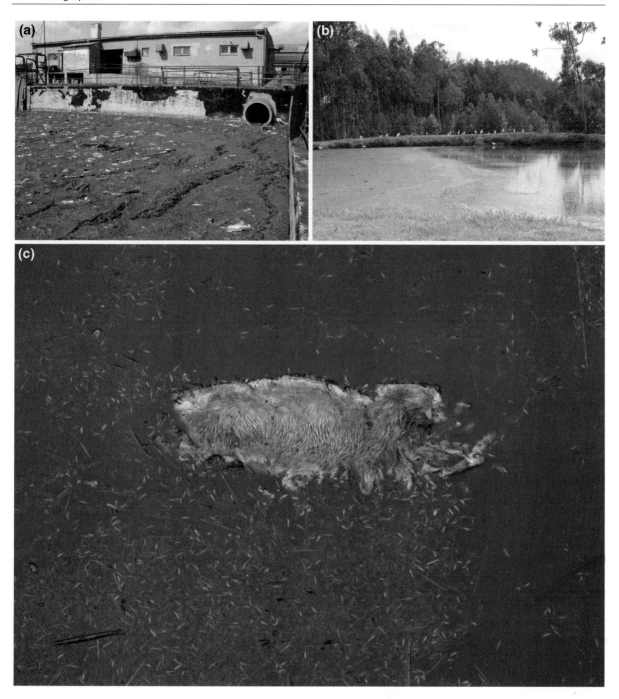

Photo 5.68 The following factors should be regarded as contributing to the occurrence of mycobacteria in liquid manure. Insufficient homogenization of liquid manure hinders effective fermentation. If the layer of materials present on the surface (especially bedding) is thick it does not allow air permeation and consequently, the liquid manure fermentation process is slowed down (**a**). Free living birds (in this case cattle egrets; *Bubul-* *cus ibis*) hunting in the dung reservoir can be involved in the transmission of various causative agents of diseases including mycobacteria (**b**). Disposing of piglet carcasses by throwing them into liquid manure is absolutely inappropriate and highly risky (**c**). The small *whitish* objects floating on the surface are straw stalks and the larvae of syrphid flies (*E. tenax*; Photo I. Pavlik)

Photo 6.1 In aquatic environments, mycobacteria are most often found in the organic sediment formed by, among others, unicellular organisms of various species. Sediment in aquaria comprises feed remains, fish faeces, plant remnants, etc. Dead snail bodies which rapidly decompose in warm aquarium water (**a**) and the bodies of dead fish (shown by the arrow) may also be present there (**b**). A lot of organic material ends up in the sediment. Such sediment can constitute a major part of the bottom layer in aquaria. Bubbles rise from the gas released by the decomposition of organic matter (**c**). Sediment is the most frequent source of PPM, which may infect both animals and people. Protozoa are a significant component of the sediment due to the fact that mycobacteria can survive inside their organisms for a long time or even multiply there. The regular removal of sediment containing various microorganisms including mycobacteria is necessary. During this activity, breeders must efficiently protect themselves from infection with mycobacteria, which can easily penetrate skin injuries, especially on hands (Sections 3.1.7 and 6.5.1; Photo I. Pavlik)

Photo 6.2 In aquaria, mycobacteria can be found in the organic material present on the surface of plants, on the walls of the aquaria and on stones. Protozoa that graze on this organic material are also present there (I. Pavlik, unpublished observation). However, this material is also grazed by higher organisms (freshwater snail; *Planobarius corneus*) because it is an important and often the only source of their food (**a**). For the intake of this material, snails (*Pomacea bridgesi* syn. *Ampullaria australis*) have an adapted tongue designated as a radula, which allows them to effectively intake this organic matter (**b**). The mouths of fishes (e.g. male of *Ancistrus* sp.) have adapted to intake food from the bottom and from surfaces covered with organic matter including growing algae, protozoa and other components of biofilms (**c**). Mycobacteria, which can infect the animals present, are often found in these components as well (Section 6.5; Photo I. Pavlik)

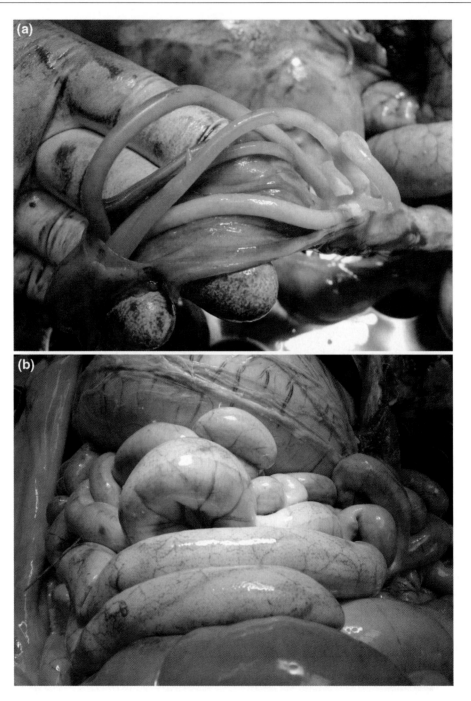

Photo 6.3 Mycobacteria have been detected inside different nematode species (Section 6.3.1.3). These intestinal parasites can be present in abundance in the bowels of both young and adult animals; they can often fill the entire lumen of the intestines (**a**). This weaned piglet was fed by a person shedding *M. tuberculosis* in their sputum. Surprising results were obtained by the study of *M. tuberculosis* distribution in the bodies of this and seven other piglets. It seems that the presence of a high amount of nematodes (the highest amounts were detected in two piglets) provoked an immune system response; the enlargement of Peyer's patches was apparent by gross examination (small, white poppy seed-sized nodules are apparent in the walls of the small intestine) (**b**). *M. tuberculosis* was detected by the IS*6110* PCR method just in these parts of the intestinal mucosa of these two highly infested piglets. It is likely that the presence of nematodes in the intestines contributed to an easier infection of the mucosa by a pathogen that is non-typical for this tissue (V. Mrlik, M. Moravkova, P. Kralik, I. Pavlik, unpublished data; Photo I. Pavlik)

Photo 6.4 *M. marinum* was detected in the ova of brine shrimps (*Artemia salina*), often used by professional and sometimes also by amateur aquarists to feed aquarium fish. Under very simple conditions, they immerse the ova in water in aerated bottles for several days (intensive aeration is necessary for the successful hatching of brine shrimps) (**a**). The hatched brine shrimps are considered high-quality feed for the young of many fish species. The use of a magnifying glass allows the observation of small eggs, which, after some days, develop into brine shrimps (several millimetres in length) (**b**). However, eggs of the snail *Pseudosuccinea columella* can be introduced into an aquarium together with brine shrimp eggs; *M. marinum* was also detected in these snails (Section 6.3.2.2; Photo I. Pavlik and J. Svobodova)

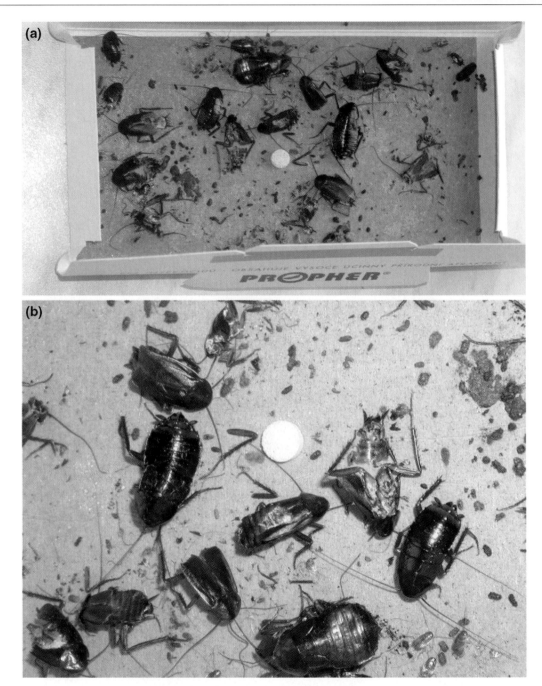

Photo 6.5 Cockroaches may also be involved in the spread of various species of mycobacteria (Section 6.3.5.2). An abundance of cockroaches may be present in warm buildings and store-rooms. Their density can be assessed by different methods. One of them is the laying of boxes with sticky bottoms to catch the cockroaches (Blattodea) in places where they are often found, i.e. in the corners of rooms, close to walls or, e.g. near stored goods (**a**). The cockroaches are drawn to attractants (it is usu-ally a pheromone), which are placed in the middle of a sticky surface; this is a round piece of filter paper impregnated with the effective agent shown in the upper part of the photograph of trapped cockroaches (**b**). If a trap is left for a long time, the dead bodies are ingested by other cockroaches, which do not come into contact with the sticky surface. This is one way in which mycobacteria can also be spread in the environment, i.e. through the infection of the hitherto non-infected cockroaches. According to the amount of trapped cockroaches, the person-nel of specialized companies opt for different procedures to con-trol them. The surfaces (floors, walls or ceilings) of buildings infested by cockroaches, their equipment, and stored raw mate-rials, etc. require different procedures (Photo I. Pavlik)

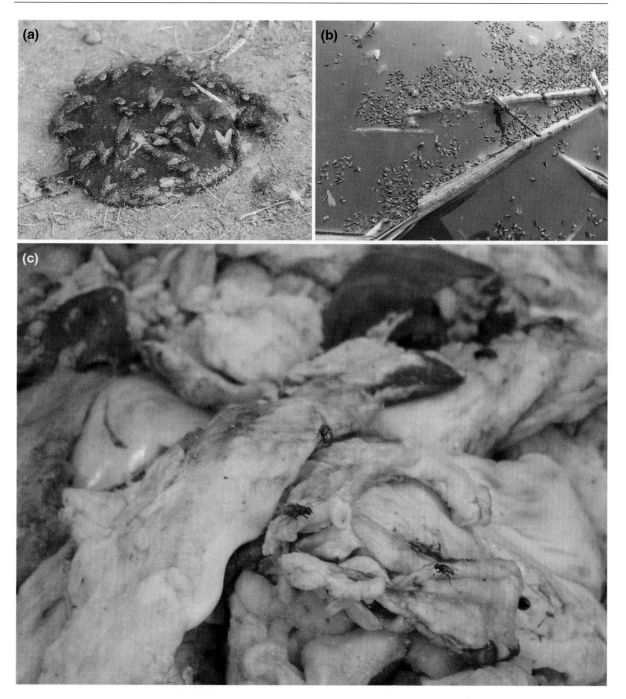

Photo 6.6 Various Diptera species number among the significant decomposers of organic waste materials in the environment. At the same time, they come into contact with a number of pathogens including mycobacteria. They can spread them both on their surface and inside their body cavities (Section 6.3.5.4). Large amounts of different species of Diptera can be seen in the summer months (the time of their occurrence in the mild climatic zone) on various excrements. The fresh faeces of a clinically ill hen infected with *M. a. avium* (**a**) (Photo J. Lamka). High numbers of Diptera are found near dunghills, where they sit on both the dung surface and leaking fluid (**b**) (Photo J. Lamka). A variety of Diptera species seek out the tissue fluids of slaughtered animals (**c**). Diptera pose a high risk for the spread of the causative agents of various diseases including mycobacterial infections (e.g. bovine tuberculosis and paratuberculosis) in slaughterhouses and their close surroundings. The situation may be often critical in abattoirs and other places of animal slaughter in less-developed subtropical and tropical countries (Photo I. Pavlik)

Photo 6.7 It may be very difficult to prevent Diptera from gaining access to the various foodstuffs present in the rooms and halls of processing plants. The protection of a breakfast served in Cape Town (the Republic of South Africa) from invertebrates, usually various Diptera species, may be an example (**a**). Various types of nets are applied to the windows of dairy rooms, forage stores, etc. on domestic animal farms (**b**). Nevertheless, it is necessary to keep in mind that in the windowless halls of slaughterhouses with temperatures below 10 °C, a variety of Diptera have also been seen to be sitting on abattoir waste (**c**) (Photo I. Pavlik)

Photo 6.8   Sometimes only empty pupae may attest to the presence of Diptera. In this bull insemination station, the massive shedding of *M. a. paratuberculosis* through faeces of a bull of Belgian Blue Beef Breed was diagnosed (**a**). Even though the nutritional state of this bull was good, the animal was an important source of the causative agent of paratuberculosis for the surroundings. *M. a. paratuberculosis* was detected in the soil under the faeces, in dirty grass and in Diptera sitting on the faeces in the pasture (**b**). A few weeks later, in repeatedly collected samples of faecal remnants from the surroundings of the pen where the bull was temporarily stabled, *M. a. paratuberculosis* was isolated from the empty pupae of various Diptera species (**c**). Due to the difficulty of preventing the spread of *M. a. paratuberculosis* in cattle herds infected with the causative agent of paratuberculosis by various Diptera species, the prevention of transmission of the causative agent of paratuberculosis into herds of healthy animals is essential (I. Pavlik, I. Trcka, A. Horvathova, unpublished data; Photo I. Pavlik)

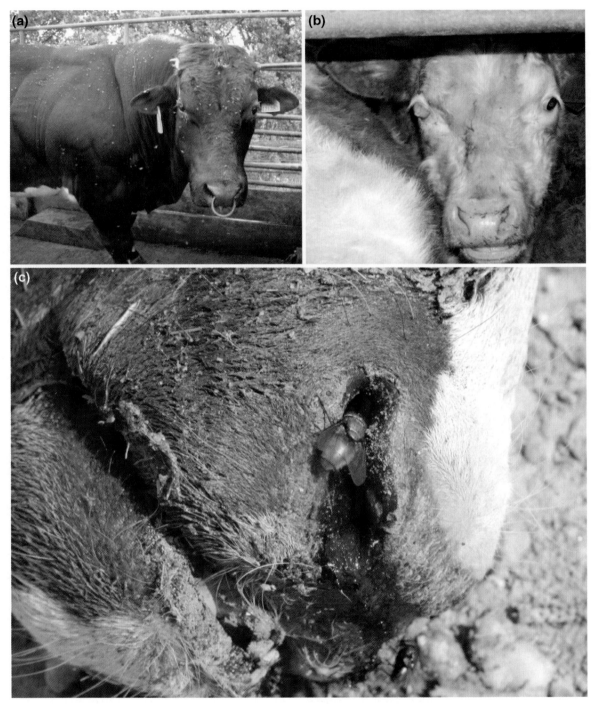

Photo 6.9 A variety of mycobacteria species have been detected in secretophagous Diptera species. Their role in the ecology of mycobacteria is not quite clear; nevertheless, these Diptera species pose certain risks for the transmission of mycobacteria both in the environment and among individual hosts (Section 6.3.5.4). Secretophagous Diptera sit on cattle, in particular, while they are at pasture, or other places outside the stables (**a**). Moreover, a variety of secretophagous Diptera are also present in stables, where they can, along with others, spread the causative agent of infectious keratoconjunctivitis *Moraxella bovis* (the right eye of a young bull has been infected) (**b**). Various Diptera species search for secretions on the mucosa of both live and dead animals; this photograph shows a drone fly on the nose of a dead calf (**c**). Photo I. Pavlik

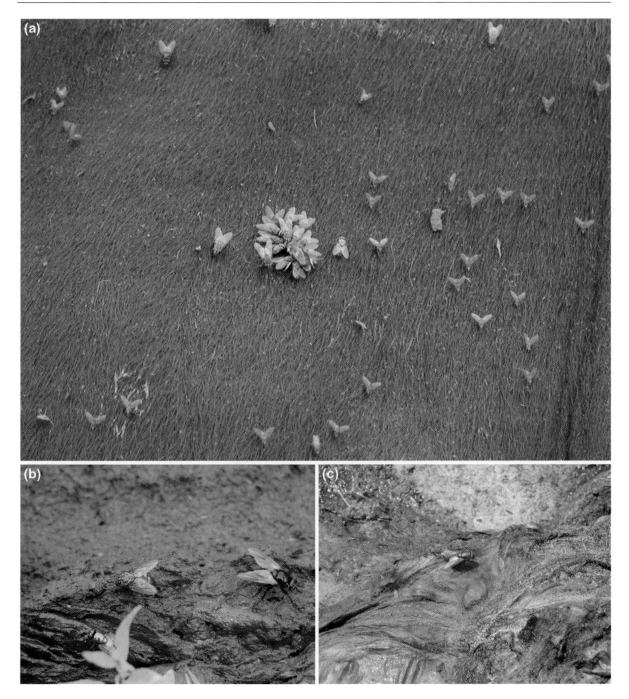

Photo 6.10 Various mycobacterial species have also been detected among hematophagous, coprophagous and necrophagous Diptera species. Their role in the spread of mycobacteria among different animals and their significance in the ecology of mycobacteria is unknown (Section 6.3.5.4). Various species of Diptera can especially be seen on the skin of grazing cattle; they feed on their blood and on the organic rem- nants of their hair (**a**). Different species of Diptera can often be sitting on faeces (**b**). Dead animals provide Diptera with organic matter; the photograph shows a drone fly sitting on the body of a mummified dead piglet (**c**). It is most likely that Diptera are one of the factors that play a very significant role in the ecology of mycobacteria due to their species variety, high adaptability to the environment and high numbers (Photo I. Pavlik)

Photo 6.11 The significance of various Diptera species in the spread of mycobacteria can be demonstrated in the detections of the causative agent of paratuberculosis in imagos and different developmental stages of Diptera. In the scope of a control programme performed in a cattle herd over a long period of time, each calf was given colostrum from its own mother and subsequently bulk milk, which had been heat treated at 80 °C for 20–30 min. If *M. a. paratuberculosis* was diagnosed in the mother, the first preventative measure, ensuring that the offspring would not be included in the breeding programme, was imposed. The second measure was imposed so as to ensure that the further spread of *M. a. paratuberculosis* through contaminated milk would be prevented. However, it was found in the herd that invertebrates had access to milk during its chilling after heat treatment. The photo shows mainly houseflies (*Musca domestica*) present on the milk surface; these were attracted to warm milk when the weather was cold outdoors (below 20 °C) (**a**). During one stage of the control programme, the farmer started to acidify the heat-treated milk with formic acid. How-

ever, this measure failed to prevent the contamination of milk with *M. a. paratuberculosis* by various Diptera species. They were attracted to the precipitated protein in the milk prepared for the feeding of the calves (**b**). Another risk of transmission of *M. a. paratuberculosis* through Diptera from adult animals to calves was the frequent presence of various species of Diptera sitting on the feeding mixtures for calves (**c**) or on the milk prepared for calves (**d**). These insects were particularly attracted to the sweet smell of saccharides present in the feeding mixtures. The calves unhesitatingly ingested dead imagos together with the feeding mixture. Hence, it follows from these observations that it is best to rear non-infected calves from herds infected with paratuberculosis on a remote farm, at best several kilometres away from the infected farm. Thus, it will be ensured that the transmission of *M. a. paratuberculosis* from adult infected cattle or other infected ruminant species (e.g. sheep or goats are sometimes kept together with cattle) is not possible either directly or through a vector (Photo I. Pavlik)

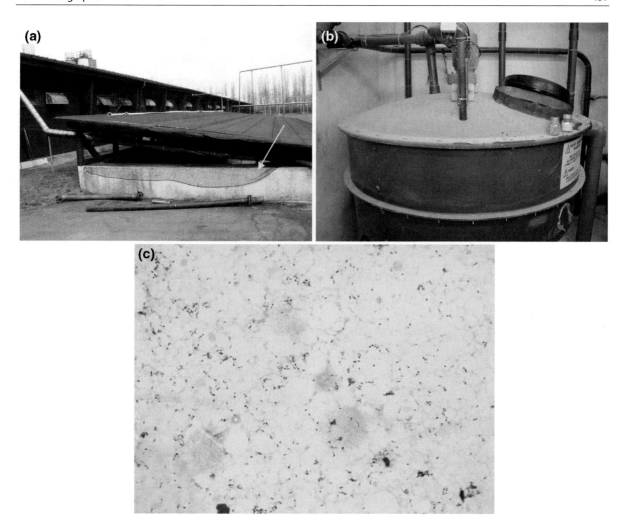

Photo 6.12 In a pig herd, the farmer fed market pigs with liquid feeds supplemented with whey, which was used throughout the year, the reservoir being continuously refilled. Every 2–3 months, it was completely emptied and the sediment from the bottom, which was sometimes up to 50–80 cm thick, was fed to the pigs after homogenization with water. After slaughter, increased numbers of tuberculous lesions in mesenteric lymph nodes caused by *M. a. hominissuis* were detected in the finished pigs. Whey was stored in a big concrete tank, protected from rain by a roof. The sides of the reservoir were also protected by small-mesh nets from birds which used to hunt invertebrate animals on the whey surface. These nets had the aim of preventing the entry of birds and particularly against their drowning in the water which previously happened sometimes, especially in the case of young or weak birds (**a**). The farmer was concerned about the spread of the causative agent of avian tuberculosis *M. a. avium* into the whey by some of the birds. While investigating the biofilms present in the vessel for the preparation of liquid feed (**b**), no mycobacteria were detected. They were found neither in the precipitations floating on the surface nor in whey collected from depths of 50, 100, 150 and 200 cm. However, sediments from a depth of 250–300 cm were abundant in mycobacteria, as detected by microscopy after Ziehl–Neelsen staining (**c**). That explained the irregularity of tuberculous lesions found in pigs fed with sediment from the bottom of the tank at the time of their delivery (i.e. 3–4 months before the end of fattening and slaughter). Regarding the spread of mycobacteria in the environment, it is necessary to emphasize that birds did not have access to the whey owing to the protective nets. However, lots of Diptera frequented the sediment in the summer months and most likely spread *M. a. hominissuis* to the surrounding environment; that pathogen was later detected in the whey sediment (I. Pavlik, I. Trcka, L. Matlova, M. Moravkova, J. Volavka, unpublished data; Photo I. Pavlik)

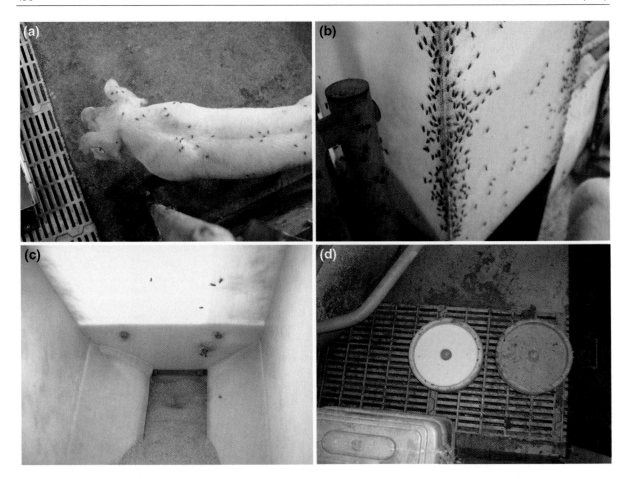

Photo 6.13 Diptera can most probably play a role in the spread of mycobacteria in pig herds by the contamination of dry feeds. Various species of Diptera (especially house flies) were present in a pig herd throughout the year, even during a few cold weeks, when the outdoor temperatures ranged between –10 and –20 °C. At that time, the flies were usually to be found on the warm bodies of the pigs (**a**). The examination of trapped flies revealed mycobacteria of species *M. fortuitum* and *M. intracellulare*. Due to the fact that their whole bodies were examined, it was not possible to distinguish whether mycobacteria contaminated their surfaces or were found inside their bodies. Large numbers of flies were also present on various parts of equipment inside the stables and defaecated on their surfaces (**b**). They were usually present on the outside of plastic reservoirs searching for feed particles, etc. Moreover, 90% of the feed reservoirs were open and flies had free access to the feed (**c**). They were also swept together with the feed into the pig troughs. Flies were also sitting on feed given to newborn piglets in the farrowing pens (**d**): on the left of the photo, a dish containing liquid milk and on the right, a dish containing a dry feeding mixture can be seen. It is likely that the above-mentioned two mycobacterial species which were also detected in the intestinal lymph nodes of pigs with tuberculous lesions in this herd were spread through this route ([1]; Photo I. Pavlik)

Source: [1] Pavlik I, Matlova L, Gilar M, Bartl J, Parmova I, Lysak F, Alexa M, Dvorska-Bartosova L, Svec V, Vrbas V, Horvathova A (2007) Veterinarni Medicina 52:392–404.

Photo 6.14 Syrphid flies (*E. tenax*) are often involved in the spread of mycobacteria from sewage on farms with cattle or other animals [1, 2]. During warm days, timid females sit on the edges of liquid manure reservoirs, where they lay their ova (**a, c**). Larvae that hatch from the ova live in liquid parts of the manure and are almost invisible on the surface. They sink with their mouth opened and facing the bottom and obtain the oxygen essential for life from the surface by means of a several-centimetre-long breathing tube. The photo shows the third larval stage in a Petri dish (**b**). Larvae live on organic remnants from the liquid environment, which include different species of mycobacteria; these, however, cannot be destroyed in their intestinal tract and subsequently digested. Various mycobacterial species, including *M. a. avium* and *M. a. paratuberculosis*, have been isolated from different developmental stages of larvae by culture (Section 6.3.5.4; Photo I. Pavlik)

Source: [1] Fischer OA, Matlova L, Dvorska L, Svastova P, Bartos M, Weston RT, Kopecna M, Trcka I, Pavlik I (2005) Med.Vet.Entomol. 19:360–366. [2] Fischer OA, Matlova L, Dvorska L, Svastova P, Bartos M, Weston RT, Pavlik I (2006) Folia Microbiol.(Praha) 51:147–153.

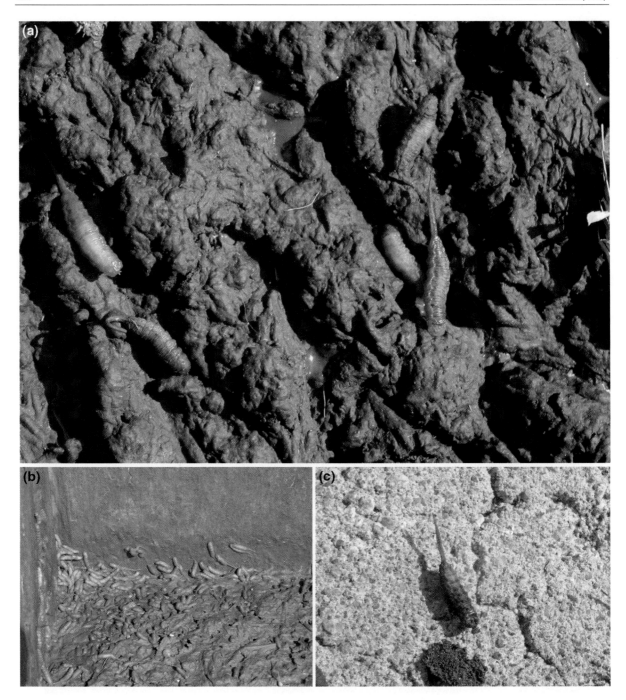

Photo 6.15 Syrphid flies (*E. tenax*) in the puparium stage are mainly involved in the spread of mycobacteria from reservoirs of liquid manure. At this time, larvae leave the liquid of the reservoirs and migrate to the surface, eventually to the particles which float there (**a**). They try to get to solid ground by climbing down the reservoir walls (**b**). If successful, they slowly crawl on the ground for up to several tens of meters to a shady and dry place (**c**). Due to the fact that they drag the tube behind them, the larvae look like small mice and are therefore also designated as such. The emergence of the imago is preceded by the appearance of colour pigment. The adults live on plant nectar and only females approach the liquid manure reservoirs to lay their eggs there (Section 6.3.5.4; Photo I. Pavlik)

(a)

(b)

Photo 6.16 During the migratory period before moving into the pupa stage, the larvae of syrphid flies (*E. tenax*) are most often preyed upon by birds. The photo shows sparrows (*Passer* sp.) flying over the reservoir and sitting on its surface engaged in catching the larvae of syrphid flies (**a**). They also caught larvae on the ground during their migration to the place of pupation. Sparrows subsequently fed the larvae to their chicks in the nests (I. Pavlik, O. A. Fischer, I. Trcka, unpublished data). Sparrow faeces were collected from the ground underneath their overnight nests, usually found on the steel constructions under the roofs of stables (**b**); *M. a. paratuberculosis* was isolated from both sparrow faeces and the larvae of syrphid flies. It was noteworthy that the larvae of syrphid flies also developed outside the liquid manure reservoir, i.e. in the shallow pools of dung water (at left shown by the arrow) found among the manure conveyors in the stable for several months and which did not dry up (I. Pavlik and D. L. Peral, unpublished data; Photo I. Pavlik)

Photo 6.17 The larvae of artificially bred darkling beetles (*Tenebrio molitor*) are used as feed for various animal species kept in captivity. Accordingly, artificially bred darkling beetles must be protected from mycobacterial infection/contamination from different risk substrates (e.g. old sawdust, bran, etc.). Darkling beetles are primarily fed to insectivorous birds and poikilothermic animals in zoological gardens (**a**). In this storeroom for feeding mixtures for pigs, hundreds of developing larvae of darkling beetles at different developmental stages were present on the floor (**b**). The risk of infection of darkling beetles with the causative agent of avian tuberculosis was high because the bodies of dead sparrows and their faeces were also found there (I. Pavlik, O. A. Fischer, I. Trcka, unpublished data; Photo I. Pavlik)

Photo 6.18   Tens to hundreds of pairs of barn swallows (*Hirundo rustica*) and house martins (*Delichon urbica*) nest on some farms; these feed on insects living in domestic animal herds (**a**). They build their nests from stalks of old grass and from mud, i.e. materials often contaminated with mycobacteria (**b**). In subtropical and tropical climatic zones, such nests made of mud can be very large (**c**) as is the case in this greater striped swallow, *Hirundo cucullata* (**d**) (Photo I. Pavlik)

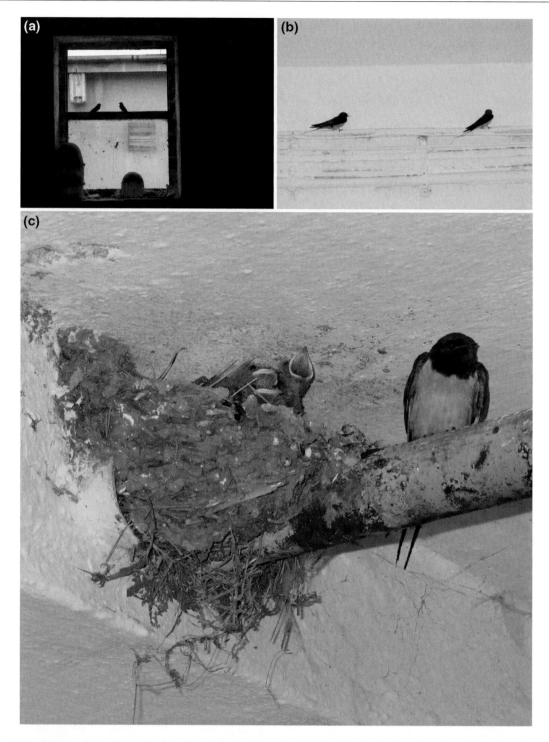

Photo 6.19 Barn swallows (*Hirundo rustica*) usually enter stables through open windows and doors (**a**). Flying insects can be present in the stables and are preyed on by the swallows (**b**). Barn swallows feed their young with invertebrate animals which they catch (**c**). The fact that mycobacteria were also detected in the imagos of invertebrates (Section 6.3.5.4) can most likely explain the isolation of mycobacteria from the faeces of swallows ([1]; Photo I. Pavlik)

Source: [1] Pavlik I, Matlova L, Gilar M, Bartl J, Parmova I, Lysak F, Alexa M, Dvorska-Bartosova L, Svec V, Vrbas V, Horvathova A (2007) Veterinarni Medicina 52:392–404.

Photo 6.20 We can only hypothesize about the circulation of mycobacteria in barn swallows (*Hirundo rustica*) and house martins (*Delichon urbica*), which live in very close contact with domestic animals (**a**). They can be heavily involved in the ecology of mycobacteria, especially in their spread in the environment, when they migrate between continents every year (**b**) (Photo I. Pavlik)

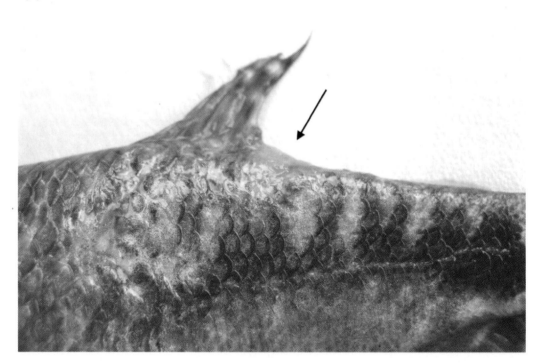

Photo 6.21 A skin ulceration of the lipoid protuberance caused by mycobacterial infection is apparent on the head of a male *Cyphotilapia frontosa* (**a**). The arrow shows skin erosions behind the dorsal fin and a loss of scales on the left under the protuberance in *Trichogaster leeri* (**b**). Both these lesions were due to mycobacterial infection caused by *M. marinum* (Photo L. Novotny)

Photo 6.22 Carcasses represent an attractive source of food for many inhabitants of aquaria. The following example of feeding on the body of a dead Angelfish (*Pterophyllum scalare*) demonstrates how mycobacteria from the body of an infected fish (*M. marinum* infection) can easily and quickly become involved in the ecology of the aquarium environment. First, the dead Angelfish was discovered by a blue gourami (*Trichogaster trichopterus sumatranus*) (**a**). Subsequently, snails (*Ampullaria australis*) started to eat it (**b**). Afterwards, the following fish came and ate the dead Angelfish: a microgeophagus or blue ram (*Mikrogeophagus ramirezi*) and female of *Ancistrus* sp. LDA16, which appropriated the whole dead body of the Angelfish (**c, d**). The muscles of the Angelfish were eaten over a few days and only the skeleton was left on the bottom (Photo I. Pavlik)

Photo 6.23 With regard to food and feed safety and mycobacterial infections, fish and other sea animals pose several risks. One of them is the risk of injury from both the sharp-edged crab shells and sharp fins and teeth of fish during kitchen processing (**a**). People can easily be infected through skin injuries with both *M. marinum* and other PPM species. The second risk is the eating of raw fish spawn which may be contaminated (infected) with *M. marinum*; the photo shows a meal with red, raw and salted salmon spawn (**b**). Ingestion of the raw meat of fish and other sea animals; the Japanese habit of eating raw fish (sushi) is gradually becoming more and more popular in other parts of the world (**c**). The risks of the spread of mycobacterial infections through this route are not exactly known. Dried or smoked fish cured with warm smoke likely constitute a low hazard, provided the smoking temperatures are higher than 72 °C for several minutes (**d**). Cold smoking (e.g. of salmon meat) may also pose a certain risk of transmission of mycobacteria (Photo I. Pavlik)

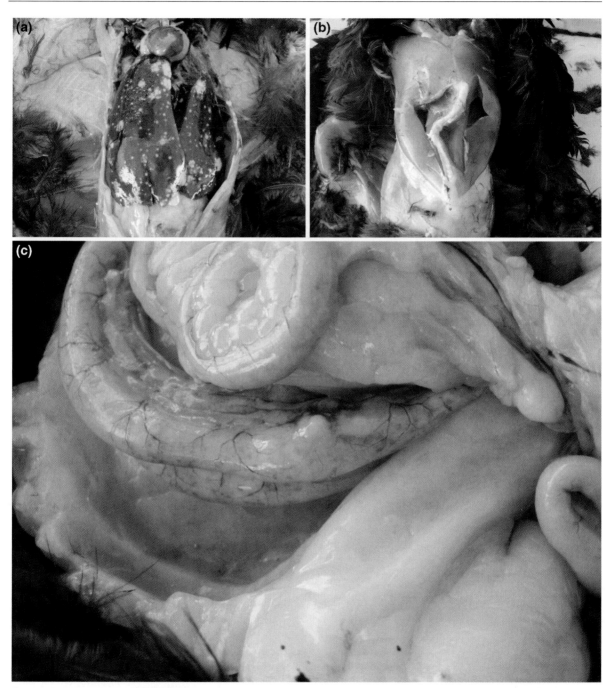

Photo 6.24 In domestic fowl, *M. a. avium* most frequently infects the liver where tuberculous nodules of various sizes are usually observed on necropsy. Such tuberculous lesions were found in a hen with a good health status; that hen was negative in both the avian tuberculin test and rapid slide agglutination test with prepared antigen of *M. a. avium* (**a**). Gross deformation of the sternum was apparent and was most likely associated with a disturbed metabolism of calcium (**b**). However, with respect to the ecology of mycobacteria, the tuberculous process found in the intestinal tract of the hen was most serious (**c**). It follows from the presented results that clinically healthy hens may massively shed *M. a. avium* through faeces ([1, 2]; Photo J. E. Shitaye)

Source: [1] Shitaye JE, Matlova L, Horvathova A, Moravkova M, Dvorska-Bartosova L, Trcka I, Lamka J, Treml F, Vrbas V, Pavlik I (2008) Veterinarni Medicina 53:101–110. [2] Shitaye JE, Matlova L, Horvathova A, Moravkova M, Dvorska-Bartosova L, Treml F, Lamka J, Pavlik I (2008) Veterinary Microbiology 127:155–164.

Photo 6.25 High concentrations of birds in their nesting places produce high amounts of faeces, which may contain a variety of mycobacterial species. On the shelves in the surroundings of the Cape of Good Hope in Cape Town, Republic of South Africa, faeces around the nests of cormorants sitting on eggs and on the nearby reefs can be seen (**a**). With regard to the fact that Cape cormorants (*Phalacrocorax capensis*) are also good fliers, they can be involved in the ecology of mycobacteria by spreading them over great distances (**b**) (Photo I. Pavlik)

Photo 6.26   Birds can spread mycobacteria through their feathers; the mycobacteria usually originate from the ash where the birds roll (Figure 5.32) and the water where they bathe. While bathing, a bald eagle (*Haliaeetus leucocephalus*) from a zoological garden immersed its whole body in water (**a**). Subsequently, the eagle shook out its feathers, whereby aerosol was produced and at the same time water penetrated more deeply to the skin (**b**). Drops of the surface water from a small lake were apparent on the feathers of a sitting bird after a bath (**c**). These photographs document the intensity of penetration of mycobacteria from water to the feathers of birds (Photo I. Pavlik)

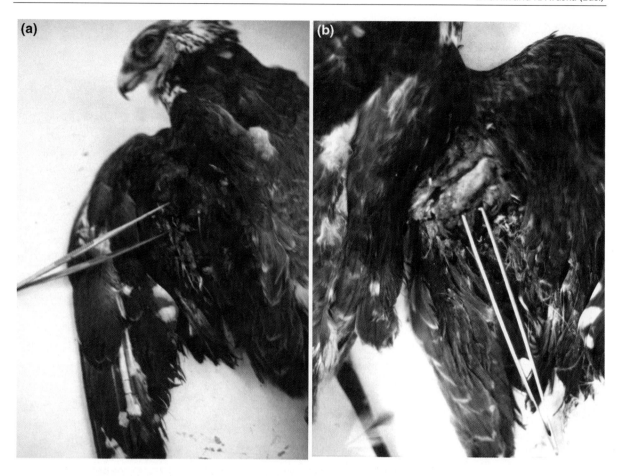

Photo 6.27 In particular, birds of prey are considerably resistant to mycobacterial infections. *M. a. avium* in these birds sometimes manifests itself as skin disease; this can first infect the skin and then subcutis, muscle or bone tissues. In a saker falcon (*Falco cherrug*) avian tuberculosis was detected on the skin and later in deeper parts of tissue on the left wing (**a**). In a peregrine falcon (*Falco peregrinus*) avian tuberculosis affected the right wing with comparable intensity; this bird also died (**b**). In such cases, *M. a. avium* spreads from the infected skin to the surrounding tissues (Photo I. Pavlik)

Photo 6.28 Runners (Paleognathae), i.e. birds that have lost the ability to fly, are in contact with soil and dust over their whole lifetime. Even though avian tuberculosis has been diagnosed in these birds, no studies concerning the ecology of other species of mycobacteria are available at present. Ostriches (*Struthio camelus*) live in the dry environment of their homeland in Africa (Cape Town region, Republic of South Africa) for the major part of the year (**a**). Nevertheless, their braun faeces are wet even in such an extremely dry environment and are attractive to Diptera as are bird faeces in a damp environment (Cape Town region, Republic of South Africa) (**b**). Runners kept in captivity, in this case the lesser rhea (*Pterocnemia pennata*) from a zoological garden (**c**) and the emu (*Dromaius novaehollandiae*) on a family farm (**d**) may be in contact with various species of ruminants or other animals. (**e**) The eating of the eggs of these animals may pose a certain risk of transmission of mycobacteria also. Cases of avian tuberculosis have been reported from such farms (Photo I. Pavlik)

Photo 6.29 Swimmers, i.e. birds that have lost the ability to fly and live in water for the duration of their life, are also occasionally in contact with soil. For example, African penguins (*Spheniscus demersus*) are excellent swimmers and they spend most of their life in the sea water (**a**). The concentration of mycobacteria is very low there and hence, the probability of their contamination with mycobacteria is also low. While resting, and in particular during the nesting season, penguins are in contact with soil and dust (Cape Town region, Republic of South Africa) (**b**). They may also dig holes in the ground where they sit on eggs and where their young hatch (**c**). In the nesting season, both the adults and the young shed their feathers and fine feathers float around the nesting colonies (**d**). No information is available about the ecology of mycobacteria in the environment around penguin colonies (Photo I. Pavlik)

Photo 6.30 The majority of birds can fly. On dry land, they number among the organisms that are obviously involved in the spread of mycobacteria in the environment. Some bird species, in this case the sacred ibis (*Threskiornis aethiopicus*) fly over the land (Fig. 3.17.), freshwater surfaces, islands (photograph from the seashore of Robin Island, Cape Town region, Republic of South Africa) and oceans (**a**). For example, the shy albatross (*Thalassarche cauta*) on the South Island of New Zealand is one of the most tireless fliers and flies hundreds of thousands of km around the southern hemisphere of our planet during the course of its life (**b**). On the other hand, e.g. the free-ranging helmeted guinea fowl (*Numida meleagris*) are territorial birds that fly only occasionally (**c**). In its natural environment, the contact of guinea fowl with mycobacteria including the causative agent of avian tuberculosis is rare. In contrast, helmeted guinea fowl kept in captivity, often in aviaries together with other birds (e.g. pheasants), are under high infectious pressure, not only with PPM (**d**). This example demonstrates a range of factors, which may considerably determine the involvement of certain species of flying birds in the ecology of mycobacteria (Photo I. Pavlik)

Photo 6.31 Some bird species, e.g. the cockatiel (*Nymphicus hollandicus*) are kept in households as pets. They often share a plate (**a**) and a cup with people in the household (**b**). It is surprising that mycobacterial infections are very rarely diagnosed in these pet bird species. Their contact with the owners, usually children, is very close (**c**). However, infections with *M. tuberculosis* transmitted from infected people are sometimes diagnosed in these birds (Photo I. Pavlik)

Photo 6.32 Due to the agility of some bird species kept in a semi-wild enclosure, unusual contact with other animals can be observed. Helmeted guineafowl (*Numida meleagris*) were observed to fly freely in paddocks for various animals including Burchell's zebra (*Equus burchellii*) in a zoological garden (**a**) (Photo I. Pavlik). A flock of wild turkeys (*Meleagris gallopavo*) visited the surroundings of a manger for fallow deer (*Dama dama*) in a game park during the winter months (**b**) (Photo J. Lamka). The Barnacle geese (*Branta leucopsis*) and swan geese (*Anser cygnoides*) were in contact with the white form of fallow deer (*D. dama*) in a hobby breed (**c**). Photo J. Lamka

Photo 6.33 The contact of Humboldt penguins (*Spheniscus humboldti*) with free-ranging grallatorial birds in a zoological garden is not usual. During the afternoon feeding time when penguins were given fish, first a night heron (*Nycticorax nycticorax*) passed over a banister and then jumped through a protective net into the paddock (**a**). Subsequently, a little egret (*Egretta garzetta*) came to the exhibition pavilion (**b**). They tried together to steel the fish remains which the Humboldt penguins left in the troughs and on the water surface. In the wild, these grallatorial birds live out of human residences and are very shy (**c**). They can be seen in the vicinity of domestic animals only along rivers, brooks, irrigation ditches or lakes; by the channel a grey heron (*Ardea cinerea*), shown by the arrow, is fishing (Photo I. Pavlik)

Photo 6.34 Semi-free living pigeons are often found on farms of cattle and other domestic animals. Owing to their parasitic life, they are usually frightened away or are even hunted on the farms and consequently they become quite timid (**a**). In the cities, however, pigeons may become used to people, because they feed them regularly throughout the year (**b**). Their feeding is prohibited in some cities, but in some others, it has become a tradition and tourists have themselves photographed with the pigeons (**c**). The pigeons may be alternately tame or aggressive while demanding food from visitors in places of their regular feeding (**d**). Their behaviour may then take the tourists by surprise. There is no doubt about the subsequent transmission of various causative agents of diseases including mycobacterial infections (Photo I. Pavlik)

Photo 6.35 In cities, pigeons (*Columba livia* f. *domestica*) that became wild are often in close contact with asocial (homeless) people (**a**). The risk of their infection with mycobacteria that are potentially spread by these pigeons is usually high due to the fact that their organism is considerably weakened. However, in big cities house sparrows (*Passer domesticus*) are also attracted to food which people bring, because sparrows are often in close contact with pigeons (**b**). Pigeons entirely lose timidity in big cities because tourists give them food regularly (**c**). Besides pigeons and sparrows, European starlings (*Sturnus vulgaris*) can also be seen in big cities (**d**). However, the latter remain cautious and do not eat out of the peoples' hands (Photo I. Pavlik)

Photo 6.36 The Indian peafowl (*Pavo cristatus*) that often live semi-wild are attractive birds kept in zoological gardens. In particular at the time of their mating call, they are in contact with adult visitors and children (**a, b**). The peafowl are highly susceptible to avian tuberculosis (Photo I. Pavlik)

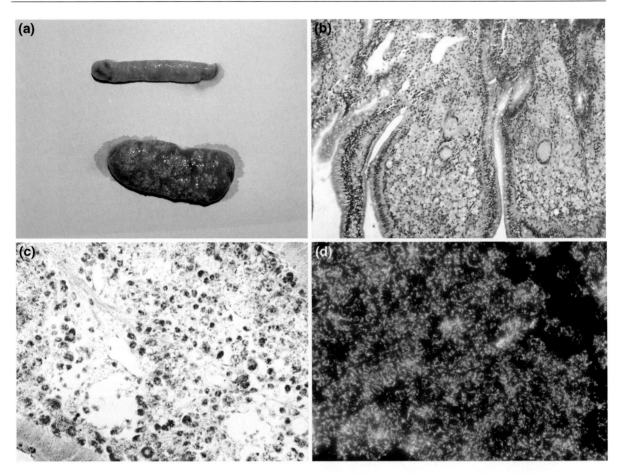

Photo 6.37 *M. genavense* is one of the rarely diagnosed causative agents of disease in birds; that disease is usually fatal. In a dead blue-headed parrot (*Pionus menstruus*) marked enlargement of a part of the small intestine was observed (**a**) (Photo V. Grymova). Diffuse intestinal mucosa infiltration by epithelioid cells, multinucleate giant cells and lymphocytes were detected after haematoxylin–eosin staining and magnification 200× (**b**). After Ziehl–Neelsen staining and 600× magnifica-

tion, the infiltration of diffuse intestinal mucosa by tuberculous granulomatous tissue was found, and abundant intracellular and extracellular mycobacteria were detected (**c**) (Photo R. Halouzka). After 5 months, using the MGIT system of liquid cultures (Becton Dickinson) *M. genavense* was identified; the morphology of the small rods after fluorescent staining resembled *M. a. avium* (**d**) (Photo J. Svobodova)

Photo 6.38 *M. pinnipedii* has also been detected in breeders of water mammals. The lungs of the breeders were infected with this species from *M. tuberculosis* complex. The pulmonary tract of the breeders was most likely infected by aspiration of *M. pinnipedii* during exercising and training (**a**), or the feeding of infected individuals (**b**) (Photo I. Pavlik)

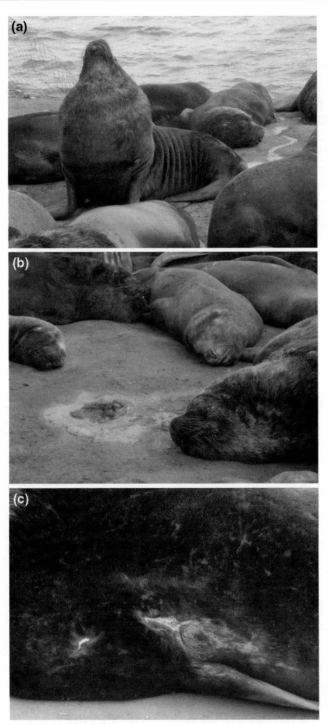

Photo 6.39 The ecology of PPM and OPM (especially *M. pinnipedii*) in South American fur seals (*Arctocephalus australis*) is unknown even though their contact on land, e.g. in the port in Mar del Plata (Argentina) is quite close. Urine from a seal colony flowing over the ground to the sea can be seen in the middle of the photo (**a**). The contact of the seals with faeces is also very close because faeces are present on the ground among the colony (**b**). Numerous skin injuries have been observed on the bodies of these free-ranging seals (**c**). The injuries can become the entrance gateways of a number of bacteria including various species of PPM (Photo I. Pavlik)

Photo 6.40 *M. tuberculosis, M. bovis, M. avium* subsp. *paratuberculosis* have been isolated from wild boar (*Sus scrofa*) and their faeces [1, 2]. These animals, abundant in Central Europe, and capable of transmitting classical swine fever (hog cholera) to domestic pigs, can also serve as vectors of mycobacterial diseases [3]. Wild boar like to search for wet places to lie and reduce their temperature in the summer months (**a**) (Photo I. Trcka). When they are in search of food, the water surface does not represent a barrier for them (**b**). Wild boar are good swimmers and can overcome water bodies that are up to several hundred metres in width (Photo I. Pavlik). Even though wild boar are quite often in contact with PPM, tuberculous lesions in their lymph nodes are only occasionally observed in contrast to domestic pigs (**c**) (Photo I. Trcka)

Source: [1] Machackova M, Matlova L, Lamka J, Smolik J, Melicharek I, Hanzlikova M, Docekal J, Cvetnic Z, Nagy G, Lipiec M, Ocepek M, Pavlik I (2003) Veterinarni Medicina 48:51–65. [2] Trcka I, Lamka J, Suchy R, Kopecna M, Beran V, Moravkova M, Horvathova A, Bartos M, Parmova I, Pavlik I (2006) Veterinarni Medicina 51:320–332. [3] Naranjo V, Gortazar C, Vicente J, de la FJ (2008) Vet.Microbiol 127:1–9.

Photo 6.41 Large herbivores, living both on land and in the aquatic environment, produce enormous amounts of faeces, which can be massively contaminated with mycobacteria. Tuberculosis has been described in white rhinoceroses (*Ceratotherium simum*) as well as in elephants and giraffes in zoological gardens [1, 2]. In the wild white rhinoceroses (**a**) are endangered not only by poachers but also by diseases, including tuberculosis. These animals defaecate regularly in a particular place, whereby they also designate their territory (**b**). Common hip-popotamuses (*Hippopotamus amphibius*) can defaecate ashore, but do so mainly in water where they spend most of their lives (**c**) (Photo I. Pavlik)

Source: [1] Lewerin SS, Olsson SL, Eld K, Roken B, Ghebremichael S, Koivula T, Kallenius G, Bolske G (2005) Vet.Rec. 156:171–175. [2] Stetter MD, Mikota SK, Gutter AF, Monterroso ER, Dalovisio JR, Degraw C, Farley T (1995) J Am Vet.Med Assoc. 207:1618–1621.

Photo 6.42 Herbivores living in the steppes of Africa are exposed to a dry arid climate for the majority of the year (Cape Town region, Republic of South Africa). For instance, free-ranging springboks (*Antidorcas marsupialis*) live on the withering vegetation of various species during the dry season in Africa (**a**). Common elands (*Taurotragus oryx*) also feed on various vegetation including shrub leaves (**b**). Their faeces are also very dry (**c**) and the question of how long mycobacteria can survive in faeces and how significant their involvement in the spread of mycobacteria in this type of environment is, remains open (Photo I. Pavlik)

Photo 6.43 Moreover, the following gigantic herbivores that also live in the steppes of Africa produce copious amounts of faeces that are potentially significant sources of mycobacteria in the environment: African savanna elephants (*Loxodonta africana*) which are the largest mammals living on land (**a**) and giraffes (*Giraffa camelopardalis*), the tallest land mammals (**b**). These two herbivorous species also produce dry faeces in periods of drought, which do not constitute a favourable environment for the survival of mycobacteria (**c**). Regarding the fact that, e.g. elephants frequent watering places, where they also like to bathe, they may spread mycobacteria through their skin, on which a thick layer of mud may be present (Photo I. Pavlik)

Photo 6.44 In zoological gardens, giraffes (*Giraffa camelopardalis*) live under completely different ecological conditions than in the steppes of Africa (Fig. 6.43). They are offered feed in special baskets and thus mycobacterial contamination is prevented (**a**). In special pavilions, their contact with the environment and the PPM present there is substantially limited, especially during the cold months. However, such a close contact with visitors poses some risks (**b**). Their contact with other animals in zoological gardens may often be unusual and sometimes even unexpected (**c**). Some contacts may be fatal for the giraffes, e.g. in a zoological garden in Sweden, the transmission of *M. tuberculosis* from infected elephants to giraffes was confirmed by IS*6110* RFLP method ([1]; Photo I. Pavlik)

Source: [1] Lewerin SS, Olsson SL, Eld K, Roken B, Ghebremichael S, Koivula T, Kallenius G, Bolske G (2005) Vet.Rec. 156:171–175.

Photo 6.45 Carnivores often come into contact with biological material which may be contaminated with mycobacteria. Badgers (*Meles meles*) in Great Britain and in Ireland most probably become infected with *M. bovis* by the ingestion of placentae, e.g. left after calving in the pastures on cattle farms. Dogs are also attracted to placentae. During parturition in a tie stall, a dog was attracted to a placenta hanging out of the cow's uterus, quickly bit off a part and promptly left the stable (**a**). The dog hesitated for a while and carrying the "haul" he quickly ran into high grass next to the stable (**b**). Several seconds later, he rushed out of the grass where he had hidden the haul (**c**). It is possible that these routes may contribute to the spread of *M. bovis* and also, e.g. *M. a. paratuberculosis* (Photo I. Pavlik)

Photo 6.46 Cats are one of the most successful free-ranging carnivores at present. After becoming wild, they may live on many different foods. In this port, fishermen gave them, e.g. waste fish, which the cats liked to eat (**a**). Their nutritional status was very good (**b**). Their breeding was also successful (**c**). We can only hypothesize about the ecology of mycobacteria, which are found, e.g. in fish and other poikilothermic animals fed to the cats in these cases (Photo I. Pavlik)

Photo 6.47 In stables with infected cattle or other ruminants, *M. a. paratuberculosis* is found in both the faeces and milk of these animals. In this herd of cattle infected with paratuberculosis, the cows were placed into separate pens before calving. Cats were often present in these pens together with the cows (**a**). The breeder often fed the cats with freshly drawn milk in which *M. a. paratuberculosis* was detected (**b**). While investigating three euthanized cats, *M. a. paratuberculosis* was detected in the mesenteric lymph nodes of two of them. The IS*900* RFLP method confirmed the identity of these isolates (I. Pavlik, W. Y. Ayele, O. A. Fischer, L. Matlova, unpublished data; Photo I. Pavlik)

Photo 7.1 Original old peat dwelling occupied by poor settlers during the unsuccessful colonization of moorland in north-west Germany (Photo K. Muller)

Photo 7.2 Various components of the environment can become sources of mycobacteria, especially *M. a. hominissuis*, for pigs. From a young age pigs like to nibble various parts of pens where they are kept (**a**), above all wooden parts, which however, are relatively quickly decomposed in a damp environ- ment (**b**). *M. a. hominissuis* was detected in such decomposing wooden parts of pens in a pig herd. After their replacement, the incidence of tuberculous lesions in mesenteric lymph nodes markedly declined (Photo I. Pavlik)

Photo 7.3 Other wooden parts of pens became sources of mycobacteria for different herds. Even though these troughs were cleaned with hot water every day after feeding, *M. a. hominissuis* was detected in the decomposing wood (**a**). In another herd, wooden boards that separated different litters (pens) were the sources of *M. a. hominissuis* for piglets (**b**). Moulding wood lying on the pen floor was the source of *M. a. hominissuis* for boars (**c**). Hence, it was shown that regular liming failed to prevent the multiplication of mycobacteria in decomposing wood which was damp for a long time (Photo I. Pavlik)

Photo 7.4 The surrounding environment constitutes a risk for pig herds, above all as a result of the seasonal migration of small terrestrial mammals. A significant risk factor from this aspect may be the bosks in fields where small rodents and insectivores often live and from where they migrate to buildings on pig farms at times of food shortages (**a**). The situation on pig farms situated in the proximity of deciduous (**b**) or mixed forests (**c**) is similar (Photo I. Pavlik)

Photo 7.5 Faeces and the characteristic smell of small terrestrial mammals are often the only evidence which attest to their presence on pig farms. Mycobacteria are rarely isolated from them because faeces often become desiccated in the dry environment of the stables (**a**). Regular disinfestations – laying traps containing poison which are usually dyed for safety reasons – appear to be an effective preventative measure (**b**). After the laying of traps, poisoned small terrestrial mammals can be found in some parts of the stables (**c**). A variety of mycobacteria including *M. a. avium* and *M. a. hominissuis* have been isolated from their organs. After being caught by pigs (these animals lose their timidity *ante finam*), they become sources of their infection. The dead bodies of small terrestrial mammals become part of the environment in pig herds and also, e.g. herds of ruminants; the photo shows a poisoned individual in silage (**d**) (Photo I. Pavlik)

Photo 7.6 On many pig farms, various types of applicators of vitamins and medicines are used. Solutions of medicaments (vitamins) are pumped into the water supply main from barrels on the floor. On the left of the photo, we can see a metal filter after the removal of rust present in water and water biofilm (**a**). This metal filter (strainer) is usually covered with a plastic case (**b**), which can be easily removed and the sieve taken out (**c**). Mycobacteria were detected in the sediment from this metal sieve by direct microscopy after Ziehl–Neelsen staining. Even though this metal filter reduces the circulation of bigger parts of biofilm, it does not prevent the penetration of mycobacteria, which subsequently cause infections in pigs (Photo I. Pavlik)

Photo 7.7 Windows on pig farms are the most common means by which birds penetrate farm buildings. The risk is constituted by both open windows (without protective nets) (**a**) and broken windows (**b**). In this cattle herd, the pens for calves were situated in a hall protected by a net which prevented the penetration of flying insects and birds. Nevertheless, they learnt to penetrate into the interior of the building through an opening at a metal column (shown by the arrow) (**c**). These birds were common redstarts (*Phoenicurus phoenicurus*) as shown in the photograph. They constitute a high risk for the spread of the causative agent of avian tuberculosis (Photo I. Pavlik)

Photo 7.8   Soft paddocks pose a high risk of infection with both mycobacteria and *Rhodococcus equi* to pigs of all age categories [1]. *M. fortuitum* and *M. a. hominissuis* were isolated from this water and mud; these pathogens caused nonspecific reactions to avian tuberculin in response to preventative skin testing (I. Pavlik, L. Matlova, L. Dvorska-Bartosova, unpublished data; Photo I. Pavlik)
Source: [1] Dvorska L, Parmova I, Lavickova M, Bartl J, Vrbas V, Pavlik I (1999) Veterinarni Medicina 44:321–330.

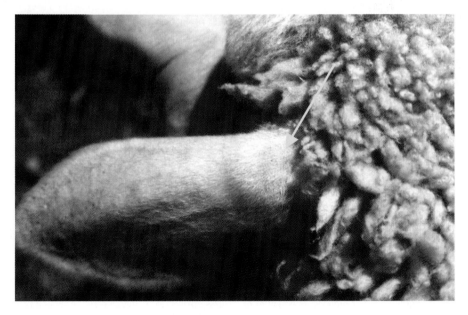

Photo 7.9   ESM and PPM may cause nonspecific responses to bovine tuberculin in sheep (shown by the *arrow*). At the base of the left ear, a marked reaction to bovine tuberculin is apparent even though bovine tuberculosis was not detected in any of the parenchymatous organs of this sheep either by gross examination or culture (Photo I. Pavlik)

Photo 7.10 Various types of watering places on pastures constitute a risk of different mycobacterial infections. Water is supplied through a pipe from the spring to some watering places (troughs). This way of supplying water reduces its potential of being contaminated with both mycobacteria and, e.g. parasites shed in animal faeces (**a**). Water was pumped to this watering place from a nearby well. The water was cold throughout the year and neither ESM nor PPM were detected by repeated test-ing of a small amount of sediment present there (**b**). Water was supplied to this barrel from a cistern every second or third day. Particularly in the summer months, the water reached temperatures of over 25 °C which resulted in a considerable multiplication of ESM and PPM (**c**). These subsequently caused the occurrence of nonspecific responses to bovine tuberculin during skin testing in cattle (Photo I. Pavlik)

Photo 7.11 Typical long-term contamination of certain components is found in the environment of animals kept in captivity. In game parks as well as in the wild, wild ruminants like to cool themselves and get rid of ectoparasites in sumps (**a**). Mud from a sump can be seen on the hair of a white red deer (*Cervus elaphus*; Photo J. Lamka). In zoological gardens, environmental fatigue is apparent in the paddocks for ruminants especially after periods of rainfall (**b**). Muddy soil contaminates various components of the paddock environment. Mud present on the hooves of East Caucasian turs (*Capra cylindricornis*) infected with paratuberculosis polluted the equipment in the paddocks, e.g. artificially arranged rocks (Photo I. Pavlik). The faeces of a clinically ill animal infected with paratuberculosis (watery faeces on the upper part of the photo) are a massive source of *M. a. paratuberculosis* for the environment and kept animals (**c**) (Photo J. Lamka)

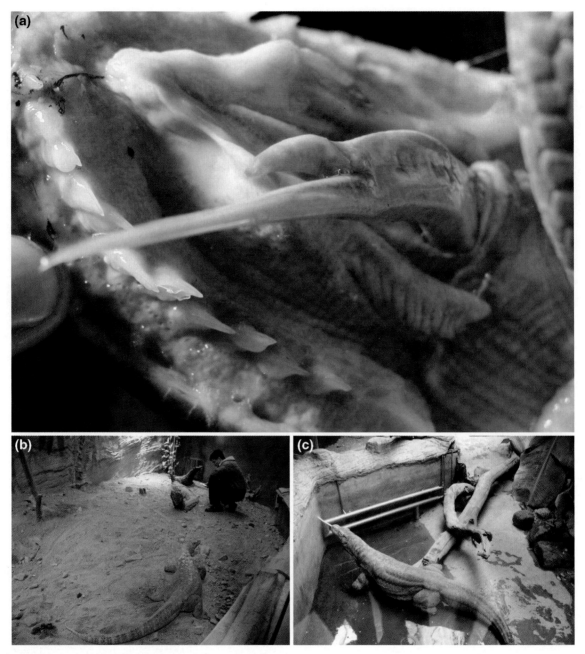

Photo 7.12 Various decorative items constitute sources of mycobacteria for animals in different parts of zoological gardens. A sample of the forked tongue of a female Komodo dragon (*Varanus komodoensis*) was excised and examined and a granulomatous inflammation of mycobacterial origin was subsequently diagnosed (**a**). This female died 4 years later and during the necropsy, massive mycobacterial infection of various organs was found. While investigating the infection sources in the environment, a variety of mycobacterial species were isolated from the faeces of a living female (at bottom left in the photo) and from sand around the feeder (in the background of the photo on the right). The temperature of the sand at a depth of about 5 cm near a decorative tree trunk was measured to be 27 °C (situated in the upper right-hand corner of the photo) (**b**). Various mycobacterial species were isolated from the terrarium of a single male. Mycobacteria were isolated from the sediment in a small lake (the lower left part of the photo), from a rotting decorative tree trunk near the water surface, and from soil and leaves of decorative plants presented on the right (**c**) (I. Pavlik, J. Pokorny, L. Matlova, D. L. Peral, J. Svobodova, J. L. Khol, V. Mrlik, unpublished data; Photo I. Pavlik)

Photo 7.13 Clean and properly stored straw may be used as an adequate alternative bedding to sawdust. Pigs enjoy running in straw whereby they mix it up and their faeces end up in deeper layers of the bedding (**a**). Lying on straw is attractive for pigs (**b**). The use of sawdust as bedding for cattle should also be considered risky. In this stable, old unsuitable straw was first spread in the pens and sawdust was gradually scattered over it (**c**). Sawdust in soil decomposes slowly and is often contaminated with mycobacteria that can cause mastitis of mycobacterial origin in milking cows (Photo I. Pavlik)

Photo 7.14 Pigs usually designate a place in the pen where they lie and rest, and another place where they defecate and urinate. However, faeces and urine should be regularly removed from such areas as shown on the right and at the back of the photo (**a**); if a grid is installed the manure automatically falls into a reservoir. If manure is left in a pen for any length of time, a variety of Diptera lay their ova in it, from which larvae emerge. These can develop in faeces over several weeks. If a herd is infected with mycobacteria, these can pass through the intestinal tract, be shed through faeces and subsequently spread by invertebrate animals. If the density of pigs in a pen is high, they can become highly contaminated again. In the event of the occurrence of mycobacteria in feeds or drinking water, animals can become repeatedly infected with mycobacteria shed with faeces (**b**). Farmers scatter feed, sawdust, wood shavings, peat or bran on the ground of pig pens with the aim of changing the behaviour of pigs (pigs prefer lying in dry places). However, experienced farmers consider such an addition of materials to a pen as risky for various reasons including mycobacterial infections (**c**). Therefore, alternative technology would be more suitable for inexperienced farmers (Photo I. Pavlik)

Photo 7.15 In arid regions (e.g. in the Mediterranean) straw can be stored in the open, e.g. packed in bales (**a**). However, in areas with a higher occurrence of precipitation, straw stored in this way would get wet (**b**). Above all in the warm months, the conditions in such straw are ideal for the multiplication of various mycobacterial species. Moreover, feeding straw, which has been in contact with soil, should be considered as most hazardous with regard to the presence of mycobacteria. The storage of straw under a roof (**c**) is therefore best (Photo I. Pavlik)

Photo 7.16 An increased occurrence of tuberculous lesions in mesenteric and cephalic lymph nodes was detected in market pigs from this herd. The breeder fed only the pigs with a purchased diet and gave them drinking water from an overpressure system. Therefore, the source of mycobacterial infections could not be identified for a long time. Even though paddocks had been constructed outside the stables, they were not used due to the pigs' repeated infections with *R. equi*. During local investigation, a peat pile was found in front of the stable (the *black pile on the *right* shown by the *arrow*) (**a**). This peat was used as bedding in the pens for piglets (**b**). Due to the fact that the breeder was convinced that the piglets did not ingest peat, he was not concerned about the proper storage of the peat. During the investigation of peat samples from the *top* and *bottom* of the pile, both microscopy and culture detected a massive contamination of the peat with *M. a. hominissuis*, the same causative agent which was isolated from the infected lymph nodes (**c**) (Photo I. Pavlik)

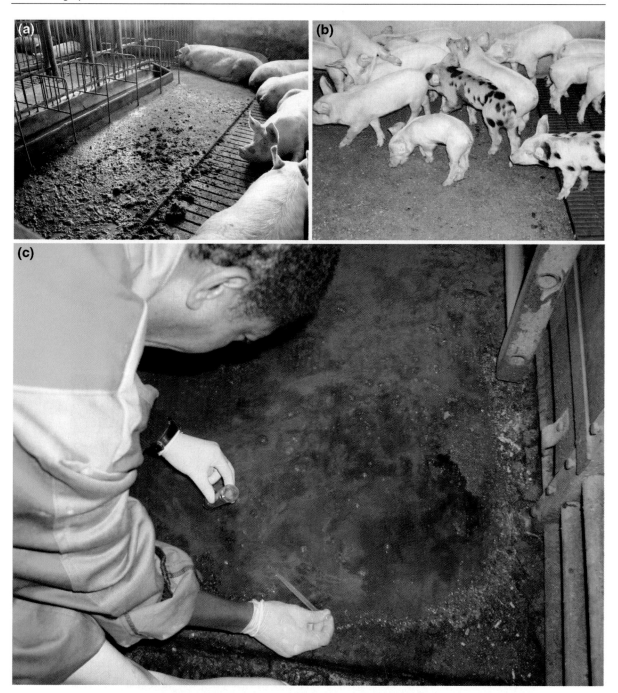

Photo 7.17 Dry concrete floors with intact surfaces are the safest option in pens for pigs with respect to their potential contamination with mycobacteria from the environment. The pigs (in this case breeding sows) usually lie in one place and defecate and urinate in a different spot in the opposite corner of the pen (**a**). The best situation is when urine and faeces fall into a liquid dung reservoir. A combination of a good-quality concrete floor with plastic grids is most suitable for piglets (**b**). Various mycobacterial species have been isolated from those places in pens where urine and faeces are left for any length of time (**c**). These mycobacteria are responsible for the formation of tuberculous lesions in the lymph nodes of pigs (I. Pavlik, J. E. Shitaye, A. Horvathova, unpublished data; Photo I. Pavlik)

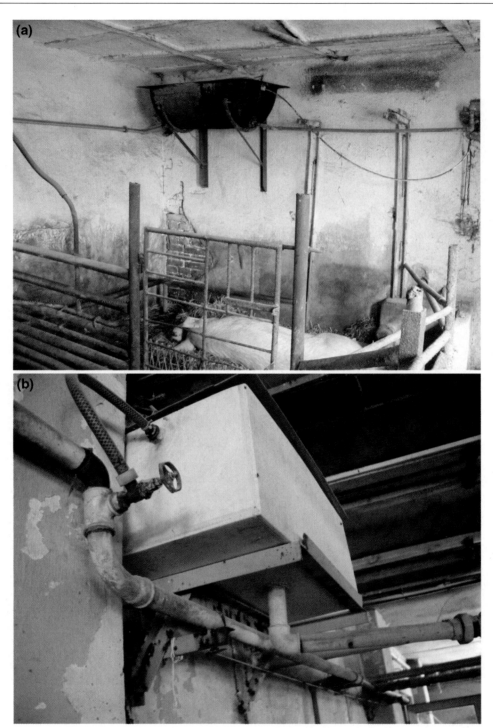

Photo 7.18 Expansion tanks containing drinking water pose the main risk to animals with regard to their potential infection with various species of mycobacteria. The water gathers in these tanks and flows to the drinkers due to its own weight. The emplacement of a bath used as an expansion tank in this herd should be considered as completely inappropriate and highly risky (**a**). Mycobacteria detected in the bath sediment most probably originated from dust, which continuously fell onto the water surface. Plastic expansion tanks with a bottom drain can be regarded as relatively safe (**b**). This bottom drain continuously prevents the formation of sediment. These tanks should also be covered from above with a removable lid which is regularly cleaned (Photo I. Pavlik)

Photo 7.19   If an overpressure system of water supply is used, it is safer to use big water tanks. Large volumes of water warm up much more slowly and the detection of mycobacteria in sediments from such tanks is less frequent and in lower concentrations in comparison with small expansion tanks. These tanks are usually constructed to hold several m^3 to several tens of m^3 of water. A concrete tank holding 30 m^3 of water was situated in the basement of this small object protecting the pumps (**a**). The water temperature did not exceed 6 °C at any point of the year, which is an absolutely unfavourable environment for the multiplication of most species of mycobacteria; therefore, mycobacteria were only occasionally detected in the sediment from this tank. Similar reservoirs for larger volumes of water that are partly situated above ground level are used on some farms. Water in this tank reached about 1 m above ground level (**b**). While looking inside it was evident that the water was clean and the sediment layer was very thin (**c**). The detection rate of mycobacteria was very low (Photo I. Pavlik)

Photo 7.20 All preventative measures were adopted in a particular pig herd. Before the weaned piglets were introduced into the pens, the latter were cleaned mechanically with high-pressure hot water and disinfected (**a**). Nevertheless, tuberculous lesions were detected in the mesenteric and cephalic lymph nodes of numerous pigs after fattening. While investigating the sources of mycobacterial infections, abundant sediment, and subsequently biofilm, ran out of the unscrewed left and right drinker (**b**). Mycobacteria were detected in it after Ziehl–Neelsen staining and were subsequently isolated *in vitro*. Steaming of the whole water main by pumped water of temperatures of around 80 °C for 20 min was found to be most effective at eliminating mycobacteria (**c**). Four to five months later, a marked decrease in the occurrence of tuberculous lesions was observed in the slaughtered pigs. Due to the fact that the water supply main piping was made of plastic material, the use of hot water caused its deformations. This was prevented by loops in each branch of the water piping; these controlled the dilatation and contraction of the water piping (Photo I. Pavlik)

Photo 7.21   If paratuberculosis occurs in a herd of ruminants, the liming of pastures when vegetation is sparse is advised (**a**). However, preventative liming is non-effective if infected, clinically ill animals suffering from chronic enteritis are present (**b**). The liming of pastures can only be effective if the infected animals do not graze there for a certain period of time or if it is used for the grazing of young, non-infected animals. The liming of, e.g. (sub)alpine pastures is impossible due to the inaccessibility of the terrain for agricultural machines (**c**). Liming with the use of aeroplanes is too expensive and therefore it is not applied in infected herds. Accordingly, the most important step for the prevention of further contamination of the environment with the causative agent of paratuberculosis *M. a. paratuberculosis* is the eradication of a herd of infected ruminants or the gradual culling of animals which shed *M. a. paratuberculosis* through faeces (Photo I. Pavlik)

Photo 7.22 Feeds given to animals and birds of prey may also be sources of mycobacteria for the environment of zoological gardens. Pieces of meat intended for a Eurasian griffon vulture (*Gyps fulvus*) thrown on the ground of the aviaries attracted common starlings (*Sturnus vulgaris*) as shown by the arrows (**a**). One-day-old chicken carcasses (shown by the arrow) put into another aviary for birds of prey decomposed quickly during warm weather and thereby attracted various invertebrate animals (**b**). Residues of feed for scavenger birds are also involved in the ecology of mycobacteria (**c**). If the feed is infected with mycobacteria, it itself becomes a source of mycobacteria for the environment. Regarding the risks of the transmission of ESM and PPM from the environment, this feed can moreover become contaminated with mycobacteria that are spread to them by flying invertebrates or birds (Photo I. Pavlik)

Photo 7.23 In the temperate climatic zone, free-ranging ruminant species do not usually graze on pastures for cattle. More than 20 European roe deer (*Capreolus capreolus*) can be seen on the cereal stubble field in the upper background of the photo (**a**). Each of the ruminant species grazed separately during the day. However, European roe deer came to the pastures for cattle in the evening when cattle were fed water and corn near the central stable. The contact of ruminants kept in captivity, e.g. white fallow deer (*Dama dama*) with other free-ranging animals has also been described. However, their contact with a white stork (*Ciconia ciconia*) on a pasture (**b**) is a rare event (Photo I. Pavlik)

Photo 7.24 Besides cattle, other animals can become sources of *M. bovis*. In a zoological garden, *M. bovis* was diagnosed in Vietnamese pigs, which were allowed to run free on the paths (**a**). They were a great attraction for children. The brush-tailed possum (*Trichosurus vulpecula*) is a significant reservoir of *M. bovis* in New Zealand (**b**). At first sight, these marsupials look cute and friendly, but in fact they are bad-tempered animals whose bite is relatively dangerous (Photo I. Pavlik)

Photo 8.1 A juvenile example of the moorland dragonfly *Leucorrhinia rubicunda* displays a mosaic composed of yellow and a darker colour (**a**). An adult moorland dragonfly *L. rubicunda* is shown with characteristic red/pink spots on the wings (**b**). Their colonization of moorland waters is dependent on mycobacteria (Photo K. Soeffing)

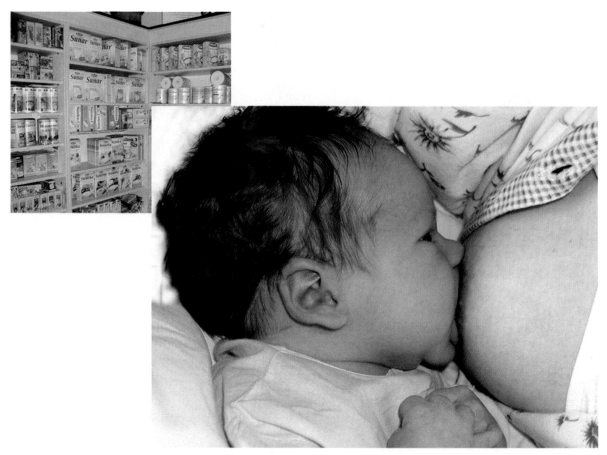

Photo 9.1   *M. leprae, M. tuberculosis* and *M. bovis* have represented a threat from ancient times right up to the present. Other mycobacteria are also considered to be pathogenic for animals or for immunodeficient people. Recently, a report of the American Academy of Microbiology has tentatively identified *M. avium paratuberculosis* as a possible public health threat. Nevertheless, the contamination of food and water by this bacterium is not yet under legal control and humans come into contact with it and different other mycobacteria also in the air and soil. Peptidoglycans, the components of the mycobacterial cell wall, and their components, muramylpeptides, have in precise experiments been confirmed as capable of provoking autoimmune and autoinflammatory reactions in some people. Nevertheless, the possible participation of *M. avium* subsp. *paratuberculosis* in the development of type I diabetes, rheumatic arthritis, multiple sclerosis, Crohn's disease and other autoimmune or autoinflammatory diseases is not yet being seriously considered. "Pasteurization and cooking devitalize all bacteria", or "contact with microorganisms increases immunity", or "non-cultivable microorganisms equals no microorganisms", are statements which reflect the views of some in the research community. However, any risk, even if hypothetical, must need to be assessed and dealt with as a real risk. Certainly, the elimination of mycobacteria from municipal water distribution systems and from dairy and beef products will not be easy. It will take dozens of years and will need enormous financial support. However, there exists a great opportunity to stop the increase of autoimmune and autoinflammatory diseases, which is occurring worldwide. These diseases are a cause of considerable suffering for patients and their treatment is very expensive. It is impossible to achieve victory if the battle is not even started, and it is unlikely that victory will be won on all fronts at once. But why not start in a very limited area at least and ban the sale of baby food containing dead mycobacteria in pharmacies? Breastfeeding is recommended for many good reasons, including avoiding the risk of the consumption of dead *M. avium* subsp. *paratuberculosis* cells together with infant formula. Nevertheless, mothers suffering from Crohn's disease should be informed that their milk can be contaminated with *MAP*. Babies should be breastfed, if possible, and baby food must be *MAP* free, sooner rather than later (Photo I Trneckova and V. Vrbas)

Index